Study and Solutions Guide

Precalculus Functions and Graphs: A Graphing Approach

and

Precalculus with Limits: A Graphing Approach

Fourth Edition

Larson/Hostetler/Edwards

Bruce H. Edwards

University of Florida
Gainesville, Florida

Houghton Mifflin Company   Boston   New York

Publisher: Jack Shira
Associate Sponsoring Editor: Cathy Cantin
Development Manager: Maureen Ross
Assistant Editor: Lisa Pettinato
Supervising Editor: Karen Carter
Senior Project Editor: Patty Bergin
Editorial Assistant: Allison Seymour
Art and Design Manager: Gary Crespo
Senior Marketing Manager: Danielle Potvin
Marketing Associate: Nicole Mollica
Senior Manufacturing Coordinator: Priscilla Bailey
Composition and Art: Meridian Creative Group

Printed in the United States of America

ISBN: 0-618-39481-8

123456789-CRS-08 07 06 05 04

Preface

This *Study and Solutions Guide* is a supplement to *Precalculus Functions and Graphs: A Graphing Approach*, Fourth Edition, and *Precalculus with Limits: A Graphing Approach*, Fourth Edition, by Ron Larson, Robert P. Hostetler, and Bruce H. Edwards.

Solutions to the exercises in the text are given in two parts. Part I contains solutions to odd-numbered Section and Review Exercises; summaries of the chapters; and Practice Tests with solutions. Part II contains solutions to the Chapter and Cumulative Tests from the textbook.

This *Study and Solutions Guide* is the result of the efforts of Larson Texts, Inc., and Meridian Creative Group. If you have any corrections or suggestions for improving this guide, we would appreciate hearing from you.

Bruce H. Edwards
358 Little Hall
University of Florida
Gainesville, FL 32611
Be@math.ufl.edu

Contents

PART I

CHAPTER 1
Functions and Their Graphs

CHAPTER 1
Functions and Their Graphs

Section 1.1   Lines in the Plane

You should know the following important facts about lines.

- The graph of $y = mx + b$ is a straight line. It is called a linear equation.
- The slope of the line through (x_1, y_1) and (x_2, y_2) is

    $$m = \frac{y_2 - y_1}{x_2 - x_1}.$$

- (a) If $m > 0$ the line rises from left to right.

    (b) If $m = 0$, the line is horizontal.

    (c) If $m < 0$,, the line falls from left to right.

    (d) If m is undefined, the line is vertical.

- Equations of Lines

    (a) Slope-Intercept: $y = mx + b$

    (b) Point-Slope: $y - y_1 = m(x - x_1)$

    (c) Two-Point: $y - y_1 = \dfrac{y_2 - y_1}{x_2 - x_1}(x - x_1)$

    (d) General: $Ax + By + c = 0$

    (e) Vertical: $x = a$

    (f) Horizontal: $y = b$

- Given two distinct nonvertical lines

    $$L_1: y = m_1 x + b_1 \quad \text{and} \quad L_2: y = m_2 x + b_2$$

    (a) L_1 is parallel to L_2 if and only if $m_1 = m_2$ and $b_1 \neq b_2$.

    (b) L_1 is perpendicular to L_2 if and only if $m_1 = -1/m_2$.

Solutions to Odd-Numbered Exercises

1. (a) $m = \frac{2}{3}$. Since the slope is positive, the line rises. Matches L_2.

    (b) m is undefined. The line is vertical. Matches L_3.

    (c) $m = -2$. The line falls. Matches L_1.

3.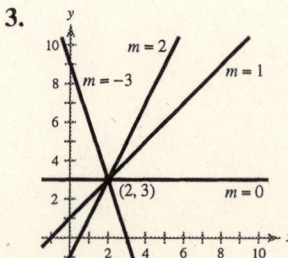

5. Slope $= \dfrac{\text{rise}}{\text{run}} = \dfrac{3}{2}$

7. slope $= \dfrac{0 - (-10)}{-4 - 0} = \dfrac{10}{-4} = -\dfrac{5}{2}$

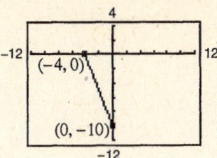

9.

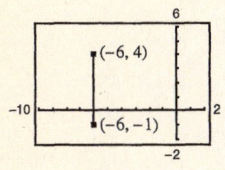

Slope is undefined.

11. Since $m = 0$, y does not change. Three points are $(0, 1)$, $(3, 1)$, and $(-1, 1)$.

13. Since m is undefined, x does not change and the line is vertical. Three points are $(1, 1)$, $(1, 2)$, and $(1, 3)$.

15. Since $m = -2$, y decreases 2 for every unit increase in x. Three points are $(1, -11)$, $(2, -13)$, and $(3, -15)$.

17. Since $m = \frac{1}{2}$, y increases 1 for every increase of 2 in x. Three points are $(9, -1)$, $(11, 0)$, $(13, 1)$.

19. $5x - y + 3 = 0$

$\qquad y = 5x + 3$

(a) Slope: $m = 5$

$\quad$ y-intercept: $(0, 3)$

(b)

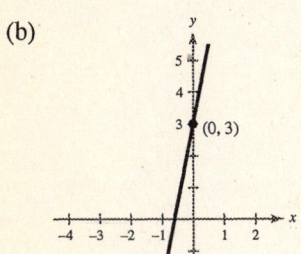

21. $5x - 2 = 0$

$\qquad x = \frac{2}{5}$

(a) Slope: undefined

$\quad$ No y-intercept

(b)

23. $3y + 5 = 0$

(a) $y = -\frac{5}{3}$

$\quad$ Slope: $m = 0$

$\quad$ y-intercept: $\left(0, -\frac{5}{3}\right)$

(b)

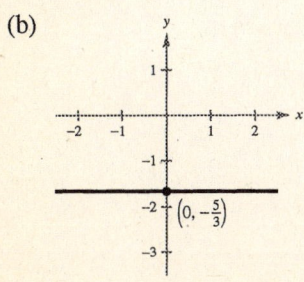

25. $y + 2 = 3(x - 0)$

$\qquad y = 3x - 2 \implies 3x - y - 2 = 0$

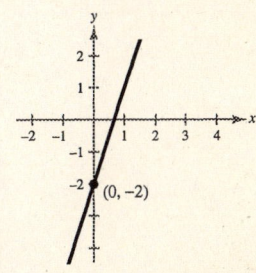

27. $y - 0 = 4(x - 0)$

$y = 4x$

$4x - y = 0$

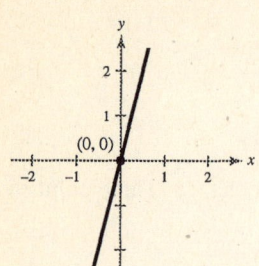

29. $x = 6$

$x - 6 = 0$

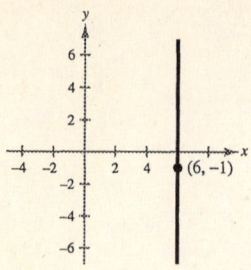

31. $y - \frac{3}{2} = 0\left(x + \frac{1}{2}\right)$

$y - \frac{3}{2} = 0$ horizontal line

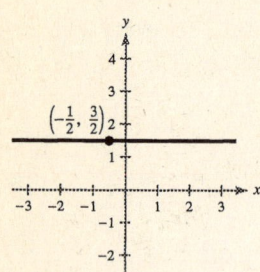

33. $y + 1 = \frac{5 + 1}{-5 - 5}(x - 5)$

$y = -\frac{3}{5}(x - 5) - 1$

$y = -\frac{3}{5}x + 2$

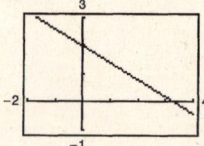

35. Since both points have $x = -8$, the slope is undefined.

$x = -8$

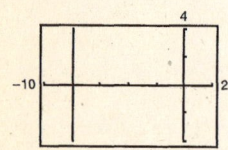

37. $y - \frac{1}{2} = \frac{\frac{5}{4} - \frac{1}{2}}{\frac{1}{2} - 2}(x - 2)$

$-y = -\frac{1}{2}(x - 2) + \frac{1}{2}$

$y = -\frac{1}{2}x + \frac{3}{2}$

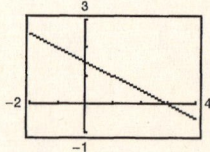

39. $y + \frac{3}{5} = \frac{-\frac{9}{5} + \frac{3}{5}}{\frac{9}{10} + \frac{1}{10}}\left(x + \frac{1}{10}\right)$

$y + \frac{3}{5} = -\frac{6}{5}\left(x + \frac{1}{10}\right)$

$y = -\frac{6}{5}x - \frac{18}{25}$

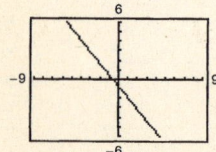

41. $y - 0.6 = \frac{-0.6 - 0.6}{-2 - 1}(x - 1)$

$y = 0.4(x - 1) + 0.6$

$y = 0.4x + 0.2$

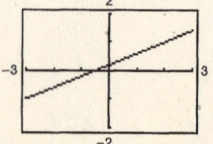

43. Using the points $(2000, 28500)$ and $(2002, 32900)$ we have

$$m = \frac{32900 - 28500}{2002 - 2000} = \frac{4400}{2} = 2200$$

$$S - 28500 = 2200(t - 2000)$$

$$S = 2200t - 4,371,500$$

When $t = 2006$,
$$S = 2200(2006) - 4,371,500 = \$41,700$$

45. $x - 2y = 4$

$$-2y = -x + 4$$

$$y = \tfrac{1}{2}x - 2$$

Slope: $\tfrac{1}{2}$

y-intercept: $(0, -2)$

The graph passes through $(0, -2)$ and rises 1 unit for each horizontal increase of 2.

47. $x = -6$

slope is undefined

no y-intercept

The line is vertical and passes through $(-6, 0)$.

49. $y = 0.5x - 3$

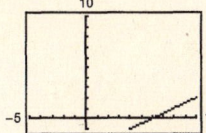 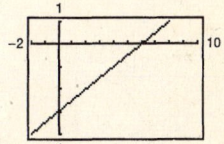

The second setting shows the x- and y-intercepts more clearly.

51. $m_{L_1} = \dfrac{9 + 1}{5 - 0} = 2$

$$m_{L_2} = \frac{1 - 3}{4 - 0} = -\frac{1}{2} = -\frac{1}{m_{L_1}}$$

L_1 and L_2 are perpendicular.

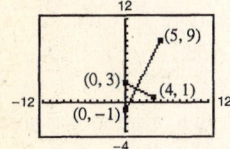

53. $m_{L_1} = \dfrac{0 - 6}{-6 - 3} = \dfrac{2}{3}$

$$m_{L_2} = \frac{\tfrac{7}{3} + 1}{5 - 0} = \frac{2}{3} = m_{L_1}$$

L_1 and L_2 are parallel.

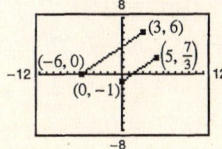

55. $4x - 2y = 3$

$$y = 2x - \tfrac{3}{2}$$

Slope: $m = 2$

(a) $y - 1 = 2(x - 2)$

$$y = 2x - 3$$

(b) $y - 1 = -\tfrac{1}{2}(x - 2)$

$$y = -\tfrac{1}{2}x + 2$$

57. $3x + 4y = 7$

$$y = -\tfrac{3}{4}x + \tfrac{7}{4}$$

Slope: $m = -\tfrac{3}{4}$

(a) $y - \tfrac{7}{8} = -\tfrac{3}{4}\left(x + \tfrac{2}{3}\right)$

$$y = -\tfrac{3}{4}x + \tfrac{3}{8}$$

(b) $y - \tfrac{7}{8} = \tfrac{4}{3}\left(x + \tfrac{2}{3}\right)$

$$y = \tfrac{4}{3}x + \tfrac{127}{72}$$

59. $x - 4 = 0$ vertical line

slope not defined

(a) $x - 3 = 0$ passes through $(3, -2)$

(b) $y + 2 = 0$ passes through $(3, -2)$ and is horizontal

61. (a) $y = 2x$ (b) $y = -2x$ (c) $y = \frac{1}{2}x$

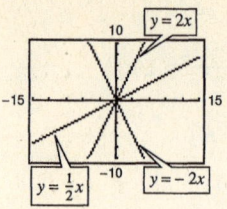

(b) and (c) are perpendicular.

63. (a) $y = -\frac{1}{2}x$ (b) $y = -\frac{1}{2}x + 3$

(c) $y = 2x - 4$

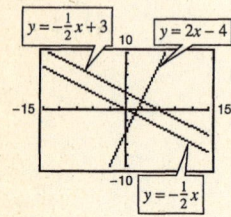

(a) and (b) are parallel.

(c) is perpendicular to (a) and (b).

65. (a) $m = 135$. The sales are increasing 135 units per year.

(b) $m = 0$. There is no change in sales.

(c) $m = -40$. The sales are decreasing 40 units per year.

67. (a)

Years	Slope
1992–1993	$0.68 - 0.58 = 0.1$
1993–1994	$0.86 - 0.68 = 0.18$
1994–1995	$0.91 - 0.86 = 0.05$
1995–1996	$0.69 - 0.91 = -0.22$
1996–1997	$0.57 - 0.69 = -0.12$
1997–1998	$0.74 - 0.57 = 0.17$
1998–1999	$1.60 - 0.74 = 0.86$
1999–2000	$0.82 - 1.60 = -0.78$
2000–2001	$0.92 - 0.82 = 0.1$
2001–2002	$0.08 - 0.92 = -0.84$

Greatest increase: 1998–1999   (0.86)

Greatest decrease: 2001–2002   (−0.84)

(b) $(2, 0.58)$, $(12, 0.08)$:

$$y - 0.58 = \frac{0.08 - 0.58}{12 - 2}(x - 2)$$

$$y = -0.05(x - 2) + 0.58$$

$$y = -0.05x + 0.68$$

(c) Between 1992 and 2002, the earnings per share decreased at a rate of 0.05 per year.

(d) For 2006, $x = 16$ and $y = -0.05(16) + 0.68 = -0.12$, which is probably reasonable.

69. $\dfrac{\text{rise}}{\text{run}} = \dfrac{3}{4} = \dfrac{x}{\frac{1}{2}(32)}$

$$\frac{3}{4} = \frac{x}{16}$$

$$4x = 48$$

$$x = 12$$

The maximum height in the attic is 12 feet.

71. $(4, 2540), m = 125$

$$V - 2540 = 125(t - 4)$$

$$V - 2540 = 125t - 500$$

$$V = 125t + 2040$$

73. $(4, 20400), m = -2000$

$$V - 20,400 = -2000(t - 4)$$

$$V - 20,400 = -2000t + 8000$$

$$V = -2000t + 28,400$$

75. The slope is $m = -10$. This represents the decrease in the amount of the loan each week.

Matches graph (b).

77. The slope is $m = 0.35$ This represents the increase in travel cost for each mile driven.

Matches graph (a).

79. Using the points $(0, 32)$ and $(100, 212)$, we have

$$m = \frac{212 - 32}{100 - 0} = \frac{180}{100} = \frac{9}{5}$$

$$F - 32 = \frac{9}{5}(C - 0)$$

$$F = \frac{9}{5}C + 32.$$

81. (a) Using the points $(0, 875)$ and $(5, 0)$, where the first coordinate represents the year t and the second coordinate represents the value V, we have

$$m = \frac{0 - 875}{5 - 0} = -175$$

$$V = -175t + 875, \ 0 \le t \le 5.$$

(c) $t = 0$: $V = -175(0) + 875 = 875$

$t = 1$: $V = -175(1) + 875 = 700$

$t = 2$: $V = -175(2) + 875 = 525$

$t = 3$: $V = -175(3) + 875 = 350$

$t = 4$: $V = -175(4) + 875 = 175$

$t = 5$: $V = -175(5) + 875 = 0$

(b)

t	0	1	2	3	4	5
V	875	700	525	350	175	0

83. (a) $C = 36,500 + 5.25t + 11.50t$

$= 16.75t + 36,500$

(c) $P = R - C$

$= 27t - (16.75t + 36,500)$

$= 10.25t - 36,500$

(b) $R = 27t$

(d) $\quad 0 = 10.25t - 36,500$

$36,500 = 10.25t$

$t \approx 3561$ hours

85. (a) $\dfrac{83,038 - 75,365}{2002 - 1990} = \dfrac{7673}{12} \approx 639$ student/year

(b) 1984: $75,365 - 6(639) \approx 71,531$ students

1997: $75,365 + 7(639) \approx 79,838$ students

2000: $75,365 + 10(639) \approx 81,755$ students

(Answer could vary slightly)

(c) Let $t = 0$ represent 1990.

$(0, 75,365), (12, 83,038)$

$$y - 75,365 = \frac{83,038 - 75,365}{12 - 0} \ t - 0$$

$$y = \frac{7673}{12}t + 75,365 \approx 639t + 75,365$$

The slope is the annual increase in students. It is positive, indicating that Penn State University increased its students from 1990 to 2002.

87. False. The slopes are different:

$$\frac{4 - 2}{-1 + 8} = \frac{2}{7}$$

$$\frac{7 + 4}{-7 - 0} = -\frac{11}{7}$$

89.
$$\frac{x}{5} + \frac{y}{-3} = 1$$

$$-3x + 5y + 15 = 0$$

$a = 5$ and $b = -3$ are the x- and y-intercepts.

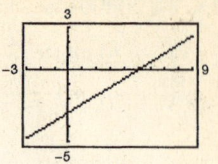

91.
$$\frac{x}{2} + \frac{y}{3} = 1$$

$$3x + 2y - 6 = 0$$

93.
$$\frac{x}{-\frac{1}{6}} + \frac{y}{-\frac{2}{3}} = 1$$

$$-6x - \frac{3}{2}y = 1$$

$$12x + 3y + 2 = 0$$

95. The line with slope -3 is steeper.

97. One way is to calculate the lengths of the sides.

$$d(A, B) = \sqrt{(2 - 2)^2 + (9 - 3)^2} = 6$$

$$d(B, C) = \sqrt{(7 - 2)^2 + (3 - 9)^2} = \sqrt{25 + 36} = \sqrt{61}$$

$$d(A, C) = \sqrt{(7 - 2)^2 + (3 - 3)^2} = 5$$

Then, $[d(A, B)]^2 + [d(A, C)]^2 = [d(B, C)]^2$, and the triangle is a right triangle.

Another way is to calculate the slopes of the lines joining AB and AC.

slope of AB: undefined   (vertical line)

slope of AC: 0   (horizontal line)

99. Yes. $x + 20$

101. No. The term $x^{-1} = \frac{1}{x}$ causes the expression to not be a polynomial.

103. No. This expression is not defined for $x = \pm 3$.

105. $x^2 - 6x - 27 = (x - 9)(x + 3)$

107. $2x^2 + 11x - 40 = (2x - 5)(x + 8)$

Section 1.2   Functions

> ■ Given a set or an equation, you should be able to determine if it represents a function.
> ■ Given a function, you should be able to do the following.
>   (a) Find the domain.
>   (b) Evaluate it at specific values.

Solutions to Odd-Numbered Exercises

1. Yes, it does represent a function. Each domain value is matched with only one range value.

3. No, it does not represent a function. The domain values are each matched with three range values.

5. Yes, it does represent a function. Each input value is matched with only one output value.

7. No, it does not represent a function. The input values of 10 and 7 are each matched with two output values.

9. (a) Each element of A is matched with exactly one element of B, so it does represent a function.

   (b) The element 1 in A is matched with two elements, -2 and 1 of B, so it does not represent a function.

   (c) Each element of A is matched with exactly one element of B, so it does represent a function.

   (d) The element 2 of A is not matched to any element of B, so it does not represent a function.

11. Each are functions. For each year there corresponds one and only one circulation.

13. $x^2 + y^2 = 4 \implies y = \pm\sqrt{4 - x^2}$

   Thus, y *is not* a function of x. For instance, the values $y = 2$ and -2 both correspond to $x = 0$.

15. $x^2 + y = -1$

   $$y = -x^2 - 1$$

   Thus, y *is* a function of x.

17. $2x + 3y = 4 \implies y = \frac{1}{3}(4 - 2x)$

   Thus, y *is* a function of x.

19. $y^2 = x^2 - 1 \implies y = \pm\sqrt{x^2 - 1}$

   Thus, y *is not* a function of x. For instance, the values $y = \sqrt{3}$ and $-\sqrt{3}$ both correspond to $x = 2$.

21. $y = |4 - x|$

   This is a function of x.

23. $x = -7$ does not represent y as a function of x. All values of y correspond to $x = -7$.

25. $f(x) = \dfrac{1}{x + 1}$

   (a) $f(4) = \dfrac{1}{(4) + 1} = \dfrac{1}{5}$

   (b) $f(0) = \dfrac{1}{(0) + 1} = 1$

   (c) $f(4t) = \dfrac{1}{(4t) + 1} = \dfrac{1}{4t + 1}$

   (d) $f(x + c) = \dfrac{1}{(x + c) + 1} = \dfrac{1}{x + c + 1}$

27. $f(x) = 2x - 3$

   (a) $f(1) = 2(1) - 3 = -1$

   (b) $f(-3) = 2(-3) - 3 = -9$

   (c) $f(x - 1) = 2(x - 1) - 3 = 2x - 5$

29. $h(t) = t^2 - 2t$

   (a) $h(2) = 2^2 - 2(2) = 0$

   (b) $h(1.5) = (1.5)^2 - 2(1.5) = -0.75$

   (c) $h(x + 2) = (x + 2)^2 - 2(x + 2) = x^2 + 2x$

31. $f(y) = 3 - \sqrt{y}$

   (a) $f(4) = 3 - \sqrt{4} = 1$

   (b) $f(0.25) = 3 - \sqrt{0.25} = 2.5$

   (c) $f(4x^2) = 3 - \sqrt{4x^2} = 3 - 2|x|$

33. $q(x) = \dfrac{1}{x^2 - 9}$

   (a) $q(0) = \dfrac{1}{0^2 - 9} = -\dfrac{1}{9}$

   (b) $q(3) = \dfrac{1}{3^2 - 9}$ is undefined.

   (c) $q(y + 3) = \dfrac{1}{(y + 3)^2 - 9} = \dfrac{1}{y^2 + 6y}$

35. $f(x) = \dfrac{|x|}{x}$

(a) $f(2) = \dfrac{|2|}{2} = 1$

(b) $f(-2) = \dfrac{|-2|}{-2} = -1$

(c) $f(x^2) = \dfrac{|x^2|}{x^2} = 1, \ x \neq 0$

37. $f(x) = \begin{cases} 2x + 1, & x < 0 \\ 2x + 2, & x \geq 0 \end{cases}$

(a) $f(-1) = 2(-1) + 1 = -1$

(b) $f(0) = 2(0) + 2 = 2$

(c) $f(2) = 2(2) + 2 = 6$

39. $h(t) = \frac{1}{2}|t + 3|$

t	-5	-4	-3	-2	-1
$h(t)$	1	$\frac{1}{2}$	0	$\frac{1}{2}$	1

41. $f(x) = \begin{cases} -\frac{1}{2}x + 4, & x \leq 0 \\ (x - 2)^2, & x > 0 \end{cases}$

x	-2	-1	0	1	2
$f(x)$	5	$\frac{9}{2}$	4	1	0

43. $f(x) = 15 - 3x = 0$

$\qquad 3x = 15$

$\qquad\quad x = 5$

45. $f(x) = \dfrac{3x - 4}{5} = 0$

$\quad 3x - 4 = 0$

$\qquad 3x = 4$

$\qquad\ x = \frac{4}{3}$

47. $\qquad\quad f(x) = g(x)$

$\qquad\quad x^2 = x + 2$

$\quad x^2 - x - 2 = 0$

$(x + 1)(x - 2) = 0$

$x = -1 \ \text{ or } \ x = 2$

49. $f(x) = 5x^2 + 2x - 1$

Since $f(x)$ is a polynomial, the domain is all real numbers x.

51. $h(t) = \dfrac{4}{t}$

Domain: All real numbers except $t = 0$

53. $f(x) = \sqrt[3]{x - 4}$

Domain: all real numbers

55. $g(x) = \dfrac{1}{x} - \dfrac{3}{x + 2}$

Domain: All real numbers except

$x = 0, \ x = -2$

57. $g(y) = \dfrac{y + 2}{\sqrt{y - 10}}$

$y - 10 > 0$

$\qquad y > 10$

Domain: all $y > 10$.

59. $f(x) = \sqrt{4 - x^2}$

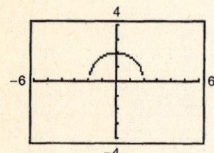

Domain: $[-2, 2]$

Range: $[0, 2]$

61. $g(x) = |2x + 3|$

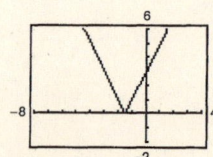

Domain: $(-\infty, \infty)$

Range: $[0, \infty)$

63. $f(x) = x^2$

$\{(-2, 4), (-1, 1), (0, 0), (1, 1), (2, 4)\}$

65. $f(x) = |x| + 2$

$\{(-2, 4), (-1, 3), (0, 2), (1, 3), (2, 4)\}$

67. $A = \pi r^2, \quad C = 2\pi r$

$r = \dfrac{C}{2\pi}$

$A = \pi\left(\dfrac{C}{2\pi}\right)^2 = \dfrac{C^2}{4\pi}$

69. (a) According to the table, the maximum profit is 3375 for $x = 150$

(b) 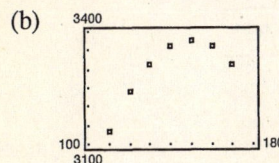   Yes, P is a function of x.

(c) Profit = Revenue − Cost

$\quad = $ (price per unit)(number of units) − (cost)(number of units)

$\quad = [90 - (x - 100)(0.15)]x - 60x$

$\quad = (105 - 0.15x)x - 60x$

$\quad = 45x - 0.15x^2, \quad x > 100$

$$P = \begin{cases} 30x, & x \le 100 \\ 45x - 0.15x^2, & x > 100 \end{cases}$$

71. $A = \frac{1}{2}(\text{base})(\text{height}) = \frac{1}{2}xy.$

Since $(0, y)$, $(2, 1)$ and $(x, 0)$ all lie on the same line, the slopes between any pair of points are equal.

$\dfrac{1 - y}{2 - 0} = \dfrac{1 - 0}{2 - x}$

$1 - y = \dfrac{2}{2 - x}$

$y = 1 - \dfrac{2}{2 - x} = \dfrac{x}{x - 2}$

Therefore, $A = \dfrac{1}{2}xy = \dfrac{1}{2}x\left(\dfrac{x}{x - 2}\right) = \dfrac{x^2}{2x - 4}$

The domain is $x > 2$, since $A > 0$.

73. (a) $V = (\text{length})(\text{width})(\text{height}) = yx^2$

But, $y + 4x = 108$, or $y = 108 - 4x$.

Thus, $V = (108 - 4x)x^2$.

(b) Since $y = 108 - 4x > 0$

$\qquad\qquad 4x < 108$

$\qquad\qquad x < 27$

Domain: $0 < x < 27$

(c)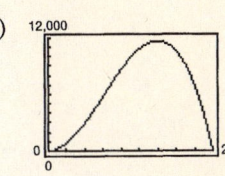

(d) The highest point on the graph occurs at $x = 18$. The dimensions that maximize the volume are $18 \times 18 \times 36$ inches.

75. The domain of $-1.97x + 26.3$ is $7 \le x \le 12$.

The domain of $0.505x^2 - 1.47x + 6.3$ is $1 \le x \le 6$.

You can tell by comparing the models to the given data. The models fit the data well on the domains above.

77. $f(11) = -1.97(11) + 26.3 = 4.63$

\$4,630 in monthly revenue for November.

79. $n(t) = \begin{cases} -9.2t^2 + 84.5t + 575, & 0 \le t \le 4 \\ 26.8t + 657, & 5 \le t \le 10 \end{cases}$

$t = 0$ corresponds to 1990

t	0	1	2	3	4	5	6	7	8	9	10
$n(t)$ (in billions)	575	650	707	746	766	791	818	845	871	898	925

81. (a) $F(y) = 149.76\sqrt{10}\, y^{5/2}$

y	5	10	20	30	40
$F(y)$	2.65×10^4	1.50×10^5	8.47×10^5	2.33×10^6	4.79×10^6

(Answers will vary.)

F increases very rapidly as y increases.

(b)

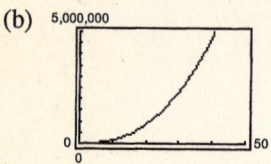

(c) From the table, $y \approx 22$ ft (slightly above 20). You could obtain a better approximation by completing the table for values of y between 20 and 30.

(d) By graphing $F(y)$ together with the horizontal line $y_2 = 1,000,000$, you obtain $y \approx 21.37$ feet.

83. $f(x) = 2x$

$$\frac{f(x + c) - f(x)}{c} = \frac{2(x + c) - 2x}{c}$$

$$= \frac{2c}{c} = 2, \quad c \ne 0$$

85. $f(x) = x^2 - x + 1, \quad f(2) = 3$

$$\frac{f(2 + h) - f(2)}{h} = \frac{(2 + h)^2 - (2 + h) + 1 - 3}{h}$$

$$= \frac{4 + 4h + h^2 - 2 - h + 1 - 3}{h}$$

$$= \frac{h^2 + 3h}{h} = h + 3, \quad h \ne 0$$

87. $f(t) = \dfrac{1}{t}, \quad f(1) = 1$

$$\frac{f(t) - f(1)}{t - 1} = \frac{\dfrac{1}{t} - 1}{t - 1} = \frac{1 - t}{t(t - 1)} = \frac{-1}{t}, \quad t \ne 1$$

89. False. The range of $f(x)$ is $[-1, \infty)$.

91. Since the function is undefined at 0, we have $r(x) = \dfrac{c}{x}$. Since $(-4, -8)$ is on the graph, we have

$$-8 = \frac{c}{-4} \implies c = 32. \text{ Thus, } r(x) = \frac{32}{x}.$$

93. The domain is the set of inputs of the function and the range is the set of corresponding outputs.

95. $12 - \dfrac{4}{x+2} = \dfrac{12(x+2) - 4}{x+2} = \dfrac{12x + 20}{x+2}$

97. $\dfrac{2x^3 + 11x^2 - 6x}{5x} \cdot \dfrac{x + 10}{2x^2 + 5x - 3} = \dfrac{x(2x^2 + 11x - 6)(x + 10)}{5x(2x - 1)(x + 3)}$

$\qquad\qquad\qquad\qquad\qquad\qquad\quad = \dfrac{(2x - 1)(x + 6)(x + 10)}{5(2x - 1)(x + 3)}$

$\qquad\qquad\qquad\qquad\qquad\qquad\quad = \dfrac{(x + 6)(x + 10)}{5(x + 3)}, x \neq 0, \dfrac{1}{2}$

Section 1.3   Graphs of Functions

■ You should be able to determine the domain and range of a function from its graph.

■ You should be able to use the vertical line test for functions.

■ You should be able to determine when a function is constant, increasing, or decreasing.

■ You should be able to find relative maximum and minimum values of a function.

■ You should know that f is

(a) Odd if $f(-x) = -f(x)$.

(b) Even if $f(-x) = f(x)$.

Solutions to Odd-Numbered Exercises

1. $f(x) = 1 - x^2$

Domain: All real numbers

Range: $(-\infty, 1]$

$f(0) = 1$

3. $f(x) = \sqrt{16 - x^2}$

Domain: $[-4, 4]$

Range: $[0, 4]$

$f(0) = 4$

5. $f(x) = 2x^2 + 3$

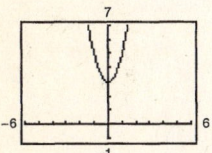

Domain: All real numbers

Range: $[3, \infty)$

7. $f(x) = \sqrt{x - 1}$

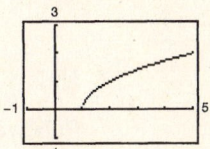

Domain: $x - 1 \geq 0 \Rightarrow x \geq 1$
or $[1, \infty)$

Range: $[0, \infty)$

9. $f(x) = |x + 3|$

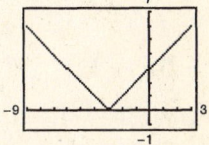

Domain: All real numbers

Range: $[0, \infty)$

11. $y = \frac{1}{2}x^2$

A vertical line intersects the graph just once, so y is a function of x.

13. $x - y^2 = 1 \Rightarrow y = \pm\sqrt{x - 1}$

y is not a function of x. Graph

$y_1 = \sqrt{x - 1}$ and $y_2 = -\sqrt{x - 1}$.

15. $x^2 = 2xy - 1$

A vertical line intersects the graph just once, so y is a function of x. Solve for y and graph

$$y = \frac{x^2 + 1}{2x}.$$

17. $f(x) = \frac{3}{2}x$

f is increasing on $(-\infty, \infty)$.

19. $f(x) = x^3 - 3x^2 + 2$

f is increasing on $(-\infty, 0)$ and $(2, \infty)$.

f is decreasing on $(0, 2)$.

21. $f(x) = 3$

(a)

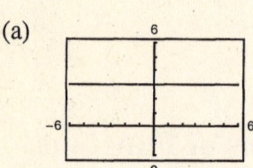

(b) f is constant on $(-\infty, \infty)$

23. $f(x) = x^{2/3}$

(a)

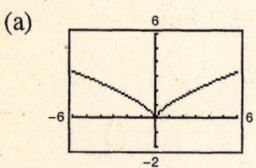

(b) Increasing on $(0, \infty)$

Decreasing on $(-\infty, 0)$

25. $f(x) = x\sqrt{x + 3}$

(a)

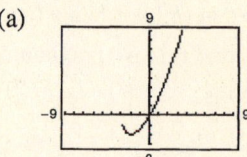

(b) Increasing on $(-2, \infty)$

Decreasing on $(-3, -2)$

27. $f(x) = |x + 1| + |x - 1|$

(a)

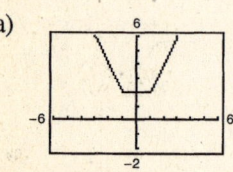

(b) Increasing on $(1, \infty)$, constant on $(-1, 1)$, decreasing on $(-\infty, -1)$

29. $f(x) = x^2 - 6x$

Relative minimum: $(3, -9)$

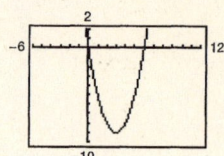

31. $y = 2x^3 + 3x^2 - 12x$

Relative minimum: $(1, -7)$

Relative maximum: $(-2, 20)$

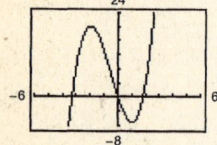

33. $h(x) = (x - 1)\sqrt{x}$

Relative minimum: $(0.33, -0.38)$

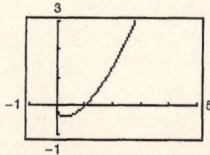

$(0, 0)$ is not a relative maximum because it occurs at the endpoint of the domain $[0, \infty)$.)

35. $f(x) = x^2 - 4x - 5$

(a)

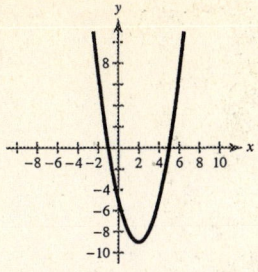

Minimum: $(2, -9)$

(b)

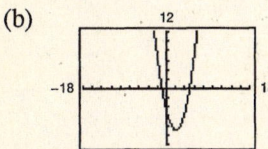

Minimum: $(2, -9)$

(c) Answers are the same

37. $f(x) = x^3 - 8x$

(a)

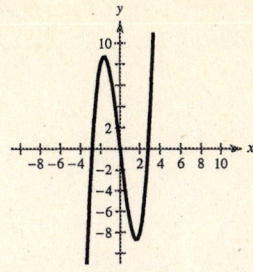

Maximum: Approximately $(-2, 9)$

Minimum: Approximately $(2, -9)$

(b)

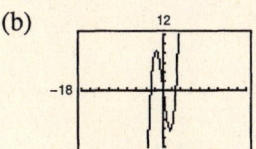

Maximum: $(-1.63, 8.71)$

Minimum: $(1.63, -8.71)$

(c) The answers are similar.

39. $f(x) = (x - 4)^{2/3}$

(a)

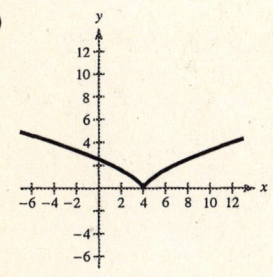

Minimum: $(4, 0)$

(b)

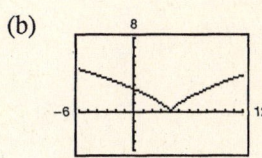

Minimum: $(4, 0)$

(c) The answers are the same.

41. $f(x) = \begin{cases} 2x + 3, & x < 0 \\ 3 - x, & x \geq 0 \end{cases}$

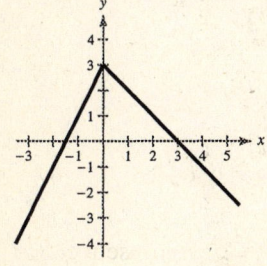

43. $f(x) = \begin{cases} \sqrt{x + 4}, & x < 0 \\ \sqrt{4 - x}, & x \geq 0 \end{cases}$

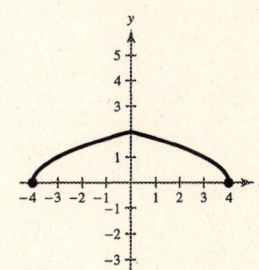

45. $f(x) = \begin{cases} x + 3, & x \le 0 \\ 3, & 0 < x \le 2 \\ 2x - 1, & x > 2 \end{cases}$

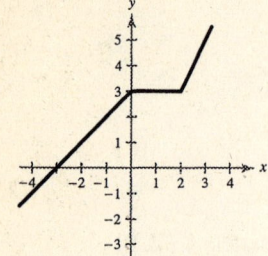

47. $f(x) = \begin{cases} 2x + 1, & x \le -1 \\ x^2 - 2, & x > -1 \end{cases}$

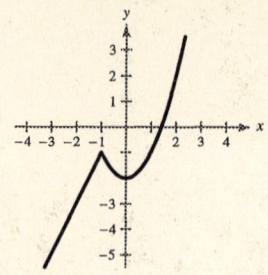

49. $f(-t) = (-t)^2 + 2(-t) - 3$

$= t^2 - 2t - 3$

$\ne f(t) \ne -f(t)$

f is neither even nor odd.

51. $g(-x) = (-x)^3 - 5(-x)$

$= -x^3 + 5x$

$= -g(x)$

g is odd.

53. $f(-x) = (-x)\sqrt{1 - (-x)^2} = -x\sqrt{1 - x^2} = -f(x)$

The function is odd.

55. $g(-s) = 4(-s)^{2/3} = 4s^{2/3} = g(s)$

The function is even.

57. $\left(-\frac{3}{2}, 4\right)$

(a) If f is even, another point is $\left(\frac{3}{2}, 4\right)$.

(b) If f is odd, another point is $\left(\frac{3}{2}, -4\right)$.

59. $(4, 9)$

(a) If f is even, another point is $(-4, 9)$.

(b) If f is odd, another point is $(-4, -9)$.

61. $(x, -y)$

(a) If f is even, another point is $(-x, -y)$.

(b) If f is odd, another point is $(-x, y)$.

63. $f(x) = 5$, even

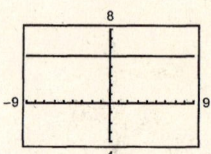

65. $f(x) = 3x - 2$, neither even nor odd

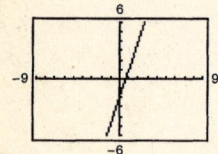

67. $h(x) = x^2 - 4$, even

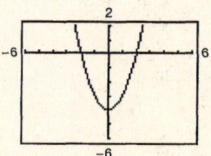

69. $f(x) = \sqrt{1 - x}$, neither even nor odd

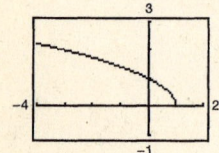

71. $f(x) = |x + 2|$, neither even nor odd

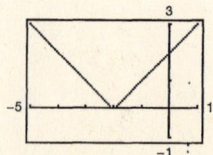

73. $f(x) = 4 - x \geq 0$

$4 \geq x$

$(-\infty, 4]$

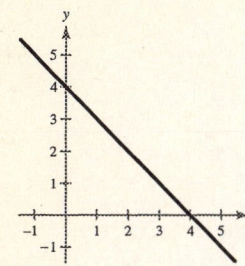

75. $f(x) = x^2 - 9 \geq 0$

$x^2 \geq 9$

$x \geq 3$ or $x \leq -3$

$[3, \infty)$ or $(-\infty, -3]$

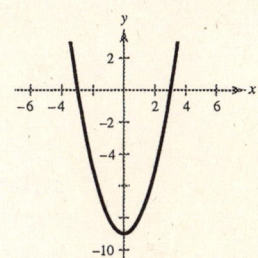

77. $s(x) = 2\left(\frac{1}{4}x - \left[\!\left[\frac{1}{4}x\right]\!\right]\right)$

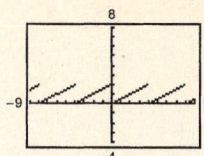

Domain: $(-\infty, \infty)$

Range: $[0, 2)$

Sawtooth pattern

79. (a) Let x and y be the length and width of the rectangle. Then $100 = 2x + 2y$ or $y = 50 - x$. Thus, the area is $A = xy = x(50 - x)$.

(b)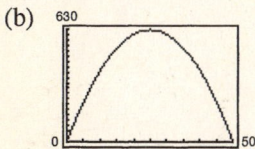

(c) The maximum area is 625 m² when $x = y = 25$ m. That is, the rectangle is a square.

81. (a) The second model is correct. For instance,

$$C_2\left(\tfrac{1}{2}\right) = 1.05 - 0.38\left[\!\left[-\left(\tfrac{1}{2} - 1\right)\right]\!\right]$$

$$= 1.05 - 0.38\left[\!\left[\tfrac{1}{2}\right]\!\right] = 1.05.$$

(b)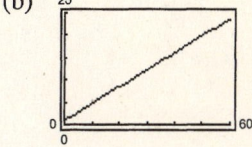

The cost of an 18-minute 45-second call is

$$C_2\left(18\tfrac{45}{60}\right)$$

$$= C_2(18.75) = 1.05 - 0.38[[-(18.75 - 1)]]$$

$$= 1.05 - 0.38[[-17.75]] = 1.05 - 0.38(-18)$$

$$= 1.05 + 0.38(18) = \$7.89$$

83. $h = \text{top} - \text{bottom}$

$= (-x^2 + 4x - 1) - 2$

$= -x^2 + 4x - 3, 1 \leq x \leq 3$

85. $L = \text{right} - \text{left}$

$= \tfrac{1}{2}y^2 - 0$

$= \tfrac{1}{2}y^2, 0 \leq y \leq 4$

87. $P = -0.76t^2 + 9.9t + 618$, $5 \leq t \leq 11$ where $t = 5$ corresponds to 1995

(a)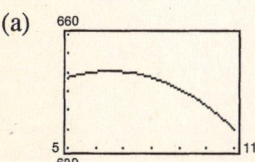

(b) Increasing on $(5, 6.5)$ or 1995 to middle of 1996

Decreases on $(6.5, 11)$ or middle of 1996 to 2001

(c) Maximum population was 650,200 for $t \approx 6.5$.

89. False. The domain of $f(x) = \sqrt{x^2}$ is the set of all real numbers.

91. $f(x) = a_{2n+1}x^{2n+1} + a_{2n-1}x^{2n-1} + \cdots + a_3x^3 + a_1x$

$f(-x) = a_{2n+1}(-x)^{2n+1} + a_{2n-1}(-x)^{2n-1} + \cdots + a_3(-x)^3 + a_1(-x)$

$\qquad = -a_{2n+1}x^{2n+1} - a_{2n-1}x^{2n-1} - \cdots - a_3x^3 - a_1x = -f(x)$

Therefore, $f(x)$ is odd.

93. f is an even function.

(a) $g(x) = -f(x)$ is even because
$g(-x) = -f(-x) = -f(x) = g(x).$

(b) $g(x) = f(-x)$ is even because
$g(-x) = f(-(-x)) = f(x) = f(-x) = g(x).$

(c) $g(x) = f(x) - 2$ is even because
$g(-x) = f(-x) - 2 = f(x) - 2 = g(x).$

(d) $g(x) = -f(x - 2)$ is neither even nor odd because
$g(-x) = -f(-x - 2) = -f(x + 2) \neq g(x)$ nor
$-g(x).$

95. No, $x^2 + y^2 = 25$ does not represent x as a function of y. For instance, $(-3, 4)$ and $(3, 4)$ both lie on the graph.

97. $-2x^2 + 8x$

Terms: $-2x^2, 8x$

Coefficients: $-2, 8$

99. $\dfrac{x}{3} - 5x^2 + x^3$

Terms: $\dfrac{x}{3}, -5x^2, x^3$

Coefficients: $\dfrac{1}{3}, -5, 1$

101. (a) $d = \sqrt{(6 - (-2))^2 + (3 - 7)^2}$

$\qquad = \sqrt{64 + 16} = \sqrt{80} = 4\sqrt{5}$

(b) midpoint $= \left(\dfrac{-2 + 6}{2}, \dfrac{7 + 3}{2}\right) = (2, 5)$

103. (a) $d = \sqrt{\left(-\dfrac{3}{2} - \dfrac{5}{2}\right)^2 + (4 - (-1))^2}$

$\qquad = \sqrt{16 + 25} = \sqrt{41}$

(b) midpoint $= \left(\dfrac{\dfrac{5}{2} - \dfrac{3}{2}}{2}, \dfrac{-1 + 4}{2}\right) = \left(\dfrac{1}{2}, \dfrac{3}{2}\right)$

105. $f(x) = 5x - 1$

(a) $f(6) = 5(6) - 1 = 29$

(b) $f(-1) = 5(-1) - 1 = -6$

(c) $f(x - 3) = 5(x - 3) - 1 = 5x - 16$

107. $f(x) = x\sqrt{x - 3}$

(a) $f(3) = 3\sqrt{3 - 3} = 0$

(b) $f(12) = 12\sqrt{12 - 3}$

$\qquad = 12\sqrt{9} = 12(3) = 36$

(c) $f(6) = 6\sqrt{6 - 3} = 6\sqrt{3}$

109. $f(x) = x^2 - 2x + 9$

$f(3 + h) = (3 + h)^2 - 2(3 + h) + 9 = 9 + 6h + h^2 - 6 - 2h + 9$

$\qquad\qquad = h^2 + 4h + 12$

$f(3) = 3^2 - 2(3) + 9 = 12$

$\dfrac{f(3 + h) - f(3)}{h} = \dfrac{(h^2 + 4h + 12) - 12}{h} = \dfrac{h(h + 4)}{h} = h + 4, \, h \neq 0$

Section 1.4   Shifting, Reflecting, and Stretching Graphs

■ You should know the graphs of the most commonly used functions in algebra, and be able to reproduce them on your graphing utility.

    (a) Constant function: $f(x) = c$ (b) Identity function: $f(x) = x$

    (c) Absolute value function: $f(x) = |x|$ (d) Square root function: $f(x) = \sqrt{x}$

    (e) Squaring function: $f(x) = x^2$ (f) Cubing function: $f(x) = x^3$

■ You should know how the graph of a function is changed by vertical and horizontal shifts.

■ You should know how the graph of a function is changed by reflection.

■ You should know how the graph of a function is changed by nonrigid transformations, like stretches and shrinks.

■ You should know how the graph of a function is changed by a sequence of transformations.

Solutions to Odd-Numbered Exercises

1.
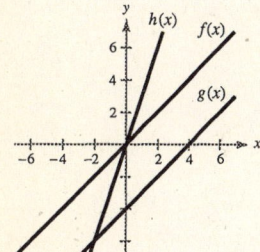

3.
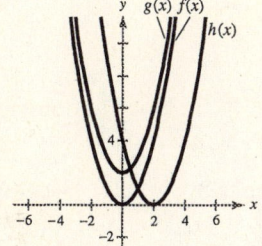

5.
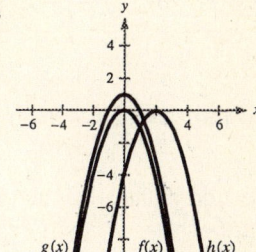

7.
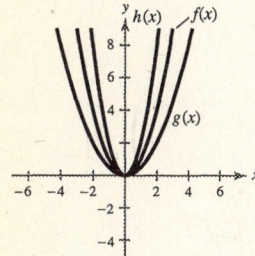

9.
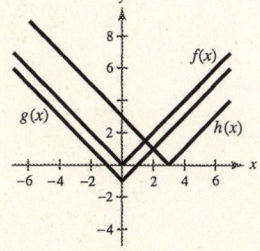

11.
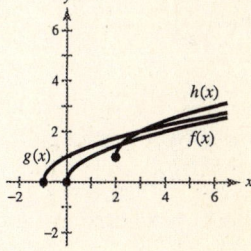

13. (a) $y = f(x) + 2$ (b) $y = -f(x)$ (c) $y = f(x - 2)$

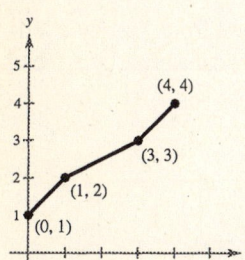
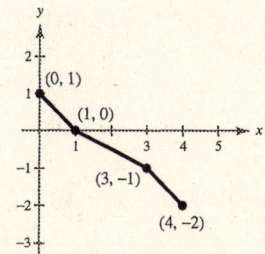
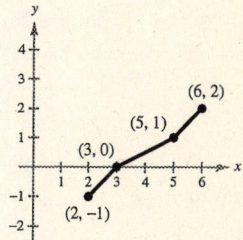

—CONTINUED—

13. —**CONTINUED**—

(d) $y = f(x + 3)$

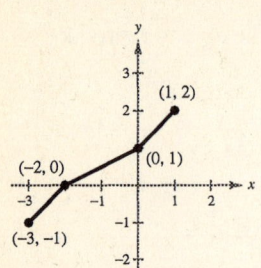

(e) $y = 2f(x)$

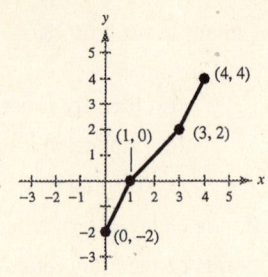

(f) $y = f(-x)$

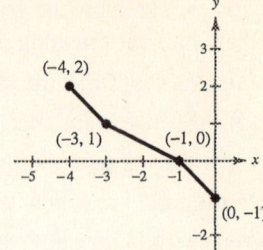

(g) Let $g(x)$ and $f\left(\frac{1}{2}x\right)$. Then from the graph,

$$g(0) = f\left(\frac{1}{2}(0)\right) = f(0) = -1$$

$$g(2) = f\left(\frac{1}{2}(2)\right) = f(1) = 0$$

$$g(6) = f\left(\frac{1}{2}(6)\right) = f(3) = 1$$

$$g(8) = f\left(\frac{1}{2}(8)\right) = f(4) = 2$$

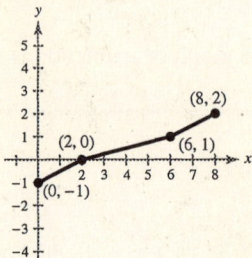

15. Vertical shrink of $y = x$: $y = \frac{1}{2}x$

17. Constant function: $y = 7$

19. Reflection in the x-axis and a vertical shift one unit upward of $y = \sqrt{x}$: $y = 1 - \sqrt{x}$

21. Horizontal shift of $y = |x|$: $y = |x + 2|$

23. Vertical shift one unit downward of $y = x^2$

$$y = x^2 - 1$$

25. Reflection in the x-axis and a vertical shift one unit upward of $y = x^3$: $y = 1 - x^3$

27. $y = -\sqrt{x} - 1$ is $f(x)$ reflected in the x-axis, followed by a vertical shift 1 unit downward.

29. $y = \sqrt{x - 2}$ is $f(x)$ shifted right two units.

31. $y = \sqrt{2x}$ is a horizontal shrink of $f(x)$ by 2.

33. $y = |x + 5|$ is $f(x)$ shifted left five units.

35. $y = -|x|$ is $f(x)$ reflected in the x-axis.

37. $y = 4|x|$ is a vertical stretch of $f(x)$.

39. $g(x) = 4 - x^3$ is obtained from $f(x)$ by a reflection in the x-axis followed by a vertical shift upward of four units.

41. $h(x) = \frac{1}{4}(x + 2)^3$ is obtained from $f(x)$ by a left shift of two units and a vertical shrink by a factor of $\frac{1}{4}$.

43. $p(x) = \left(\frac{1}{3}x\right)^3 + 2$ is obtained from $f(x)$ by a horizontal stretch followed by a vertical shift 2 units upward.

45. $f(x) = x^3 - 3x^2$

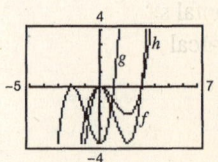

$g(x) = f(x + 2) = (x + 2)^3 - 3(x + 2)^2$ is a horizontal shift 2 units to left

$h(x) = \frac{1}{2}f(x) = \frac{1}{2}(x^3 - 3x^2)$ is a vertical shrink.

47. $f(x) = x^3 - 3x^2$

$g(x) = -\frac{1}{3}f(x) = -\frac{1}{3}(x^3 - 3x^2)$ reflection in the *x*-axis and vertical shrink

$h(x) = f(-x) = (-x)^3 - 3(-x)^2$ reflection in the *y*-axis

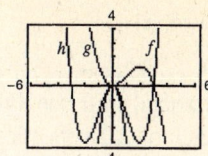

49. The graph of *g* is obtained from that of *f* by first negating *f*, and then shifting vertically one unit upward: $g(x) = -x^3 + 3x^2 + 1$.

51. (a) $f(x) = x^2$

(b) $g(x) = 2 - (x + 5)^2$ is obtained from *f* by a horizontal shift to the left 5 units, a reflection in the *x*-axis, and a vertical shift upward 2 units.

(c)

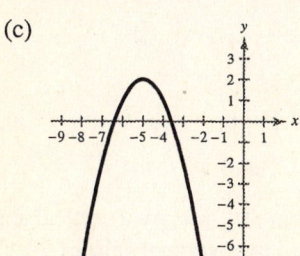

(d) $g(x) = 2 - f(x + 5)$

53. (a) $f(x) = x^2$

(b) $g(x) = 3 + 2(x - 4)^2$ is obtained from *f* by a horizontal shift 4 units to the right, a vertical stretch of 2, and a vertical shift upward 3 units.

(c)

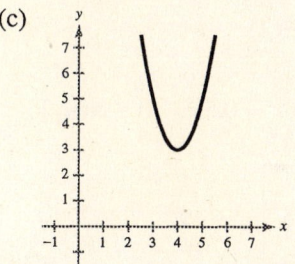

(d) $g(x) = 3 + 2f(x - 4)$

55. (a) $f(x) = x^3$

(b) $g(x) = 3(x - 2)^3$ is obtained from *f* by a horizontal shift 2 units to the right followed by a vertical stretch of 3.

(c)

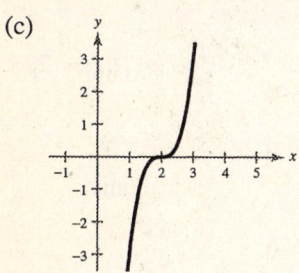

(d) $g(x) = 3f(x - 2)$

57. (a) $f(x) = x^3$

(b) $g(x) = (x - 1)^3 + 2$ is obtained from *f* by a horizontal shift 1 unit to the right, and a vertical shift upward 2 units.

(c)

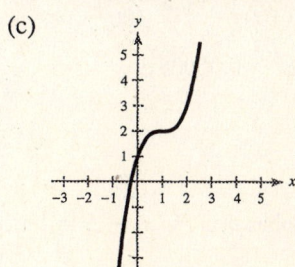

(d) $g(x) = f(x - 1) + 2$

59. (a) $f(x) = |x|$

(b) $g(x) = |x + 4| + 8$ is obtained from *f* by a horizontal shift 4 units to the left, followed by a vertical shift 8 units upward.

(c)

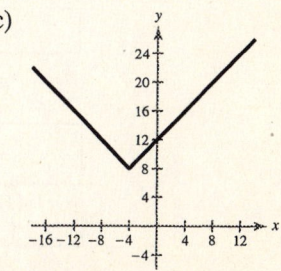

(d) $g(x) = f(x + 4) + 8$

61. (a) $f(x) = |x|$

(b) $g(x) = -2|x - 1| - 4$ is obtained from f by a horizontal shift one unit to the right, a vertical stretch of 2, a reflection in the x-axis, and a vertical shift downward 4 units.

(c)

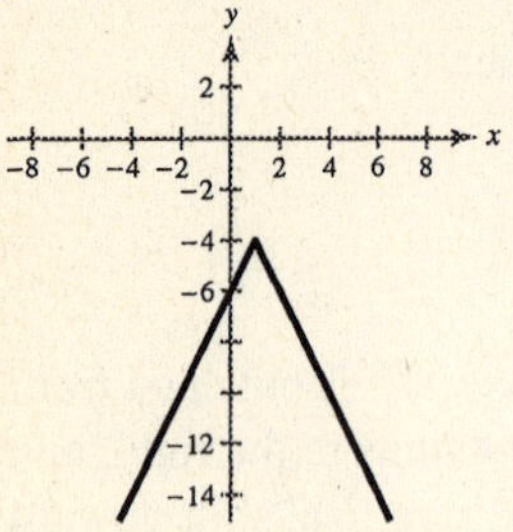

(d) $g(x) = -2f(x - 1) - 4$

63. (a) $f(x) = \sqrt{x}$

(b) $g(x) = -\frac{1}{2}\sqrt{x + 3} - 1$ is obtained from f by a horizontal shift 3 units to the left, a vertical shrink, a reflection in the x-axis, and a vertical shift 1 unit downward.

(c)

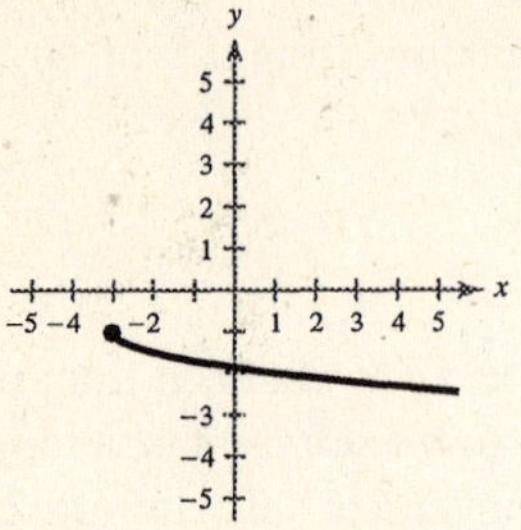

(d) $g(x) = -\frac{1}{2}f(x + 3) - 1$

65. (a) $P(x) = 80 + 20x - 0.5x^2, \quad 0 \le x \le 20$

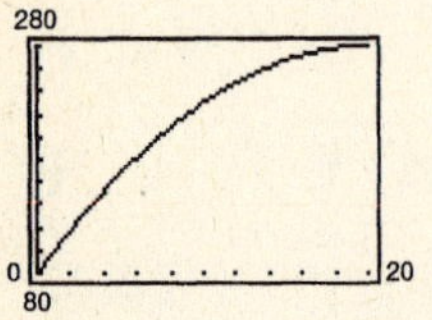

(b) $P(x)$ is shifted downward by a vertical shift of 25.

$P(x) = 55 + 20x - 0.5x^2, \quad 0 \le x \le 20$

(c) $P(x)$ is changed by a horizontal stretch.

$$P(x) = 80 + 20\left(\frac{x}{100}\right) - 0.5\left(\frac{x}{100}\right)^2$$

$$= 80 + 0.2x - 0.00005x^2$$

67. $F(t) = 0.036t^2 + 20.1, \quad 0 \le t \le 20$

$t = 0$ corresponds to 1980

(a) F is obtained from $f(t) = t^2$ by a vertical shrink of 0.036 followed by a vertical shift of 20.1 units upward

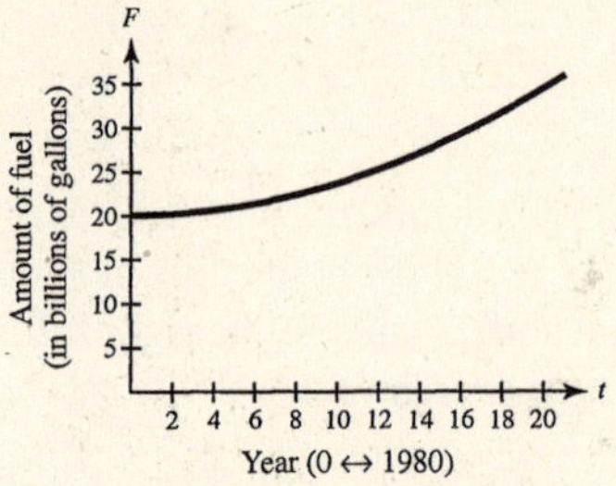

(b) $G(t) = F(t + 10) = 0.036(t + 10)^2 + 20.1$
$-10 \le t \le 10$

$G(0) = F(10)$ corresponds to 1990.

69. True. $|x| = |-x|$ implies $f(x) = |x| - 5 = |-x| - 5 = g(x)$

71. (a)

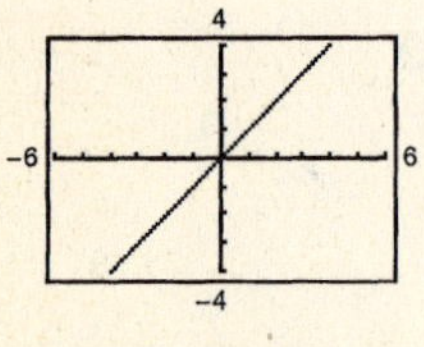

(b)

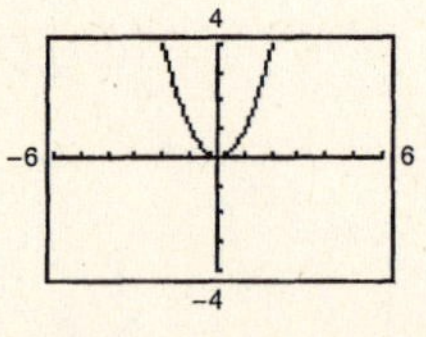

(c)

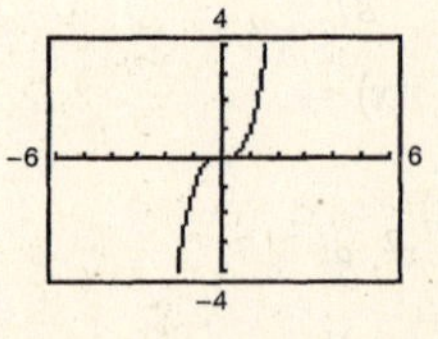

(d)

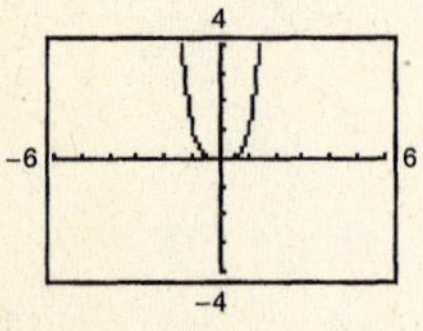

(e)

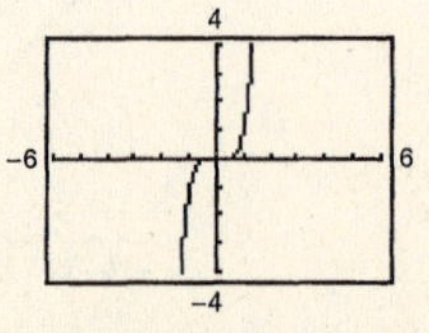

(f)

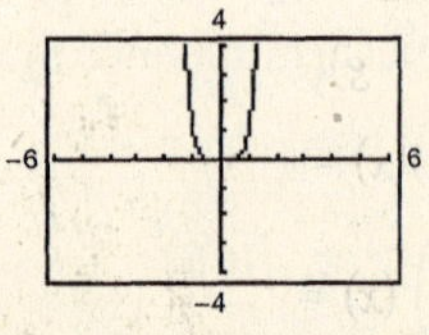

All the graphs pass through the origin. The graphs of the odd powers of x are symmetric to the origin and the graphs of the even powers are symmetric to the y-axis. As the powers increase, the graphs become flatter in the interval $-1 < x < 1$.

73. slope L_1: $\dfrac{10 + 2}{2 + 2} = 3$

slope L_2: $\dfrac{9 - 3}{3 + 1} = \dfrac{3}{2}$

Neither parallel nor perpendicular

75. Domain: All $x \neq 9$

77. Domain: $100 - x^2 \geq 0 \Rightarrow x^2 \leq 100 \Rightarrow -10 \leq x \leq 10$

Section 1.5   Combinations of Functions

■ Given two functions, f and g, you should be able to form the following functions (if defined):
1. Sum: $(f + g)(x) = f(x) + g(x)$
2. Difference: $(f - g)(x) = f(x) - g(x)$
3. Product: $(fg)(x) = f(x)g(x)$
4. Quotient: $(f/g)(x) = f(x)/g(x)$, $g(x) \neq 0$
5. Composition of f with g: $(f \circ g)(x) = f(g(x))$
6. Composition of g with f: $(g \circ f)(x) = g(f(x))$

Solutions to Odd-Numbered Exercises

1.

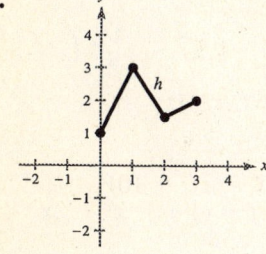

3.

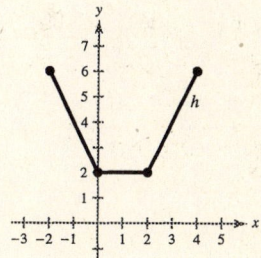

5. $f(x) = x + 3$, $\quad g(x) = x - 3$

(a) $(f + g)(x) = f(x) + g(x) = (x + 3) + (x - 3) = 2x$

(b) $(f - g)(x) = f(x) - g(x) = (x + 3) - (x - 3) = 6$

(c) $(fg)(x) = f(x)g(x) = (x + 3)(x - 3) = x^2 - 9$

(d) $\left(\dfrac{f}{g}\right)(x) = \dfrac{f(x)}{g(x)} = \dfrac{x + 3}{x - 3}$, $\quad x \neq 3$

Domain: all $x \neq 3$

7. $f(x) = x^2$, $g(x) = 1 - x$

(a) $(f + g)(x) = f(x) + g(x) = x^2 + (1 - x) = x^2 - x + 1$

(b) $(f - g)(x) = f(x) - g(x) = x^2 - (1 - x) = x^2 + x - 1$

(c) $(fg)(x) = f(x) \cdot g(x) = x^2(1 - x) = x^2 - x^3$

(d) $\left(\dfrac{f}{g}\right)(x) = \dfrac{f(x)}{g(x)} = \dfrac{x^2}{1 - x}$, $x \neq 1$

Domain: all $x \neq 1$.

9. $f(x) = x^2 + 5, g(x) = \sqrt{1 - x}$

    (a) $(f + g)(x) = f(x) + g(x) = (x^2 + 5) + \sqrt{1 - x}$

    (b) $(f - g)(x) = f(x) - g(x) = (x^2 + 5) - \sqrt{1 - x}$

    (c) $(fg)(x) = f(x) \cdot g(x) = (x^2 + 5)\sqrt{1 - x}$

    (d) $\left(\dfrac{f}{g}\right)(x) = \dfrac{f(x)}{g(x)} = \dfrac{x^2 + 5}{\sqrt{1 - x}}, \; x < 1$

    Domain: $x < 1$.

11. $f(x) = \dfrac{1}{x}, g(x) = \dfrac{1}{x^2}$

    (a) $(f + g)(x) = f(x) + g(x) = \dfrac{1}{x} + \dfrac{1}{x^2} = \dfrac{x + 1}{x^2}$

    (b) $(f - g)(x) = f(x) - g(x) = \dfrac{1}{x} - \dfrac{1}{x^2} = \dfrac{x - 1}{x^2}$

    (c) $(fg)(x) = f(x) \cdot g(x) = \dfrac{1}{x}\left(\dfrac{1}{x^2}\right) = \dfrac{1}{x^3}$

    (d) $\left(\dfrac{f}{g}\right)(x) = \dfrac{f(x)}{g(x)} = \dfrac{1/x}{1/x^2} = \dfrac{x^2}{x} = x, \; x \neq 0$

    Domain: $x \neq 0$.

13. $(f + g)(3) = f(3) + g(3) = (3^2 + 1) + (3 - 4) = 9$

15. $(f - g)(0) = f(0) - g(0) = [0^2 + 1] - (0 - 4) = 5$

17. $(fg)(4) = f(4)g(4) = (4^2 + 1)(4 - 4) = 0$

19. $\left(\dfrac{f}{g}\right)(-5) = \dfrac{f(-5)}{g(-5)} = \dfrac{(-5)^2 + 1}{(-5 - 4)} = \dfrac{26}{-9} = -\dfrac{26}{9}$

21. $(f - g)(2t) = f(2t) - g(2t) = [(2t)^2 + 1] - (2t - 4) = 4t^2 - 2t + 5$

23. $(fg)(-5t) = f(-5t)g(-5t) = [(-5t)^2 + 1][(-5t) - 4]$

            $= (25t^2 + 1)(-5t - 4) = -125t^3 - 100t^2 - 5t - 4$

25. $\left(\dfrac{f}{g}\right)(-t) = \dfrac{f(-t)}{g(-t)} = \dfrac{(-t)^2 + 1}{-t - 4} = \dfrac{t^2 + 1}{-t - 4}, \; t \neq -4$

27. $f(x) = \frac{1}{2}x, g(x) = x - 1, (f + g)(x) = \frac{3}{2}x - 1$

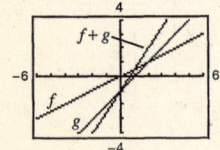

29. $f(x) = x^2, g(x) = -2x, (f + g)(x) = x^2 - 2x$

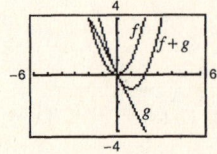

31. $f(x) = 3x, g(x) = -\dfrac{x^3}{10}, (f + g)(x) = 3x - \dfrac{x^3}{10}$

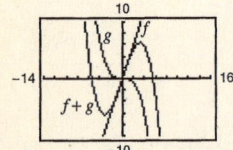

For $0 \leq x \leq 2, f(x)$ contributes more to the magnitude.

For $x > 6, g(x)$ contributes more to the magnitude.

33. $f(x) = 3x + 2, g(x) = -\sqrt{x + 5},$

    $(f + g)(x) = 3x + 2 - \sqrt{x + 5}$

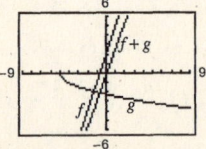

$f(x) = 3x + 2$ contributes more to the magnitude in both intervals.

35. $f(x) = x^2, g(x) = x - 1$

(a) $(f \circ g)(x) = f(g(x)) = f(x - 1) = (x - 1)^2$ (c) $(f \circ g)(0) = (0 - 1)^2 = 1$

(b) $(g \circ f)(x) = g(f(x)) = g(x^2) = x^2 - 1$

37. $f(x) = 3x + 5, g(x) = 5 - x$

(a) $(f \circ g)(x) = f(g(x)) = f(5 - x) = 3(5 - x) + 5 = 20 - 3x$

(b) $(g \circ f)(x) = g(f(x)) = g(3x + 5) = 5 - (3x + 5) = -3x$

(c) $(f \circ g)(0) = 20$

39. (a) $(f \circ g)(x) = f(g(x)) = f(x^2) = \sqrt{x^2 + 4}$

$(g \circ f)(x) = g(f(x)) = g(\sqrt{x + 4}) = (\sqrt{x + 4})^2$

$= x + 4, \ x \geq -4$

(b) 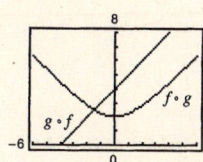   They are not equal.

41. (a) $(f \circ g)(x) = f(g(x)) = f(3x + 1)$

$= \frac{1}{3}(3x + 1) - 3 = x - \frac{8}{3}$

$(g \circ f)(x) = g(f(x)) = g(\frac{1}{3}x - 3) \circ$

$= 3(\frac{1}{3}x - 3) + 1 = x - 8$

(b) 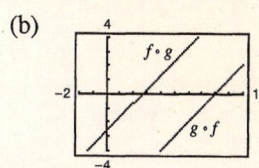   They are not equal.

43. (a) $(f \circ g)(x) = f(g(x)) = f(x^6) = (x^6)^{2/3} = x^4$

$(g \circ f)(x) = g(f(x)) = g(x^{2/3}) = (x^{2/3})^6 = x^4$

(b) 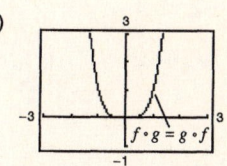   They are equal.

45. (a) $(f \circ g)(x) = f(g(x)) = f(4 - x) = 5(4 - x) + 4 = 24 - 5x$

$(g \circ f)(x) = g(f(x)) = g(5x + 4) = 4 - (5x + 4) = -5x$

(b) No, $(f \circ g)(x) \neq (g \circ f)(x)$ because $24 - 5x \neq -5x$.

(c)

x	$f(g(x))$	$g(f(x))$
0	24	0
1	19	-5
2	14	-10
3	9	-15

47. (a) $(f \circ g)(x) = f(g(x)) = f(x^2 - 5) = \sqrt{(x^2 - 5) + 6} = \sqrt{x^2 + 1}$

$(g \circ f)(x) = g(f(x)) = g(\sqrt{x + 6}) = (\sqrt{x + 6})^2 - 5 = (x + 6) - 5 = x + 1, \ x \geq -6$

(b) No, $(f \circ g)(x) \neq (g \circ f)(x)$ because $\sqrt{x^2 + 1} \neq x + 1$.

(c)

x	$f(g(x))$	$g(f(x))$
0	1	1
-2	$\sqrt{5}$	-1
3	$\sqrt{10}$	4

49. (a) $(f \circ g)(x) = f(g(x)) = f(2x - 1) = |(2x - 1) + 3| = |2x + 2| = 2|x + 1|$

   $(g \circ f)(x) = g(f(x)) = g(|x + 3|) = 2|x + 3| - 1$

(b) No, $(f \circ g)(x) \neq (g \circ f)(x)$ because $2|x + 1| \neq 2|x + 3| - 1$.

(c)

x	$f(g(x))$	$g(f(x))$
-1	0	3
0	2	5
1	4	7

51. (a) $(f + g)(3) = f(3) + g(3) = 2 + 1 = 3$

(b) $\left(\dfrac{f}{g}\right)(2) = \dfrac{f(2)}{g(2)} = \dfrac{0}{2} = 0$

53. (a) $(f \circ g)(2) = f(g(2)) = f(2) = 0$

(b) $(g \circ f)(2) = g(f(2)) = g(0) = 4$

55. Let $f(x) = x^2$ and $g(x) = 2x + 1$, then $(f \circ g)(x) = h(x)$. This is not a unique solution. For example, if $f(x) = (x + 1)^2$ and $g(x) = 2x$, then $(f \circ g)(x) = h(x)$ as well.

57. Let $f(x) = \sqrt[3]{x}$ and $g(x) = x^2 - 4$, then $(f \circ g)(x) = h(x)$. This answer is not unique. Other possibilities may be:

   $f(x) = \sqrt[3]{x - 4}$ and $g(x) = x^2$ or

   $f(x) = \sqrt[3]{-x}$ and $g(x) = 4 - x^2$ or

   $f(x) = \sqrt[9]{x}$ and $g(x) = (x^2 - 4)^3$

59. Let $f(x) = 1/x$ and $g(x) = x + 2$, then $(f \circ g)(x) = h(x)$. Again, this is not a unique solution. Other possibilities may be:

   $f(x) = \dfrac{1}{x + 2}$ and $g(x) = x$

   or $f(x) = \dfrac{1}{x + 1}$ and $g(x) = x + 1$

61. Let $f(x) = x^2 + 2x$ and $g(x) = x + 4$. Then $(f \circ g)(x) = h(x)$. (Answer is not unique.)

63. (a) The domain of $f(x) = \sqrt{x + 4}$ is $x + 4 \geq 0$ or $x \geq -4$

(b) The domain of $g(x) = x^2$ is all real numbers.

(c) $(f \circ g)(x) = f(g(x)) = f(x^2) = \sqrt{x^2 + 4}$.

The domain of $(f \circ g)$ is all real numbers.

65. (a) The domain of $f(x) = x^2 + 1$ is all real numbers.

(b) The domain of $g(x) = \sqrt{x}$ is all $x \geq 0$

(c) $(f \circ g)(x) = f(g(x)) = f(\sqrt{x})$

   $= (\sqrt{x})^2 + 1 = x + 1, \quad x \geq 0$

The domain of $f \circ g$ is $x \geq 0$

67. (a) The domain of $f(x) = \dfrac{1}{x}$ is all $x \neq 0$.

(b) The domain of $g(x) = x + 3$ is all real numbers.

(c) The domain of $(f \circ g)(x) = f(x + 3) = \dfrac{1}{x + 3}$ is all $x \neq -3$.

69. (a) The domain of $f(x) = |x - 4|$ is all real numbers.

(b) The domain of $g(x) = 3 - x$ is all real numbers.

(c) $(f \circ g)(x) = f(g(x)) = f(3 - x) = |(3 - x) - 4| = |-x - 1| = |x + 1|$

Domain: all real numbers

71. (a) The domain of $f(x) = x + 2$ is all real numbers.

(b) The domain of $g(x) = \dfrac{1}{x^2 - 4}$ is all $x \neq \pm 2$

(c) $(f \circ g)(x) = f(g(x)) = f\left(\dfrac{1}{x^2 - 4}\right) = \dfrac{1}{x^2 - 4} + 2$

Domain: $x \neq \pm 2$

73. (a) $T(x) = R(x) + B(x) = \frac{3}{4}x + \frac{1}{15}x^2$

(b)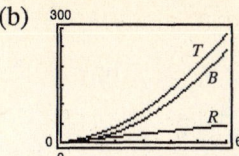

(c) $B(x)$ contributes more to $T(x)$ at higher speeds.

75. Let $t = 4$ represent 1994

$y_1 = 8.93t + 103.0$

$y_2 = 1.886t^2 - 5.24t + 305.7$

$y_3 = -0.361t^2 + 7.97t + 14.2$

Year	1994	1995	1996	1997	1998	1999	2000
y_1	138.7	147.7	156.6	165.5	174.4	183.4	192.3
y_2	314.9	326.7	342.2	361.4	384.5	411.3	441.9
y_3	40.3	45.0	49.0	52.3	54.9	56.7	57.8

77. $(A \circ r)(t)$ gives the area of the circle as a function of time.

$(A \circ r)(t) = A(r(t))$

$= A(0.6t)$

$= \pi(0.6t)^2 = 0.36\pi t^2$

79. (a) $(C \circ x)(t) = C(x(t))$

$= 60(50t) + 750$

$= 3000t + 750$

$(C \circ x)(t)$ represents the cost after t production hours.

(b)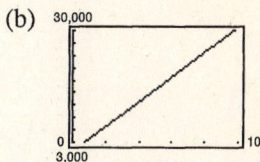

The cost increases to $15,000 when $t = 4.75$ hours.

81. $g(f(x)) = g(x - 500,000) = 0.03(x - 500,000)$ represents 3 percent of the amount over $500,000.

83. False. $(f \circ g)(x) = f(6x) = 6x + 1$, but $(g \circ f)(x) = g(x + 1) = 6(x + 1)$

85. Let $f(x)$ and $g(x)$ be odd functions, and define $h(x) = f(x)g(x)$. Then,

$h(-x) = f(-x)g(-x)$

$= [-f(x)][-g(x)]$ since f and g are both odd

$= f(x)g(x) = h(x)$.

Thus, h is even.

Let $f(x)$ and $g(x)$ be even functions, and define $h(x) = f(x)g(x)$. Then,

$h(-x) = f(-x)g(-x)$

$= f(x)g(x)$ since f and g are both even

$= h(x)$.

Thus, h is even.

87. $g(-x) = \frac{1}{2}[f(-x) + f(-(-x))] = \frac{1}{2}[f(-x) + f(x)] = g(x),$

which shows that g is even.

$$h(-x) = \frac{1}{2}[f(-x) - f(-(-x))] = \frac{1}{2}[f(-x) - f(x)]$$
$$= -\frac{1}{2}[f(x) - f(-x)] = -h(x),$$

which shows that h is odd.

89. $(0, -5), (1, -5), (2, -7)$

(other answers possible)

91. $\left(\sqrt{24}, 0\right), \left(-\sqrt{24}, 0\right), \left(0, \sqrt{24}\right)$

(other answers possible)

93. $y - (-2) = \dfrac{8 - (-2)}{-3 - (-4)}(x - (-4))$

$y + 2 = 10(x + 4)$

$y - 10x - 38 = 0$

95. $y - (-1) = \dfrac{4 - (-1)}{-\frac{1}{3} - \frac{3}{2}}\left(x - \frac{3}{2}\right)$

$y + 1 = \dfrac{5}{-\frac{11}{6}}\left(x - \frac{3}{2}\right) = -\frac{30}{11}\left(x - \frac{3}{2}\right)$

$11y + 11 = -30x + 45$

$30x + 11y - 34 = 0$

97. Figure shifts 4 units to the right

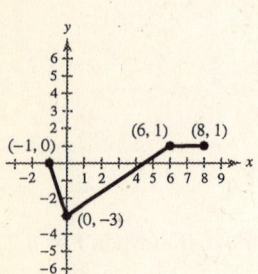

99. Figure shifts 4 units upward

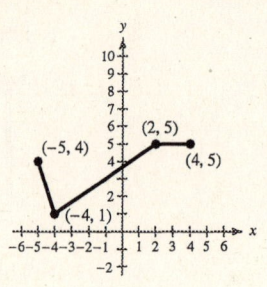

101. Vertical stretch by 2

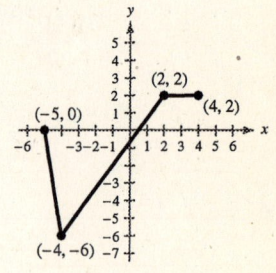

Section 1.6   Inverse Functions

- ■ Two functions f and g are inverses of each other if $f(g(x)) = x$ for every x in the domain of g and $g(f(x)) = x$ for every x in the domain of f.

- ■ Be able to find the inverse of a function, if it exists.

  1. Replace $f(x)$ with y.

  2. Interchange x and y.

  3. Solve for y. If this equation represents y as a function of x, then you have found $f^{-1}(x)$. If this equation does not represent y as a function of x, then f does not have an inverse function.

- ■ A function f has an inverse function if and only if no **horizontal** line crosses the graph of f at more than one point.

- ■ A function f has an inverse function if and only if f is one-to-one.

Solutions to Odd-Numbered Exercises

1. $f(x) = 6x$

$f^{-1}(x) = \frac{1}{6}x$

$f(f^{-1}(x)) = f(\frac{1}{6}x) = 6(\frac{1}{6}x) = x$

$f^{-1}(f(x)) = f^{-1}(6x) = \frac{1}{6}(6x) = x$

3. $f(x) = x + 7$

$f^{-1}(x) = x - 7$

$f(f^{-1}(x)) = f(x - 7) = (x - 7) + 7 = x$

$f^{-1}(f(x)) = f^{-1}(x + 7) = (x + 7) - 7 = x$

5. $f^{-1}(x) = \dfrac{x - 1}{2}$

$f(f^{-1}(x)) = f\left(\dfrac{x - 1}{2}\right) = 2\left(\dfrac{x - 1}{2}\right) + 1 = (x - 1) + 1 = x$

$f^{-1}(f(x)) = f^{-1}(2x + 1) = \dfrac{(2x + 1) - 1}{2} = \dfrac{2x}{2} = x$

7. $f^{-1}(x) = x^3$

$f(f^{-1}(x)) = f(x^3) = \sqrt[3]{x^3} = x$

$f^{-1}(f(x)) = f^{-1}(\sqrt[3]{x}) = (\sqrt[3]{x})^3 = x$

9. (a) $f(g(x)) = f\left(-\dfrac{2x + 6}{7}\right) = -\dfrac{7}{2}\left(-\dfrac{2x + 6}{7}\right) - 3 = \dfrac{2x + 6}{2} - 3 = (x + 3) - 3 = x$

$g(f(x)) = g\left(-\dfrac{7}{2}x - 3\right) = -\dfrac{2\left(-\frac{7}{2}x - 3\right) + 6}{7} = -\dfrac{-7x - 6 + 6}{7} = \dfrac{7x}{7} = x$

(b)

x	2	0	-2	-4	-6
$f(x)$	-10	-3	4	11	18

x	-10	-3	4	11	18
$g(x)$	2	0	-2	-4	-6

Note that the entries in the tables are the same except that the rows are interchanged.

11. (a) $f(g(x)) = f(\sqrt[3]{x - 5}) = [\sqrt[3]{x - 5}]^3 + 5 = (x - 5) + 5 = x$

$g(f(x)) = g(x^3 + 5) = \sqrt[3]{(x^3 + 5) - 5} = \sqrt[3]{x^3} = x$

(b)

x	-3	-2	-1	0	1
$f(x)$	-22	-3	4	5	6

x	-22	-3	4	5	6
$g(x)$	-3	-2	-1	0	1

Note that the entries in the tables are the same except that the rows are interchanged.

13. (a) $f(g(x)) = f(8 + x^2) = -\sqrt{(8 + x^2) - 8} = -\sqrt{x^2} = -(-x) = x \quad x \le 0$

[Since $x \le 0$, $\sqrt{x^2} = -x$]

$g(f(x)) = g\left(-\sqrt{x - 8}\right) = 8 + \left[-\sqrt{x - 8}\right]^2 = 8 + (x - 8) = x$

(b)

x	8	9	12	17	24
$f(x)$	0	-1	-2	-3	-4

x	0	-1	-2	-3	-4
$g(x)$	8	9	12	17	24

Note that the entries in the tables are the same except that the rows are interchanged.

15. (a) $f(g(x)) = f\left(\sqrt[3]{x}\right) = \left(\sqrt[3]{x}\right)^3 = x$

$g(f(x)) = g(x^3) = \sqrt[3]{x^3} = x$

(b)

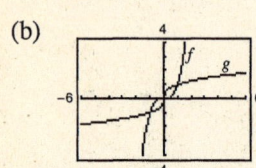

Reflections in the line $y = x$

17. (a) $f(g(x)) = f(x^2 + 4), \ x \ge 0$

$= \sqrt{(x^2 + 4) - 4} = x$

$g(f(x)) = g\left(\sqrt{x - 4}\right)$

$= \left(\sqrt{x - 4}\right)^2 + 4 = x$

(b)

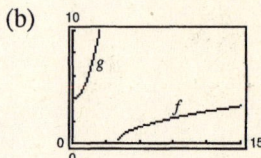

Reflections in the line $y = x$

19. (a) $f(g(x)) = f\left(\sqrt[3]{1 - x}\right) = 1 - \left(\sqrt[3]{1 - x}\right)^3 = 1 - (1 - x) = x$

$g(f(x)) = g(1 - x^3) = \sqrt[3]{1 - (1 - x^3)} = \sqrt[3]{x^3} = x$

(b)

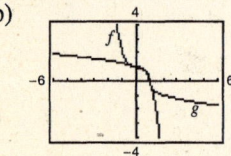

Reflections in the line $y = x$

21. The inverse is a line through $(-1, 0)$.

Matches graph (c).

23. The inverse is half a parabola starting at $(1, 0)$.

Matches graph (a).

25. $f(x) = 2x, \quad g(x) = \dfrac{x}{2}$.

(a)

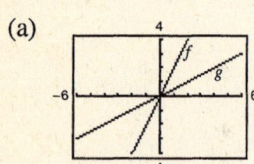

Reflection in the line $y = x$

(b)

x	-2	-1	0	1	2
$f(x)$	-4	-2	0	2	4

x	-4	-2	0	2	4
$g(x)$	-2	-1	0	1	2

The entries in the tables are the same, except that the rows are interchanged.

27. $f(x) = \dfrac{x-1}{x+5}$, $g(x) = -\dfrac{5x+1}{x-1} = \dfrac{5x+1}{1-x}$

(a)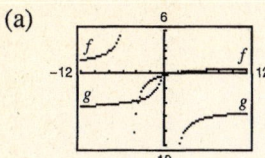

Reflection in the line $y = x$

(b)

x	-2	-1	0	3	5
$f(x)$	-1	$-\frac{1}{2}$	$-\frac{1}{5}$	$\frac{1}{4}$	$\frac{2}{5}$

x	-1	$-\frac{1}{2}$	$-\frac{1}{5}$	$\frac{1}{4}$	$\frac{2}{5}$
$g(x)$	-2	-1	0	3	5

The entries in the tables are the same, except that the rows are interchanged.

29. $f(x) = 3 - \dfrac{1}{2}x$

f is one-to-one because a horizontal line will intersect the graph at most once.

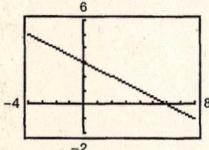

31. $h(x) = \dfrac{x^2}{x^2 + 1}$

h is not one-to-one because some horizontal lines intersect the graph twice.

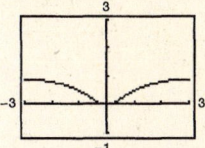

33. $h(x) = \sqrt{16 - x^2}$

h is not one-to-one because some horizontal lines intersect the graph twice.

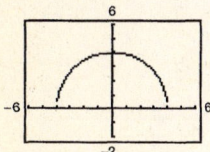

35. $f(x) = 10$

f is not one-to-one because the horizontal line $y = 10$ intersects the graph at every point on the graph.

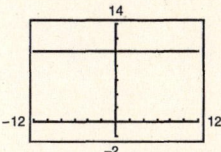

37. $g(x) = (x + 5)^3$

g is one-to-one because a horizontal line will intersect the graph at most once.

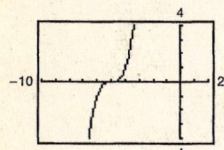

39. $h(x) = |x + 4| - |x - 4|$

h is not one-to-one because some horizontal lines intersect the graph more than once.

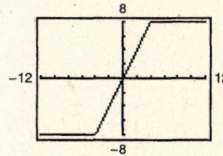

41. $f(x) = x^4$

$y = x^4$

$x = y^4$

$y = \pm\sqrt[4]{x}$

f is not one-to-one.

This does not represent y as a function of x. f does not have an inverse.

43. $f(x) = \dfrac{3x + 4}{5}$

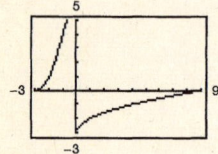

$y = \dfrac{3x + 4}{5}$

$x = \dfrac{3y + 4}{5}$

$5x = 3y + 4$

$5x - 4 = 3y$

$(5x - 4)/3 = y$

$f^{-1}(x) = \dfrac{5x - 4}{3}$

f is one-to-one and has an inverse.

45. $f(x) = \dfrac{1}{x^2}$ is not one-to-one, and does not have an inverse. For example, $f(1) = f(-1) = 1$.

47. $f(x) = (x + 3)^2, \ x \ge -3, \ y \ge 0$

$\quad y = (x + 3)^2, \ x \ge -3, \ y \ge 0$

$\quad x = (y + 3)^2, \ y \ge -3, \ x \ge 0$

$\quad \sqrt{x} = y + 3, \ y \ge -3, \ x \ge 0$

$\quad y = \sqrt{x} - 3, \ x \ge 0, \ y \ge -3$

f is one-to-one.

This is a function of x, so f has an inverse.

$f^{-1}(x) = \sqrt{x} - 3, \ x \ge 0$

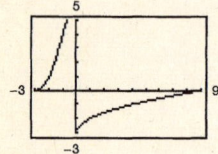

49. $f(x) = \sqrt{2x + 3} \ \implies \ x \ge -\dfrac{3}{2}, \ y \ge 0$

$\quad y = \sqrt{2x + 3}, \ x \ge -\dfrac{3}{2}, \ y \ge 0$

$\quad x = \sqrt{2y + 3}, \ y \ge -\dfrac{3}{2}, \ x \ge 0$

$\quad x^2 = 2y + 3, \ x \ge 0, \ y \ge -\dfrac{3}{2}$

$\quad y = \dfrac{x^2 - 3}{2}, \ x \ge 0, \ y \ge -\dfrac{3}{2}$

f is one to one.

This is a function of x, so f has an inverse.

$f^{-1}(x) = \dfrac{x^2 - 3}{2}, \ x \ge 0$

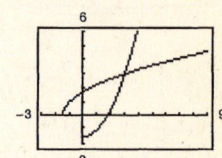

51. $f(x) = |x - 2|, \ x \le 2, \ y \ge 0$

$\quad y = |x - 2|$

$\quad x = |y - 2| \ \ y \le 2, \ x \ge 0$

$\quad x = -(y - 2) \ \text{since } y - 2 \le 0$

$\quad x = -y + 2$

$\quad y = -x + 2, \ x \ge 0, \ y \le 2$

$\quad f^{-1}(x) = -x + 2, \ x \ge 0$

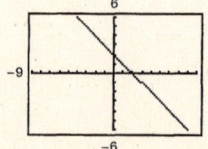

53. $f(x) = 2x - 3$

$y = 2x - 3$

$x = 2y - 3$

$y = \dfrac{x + 3}{2}$

$f^{-1}(x) = \dfrac{x + 3}{2}$

Reflections in the line $y = x$

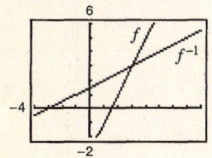

55. $f(x) = x^5$

$y = x^5$

$x = y^5$

$y = \sqrt[5]{x}$

$f^{-1}(x) = \sqrt[5]{x}$

Reflections in the line $y = x$

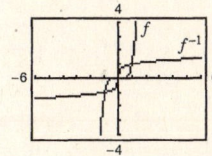

57. $f(x) = x^{3/5}$

$y = x^{3/5}$

$x = y^{3/5}$

$y = x^{5/3}$

$f^{-1}(x) = x^{5/3}$

Reflections in the line $y = x$

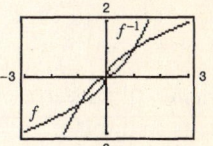

59. $f(x) = \sqrt{4 - x^2}, 0 \le x \le 2$

$y = \sqrt{4 - x^2}$

$x = \sqrt{4 - y^2}$

$x^2 = 4 - y^2$

$y^2 = 4 - x^2$

$y = \sqrt{4 - x^2}$

$f^{-1}(x) = \sqrt{4 - x^2}, 0 \le x \le 2$

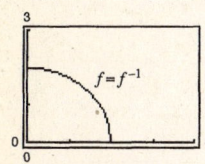

Reflections in the line $y = x$

61. $f(x) = \dfrac{4}{x}$

$y = \dfrac{4}{x}$

$x = \dfrac{4}{y}$

$xy = 4$

$y = \dfrac{4}{x}$

$f^{-1}(x) = \dfrac{4}{x}$

Reflections in the line $y = x$

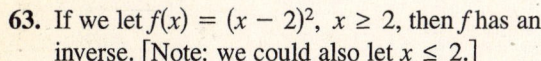

63. If we let $f(x) = (x - 2)^2$, $x \ge 2$, then f has an inverse. [Note: we could also let $x \le 2$.]

$f(x) = (x - 2)^2, \ x \ge 2, \ y \ge 0$

$y = (x - 2)^2, \ x \ge 2, \ y \ge 0$

$x = (y - 2)^2, \ x \ge 0, \ y \ge 2$

$\sqrt{x} = y - 2, \ x \ge 0, \ y \ge 2$

$\sqrt{x} + 2 = y, \ x \ge 0, \ y \ge 2$

Thus, $f^{-1}(x) = \sqrt{x} + 2, \ x \ge 0$.

65. If we let $f(x) = |x + 2|$, $x \ge -2$, then f has an inverse. [[Note: we could also let $x \le -2$.]

$f(x) = |x + 2|, \ x \ge -2$

$f(x) = x + 2$ when $x \ge -2$

$y = x + 2, \ x \ge -2, \ y \ge 0$

$x = y + 2, \ x \ge 0, \ y \ge -2$

$x - 2 = y, \ x \ge 0, \ y \ge -2$

Thus, $f^{-1}(x) = x - 2, \ x \ge 0$.

67.

x	$f(x)$
-2	-4
-1	-2
1	2
3	3

x	$f^{-1}(x)$
-4	-2
-2	-1
2	1
3	3

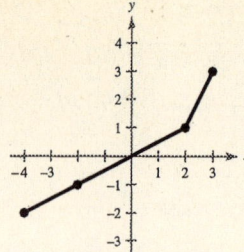

69. $f(x) = x^3 + x + 1$

The graph of the inverse relation is an inverse function since it satisfies the vertical line test.

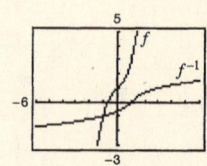

71. $g(x) = \dfrac{3x^2}{x^2 + 1}$

The graph of the inverse relation is not an inverse function since it does not satisfy the vertical line test.

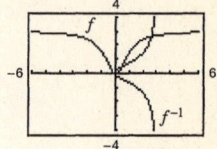

In Exercises 73, 75, and 77, $f(x) = \frac{1}{8}x - 3$, $f^{-1}(x) = 8(x + 3)$, $g(x) = x^3$, $g^{-1}(x) = \sqrt[3]{x}$.

73. $(f^{-1} \circ g^{-1})(1) = f^{-1}(g^{-1}(1)) = f^{-1}(\sqrt[3]{1}) = 8(\sqrt[3]{1} + 3) = 8(1 + 3) = 32$

75. $(f^{-1} \circ f^{-1})(6) = f^{-1}(f^{-1}(6)) = f^{-1}(8[6 + 3]) = f^{-1}(72) = 8(72 + 3) = 600$

77. $(f \circ g)(x) = f(g(x)) = f(x^3) = \frac{1}{8}x^3 - 3$. Now find the inverse of $(f \circ g)(x) = \frac{1}{8}x^3 - 3$:

$$y = \tfrac{1}{8}x^3 - 3$$
$$x = \tfrac{1}{8}y^3 - 3$$
$$x + 3 = \tfrac{1}{8}y^3$$
$$8(x + 3) = y^3$$
$$\sqrt[3]{8(x + 3)} = y$$
$$(f \circ g)^{-1}(x) = 2\sqrt[3]{x + 3}$$

Note: $(f \circ g)^{-1} = g^{-1} \circ f^{-1}$

In Exercises 79 and 81, $f(x) = x + 4$, $f^{-1}(x) = x - 4$, $g(x) = 2x - 5$, $g^{-1}(x) = \dfrac{x + 5}{2}$.

79. $(g^{-1} \circ f^{-1})(x) = g^{-1}(f^{-1}(x)) = g^{-1}(x - 4) = \dfrac{(x - 4) + 5}{2} = \dfrac{x + 1}{2}$

81. $(f \circ g)(x) = f(g(x)) = f(2x - 5) = (2x - 5) + 4 = 2x - 1$. Now find the inverse of $(f \circ g)(x) = 2x - 1$:

$$y = 2x - 1$$
$$x = 2y - 1$$
$$x + 1 = 2y$$
$$y = \dfrac{x + 1}{2}$$
$$(f \circ g)^{-1}(x) = \dfrac{x + 1}{2}$$

Note that $(f \circ g)^{-1}(x) = (g^{-1} \circ f^{-1})(x)$; see Exercise 87.

83. (a) Yes, f is one-to-one

(b) f^{-1} gives the year corresponding to the 7 values in the second column

(c) $f^{-1}(650.3) = 10$ because $f(10) = 650.3$

(d) No, because $f(8) = f(12) = 546.3$.

85. False. $f(x) = x^2$ is even, but f^{-1} does not exist

87. We will show that $(f \circ g)^{-1}(x) = (g^{-1} \circ f^{-1})(x)$ for all x in their domains.
Let $y = (f \circ g)^{-1}(x) \implies (f \circ g)(y) = x$ then $f(g(y)) = x \implies f^{-1}(x) = g(y)$.

Hence, $(g^{-1} \circ f^{-1})(x) = g^{-1}(f^{-1}(x)) = g^{-1}(g(y)) = y = (f \circ g)^{-1}(x)$.

Thus, $g^{-1} \circ f^{-1} = (f \circ g)^{-1}$.

89. $\dfrac{27x^3}{3x^2} = 9x, x \neq 0$

91. $\dfrac{x^2 - 36}{6 - x} = \dfrac{(x-6)(x+6)}{-(x-6)} = \dfrac{x+6}{-1} = -x - 6, x \neq 6$

93. $4x - y = 3$

$y = 4x - 3$

Yes, y is a function of x.

95. $x^2 + y^2 = 9$

$y = \pm\sqrt{9 - x^2}$

No, y is not a function of x.

97. $y = \sqrt{x + 2}$

Yes, y is a function of x

Section 1.7   Exploring Data: Linear Models and Scatter Plots

Solutions to Odd-Numbered Exercises

1. (a)

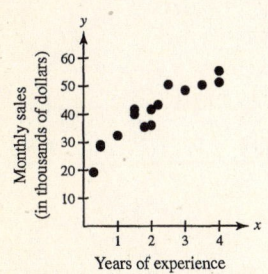

(b) Yes, the data appears somewhat linear. The more experience, x, corresponds to higher sales, y.

3. Negative correlation—y decreases as x increases.

5. No correlation.

7. (a)

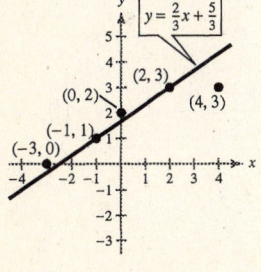

(b) $y = 0.46x + 1.62$

(c)

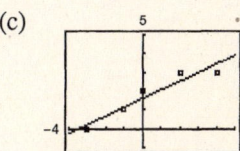

(d) Yes, the model appears valid.

9. (a)

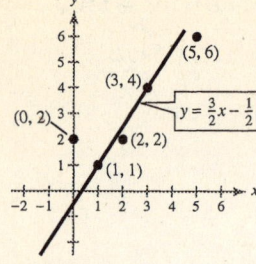
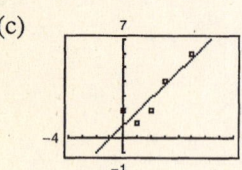

(b) $y = 0.95x + 0.92$

(c)

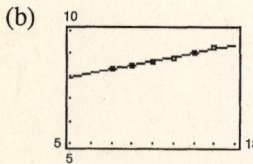

(d) Yes, the model appears valid.

11. (a)

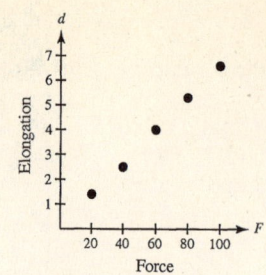

(b) $d = 0.07F - 0.3$

(c) $d = 0.066F$ or
$F = 15.13d + 0.096$

(d) If $F = 55$,
$d = 0.066(55) \approx 3.63$ cm

13. (a) $S = 0.2t - 0.14$

(b)

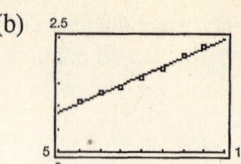

(c) The slope 0.2 is the average annual increase of salary in millions of dollars.

(d) For 2006, $t = 16$ and $S \approx 3.06$ or about $3.1 million.

15. (a) $S = 0.183t + 7.013$

(b)

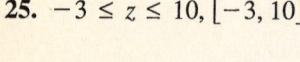

(c) The slope 0.183 is the average annual increase in monthly spending.

(d) For 2008, $t = 18$ and $S \approx \$10.31$

(e)

Year	1997	1998	1999	2000	2001	2002
S	8.30	8.50	8.65	8.80	9.00	9.25
Model	8.29	8.48	8.66	8.84	9.03	9.21

The model fits well

17. (a) $y = -0.024x + 5.06$

(b) The negative slope indicates that the times are decreasing

(c)

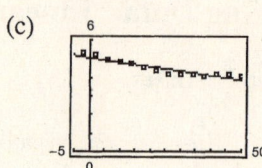

(d) The model is reasonably close

(e) Answers will vary

19. True. To have positive correlation, the y-values tend to increase as x increases

21. Answers will vary

23. $P \leq 2, (-\infty, 2]$

25. $-3 \leq z \leq 10, [-3, 10]$

27. $\dfrac{x^2 - 4}{\left(\dfrac{x + 2}{5}\right)} = \dfrac{(x + 2)(x - 2)5}{x + 2} = 5x - 10, x \neq -2$

29. $f(x) = 2x^2 - 3x + 5$

(a) $f(-1) = 2 + 3 + 5 = 10$

(b) $f(w + 2) = 2(w + 2)^2 - 3(w + 2) + 5$
$= 2w^2 + 5w + 7$

31. $h(x) = \begin{cases} 1 - x^2, & x \leq 0 \\ 2x + 3, & x > 0 \end{cases}$

(a) $h(1) = 2(1) + 3 = 5$

(b) $h(0) = 1 - 0 = 1$

33. $6x + 1 = -9x - 8$

$15x = -9$

$x = -\dfrac{9}{15} = -\dfrac{3}{5}$

35. $8x^2 - 10x - 3 = 0$

$(4x + 1)(2x - 3) = 0$

$x = -\dfrac{1}{4}, \dfrac{3}{2}$

37. $2x^2 - 7x + 4 = 0$

$x = \dfrac{7 \pm \sqrt{49 - 4(4)(2)}}{4}$

$= \dfrac{7 \pm \sqrt{17}}{4}$

Review Exercises for Chapter 1

Solutions to Odd-Numbered Exercises

1. $m = \dfrac{2 - 2}{8 - (-3)} = \dfrac{0}{11} = 0$

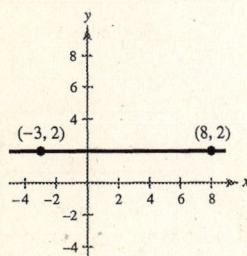

3. $m = \dfrac{5/2 - 1}{5 - 3/2} = \dfrac{3/2}{7/2} = \dfrac{3}{7}$

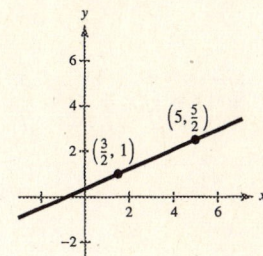

5. $(-4.5, 6)$, $(2.1, 3)$

$m = \dfrac{3 - 6}{2.1 - (-4.5)} = \dfrac{-3}{6.6} = -\dfrac{30}{66} = -\dfrac{5}{11}$

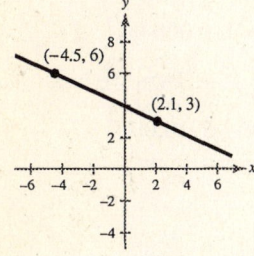

7. (a) $\qquad y + 1 = \frac{1}{4}(x - 2)$

$4y + 4 = x - 2$

$-x + 4y + 6 = 0$

(b) Three additional points:

$(2 + 4, -1 + 1) = (6, 0)$

$(6 + 4, 0 + 1) = (10, 1)$

$(10 + 4, 1 + 1) = (14, 2)$

(other answers possible)

9. (a) $\qquad y + 5 = \frac{3}{2}(x - 0)$

$2y + 10 = 3x$

$-3x + 2y + 10 = 0$

(b) Three additional points:

$(0 + 2, -5 + 3) = (2, -2)$

$(2 + 2, -2 + 3) = (4, 1)$

$(4 + 2, 1 + 3) = (6, 4)$

(other answers possible)

11. (a) $\quad y + 5 = -1\left(x - \frac{1}{5}\right)$

$y + 5 = -x + \frac{1}{5}$

$5y + 25 = -5x + 1$

$5x + 5y + 24 = 0$

(b) Three additional points:

$\left(\frac{1}{5} + 1, -5 - 1\right) = \left(\frac{6}{5}, -6\right)$

$\left(\frac{6}{5} + 1, -6 - 1\right) = \left(\frac{11}{5}, -7\right)$

$\left(\frac{11}{5} + 1, -7 - 1\right) = \left(\frac{16}{5}, -8\right)$

(other answers possible)

13. (a) $y - 6 = 0(x + 2)$

$y - 6 = 0$

(b) Three additional points:

$(0, 6), (1, 6), (2, 6)$

(other answers possible)

15. (a) m is undefined means that the line is vertical.

$x - 10 = 0$

(b) Three additional points: $(10, 0), (10, 1), (10, 2)$

(other answers possible)

17. (a) $y + 1 = \dfrac{-1 + 1}{4 - 2}(x - 2) = 0(x - 2) = 0 \implies y = -1$ (slope = 0)

(b)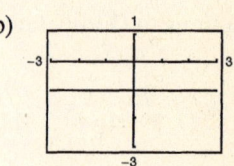

19. (a) $y - 0 = \dfrac{2 - 0}{6 - (-1)}(x + 1) = \dfrac{2}{7}(x + 1) = \dfrac{2}{7}x + \dfrac{2}{7} \implies y = \dfrac{2}{7}x + \dfrac{2}{7}$

(b)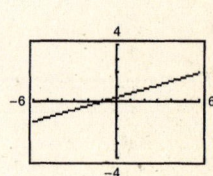

21. $t = 5$ corresponds to 2005.

Point: $(5, 12500)$, slope: 850

$y - 12,500 = 850(t - 5)$

$y = 850t + 8250$

23. $(2, 160,000), \ (3, 185,000)$

$m = \dfrac{185,000 - 160,000}{3 - 2} = 25,000$

$S - 160,000 = 25,000(t - 2)$

$S = 25,000t + 110,000$

For the fourth quarter let $t = 4$. Then we have

$S = 25,000(4) + 110,000 = \$210,000.$

25. $5x - 4y = 8 \implies y = \dfrac{5}{4}x - 2$ and $m = \dfrac{5}{4}$

(a) Parallel slope: $m = \dfrac{5}{4}$

$y - (-2) = \dfrac{5}{4}(x - 3)$

$4y + 8 = 5x - 15$

$0 = 5x - 4y - 23$

$y = \dfrac{5}{4}x - \dfrac{23}{4}$

(b) Perpendicular slope: $m = -\dfrac{4}{5}$

$y - (-2) = -\dfrac{4}{5}(x - 3)$

$5y + 10 = -4x + 12$

$4x + 5y - 2 = 0$

$y = -\dfrac{4}{5}x + \dfrac{2}{5}$

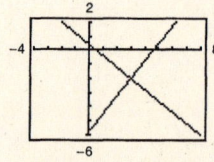

27. $x = 4$ is a vertical line; the slope is not defined.

(a) Parallel line: $x = -6$

(b) Perpendicular slope: $m = 0$

Perpendicular line: $y - 2 = 0(x + 6) = 0 \implies y = 2$

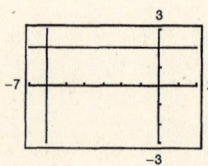

29. (a) Not a function. 20 is assigned two different values.

  (b) Function

  (c) Function

  (d) Not a function. No value is assigned to 30.

31. $16x - y^4 = 0$

$$y^4 = 16x$$

$$y = \pm 2\sqrt[4]{x}$$

y is *not* a function of x. Some x-values correspond to two y-values.

For example, $x = 1$ corresponds to $y = 2$ and $y = -2$.

33. $y = \sqrt{1 - x}$

Each x value, $x \leq 1$, corresponds to only one y value so y is a function of x.

35. $f(x) = x^2 + 1$

  (a) $f(2) = 2^2 + 1 = 5$

  (b) $f(-4) = (-4)^2 + 1 = 17$

  (c) $f(t^2) = (t^2)^2 + 1 = t^4 + 1$

  (d) $-f(x) = -(x^2 + 1) = -x^2 - 1$

37. $h(x) = \begin{cases} 2x + 1, & x \leq -1 \\ x^2 + 2, & x > -1 \end{cases}$

  (a) $h(-2) = 2(-2) + 1 = -3$

  (b) $h(-1) = 2(-1) + 1 = -1$

  (c) $h(0) = 0^2 + 2 = 2$

  (d) $h(2) = 2^2 + 2 = 6$

39. $f(x) = (x - 1)(x + 2)$ is defined for all real numbers.

  Domain: $(-\infty, \infty)$

41. $f(x) = \sqrt{25 - x^2}$

  Domain: $25 - x^2 \geq 0$

  $$(5 + x)(5 - x) \geq 0$$

  Domain: $[-5, 5]$

43. $g(s) = \dfrac{5}{3s - 9} = \dfrac{5}{3(s - 3)}$

  Domain: All real numbers except $s = 3$

45. (a) $C(x) = 16{,}000 + 5.35x$

  (b) $P(x) = R(x) - C(x)$

$$= 8.20x - (16{,}000 + 5.35x)$$

$$= 2.85x - 16{,}000$$

47.
$$f(x) = 2x^2 + 3x - 1$$

$$f(x + h) = 2(x + h)^2 + 3(x + h) - 1$$

$$= 2x^2 + 4xh + 2h^2 + 3x + 3h - 1$$

$$\frac{f(x + h) - f(x)}{h} = \frac{(2x^2 + 4xh + 2h^2 + 3x + 3h - 1) - (2x^2 + 3x - 1)}{h}$$

$$= \frac{4xh + 2h^2 + 3h}{h}$$

$$= 4x + 2h + 3, \quad h \neq 0$$

49. Domain: All real numbers

Range: $y \leq 3$

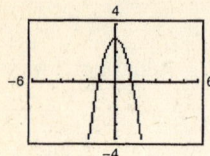

51. Domain: $36 - x^2 \geq 0 \Rightarrow x^2 \leq 36 \Rightarrow -6 \leq x \leq 6$

Range: $0 \leq y \leq 6$

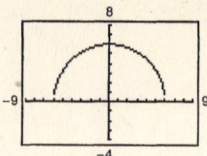

53. (a) $y = \dfrac{x^2 + 3x}{6}$

(b) y is a function of x.

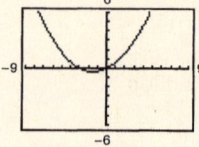

55. (a) $3x + y^2 = 2$

$$y^2 = 2 - 3x$$

$$y = \pm\sqrt{2 - 3x}$$

(b) y is not a function of x.

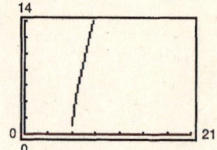

57. $f(x) = x^3 - 3x$

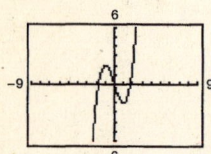

Increasing on $(-\infty, -1)$ and $(1, \infty)$. Decreasing on $(-1, 1)$.

59. $f(x) = x\sqrt{x - 6}$

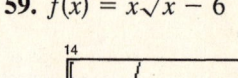

Increasing on $(6, \infty)$

61. $f(x) = (x^2 - 4)^2$. Relative minimums at $(-2, 0)$ and $(2, 0)$. Relative maximum at $(0, 16)$.

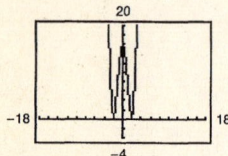

63. $h(x) = 4x^3 - x^4$. Relative maximum $(3, 27)$

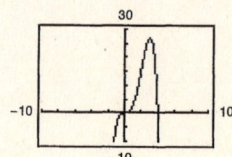

65. $f(x) = \begin{cases} 3x + 5 & , x < 0 \\ x - 4 & , x \geq 0 \end{cases}$

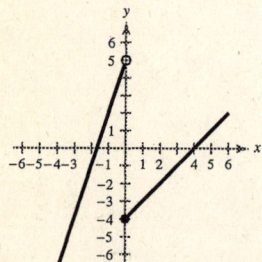

67. $f(-x) = ((-x)^2 - 8)^2 = (x^2 - 8)^2 = f(x)$.

f is even.

69. $f(x) = -2$ is a constant function.

71. $g(x) = -x^3 - 2$ is obtained from $f(x) = x^3$ by a reflection in the y-axis, followed by a vertical shift 2 units downward. $g(x) = -f(x) - 2$.

73. $h(x) = x^2 - 6$

(a) $f(x) = x^2$

(b) The graph of h is a vertical shift of f 6 units downward.

(c)

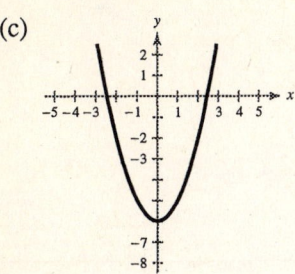

(d) $h(x) = x^2 - 6 = f(x) - 6$

77. $h(x) = \sqrt{x} - 5$

(a) $f(x) = \sqrt{x}$

(b) The graph of h is a vertical shift of f 5 units downward.

(c)

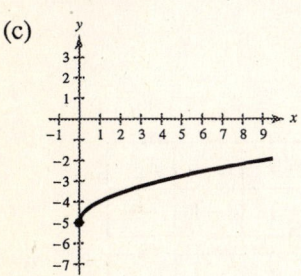

(d) $h(x) = \sqrt{x} - 5 = f(x) - 5$

81. $h(x) = -2x^2 + 3$

(a) $f(x) = x^2$

(b) The graph of h is obtained from f by a vertical stretch of 2, a reflection in the x-axis, and a vertical shift 3 units upward.

(c)

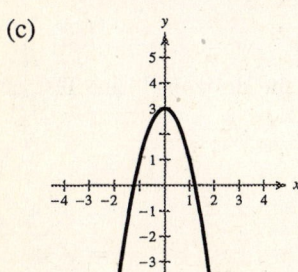

(d) $h(x) = -2x^2 + 3 = -2f(x) + 3$

75. $h(x) = (x - 1)^3 + 7$

(a) $f(x) = x^3$

(b) The graph of h is obtained from f by a horizontal shift 1 unit to the right, followed by a vertical shift 7 units upward.

(c)

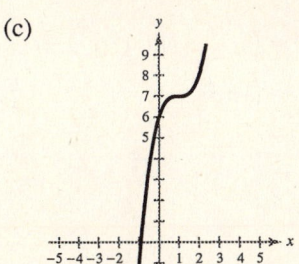

(d) $h(x) = (x - 1)^3 + 7 = f(x - 1) + 7$

79. $h(x) = -x^2 - 3$

(a) $f(x) = x^2$

(b) The graph of h is a reflection in the x-axis, followed by a vertical shift downward 3 units of the graph of f.

(c)

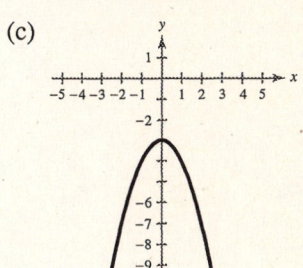

(d) $h(x) = -x^2 - 3 = -f(x) - 3$

83. $h(x) = -\frac{1}{2}|x| + 9$

(a) $f(x) = |x|$

(b) The graph of h is obtained from f by a vertical shrink of $\frac{1}{2}$, a reflection in the x-axis, and a vertical shift 9 units upward.

(c)

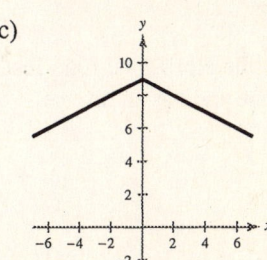

(d) $h(x) = -\frac{1}{2}|x| + 9 = -\frac{1}{2}f(x) + 9$

85. $(f - g)(4) = f(4) - g(4)$

$$= [3 - 2(4)] - \sqrt{4}$$
$$= -5 - 2$$
$$= -7$$

87. $(f + g)(25) = f(25) + g(25) = -47 + 5 = -42$

89. $(fh)(1) = f(1)h(1) = (3 - 2(1))(3(1)^2 + 2)$

$$= (1)(5) = 5$$

91. $(h \circ g)(7) = h(g(7))$

$$= h(\sqrt{7})$$
$$= 3(\sqrt{7})^2 + 2$$
$$= 23$$

93. $(f \circ h)(-4) = f(h(-4)) = f(50) = -97$

95. Let $t = 6$ represent 1996

$$y_1 = -2.75t^2 + 86.8t + 659$$
$$y_2 = -1.88t^2 + 62.4t + 616$$

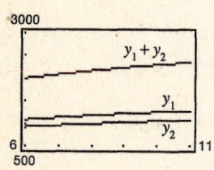

97. $f(x) = 6x \implies f^{-1}(x) = \frac{1}{6}x$

$$f(f^{-1}(x)) = f(\tfrac{1}{6}x) = 6(\tfrac{1}{6}x) = x$$
$$f^{-1}(f(x)) = f^{-1}(6x) = \tfrac{1}{6}(6x) = x$$

99. (a)

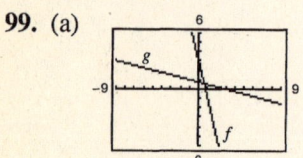

Reflection in the line $y = x$

(b)

x	-5	-1	0	1	3
$f(x)$	23	7	3	-1	-9

x	23	7	3	-1	-9
$g(x)$	-5	-1	0	1	3

The entries in the table are the same except that their rows are interchanged.

101.

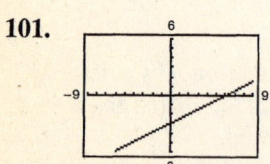

$f(x) = \frac{1}{2}x - 3$ passes the Horizontal Line Test, and hence is one-to-one and has an inverse $(f^{-1}(x) = 2(x + 3))$.

103.

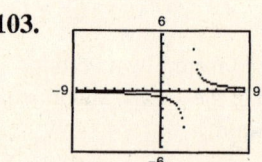

$h(t) = \dfrac{2}{t - 3}$ passes the Horizontal Line Test, and hence is one-to-one.

105. $f(x) = \dfrac{x}{12}$

$y = \dfrac{x}{12}$

$x = \dfrac{y}{12}$

$12x = y$

$f^{-1}(x) = 12x$

107. $f(x) = 4x^3 - 3$

$y = 4x^3 - 3$

$x = 4y^3 - 3$

$x + 3 = 4y^3$

$\dfrac{x + 3}{4} = y^3$

$f^{-1}(x) = \sqrt[3]{\dfrac{x + 3}{4}}$

109. (a)

(b) Yes, the relationship is approximately linear. Higher entrance exam scores x are associated with higher grade-point averages, y.

111. (a)

(b) $S \approx 10t$ (approximations will vary)

(c) $S = 9.7t + 0.4$

(d) For $t = 2.5$, $S \approx 24.7$ m/sec

113. $y = 95.174x - 458.423$

115. The model does not fit well.

117. False. $g(x) = -[(x - 6)^2 + 3] = -(x - 6)^2 - 3$
and $g(-1) = -52 \neq 28$

119. False. $f(x) = \dfrac{1}{x}$ or $f(x) = x$ satisfy $f = f^{-1}$.

Chapter 1   Practice Test

1. Find the slope of the line passing through the points $(-2, 2)$ and $(1, 3)$.

2. Find an equation for the line passing through the points $(3, -2)$ and $(4, -5)$. Use a graphing utility to sketch a graph of the line.

3. Find an equation of the line that passes through the point $(-1, 5)$ and has slope -3. Use a graphing utility to sketch a graph of the line.

4. Find the slope-intercept form of the line that passes through the point $(-3, 2)$ and is perpendicular to $3x + 5y = 7$.

5. Does the equation $x^4 + y^4 = 16$ represent y as a function of x?

6. Evaluate the function $f(x) = |x - 2|/(x - 2)$ at the points $x = 0$, $x = 2$, and $x = 4$.

7. Find the domain of the function $f(x) = 5/(x^2 - 16)$.

8. Find the domain of the function $g(t) = \sqrt{4 - t}$.

9. Use a graphing utility to sketch the graph of the function $f(x) = 3 - x^6$ and determine if the function is even, odd, or neither.

10. Determine the open interval(s) on which the function $f(x) = 12x - x^3$ is increasing.

11. Use a graphing utility to approximate any relative minimum or maximum values of the function $y = 4 - x + x^3$.

12. Compare the graph of $f(x) = x^3 - 3$ with the graph of $y = x^3$.

13. Compare the graph of $f(x) = \sqrt{x - 6}$ with the graph of $y = \sqrt{x}$.

14. Find $g \circ f$ if $f(x) = \sqrt{x}$ and $g(x) = x^2 - 2$. What is the domain of $g \circ f$?

15. Find f/g if $f(x) = 3x^2$ and $g(x) = 16 - x^4$. What is the domain of f/g?

16. Show that $f(x) = 3x + 1$ and $g(x) = \dfrac{x - 1}{3}$ are inverse functions algebraically and graphically.

17. Find the inverse of $f(x) = \sqrt{9 - x^2}$, $0 \leq x \leq 3$. Graph f and f^{-1} in the same viewing rectangle.

18. Use a graphing utility to find the least squares regression line for the points $(-1, 0)$, $(0, 1)$, $(3, 3)$, $(4, 5)$. Graph the points and the line.

CHAPTER 2
Polynomial and Rational Functions

CHAPTER 2
Polynomial and Rational Functions

Section 2.1   Quadratic Functions

You should know the following facts about parabolas.

- $f(x) = ax^2 + bx + c, \ a \neq 0$, is a quadratic function, and its graph is a parabola.
- If $a > 0$, the parabola opens upward and the vertex is the minimum point. If $a < 0$, the parabola opens downward and the vertex is the maximum point.
- The vertex is $(-b/2a, f(-b/2a))$.
- To find the x-intercepts (if any), solve
    $$ax^2 + bx + c = 0.$$
- The standard form of the equation of a parabola is
    $$f(x) = a(x - h)^2 + k$$
  where $a \neq 0$.

  (a) The vertex is (h, k).

  (b) The axis is the vertical line $x = h$.

Solutions to Odd-Numbered Exercises

1. $f(x) = (x - 2)^2$ opens upward and has vertex $(2, 0)$.

Matches graph (g).

3. $f(x) = x^2 - 2$ opens upward and has vertex $(0, -2)$.

Matches graph (b).

5. $f(x) = 4 - (x - 2)^2 = -(x - 2)^2 + 4$ opens downward and has vertex $(2, 4)$.

Matches graph (f).

7. $f(x) = x^2 + 3$ opens upward and has vertex $(0, 3)$.

Matches graph (e)

9. (a) $y = \frac{1}{2}x^2$ vertical shrink

 (b) $y = \frac{1}{2}x^2 - 1$ vertical shrink and vertical shift 1 unit downward

 (c) $y = \frac{1}{2}(x + 3)^2$ vertical shrink and horizontal shift 3 units to the left

 (d) $y = -\frac{1}{2}(x + 3)^2 - 1$ horizontal shift 3 units to the left, vertical shrink, reflection in x-axis, and vertical shift 1 unit downward

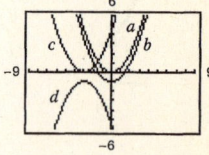

11. (a) $y = -2x^2$ vertical stretch and reflection in the x-axis

 (b) $y = -2x^2 - 1$ vertical stretch, reflection in the x-axis, and vertical shift 1 unit downward

 (c) $y = -2(x - 3)^2$ horizontal shift 3 units to the right, vertical stretch, and reflection in the x-axis

 (d) $y = 2(x - 3)^2 - 1$ horizontal shift 3 units to the right, vertical stretch, and vertical shift 1 unit downward

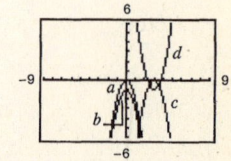

13. $f(x) = 25 - x^2$

Vertex: $(0, 25)$

x-intercepts: $(-5, 0), (5, 0)$

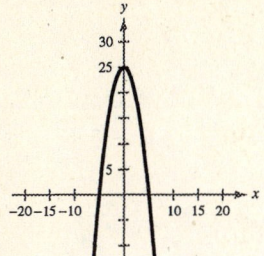

15. $f(x) = \frac{1}{2}x^2 - 4$

Vertex: $(0, -4)$

x-intercepts: $(\pm 2\sqrt{2}, 0)$

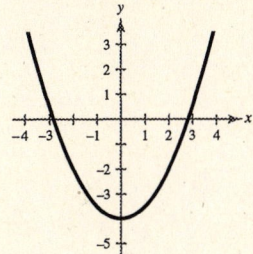

17. $f(x) = (x + 4)^2 - 3$

Vertex: $(-4, -3)$

x-intercepts: $(-4 \pm \sqrt{3}, 0)$

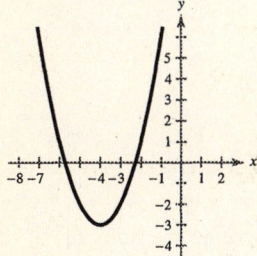

19. $h(x) = x^2 - 8x + 16 = (x - 4)^2$

Vertex: $(4, 0)$

x-intercepts: $(4, 0)$

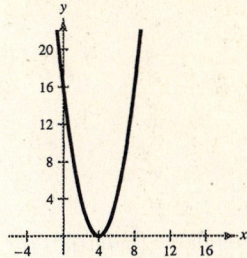

21. $f(x) = x^2 - x + \frac{5}{4} = \left(x - \frac{1}{2}\right)^2 + 1$

Vertex: $\left(\frac{1}{2}, 1\right)$

x-intercepts: None

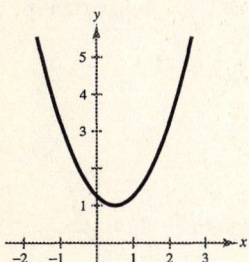

23. $f(x) = -x^2 + 2x + 5 = -(x - 1)^2 + 6$

Vertex: $(1, 6)$

x-intercepts: $\left(1 - \sqrt{6}, 0\right), \left(1 + \sqrt{6}, 0\right)$

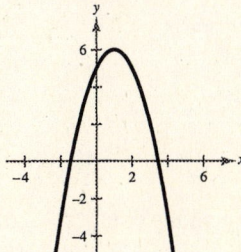

25. $h(x) = 4x^2 - 4x + 21 = 4\left(x - \frac{1}{2}\right)^2 + 20$

Vertex: $\left(\frac{1}{2}, 20\right)$

x-intercepts: None

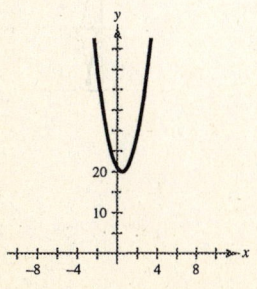

27. $f(x) = -(x^2 + 2x - 3) = -(x + 1)^2 + 4$

Vertex: $(-1, 4)$

x-intercepts: $(-3, 0), (1, 0)$

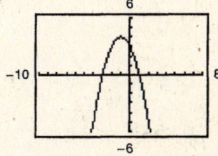

29. $g(x) = x^2 + 8x + 11 = (x + 4)^2 - 5$

Vertex: $(-4, -5)$

x- intercepts: $\left(-4 \pm \sqrt{5}, 0\right)$

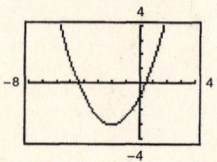

31. $f(x) = -2x^2 + 16x - 31$

$\quad = -2\left(x^2 - 8x + \frac{31}{2}\right)$

$\quad = -2\left(x^2 - 8x + 16 - \frac{1}{2}\right)$

$\quad = -2(x - 4)^2 + 1$

Vertex: $(4, 1)$

x- intercepts: $\left(4 \pm \frac{1}{2}\sqrt{2}, 0\right)$

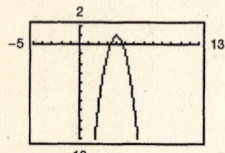

33. $g(x) = \frac{1}{2}(x^2 + 4x - 2) = \frac{1}{2}(x^2 + 4x + 4 - 6)$

$\qquad\qquad\qquad\qquad = \frac{1}{2}(x + 2)^2 - 3$

Vertex: $(-2, -3)$

x- intercepts: $\left(-2 \pm \sqrt{6}, 0\right)$

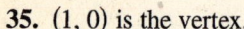

35. $(1, 0)$ is the vertex.

$f(x) = a(x - 1)^2 + 0 = a(x - 1)^2$

Since the graph passes through the point $(0, 1)$ we have:

$\quad 1 = a(0 - 1)^2$

$\quad 1 = a$

$f(x) = 1(x - 1)^2 = (x - 1)^2$

37. $(-1, 4)$ is the vertex.

$f(x) = a(x + 1)^2 + 4$

Since the graph passes through the point $(1, 0)$ we have

$\quad 0 = a(1 + 1)^2 + 4$

$\quad 0 = 4a + 4$

$\quad -1 = a$

Thus, $f(x) = -(x + 1)^2 + 4$. Note that $(-3, 0)$ is on the parabola.

39. $(-2, 5)$ is the vertex.

$f(x) = a(x + 2)^2 + 5$

Since the graph passes through the point $(0, 9)$, we have:

$\quad 9 = a(0 + 2)^2 + 5$

$\quad 4 = 4a$

$\quad 1 = a$

$f(x) = 1(x + 2)^2 + 5 = (x + 2)^2 + 5$

41. $\left(\frac{5}{2}, -\frac{3}{4}\right)$ is the vertex.

$f(x) = a\left(x - \frac{5}{2}\right)^2 - \frac{3}{4}$

Since the graph passes through $(-2, 4)$,

$\quad 4 = a\left(-2 - \frac{5}{2}\right)^2 - \frac{3}{4}$

$\quad \frac{19}{4} = a\left(-\frac{9}{2}\right)^2$

$\quad 19 = 81a$

$\quad a = \frac{19}{81}$

Thus, $f(x) = \frac{19}{81}\left(x - \frac{5}{2}\right)^2 - \frac{3}{4}$

43. $y = x^2 - 4x - 5 \qquad\qquad 0 = x^2 - 4x - 5$

x-intercepts: $(5, 0), (-1, 0) \quad 0 = (x - 5)(x + 1)$

$\qquad\qquad\qquad\qquad\qquad\quad x = 5 \text{ or } x = -1$

45. $y = x^2 + 8x + 16$

x-intercept: $(-4, 0)$

$0 = x^2 + 8x + 16$

$0 = (x + 4)^2$

$x = -4$

x-intercept: $(-4, 0)$

47. $y = x^2 - 4x$ $0 = x^2 - 4x$

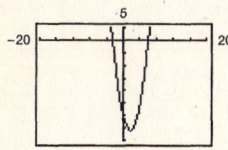

$0 = x(x - 4)$

$x = 0 \text{ or } x = 4$

x-intercepts: $(0, 0), (4, 0)$

49. $y = 2x^2 - 7x - 30$ $0 = 2x^2 - 7x - 30$

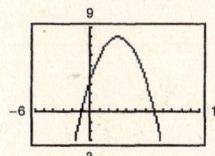

$0 = (2x + 5)(x - 6)$

$x = -\frac{5}{2} \text{ or } x = 6$

x-intercepts: $\left(-\frac{5}{2}, 0\right), (6, 0)$

51. $y = -\frac{1}{2}(x^2 - 6x - 7)$

$0 = -\frac{1}{2}(x^2 - 6x - 7)$

$0 = x^2 - 6x - 7$

$0 = (x + 1)(x - 7)$

$x = -1, 7$

x-intercepts: $(-1, 0), (7, 0)$

53. $f(x) = [x - (-1)](x - 3)$ opens upward

$= (x + 1)(x - 3)$

$= x^2 - 2x - 3$

$g(x) = -[x - (-1)](x - 3)$ opens downward

$= -(x + 1)(x - 3)$

$= -(x^2 - 2x - 3)$

$= -x^2 + 2x + 3$

Note: $f(x) = a(x + 1)(x - 3)$ has *x*-intercepts $(-1, 0)$ and $(3, 0)$ for all real numbers $a \neq 0$.

55. $f(x) = [x - (-3)]\left[x - \left(-\frac{1}{2}\right)\right](2)$ opens upward

$= (x + 3)\left(x + \frac{1}{2}\right)(2)$

$= (x + 3)(2x + 1)$

$= 2x^2 + 7x + 3$

$g(x) = -(2x^2 + 7x + 3)$ opens downward

$= -2x^2 - 7x - 3$

Note: $f(x) = a(x + 3)(2x + 1)$ has *x*-intercepts $(-3, 0)$ and $\left(-\frac{1}{2}, 0\right)$ for all real numbers $a \neq 0$.

57. Let $x =$ the first number and $y =$ the second number. Then the sum is

$$x + y = 110 \implies y = 110 - x.$$

The product is

$$P(x) = xy = x(110 - x) = 110x - x^2.$$

$$P(x) = -x^2 + 110x$$

$$= -(x^2 - 110x + 3025 - 3025)$$

$$= -[(x - 55)^2 - 3025]$$

$$= -(x - 55)^2 + 3025$$

The maximum value of the product occurs at the vertex of $P(x)$ and is 3025. This happens when $x = y = 55$.

59. Let x be the first number and y be the second number.

Then $x + 2y = 24 \implies x = 24 - 2y.$

The product is $P = xy = (24 - 2y)y = 24y - 2y^2.$

Completing the square,

$$P = -2y^2 + 24y$$

$$= -2(y^2 - 12y + 36) + 72$$

$$= -2(y - 6)^2 + 72.$$

The maximum value of the product P occurs at the vertex of the parabola and equals 72. This happens when $y = 6$ and $x = 24 - 2(6) = 12.$

61. (a)

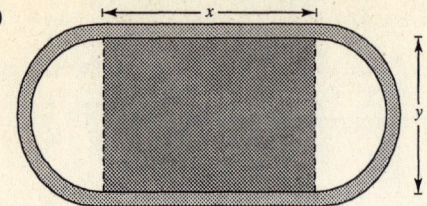

(b) Radius of semicircular ends of track: $r = \frac{1}{2}y$ distance around two semicircular parts of track:

$$d = 2\pi r = 2\pi\left(\frac{1}{2}y\right) = \pi y$$

(c) Distance traveled around track in one lap:

$$d = \pi y + 2x = 200$$

$$\pi y = 200 - 2x$$

$$y = \frac{200 - 2x}{\pi}$$

(e)

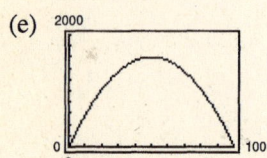

The area is maximum when $x = 50$ and

$$y = \frac{200 - 2(50)}{\pi} = \frac{100}{\pi}.$$

(d) Area of rectangular region:

$$A = xy = x\left(\frac{200 - 2x}{\pi}\right)$$

$$= \frac{1}{\pi}(200x - 2x^2)$$

$$= -\frac{2}{\pi}(x^2 - 100x)$$

$$= -\frac{2}{\pi}(x^2 - 100x + 2500 - 2500)$$

$$= -\frac{2}{\pi}(x - 50)^2 + \frac{5000}{\pi}$$

The area is maximum when $x = 50$ and

$$y = \frac{200 - 2(50)}{\pi} = \frac{100}{\pi}.$$

63. $y = -\dfrac{1}{12}x^2 + 2x + 4$

(a)

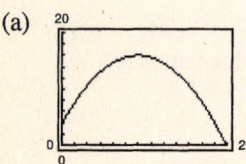

(b) When $x = 0$, $y = 4$ feet.

(c) The vertex occurs at

$$x = -\frac{b}{2a} = -\frac{2}{2(-1/12)} = 12.$$

The maximum height is

$$y = -\frac{1}{12}(12)^2 + 2(12) + 4$$

$$= 16 \text{ feet.}$$

(d) You can solve this part graphically by finding the x-intercept of the graph:

$$x \approx 25.856.$$

Algebraically,

$$0 = -\frac{1}{12}x^2 + 2x + 4$$

$$0 = x^2 - 24x - 48 \quad \text{(Multiply both sides by } -12.\text{)}$$

$$x = \frac{-(-24) \pm \sqrt{(-24)^2 - 4(1)(-48)}}{2(1)}$$

$$= \frac{24 \pm \sqrt{768}}{2} = \frac{24 \pm 16\sqrt{3}}{2} = 12 \pm 8\sqrt{3}$$

Using the positive value for x, we have

$$x = 12 + 8\sqrt{3} \approx 25.86 \text{ feet.}$$

65. $C = 800 - 10x + 0.25x^2$

x	10	15	20	25	30
C	725	706.25	700	706.25	725

From the table, the minimum cost seems to be at $x = 20$.

The minimum cost occurs at the vertex.

$$x = -\frac{b}{2a} = -\frac{(-10)}{2(0.25)} = \frac{10}{.5} = 20$$

$C(20) = 700$ is the minimum cost.

Graphically, you could graph
$C = 800 - 10x + 0.25x^2$ in the window
$[0, 40] \times [0, 1000]$ and find the vertex $(20, 700)$.

67. (a) $C = 4274 + 3.4t - 1.52t^2$ $0 \le t \le 41$

($t = 0$ corresponds to 1960)

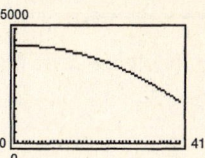

(b) Using a graphing utility, the maximum is about 4276 for $t = 1.12$, or 1961.

Yes the per capita consumption is decreasing. (Answers will vary.)

(c) For 2000 ($t = 40$), $C = 1978$ cigarettes per person. The annual consumption per smoker was

$$\frac{1978(209,128,000)}{48,300,00} \approx 8564 \text{ per smoker per year.}$$

The daily consumption was

$$\frac{8564}{365} \approx 23.5 \text{ cigarettes per smoker per day.}$$

69. True

$$-12x^2 - 1 = 0$$

$$12x^2 = -1 \text{ impossible}$$

71. Model (a) is preferable. $a > 0$ means the parabola opens upward and profits are increasing for t to the right of the vertex,

$$t \ge -\frac{b}{(2a)}.$$

73. $x + y = 8 \implies y = 8 - x$.

Then $-\frac{2}{3}x + y = -\frac{2}{3}x + (8 - x) = 6 \implies -\frac{5}{3}x = -2 \implies x = \frac{6}{5}$ and $y = 8 - \frac{6}{5} = \frac{34}{5}$

$(1.2, 6.8)$

75. $y = x + 3 = 9 - x^2$

$x^2 + x - 6 = 0$

$(x + 3)(x - 2) = 0$

$x = -3, x = 2$

Thus, $(-3, 0)$ and $(2, 5)$ are the points of intersection.

Section 2.2   Polynomial Functions of Higher Degree

■ You should know the following basic principles about polynomials.

■ $f(x) = a_n x^n + a_{n-1} x^{n-1} + \cdots + a_2 x^2 + a_1 x + a_0$, $a_n \neq 0$, is a polynomial function of degree n.

■ If f is of odd degree and

(a) $a_n > 0$, then

1. $f(x) \to \infty$ as $x \to \infty$.
2. $f(x) \to -\infty$ as $x \to -\infty$.

(b) $a_n < 0$, then

1. $f(x) \to -\infty$ as $x \to \infty$.
2. $f(x) \to \infty$ as $x \to -\infty$.

■ If f is of even degree and

(a) $a_n > 0$, then

1. $f(x) \to \infty$ as $x \to \infty$.
2. $f(x) \to \infty$ as $x \to -\infty$.

(b) $a_n < 0$, then

1. $f(x) \to -\infty$ as $x \to \infty$.
2. $f(x) \to -\infty$ as $x \to -\infty$.

■ The following are equivalent for a polynomial function.

(a) $x = a$ is a zero of a function.

(b) $x = a$ is a solution of the polynomial equation $f(x) = 0$.

(c) $(x - a)$ is a factor of the polynomial.

(d) $(a, 0)$ is an x-intercept of the graph of f.

■ A polynomial of degree n has at most n distinct zeros.

■ If f is a polynomial function such that $a < b$ and $f(a) \neq f(b)$, then f takes on every value between $f(a)$ and $f(b)$ in the interval $[a, b]$.

■ If you can find a value where a polynomial is positive and another value where it is negative, then there is at least one real zero between the values.

Solutions to Odd-Numbered Exercises

1. $f(x) = -2x + 3$ is a line with y-intercept $(0, 3)$. Matches graph (f).

3. $f(x) = -2x^2 - 5x$ is a parabola with x-intercepts $(0, 0)$ and $\left(-\frac{5}{2}, 0\right)$ and opens downward. Matches graph (c).

5. $f(x) = -\frac{1}{4}x^4 + 3x^2$ has intercepts $(0, 0)$ and $\left(\pm 2\sqrt{3}, 0\right)$. Matches graph (e).

7. $f(x) = x^4 + 2x^3$ has intercepts $(0, 0)$ and $(-2, 0)$. Matches graph (g).

9. $y = x^3$

(a) $f(x) = (x - 2)^3$

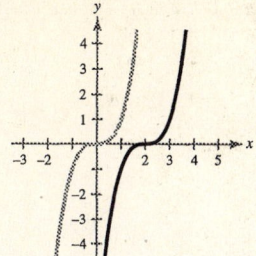

Horizontal shift two units to the right

(b) $f(x) = x^3 - 2$

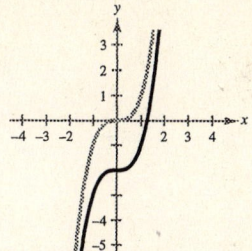

Vertical shift two units downward

(c) $f(x) = -\frac{1}{2}x^3$

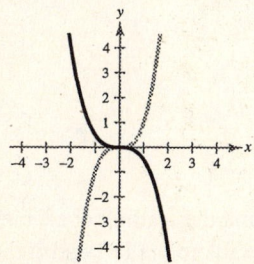

Reflection in the x-axis and a vertical shrink

(d) $f(x) = (x - 2)^3 - 2$

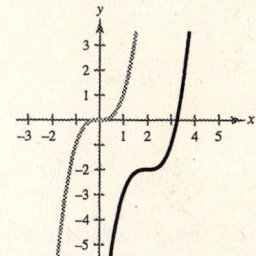

Horizontal shift two units to the right and a vertical shift two units downward

11. $y = x^4$

(a) $f(x) = (x + 5)^4$

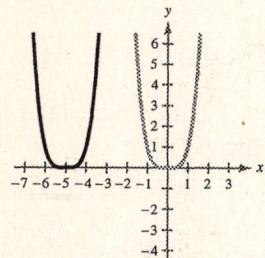

Horizontal shift five units to the left

(b) $f(x) = x^4 - 5$

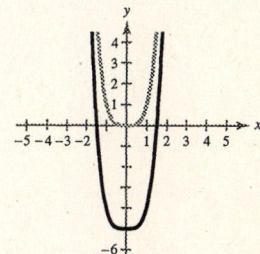

Vertical shift five units downward

(c) $f(x) = 4 - x^4$

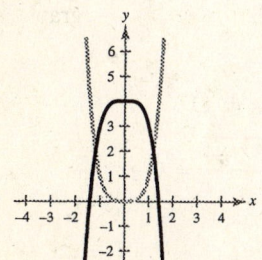

Reflection in the x-axis and then a vertical shift four units upward

(d) $f(x) = \frac{1}{2}(x - 1)^4$

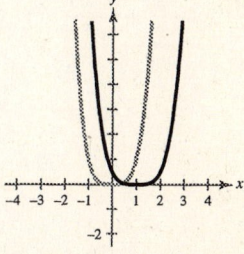

Horizontal shift one unit to the right and a vertical shrink

13. $f(x) = 3x^3 - 9x + 1$; $g(x) = 3x^3$

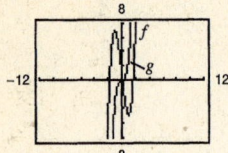

15. $f(x) = -(x^4 - 4x^3 + 16x)$; $g(x) = -x^4$

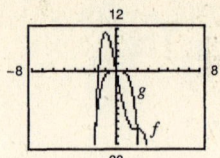

17. $f(x) = 2x^2 - 3x + 1$

Degree: 2

Leading coefficient: 2

The degree is even and the leading coefficient is positive. The graph rises to the left and right.

19. $g(x) = 5 - \frac{7}{2}x - 3x^2$

Degree: 2

Leading coefficient: -3

The degree is even and the leading coefficient is negative. The graph falls to the left and right.

21. $f(x) = \dfrac{6 - 2x + 4x^2 - 5x^3}{3}$

Degree: 3

Leading coefficient: $-\frac{5}{3}$

The degree is odd and the leading coefficient is negative. The graph rises to the left and falls to the right.

23. $h(t) = -\frac{2}{3}(t^2 - 5t + 3)$

Degree: 2

Leading coefficient: $-\frac{2}{3}$

The degree is even and the leading coefficient is negative. The graph falls to the left and right.

25. $f(x) = x^2 - 25$

$\quad = (x + 5)(x - 5)$

$\quad x = \pm 5$

27. $h(t) = t^2 - 6t + 9$

$\quad = (t - 3)^2$

$\quad t = 3 \quad \text{(multiplicity 2)}$

29. $f(x) = x^2 + x - 2$

$\quad = (x + 2)(x - 1)$

$\quad x = -2, 1$

31. $f(t) = t^3 - 4t^2 + 4t$

$\quad = t(t - 2)^2$

$\quad t = 0, 2 \quad \text{(multiplicity 2)}$

33. $f(x) = \dfrac{1}{2}x^2 + \dfrac{5}{2}x - \dfrac{3}{2}$

$\quad = \dfrac{1}{2}(x^2 + 5x - 3)$

$\quad x = \dfrac{-5 \pm \sqrt{25 - 4(-3)}}{2} = -\dfrac{5}{2} \pm \dfrac{\sqrt{37}}{2}$

$\quad \approx 0.5414, -5.5414$

35. (a)

(b) $x \approx 3.732, 0.268$

(c) $f(x) = 3x^2 - 12x + 3$

$\quad = 3(x^2 - 4x + 1)$

$\quad x = \dfrac{4 \pm \sqrt{16 - 4}}{2} = 2 \pm \sqrt{3}$

37. (a)

(b) $t = \pm 1$

(c) $g(t) = \frac{1}{2}t^4 - \frac{1}{2}$

$\quad = \frac{1}{2}(t + 1)(t - 1)(t^2 + 1)$

$\quad t = \pm 1$

39. (a)

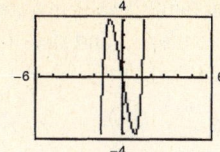

(b) $x = 0, 1.414, -1.414$

(c) $f(x) = x^5 + x^3 - 6x$

$\qquad = x(x^4 + x^2 - 6)$

$\qquad = x(x^2 + 3)(x^2 - 2)$

$x = 0, \pm\sqrt{2}$

41. (a)

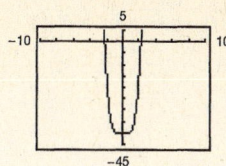

(b) $2.236, -2.236$

(c) $f(x) = 2x^4 - 2x^2 - 40$

$\qquad = 2(x^2 + 4)\left(x + \sqrt{5}\right)\left(x - \sqrt{5}\right)$

$x = \pm\sqrt{5}$

43. (a)

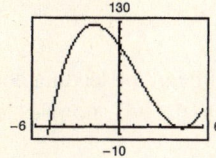

(b) $x = 4, 5, -5$

(c) $f(x) = x^3 - 4x^2 - 25x + 100$

$\qquad = x^2(x - 4) - 25(x - 4)$

$\qquad = (x^2 - 25)(x - 4)$

$\qquad = (x - 5)(x + 5)(x - 4)$

$x = \pm 5, 4$

45. (a)

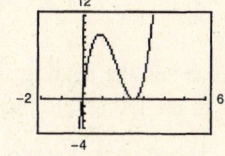

(b) $x = 0, \frac{5}{2}$

(c) $y = 4x^3 - 20x^2 + 25x$

$\quad 0 = 4x^3 - 20x^2 + 25x$

$\quad 0 = x(2x - 5)^2$

$\quad x = 0$ or $x = \frac{5}{2}$ (multiplicity 2)

47.

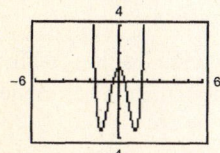

Zeros: $x \approx \pm\, 0.421, \pm 1.680$

Relative maximum: $(0, 1)$

Relative minimums: $(1.225, -3.5), (-1.225, -3.5)$

49.

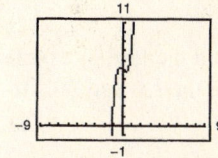

Zeros: $x \approx -1.178$

Relative maximum: $(-0.324, 6.218)$

Relative minimum: $(0.324, 5.782)$

51. $f(x) = (x - 0)(x - 4) = x^2 - 4x$

Note: $f(x) = a(x - 0)(x - 4) = ax(x - 4)$ has zeros 0 and 4 for all nonzero real numbers a.

53. $f(x) = (x - 0)(x + 2)(x + 3) = x^3 + 5x^2 + 6x$

Note: $f(x) = ax(x + 2)(x + 3)$ has zeros 0, -2, and -3 for all nonzero real numbers a.

55. $f(x) = (x - 4)(x + 3)(x - 3)(x - 0)$

$\qquad = (x - 4)(x^2 - 9)x$

$\qquad = x^4 - 4x^3 - 9x^2 + 36x$

Note: $f(x) = a(x^4 - 4x^3 - 9x^2 + 36x)$ has zeros 4, -3, 3, and 0 for all nonzero real numbers a.

57. $f(x) = \left[x - \left(1 + \sqrt{3}\right)\right]\left[x - \left(1 - \sqrt{3}\right)\right]$

$\qquad = \left[(x - 1) - \sqrt{3}\right]\left[(x - 1) + \sqrt{3}\right]$

$\qquad = (x - 1)^2 - \left(\sqrt{3}\right)^2$

$\qquad = x^2 - 2x + 1 - 3$

$\qquad = x^2 - 2x - 2$

Note: $f(x) = a(x^2 - 2x - 2)$ has zeros $1 + \sqrt{3}$ and $1 - \sqrt{3}$ for all nonzero real numbers a.

59. $f(x) = (x - 2)[x - (4 + \sqrt{5})][x - (4 - \sqrt{5})]$

$= (x - 2)[(x - 4) - \sqrt{5}][(x - 4) + \sqrt{5}]$

$= (x - 2)[(x - 4)^2 - 5]$

$= x^3 - 10x^2 + 27x - 22$

Note: $f(x) = a(x - 2)[(x - 4)^2 - 5]$ has zeros
$2, 4 + \sqrt{5},$ and $4 - \sqrt{5}$ for all nonzero real
numbers a.

61. (a) The degree of f is odd and the leading coefficient
is 1. The graph falls to the left and rises to the
right.

(b) $f(x) = x^3 - 9x = x(x^2 - 9) = x(x - 3)(x + 3)$

zeros: $0, 3, -3$

(c), (d)

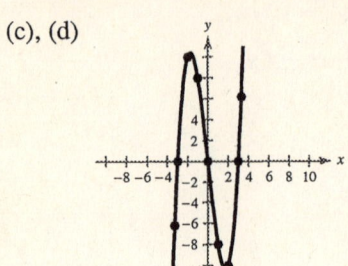

63. (a) The degree of f is even and the leading coefficient
is $\frac{1}{4}$. The graph rises to the left and to the right.

(b) $f(t) = \frac{1}{4}(t^2 - 2t + 15)$ has no real zeros.

(c), (d)

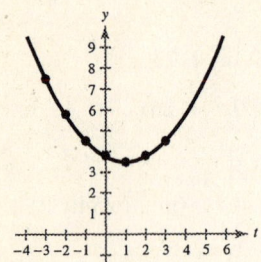

65. (a) The degree of f is odd and the leading coefficient
is 1. The graph falls to the left and rises to the
right.

(b) $f(x) = x^3 - 3x^2 = x^2(x - 3)$; zeros: $0, 3$

(c), (d)

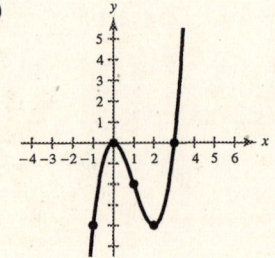

67. (a) The degree of f is odd and the leading coefficient
is -1. The graph rises to the left and falls to the
right.

(b) $f(x) = -x^3 - 5x^2 = x^2(-x - 5)$; zeros: $0, -5$

(c), (d)

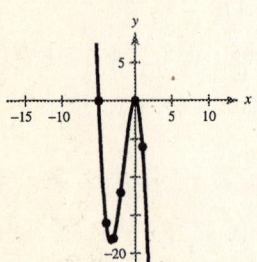

69. (a) The degree of f is odd and the leading coefficient
is 1. The graph falls to the left and rises to the
right.

(b) $f(x) = x^2(x - 4)$; zeros: $0, 4$

(c), (d)

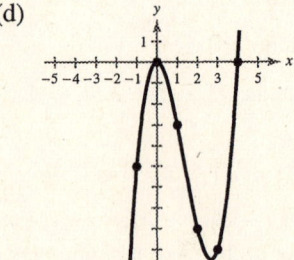

71. (a) The degree of g is even (4) and the leading
coefficient is $-\frac{1}{4}$. The graph falls to the left
and to the right.

(b) $g(t) = -\frac{1}{4}(t - 2)^2(t + 2)^2$; zeros: $2, -2$

(c), (d)

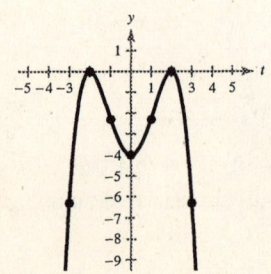

73. $f(x) = x^3 - 3x^2 + 3$

(a)

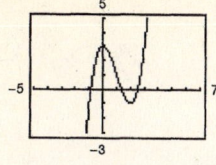

The function has three zeros. They are in the intervals $(-1, 0)$, $(1, 2)$ and $(2, 3)$.

(b) $-0.879, 1.347, 2.532$

(c)

x	$y1$	x	$y1$	x	$y1$
-0.9	-0.159	1.3	0.127	2.5	-0.125
-0.89	-0.0813	1.31	0.09979	2.51	-0.087
-0.88	-0.0047	1.32	0.07277	2.52	-0.0482
-0.87	0.0708	1.33	0.04594	2.53	-0.0084
-0.86	0.14514	1.34	0.0193	2.54	0.03226
-0.85	0.21838	1.35	-0.0071	2.55	0.07388
-0.84	0.2905	1.36	-0.0333	2.56	0.11642

75. $g(x) = 3x^4 + 4x^3 - 3$

(a)

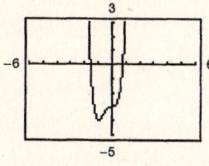

The function has two zeros. They are in the intervals $(-2, -1)$ and $(0, 1)$.

(b) $-1.585, 0.779$

(c)

x	$y1$	x	$y1$
-1.6	0.2768	0.75	-0.3633
-1.59	0.09515	0.76	-0.2432
-1.58	-0.0812	0.77	-0.1193
-1.57	-0.2524	0.78	0.00866
-1.56	-0.4184	0.79	0.14066
-1.55	-0.5795	0.80	0.2768
-1.54	-0.7356	0.81	0.41717

77.

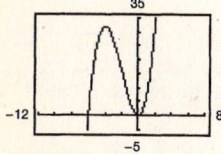

$f(x) = x^2(x + 6)$

No symmetry

Two x-intercepts

79.

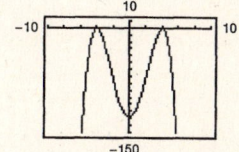

$g(t) = -\frac{1}{2}(t - 4)^2(t + 4)^2$

Symmetric about the y-axis

Two x-intercepts

81.

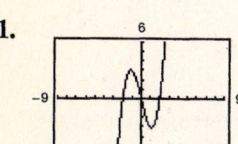

$f(x) = x^3 - 4x = x(x + 2)(x - 2)$

Symmetric to origin

Three x-intercepts

83.

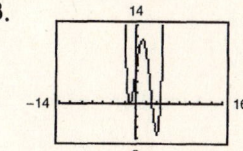

$g(x) = \frac{1}{5}(x + 1)^2(x - 3)(2x - 9)$

Three x-intercepts

No symmetry

85. (a) Volume = length × width × height

Because the box is made from a square, length = width.

Thus:

Volume = (length)² × height

$$= (36 - 2x)^2 x$$

(b) Domain: $0 < 36 - 2x < 36$

$$-36 < -2x < 0$$

$$18 > x > 0$$

(c)

Height, x	Length and Width	Volume, V
1	$36 - 2(1)$	$1[36 - 2(1)]^2 = 1156$
2	$36 - 2(2)$	$2[36 - 2(2)]^2 = 2048$
3	$36 - 2(3)$	$3[36 - 2(3)]^2 = 2700$
4	$36 - 2(4)$	$4[36 - 2(4)]^2 = 3136$
5	$36 - 2(5)$	$5[36 - 2(5)]^2 = 3380$
6	$36 - 2(6)$	$6[36 - 2(6)]^2 = 3456$
7	$36 - 2(7)$	$7[36 - 2(7)]^2 = 3388$

Maximum volume 3456 for $x = 6$

(d)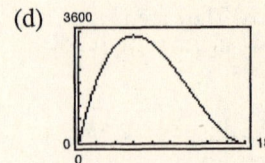

$x = 6$ when $V(x)$ is maximum.

87. The point of diminishing returns (where the graph changes from curving upward to curving downward) occurs when $x = 200$. The point is (200, 160) which corresponds to spending $2,000,000 on advertising to obtain a revenue of $160 million.

89. $y_1 = 0.1250t^2 - 1.446t^2 + 9.07t + 155.5$

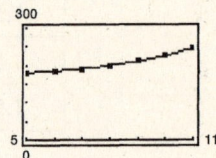

The model is a good fit.

91. For 2007, $t = 17$ and

$$y_1 = 1250(17)^3 - 1.446(17)^2 + 9.07(17) + 155.5$$

$$\approx 505.92, \quad \text{or} \quad \$505,920$$

$$y_2 = -0.2(17)^3 + 5.155(17)^2 - 37.23(17) + 206.8$$

$$\approx 81.085, \quad \text{or} \quad \$81,085$$

The answer for y_2 does not seem reasonable.

93. False. A sixth-degree polynomial can have at most five turning points.

95. $f(x) = -x^3$ matches graph (b)

No, a polynomial of odd degree must cross the x-axis.

97. $f(x) = x^3$ matches graph (a)

No, a polynomial of odd degree must cross the x-axis.

99. $(f + g)(-4) = f(-4) + g(-4)$

$$= -59 + 128 = 69$$

101. $(fg)\left(-\dfrac{4}{7}\right) = f\left(-\dfrac{4}{7}\right)g\left(-\dfrac{4}{7}\right)$

$$= (-11)\left(\dfrac{8 \cdot 16}{49}\right)$$

$$= -\dfrac{1408}{49} \approx -28.7347$$

103. $(fg)(-1) = f(g(-1)) = f(8) = 109$

105. $3(x - 5) < 4x - 7$

$3x - 15 < 4x - 7$

$-8 < x$

107.
$$\frac{5x - 2}{x - 7} \le 4$$

$$\frac{5x - 2}{x - 7} - 4 \le 0$$

$$\frac{5x - 2 - 4(x - 7)}{x - 7} \le 0$$

$$\frac{x + 26}{x - 7} \le 0$$

$[x + 26 \ge 0 \text{ and } x - 7 < 0]$ or $[x + 26 \le 0 \text{ and } x - 7 > 0]$

$[x \ge -26 \text{ and } x < 7]$ or $[x \le -26 \text{ and } x > 7]$

$\qquad\qquad\qquad\qquad$ impossible

$-26 \le x < 7$

Section 2.3   Real Zeros of Polynomial Functions

You should know the following basic techniques and principles of polynomial division.
- The Division Algorithm (Long Division of Polynomials)
- Synthetic Division
- $f(k)$ is equal to the remainder of $f(x)$ divided by $(x - k)$.
- $f(k) = 0$ if and only if $(x - k)$ is a factor of $f(x)$.
- The Rational Zero Test
- The Upper and Lower Bound Rule

Solutions to Odd-Numbered Exercises

1.
$$\begin{array}{r} 2x + 4 \\ x + 3 \overline{) 2x^2 + 10x + 12} \\ -(2x^2 + 6x) \\ \hline 4x + 12 \\ -(4x + 12) \\ \hline 0 \end{array}$$

$$\frac{2x^2 + 10x + 12}{x + 3} = 2x + 4, x \ne -3$$

3.
$$\begin{array}{r} x^2 - 3x + 1 \\ 4x + 5 \overline{) 4x^3 - 7x^2 - 11x + 5} \\ -(4x^3 + 5x^2) \\ \hline -12x^2 - 11x \\ -(-12x^2 - 15x) \\ \hline 4x + 5 \\ -(4x + 5) \\ \hline 0 \end{array}$$

$$\frac{4x^3 - 7x^2 - 11x + 5}{4x + 5} = x^2 - 3x + 1, \quad x \ne -\frac{5}{4}$$

5.
$$\begin{array}{r} 7 \\ x + 2 \overline{) 7x + 3} \\ -(7x + 14) \\ \hline -11 \end{array}$$

$$\frac{7x + 3}{x + 2} = 7 - \frac{11}{x + 2}$$

7.
$$\begin{array}{r} 3x + 5 \\ 2x^2 + 0x + 1 \overline{) 6x^3 + 10x^2 + x + 8} \\ -(6x^3 + 0x^2 + 3x) \\ \hline 10x^2 - 2x + 8 \\ -(10x^2 + 0x + 5) \\ \hline -2x + 3 \end{array}$$

$$\frac{6x^3 + 10x^2 + x + 8}{2x^2 + 1} = 3x + 5 - \frac{2x - 3}{2x^2 + 1}$$

9.
$$\begin{array}{r} x \\ x^2 + 1 \overline{) x^3 + 0x^2 + 0x - 9} \\ -(x^3 + x) \\ \hline - x - 9 \end{array}$$

$$\frac{x^3 - 9}{x^2 + 1} = x - \frac{x + 9}{x^2 + 1}$$

11.
$$\begin{array}{r} 2x \\ x^2 - 2x + 1 \overline{) 2x^3 - 4x^2 - 15x + 5} \\ -(2x^3 - 4x^2 + 2x) \\ \hline - 17x + 5 \end{array}$$

$$\frac{2x^3 - 4x^2 - 15x + 5}{(x - 1)^2} = 2x - \frac{17x - 5}{(x - 1)^2}$$

13.
$$\begin{array}{r|rrrr} 5 & 3 & -17 & 15 & -25 \\ & & 15 & -10 & 25 \\ \hline & 3 & -2 & 5 & 0 \end{array}$$

$$\frac{3x^3 - 17x^2 + 15x - 25}{x - 5} = 3x^2 - 2x + 5, \quad x \neq 5$$

15.
$$\begin{array}{r|rrrr} 3 & 6 & 7 & -1 & 26 \\ & & 18 & 75 & 222 \\ \hline & 6 & 25 & 74 & 248 \end{array}$$

$$\frac{6x^3 + 7x^2 - x + 26}{x - 3} = 6x^2 + 25x + 74 + \frac{248}{x - 3}$$

17.
$$\begin{array}{r|rrrr} 2 & 9 & -18 & -16 & 32 \\ & & 18 & 0 & -32 \\ \hline & 9 & 0 & -16 & 0 \end{array}$$

$$\frac{9x^3 - 18x^2 - 16x + 32}{x - 2} = 9x^2 - 16, \quad x \neq 2$$

19.
$$\begin{array}{r|rrrr} -8 & 1 & 0 & 0 & 512 \\ & & -8 & 64 & -512 \\ \hline & 1 & -8 & 64 & 0 \end{array}$$

$$\frac{x^3 + 512}{x + 8} = x^2 - 8x + 64, \quad x \neq -8$$

21.
$$\begin{array}{r|rrrr} -\tfrac{1}{2} & 4 & 16 & -23 & -15 \\ & & -2 & -7 & 15 \\ \hline & 4 & 14 & -30 & 0 \end{array}$$

$$\frac{4x^3 + 16x^2 - 23x - 15}{x + \tfrac{1}{2}} = 4x^2 + 14x - 30,$$

$$x \neq -\frac{1}{2}$$

23.
$$y_2 = x - 2 + \frac{4}{x + 2}$$

$$= \frac{(x - 2)(x + 2) + 4}{x + 2}$$

$$= \frac{x^2 - 4 + 4}{x + 2}$$

$$= \frac{x^2}{x + 2}$$

$$= y_1$$

25. $f(x) = x^3 - x^2 - 14x + 11, \quad k = 4$

$$\begin{array}{r|rrrr} 4 & 1 & -1 & -14 & 11 \\ & & 4 & 12 & -8 \\ \hline & 1 & 3 & -2 & 3 \end{array}$$

$f(x) = (x - 4)(x^2 + 3x - 2) + 3$

$f(4) = (0)(26) + 3 = 3$

27. $\begin{array}{r|rrrr} \sqrt{2} & 1 & 3 & -2 & -14 \\ & & \sqrt{2} & 2 + 3\sqrt{2} & 6 \\ \hline & 1 & 3 + \sqrt{2} & 3\sqrt{2} & -8 \end{array}$

$f(x) = (x - \sqrt{2})(x^2 + (3 + \sqrt{2})x + 3\sqrt{2}) - 8$

$f(\sqrt{2}) = 0(4 + 6\sqrt{2}) - 8 = -8$

29. $\begin{array}{r|rrrr} 1 - \sqrt{3} & 4 & -6 & -12 & -4 \\ & & 4 - 4\sqrt{3} & 10 - 2\sqrt{3} & 4 \\ \hline & 4 & -2 - 4\sqrt{3} & -2 - 2\sqrt{3} & 0 \end{array}$

$f(x) = (x - 1 + \sqrt{3})[4x^2 - (2 + 4\sqrt{3})x - (2 + 2\sqrt{3})]$

$f(1 - \sqrt{3}) = 0$

31. $f(x) = 4x^3 - 13x + 10$

(a) $\begin{array}{r|rrrr} 1 & 4 & 0 & -13 & 10 \\ & & 4 & 4 & -9 \\ \hline & 4 & 4 & -9 & \underline{1} = f(1) \end{array}$

(b) $\begin{array}{r|rrrr} -2 & 4 & 0 & -13 & 10 \\ & & -8 & 16 & -6 \\ \hline & 4 & -8 & 3 & \underline{4} = f(-2) \end{array}$

(c) $\begin{array}{r|rrrr} \frac{1}{2} & 4 & 0 & -13 & 10 \\ & & 2 & 1 & -6 \\ \hline & 4 & 2 & -12 & \underline{4} = f(\frac{1}{2}) \end{array}$

(d) $\begin{array}{r|rrrr} 8 & 4 & 0 & -13 & 10 \\ & & 32 & 256 & 1944 \\ \hline & 4 & 32 & 243 & \underline{1954} = f(8) \end{array}$

33. $h(x) = 3x^3 + 5x^2 - 10x + 1$

(a) $\begin{array}{r|rrrr} 3 & 3 & 5 & -10 & 1 \\ & & 9 & 42 & 96 \\ \hline & 3 & 14 & 32 & \underline{97} = h(3) \end{array}$

(b) $\begin{array}{r|rrrr} \frac{1}{3} & 3 & 5 & -10 & 1 \\ & & 1 & 2 & -\frac{8}{3} \\ \hline & 3 & 6 & -8 & \underline{-\frac{5}{3}} = h(\frac{1}{3}) \end{array}$

(c) $\begin{array}{r|rrrr} -2 & 3 & 5 & -10 & 1 \\ & & -6 & 2 & 16 \\ \hline & 3 & -1 & -8 & \underline{17} = h(-2) \end{array}$

(d) $\begin{array}{r|rrrr} -5 & 3 & 5 & -10 & 1 \\ & & -15 & 50 & -200 \\ \hline & 3 & -10 & 40 & \underline{-199} = h(-5) \end{array}$

35. $\begin{array}{r|rrrr} 2 & 1 & 0 & -7 & 6 \\ & & 2 & 4 & -6 \\ \hline & 1 & 2 & -3 & 0 \end{array}$

$x^3 - 7x + 6 = (x - 2)(x^2 + 2x - 3)$

$\qquad\qquad\quad = (x - 2)(x + 3)(x - 1)$

Zeros: $2, -3, 1$

37. $\begin{array}{r|rrrr} \frac{1}{2} & 2 & -15 & 27 & -10 \\ & & 1 & -7 & 10 \\ \hline & 2 & -14 & 20 & 0 \end{array}$

$2x^3 - 15x^2 + 27x - 10$

$\qquad = (x - \frac{1}{2})(2x^2 - 14x + 20)$

$\qquad = (2x - 1)(x - 2)(x - 5)$

Zeros: $\frac{1}{2}, 2, 5$

39. (a)
$$
\begin{array}{r|rrrr}
-2 & 2 & 1 & -5 & 2 \\
   &   & -4 & 6 & -2 \\
\hline
   & 2 & -3 & 1 & 0
\end{array}
$$

$$
\begin{array}{r|rrr}
1 & 2 & -3 & 1 \\
  &   & 2 & -1 \\
\hline
  & 2 & -1 & 0
\end{array}
$$

(b) Remaining factor: $(2x - 1)$

(c) $f(x) = (x + 2)(x - 1)(2x - 1)$

(d) Real zeros: $-2, 1, \frac{1}{2}$

(e)

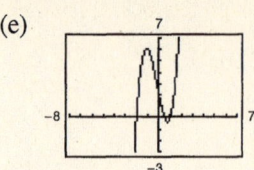

41. (a)
$$
\begin{array}{r|rrrrr}
5 & 1 & -4 & -15 & 58 & -40 \\
  &   & 5 & 5 & -50 & 40 \\
\hline
  & 1 & 1 & -10 & 8 & 0
\end{array}
$$

$$
\begin{array}{r|rrrr}
-4 & 1 & 1 & -10 & 8 \\
   &   & -4 & 12 & -8 \\
\hline
   & 1 & -3 & 2 & 0
\end{array}
$$

(b) $x^2 - 3x + 2 = (x - 2)(x - 1)$,

     Remaining factors: $(x - 2), (x - 1)$

(c) $f(x) = (x - 5)(x + 4)(x - 2)(x - 1)$

(d) Real zeros: $5, -4, 2, 1$

(e)

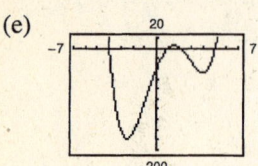

43. (a)
$$
\begin{array}{r|rrrr}
-\frac{1}{2} & 6 & 41 & -9 & -14 \\
             &   & -3 & -19 & 14 \\
\hline
             & 6 & 38 & -28 & 0
\end{array}
$$

$$
\begin{array}{r|rrr}
\frac{2}{3} & 6 & 38 & -28 \\
            &   & 4 & 28 \\
\hline
            & 6 & 42 & 0
\end{array}
$$

(b) $6x + 42$ Remaining factor $\big($or $6(x + 7)\big)$

(c) $f(x) = (2x + 1)(3x - 2)(x + 7)$

     Note: Use $\frac{1}{6}(6x + 42) = x + 7$

(d) Real zeros: $-\frac{1}{2}, \frac{2}{3}, -7$

(e)

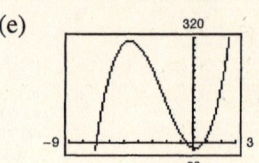

45. $f(x) = x^3 + 3x^2 - x - 3$

Possible rational zeros: $\pm 1, \pm 3$

Zeros shown on graph: $-3, -1, 1$

47. $f(x) = 2x^4 - 17x^3 + 35x^2 + 9x - 45$

Possible rational zeros: $\pm 1, \pm 3, \pm 5, \pm 9, \pm 15, \pm 45,$

$$\pm \tfrac{1}{2}, \pm \tfrac{3}{2}, \pm \tfrac{5}{2}, \pm \tfrac{9}{2}, \pm \tfrac{15}{2}, \pm \tfrac{45}{2}$$

Zeros shown of graph: $-1, \frac{3}{2}, 3, 5$

49. $z^4 - z^3 - 2z - 4 = 0$

Possible rational zeros: $\pm 1, \pm 2, \pm 4$

$$
\begin{array}{r|rrrrr}
-1 & 1 & -1 & 0 & -2 & -4 \\
   &   & -1 & 2 & -2 & 4 \\
\hline
   & 1 & -2 & 2 & -4 & 0
\end{array}
$$

$$
\begin{array}{r|rrrr}
2 & 1 & -2 & 2 & -4 \\
  &   & 2 & 0 & 4 \\
\hline
  & 1 & 0 & 2 & 0
\end{array}
$$

$z^4 - z^3 - 2z - 4 = (z + 1)(z - 2)(z^2 + 2) = 0$

The only real zeros are -1 and 2. You can verify this by graphing the function $f(z) = z^4 - z^3 - 2z - 4$.

51. $2y^4 + 7y^3 - 26y^2 + 23y - 6 = 0$

Using a graphing utility and synthetic division, $1/2, 1,$ and -6 are rational zeros. Hence,

$$(y + 6)(y - 1)^2\big(y - \tfrac{1}{2}\big) = 0 \implies y = -6, 1, \tfrac{1}{2}.$$

53. $h(t) = t^3 - 2t^2 - 7t + 2$

(a) zeros: -2, 3.732, 0.268

(b)
$$
\begin{array}{r|rrrr}
-2 & 1 & -2 & -7 & 2 \\
   &   & -2 & 8 & -2 \\
\hline
   & 1 & -4 & 1 & 0
\end{array}
$$
$t = -2$ is a zero

(c) $h(t) = (t + 2)(t^2 - 4t + 1)$

$\quad = (t + 2)\big[t - (\sqrt{3} + 2)\big]\big[t + (\sqrt{3} - 2)\big]$

55. $h(x) = x^5 - 7x^4 + 10x^3 + 14x^2 - 24x$

(a) $h(x) = x(x^4 - 7x^3 + 10x^2 + 14x - 24)$

From the calculator we have $x = 0, 3, 4$ and $x \approx \pm 1.414$.

(b)
$$
\begin{array}{r|rrrrr}
3 & 1 & -7 & 10 & 14 & -24 \\
  &   & 3 & -12 & -6 & 24 \\
\hline
  & 1 & -4 & -2 & 8 & 0
\end{array}
$$

$$
\begin{array}{r|rrrr}
4 & 1 & -4 & -2 & 8 \\
  &   & 4 & 0 & -8 \\
\hline
  & 1 & 0 & -2 & 0
\end{array}
$$

(c) $h(x) = x(x - 3)(x - 4)(x^2 - 2)$

$\quad = x(x - 3)(x - 4)\big(x - \sqrt{2}\big)\big(x + \sqrt{2}\big)$

The exact roots are $x = 0, \ 3, \ 4, \ \pm\sqrt{2}$.

57. $f(x) = 2x^4 - x^3 + 6x^2 - x + 5$

4 variations in sign $\implies$ 4, 2 or 0 positive real zeros

$f(-x) = 2x^4 + x^3 + 6x^2 + x + 5$

0 variations in sign $\implies$ 0 negative real zeros

59. $g(x) = 4x^3 - 5x + 8$

2 variations in sign $\implies$ 2 or 0 positive real zeros

$g(-x) = -4x^3 + 5x + 8$

1 variation in sign $\implies$ 1 negative real zero

61. $f(x) = x^3 + x^2 - 4x - 4$

(a) $f(x)$ has 1 variation in sign $\implies$ 1 positive real zero

$f(-x) = -x^3 + x^2 + 4x - 4$ has 2 variations in sign $\implies$ 2 or 0 negative real zeros

(b) Possible rational zeros: $\pm 1, \pm 2, \pm 4$

(c)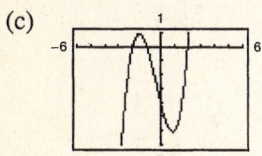

(d) Real zeros: $-2, -1, 2$

63. $f(x) = -2x^4 + 13x^3 - 21x^2 + 2x + 8$

(a) $f(x)$ has 3 variations in sign $\implies$ 3 or 1 positive real zeros

$f(-x) = -2x^4 - 13x^3 - 21x^2 - 2x + 8$ has 1 variation in sign $\implies$ 1 negative real zero

(b) Possible rational zeros: $\pm\frac{1}{2}, \pm 1, \pm 2, \pm 4, \pm 8$

(c)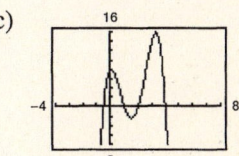

(d) Real zeros: $-\frac{1}{2}, 1, 2, 4$

65. $f(x) = 32x^3 - 52x^2 + 17x + 3$

(a) $f(x)$ has 2 variations in sign $\implies$ 2 or 0 positive real zeros

$f(-x) = -32x^3 - 52x^2 - 17x + 3$ has 1 variation in sign $\implies$ 1 negative real zero

(b) Possible rational zeros: $\pm\frac{1}{32}, \pm\frac{1}{16}, \pm\frac{1}{8}, \pm\frac{1}{4}, \pm\frac{1}{2},$
$\pm 1, \pm\frac{3}{32}, \pm\frac{3}{16}, \pm\frac{3}{8}, \pm\frac{3}{4}, \pm\frac{3}{2}, \pm 3$

(c)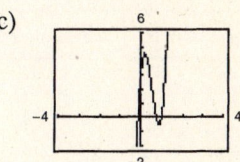

(d) Real zeros: $1, \frac{3}{4}, -\frac{1}{8}$

67. $f(x) = x^4 - 4x^3 + 15$

(a)
$$
\begin{array}{r|rrrrr}
4 & 1 & -4 & 0 & 0 & 15 \\
  &   & 4 & 0 & 0 & 0 \\
\hline
  & 1 & 0 & 0 & 0 & 15
\end{array}
$$

4 is an upper bound.

(b)
$$
\begin{array}{r|rrrrr}
-1 & 1 & -4 & 0 & 0 & 15 \\
   &   & -1 & 5 & -5 & 5 \\
\hline
   & 1 & -5 & 5 & -5 & 20
\end{array}
$$

-1 is a lower bound.

71. $P(x) = x^4 - \frac{25}{4}x^2 + 9$

$\qquad = \frac{1}{4}(4x^4 - 25x^2 + 36)$

$\qquad = \frac{1}{4}(4x^2 - 9)(x^2 - 4)$

$\qquad = \frac{1}{4}(2x + 3)(2x - 3)(x + 2)(x - 2)$

The rational zeros are $\pm\frac{3}{2}$ and ± 2.

75. $f(x) = x^3 - 1 = (x - 1)(x^2 + x + 1)$

Rational zeros: 1 $(x = 1)$

Irrational zeros: 0

Matches (d).

69. $f(x) = x^4 - 4x^3 + 16x - 16$

(a)
$$
\begin{array}{r|rrrrr}
5 & 1 & -4 & 0 & 16 & -16 \\
  &   & 5 & 5 & 25 & 205 \\
\hline
  & 1 & 1 & 5 & 41 & 189
\end{array}
$$

5 is an upper bound.

(b)
$$
\begin{array}{r|rrrrr}
-3 & 1 & -4 & 0 & 16 & -16 \\
   &   & -3 & 21 & -63 & 141 \\
\hline
   & 1 & -7 & 21 & -47 & 125
\end{array}
$$

-3 is a lower bound

73. $f(x) = x^3 - \frac{1}{4}x^2 - x + \frac{1}{4}$

$\qquad = \frac{1}{4}(4x^3 - x^2 - 4x + 1)$

$\qquad = \frac{1}{4}[x^2(4x - 1) - 1(4x - 1)]$

$\qquad = \frac{1}{4}(4x - 1)(x^2 - 1)$

$\qquad = \frac{1}{4}(4x - 1)(x + 1)(x - 1)$

The rational zeros are $\frac{1}{4}$ and ± 1.

77. $f(x) = x^3 - x = x(x + 1)(x - 1)$

Rational zeros: 3 $(x = 0, \pm 1)$

Irrational zeros: 0

Matches (b).

79. $R = 0.03889t^3 - 0.9064t^2 + 8.327 - 0.92$

(a)

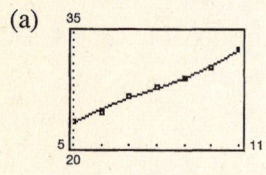

The model is a good approximation.

(b)

Year	1995	1996	1997	1998	1999	2000	2001
R	23.07	24.41	26.48	27.81	28.92	30.37	32.87
Model	22.92	24.81	26.29	27.60	28.96	30.60	32.77

(c) For 2008, $t = 18$ and

$$
\begin{array}{r|rrrr}
18 & 0.03889 & -0.9064 & 8.327 & -0.92 \\
   &   & 0.70002 & -3.7148 & 83.0189 \\
\hline
   & .03889 & -0.20638 & 4.6122 & 82.099
\end{array}
$$

$R(18) \approx 82.099$

No, the model has a cubic term which will grow rapidly.

81. (a) Combined length and width:

$$4x + y = 120 \implies y = 120 - 4x$$

$$\text{Volume} = l \cdot w \cdot h = x^2y$$

$$= x^2(120 - 4x)$$

$$= 4x^2(30 - x)$$

(b)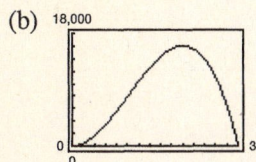

Dimensions with maximum volume: $20 \times 20 \times 40$

(c) $13,500 = 4x^2(30 - x)$

$$4x^3 - 120x^2 + 13,500 = 0$$

$$x^3 - 30x^2 + 3375 = 0$$

$$\begin{array}{r|rrrr} 15 & 1 & -30 & 0 & 3375 \\ & & 15 & -225 & -3375 \\ \hline & 1 & -15 & -225 & 0 \end{array}$$

$$(x - 15)(x^2 - 15x - 225) = 0$$

Using the Quadratic equation, $x = 15, \dfrac{15 \pm 15\sqrt{5}}{2}$

The value of $\dfrac{15 - 15\sqrt{5}}{2}$ is not possible because it is negative.

83. False, $-\frac{4}{7}$ is a zero of f.

85.
$$\begin{array}{r} x^{2n} + 6x^n + 9 \\ x^n + 3 \overline{)\, x^{3n} + 9x^{2n} + 27x^n + 27} \\ \underline{x^{3n} + 3x^{2n}} \\ 6x^{2n} + 27x^n \\ \underline{6x^{2n} + 18x^n} \\ 9x^n + 27 \\ \underline{9x^n + 27} \\ 0 \end{array}$$

$$\frac{x^{3n} + 9x^{2n} + 27x^n + 27}{x^n + 3} = x^{2n} + 6x^n + 9,$$

$$x^n \neq -3$$

[Note: let $y = x^n$ and calculate $(y^3 + 9y^2 + 27y + 27) \div (y + 3)$]

87. (a) $\dfrac{x^2 - 1}{x - 1} = x + 1, \quad x \neq 1$

(b) $\dfrac{x^3 - 1}{x - 1} = x^2 + x + 1, \quad x \neq 1$

(c) $\dfrac{x^4 - 1}{x - 1} = x^3 + x^2 + x + 1, \quad x \neq 1$

In general,

$$\frac{x^n - 1}{x - 1} = x^{n-1} + x^{n-2} + \cdots + x + 1, \quad x \neq 1$$

89. $9x^2 - 25 = 0$

$$(3x + 5)(3x - 5) = 0$$

$$x = -\frac{5}{3}, \frac{5}{3}$$

91. $2x^2 + 6x + 3 = 0$

$$x = \frac{-6 \pm \sqrt{6^2 - 4(2)(3)}}{2(2)}$$

$$= \frac{-6 \pm \sqrt{12}}{4}$$

$$= \frac{-3 \pm \sqrt{3}}{2}$$

$$x = -\frac{3}{2} + \frac{\sqrt{3}}{2}, \quad -\frac{3}{2} - \frac{\sqrt{3}}{2}$$

93. $(x - 0)(x + 12) = x^2 + 12x$

95. $(x - 0)(x + 1)(x - 2)(x - 5)$

$$= (x^2 + x)(x^2 - 7x + 10)$$

$$= x^4 - 6x^3 + 3x^2 + 10x$$

Section 2.4   Complex Numbers

■ You should know how to work with complex numbers.

■ Operations on complex numbers

   (a) Addition: $(a + bi) + (c + di) = (a + c) + (b + d)i$

   (b) Subtraction: $(a + bi) - (c + di) = (a - c) + (b - d)i$

   (c) Multiplication: $(a + bi)(c + di) = (ac - bd) + (ad + bc)i$

   (d) Division: $\dfrac{a + bi}{c + di} = \dfrac{a + bi}{c + di} \cdot \dfrac{c - di}{c - di} = \dfrac{ac + bd}{c^2 + d^2} + \dfrac{bc - ad}{c^2 + d^2}i$

■ The complex conjugate of $a + bi$ is $a - bi$:

   $(a + bi)(a - bi) = a^2 + b^2$

■ The additive inverse of $a + bi$ is $-a - bi$.

■ The multiplicative inverse of $a + bi$ is

   $\dfrac{a - bi}{a^2 + b^2}.$

■ $\sqrt{-a} = \sqrt{a}\,i$ for $a > 0$.

Solutions to Odd-Numbered Exercises

1. $a + bi = -9 + 4i$

   $a = -9$

   $b = 4$

3. $(a - 1) + (b + 3)i = 5 + 8i$

   $a - 1 = 5 \implies a = 6$

   $b + 3 = 8 \implies b = 5$

5. $4 + \sqrt{-25} = 4 + 5i$

7. $7 = 7 + 0i$

9. $-5i + i^2 = -5i - 1 = -1 - 5i$

11. $\left(\sqrt{-75}\right)^2 = -75$

13. $\sqrt{-0.09} = \sqrt{0.09}\,i = 0.3i$

15. $(4 + i) + (7 - 2i) = 11 - i$

17. $\left(-1 + \sqrt{-8}\right) + \left(8 - \sqrt{-50}\right) = 7 + 2\sqrt{2}i - 5\sqrt{2}i = 7 - 3\sqrt{2}i$

19. $13i - (14 - 7i) = 13i - 14 + 7i = -14 + 20i$

21. $\left(\dfrac{3}{2} + \dfrac{5}{2}i\right) + \left(\dfrac{5}{3} + \dfrac{11}{3}i\right) = \left(\dfrac{3}{2} + \dfrac{5}{3}\right) + \left(\dfrac{5}{2} + \dfrac{11}{3}\right)i$

          $= \dfrac{9 + 10}{6} + \dfrac{15 + 22}{6}i$

          $= \dfrac{19}{6} + \dfrac{37}{6}i$

23. $(1.6 + 3.2i) + (-5.8 + 4.3i) = -4.2 + 7.5i$

25. $\sqrt{-6} \cdot \sqrt{-2} = \left(\sqrt{6}i\right)\left(\sqrt{2}i\right) = \sqrt{12}i^2 = \left(2\sqrt{3}\right)(-1) = -2\sqrt{3}$

27. $\left(\sqrt{-10}\right)^2 = \left(\sqrt{10}\,i\right)^2 = 10i^2 = -10$

29. $(1 + i)(3 - 2i) = 3 - 2i + 3i - 2i^2$
$$= 3 + i + 2$$
$$= 5 + i$$

31. $4i(8 + 5i) = 32i + 20i^2 = 32i + 20(-1) = -20 + 32i$

33. $\left(\sqrt{14} + \sqrt{10}\,i\right)\left(\sqrt{14} - \sqrt{10}\,i\right) = 14 - 10i^2 = 14 + 10 = 24$

35. $(4 + 5i)^2 - (4 - 5i)^2 = [(4 + 5i) + (4 - 5i)][(4 + 5i) - (4 - 5i)] = 8(10i) = 80i$

37. $4 - 3i$ is the complex conjugate of $4 + 3i$
$(4 + 3i)(4 - 3i) = 16 + 9 = 25$

39. $-6 + \sqrt{5}\,i$ is the complex conjugate of $-6 - \sqrt{5}\,i$
$\left(-6 - \sqrt{5}\,i\right)\left(-6 + \sqrt{5}\,i\right) = 36 + 5 = 41$

41. $-\sqrt{20}\,i$ is the complex conjugate of
$\sqrt{-20} = \sqrt{20}\,i$
$\left(\sqrt{20}\,i\right)\left(-\sqrt{20}\,i\right) = 20$

43. $3 + \sqrt{2}\,i$ is the complex conjugate of
$3 - \sqrt{-2} = 3 - \sqrt{2}\,i$
$\left(3 - \sqrt{2}\,i\right)\left(3 + \sqrt{2}\,i\right) = 9 + 2 = 11$

45. $\dfrac{6}{i} = \dfrac{6}{i} \cdot \dfrac{-i}{-i} = \dfrac{-6i}{-i^2} = \dfrac{-6i}{1} = -6i$

47. $\dfrac{2}{4 - 5i} = \dfrac{2}{4 - 5i} \cdot \dfrac{4 + 5i}{4 + 5i} = \dfrac{8 + 10i}{16 + 25} = \dfrac{8}{41} + \dfrac{10}{41}i$

49. $\dfrac{2 + i}{2 - i} = \dfrac{2 + i}{2 - i} \cdot \dfrac{2 + i}{2 + i}$
$$= \dfrac{4 + 4i + i^2}{4 + 1}$$
$$= \dfrac{3 + 4i}{5}$$
$$= \dfrac{3}{5} + \dfrac{4}{5}i$$

51. $\dfrac{i}{(4 - 5i)^2} = \dfrac{i}{16 - 25 - 40i}$
$$= \dfrac{i}{-9 - 40i} \cdot \dfrac{-9 + 40i}{-9 + 40i}$$
$$= \dfrac{-40 - 9i}{81 + 40^2}$$
$$= \dfrac{-40}{1681} - \dfrac{9}{1681}i$$

53. $\dfrac{2}{1 + i} - \dfrac{3}{1 - i} = \dfrac{2(1 - i) - 3(1 + i)}{(1 + i)(1 - i)}$
$$= \dfrac{2 - 2i - 3 - 3i}{1 + 1}$$
$$= \dfrac{-1 - 5i}{2}$$
$$= -\dfrac{1}{2} - \dfrac{5}{2}i$$

55. $\dfrac{i}{3 - 2i} + \dfrac{2i}{3 + 8i} = \dfrac{3i + 8i^2 + 6i - 4i^2}{(3 - 2i)(3 + 8i)}$
$$= \dfrac{-4 + 9i}{9 + 18i + 16}$$
$$= \dfrac{-4 + 9i}{25 + 18i} \cdot \dfrac{25 - 18i}{25 - 18i}$$
$$= \dfrac{-100 + 72i + 225i + 162}{25^2 + 18^2}$$
$$= \dfrac{62 + 297i}{949}$$
$$= \dfrac{62}{949} + \dfrac{297}{949}i$$

57. $-6i^3 + i^2 = -6i^2i + i^2$

$$= -6(-1)i + (-1)$$

$$= 6i - 1$$

$$= -1 + 6i$$

59. $\left(\sqrt{-75}\right)^3 = \left(5\sqrt{3}i\right)^3 = 5^3\left(\sqrt{3}\right)^3 i^3$

$$= 125\left(3\sqrt{3}\right)(-i)$$

$$= -375\sqrt{3}i$$

61. $\dfrac{1}{i^3} = \dfrac{1}{i^3} \cdot \dfrac{i}{i} = \dfrac{i}{i^4} = \dfrac{i}{1} = i$

63. $(2)^3 = 8$

$$\left(-1 + \sqrt{3}i\right)^3 = (-1)^3 + 3(-1)^2\left(\sqrt{3}i\right) + 3(-1)\left(\sqrt{3}i\right)^2 + \left(\sqrt{3}i\right)^3$$

$$= -1 + 3\sqrt{3}i - 9i^2 + 3\sqrt{3}i^3$$

$$= -1 + 3\sqrt{3}i + 9 - 3\sqrt{3}i$$

$$= 8$$

$$\left(-1 - \sqrt{3}i\right)^3 = (-1)^3 + 3(-1)^2\left(-\sqrt{3}i\right) + 3(-1)\left(-\sqrt{3}i\right)^2 + \left(-\sqrt{3}i\right)^3$$

$$= -1 - 3\sqrt{3}i - 9i^2 - 3\sqrt{3}i^3$$

$$= -1 - 3\sqrt{3}i + 9 + 3\sqrt{3}i$$

$$= 8$$

The three numbers are cube roots of 8.

65. $4 + 3i$

67. $4 - 5i$

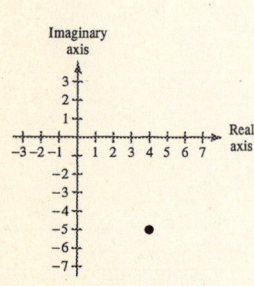

69. $3i$

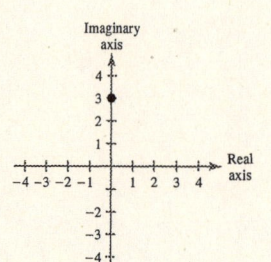

71. 1

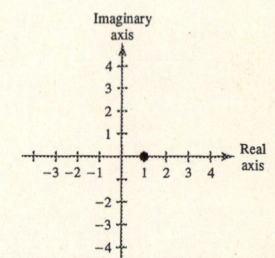

73. The complex number $\frac{1}{2}i$, is in the Mandelbrot Set since for $c = \frac{1}{2}i$, the corresponding Mandelbrot sequence is

$$\frac{1}{2}i, \ -\frac{1}{4} + \frac{1}{2}i, \ -\frac{3}{16} + \frac{1}{4}i, \ -\frac{7}{256} + \frac{13}{32}i, \ -\frac{10,767}{65,536} + \frac{1957}{4096}i, \ -\frac{864,513,055}{4,294,967,296} + \frac{46,037,845}{134,217,728}i$$

which is bounded. Or in decimal form

$0.5i, \ -0.25 + 0.5i, \ -0.1875 + 0.25i, \ -0.02734 + 0.40625i,$

$-0.164291 + 0.477783i, \ -0.201285 + 0.343009i.$

75. The complex number 1 is not in the Mandelbrot Set since for $c = 1$, the corresponding Mandelbrot sequence is 1, 2, 5, 26, 677, 458,330 which is unbounded.

77. $z_1 = 5 + 2i$

$z_2 = 3 - 4i$

$$\frac{1}{z} = \frac{1}{z_1} + \frac{1}{z_2} = \frac{1}{5 + 2i} + \frac{1}{3 - 4i} = \frac{(3 - 4i) + (5 + 2i)}{(5 + 2i)(3 - 4i)} = \frac{8 - 2i}{23 - 14i}$$

$$z = \frac{23 - 14i}{8 - 2i}\left(\frac{8 + 2i}{8 + 2i}\right) = \frac{212 - 66i}{68} \approx 3.118 - 0.971i$$

79. False. A real number $a + 0i = a$ is equal to its conjugate.

81. $(4x - 5)(4x + 5) = 16x^2 - 20x + 20x - 25$

$= 16x^2 - 25$

83. $\left(3x - \frac{1}{2}\right)(x + 4) = 3x^2 - \frac{1}{2}x + 12x - 2 = 3x^2 + \frac{23}{2}x - 2$

Section 2.5   The Fundamental Theorem of Algebra

■   You should know that if f is a polynomial of degree $n > 0$, then f has at least one zero in the complex number system. (Fundamental Theorem of Algebra)

■   You should know that if $a + bi$ is a complex zero of a polynomial f, with real coefficients, then $a - bi$ is also a complex zero of f.

■   You should know the difference between a factor that is irreducible over the rationals (such as $x^2 - 7$) and a factor that is irreducible over the reals (such as $x^2 + 9$).

Solutions to Odd-Numbered Exercises

1. $f(x) = x^2(x + 3)$

The three zeros are $x = 0$, $x = 0$ and $x = -3$

3. $f(x) = (x + 9)(x + 2i)(x - 2i)$

The three zeros are $x = -9$, $x = -2i$ and $x = 2i$

5. $f(x) = x^3 - 4x^2 + x - 4 = x^2(x - 4) + 1(x - 4)$

$= (x - 4)(x^2 + 1)$ zeros: $4, \pm i$

The only real zero of $f(x)$ is $x = 4$. This corresponds to the x-intercept of $(4, 0)$ on the graph.

7. $f(x) = x^4 + 4x^2 + 4 = (x^2 + 2)^2$

zeros: $\pm\sqrt{2}i, \pm\sqrt{2}i$

$f(x)$ has no real zeros and the graph of $f(x)$ has no x-intercepts.

9. $h(x) = x^2 - 4x + 1$

h has no rational zeros. By the Quadratic Formula, the zeros are $x = \dfrac{4 \pm \sqrt{16 - 4}}{2} = 2 \pm \sqrt{3}$.

$h(x) = \left[x - \left(2 + \sqrt{3}\right)\right]\left[x - \left(2 - \sqrt{3}\right)\right] = \left(x - 2 - \sqrt{3}\right)\left(x - 2 + \sqrt{3}\right)$

11. $f(x) = x^2 - 12x + 26$

f has no rational zeros. By the Quadratic Formula, the zeros are

$$x = \frac{12 \pm \sqrt{(-12)^2 - 4(26)}}{2} = 6 \pm \sqrt{10}.$$

$$f(x) = \left[x - \left(6 + \sqrt{10}\right)\right]\left[x - \left(6 - \sqrt{10}\right)\right]$$
$$= \left(x - 6 - \sqrt{10}\right)\left(x - 6 + \sqrt{10}\right)$$

13. $f(x) = x^2 + 25$

$$= (x + 5i)(x - 5i)$$

The zeros of $f(x)$ are $x = \pm 5i$.

15. $f(x) = x^4 - 81$

$$= (x^2 - 9)(x^2 + 9)$$
$$= (x + 3)(x - 3)(x + 3i)(x - 3i)$$

The zeros of $f(x)$ are $x = \pm 3$ and $x = \pm 3i$.

17. $f(z) = z^2 - z + 56$

$$z = \frac{1 \pm \sqrt{1 - 4(56)}}{2}$$
$$= \frac{1 \pm \sqrt{-223}}{2}$$
$$= \frac{1}{2} \pm \frac{\sqrt{223}}{2}i$$

$$f(z) = \left(z - \frac{1}{2} + \frac{\sqrt{223}\,i}{2}\right)\left(z - \frac{1}{2} - \frac{\sqrt{223}\,i}{2}\right)$$

19. $f(t) = t^3 - 3t^2 - 15t + 125$

Possible rational zeros: $\pm 1, \pm 5, \pm 25, \pm 125$

$$
\begin{array}{r|rrrr}
-5 & 1 & -3 & -15 & 125 \\
   &   & -5 & 40 & -125 \\
\hline
   & 1 & -8 & 25 & 0
\end{array}
$$

By the Quadratic Formula, the zeros of

$$t^2 - 8t + 25 \text{ are } t = \frac{8 \pm \sqrt{64 - 100}}{2} = 4 \pm 3i.$$

The zeros of $f(t)$ are $t = -5$ and $t = 4 \pm 3i$.

$$f(t) = [t - (-5)][t - (4 + 3i)][t - (4 - 3i)]$$
$$= (t + 5)(t - 4 - 3i)(t - 4 + 3i)$$

21. $f(x) = 5x^3 - 9x^2 + 28x + 6$

Possible rational zeros: $\pm 6, \pm \frac{6}{5}, \pm 3, \pm \frac{3}{5}, \pm 2, \pm \frac{2}{5},$ $\pm 1, \pm \frac{1}{5}$

$$
\begin{array}{r|rrrr}
-\frac{1}{5} & 5 & -9 & 28 & 6 \\
             &   & -1 & 2 & -6 \\
\hline
             & 5 & -10 & 30 & 0
\end{array}
$$

By the Quadratic Formula, the zeros of $5x^2 - 10x + 30$ are those of $x^2 - 2x + 6$:

$$x = \frac{2 \pm \sqrt{4 - 4(6)}}{2} = 1 \pm \sqrt{5}i$$

zeros: $-\frac{1}{5}, 1 \pm \sqrt{5}i$

$$f(x) = 5\left(x + \frac{1}{5}\right)\left(x - \left(1 + \sqrt{5}i\right)\right)\left(x - \left(1 - \sqrt{5}i\right)\right)$$
$$= (5x + 1)\left(x - 1 - \sqrt{5}i\right)\left(x - 1 + \sqrt{5}i\right)$$

23. $f(x) = x^4 + 10x^2 + 9$

$$= (x^2 + 1)(x^2 + 9)$$
$$= (x + i)(x - i)(x + 3i)(x - 3i)$$

The zeros of $f(x)$ are $x = \pm i$ and $x = \pm 3i$.

25. $g(x) = x^4 - 4x^3 + 8x^2 - 16x + 16$

Possible rational zeros: $\pm 1, \pm 2, \pm 4, \pm 8, \pm 16$

$$
\begin{array}{r|rrrrr}
2 & 1 & -4 & 8 & -16 & 16 \\
  &   & 2 & -4 & 8 & -16 \\
\hline
2 & 1 & -2 & 4 & -8 & 0 \\
  &   & 2 & 0 & 8 &   \\
\hline
  & 1 & 0 & 4 & 0 &
\end{array}
$$

$$g(x) = (x - 2)(x - 2)(x^2 + 4)$$
$$= (x - 2)^2(x + 2i)(x - 2i)$$

The zeros of g are 2, 2, and $\pm 2i$.

27. (a) $f(x) = x^2 - 14x + 46$. By the Quadratic Formula

$$x = \frac{14 \pm \sqrt{(-14)^2 - 4(46)}}{2} = 7 \pm \sqrt{3}.$$

The zeros are $7 + \sqrt{3}$ and $7 - \sqrt{3}$.

(b) $f(x) = \left[x - \left(7 + \sqrt{3}\right)\right]\left[x - \left(7 - \sqrt{3}\right)\right]$

$\quad = \left(x - 7 - \sqrt{3}\right)\left(x - 7 + \sqrt{3}\right)$

(c) x-intercepts: $\left(7 + \sqrt{3}, 0\right)$ and $\left(7 - \sqrt{3}, 0\right)$

(d)
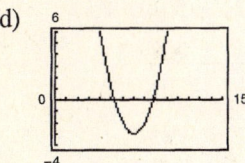

29. (a) $f(x) = x^2 + 14x + 44$. By the Quadratic Formula,

$$x = \frac{-14 \pm \sqrt{14^2 - 4(44)}}{2} = -7 \pm \sqrt{5}$$

The zeros are $-7 + \sqrt{5}$ and $-7 - \sqrt{5}$.

(b) $f(x) = \left[x - \left(-7 + \sqrt{5}\right)\right]\left[x - \left(-7 - \sqrt{5}\right)\right]$

$\quad = \left(x + 7 - \sqrt{5}\right)\left(x + 7 + \sqrt{5}\right)$

(c) x-intercepts: $\left(-7 + \sqrt{5}, 0\right)$, $\left(-7 - \sqrt{5}, 0\right)$

(d)
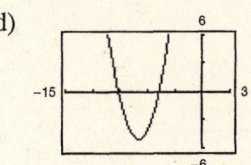

31. (a) $f(x) = x^3 - 11x + 150$

$\quad = (x + 6)(x^2 - 6x + 25).$

Use the Quadratic Formula to find the zeros of $x^2 - 6x + 25$:

$$x = \frac{6 \pm \sqrt{(-6)^2 - 4(25)}}{2} = 3 \pm 4i.$$

The zeros are -6, $3 + 4i$, and $3 - 4i$.

(b) $f(x) = (x + 6)(x - 3 + 4i)(x - 3 - 4i)$

(c) x-intercept: $(-6, 0)$

(d)
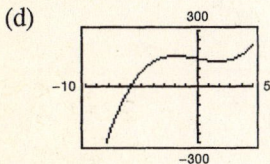

33. (a) $f(x) = x^4 + 25x^2 + 144$

$\quad = (x^2 + 9)(x^2 + 16)$

The zeros are $\pm 3i, \pm 4i$.

(b) $f(x) = (x^2 + 9)(x^2 + 16)$

$\quad = (x + 3i)(x - 3i)(x + 4i)(x - 4i)$

(c) No x-intercepts

(d)
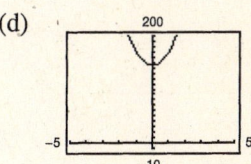

35. $f(x) = (x - 3)(x - i)(x + i)$

$\quad = (x - 3)(x^2 + 1)$

$\quad = x^3 - 3x^2 + x - 3$

Note: $f(x) = a(x^3 - 3x^2 + x - 3)$, where a is any nonzero real number, has zeros $3, \pm i$

37. $f(x) = (x - 2)(x + 4 - i)(x + 4 + i)$

$\quad = (x - 2)[(x + 4)^2 + 1]$

$\quad = (x - 2)(x^2 + 8x + 17)$

$\quad = x^3 + 6x^2 + x - 34$

39. If $1 + \sqrt{3}i$ is a zero, so is its conjugate $1 - \sqrt{3}i$.

$f(x) = (x + 5)^2\left(x - 1 + \sqrt{3}i\right)\left(x - 1 - \sqrt{3}i\right)$

$\quad = (x^2 + 10x + 25)(x^2 - 2x + 4)$

$\quad = x^4 + 8x^3 + 9x^2 - 10x + 100$

Note: $f(x) = a(x^4 + 8x^3 + 9x^2 - 10x + 100)$, where a is any nonzero real number, has these zeros.

41. $f(x) = x^4 - 6x^2 - 7$

(a) $f(x) = (x^2 - 7)(x^2 + 1)$

(b) $f(x) = \left(x - \sqrt{7}\right)\left(x + \sqrt{7}\right)(x^2 + 1)$

(c) $f(x) = \left(x - \sqrt{7}\right)\left(x + \sqrt{7}\right)(x + i)(x - i)$

43. $f(x) = x^4 - 2x^3 - 3x^2 + 12x - 18$

(a) $f(x) = (x^2 - 6)(x^2 - 2x + 3)$　　　　(b) $f(x) = (x + \sqrt{6})(x - \sqrt{6})(x^2 - 2x + 3)$

(c) $f(x) = (x + \sqrt{6})(x - \sqrt{6})(x - 1 - \sqrt{2}i)(x - 1 + \sqrt{2}i)$

45. $f(x) = 2x^3 + 3x^2 + 50x + 75$

Since $5i$ is a zero, so is $-5i$.

$$
\begin{array}{r|rrrr}
5i & 2 & 3 & 50 & 75 \\
   &   & 10i & -50 + 15i & -75 \\
\hline
   & 2 & 3 + 10i & 15i & 0 \\
\end{array}
$$

$$
\begin{array}{r|rrr}
-5i & 2 & 3 + 10i & 15i \\
    &   & -10i & -15i \\
\hline
    & 2 & 3 & 0 \\
\end{array}
$$

The zero of $2x + 3$ is $x = -\frac{3}{2}$. The zeros of f are $x = -\frac{3}{2}$ and $x = \pm 5i$.

Alternate Solution

Since $x = \pm 5i$ are zeros of $f(x)$, $(x + 5i)(x - 5i) = x^2 + 25$ is a factor of $f(x)$. By long division we have:

$$
\require{enclose}
\begin{array}{r}
2x + 3 \\
x^2 + 0x + 25 \enclose{longdiv}{2x^3 + 3x^2 + 50x + 75} \\
\underline{2x^3 + 0x^2 + 50x } \\
3x^2 + 0x + 75 \\
\underline{3x^2 + 0x + 75} \\
0
\end{array}
$$

Thus, $f(x) = (x^2 + 25)(2x + 3)$ and the zeros of f are $x = \pm 5i$ and $x = -\frac{3}{2}$.

47. $g(x) = x^3 - 7x^2 - x + 87$. Since $5 + 2i$ is a zero, so is $5 - 2i$.

$$
\begin{array}{r|rrrr}
5 + 2i & 1 & -7 & -1 & 87 \\
       &   & 5 + 2i & -14 + 6i & -87 \\
\hline
       & 1 & -2 + 2i & -15 + 6i & 0 \\
\end{array}
$$

$$
\begin{array}{r|rrr}
5 - 2i & 1 & -2 + 2i & -15 + 6i \\
       &   & 5 - 2i & 15 - 6i \\
\hline
       & 1 & 3 & 0 \\
\end{array}
$$

The zero of $x + 3$ is $x = -3$.

The zeros of f are $-3, 5 \pm 2i$.

49. $h(x) = 3x^3 - 4x^2 + 8x + 8$. Since $1 - \sqrt{3}i$ is a zero, so is $1 + \sqrt{3}i$.

$$
\begin{array}{r|rrrr}
1 - \sqrt{3}i & 3 & -4 & 8 & 8 \\
              &   & 3 - 3\sqrt{3}i & -10 - 2\sqrt{3}i & -8 \\
\hline
              & 3 & -1 - 3\sqrt{3}i & -2 - 2\sqrt{3}i & 0 \\
\end{array}
$$

$$
\begin{array}{r|rrr}
1 + \sqrt{3}i & 3 & -1 - 3\sqrt{3}i & -2 - 2\sqrt{3}i \\
              &   & 3 + 3\sqrt{3}i & 2 + 2\sqrt{3}i \\
\hline
              & 3 & 2 & 0 \\
\end{array}
$$

The zero of $3x + 2$ is $x = -\frac{2}{3}$.

The zeros of h are $x = -\frac{2}{3}, 1 \pm \sqrt{3}i$.

51. $h(x) = 8x^3 - 14x^2 + 18x - 9$. Since $\frac{1}{2}(1 - \sqrt{5}i)$ is a zero, so is $\frac{1}{2}(1 + \sqrt{5}i)$.

$$
\begin{array}{r|rrrr}
\frac{1}{2}(1 - \sqrt{5}i) & 8 & -14 & 18 & -9 \\
                          &   & 4 - 4\sqrt{5}i & -15 + 3\sqrt{5}i & 9 \\
\hline
                          & 8 & -10 - 4\sqrt{5}i & 3 + 3\sqrt{5}i & 0 \\
\end{array}
$$

$$
\begin{array}{r|rrr}
\frac{1}{2}(1 + \sqrt{5}i) & 8 & -10 - 4\sqrt{5}i & 3 + 3\sqrt{5}i \\
                          &   & 4 + 4\sqrt{5}i & -3 - 3\sqrt{5}i \\
\hline
                          & 8 & -6 & 0 \\
\end{array}
$$

The zero of $8x - 6$ is $x = \frac{3}{4}$.

The zeros of h are $x = \frac{3}{4}, \frac{1}{2}(1 \pm \sqrt{5}i)$.

53. $f(x) = x^4 + 3x^3 - 5x^2 - 21x + 22$

(a) The root feature yields the real roots 1 and 2, and the complex roots $-3 \pm 1.414i$.

(b) By synthetic division,

$$
\begin{array}{r|rrrrr}
1 & 1 & 3 & -5 & -21 & 22 \\
  &   & 1 & 4 & -1 & -22 \\
\hline
  & 1 & 4 & -1 & -22 & 0
\end{array}
$$

$$
\begin{array}{r|rrrr}
2 & 1 & 4 & -1 & -22 \\
  &   & 2 & 12 & 22 \\
\hline
  & 1 & 6 & 11 & 0
\end{array}
$$

The complex roots of $x^2 + 6x + 11$ are

$$x = \frac{-6 \pm \sqrt{6^2 - 4(11)}}{2} = -3 \pm \sqrt{2}i.$$

55. $h(x) = 8x^3 - 14x^2 + 18x - 9$

(a) The root feature yields the real root 0.75, and the complex roots $0.5 \pm 1.118i$.

(b) By synthetic division,

$$
\begin{array}{r|rrrr}
\frac{3}{4} & 8 & -14 & 18 & -9 \\
  &   & 6 & -6 & 9 \\
\hline
  & 8 & -8 & 12 & 0
\end{array}
$$

The complex roots of $8x^2 - 8x + 12$ are

$$x = \frac{8 \pm \sqrt{64 - 4(8)(12)}}{2(8)} = \frac{1}{2} \pm \frac{\sqrt{5}}{2}i.$$

57.

$$-16t^2 + 48t = 64, \quad 0 \le t \le 3$$

$$-16t^2 + 48t - 64 = 0$$

$$t = \frac{-48 \pm \sqrt{1792}i}{-32}$$

Since the roots are imaginary, the ball never will reach a height of 64 feet. You can verify this graphically by observing that $y_1 = -16t^2 + 48t$ and $y_2 = 64$ do not intersect.

59. False, a third degree polynomial must have at least one real zero.

61. $f(x) = x^4 - 4x^2 + k$

(a) f has two real zeros each of multiplicity 2 for $k = 4$: $f(x) = x^4 - 4x^2 + 4 = (x^2 - 2)^2$.

(b) f has two real zeros and two complex zeros if $k < 0$.

63. $f(x) = x^2 - 7x - 8 = \left(x^2 - 7x + \frac{49}{4}\right) - 8 - \frac{49}{4}$

$$= \left(x - \frac{7}{2}\right)^2 - \frac{81}{4}$$

Vertex: $\left(\frac{7}{2}, -\frac{81}{4}\right)$

$f(x) = (x - 8)(x + 1)$

Intercepts: $(8, 0), (-1, 0), (0, -8)$

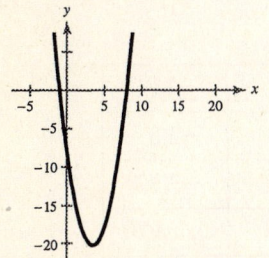

65. $f(x) = 6x^2 + 5x - 6 = (3x - 2)(2x + 3)$

Intercepts: $\left(\frac{2}{3}, 0\right), \left(-\frac{3}{2}, 0\right), (0, -6)$

$f(x) = 6x^2 + 5x - 6$

$$= 6\left(x^2 + \frac{5}{6}x + \frac{25}{144}\right) - 6 - \frac{25}{24}$$

$$= 6\left(x + \frac{5}{12}\right)^2 - \frac{169}{24}$$

Vertex: $\left(-\frac{5}{12}, -\frac{169}{24}\right)$

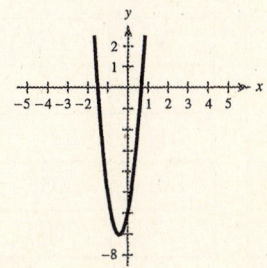

Section 2.6 Rational Functions and Asymptotes

■ You should know the following basic facts about rational functions.

(a) A function of the form $f(x) = P(x)/Q(x)$, $Q(x) \neq 0$, where $P(x)$ and $Q(x)$ are polynomials, is called a rational function.

(b) The domain of a rational function is the set of all real numbers except those which make the denominator zero.

(c) If $f(x) = P(x)/Q(x)$ is in reduced form, and a is a value such that $Q(a) = 0$, then the line $x = a$ is a vertical asymptote of the graph of f. $f(x) \to \infty$ or $f(x) \to -\infty$ as $x \to a$.

(d) The line $y = b$ is a horizontal asymptote of the graph of f if $f(x) \to b$ as $x \to \infty$ or $x \to -\infty$.

(e) Let $f(x) = \dfrac{P(x)}{Q(x)} = \dfrac{a_n x^n + a_{n-1} x^{n-1} + \cdots + a_1 x + a_0}{b_m x^m + b_{m-1} x^{m-1} + \cdots + b_1 x + b_0}$ where $P(x)$ and $Q(x)$ have no common factors.

  1. If $n < m$, then the x-axis $(y = 0)$ is a horizontal asymptote.

  2. If $n = m$, then $y = \dfrac{a_n}{b_m}$ is a horizontal asymptote.

  3. If $n > m$, then there are no horizontal asymptotes.

Solutions to Odd-Numbered Exercises

1. $f(x) = \dfrac{1}{x - 1}$

(a)

x	$f(x)$
0.5	-2
0.9	-10
0.99	-100
0.999	-1000

x	$f(x)$
1.5	2
1.1	10
1.01	100
1.001	1000

x	$f(x)$
5	0.25
10	$0.\overline{1}$
100	$0.\overline{01}$
1000	$0.\overline{001}$

x	$f(x)$
-5	-0.167
-10	-0.0909
-100	-0.0099
-1000	-0.001

(b) The zero of the denominator is $x = 1$, so $x = 1$ is a vertical asymptote. The degree of the numerator is less than the degree of the denominator so the x-axis, or $y = 0$ is a horizontal asymptote.

(c) The domain is all real numbers except $x = 1$.

3. $f(x) = \dfrac{3x}{|x - 1|}$

(a)

x	$f(x)$
0.5	3
0.9	27
0.99	297
0.999	2997

x	$f(x)$
1.5	9
1.1	33
1.01	303
1.001	3003

x	$f(x)$
5	3.75
10	$3.\overline{33}$
100	$3.\overline{03}$
1000	$3.\overline{003}$

x	$f(x)$
-5	-2.5
-10	-2.727
-100	-2.970
-1000	-2.997

(b) The zero of the denominator is $x = 1$, so $x = 1$ is a vertical asymptote. Since $f(x) \to 3$ as $x \to \infty$ and $f(x) \to -3$ as $x \to -\infty$, both $y = 3$ and $y = -3$ are horizontal asymptotes.

(c) The domain is all real numbers except $x = 1$.

5. $f(x) = \dfrac{3x^2}{x^2 - 1}$

(a)

x	$f(x)$
0.5	-1
0.9	-12.79
0.99	-148.79
0.999	-1498

x	$f(x)$
1.5	5.4
1.1	17.29
1.01	152.3
1.001	1502.3

x	$f(x)$
5	3.125
10	$3.\overline{03}$
100	$3.\overline{0003}$
1000	3

x	$f(x)$
-5	3.125
-10	$3.\overline{03}$
-100	$3.\overline{0003}$
-1000	3

(b) The zeros of the denominator are $x = \pm 1$ so both $x = 1$ and $x = -1$ are vertical asymptotes.

Since the degree of the numerator equals the degree of the denominator, $y = \frac{3}{1} = 3$ is a horizontal asymptote.

(c) The domain is all real numbers except $x = \pm 1$.

7. $f(x) = \dfrac{2}{x + 2}$

Vertical asymptote: $x = -2$

Horizontal asymptote: $y = 0$

Matches graph (a)

9. $f(x) = \dfrac{4x + 1}{x}$

Vertical asymptote: $x = 0$

Horizontal asymptote: $y = 4$

Matches graph (c)

11. $f(x) = \dfrac{x - 2}{x - 4}$

Vertical asymptote: $x = 4$

Horizontal asymptote: $y = 1$

Matches graph (b)

13. $f(x) = \dfrac{1}{x^2}$

(a) Domain: all real numbers except $x = 0$

(b) Vertical asymptote: $x = 0$

Horizontal asymptote: $y = 0$

[Degree of $p(x)$ < degree of $q(x)$]

(c)

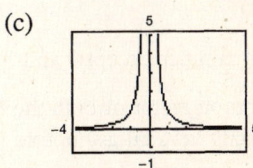

15. $f(x) = \dfrac{2 + x}{2 - x}$

(a) Domain: all real numbers except $x = 2$

(b) Vertical asymptote: $x = 2$

Horizontal asymptote: $y = -1$

(c)

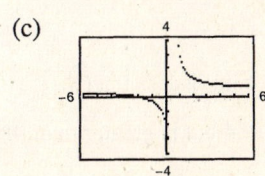

17. $f(x) = \dfrac{x^2 + 2x}{2x^2 - x} = \dfrac{x + 2}{2x - 1}, \quad x \neq 0$

(a) Domain: all real numbers except $x = 0, \ \frac{1}{2}$

(b) Vertical asymptote: $x = \frac{1}{2}$

Horizontal asymptote: $y = \frac{1}{2}$

[$x = 0$ is not a vertical asymptote]

(c)

19. $f(x) = \dfrac{3x^2 + x - 5}{x^2 + 1}$

   (a) Domain: all real numbers

   (b) Vertical asymptote: none

      Horizontal asymptote: $y = 3$

   (c)

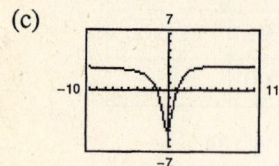

21. $f(x) = \dfrac{x - 3}{|x|}$

   (a) Domain: all real numbers except $x = 0$

   (b) Vertical asymptote: $x = 0$

      Horizontal asymptote:

         $y = 1$ to the right

         $y = -1$ (to the left)

   (c)

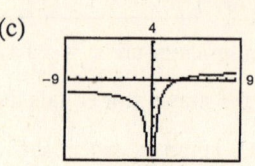

23. $f(x) = \dfrac{x^2 - 4}{x + 2}$, $g(x) = x - 2$

   (a) Domain of f: all real numbers except -2 Domain of g: all real numbers

   (b) Since $x + 2$ is a common factor of both the numerator and the denominator of $f(x)$, $x = -2$ is not a vertical asymptote of f. f has no vertical asymptotes.

   (c)

x	-4	-3	-2.5	-2	-1.5	-1	0
$f(x)$	-6	-5	-4.5	undef.	-3.5	-3	-2
$g(x)$	-6	-5	-4.5	-4	-3.5	-3	-2

   (d) f and g differ only where f is undefined.

25. $f(x) = \dfrac{x - 3}{x^2 - 3x}$, $g(x) = \dfrac{1}{x}$

   (a) Domain of f: all real number except 0 and 3 Domain of g: all real numbers except 0

   (b) Since $x - 3$ is a common factor of both the numerator and the denominator of f, $x = 3$ is not a vertical asymptote of f. The only vertical asymptote is $x = 0$.

   (c)

x	-1	-0.5	0	0.5	2	3	4
$f(x)$	-1	-2	undef.	2	$\frac{1}{2}$	undef.	$\frac{1}{4}$
$g(x)$	-1	-2	undef.	2	$\frac{1}{2}$	$\frac{1}{3}$	$\frac{1}{4}$

   (d) They differ only at $x = 3$, where f is undefined and g is defined.

27. $f(x) = 4 - \dfrac{1}{x}$

   (a) As $x \to \pm\infty$, $f(x) \to 4$

   (b) As $x \to \infty$, $f(x) \to 4$ but is less than 4

   (c) As $x \to -\infty$, $f(x) \to 4$ but is greater than 4

29. $f(x) = \dfrac{2x - 1}{x - 3}$

   (a) As $x \to \pm\infty$, $f(x) \to 2$

   (b) As $x \to \infty$, $f(x) \to 2$ but is greater than 2

   (c) As $x \to -\infty$, $f(x) \to 2$ but is less than 2

31. $g(x) = \dfrac{x^2 - 4}{x + 3} = \dfrac{(x - 2)(x + 2)}{x + 3}$

The zeros of g are the zeros of the numerator:
$x = \pm 2$

33. $f(x) = 1 - \dfrac{2}{x - 5} = \dfrac{x - 7}{x - 5}$

The zero of f corresponds to the zero of the numerator and is $x = 7$.

35. $C = \dfrac{255p}{100 - p}, \quad 0 \le p < 100$

(a) $C(10) = \dfrac{255(10)}{100 - 10} \approx 28.33$ million dollars

(b) $C(40) = \dfrac{255(40)}{100 - 40} = 170$ million dollars

(c) $C(75) = \dfrac{255(75)}{100 - 75} = 765$ million dollars

(d)
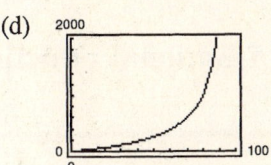

(e) $C \to \infty$ as $x \to 100$. No, it would not be possible to remove 100% of the pollutants.

37. (a) Use data $\left(16, \dfrac{1}{3}\right), \left(32, \dfrac{1}{4.7}\right), \left(44, \dfrac{1}{9.8}\right),$

$\left(50, \dfrac{1}{19.7}\right), \left(60, \dfrac{1}{39.4}\right)$

$\dfrac{1}{y} = -0.00742x \times 0.445$

$y = \dfrac{1}{0.445 - 0.007x}$

(b)

x	16	32	44	50	60
y	3.0	4.5	7.3	10.5	40

(Answers will vary.)

(c) No, the function is negative for $x = 70$

39. $N = \dfrac{20(5 + 3t)}{1 + 0.04t}, \quad 0 \le t$

(a)
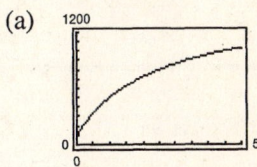

(b) $N(5) \approx 333$ deer

$N(10) = 500$ deer

$N(25) = 800$ deer

(c) The herd is limited by the horizontal asymptote:

$N = \dfrac{60}{0.04} = 1500$ deer

41. False. A rational function can have at most n vertical asymptotes, where n is the degree of the denominator.

43. One possible answer:

$f(x) = \dfrac{1}{(x + 2)(x - 1)}$

$= \dfrac{1}{x^2 + x - 2}$

45. One possible answer:

$f(x) = \dfrac{2x^2}{x^2 + 1}$

47. $y - 2 = \dfrac{-1 - 2}{0 - 3}(x - 3) = 1(x - 3)$

$y = x - 1$

$y - x + 1 = 0$

49. $y - 7 = \dfrac{10 - 7}{3 - 2}(x - 2) = 3(x - 2)$

$y = 3x + 1$

$3x - y + 1 = 0$

51.
$$x - 4 \overline{)\, x^2 + 5x + 6} \qquad \frac{x + 9}{}$$

$$\frac{x^2 - 4x}{}$$
$$9x + 6$$
$$\frac{9x - 36}{}$$
$$42$$

$$\frac{x^2 + 5x + 6}{x - 4} = x + 9 + \frac{42}{x - 4}$$

53.
$$x + 5 \overline{)\, 2x^2 + x - 11} \qquad \frac{2x - 9}{}$$

$$\frac{2x^2 + 10x}{}$$
$$-9x - 11$$
$$\frac{-9x - 45}{}$$
$$34$$

$$\frac{2x^2 + x - 11}{x + 5} = 2x - 9 + \frac{34}{x + 5}$$

Section 2.7   Graphs of Rational Functions

■ You should be able to graph $f(x) = \dfrac{p(x)}{q(x)}$.

(a) Find the x- and y-intercepts.

(b) Find any vertical or horizontal asymptotes.

(c) Plot additional points.

(d) If the degree of the numerator is one more than the degree of the denominator, use long division to find the slant asymptote.

Solutions to Odd-Numbered Exercises

1. $g(x) = \dfrac{2}{x} + 1$

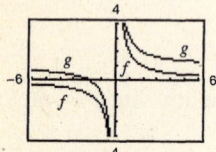

Vertical shift one unit upward

3. $g(x) = -\dfrac{2}{x}$

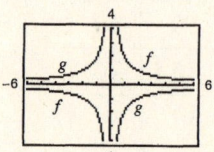

Reflection in the x-axis

5. $g(x) = \dfrac{2}{x^2} - 2$

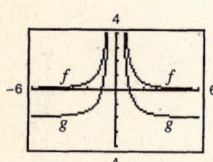

Vertical shift two units downward

7. $g(x) = \dfrac{2}{(x - 2)^2}$

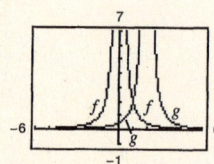

Horizontal shift two units to the right

9. $f(x) = \dfrac{1}{x + 2}$

y-intercept: $\left(0, \dfrac{1}{2}\right)$

Vertical asymptote: $x = -2$

Horizontal asymptote: $y = 0$

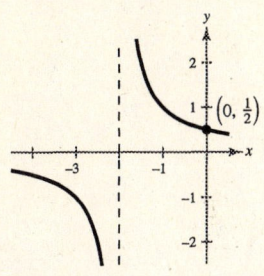

x	-4	-3	-1	0	1
y	$-\frac{1}{2}$	-1	1	$\frac{1}{2}$	$\frac{1}{3}$

11. $C(x) = \dfrac{5 + 2x}{1 + x} = \dfrac{2x + 5}{x + 1}$

x-intercept: $\left(-\dfrac{5}{2}, 0\right)$

y-intercept: $(0, 5)$

Vertical asymptote: $x = -1$

Horizontal asymptote: $y = 2$

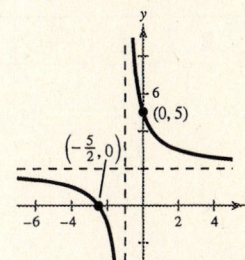

x	-4	-3	-2	0	1	2
$C(x)$	1	$\frac{1}{2}$	-1	5	$\frac{7}{2}$	3

13. $f(t) = \dfrac{1 - 2t}{t} = -\dfrac{2t - 1}{t}$

t-intercept: $\left(\dfrac{1}{2}, 0\right)$

Vertical asymptote: $t = 0$

Horizontal asymptote: $y = -2$

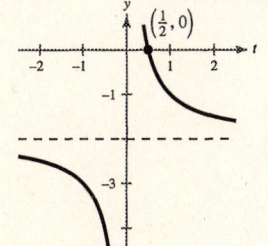

x	-2	-1	$\frac{1}{2}$	1	2
y	$-\frac{5}{2}$	-3	0	-1	$-\frac{3}{2}$

15. $f(x) = \dfrac{x^2}{x^2 - 4}$

Intercept: $(0, 0)$

Vertical asymptotes: $x = 2$, $x = -2$

Horizontal asymptote: $y = 1$

y-axis symmetry

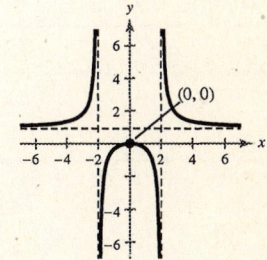

x	± 4	± 3	± 1	0
y	$\frac{4}{3}$	$\frac{9}{5}$	$-\frac{1}{3}$	0

17. $f(x) = \dfrac{x}{x^2 - 1} = \dfrac{x}{(x + 1)(x - 1)}$

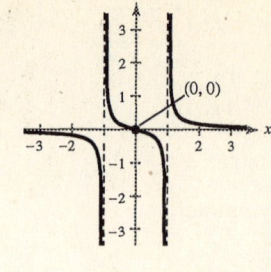

Intercept: $(0, 0)$

Vertical asymptotes: $x = 1$ and $x = -1$

Horizontal asymptote: $y = 0$

Origin symmetry

x	-3	-2	$-\frac{1}{2}$	0	$\frac{1}{2}$	2	3	4
y	$-\frac{3}{8}$	$-\frac{2}{3}$	$\frac{2}{3}$	0	$-\frac{2}{3}$	$\frac{2}{3}$	$\frac{3}{8}$	$\frac{4}{15}$

19. $g(x) = \dfrac{4(x + 1)}{x(x - 4)}$

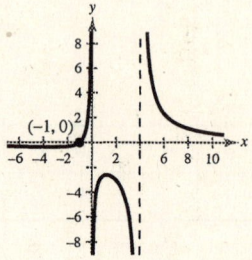

Intercept: $(-1, 0)$

Vertical asymptotes: $x = 0$ and $x = 4$

Horizontal asymptote: $y = 0$

x	-2	-1	1	2	3	5	6
y	$-\frac{1}{3}$	0	$-\frac{8}{3}$	-3	$-\frac{16}{3}$	$\frac{24}{5}$	$\frac{7}{3}$

21. $f(x) = \dfrac{3x}{x^2 - x - 2} = \dfrac{3x}{(x + 1)(x - 2)}$

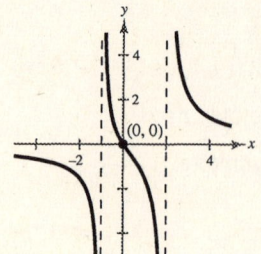

Intercept: $(0, 0)$

Vertical asymptotes: $x = -1, 2$

Horizontal asymptote: $y = 0$

x	-3	0	1	3	4
y	$-\frac{9}{10}$	0	$-\frac{3}{2}$	$\frac{9}{4}$	$\frac{6}{5}$

23. $f(x) = \dfrac{x^2 + 3x}{x^2 + x - 6} = \dfrac{x(x + 3)}{(x - 2)(x + 3)} = \dfrac{x}{x - 2},$

$x \neq -3$

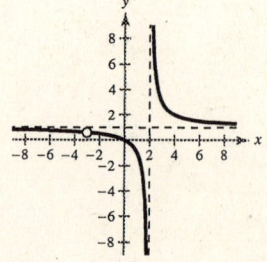

Intercept: $(0, 0)$

Vertical asymptote: $x = 2$

   (There is a hole at $x = -3$)

Horizontal asymptote: $y = 1$

x	-2	-1	0	1	2	3
y	$\frac{1}{2}$	$\frac{1}{3}$	0	-1	undef.	3

25. $f(x) = \dfrac{x^2 - 1}{x + 1} = \dfrac{(x + 1)(x - 1)}{x + 1} = x - 1,$

$x \neq -1$

The graph is a line, with a hole at $x = -1$.

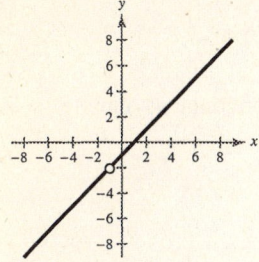

27. $f(x) = \dfrac{2 + x}{1 - x} = -\dfrac{x + 2}{x - 1}$

Vertical asymptote: $x = 1$

Horizontal asymptote: $y = -1$

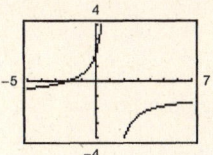

Domain: $x \neq 1$ or $(-\infty, 1) \cup (1, \infty)$

29. $f(t) = \dfrac{3t + 1}{t}$

Vertical asymptote: $t = 0$

Horizontal asymptote: $y = 3$

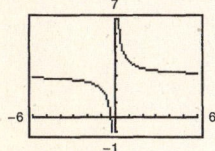

Domain: $t \neq 0$ or $(-\infty, 0) \cup (0, \infty)$

31. $h(t) = \dfrac{4}{t^2 + 1}$

Domain: all real numbers OR $(-\infty, \infty)$

Horizontal asymptote: $y = 0$

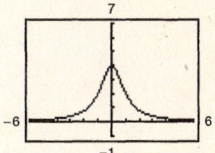

33. $f(x) = \dfrac{x + 1}{x^2 - x - 6} = \dfrac{x + 1}{(x - 3)(x + 2)}$

Domain: all real numbers except $x = 3, -2$

Vertical asymptotes: $x = 3$, $x = -2$

Horizontal asymptote: $y = 0$

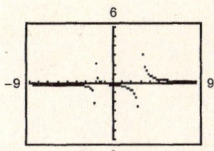

35. $f(x) = \dfrac{20x}{x^2 + 1} - \dfrac{1}{x} = \dfrac{19x^2 - 1}{x(x^2 + 1)}$

Domain: all real numbers except 0,
OR $(-\infty, 0) \cup (0, \infty)$

Vertical asymptote: $x = 0$

Horizontal asymptote: $y = 0$

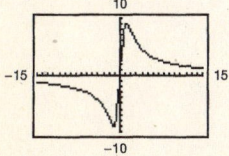

37. $h(x) = \dfrac{6x}{\sqrt{x^2 + 1}}$

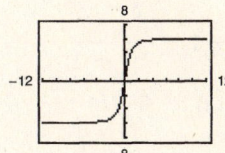

There are two horizontal asymptotes: $y = \pm 6$.

39. $g(x) = \dfrac{4|x - 2|}{x + 1}$

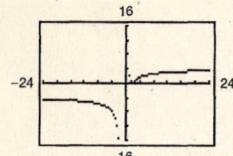

There are two horizontal asymptotes: $y = \pm 4$.

One vertical asymptote: $x = -1$.

41. $f(x) = \dfrac{4(x - 1)^2}{x^2 - 4x + 5}$

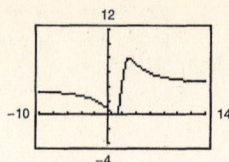

The graph crosses its horizontal asymptote: $y = 4$.

43. $f(x) = \dfrac{2x^2 + 1}{x} = 2x + \dfrac{1}{x}$

Vertical asymptote: $x = 0$

Slant asymptote: $y = 2x$

Origin symmetry

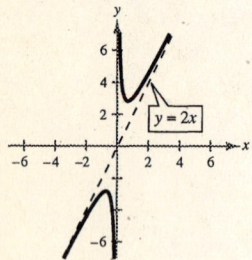

45. $h(x) = \dfrac{x^2}{x - 1} = x + 1 + \dfrac{1}{x - 1}$

Intercept: $(0, 0)$

Vertical asymptote: $x = 1$

Slant asymptote: $y = x + 1$

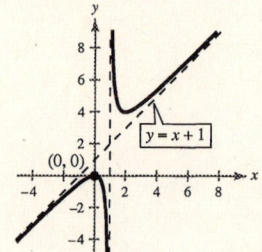

47. $g(x) = \dfrac{x^3}{2x^2 - 8} = \dfrac{1}{2}x + \dfrac{4x}{2x^2 - 8}$

Intercept: $(0, 0)$

Vertical asymptotes: $x = \pm 2$

Slant asymptote: $y = \dfrac{1}{2}x$

Origin symmetry

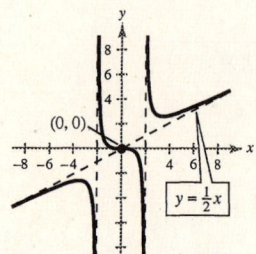

49. $f(x) = \dfrac{x^3 + 2x^2 + 4}{2x^2 + 1} = \dfrac{x}{2} + 1 + \dfrac{3 - \dfrac{x}{2}}{2x^2 + 1}$

Intercepts: $(-2.594, 0)$, $(0, 4)$

Slant asymptote: $y = \dfrac{x}{2} + 1$

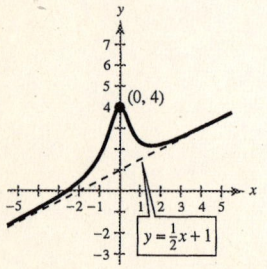

51. $y = \dfrac{x + 1}{x - 3}$

(a) x-intercept: $(-1, 0)$

(b) $0 = \dfrac{x + 1}{x - 3}$

$0 = x + 1$

$-1 = x$

53. $y = \dfrac{1}{x} - x$

(a) x-intercepts: $(\pm 1, 0)$

(b) $0 = \dfrac{1}{x} - x$

$x = \dfrac{1}{x}$

$x^2 = 1$

$x = \pm 1$

55. $y = \dfrac{2x^2 + x}{x + 1} = 2x - 1 + \dfrac{1}{x + 1}$

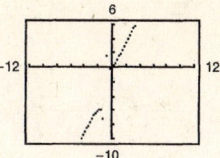

Domain: all real numbers except $x = -1$

Vertical asymptote: $x = -1$

Slant asymptote: $y = 2x - 1$

57. $y = \dfrac{1 + 3x^2 - x^3}{x^2} = \dfrac{1}{x^2} + 3 - x = -x + 3 + \dfrac{1}{x^2}$

Domain: all real numbers except 0

OR $(-\infty, 0) \cup (0, \infty)$

Vertical asymptote: $x = 0$

Slant asymptote: $y = -x + 3$

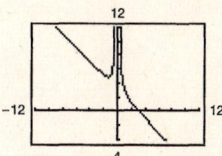

59. $y = \dfrac{1}{x + 5} + \dfrac{4}{x}$

(a)

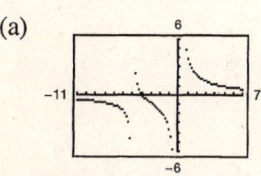

x-intercept: $(-4, 0)$

(b) $0 = \dfrac{1}{x + 5} + \dfrac{4}{x}$

$-\dfrac{4}{x} = \dfrac{1}{x + 5}$

$-4(x + 5) = x$

$-4x - 20 = x$

$-5x = 20$

$x = -4$

61. $y = x - \dfrac{6}{x - 1}$

(a)

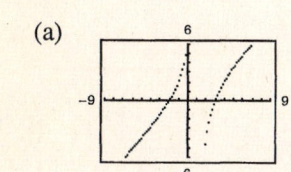

x-intercept: $(-2, 0), (3, 0)$

(b) $0 = x - \dfrac{6}{x - 1}$

$\dfrac{6}{x - 1} = x$

$6 = x(x - 1)$

$0 = x^2 - x - 6$

$0 = (x + 2)(x - 3)$

$x = -2, \quad x = 3$

63. (a) $.25(50) + .75(x) = C(50 + x)$

$$\frac{12.5 + .75x}{50 + x} = C$$

$$\frac{50 + 3x}{200 + 4x} = C$$

$$C = \frac{3x + 50}{4(x + 50)}$$

(b) Domain: $x \geq 0$ and $x \leq 1000 - 50 = 950$

Thus, $0 \leq x \leq 950$

(c)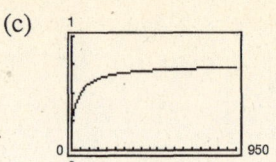

As the tank fills, the rate that the concentration is increasing slows down. It approaches the horizontal asymptote $C = \frac{3}{4} = 0.75$. When the tank is full $(x = 950)$, the concentration is $C = 0.725$.

65. (a) $A = xy$ and

$$(x - 2)(y - 4) = 30$$

$$y - 4 = \frac{30}{x - 2}$$

$$y = 4 + \frac{30}{x - 2} = \frac{4x + 22}{x - 2}$$

Thus, $A = xy = x\left(\frac{4x + 22}{x - 2}\right) = \frac{2x(2x + 11)}{x - 2}$.

(b) Domain: Since the margins on the left and right are each 1 inch, $x > 2$, OR $(2, \infty)$.

(c)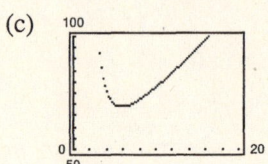

The area is minimum when $x \approx 5.87$ in. and $y \approx 11.75$ in.

67. $C = 100\left(\dfrac{200}{x^2} + \dfrac{x}{x + 30}\right), \ 1 \leq x$

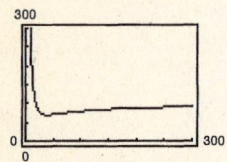

The minimum occurs when $x \approx 40.4 \approx 40$.

69. $C = \dfrac{3t^2 + t}{t^3 + 50}, \ 0 \leq t$

(a) The horizontal asymptote is the t-axis, or $C = 0$. This indicates that the chemical eventually dissipates.

(b) 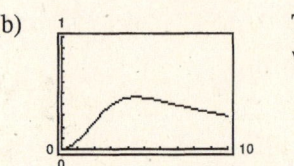 The maximum occurs when $t \approx 4.5$.

(c) Graph C together with $y = 0.345$. The graphs intersect at $t \approx 2.65$ and $t \approx 8.32$. $C < 0.345$ when $0 \leq t < 2.65$ hours and when $t > 8.32$ hours

71. (a) $A = 0.44t + 1.8$

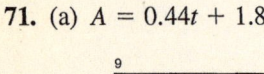

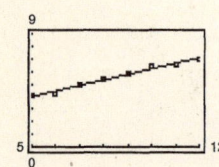

(b) $\dfrac{1}{A} = -0.016t + 0.32$

$$A = \frac{1}{0.32 - 0.016t}$$

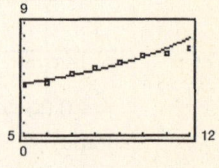

(c)
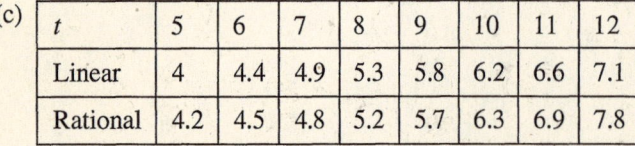

t	5	6	7	8	9	10	11	12
Linear	4	4.4	4.9	5.3	5.8	6.2	6.6	7.1
Rational	4.2	4.5	4.8	5.2	5.7	6.3	6.9	7.8

The linear model is closer to the actual data.

73. False, you will have to lift your pencil to cross the vertical asymptote.

75. $h(x) = \dfrac{6 - 2x}{3 - x} = \dfrac{2(3 - x)}{3 - x} = 2, \quad x \neq 3$

Since $h(x)$ is not reduced and $(3 - x)$ is a factor of both the numerator and the denominator, $x = 3$ is not a horizontal asymptote.

There is a hole in the graph at $x = 3$.

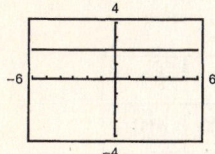

77. $y = x + 1 + \dfrac{a}{x - 2}$ has slant asymptote $y = x + 1$ and vertical asymptote $x = 2$.

$0 = -2 + 1 + \dfrac{a}{-2 - 2}$

$1 = \dfrac{a}{-4}$

$a = -4$

Hence, $y = x + 1 + \dfrac{-4}{x - 2} = \dfrac{x^2 - x - 6}{x - 2}$

79. $\left(\dfrac{x}{8}\right)^{-3} = \left(\dfrac{8}{x}\right)^3 = \dfrac{512}{x^3}$

81. $\dfrac{3x^3 y^2}{15xy^4} = \dfrac{x^2}{5y^2}, x \neq 0$

83. $\dfrac{3^{7/6}}{3^{1/6}} = 3^{6/6} = 3$

85.

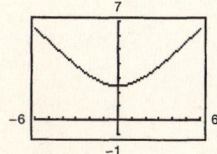

Domain: all x

Range: $y \geq \sqrt{6}$

87.

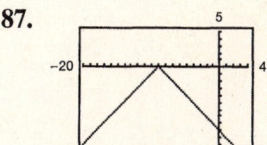

Domain: all x

Range: $y \leq 0$

Section 2.8   Exploring Data: Quadratic Models

Solutions to Odd-Numbered Exercises

1. A quadratic model is better

3. A linear model is better

5. Neither linear nor quadratic

7. (a)

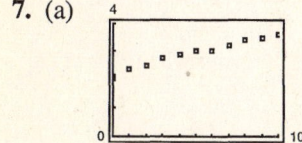

(b) Linear model is better

(c) $y = 0.14x + 2.24$ linear

[$y = -0.00478x^2 + 0.1887x + 2.1692$ quadratic]

(d)

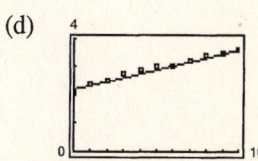

(e)

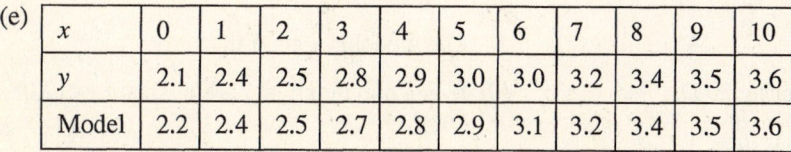

x	0	1	2	3	4	5	6	7	8	9	10
y	2.1	2.4	2.5	2.8	2.9	3.0	3.0	3.2	3.4	3.5	3.6
Model	2.2	2.4	2.5	2.7	2.8	2.9	3.1	3.2	3.4	3.5	3.6

9. (a)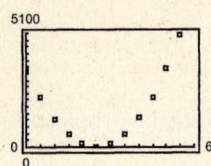

(b) Quadratic model is better

(c) $y = 5.55x^2 - 277.5x + 3478$

(d)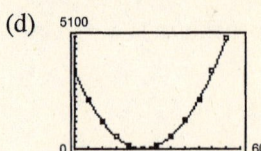

(e)

x	0	5	10	15	20	25	30	35	40	45	50	55
y	3480	2235	1250	565	150	12	145	575	1275	2225	3500	5010
Model	3478	2229	1258	564	148	9	140	564	1258	2229	3478	5004

11. (a)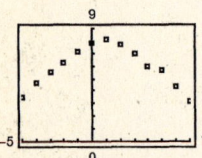

(b) Quadratic model is better

(c) $y = -0.1203x^3 + 0.208x + 7.53$

(d)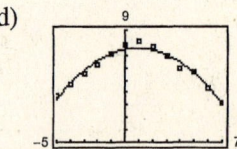

(e)

x	-5	-4	-3	-2	-1	0	1	2	3	4	5	6	7
y	3.8	4.7	5.5	6.2	7.1	7.9	8.1	7.7	6.9	6	5.6	4.4	3.2
Model	3.5	4.8	5.8	6.6	7.2	7.5	7.6	7.5	7.1	6.4	5.6	4.4	3.1

13. (a)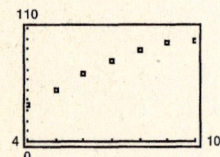

(b) $P = -1.298t^2 + 28.99t - 61.3$

(c)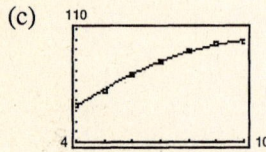

The model is a good fit.

(d) $P = 100$ when $t \approx 10.48$, or during 2000.

(e) No, the model turns downward as time increases.

15. (a)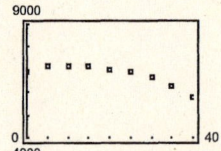

(b) $H = -1.682t^2 + 40.63t + 6903.2$

(c)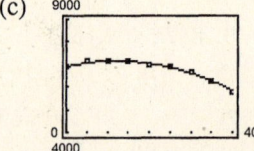

The model is a good fit.

(d) H is a maximum of 7148.6 when $t \approx 12.08$, or 1972.

(e) No, the model decreases as time increases.

17. (a) $y = 2.477x + 1.13$ Linear

      $y = 0.071x^2 + 1.69x + 2.7$ Quadratic

  (b) 0.989 for linear model

      0.9952 for quadratic model

  (c) Quadratic fits better

19. (a) $y = -0.892x + 0.24$ Linear

      $y = 0.0012x^2 - 0.895x + 0.203$ Quadratic

  (b) 0.999824 for linear model

      0.999873 for quadratic model

  (c) Quadratic fits slightly better

21. (a)

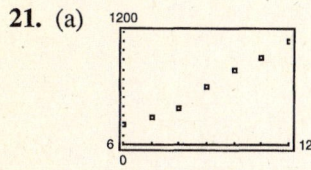

  (b) $S = 155.014t - 774.16$

  (c)

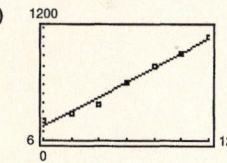

  (d) $S = 6.66t^2 + 35.1t - 261$

  (e)

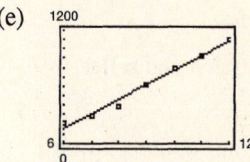

  (f) Both models fit well. For the linear model, 2007 corresponds to $t = 17$ and $S \approx 1861$.

23. True.

25. (a) $(f \circ g)(x) = f(x^2 + 3) = 2(x^2 + 3) - 1 = 2x^2 + 5$

  (b) $(g \circ f)(x) = g(2x - 1) = (2x - 1)^2 + 3 = 4x^2 - 4x + 4$

27. (a) $(f \circ g)(x) = f(\sqrt[3]{x + 1}) = x + 1 - 1 = x$

  (b) $(g \circ f)(x) = g(x^3 - 1) = \sqrt[3]{x^3 - 1 + 1} = x$

29. f is one-to-one

     $y = 2x + 5$

     $x = 2y + 5$

     $2y = x - 5$

     $y = (x - 5)/2 \Rightarrow f^{-1}x = \dfrac{x - 5}{2}$

31. f is one-to-one on $[0, \infty)$

     $y = x^2 + 5$ $x \geq 0$

     $x = y^2 + 5$ $y \geq 0$

     $y^2 = x - 5$

     $y = \sqrt{x - 5} \Rightarrow f^{-1}(x) = \sqrt{x - 5}$

33.

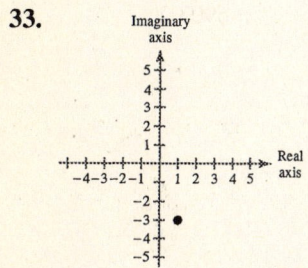

35.

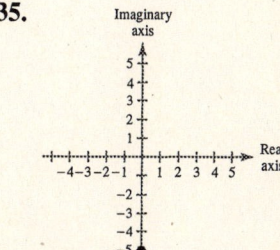

Review Exercises for Chapter 2

Solutions to Odd-Numbered Exercises

1.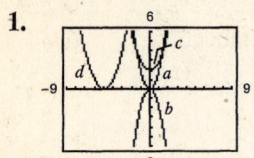

(a) $y = 2x^2$ is a vertical stretch

(b) $y = -2x^2$ is a vertical stretch and reflection in the x-axis

(c) $y = x^2 + 2$ is a vertical shift 2 units upward

(d) $y = (x + 5)^2$ is a horizontal shift 5 units to the left

3. $f(x) = \left(x + \frac{3}{2}\right)^2 + 1$

Vertex: $\left(-\frac{3}{2}, 1\right)$

y-intercept: $\left(0, \frac{13}{4}\right)$

No x-intercepts

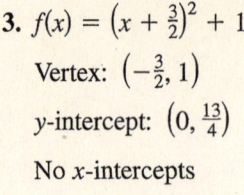

5. $f(x) = \frac{1}{3}(x^2 + 5x - 4)$

$= \frac{1}{3}\left(x^2 + 5x + \frac{25}{4} - \frac{25}{4} - 4\right)$

$= \frac{1}{3}\left[\left(x + \frac{5}{2}\right)^2 - \frac{41}{4}\right]$

$= \frac{1}{3}\left(x + \frac{5}{2}\right)^2 - \frac{41}{12}$

Vertex: $\left(-\frac{5}{2}, -\frac{41}{12}\right)$

y-intercepts: $\left(0, -\frac{4}{3}\right)$

x-intercepts: $0 = \frac{1}{3}(x^2 + 5x - 4)$

$0 = x^2 + 5x - 4$

$x = \frac{-5 \pm \sqrt{41}}{2}$ Use the Quadratic Formula.

$\left(\frac{-5 \pm \sqrt{41}}{2}, 0\right)$

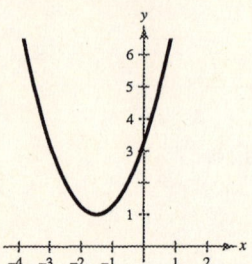

7. Vertex: $(1, -4) \implies f(x) = a(x - 1)^2 - 4$

Point: $(2, -3) \implies -3 = a(2 - 1)^2 - 4$

$1 = a$

Thus, $f(x) = (x - 1)^2 - 4$.

9. Vertex: $(-2, -2) \implies f(x) = a(x + 2)^2 - 2$

Point: $(-1, 0) \implies 0 = a(-1 + 2)^2 - 2$

$a = 2$

Thus, $f(x) = 2(x + 2)^2 - 2$.

11. (a) $A = xy = x\left(\dfrac{8-x}{2}\right)$, since $x + 2y - 8 = 0 \implies y = \dfrac{8-x}{2}$.

Since the figure is in the first quadrant and x and y must be positive, the domain of

$A = x\left(\dfrac{8-x}{2}\right)$ is $0 < x < 8$.

(b)

x	y	Area
1	$4 - \frac{1}{2}(1)$	$(1)\left[4 - \frac{1}{2}(1)\right] = \frac{7}{2}$
2	$4 - \frac{1}{2}(2)$	$(2)\left[4 - \frac{1}{2}(2)\right] = 6$
3	$4 - \frac{1}{2}(3)$	$(3)\left[4 - \frac{1}{2}(3)\right] = \frac{15}{2}$
4	$4 - \frac{1}{2}(4)$	$(4)\left[4 - \frac{1}{2}(4)\right] = 8$
5	$4 - \frac{1}{2}(5)$	$(5)\left[4 - \frac{1}{2}(5)\right] = \frac{15}{2}$
6	$4 - \frac{1}{2}(6)$	$(6)\left[4 - \frac{1}{2}(6)\right] = 6$

(c)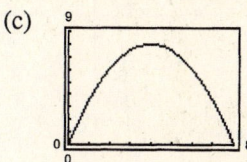

The maximum area of 8 occurs at the vertex when $x = 4$ and $y = \dfrac{8-4}{2} = 2$.

The dimensions that will produce a maximum area seem to be $x = 4$ and $y = 2$.

(d) $A = x\left(\dfrac{8-x}{2}\right)$

$= \dfrac{1}{2}(8x - x^2)$

$= -\dfrac{1}{2}(x^2 - 8x)$

$= -\dfrac{1}{2}(x^2 - 8x + 16 - 16)$

$= -\dfrac{1}{2}[(x-4)^2 - 16]$

$= -\dfrac{1}{2}(x-4)^2 + 8$ The maximum area of 8 occurs when $x = 4$ and $y = \dfrac{8-4}{2} = 2$.

(e) The answers are the same.

13. $y = x^4$

(a)   (b)   (c)   (d)

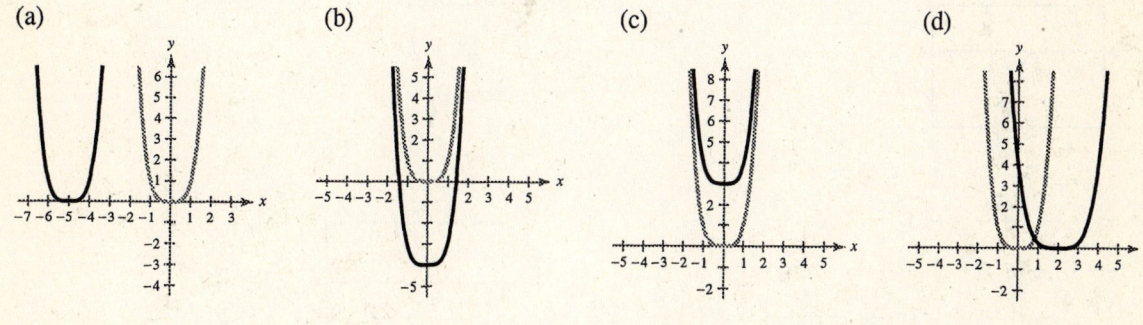

15. $y = x^6$

(a)

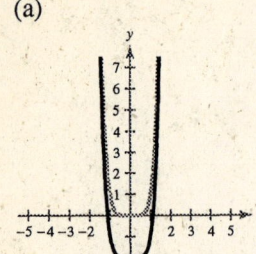

(b)

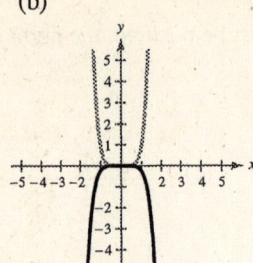

(c)

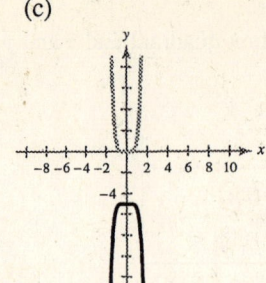

(d)

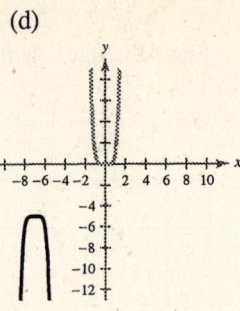

17. $f(x) = \frac{1}{2}x^3 - 2x + 1$; $g(x) = \frac{1}{2}x^3$

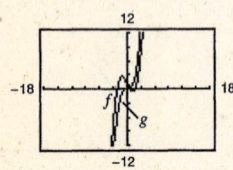

19. $f(x) = -x^2 + 6x + 9$

The degree is even and the leading coefficient is negative. The graph falls to the left and right.

21. $f(x) = \frac{3}{4}(x^4 + 3x^2 + 2)$

The degree is even and the leading coefficient is positive. The graph rises to the left and right.

23. $g(x) = x^4 - x^3 - 2x^2$

(a)

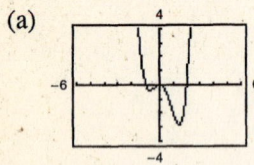

(b) zeros: $0, 2, -1$

(c) $0 = x^4 - x^3 - 2x^2$

$0 = x^2(x^2 - x - 2)$

$0 = x^2(x - 2)(x + 1)$

zeros: $0, 0, 2, -1$

25. $f(t) = t^3 - 3t$

(a)

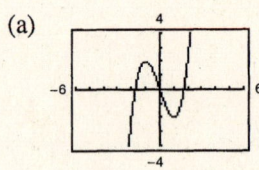

(b) zeros: $\pm 1.732, 0$

(c) $0 = t^3 - 3t$

$0 = t(t^2 - 3)$

zeros: $0, \pm\sqrt{3}$

27. $f(x) = x(x + 3)^2$

(a)

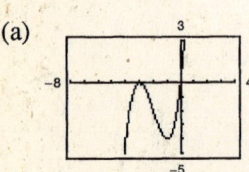

(b) zeros: $0, -3$

(c) $0 = (x + 3)^2$

zeros: $0, -3, -3$

29. $f(x) = x^3 + 2x^2 - x - 1$

(a) $f(-3) < 0, f(-2) > 0 \implies$ zero in $[-3, -2]$

$f(-1) > 0, f(0) < 0 \implies$ zero in $[-1, 0]$

$f(0) < 0, f(1) > 0 \implies$ zero in $[0, 1]$

(b) zeros: $-2.247, -0.555, 0.802$

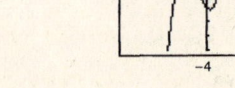

31. $f(x) = x^4 - 6x^2 - 4$

(a) $f(-3) > 0$, $f(-2) < 0 \implies$ zero in $[-3, -2]$

$\quad f(2) < 0$, $f(3) > 0 \implies$ zero in $[2, 3]$

(b) zeros: ± 2.570

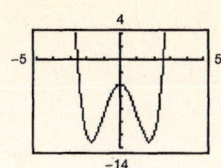

33. $y_1 = \dfrac{x^2}{x - 2}$

$y_2 = x + 2 + \dfrac{4}{x - 2}$

$\quad = \dfrac{(x + 2)(x - 2)}{x - 2} + \dfrac{4}{x - 2}$

$\quad = \dfrac{x^2 - 4}{x - 2} + \dfrac{4}{x - 2}$

$\quad = \dfrac{x^2}{x - 2}$

$\quad = y_1$

35.

$$
\begin{array}{r}
8x + 5 \\
3x - 2 \overline{\smash{)}\ 24x^2 -\ \ x - 8} \\
\underline{24x^2 - 16x} \\
15x -\ 8 \\
\underline{15x - 10} \\
2
\end{array}
$$

Thus, $\dfrac{24x^2 - x - 8}{3x - 2} = 8x + 5 + \dfrac{2}{3x - 2}$.

37.

$$
\begin{array}{r}
x^2 - 2 \\
x^2 - 1 \overline{\smash{)}\ x^4 - 3x^2 + 2} \\
\underline{x^4 -\ \ x^2} \\
-2x^2 + 2 \\
\underline{-2x^2 + 2} \\
0
\end{array}
$$

Thus, $\dfrac{x^4 - 3x^2 + 2}{x^2 - 1} = x^2 - 2 \qquad (x \neq \pm 1)$

39.

$$
\begin{array}{r}
5x + 2 \\
x^2 - 3x + 1 \overline{\smash{)}\ 5x^3 - 13x^2 -\ \ x + 2} \\
\underline{5x^3 - 15x^2 + 5x} \\
2x^2 - 6x + 2 \\
\underline{2x^2 - 6x + 2} \\
0
\end{array}
$$

Thus, $\dfrac{5x^3 - 13x^2 - x + 2}{x^2 - 3x + 1} = 5x + 2$,

$x \neq \dfrac{1}{2}\left(3 \pm \sqrt{5}\right)$

41.

$$
\begin{array}{r}
3x^2 + 5x + 8 \\
2x^2 + 0x - 1 \overline{\smash{)}\ 6x^4 + 10x^3 + 13x^2 - 5x + 2} \\
\underline{6x^4 +\ \ 0x^3 -\ \ 3x^2} \\
10x^3 + 16x^2 - 5x \\
\underline{10x^3 +\ \ 0x^2 - 5x} \\
16x^2 -\ \ 0 + 2 \\
\underline{16x^2 +\ \ 0 - 8} \\
10
\end{array}
$$

$\dfrac{6x^4 + 10x^3 + 13x^2 - 5x + 2}{2x^2 - 1} = 3x^2 + 5x + 8 + \dfrac{10}{2x^2 - 1}$

43.

$$
\begin{array}{r|rrrrr}
-2 & 0.25 & -4 & 0 & 0 & 0 \\
   &   & -\frac{1}{2} & 9 & -18 & 36 \\
\hline
   & \frac{1}{4} & -\frac{9}{2} & 9 & -18 & 36
\end{array}
$$

Hence,

$$\frac{0.25x^4 - 4x^3}{x + 2} = \frac{1}{4}x^3 - \frac{9}{2}x^2 + 9x - 18 + \frac{36}{x + 2}$$

45.

$$
\begin{array}{r|rrrrr}
\frac{2}{3} & 6 & -4 & -27 & 18 & 0 \\
   &   & 4 & 0 & -18 & 0 \\
\hline
   & 6 & 0 & -27 & 0 & 0
\end{array}
$$

Thus,

$$\frac{6x^4 - 4x^3 - 27x^2 + 18x}{x - \left(\dfrac{2}{3}\right)} = 6x^3 - 27x, \ x \neq \frac{2}{3}$$

47.

$$
\begin{array}{r|rrrr}
4 & 3 & -10 & 12 & -22 \\
  &   & 12 & 8 & 80 \\
\hline
  & 3 & 2 & 20 & 58
\end{array}
$$

Thus, $\dfrac{3x^3 - 10x^2 + 12x - 22}{x - 4} = 3x^2 + 2x + 20 + \dfrac{58}{x - 4}$

49. (a)

$$
\begin{array}{r|rrrrr}
-3 & 1 & 10 & -24 & 20 & 44 \\
   &   & -3 & -21 & 135 & -465 \\
\hline
   & 1 & 7 & -45 & 155 & -421
\end{array}
$$

$f(-3) = -421$

(b)

$$
\begin{array}{r|rrrrr}
-1 & 1 & 10 & -24 & 20 & 44 \\
   &   & -1 & -9 & 33 & -53 \\
\hline
   & 1 & 9 & -33 & 53 & \underline{-9}
\end{array}
$$

$f(-1) = -9$

51. $f(x) = x^3 + 4x^2 - 25x - 28$

(a)

$$
\begin{array}{r|rrrr}
4 & 1 & 4 & -25 & -28 \\
  &   & 4 & 32 & 28 \\
\hline
  & 1 & 8 & 7 & 0
\end{array}
$$

$(x - 4)$ is a factor

(b) $x^2 + 8x + 7 = (x + 1)(x + 7)$

Remaining factors: $(x + 1), (x + 7)$

(c) $f(x) = (x - 4)(x + 1)(x + 7)$

(d) Zeros: $4. -1, -7$

(e)

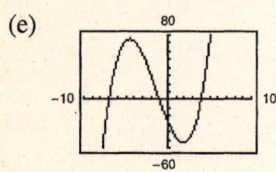

53. $f(x) = x^4 - 4x^3 - 7x^2 + 22x + 24$

(a)

$$
\begin{array}{r|rrrrr}
-2 & 1 & -4 & -7 & 22 & 24 \\
   &   & -2 & 12 & -10 & -24 \\
\hline
   & 1 & -6 & 5 & 12 & 0
\end{array}
$$

$(x + 2)$ is a factor

$$
\begin{array}{r|rrrr}
3 & 1 & -6 & 5 & 12 \\
  &   & 3 & -9 & -12 \\
\hline
  & 1 & -3 & -4 & 0
\end{array}
$$

$(x - 3)$ is a factor

(b) $x^2 - 3x - 4 = (x - 4)(x + 1)$

Remaining factors: $(x - 4), (x + 1)$

(c) $f(x) = (x + 2)(x - 3)(x - 4)(x + 1)$

(d) zeros: $-2, 3, 4, -1$

(e)

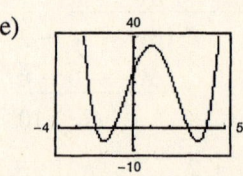

55. Possible rational zeros: $\pm 3, \pm\frac{3}{2}, \pm\frac{3}{4}, \pm 1, \pm\frac{1}{2}, \pm\frac{1}{4}$

zeros: $1, 1, \frac{3}{4}$

57. $f(x) = 6x^3 - 5x^2 + 24x - 20$

Graphing $f(x)$ with a graphing utility suggests that $x = \frac{5}{6}$ is a zero.

$$
\begin{array}{r|rrrr}
\frac{5}{6} & 6 & -5 & 24 & -20 \\
 & & 5 & 0 & 20 \\
\hline
 & 6 & 0 & 24 & 0
\end{array}
$$

The quadratic $6x^2 + 24 = 0$ has complex zeros $x = \pm 2i$. Thus, the zeros are $\frac{5}{6}, 2i, -2i$.

59. $f(x) = 6x^4 - 25x^3 + 14x^2 + 27x - 18$

Possible Rational Zeros: $\pm 1, \pm 2, \pm 3, \pm 6, \pm 9, \pm 18, \pm\frac{1}{2}, \pm\frac{3}{2}, \pm\frac{9}{2}, \pm\frac{1}{3}, \pm\frac{2}{3}, \pm\frac{1}{6}$. Use a graphing utility to see that $x = -1$ and $x = 3$ are probably zeros.

$$
\begin{array}{r|rrrrr}
-1 & 6 & -25 & 14 & 27 & -18 \\
 & & -6 & 31 & -45 & 18 \\
\hline
 & 6 & -31 & 45 & -18 & 0
\end{array}
$$

$$
\begin{array}{r|rrrr}
3 & 6 & -31 & 45 & -18 \\
 & & 18 & -39 & 18 \\
\hline
 & 6 & -13 & 6 & 0
\end{array}
$$

$$
\begin{aligned}
6x^4 - 25x^3 + 14x^2 + 27x - 18 &= (x + 1)(x - 3)(6x^2 - 13x + 6) \\
&= (x + 1)(x - 3)(3x - 2)(2x - 3)
\end{aligned}
$$

Thus, the zeros of f are $x = -1$, $x = 3$, $x = \frac{2}{3}$, and $x = \frac{3}{2}$.

61. $g(x) = 5x^3 + 3x^2 - 6x + 9$ has two variations in sign $\implies$ 0 or 2 positive real zeros.

$g(-x) = -5x^3 + 3x^2 + 6x + 9$ has one variation in sign $\implies$ 1 negative real zero.

63.
$$
\begin{array}{r|rrrr}
1 & 4 & -3 & 4 & -3 \\
 & & 4 & 1 & 5 \\
\hline
 & 4 & 1 & 5 & 2
\end{array}
$$

All entries positive. $x = 1$ is upper bound.

$$
\begin{array}{r|rrrr}
-\frac{1}{4} & 4 & -3 & 4 & -3 \\
 & & -1 & 1 & -\frac{5}{4} \\
\hline
 & 4 & -4 & 5 & -\frac{17}{4}
\end{array}
$$

Alternating signs. $x = -\frac{1}{4}$ is lower bound.

65. $6 + \sqrt{-25} = 6 + 5i$

67. $-2i^2 + 7i = 2 + 7i$

69. $(7 + 5i) + (-4 + 2i) = (7 - 4) + (5i + 2i)$
$$= 3 + 7i$$

71. $5i(13 - 8i) = 65i - 40i^2 = 40 + 65i$

73. $(10 - 8i)(2 - 3i) = 20 - 30i - 16i + 24i^2 = -4 - 46i$

75. $(3 + 7i)^2 + (3 - 7i)^2 = (9 + 42i - 49) + (9 - 42i - 49)$
$$= -80$$

77. $\dfrac{6+i}{i} = \dfrac{6+i}{i} \cdot \dfrac{-i}{-i} = \dfrac{-6i-i^2}{-i^2}$

$= \dfrac{-6i+1}{1} = 1-6i$

79. $\dfrac{3+2i}{5+i} \cdot \dfrac{5-i}{5-i} = \dfrac{15+10i-3i+2}{25+1}$

$= \dfrac{17}{26} + \dfrac{7}{26}i$

81. $2-5i$

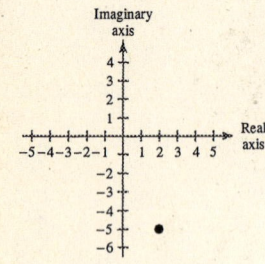

83. $-6i$

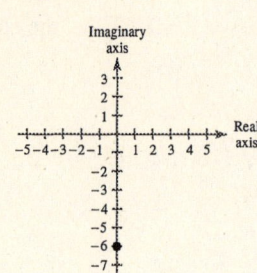

85. 3

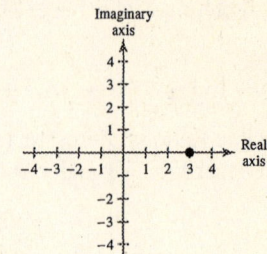

87. $f(x) = 3x(x-2)^2$

zeros: $0, 2, 2$

89. $f(x) = (x+4)(x-6)(x-2i)(x+2i)$

zeros: $-4, 6, 2i, -2i$

91. $f(x) = 2x^4 - 5x^3 + 10x - 12$

$$\begin{array}{r|rrrrr} 2 & 2 & -5 & 0 & 10 & -12 \\ & & 4 & -2 & -4 & 12 \\ \hline & 2 & -1 & -2 & 6 & 0 \end{array}$$

$x = 2$ is a zero

$f(x) = (x-2)\left(x+\tfrac{3}{2}\right)(2x^2 - 4x + 4)$

$\quad = (x-2)(2x+3)(x^2 - 2x + 2)$

By the Quadratic Formula, applied to $x^2 - 2x + 2$,

$x = \dfrac{2 \pm \sqrt{4 - 4(2)}}{2} = 1 \pm i$

zeros: $2, -\tfrac{3}{2}, 1 \pm i$

$f(x) = (x-2)(2x+3)(x-1+i)(x-1-i)$

$$\begin{array}{r|rrrr} -\frac{3}{2} & 2 & -1 & -2 & 6 \\ & & -3 & 6 & -6 \\ \hline & 2 & -4 & 4 & 0 \end{array}$$

$x = -\tfrac{3}{2}$ is a zero

93. $h(x) = x^3 - 7x^2 + 18x - 24$

$$\begin{array}{r|rrrr} 4 & 1 & -7 & 18 & -24 \\ & & 4 & -12 & 24 \\ \hline & 1 & -3 & 6 & 0 \end{array}$$

$x = 4$ is a zero. Applying the Quadratic Formula on $x^2 - 3x + 6$,

$x = \dfrac{3 \pm \sqrt{9 - 4(6)}}{2} = \dfrac{3}{2} \pm \dfrac{\sqrt{15}}{2}i$

zeros: $4, \dfrac{3}{2} + \dfrac{\sqrt{15}}{2}i, \dfrac{3}{2} - \dfrac{\sqrt{15}}{2}i$

$h(x) = (x-4)\left(x - \dfrac{3+\sqrt{15}i}{2}\right)\left(x - \dfrac{3-\sqrt{15}i}{2}\right)$

95. $f(x) = x^3 - 4x^2 + 6x - 4$

(a) $x^3 - 4x^2 + 6x - 4 = (x-2)(x^2 - 2x + 2)$

By the Quadratic Formula, for $x^2 - 2x + 2$,

$x = \dfrac{2 \pm \sqrt{(-2)^2 - 4(2)}}{2} = 1 \pm i$

zeros: $2, 1+i, 1-i$

(b) $f(x) = (x-2)(x-1-i)(x-1+i)$

(c) x-intercept: $(2, 0)$

(d)

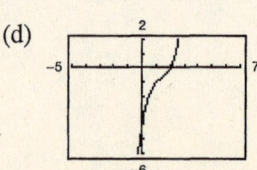

97. $f(x) = x^3 + 6x^2 + 11x + 12$

(a) $x^3 + 6x^2 + 11x + 12 = (x + 4)(x^2 + 2x + 3)$

By the Quadratic Formula for $x^2 + 2x + 3$,

$$x = \frac{-2 \pm \sqrt{4 - 4(3)}}{2} = -1 \pm \sqrt{2}i$$

zeros: $-4, -1 \pm \sqrt{2}i$

(b) $f(x) = (x + 4)(x + 1 - \sqrt{2}i)(x + 1 + \sqrt{2}i)$

(c) x-intercept: $(-4, 0)$

(d)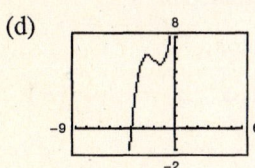

99. $f(x) = x^4 + 34x^2 + 225$

(a) $x^4 + 34x^2 + 225 = (x^2 + 9)(x^2 + 25)$

zeros: $\pm 3i, \pm 5i$

(b) $(x + 3i)(x - 3i)(x + 5i)(x - 5i)$

(c) No x-intercepts

(d)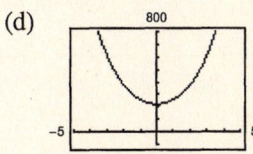

101. $f(x) = (x + 2)(x + 2)(x + 5i)(x - 5i)$

$= (x^2 + 4x + 4)(x^2 + 25)$

$= x^4 + 4x^3 + 29x^2 + 100x + 100$

103. $f(x) = (x - 1)(x + 4)(x + 3 - 5i)(x + 3 + 5i)$

$= (x^2 + 3x - 4)((x + 3)^2 + 25)$

$= (x^2 + 3x - 4)(x^2 + 6x + 34)$

$= x^4 + 9x^3 + 48x^2 + 78x - 136$

105. $f(x) = x^4 + 2x^2 - 8$

(a) $f(x) = (x^2 + 4)(x^2 - 2)$

(b) $f(x) = (x^2 + 4)(x + \sqrt{2})(x - \sqrt{2})$

(c) $f(x) = (x + 2i)(x - 2i)(x + \sqrt{2})(x - \sqrt{2})$

107. $f(x) = x^4 - 2x^3 + 8x^2 - 18x - 9$

(a) $f(x) = (x^2 + 9)(x^2 - 2x - 1)$

For the quadratic $x^2 - 2x - 1$, $x = \dfrac{2 \pm \sqrt{(-2)^2 - 4(-1)}}{2} = 1 \pm \sqrt{2}$

(b) $f(x) = (x^2 + 9)(x - 1 + \sqrt{2})(x - 1 - \sqrt{2})$

(c) $f(x) = (x + 3i)(x - 3i)(x - 1 + \sqrt{2})(x - 1 - \sqrt{2})$

109. $f(x) = \dfrac{x - 8}{1 - x}$

(a) Domain: all $x \neq 1$

(b) Horizontal asymptote: $y = -1$

Vertical asymptote: $x = 1$

111. $f(x) = \dfrac{2}{x^2 - 3x - 18} = \dfrac{2}{(x - 6)(x + 3)}$

(a) Domain: all $x \neq 6, -3$

(b) Horizontal asymptote: $y = 0$

Vertical asymptotes: $x = 6$, $x = -3$

113. $f(x) = \dfrac{7 + x}{7 - x}$

    (a) Domain: all $x \neq 7$

    (b) Horizontal asymptote: $y = -1$

        Vertical asymptote: $x = 7$

115. $f(x) = \dfrac{4x^2}{2x^2 - 3}$

    (a) Domain: all $x \neq \pm\sqrt{\frac{3}{2}}$

    (b) Horizontal asymptote: $y = 2$

        Vertical asymptote: $x = \pm\sqrt{\frac{3}{2}}$

117. $f(x) = \dfrac{2x - 10}{x^2 - 2x - 15} = \dfrac{2(x - 5)}{(x - 5)(x + 3)} = \dfrac{2}{x + 3}$,

$x \neq 5$

    (a) Domain: all $x \neq 5, -3$

    (b) Vertical asymptote: $x = -3$

        (There is a hole at $x = 5$)

        Horizontal asymptote: $y = 0$

119. $f(x) = \dfrac{x - 2}{|x| + 2}$

    (a) Domain: all real numbers

    (b) No vertical asymptotes

        Horizontal asymptotes: $y = 1, y = -1$

121. $C = \dfrac{528p}{100 - p}$, $0 \leq p < 100$

    (a) When $p = 25$, $C = \dfrac{528(25)}{100 - 25} = 176$ million

        When $p = 50$, $C = \dfrac{528(50)}{100 - 50} = 528$ million

        When $p = 75$, $C = \dfrac{528(75)}{100 - 75} = 1584$ million

(b)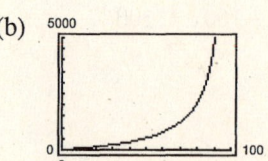

(c) No. As $p \to 100$, C tends to infinity.

123. $f(x) = \dfrac{2x - 1}{x - 5}$

    Intercepts: $\left(0, \frac{1}{5}\right), \left(\frac{1}{2}, 0\right)$

    Vertical asymptote: $x = 5$

    Horizontal asymptote: $y = 2$

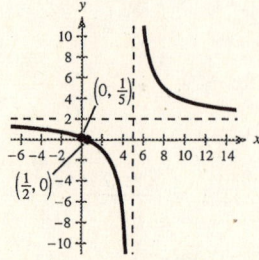

125. $f(x) = \dfrac{2x}{x^2 + 4}$

    Intercept: $(0, 0)$

    Origin symmetry

    Horizontal asymptote: $y = 0$

x	-2	-1	0	1	2
y	$-\frac{1}{2}$	$-\frac{2}{5}$	0	$\frac{2}{5}$	$\frac{1}{2}$

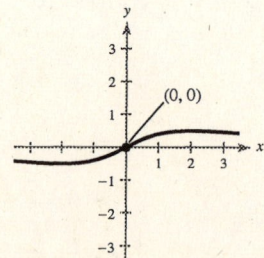

127. $f(x) = \dfrac{x^2}{x^2 + 1}$

Intercept: $(0, 0)$

y-axis symmetry

Horizontal asymptote: $y = 1$

x	± 3	± 2	± 1	0
y	$\frac{9}{10}$	$\frac{4}{5}$	$\frac{1}{2}$	0

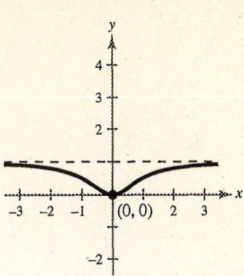

129. $f(x) = \dfrac{2(x^2 - 16)}{x^2 + 2x - 8} = \dfrac{2(x + 4)(x - 4)}{(x + 4)(x - 2)} = \dfrac{2(x - 4)}{x - 2},$

$x \neq -4$

Intercepts: $(0, 4), (4, 0)$

Horizontal asymptote: $y = 2$

Vertical asymptote: $x = 2$

Hole at $x = -4$

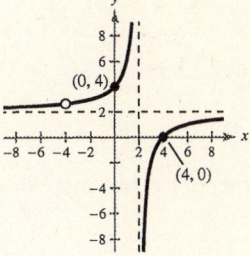

131. $f(x) = \dfrac{2x^3}{x^2 + 1} = 2x - \dfrac{2x}{x^2 + 1}$

Intercept: $(0, 0)$

Origin symmetry

Slant asymptote: $y = 2x$

x	-2	-1	0	1	2
y	$-\frac{16}{5}$	-1	0	1	$\frac{16}{5}$

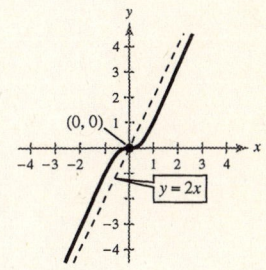

133. $f(x) = \dfrac{x^2 - x + 1}{x - 3}$

$= x + 2 + \dfrac{7}{x - 3}$

Intercept: $\left(0, -\frac{1}{3}\right)$

Vertical asymptote: $x = 3$

Slant asymptote: $y = x + 2$

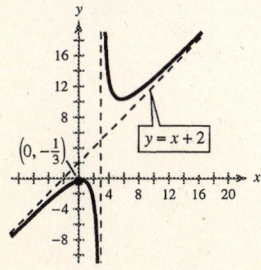

135. $N = \dfrac{20(4 + 3t)}{1 + 0.05t}, \quad t \geq 0$

(a)
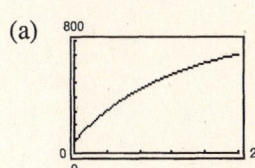

(b) $N(5) \; = 304{,}000$ fish

$N(10) \approx 453{,}333$ fish

$N(25) \approx 702{,}222$ fish

(c) The limit is

$\dfrac{60}{0.05} = 1{,}200{,}000$ fish,

the horizontal asymptote.

137. Quadratic model

139. Linear model

141. (a)

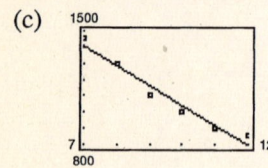

(d) $P = 16.518t^2 - 435.982t + 3703.75$

(e)

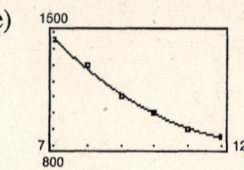

(b) $P = -122.143t + 2261.19$

(c)

The quadratic model fits the data better. For 2008, $t = 18$ and $P \approx \$1208$.

The answer does not seem reasonable.

143. False. The degree of the numerator is two more than the degree of the denominator.

145. It means that the divisor is a factor of the dividend.

Chapter 2   Practice Test

1. Sketch the graph of $f(x) = x^2 - 6x + 5$ by hand and identify the vertex and the intercepts.

2. Find the number of units x that produce a minimum cost C if $C = 0.01x^2 - 90x + 15,000$.

3. Find the quadratic function that has a maximum at $(1, 7)$ and passes through the point $(2, 5)$.

4. Find two quadratic functions that have x-intercepts $(2, 0)$ and $\left(\frac{4}{3}, 0\right)$.

5. Use the leading Coefficient Test to determine the right-hand and left-hand behavior of the graph of the polynomial function $f(x) = -3x^5 + 2x^3 - 17$.

6. Find all the real zeros of $f(x) = x^5 - 5x^3 + 4x$. Verify your answer with a graphing utility.

7. Find a polynomial function with 0, 3, and -2 as zeros.

8. Sketch $f(x) = x^3 - 12x$ by hand.

9. Divide $3x^4 - 7x^2 + 2x - 10$ by $x - 3$ using long division.

10. Divide $x^3 - 11$ by $x^2 + 2x - 1$.

11. Use synthetic division to divide $3x^5 + 13x^4 + 12x - 1$ by $x + 5$.

12. Use synthetic division to find $f(-6)$ when $f(x) = 7x^3 + 40x^2 - 12x + 15$.

13. Find the real zeros of $f(x) = x^3 - 19x - 30$.

14. Find the real zeros of $f(x) = x^4 + x^3 - 8x^2 - 9x - 9$.

15. List all possible rational zeros of the function $f(x) = 6x^3 - 5x^2 + 4x - 15$.

16. Find the rational zeros of the polynomial $f(x) = x^3 - \frac{20}{3}x^2 + 9x - \frac{10}{3}$.

17. Write $f(x) = x^4 + x^3 + 3x^2 + 5x - 10$ as a product of linear factors.

18. Write $\dfrac{2}{1 + i}$ in standard form.

19. Write $\dfrac{3 + i}{2} - \dfrac{i + 1}{4}$ in standard form.

20. Find a polynomial with real coefficients that has 2, $3 + i$, and $3 - 2i$ as zeros.

21. Use synthetic division to show that $3i$ is a zero of $f(x) = x^3 + 4x^2 + 9x + 36$.

22. Find a mathematical model for the statement, "z varies directly as the square of x and inversely as the square root of y".

23. Sketch the graph of $f(x) = \dfrac{x - 1}{2x}$ and label all intercepts and asymptotes.

24. Sketch the graph of $f(x) = \dfrac{3x^2 - 4}{x}$ and label all intercepts and asymptotes.

25. Find all the asymptotes of $f(x) = \dfrac{8x^2 - 9}{x^2 + 1}$.

26. Find all the asymptotes of $f(x) = \dfrac{4x^2 - 2x + 7}{x - 1}$.

27. Sketch the graph of $f(x) = \dfrac{x - 5}{(x - 5)^2}$.

CHAPTER 3
Exponential and Logarithmic Functions

CHAPTER 3
Exponential and Logarithmic Functions

Section 3.1   Exponential Functions and Their Graphs

- You should know that a function of the form $y = a^x$, where $a > 0$, $a \neq 1$, is called an exponential function with base a.
- You should be able to graph exponential functions.
- You should be familiar with the number e and the natural exponential function $f(x) = e^x$.
- You should know formulas for compound interest.
    - (a) For n compoundings per year: $A = P\left(1 + \dfrac{r}{n}\right)^{nt}$.
    - (b) For continuous compoundings: $A = Pe^{rt}$.

Solutions to Odd-Numbered Exercises

1. $(3.4)^{6.8} \approx 4112.033$

3. $5^{-\pi} \approx 0.006$

5. $17^{2(\sqrt{3})} \approx 18{,}297.851$

7. $g(x) = 5^x$

Asymptote: $y = 0$

Intercept: $(0, 1)$

Increasing

x	-2	-1	0	1	2
y	$\frac{1}{25}$	$\frac{1}{5}$	1	5	25

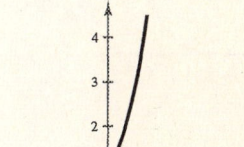

9. $f(x) = \left(\frac{1}{5}\right)^x = 5^{-x}$

Asymptote: $y = 0$

Intercepts: $(0, 1)$

Decreasing

x	-2	-1	0	1	2
y	25	5	1	$\frac{1}{5}$	$\frac{1}{25}$

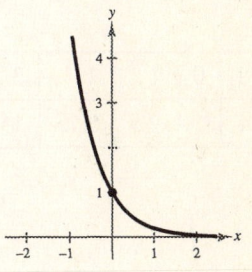

11. $h(x) = 5^{x-2}$

Asymptote: $y = 0$

Intercepts: $\left(0, \frac{1}{25}\right)$

Increasing

x	-1	0	1	2	3
y	$\frac{1}{125}$	$\frac{1}{25}$	$\frac{1}{5}$	1	5

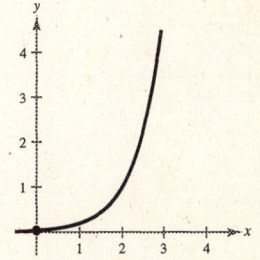

13. $g(x) = 5^{-x} - 3$

Asymptote: $y = -3$

Intercepts: $(0, -2)$, $(-0.683, 0)$

Decreasing

x	-1	0	1	2
y	2	-2	$-2\frac{4}{5}$	$-2\frac{24}{25}$

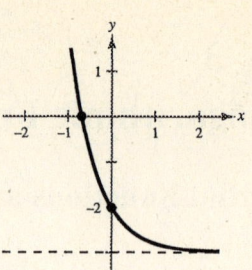

15. $f(x) = 2^{x-2}$ rises to the right.

Asymptote: $y = 0$

Intercept: $\left(0, \frac{1}{4}\right)$

Matches graph (d).

17. $f(x) = 2^x - 4$ rises to the right.

Asymptote: $y = -4$

Intercept: $(0, -3)$

Matches graph (c).

19. $f(x) = 3^x$

$g(x) = 3^{x-5} = f(x - 5)$

Horizontal shift five units to the right

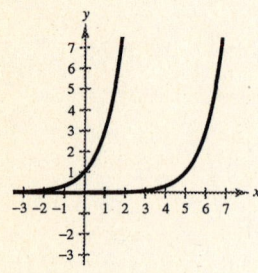

21. $f(x) = \left(\frac{3}{5}\right)^x$

$g(x) = -\left(\frac{3}{5}\right)^{x+4} = -f(x + 4)$

Horizontal shift 4 units to the left, followed by reflection in x-axis.

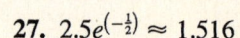

23. $e^{9.2} \approx 9897.129$

25. $50e^{4(0.02)} \approx 54.164$

27. $2.5e^{\left(-\frac{1}{2}\right)} \approx 1.516$

29. $f(x) = \left(\frac{5}{2}\right)^x$

x	-1	0	1	2	3
$f(x)$	0.4	1	2.5	6.25	15.625

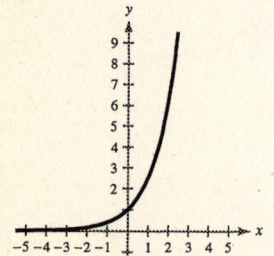

31. $f(x) = 6^x$

x	-1	0	1	2
$f(x)$	0.167	1	6	36

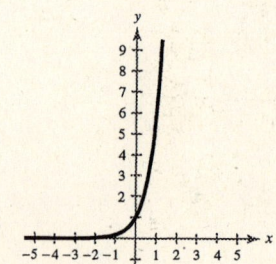

33. $f(x) = 3^{x+2} = 9 \cdot 3^x$

x	-1	-2	0	1
$f(x)$	3	1	9	27

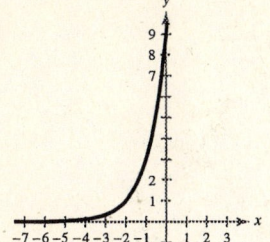

35. $f(x) = 3e^{x+4}$

x	-7	-6	-5	-4	-3
$f(x)$	0.149	0.406	1.104	3	8.155

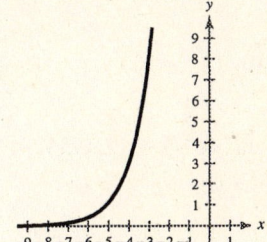

37. $f(x) = 2 + e^{x-5}$

x	2	3	4	5	6	7
$f(x)$	2.05	2.135	2.368	3	4.718	9.389

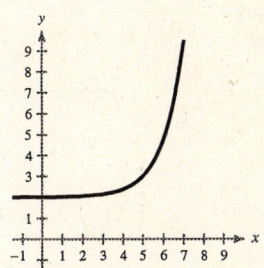

39. $y = 2^{-x^2}$

Asymptote: $y = 0$

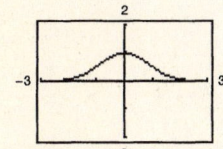

41. $f(x) = 3^{x-2} + 1$

Asymptote: $y = 1$

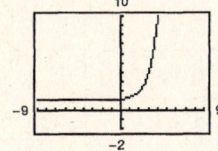

43. $g(x) = 2 - e^{-x}$

Asymptote: $y = 2$

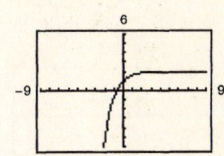

45. $s(t) = 2e^{0.12t}$

Asymptote: $y = 0$

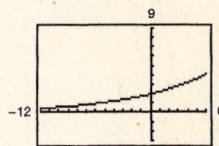

47. $f(x) = \dfrac{8}{1 + e^{-0.5x}}$

(a)

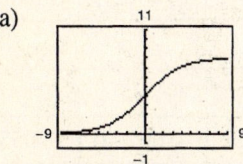

(b)

x	-30	-20	-10	0	10	20	30
$f(x)$	≈ 0	≈ 0	0.05	4	7.95	≈ 8	≈ 8

Horizontal asymptotes: $y = 0, y = 8$

49. $f(x) = \dfrac{-6}{2 - e^{0.2x}}$

(a)

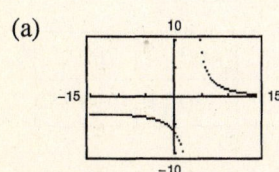

(b)

x	-20	-10	0	3	3.4	3.46
$f(x)$	-3.03	-3.22	-6	-34	-230	-2617

x	3.47	4	5	10	20
$f(x)$	3516	26.6	8.4	1.11	0.11

Horizontal asymptotes: $y = -3, y = 0$

Vertical asymptote: $x \approx 3.46$

51. $f(x) = x^2 e^{-x}$

(a)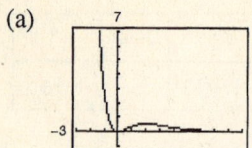

(b) Decreasing: $(-\infty, 0)$, $(2, \infty)$

Increasing: $(0, 2)$

(c) Relative maximum: $(2, 4e^{-2}) \approx (2, 0.541)$

Relative minimum: $(0, 0)$

53. $f(x) = x(2^{3-x})$

(a)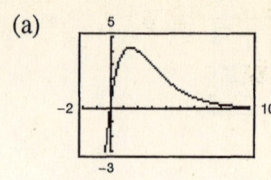

(b) Decreasing: $(1.44, \infty)$

Increasing: $(-\infty, 1.44)$

(c) Relative maximum: $(1.44, 4.25)$

55. $P = 2500, r = 2.5\% = 0.025, t = 10$

Compounded n times per year: $A = P\left(1 + \dfrac{r}{n}\right)^{nt} = 2500\left(1 + \dfrac{0.025}{n}\right)^{10n}$

Compounded continuously: $A = Pe^{rt} = 2500e^{(0.025)(10)}$

n	1	2	4	12	365	Continuos
A	3200.21	3205.09	3207.57	3209.23	3210.04	3210.06

57. $P = 2500, r = 4\% = 0.04, t = 20$

Compounded n times per year: $A = P\left(1 + \dfrac{r}{n}\right)^{nt} = 2500\left(1 + \dfrac{.04}{n}\right)^{20n}$

Compounded continuously: $A = Pe^{rt} = 2500e^{(.04)(20)}$

n	1	2	4	12	365	Continuos
A	5477.81	5520.10	5541.79	5556.46	5563.61	5563.85

59. $P = 12,000, r = 4\% = 0.04$

$A = Pe^{rt} = 12000e^{0.04t}$

t	1	10	20	30	40	50
A	12489.73	17901.90	26706.49	39841.40	59436.39	88668.67

61. $P = 12,000, r = 3.5\% = 0.035$

$A = Pe^{rt} = 12000e^{0.035t}$

t	1	10	20	30	40	50
A	12427.44	17028.81	24165.03	34291.81	48662.40	69055.23

63. $p = 5000\left(1 - \dfrac{4}{4 + e^{-0.002x}}\right)$

(a)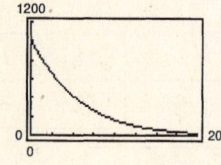

(b) If $x = 500, p \approx \$421.12$.

(c) For $x = 600, p \approx \$350.13$.

(d)

x	100	200	300	400	500	600	700
p	849.53	717.64	603.25	504.94	421.12	350.13	290.35

65. $Q = 25\left(\frac{1}{2}\right)^{t/1620}$

(a) When $t = 0$, $Q = 25\left(\frac{1}{2}\right)^{0/1620} = 25(1) = 25$ grams.

(b) When $t = 1000$, $Q = 25\left(\frac{1}{2}\right)^{1000/1620} \approx 16.30$ grams.

(c)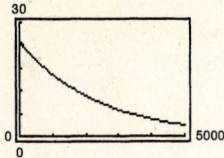

(d) Never, $Q \to 0$ as $t \to \infty$, but Q never reaches 0.

67. $P(t) = 100e^{0.2197t}$

(a)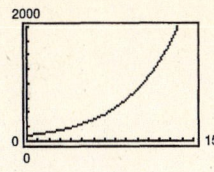

(b) $P(0) = 100$

$P(5) \approx 300$

$P(10) \approx 900$

(c) $P(0) = 100e^{0.2197(0)} = 100$

$P(5) = 100e^{0.2197(5)} = 299.966 \approx 300$

$P(10) = 100e^{0.2197(10)} = 899.798 \approx 900$

69. $C(t) = P(1.04)^t$

(a)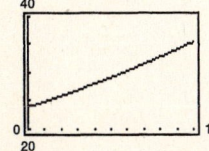

(b) $C(10) \approx 35.45$

(c) $C(10) = 23.95(1.04)^{10} \approx 35.45$

71. True. $f(x) = 1^x$ is not an exponential function.

73. $y_1 = e^x$

$y_2 = x^2$

$y_3 = x^3$

$y_4 = \sqrt{x}$

$y_5 = |x|$

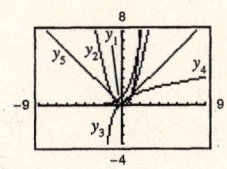

(a) $y_1 = e^x$ increases at the fastest rate.

(b) For any positive integer n, $e^x > x^n$ for x sufficiently large. That is, e^x grows faster than x^n.

(c) A quantity is growing exponentially if its growth rate is of the form $y = ce^{rx}$. This is a faster rate than any polynomial growth rate.

75. $f(x) = \left(1 + \dfrac{0.5}{x}\right)^x$ and

$g(x) = e^{0.5} \approx 1.6487$

(Horizontal line)

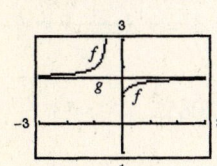

As $x \to \infty$, $f(x) \to g(x)$.

77. f has an inverse because f is one-to-one.

$y = 5x - 7$

$x = 5y - 7$

$x + 7 = 5y$

$f^{-1}(x) = \frac{1}{5}(x + 7)$

79. f has an inverse because f is one-to-one.

$y = \sqrt[3]{x + 8}$

$x = \sqrt[3]{y + 8}$

$x^3 = y + 8$

$x^3 - 8 = y$

$f^{-1}(x) = x^3 - 8$

81. $f(x) = \dfrac{2x}{x-7}$

Vertical asymptote: $x = 7$

Horizontal asymptote: $y = 2$

Intercept: $(0, 0)$

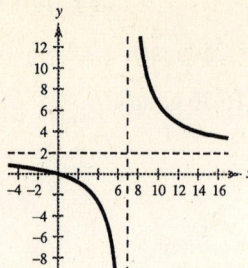

Section 3.2   Logarithmic Functions and Their Graphs

- ■ You should know that a function of the form $y = \log_a x$, where $a \geq 0$, $a \neq 1$, and $x > 0$, is called a logarithm of x to base a.

- ■ You should be able to convert from logarithmic form to exponential form and vice versa.

    $$y = \log_a x \iff a^y = x$$

- ■ You should know the following properties of logarithms.

    (a) $\log_a 1 = 0$ since $a^0 = 1$.

    (b) $\log_a a = 1$ since $a^1 = a$.

    (c) $\log_a a^x = x$ since $a^x = a^x$.

    (d) If $\log_a x = \log_a y$, then $x = y$.

- ■ You should know the definition of the natural logarithmic function.

    $$\log_e x = \ln x, \, x > 0$$

- ■ You should know the properties of the natural logarithmic function.

    (a) $\ln 1 = 0$ since $e^0 = 1$.

    (b) $\ln e = 1$ since $e^1 = e$.

    (c) $\ln e^x = x$ since $e^x = e^x$.

    (d) If $\ln x = \ln y$, then $x = y$.

- ■ You should be able to graph logarithmic functions.

Solutions to Odd-Numbered Exercises

1. $\log_4 64 = 3 \implies 4^3 = 64$

3. $\log_7 \frac{1}{49} = -2 \implies 7^{-2} = \frac{1}{49}$

5. $\log_{32} 4 = \frac{2}{5} \implies 32^{2/5} = 4$

7. $\ln 1 = 0 \implies e^0 = 1$

9. $5^3 = 125 \implies \log_5 125 = 3$

11. $81^{1/4} = 3 \implies \log_{81} 3 = \frac{1}{4}$

13. $6^{-2} = \frac{1}{36} \implies \log_6 \frac{1}{36} = -2$

15. $e^3 = 20.0855 \ldots \implies \ln 20.0855 \ldots = 3$

17. $\log_2 16 = \log_2 2^4 = 4$

19. $\log_{10} 0.01 = \log_{10} 10^{-2} = -2$

21. $\log_{10} 345 \approx 2.538$

23. $6 \log_{10} 14.8 \approx 7.022$

25. $\log_7 x = \log_7 9$

$x = 9$

27. $\log_6 6^2 = x$

$2\log_6 6 = x$

$2 = x$

29. $\log_8 x = \log_8 10^{-1}$

$x = 10^{-1} = \frac{1}{10}$

31. $f(x) = 3^x$, $g(x) = \log_3 x$

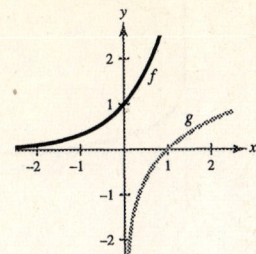

f and g are inverses. Their graphs are reflected about the line $y = x$.

33. $f(x) = e^x$, $g(x) = \ln x$

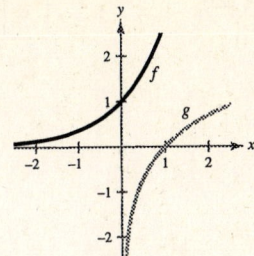

f and g are inverses. Their graphs are reflected about the line $y = x$.

35. $f(x) = \log_4 x$

Domain: $x > 0 \Rightarrow$ The domain is $(0, \infty)$.

Vertical asymptote: $x = 0$

x-intercept: $(1, 0)$

$y = \log_4 x \Rightarrow 4^y = x$

x	$\frac{1}{4}$	1	4	2
y	-1	0	1	$\frac{1}{2}$

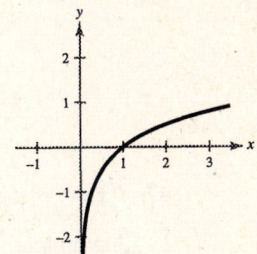

37. $f(x) = \log_{10}\left(\dfrac{x}{5}\right)$

Domain: $\dfrac{x}{5} > 0 \Rightarrow x > 0$

The domain is $(0, \infty)$.

Vertical asymptote: $\dfrac{x}{5} = 0 \Rightarrow x = 0$

The vertical asymptote is the y-axis.

x-intercept: $\log_{10}\left(\dfrac{x}{5}\right) = 0$

$$\frac{x}{5} = 10^0$$

$$\frac{x}{5} = 1 \Rightarrow x = 5$$

The x-intercept is $(5, 0)$.

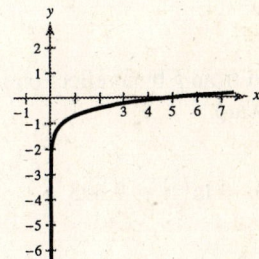

39. $h(x) = \log_4(x - 3)$

Domain: $x - 3 > 0$ or $(3, \infty)$

Vertical asymptote: $x = 3$

Intercept: $(4, 0)$

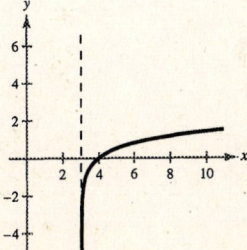

41. $y = -\log_{10} x + 2$

Domain: $x > 0$ or $(0, \infty)$

Vertical asymptote: $x = 0$

x-intercept: $(100, 0)$

x	$\frac{1}{10}$	1	10	100
y	3	2	1	0

43. $f(x) = 6 + \log_6 (x - 3)$

Domain: $(3, \infty)$

Vertical asymptote: $x = 3$

x-intercept: $\log_6(x - 3) = -6$

$6^{-6} = x - 3$

$x = 3 + 6^{-6} \approx 3$

x	4	9	$3\frac{1}{6}$
y	6	7	5

45. $f(x) = \log_3 x + 2$

Asymptote: $x = 0$

Point on graph: $(1, 2)$

Matches graph (b).

47. $f(x) = -\log_3(x + 2)$

Asymptote: $x = -2$

Point on graph: $(-1, 0)$

Matches graph (d).

49. $f(x) = \log_{10} x$

$g(x) = -\log_{10} x$ is a reflection in the x-axis of the graph of f.

51. $f(x) = \log_2 x$

$g(x) = 4 - \log_2 x$ is obtained from f by a reflection in the x-axis followed by a vertical shift 4 units upward.

53. $\ln\sqrt{42} \approx 1.869$

55. $-\ln\left(\frac{1}{2}\right) \approx 0.693$

57. $\ln e^2 = 2$

(inverse property)

59. $e^{\ln 1.8} = 1.8$

(inverse property)

61. $f(x) = \ln(x - 1)$

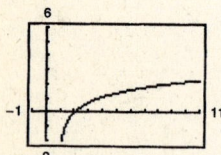

Domain: $x > 1$

Vertical asymptote: $x = 1$

x-intercept: $(2, 0)$

63. $g(x) = \ln(-x)$

Domain: $-x > 0 \implies x < 0$

The domain is $(-\infty, 0)$.

Vertical asymptote: $-x = 0 \implies x = 0$

x-intercept: $0 = \ln(-x)$

$e^0 = -x$

$-1 = x$

The x-intercept is $(-1, 0)$.

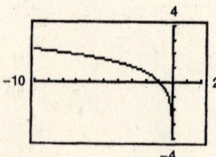

65. $f(x) = \dfrac{x}{2} - \ln \dfrac{x}{4}$

(a)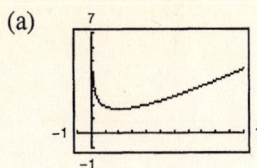

(b) Domain: $(0, \infty)$

(c) Increasing on $(2, \infty)$

    Decreasing on $(0, 2)$

(d) Relative minimum: $(2, 1.693)$

69. $f(t) = 80 - 17 \log_{10}(t + 1), \quad 0 \le t \le 12$

(a) $f(0) = 80 - 17 \log_{10}(0 + 1) = 80$

(b) $f(4) = 80 - 17 \log_{10}(4 + 1) \approx 68.1$

(c) $f(10) = 80 - 17 \log_{10}(10 + 1) \approx 62.3$

(d)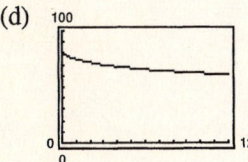

73. $\beta = 10 \log_{10}\left(\dfrac{I}{10^{-12}}\right)$

(a) $I = 1$: $\beta = 10 \log_{10}\left(\dfrac{1}{10^{-12}}\right) = 10 \cdot \log_{10}(10^{12}) = 10(12) = 120$ decibels

(b) $I = 10^{-2}$: $\beta = 10 \log_{10}\left(\dfrac{10^{-2}}{10^{-12}}\right) = 10 \log_{10}(10^{10}) = 10(10) = 100$ decibels

(c) No, this is a logarithmic scale.

75. $y = 80.4 - 11 \ln x$

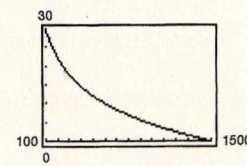

    $y(300) = 80.4 - 11 \ln 300 \approx 17.66 \text{ ft}^3/\text{min}$

79. $f(x) = \log_a x$ is the inverse of $g(x) = a^x$, where $a > 0, a \ne 1$

67. $h(x) = 4x \ln x$

(a)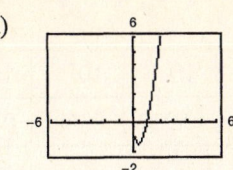

(b) Domain: $(0, \infty)$

(c) Increasing on $(0.368, \infty)$

    Decreasing on $(0, 0.368)$

(d) Relative minimum: $(0.368, -1.472)$

71. $t = \dfrac{\ln K}{0.055}$

(a)

K	1	2	4	6	8	10	12
t	0	12.6	25.2	32.6	37.8	41.9	45.2

As the amount increases, the time increases, but at a lesser rate.

(b)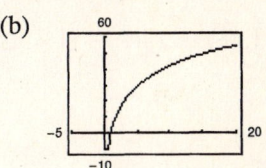

77. False. You would reflect $y = 6^x$ in the line $y = x$.

81. (a) False, y is not an exponential function of x.
      (y can never be 0.)

(b) True, y could be $\log_2 x$.

(c) True, x could be 2^y.

(d) False, y is not linear. (The points are not collinear.)

83. $f(x) = \dfrac{\ln x}{x}$

(a)

x	1	5	10	10^2	10^4	10^6
$f(x)$	0	0.322	0.230	0.046	0.00092	0.0000138

(b) As x increases without bound, $f(x)$ approaches 0.

(c)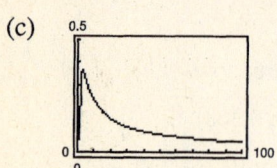

85. $x^2 + 2x - 3 = (x + 3)(x - 1)$

87. $12x^2 + 5x - 3 = (4x + 3)(3x - 1)$

89. $16x^2 - 25 = (4x + 5)(4x - 5)$

91. $2x^3 + x^2 - 45x = x(2x^2 + x - 45) = x(2x - 9)(x + 5)$

93. $(f + g)(2) = f(2) + g(2) = [3(2) + 2] + [2^3 - 1] = 8 + 7 = 15$

95. $(fg)(6) = f(6)g(6) = [3(6) + 2][6^3 - 1] = [20][215] = 4300$

97. $5x - 7 = x + 4$

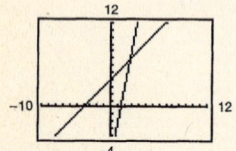

The graphs of $y = 5x - 7$ and $y_2 = x + 4$ intersect when $x = 2.75$ or $\frac{11}{4}$.

99. $\sqrt{3x - 2} = 9$

The graphs of $y_1 = \sqrt{3x - 2}$ and $y_2 = 9$ intersect when $x \approx 27.667$ or $\frac{83}{3}$.

101. $f(x) = \dfrac{4}{-8 - x} = \dfrac{-4}{x + 8}$

Vertical asymptote: $x = -8$

Horizontal asymptote: $y = 0$

103. $f(x) = \dfrac{x + 5}{(2x^2 + x - 15)} = \dfrac{x + 5}{(2x - 5)(x + 3)}$

Vertical asymptotes: $x = \dfrac{5}{2}, -3$

Horizontal asymptote: $y = 0$

105. $g(x) = \dfrac{x^2 - 4}{x^2 - 4x - 12}$

$= \dfrac{(x + 2)(x - 2)}{(x - 6)(x + 2)}$

$= \dfrac{x - 2}{x - 6}, \quad x \neq -2$

Vertical asymptote: $x = 6$

Horizontal asymptote: $y = 1$

Section 3.3   Properties of Logarithms

■ You should know the following properties of logarithms.

(a) $\log_a x = \dfrac{\log_b x}{\log_b a}$

(b) $\log_a (uv) = \log_a u + \log_a v$ $\ln (uv) = \ln u + \ln v$

(c) $\log_a (u/v) = \log_a u - \log_a v$ $\ln (u/v) = \ln u - \ln v$

(d) $\log_a u^n = n \log_a u$ $\ln u^n = n \ln u$

■ You should be able to rewrite logarithmic expressions using these properties.

Solutions to Odd-Numbered Exercises

1. (a) $\log_5 x = \dfrac{\log_{10} x}{\log_{10} 5}$

(b) $\log_5 x = \dfrac{\ln x}{\ln 5}$

3. (a) $\log_{1/5} x = \dfrac{\log_{10} x}{\log_{10} \dfrac{1}{5}} = \dfrac{\log_{10} x}{-\log_{10} 5}$

(b) $\log_{1/5} x = \dfrac{\ln x}{\ln \dfrac{1}{5}} = \dfrac{\ln x}{-\ln 5}$

5. (a) $\log_a \left(\dfrac{3}{10}\right) = \dfrac{\log_{10}\left(\dfrac{3}{10}\right)}{\log_{10} a}$

(b) $\log_a \left(\dfrac{3}{10}\right) = \dfrac{\ln\left(\dfrac{3}{10}\right)}{\ln a}$

7. (a) $\log_{2.6} x = \dfrac{\log_{10} x}{\log_{10} 2.6}$

(b) $\log_{2.6} x = \dfrac{\ln x}{\ln 2.6}$

9. $\log_3 7 = \dfrac{\ln 7}{\ln 3} \approx 1.771$

11. $\log_{1/2} 4 = \dfrac{\ln 4}{\ln \dfrac{1}{2}} = -2$

13. $\log_9(0.8) = \dfrac{\ln(0.8)}{\ln 9} \approx -0.102$

15. $\log_{15} 1460 = \dfrac{\ln 1460}{\ln 15} \approx 2.691$

17. $\log_4 8 = \log_4 2^3 = 3 \log_4 2$

$\quad = 3 \log_4 4^{1/2} = 3\left(\dfrac{1}{2}\right)\log_4 4$

$\quad = \dfrac{3}{2}$

19. $\ln(5e^6) = \ln 5 + \ln e^6 = \ln 5 + 6 = 6 + \ln 5$

21. $\log_5 \dfrac{1}{250} = \log_5 1 - \log_5 250 = 0 - \log_5 (125 \cdot 2)$

$\quad = -\log_5(5^3 \cdot 2) = -[\log_5 5^3 + \log_5 2]$

$\quad = -[3 \log_5 5 + \log_5 2] = -3 - \log_5 2$

23. $\log_{10} 5x = \log_{10} 5 + \log_{10} x$

25. $\log_{10} \dfrac{5}{x} = \log_{10} 5 - \log_{10} x$

27. $\log_8 x^4 = 4 \log_8 x$

29. $\ln \sqrt{z} = \ln z^{1/2} = \dfrac{1}{2} \ln z$

31. $\ln xyz = \ln x + \ln y + \ln z$

33. $\ln\left(a^2\sqrt{a-1}\right) = \ln a^2 + \ln(a-1)^2$
$$= 2\ln a + \tfrac{1}{2}\ln(a-1), \, a > 1$$

35. $\ln\sqrt[3]{\dfrac{x}{y}} = \dfrac{1}{3}\ln\dfrac{x}{y}$
$$= \frac{1}{3}\left[\ln x - \ln y\right]$$
$$= \frac{1}{3}\ln x - \frac{1}{3}\ln y$$

37. $\ln\left(\dfrac{x^2-1}{x^3}\right) = \ln(x^2-1) - \ln x^3$
$$= \ln[(x-1)(x+1)] - 3\ln x$$
$$= \ln(x-1) + \ln(x+1) - 3\ln x,$$
$$x > 1$$

39. $\ln\left(\dfrac{x^4\sqrt{y}}{z^5}\right) = \ln x^4\sqrt{y} - \ln z^5$
$$= \ln x^4 + \ln\sqrt{y} - \ln z^5$$
$$= 4\ln x + \frac{1}{2}\ln y - 5\ln z$$

41. $\log_b\left(\dfrac{x^2}{y^2z^3}\right) = \log_b x^2 - \log_b y^2z^3$
$$= \log_b x^2 - [\log_b y^2 + \log_b z^3]$$
$$= 2\log_b x - 2\log_b y - 3\log_b z$$

43. $y_1 = \ln[x^3(x+4)]$

$y_2 = 3\ln x + \ln(x+4)$

(a)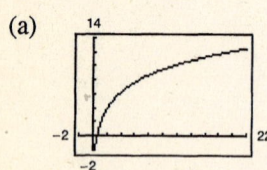

(b)

x	0.5	1	1.5	2	3	10
y_1	-0.5754	1.6094	2.9211	3.8712	5.2417	9.5468
y_2	-0.5754	1.6094	2.9211	3.8712	5.2417	9.5468

(c) The graphs and table suggest that $y_1 = y_2$ for $x > 0$. In fact,

$$y_1 = \ln[x^3(x+4)] = \ln x^3 + \ln(x+4)$$
$$= 3\ln x + \ln(x+4) = y_2$$

45. $\ln x + \ln 4 = \ln 4x$

47. $\log_4 z - \log_4 y = \log_4 \dfrac{z}{y}$

49. $2\log_2(x+3) = \log_2(x+3)^2$

51. $\tfrac{1}{3}\log_3 7x = \log_3(7x)^{1/3} = \log_3\sqrt[3]{7x}$

53. $\ln x - 3\ln(x+1) = \ln x - \ln(x+1)^3$
$$= \ln\frac{x}{(x+1)^3}$$

55. $\ln(x-2) - \ln(x+2) = \ln\left(\dfrac{x-2}{x+2}\right)$

57. $\ln x - 2[\ln(x+2) + \ln(x-2)] = \ln x - 2\ln[(x+2)(x-2)]$
$$= \ln x - 2\ln(x^2-4)$$
$$= \ln x - \ln(x^2-4)^2$$
$$= \ln\frac{x}{(x^2-4)^2}$$

59. $\frac{1}{3}[2\ln(x+3) + \ln x - \ln(x^2 - 1)] = \frac{1}{3}[\ln(x+3)^2 + \ln x - \ln(x^2 - 1)]$

$$= \frac{1}{3}[\ln[x(x+3)^2] - \ln(x^2 - 1)]$$

$$= \frac{1}{3}\ln\frac{x(x+3)^2}{x^2 - 1}$$

$$= \ln\sqrt[3]{\frac{x(x+3)^2}{x^2 - 1}}$$

61. $\frac{1}{3}[\ln y + 2\ln(y+4)] - \ln(y-1) = \frac{1}{3}[\ln y + \ln(y+4)^2] - \ln(y-1)$

$$= \frac{1}{3}\ln[y(y+4)^2] - \ln(y-1)$$

$$= \ln\sqrt[3]{y(y+4)^2} - \ln(y-1)$$

$$= \ln\frac{\sqrt[3]{y(y+4)^2}}{y-1}$$

63. $y_1 = 2[\ln 8 - \ln(x^2 + 1)]$

$y_2 = \ln\left[\frac{64}{(x^2 + 1)^2}\right]$

(a)

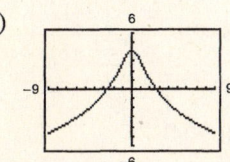

(c) The graphs and table suggest that $y_1 = y_2$. In fact,

$$y_1 = 2[\ln 8 - \ln(x^2 + 1)]$$

$$= 2\ln\frac{8}{x^2 + 1} = \ln\frac{64}{(x^2 + 1)^2} = y_2$$

(b)

x	-8	-4	-2	0	2	4	8
y_1	-4.1899	-1.5075	0.9400	4.1589	0.9400	-1.5075	-4.1899
y_2	-4.1899	-1.5075	0.9400	4.1589	0.9400	-1.5075	-4.1899

65. $y_1 = \ln x^2$

$y_2 = 2\ln x$

(a)

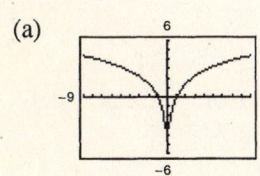

(The domain of y_2 is $x > 0$)

(b)

x	-8	-4	1	2	4
y_1	4.1589	2.7726	0	1.3863	2.7726
y_2	undefined	undefined	0	1.3863	2.7726

(c) The graphs and table suggest that $y_1 = y_2$ for $x > 0$. The functions are not equivalent because the domains are different.

67. $\log_3 9 = 2\log_3 3 = 2$

69. $\log_4 16^{3.4} = 3.4\log_4(4^2) = 6.8\log_4 4 = 6.8$

71. $\log_2(-4)$ is undefined. -4 is not in the domain of $f(x) = \log_2 x$

73. $\log_5 75 - \log_5 3 = \log_5 \frac{75}{3} = \log_5 25 = \log_5 5^2 = 2$

75. $\ln e^3 - \ln e^7 = 3 - 7 = -4$

77. $2 \ln e^4 = 2(4)\ln e = 8$

79. $\ln\left(\dfrac{1}{\sqrt{e}}\right) = \ln(1) - \ln e^{1/2} = 0 - \dfrac{1}{2} \ln e = -\dfrac{1}{2}$

81. (a) $\beta = 10 \cdot \log_{10}\left(\dfrac{I}{10^{-12}}\right) = 10[\log_{10} I - \log_{10} 10^{-12}]$

$= 10[\log_{10} I - (-12) \log_{10} 10]$

$= 10\,[\log_{10} I + 12] = 120 + 10 \cdot \log_{10} I$

(b)

I	10^{-4}	10^{-6}	10^{-8}	10^{-10}	10^{-12}	10^{-14}
β	80	60	40	20	0	-20

(c) $\beta(10^{-4}) = 120 + 10 \cdot \log_{10} 10^{-4} = 120 - 40 = 80$

$\beta(10^{-6}) = 120 + 10 \cdot \log_{10} 10^{-6} = 120 - 60 = 60$

$\beta(10^{-8}) = 120 + 10 \cdot \log_{10} 10^{-8} = 120 - 80 = 40$

$\beta(10^{-10}) = 120 + 10 \cdot \log_{10} 10^{-10} = 120 - 100 = 20$

$\beta(10^{-12}) = 120 + 10 \cdot \log_{10} 10^{-12} = 120 - 120 = 0$

$\beta(10^{-14}) = 120 + 10 \cdot \log_{10} 10^{-14} = 120 - 140 = -20$

83. (a)

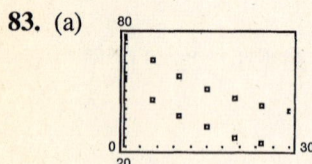

(b) $T - 21 = 54.4(0.964)^t$

$T = 21 + 54.4(0.964)^t$

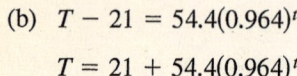

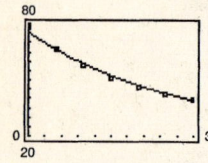

The data $(t, T - 21)$ fits the model
$T - 21 = 54.4(0.964)^t$ The model
$T = 21 + 54.4(0.964)^t$ fits the original data.

(c) $\ln(T - 21) = -0.0372t + 3.9971$ linear model

$T - 21 = e^{-0.0372t + 3.9971}$

$T = 21 + 54.4e^{-0.0372t}$

$= 21 + 54.4(.964)^t$

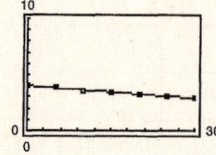

(d)

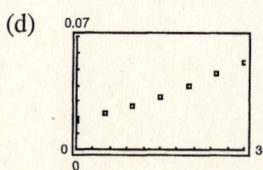

$\dfrac{1}{T - 21} = 0.00121t + 0.01615$ linear model

$T - 21 = \dfrac{1}{0.00121t + 0.01615}$

$T = 21 + \dfrac{1}{0.00121t + 0.01615}$

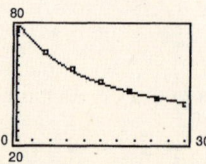

85. $f(x) = \ln x$

False, $f(0) \neq 0$ since 0 is not in the domain of $f(x)$. $f(1) = \ln 1 = 0$

87. True.

89. False. $\sqrt{\ln x} \neq \frac{1}{2} \ln x$

In fact, $\ln x^{1/2} = \frac{1}{2} \ln x$

91. True.

93. Let $x = \log_b u$.

Then $b^x = u$ and $u^n = b^{nx}$.

Hence,

$\log_b u^n = \log_b b^{nx} = nx = n \log_b u$.

95. $f(x) = \log_2 x = \dfrac{\ln x}{\ln 2}$

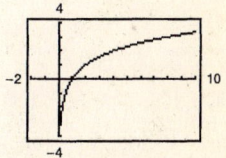

97. $f(x) = \log_3 \sqrt{x} = \dfrac{1}{2} \dfrac{\ln x}{\ln 3}$

99. $f(x) = \log_5\left(\dfrac{x}{3}\right) = \dfrac{\ln\left(\dfrac{x}{3}\right)}{\ln 5}$

101. $f(x) = \ln \dfrac{x}{2}$

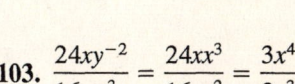

$g(x) = \dfrac{\ln x}{\ln 2}$

$h(x) = \ln x - \ln 2$

$f(x) = h(x)$ by Property 2.

103. $\dfrac{24xy^{-2}}{16x^{-3}y} = \dfrac{24xx^3}{16yy^2} = \dfrac{3x^4}{2y^3}$

105. $(18x^3y^4)^{-3}(18x^3y^4)^3 = \dfrac{(18x^3y^4)^3}{(18x^3y^4)^3} = 1$ if $x \neq 0, y \neq 0$.

107. $x^2 - 6x + 2 = 0$

$x = \dfrac{6 \pm \sqrt{36 - 4(2)}}{2} = 3 \pm \sqrt{7}$

109. $x^4 - 19x^2 + 48 = 0$

$(x^2 - 16)(x^2 - 3) = 0$

$(x - 4)(x + 4)(x - \sqrt{3})(x + \sqrt{3}) = 0$

$x = \pm 4, \pm\sqrt{3}$

111. $x^3 - 6x^2 - 4x + 24 = 0$

$x^2(x - 6) - 4(x - 6) = 0$

$(x^2 - 4)(x - 6) = 0$

$(x - 2)(x + 2)(x - 6) = 0$

$x = 2, -2, 6$

Section 3.4   Solving Exponential and Logarithmic Equations

■ To solve an exponential equation, isolate the exponential expression, then take the logarithm of both sides. Then solve for the variable.

   1. $\log_a a^x = x$

   2. $\ln e^x = x$

■ To solve a logarithmic equation, rewrite it in exponential form. Then solve for the variable.

   1. $a^{\log_a x} = x$

   2. $e^{\ln x} = x$

■ If $a > 0$ and $a \neq 1$ we have the following:

   1. $\log_a x = \log_a y \implies x = y$

   2. $a^x = a^y \implies x = y$

■ Use your graphing utility to approximate solutions.

Solutions to Odd-Numbered Exercises

1. $4^{2x-7} = 64$

   (a) $x = 5$

      $4^{2(5)-7} = 4^3 = 64$

      Yes, $x = 5$ is a solution.

   (b) $x = 2$

      $4^{2(2)-7} = 4^{-3} = \frac{1}{64} \neq 64$

      No, $x = 2$ is not a solution.

3. $3e^{x+2} = 75$

   (a) $x = -2 + e^{25}$

      $3e^{(-2+e^{25})+2} = 3e^{e^{25}} \neq 75$

      No, $x = -2 + e^{25}$ is not a solution.

   (b) $x = -2 + \ln 25$

      $3e^{(-2+\ln 25)+2} = 3e^{\ln 25} = 3(25) = 75$

      Yes, $x = -2 + \ln 25$ is a solution.

   (c) $x \approx 1.2189$

      $3e^{1.2189+2} = 3e^{3.2189} \approx 75$

      Yes, $x \approx 1.2189$ is a solution.

5. $\log_4(3x) = 3$

   $4^3 = 3x$

   $x = \frac{64}{3} \approx 21.333$

   (a) $x = 21.3560$ is an approximate solution

   (b) No, $x = -4$ is not a solution

   (c) Yes, $x = \frac{64}{3}$ is a solution

7. $\ln(x - 1) = 3.8$

   (a) $x = 1 + e^{3.8}$

      $\ln(1 + e^{3.8} - 1) = \ln e^{3.8} = 3.8$

      Yes, $x = 1 + e^{3.8}$ is a solution.

   (b) $x \approx 45.7012$

      $\ln(45.7012 - 1) = \ln(44.7012) \approx 3.8$

      Yes, $x \approx 45.7012$ is a solution.

   (c) $x = 1 + \ln 3.8$

      $\ln(1 + \ln 3.8 - 1) = \ln(\ln 3.8) \approx 0.289$

      No, $x = 1 + \ln 3.8$ is not a solution.

9.

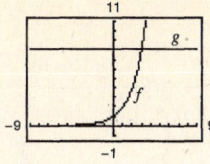

Intersection Point: $(3, 8)$

Algebraically, $2^x = 8$

$$2^x = 2^3$$

$$x = 3 \implies y = 8 \implies (3, 8)$$

11.

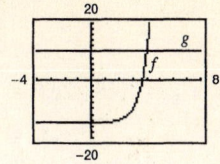

Intersection: $(4, 10)$

Algebraically, $5^{x-2} - 15 = 10$

$$5^{x-2} = 25 = 5^2$$

$$x - 2 = 2$$

$$x = 4$$

$(4, 10)$

13.

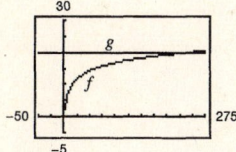

Intersection: $(243, 20)$

Algebraically, $4 \log_3 x = 20$

$$\log_3 x = 5$$

$$x = 3^5 = 243$$

$(243, 20)$

15.

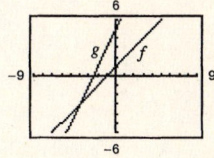

Intersection: $(-4, -3)$

Algebraically, $\ln e^{x+1} = 2x + 5$

$$x + 1 = 2x + 5$$

$$-4 = x$$

$(-4, -3)$

17. $4^x = 16$

$$4^x = 4^2$$

$$x = 2$$

19. $5^x = \dfrac{1}{625}$

$$5^x = \dfrac{1}{5^4} = 5^{-4}$$

$$x = -4$$

21. $\left(\dfrac{1}{8}\right)^x = 64$

$$8^{-x} = 8^2$$

$$-x = 2$$

$$x = -2$$

23. $\left(\dfrac{2}{3}\right)^x = \dfrac{81}{16}$

$$\left(\dfrac{3}{2}\right)^{-x} = \left(\dfrac{3}{2}\right)^4$$

$$-x = 4$$

$$x = -4$$

25. $e^x = 4$

$$x = \ln 4 \approx 1.386$$

27. $\ln x - \ln 5 = 0$

$$\ln x = \ln 5$$

$$x = 5$$

29. $\ln x = -7$

$$x = e^{-7}$$

31. $\log_x 625 = 4$

$$x^4 = 625$$

$$x^4 = 5^4$$

$$x = 5$$

33. $\log_{10} x = -1$

$$x = 10^{-1}$$

$$x = \dfrac{1}{10}$$

35. $\ln(2x - 1) = 5$

$$2x - 1 = e^5$$

$$x = \dfrac{1 + e^5}{2} \approx 74.707$$

37. $\ln e^{x^2} = x^2 \ln e = x^2$

39. $e^{\ln(5x+2)} = 5x + 2$

41. $e^{\ln x^2} = x^2$

43.
$$8^{3x} = 360$$
$$\ln 8^{3x} = \ln 360$$
$$3x \ln 8 = \ln 360$$
$$3x = \frac{\ln 360}{\ln 8}$$
$$x = \frac{1}{3}\frac{\ln 360}{\ln 8}$$
$$x \approx 0.944$$

45. $2e^{5x} = 18$
$$e^{5x} = 9$$
$$5x = \ln 9$$
$$x = \frac{1}{5}\ln 9$$
$$x \approx 0.439$$

47. $500e^{-x} = 300$
$$e^{-x} = \frac{3}{5}$$
$$-x = \ln\frac{3}{5}$$
$$x = -\ln\frac{3}{5} = \ln\frac{5}{3} \approx 0.511$$

49. $7 - 2e^x = 5$
$$-2e^x = -2$$
$$e^x = 1$$
$$x = \ln 1 = 0$$

51. $5^{-t/2} = 0.20 = \frac{1}{5}$
$$-\frac{t}{2}\ln 5 = \ln\left(\frac{1}{5}\right)$$
$$-\frac{t}{2}\ln 5 = -\ln 5$$
$$\frac{t}{2} = 1$$
$$t = 2$$

53. $2^{3-x} = 565$
$$(3 - x)\ln 2 = \ln 565$$
$$-x \ln 2 = \ln 565 - 3\ln 2$$
$$x = (3\ln 2 - \ln 565)/\ln 2 \approx -6.142$$

55. $e^{2x} - 4e^x - 5 = 0$
$$(e^x - 5)(e^x + 1) = 0$$
$$e^x = 5 \text{ or } e^x = -1$$
$$x = \ln 5 \approx 1.609$$
$(e^x = -1 \text{ is impossible.})$

57. $\dfrac{400}{1 + e^{-x}} = 350$
$$1 + e^{-x} = \frac{400}{350} = \frac{8}{7}$$
$$e^{-x} = \frac{1}{7}$$
$$-x = \ln\left(\frac{1}{7}\right) = -\ln 7$$
$$x = \ln 7 \approx 1.946$$

59. $\left(1 + \dfrac{0.10}{12}\right)^{12t} = 2$
$$\left(\frac{12.1}{12}\right)^{12t} = 2$$
$$(12t)\ln\left(\frac{12.1}{12}\right) = \ln 2$$
$$t = \frac{1}{12}\frac{\ln 2}{\ln\left(\frac{12.1}{12}\right)}$$
$$t \approx 6.960$$

61. $e^{3x} = 12$

x	0.6	0.7	0.8	0.9	1.0
$f(x)$	6.05	8.17	11.02	14.88	20.09

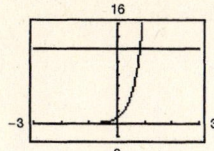

$x \approx 0.828$

63. $20(100 - e^{x/12}) = 500$

x	5	6	7	8	9
$f(x)$	1756	1598	1338	908	200

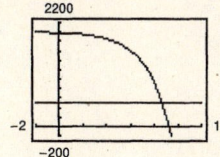

$x \approx 8.635$

65. $\left(1 + \dfrac{0.065}{365}\right)^{365t} = 4 \implies t = 21.330$

67. $\dfrac{3000}{2 + e^{2x}} = 2$

$1500 = 2 + e^{2x}$

$1498 = e^{2x}$

$\ln 1498 = \ln e^{2x}$

$\ln 1498 = 2x$

$\dfrac{\ln 1498}{2} = x \approx 3.656$

69. $g(x) = 6e^{1-x} - 25$

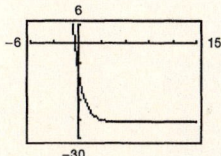

zero at $x = -0.427$

71. $g(t) = e^{0.09t} - 3$

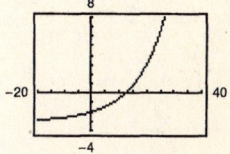

zero at $t = 12.207$

73. $\ln x = -3$

$x = e^{-3} \approx 0.050$

75. $\ln 4x = 2.1$

$4x = e^{2.1}$

$x = \frac{1}{4}e^{2.1}$

≈ 2.042

77. $-2 + 2\ln 3x = 17$

$2\ln 3x = 19$

$\ln 3x = \frac{19}{2}$

$3x = e^{19/2}$

$x = \frac{1}{3}e^{19/2}$

$x \approx 4453.242$

79. $\log_{10}(z - 3) = 2$

$z - 3 = 10^2$

$z = 10^2 + 3 = 103$

81. $7\log_4(0.6x) = 12$

$\log_4(0.6x) = \dfrac{12}{7}$

$4^{12/7} = 0.6x = \dfrac{3}{5}x$

$x = \dfrac{5}{3}4^{12/7} \approx 17.945$

83. $\ln\sqrt{x + 2} = 1$

$\sqrt{x + 2} = e^1$

$x + 2 = e^2$

$x = e^2 - 2 \approx 5.389$

85. $\ln(x + 1)^2 = 2$

$e^{\ln(x+1)^2} = e^2$

$(x + 1)^2 = e^2$

$x + 1 = e$ or $x + 1 = -e$

$x = e - 1 \approx 1.718$

or

$x = -e - 1 \approx -3.718$

87. $\log_4 x - \log_4(x - 1) = \dfrac{1}{2}$

$\log_4\left(\dfrac{x}{x - 1}\right) = \dfrac{1}{2}$

$4^{\log_4(x/x-1)} = 4^{1/2}$

$\dfrac{x}{x - 1} = 2$

$x = 2(x - 1)$

$x = 2x - 2$

$2 = x$

89. $\ln(x + 5) = \ln(x - 1) - \ln(x + 1)$

$\ln(x + 5) = \ln\left(\dfrac{x - 1}{x + 1}\right)$

$x + 5 = \dfrac{x - 1}{x + 1}$

$(x + 5)(x + 1) = x - 1$

$x^2 + 6x + 5 = x - 1$

$x^2 + 5x + 6 = 0$

$(x + 2)(x + 3) = 0$

$x = -2$ or $x = -3$

Both of these solutions are extraneous, so the equation has no solution.

91. $\log_{10} 8x - \log_{10}\left(1 + \sqrt{x}\right) = 2$

$\log_{10} \dfrac{8x}{1 + \sqrt{x}} = 2$

$\dfrac{8x}{1 + \sqrt{x}} = 10^2$

$8x = 100 + 100\sqrt{x}$

$8x - 100\sqrt{x} - 100 = 0$

$2x - 25\sqrt{x} - 25 = 0$

$\sqrt{x} = \dfrac{25 \pm \sqrt{25^2 - 4(2)(-25)}}{4}$

$= \dfrac{25 \pm 5\sqrt{33}}{4}$

Choosing the positive value, we have $\sqrt{x} \approx 13.431$ and $x \approx 180.384$.

93. $\ln 2x = 2.4$

x	2	3	4	5	6
$f(x)$	1.39	1.79	2.08	2.30	2.48

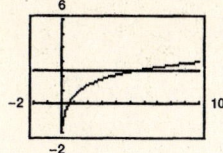

$x \approx 5.512$

95. $6 \log_3(0.5x) = 11$

x	12	13	14	15	16
$f(x)$	9.79	10.22	10.63	11.00	11.36

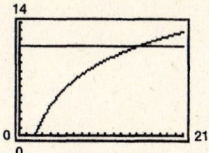

$x \approx 14.988$

97. $\log_{10} x = x^3 - 3$

Graphing $y = \log_{10} x - x^3 + 3$, you obtain 2 zeros

$x \approx 1.469$ and $x \approx 0.001$

99. $\log_3 x + \log_3(x - 3) = 1$

Graphing $y = \dfrac{\log x}{\log 3} + \dfrac{\log(x - 3)}{\log 3} - 1$,

you obtain $x \approx 3.791$

101. $\ln(x - 3) + \ln(x + 3) = 1$

Graphing $y = \ln(x - 3) + \ln(x + 3) - 1$, you obtain $x \approx 3.423$

103. $y_1 = 7$

$y_2 = 2^{x-1} - 5$

Intersection: $(4.585, 7)$

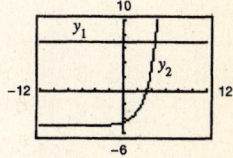

105. $y_1 = 80$

$y_2 = 4e^{-0.2x}$

Intersection:
$(-14.979, 80)$

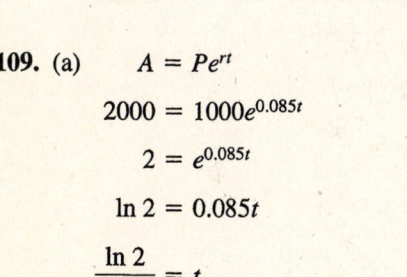

107. $y_1 = 3.25$

$y_2 = \frac{1}{2} \ln(x + 2)$

Intersection:
$(663.142, 3.25)$

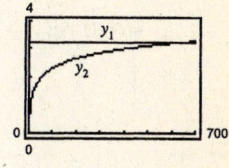

109. (a) $A = Pe^{rt}$

$2000 = 1000e^{0.085t}$

$2 = e^{0.085t}$

$\ln 2 = 0.085t$

$\dfrac{\ln 2}{0.085} = t$

$t \approx 8.2$ years

(b) $3000 = 1000e^{0.085t}$

$3 = e^{0.085t}$

$\ln 3 = 0.085t$

$\dfrac{\ln 3}{0.085} = t$

$t \approx 12.9$ years

111. $p = 500 - 0.5(e^{0.004x})$

(a) $p = 350$

$350 = 500 - 0.5(e^{0.004x})$

$300 = e^{0.004x}$

$0.004x = \ln 300$

$x \approx 1426$ units

(b) $p = 300$

$300 = 500 - 0.5(e^{0.004x})$

$400 = e^{0.004x}$

$0.004x = \ln 400$

$x \approx 1498$ units

113. $N = 68(10^{-0.04x})$

When $N = 21$:

$$21 = 68(10^{-0.04x})$$

$$\frac{21}{68} = 10^{-0.04x}$$

$$\log_{10} \frac{21}{68} = -0.04x$$

$$x = -\frac{\log(21/68)}{0.04} \approx 12.76 \text{ inches}$$

115. (a)

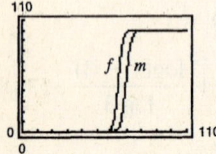

(b) From the graph we see horizontal asymptotes at $y = 0$ and $y = 100$. These represent the lower and upper percent bounds.

(c) Males:

$$50 = \frac{100}{1 + e^{-0.6114(x-69.71)}}$$

$$1 + e^{-0.6114(x-69.71)} = 2$$

$$e^{-0.6114(x-69.71)} = 1$$

$$-0.6114(x - 69.71) = \ln 1$$

$$-0.6114(x - 69.71) = 0$$

$$x = 69.71 \text{ inches}$$

Females:

$$50 = \frac{100}{1 + e^{-0.66607(x-64.51)}}$$

$$1 + e^{-0.66607(x-64.51)} = 2$$

$$e^{-0.66607(x-64.51)} = 1$$

$$-0.66607(x - 64.51) = \ln 1$$

$$-0.66607(x - 64.51) = 0$$

$$x = 64.51 \text{ inches}$$

117. $T = 20[1 + 7(2^{-h})]$

(a)

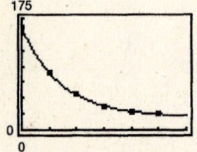

(b) We see a horizontal asymptote at $y = 20$. This represents the room temperature.

(c)

$$100 = 20[1 + 7(2^{-h})]$$

$$5 = 1 + 7(2^{-h})$$

$$4 = 7(2^{-h})$$

$$\frac{4}{7} = 2^{-h}$$

$$\ln\left(\frac{4}{7}\right) = \ln 2^{-h}$$

$$\ln\left(\frac{4}{7}\right) = -h \ln 2$$

$$\frac{\ln(4/7)}{-\ln 2} = h$$

$$h \approx 0.81 \text{ hour}$$

119. True

121. Answers will vary

123. Yes. The doubling time is given by

$$2P = Pe^{rt}$$

$$2 = e^{rt}$$

$$\ln 2 = rt$$

$$t = \frac{\ln 2}{r}$$

The time to quadruple is given by

$$4P = Pe^{rt}$$

$$4 = e^{rt}$$

$$\ln 4 = rt$$

$$t = \frac{\ln 4}{r} = \frac{\ln 2^2}{r} = \frac{2\ln 2}{r} = 2\left[\frac{\ln 2}{r}\right]$$

which is twice as long.

125. $f(x) = 3x^3 - 4$

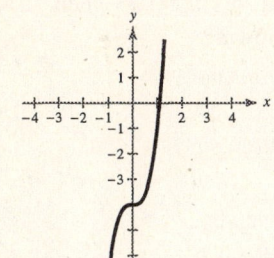

127. $f(x) = |x| + 9$

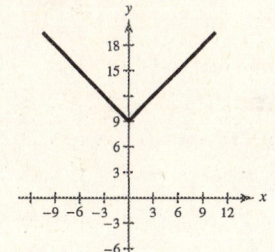

129. $f(x) = \begin{cases} 2x & x < 0 \\ -x^2 + 4 & x \geq 0 \end{cases}$

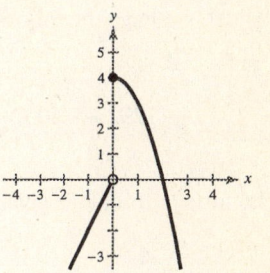

Section 3.5   Exponential and Logarithmic Models

■ You should be able to solve compound interest problems.

1. $A = P\left(1 + \dfrac{r}{n}\right)^{nt}$

2. $A = Pe^{rt}$

■ You should be able to solve growth and decay problems.

(a) Exponential growth if $b > 0$ and $y = ae^{bx}$.

(b) Exponential decay if $b > 0$ and $y = ae^{-bx}$.

■ You should be able to use the Gaussian model

$y = ae^{-(x-b)^2/c}$.

■ You should be able to use the logistics growth model

$y = \dfrac{a}{1 + be^{-(x-c)/d}}$.

■ You should be able to use the logarithmic models

$y = \ln(ax + b)$ and $y = \log_{10}(ax + b)$.

Solutions to Odd-Numbered Exercises

1. $y = 2e^{x/4}$

This is an exponential growth model.

Matches graph (c).

3. $y = 6 + \log_{10}(x + 2)$

This is a logarithmic model, and contains $(-1, 6)$.

Matches graph (b).

5. $y = \ln(x + 1)$

This is a logarithmic model.

Matches graph (d).

7. Since $A = 1000e^{0.035t}$, the time to double is given by

$$2000 = 1000e^{0.035t}$$

$$2 = e^{0.035t}$$

$$\ln 2 = 0.035t$$

$$t = \frac{1}{0.035} \ln 2 \approx 19.8 \text{ years}$$

Amount after 10 years:

$$A = 1000e^{0.035(10)} \approx \$1419.07$$

9. Since $A = 750e^{rt}$ and $A = 1500$ when $t = 7.75$, we have the following.

$$1500 = 750e^{7.75r}$$

$$r = \frac{\ln 2}{7.75} \approx 0.0894 = 8.94\%$$

Amount after 10 years:

$$A = 750e^{0.0894(10)} \approx \$1833.67$$

11. Since $A = 500e^{rt}$ and $A = 1292.85$ when $t = 10$, we have the following:

$$1292.85 = 500e^{10r}$$

$$r = \frac{\ln(1292.85/500)}{10} \approx 0.0950 = 9.5\%$$

The time to double is given by

$$1000 = 500e^{0.095t}$$

$$2 = e^{0.095t}$$

$$\ln 2 = 0.095t$$

$$t = \frac{\ln 2}{0.095} \approx 7.30 \text{ years}$$

13. Since $A = Pe^{0.045t}$ and $A = 10,000.00$ when $t = 10$, we have the following:

$$10,000.00 = Pe^{0.045(10)}$$

$$\frac{10,000.00}{e^{0.045(10)}} = P \approx 6376.28$$

The time to double is given by

$$12,752.56 = 6376.28e^{0.045t}$$

$$t = \frac{\ln 2}{0.045} \approx 15.40 \text{ years}.$$

15. $3P = Pe^{rt}$

$$3 = e^{rt}$$

$$\ln 3 = rt$$

$$\frac{\ln 3}{r} = t$$

r	2%	4%	6%	8%	10%	12%
$t = \dfrac{\ln 3}{r}$	54.93	27.47	18.31	13.73	10.99	9.16

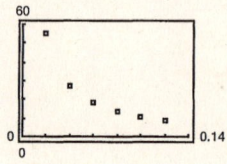

17.

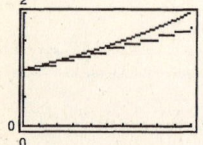

Continuous compounding results in faster growth.

$$A = 1 + 0.075[\![t]\!]$$

and $A = e^{0.07t}$

19. $\frac{1}{2}C = Ce^{k(1600)}$

$$k = \frac{\ln 0.5}{1600}$$

Given $C = 10$ and $t = 1000$,

$$y = Ce^{kt}$$

$$= 10e^{\left[\frac{\ln 0.5}{1600} 1000\right]}$$

$$\approx 6.484 \text{ grams}$$

21. $\frac{1}{2}C = Ce^{k(5730)}$

$$k = \frac{\ln 0.5}{5730}$$

Given $C = 3$ grams, after 1000 years, we have

$$y = 3e^{[(\ln 0.5)/5730](1000)}$$

$$y \approx 2.66 \text{ grams}.$$

23. $y = ae^{bx}$

$1 = ae^{b(0)} \implies 1 = a$

$10 = e^{b(3)}$

$\ln 10 = 3b$

$\dfrac{\ln 10}{3} = b \quad \implies \quad b \approx 0.7675$

Thus, $y = e^{0.7675x}$.

25. $(0, 4) \implies a = 4$

$(5, 1) \implies 1 = 4e^{b(5)} \implies b = \frac{1}{5}\ln\left(\frac{1}{4}\right)$

$\qquad\qquad\qquad\qquad = -\frac{1}{5}\ln 4 \approx -0.2773$

$y = 4e^{-0.2773x}$

27. (a)

Australia: $(0, 19.2)$ $(10, 20.9)$

$\qquad a = 19.2$ and $20.9 = 19.2e^{b(10)} \implies b = 0.008484$

$\qquad y = 19.2e^{0.008484t} \qquad$ For 2030, $y \approx 24.8$ million

Canada: $(0, 31.3)$ $(10, 34.3)$

$\qquad a = 31.3$ and $34.3 = 31.3e^{b(10)} \implies b = 0.009153$

$\qquad y = 31.3e^{0.009153t} \qquad$ For 2030, $y \approx 41.2$ million

Philippines: $(0, 81.2)$ $(10, 97.9)$

$\qquad a = 81.2$ and $97.9 = 81.2e^{b(10)} \implies b = 0.0187$

$\qquad y = 81.2e^{0.0187t} \qquad$ For 2030, $y \approx 142.3$ million

South Africa: $(0, 43.4)$ $(10, 41.1)$

$\qquad a = 43.4$ and $41.1 = 43.4e^{b(10)} \implies b = -0.005445$

$\qquad y = 43.4e^{-0.005445t} \qquad$ For 2030, $y \approx 36.9$ million

Turkey: $(0, 65.7)$ $(10, 73.3)$

$\qquad a = 65.7$ and $73.3 = 65.7e^{b(10)} \implies b = 0.01095$

$\qquad y = 65.7e^{0.01095t} \qquad$ For 2030, $y \approx 91.2$ million

(b) The constant b gives the growth rates.

(c) The constant b is negative for South Africa.

29. $N = 100e^{kt}$

$300 = 100e^{5k}$

$k = \dfrac{\ln 3}{5} \approx 0.2197$

$N = 100e^{0.2197t}$

$200 = 100e^{0.2197t}$

$t = \dfrac{\ln 2}{0.2197} \approx 3.15$ hours

31. $y = Ce^{kt}$

$\dfrac{1}{2}C = Ce^{(1620)k}$

$\ln\dfrac{1}{2} = 1620k$

$k = \dfrac{\ln(1/2)}{1620}$

When $t = 100$, we have

$y = Ce^{[\ln(1/2)/1620](100)} \approx 0.958C = 95.8\%C.$

After 100 years, approximately 95.8% of the radioactive radium will remain.

33. (a) $V = mt + b$, $V(0) = 32{,}000 \Rightarrow b = 32{,}000$.

$V(2) = 18{,}000 \Rightarrow 18{,}000 = 2m + 32{,}000 \Rightarrow m = -7000$

$V(t) = -7000t + 32{,}000$

(b) $V = ae^{kt}$, $V(0) = 32{,}000 \Rightarrow a = 32{,}000$

$V(2) = 18{,}000 \Rightarrow 18{,}000 = 32{,}000e^{2k} \Rightarrow \frac{18}{32} = e^{2k} \Rightarrow k = \frac{1}{2}\ln\left(\frac{18}{32}\right) \approx -0.2877$

$V(t) = 32{,}000e^{-0.2877t}$

(c)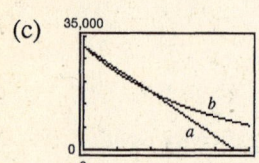

The exponential model depreciates faster in the first year.

(e) The negative slope means the car depreciates $7000 per year.

(d) Straight line: $V(1) = \$25{,}000$

$V(3) = \$11{,}000$

Exponential: $V(1) \approx \$23{,}999.57$

$V(3) \approx \$13{,}499.27$

35. $S(t) = 100(1 - e^{kt})$

(a) $15 = 100(1 - e^{k(1)})$

$-85 = -100e^k$

$k = \ln 0.85$

$k \approx -0.1625$

$S(t) = 100(1 - e^{-0.1625t})$

(b)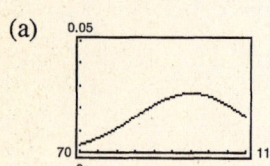

(c) $S(5) = 100(1 - e^{-0.1625(5)})$

$\approx 55.625 = 55{,}625$ units

37. $y = 0.0266e^{-(x-100)^2/450}$, $70 \le x \le 115$

(a)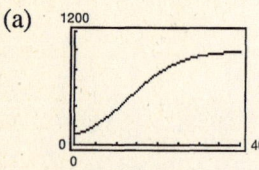

(b) Maximum point is $x = 100$, the average IQ score.

39. $p(t) = \dfrac{1000}{1 + 9e^{-0.1656t}}$

(a)

The horizontal asymptotes are $y = 0$ and $y = 1000$. The asymptote with the larger p-value, $y = 1000$, indicates that the population size will approach 1000 as time increases.

(b) $p(5) = \dfrac{1000}{1 + 9e^{-0.1656(5)}} \approx 203$ animals

(c) $500 = \dfrac{1000}{1 + 9e^{-0.1656t}}$

$1 + 9e^{-0.1656t} = 2$

$9e^{-0.1656t} = 1$

$e^{-0.1656t} = \dfrac{1}{9}$

$t = -\dfrac{\ln(1/9)}{0.1656} \approx 13$ months

41. $R = \log_{10}\left(\dfrac{I}{I_0}\right) = \log_{10}(I)$

$I = 10^R$

(a) $I = 10^{6.5} \approx 3{,}162{,}278$

(b) $I = 10^{7.9} \approx 79{,}432{,}823$

(c) $I = 10^{5.2} \approx 158{,}489$

43. $\beta(I) = 10 \log_{10}(I/I_0)$, where $I_0 = 10^{-12}$ watt per square meter.

(a) $\beta(10^{-10}) = 10 \cdot \log_{10}\left(\dfrac{10^{-10}}{10^{-12}}\right) = 10 \log_{10}10^2 = 20$ decibels

(b) $\beta(10^{-5}) = 10 \cdot \log_{10}\left(\dfrac{10^{-5}}{10^{-12}}\right) = 10 \log_{10}10^7 = 70$ decibels

(c) $\beta(10^0) = 10 \cdot \log_{10}\left(\dfrac{10^0}{10^{-12}}\right) = 10 \log_{10}10^{12} = 120$ decibels

45. $\beta = 10 \log_{10}\left(\dfrac{I}{I_0}\right)$

$10^{\beta/10} = \dfrac{I}{I_0}$

$I = I_0\, 10^{\beta/10}$

$\%\text{ decrease} = \dfrac{I_0\, 10^{8.8} - I_0\, 10^{7.2}}{I_0\, 10^{8.8}} \times 100$

$\qquad\qquad = 97.5\%$

47. $pH = -\log_{10}[\text{H}^+] = -\log_{10}[2.3 \times 10^{-5}] \approx 4.64$

49. $pH = -\log_{10}[H^+]$

$-pH = \log_{10}[H^+]$

$10^{-ph} = [H^+]$

$\dfrac{\text{Hydrogen ion concentration of grape}}{\text{Hydrogen ion concentration of milk of magnesia}} = \dfrac{10^{-3.5}}{10^{-10.5}} = 10^7$

51. (a) $P = 120{,}000,\, t = 30,\, r = 0.075,\, M = 839.06$

$u = M - \left(M - \dfrac{Pr}{12}\right)\left(1 + \dfrac{r}{12}\right)^{12t}$

$\quad = 839.06 - (839.06 - 750)(1 + 0.00625)^{12t}$

$v = (839.06 - 750)(1.00625)^{12t}$

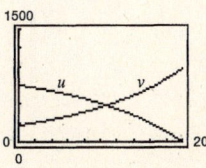

(b) In the early years, the majority of the monthly payment goes toward interest. The interest and principle are equal when $t \approx 20.729 \approx 21$ years.

(c) $P = 120{,}000,\, t = 20,\, r = 0.075,\, M = 966.71$

$u = 966.71 - (966.71 - 750)(1.00625)^{12t}$

$v = (966.71 - 750)(1.00625)^{12t}$

$u = v$ when $t \approx 10.73$ years

53. $t = -10 \ln\left(\dfrac{T - 70}{98.6 - 70}\right)$

At 9:00 A.M. we have

$t = -10 \ln(85.7 - 70/98.6 - 70) \approx 6$ hours

Thus, we can conclude that the person died 6 hours before 9 A.M., or 3:00 A.M.

55. False. The domain could be all real numbers.

57. True. For the Gaussian model, $y > 0$.

59. $4x - 3y - 9 = 0 \implies y = \frac{1}{3}(4x - 9)$

Slope: $\frac{4}{3}$

Matches (a).

Intercepts: $(0, -3), \left(\frac{9}{4}, 0\right)$

61. $y = 25 - 2.25x$

Slope: -2.25

Matches (d).

Intercepts: $(0, 25), \left(\frac{100}{9}, 0\right)$

63. $f(x) = 2x^3 - 3x^2 + x - 1$

The graph falls to the left and rises to the right.

65. $g(x) = -1.6x^5 + 4x^2 - 2$

The graph rises to the left and falls to the right.

67.

$$
\begin{array}{r|rrrr}
4 & 2 & -8 & 3 & -9 \\
  &   & 8 & 0 & 12 \\
\hline
  & 2 & 0 & 3 & 3
\end{array}
$$

$\dfrac{2x^3 - 8x^2 + 3x - 9}{x - 4} = 2x^2 + 3 + \dfrac{3}{x - 4}$

Section 3.6   Exploring Data: Nonlinear Models

Solutions to Odd-Numbered Exercises

1. Logarithmic model

3. Quadratic model

5. Exponential model

7. Quadratic model

9.

Logarithmic model

11.

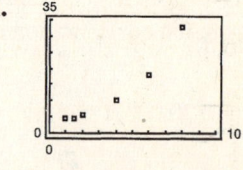

Exponential model

13.

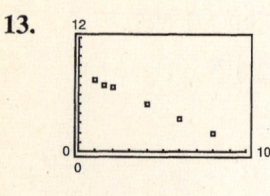

Linear model

15. $y = 3.807(1.3057)^x$

17. $y = 8.463(0.7775)^x$

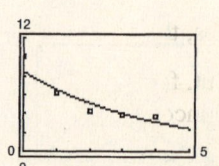

19. $y = 2.083 + 1.257 \ln x$

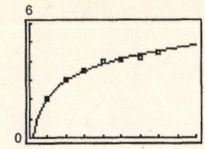

21. $y = 9.826 - 4.097 \ln x$

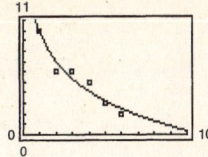

23. $y = 1.985x^{0.760}$

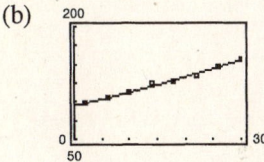

25. $y = 16.103x^{-3.174}$

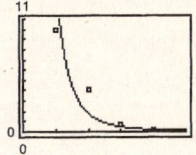

27. (a) Quadratic model: $R = 0.0125x^2 + 1.635x + 94.86$

Exponential model: $R = 95.324(1.0165)^x$

Power model: $R = 81.230x^{0.1682}$

(b)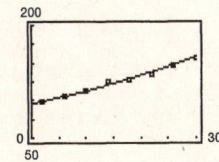

(c) The quadratic model fits best

(d) For 2004, $t = 34$ and $R \approx 164.9$ or 164.9 million

29. (a) $y = 3.127x + 250.87$

(b) $y = 251.453(1.0116)^x$

(c) No, the linear model is better.
Its r-value is closer to one.

(d) Linear model: 307 million

Exponential model: 309 million

[Answer will vary depending on number of digits.]

31. (a) $T = -1.239t + 73.02$

No, the data does not appear linear

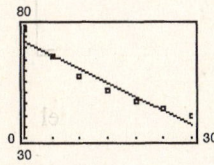

(b) $T = 0.034t^2 - 2.26t + 77.3$

Yes, the data appears quadratic.

But, for $t = 60$, the graph is increasing, which is incorrect.

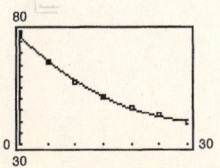

(c) Subtracting 21 from the T values, the exponential model is

$y = 54.438(0.9635)^x$

Adding back 21,

$T = 54.438(0.9635)^t + 21$

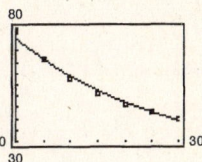

(d) Answers will vary.

33. (a) $S = 925.734(1.155)^x$

(b)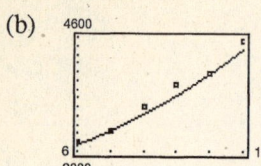

(c) The model is a good fit

(d) For 2007, $x = 17$ and $S = 10,725$ million

[Answers will vary.]

35. (a) Linear model: $y = 15.79x + 47.9$

Logarithmic model: $y = -97.5 + 131.92 \ln x$

Quadratic model: $y = -1.968x^2 + 49.24x - 88.6$

Exponential model: $y = 83.94(1.09)^x$

Power model: $y = 36.51x^{0.7525}$

(b) Linear model: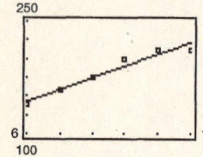

Logarithmic model: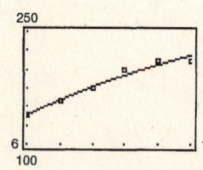

Quadratic model: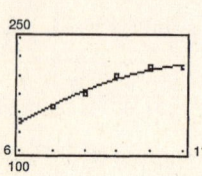

Exponential model: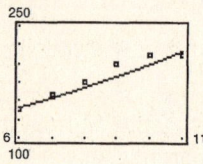

Power model: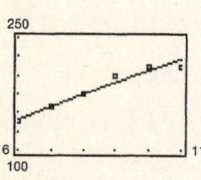

The quadratic model appears to fit best.

(c) Linear model: 217.2

Logarithmic model: 120.91

Quadratic model: 72.7

Exponential model: 532.2

Power model: 189.0

The quadratic model fits best because 72.7 is smallest.

(d) Linear model: 0.9526

Logarithmic model: 0.9736

Quadratic model: 0.9841

Exponential model: 0.9402

Power model: 0.9697

The quadratic model has the largest r^2 value.

(e) The quadratic model is best.

37. True

39. $2x + 5y = 10$

$\qquad 5y = -2x + 10$

$\qquad y = -\frac{2}{5}x + 2$

slope: $-\frac{2}{5}$

y-intercept: $(0, 2)$

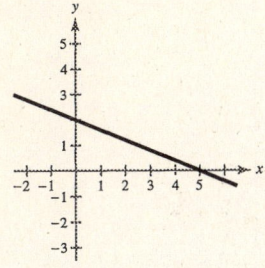

41. $1.2x + 3.5y = 10.5$

$\qquad 35y = -12x + 105$

$\qquad y = -\frac{12}{35}x + \frac{105}{35}$

$\qquad = -\frac{12}{35}x + 3$

slope: $-\frac{12}{35}$

y-intercept: $(0, 3)$

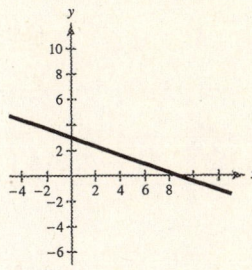

Review Exercises for Chapter 3

Solutions to Odd-Numbered Exercises

1. $(1.45)^{2\pi} \approx 10.3254$

3. $60^{2(-1.1)} = 60^{-2.2} \approx 0.0001225 \approx 0.0$

5. $f(x) = 4^x$

Intercept: $(0, 1)$

Horizontal asymptote: x-axis

Increasing on: $(-\infty, \infty)$

Matches graph (c).

7. $f(x) = -4^x$

Intercept: $(0, -1)$

Horizontal asymptote: x-axis

Decreasing on: $(-\infty, \infty)$

Matches graph (b).

9. $f(x) = 6^x$

Intercept: $(0, 1)$

Horizontal asymptote: x-axis

Increasing on: $(-\infty, \infty)$

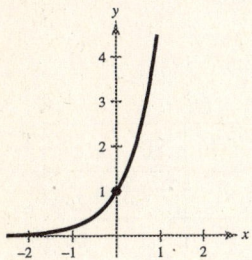

11. $g(x) = 1 + 6^{-x}$

Intercept: $(0, 2)$

Horizontal asymptote: $y = 1$

Decreasing on: $(-\infty, \infty)$

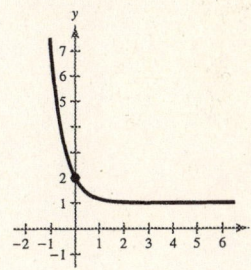

13. $e^8 \approx 2980.958$

15. $e^{-2.1} \approx 0.122$

17. $h(x) = e^{x-1}$

x	-4	-2	0	1	2	4
$h(x)$	0.0067	0.0498	0.3679	1	2.7183	20.086

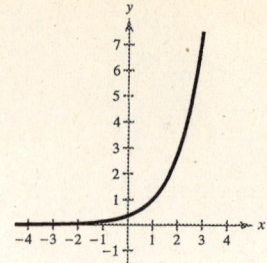

19. $h(x) = -e^x$

x	-4	-2	0	1	2	4
$h(x)$	-0.0183	-0.1353	-1	-2.718	-7.389	-54.6

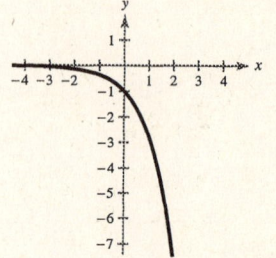

21. $f(x) = 4e^{-0.5x}$

x	-4	-2	0	1	2
$f(x)$	29.556	10.873	4	2.4261	1.4715

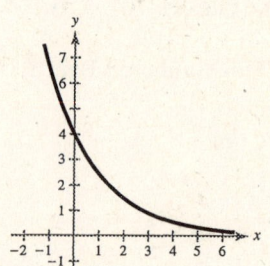

23. $g(t) = 8 - 0.5e^{-t/4}$

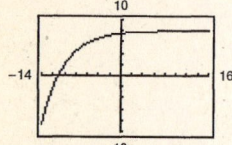

Horizontal asymptote: $y = 8$

25. $g(x) = 200e^{4/x}$

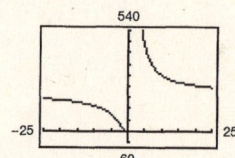

Vertical asymptote: $x = 0$

Horizontal asymptote: $y = 200$

27. $f(x) = \dfrac{10}{1 + 2^{-0.05x}}$

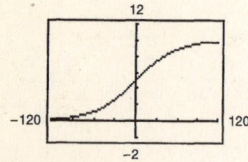

Horizontal asymptotes: $y = 0$, $y = 10$

29. $A = Pe^{rt} = 10{,}000e^{0.08t}$

t	1	10	20
A	10,832.87	22,255.41	49,530.32

t	30	40	50
A	110,231.76	245,325.30	545,981.50

31. $V(t) = 26,000\left(\frac{3}{4}\right)^t$

(a)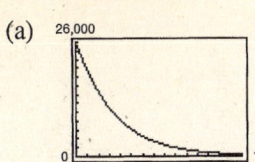

(b) For $t = 2$, $V(2) = \$14,625$

(c) The car depreciates most rapidly at the beginning, which is realistic.

33. $4^3 = 64$

$\log_4 64 = 3$

35. $25^{3/2} = 125$

$\log_{25} 125 = \frac{3}{2}$

37. $\log_6 216 = \log_6 6^3 = 3 \log_6 6 = 3$

39. $\log_4\left(\frac{1}{4}\right) = \log_4(4^{-1}) = -\log_4 4 = -1$

41. $g(x) = -\log_2 x + 5 = 5 - \dfrac{\ln x}{\ln 2}$

Domain: $x > 0$
Vertical asymptote: $x = 0$
x-intercept: $(32, 0)$

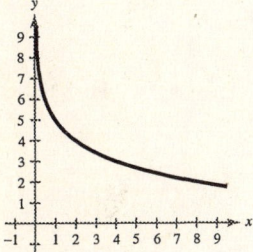

43. $f(x) = \log_2(x - 1) + 6 = 6 + \dfrac{\ln(x - 1)}{\ln (2)}$

Domain: $x > 1$
Vertical asymptote: $x = 1$
x-intercept: $(1.016, 0)$

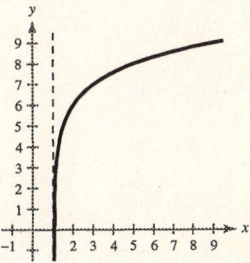

45. $\ln(21.5) \approx 3.068$

47. $\ln(e^7) = 7$

49. $\ln\sqrt{6} \approx 0.896$

51. $f(x) = \ln x + 3$

Domain: $(0, \infty)$
Vertical asymptote: $x = 0$
x-intercept: $(0.05, 0)$

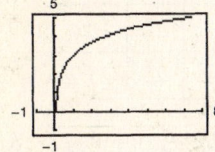

53. $h(x) = \frac{1}{2} \ln x$

Domain: $x > 0$
Vertical asymptote: $x = 0$
x-intercept: $(1, 0)$

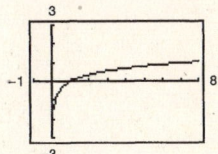

55. $t = 50 \log_{10} \dfrac{18,000}{18,000 - h}$

(a) $0 \le h < 18,000$

(c) The plane climbs at a faster rate as it approaches its absolute ceiling.

(d) If $h = 4000$, $t = 50 \log_{10} \dfrac{18,000}{18,000 - 4000} \approx 5.46$ minutes.

(b)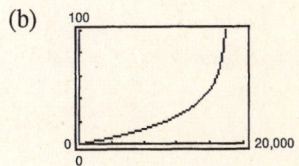

Vertical asymptote: $h = 18,000$

57. $\log_4 9 = \dfrac{\log_{10} 9}{\log_{10} 4} \approx 1.585$

$\log_4 9 = \dfrac{\ln 9}{\ln 4} \approx 1.585$

59. $\log_{12} 200 = \dfrac{\log_{10} 200}{\log_{10} 12} \approx 2.132$

$\log_{12} 200 = \dfrac{\ln 200}{\ln 12} \approx 2.132$

61. $\ln 20 = \ln[4 \cdot 5] = \ln 4 + \ln 5$

63. $\log_5\left(\tfrac{1}{15}\right) = \log_5(15)^{-1}$

$= -\log_5(3 \cdot 5)$

$= -\log_5 3 - \log_5 5$

$= -\log_5 3 - 1$

65. $\log_5 5x^2 = \log_5 5 + \log_5 x^2$

$= 1 + 2\log_5 x$

67. $\log_{10} \dfrac{5\sqrt{y}}{x^2} = \log_{10} 5\sqrt{y} - \log_{10} x^2$

$= \log_{10} 5 + \log_{10}\sqrt{y} - \log_{10} x^2$

$= \log_{10} 5 + \dfrac{1}{2}\log_{10} y - 2\log_{10} x$

69. $\ln\left(\dfrac{x+3}{xy}\right) = \ln(x+3) - \ln(xy)$

$= \ln(x+3) - \ln x - \ln y$

71. $\log_2 5 + \log_2 x = \log_2 5x$

73. $\dfrac{1}{2}\ln(2x-1) - 2\ln(x+1) = \ln\sqrt{2x-1} - \ln(x+1)^2$

$= \ln\dfrac{\sqrt{2x-1}}{(x+1)^2}$

75. $\ln 3 + \dfrac{1}{3}\ln(4-x^2) - \ln x = \ln\left[\dfrac{3(4-x^2)^{1/3}}{x}\right] = \ln\left[\dfrac{3\sqrt[3]{4-x^2}}{x}\right]$

77. $s = 25 - \dfrac{13\ln\left(\dfrac{h}{12}\right)}{\ln 3}$

(a)

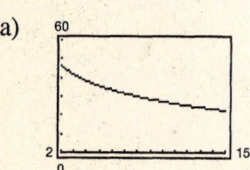

(b)

h	4	6	8	10	12	14
s	38	33.2	29.8	27.2	25	23.2

(c) As the depth increases, the number of miles of roads cleared decreases.

79. $8^x = 512 = 8^3 \implies x = 3$

81. $6^x = \dfrac{1}{216} = \dfrac{1}{6^3} = 6^{-3} \implies x = -3$

83. $\log_7 x = 4 \implies x = 7^4 = 2401$

85. $\ln x = 4$

$x = e^4 \approx 54.598$

87. $e^x = 12$

$x = \ln 12 \approx 2.485$

89. $3e^{-5x} = 132$

$e^{-5x} = 44$

$-5x = \ln 44$

$x = -\dfrac{\ln 44}{5} \approx -0.757$

91. $2^x + 13 = 35$

$2^x = 22$

$x \ln 2 = \ln 22$

$x = \dfrac{\ln 22}{\ln 2} \approx 4.459$

93. $-4(5^x) = -68$

$5^x = 17$

$x \ln 5 = \ln 17$

$x = \dfrac{\ln 17}{\ln 5} \approx 1.760$

95. $e^{2x} - 7e^x + 10 = 0$

$(e^x - 5)(e^x - 2) = 0$

$e^x = 5 \implies x = \ln 5 \approx 1.609$

$e^x = 2 \implies x = \ln 2 \approx 0.693$

97. $\ln 3x = 8.2$

$3x = e^{8.2}$

$x = \dfrac{e^{8.2}}{3} \approx 1213.650$

99. $2 \ln 4x = 15$

$\ln 4x = \dfrac{15}{2}$

$4x = e^{15/2}$

$x = \dfrac{1}{4} e^{15/2} \approx 452.011$

101. $\ln x - \ln 3 = 2$

$\ln \dfrac{x}{3} = 2$

$\dfrac{x}{3} = e^2$

$x = 3e^2 \approx 22.167$

103. $\ln \sqrt{x + 1} = 2$

$\dfrac{1}{2} \ln(x + 1) = 2$

$\ln(x + 1) = 4$

$x + 1 = e^4$

$x = e^4 - 1 \approx 53.598$

105.

$\log_{10}(x - 1) = \log_{10}(x - 2) - \log_{10}(x + 2)$

$\log_{10}(x - 1) = \log_{10}\left(\dfrac{x - 2}{x + 2}\right)$

$x - 1 = \dfrac{x - 2}{x + 2}$

$(x - 1)(x + 2) = x - 2$

$x^2 + x - 2 = x - 2$

$x^2 = 0$

$x = 0$

Since $x = 0$ is not in the domain of $\ln(x - 1)$ or of $\ln(x - 2)$, it is an extraneous solution. The equation has no solution. You can verify this by graphing each side of the equation and observing that the two curves do not intersect.

107. $\log_{10}(1 - x) = -1$

$10^{-1} = 1 - x$

$x = 1 - 10^{-1} = 0.9$

109. $3(7550) = 7550e^{0.0725t}$

$3 = e^{0.0725t}$

$\ln 3 = 0.0725t$

$t = \dfrac{\ln 3}{0.0725} \approx 15.2 \text{ years}$

111. $y = 3e^{-2x/3}$.

Decreasing exponential.

Matches graph (e).

113. $y = \ln(x + 3)$

Logarithmic function shifted to left.

Matches graph (f).

115. $y = 2e^{-(x + 4)^2/3}$

Gaussian model.

Matches graph (a).

117.

$$y = ae^{bx}$$

$$2 = ae^{b(0)} \implies a = 2$$

$$3 = 2e^{b(4)}$$

$$1.5 = e^{4b}$$

$$\ln 1.5 = 4b \quad \implies b \approx 0.1014$$

Thus, $y = 2e^{0.1014x}$.

119.

$$y = ae^{bx}$$

$$\tfrac{1}{2} = ae^{b(0)} \implies a = \tfrac{1}{2}$$

$$5 = \tfrac{1}{2}e^{b(5)}$$

$$10 = e^{5b}$$

$$\ln 10 = 5b \quad \implies b \approx 0.4605$$

Thus, $y = \tfrac{1}{2}e^{0.4605x}$.

121. $P = 361e^{kt}$

$t = 0$ corresponds to 2000

$(-20, 215)$:

$$215 = 361e^{k(-20)}$$

$$\frac{215}{361} = e^{-20k}$$

$$-20k = \ln\left(\frac{215}{361}\right)$$

$$k = -\frac{1}{20}\ln\left(\frac{215}{361}\right) = \frac{1}{20}\ln\left(\frac{361}{215}\right) \approx 0.02591$$

$P = 361e^{0.02591t}$

For 2020, $P(20) = 361e^{0.02591(20)} \approx 606.1$

606,100 population in 2020

123. (a) $20,000 = 10,000e^{r(12)}$

$$2 = e^{12r}$$

$$\ln 2 = 12r$$

$$r = \frac{\ln 2}{12} \approx 0.0578 \text{ or } 5.78\%$$

(b) $10,000e^{0.0578(1)} \approx \$10,595.03$

125. $N = \dfrac{157}{1 + 5.4e^{-0.12t}}$

(a) When $N = 50$:

$$50 = \frac{157}{1 + 5.4e^{-0.12t}}$$

$$1 + 5.4e^{-0.12t} = \frac{157}{50}$$

$$5.4e^{-0.12t} = \frac{107}{50}$$

$$e^{-0.12t} = \frac{107}{270}$$

$$-0.12t = \ln\frac{107}{270}$$

$$t = \frac{\ln(107/270)}{-0.12} \approx 7.7 \text{ weeks}$$

(b) When $N = 75$:

$$75 = \frac{157}{1 + 5.4e^{-0.12t}}$$

$$1 + 5.4e^{-0.12t} = \frac{157}{75}$$

$$5.4e^{-0.12t} = \frac{82}{75}$$

$$e^{-0.12t} = \frac{82}{405}$$

$$-0.12t = \ln\frac{82}{405}$$

$$t = \frac{\ln(82/405)}{-0.12} \approx 13.3 \text{ weeks}$$

127. Logistics model

129. Logarithmic model

131. (a) Quadratic model: $y = 0.025x^2 + 0.12x + 10.1$

Exponential model: $y = 8.73(1.05)^x$

Power model: $y = 3.466x^{0.647}$

(b) Quadratic model:

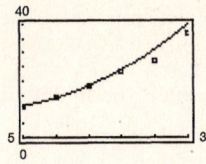

Exponential model:

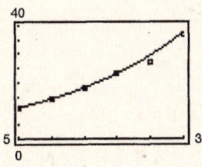

Power model:

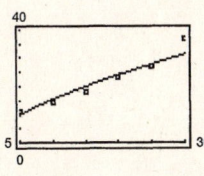

(c) The exponential model appears to fit best

(d) For 2007, $x = 37$ and $y \approx 53.1$ or 53,100 screens.

133. (a) $P = \dfrac{9999.887}{1 + 19.0e^{-0.2x}}$

(b)

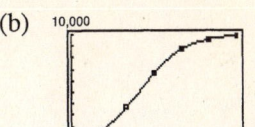

(c) The model is a good fit

(d) The limiting size is $\dfrac{9999.887}{1 + 0} \approx 10,000$ fish

135. True; by the inverse properties, $\log_b b^{2x} = 2x$.

137. False; $\ln x + \ln y = \ln(xy) \neq \ln(x + y)$

139. False. The domain of $f(x) = \ln(x)$ is $x > 0$

141. (a)

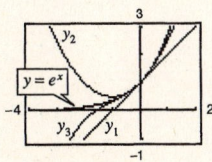

(b) Pattern $\displaystyle\sum_{i=0}^{n} \dfrac{x^i}{i!}$

$$y_4 = 1 + x + \dfrac{x^2}{2!} + \dfrac{x^3}{3!} + \dfrac{x^4}{4!}$$

The graph of y_4 closely approximates $y = e^x$ near $(0, 1)$.

Chapter 3 Practice Test

1. Solve for x: $x^{3/5} = 8$

2. Solve for x: $3^{x-1} = \frac{1}{81}$

3. Graph $f(x) = 2^{-x}$ by hand.

4. Graph $g(x) = e^x + 1$ by hand.

5. If \$5000 is invested at 9% interest, find the amount after three years if the interest is compounded

   (a) monthly   (b) quarterly   (c) continuously.

6. Write the equation in logarithmic form: $7^{-2} = \frac{1}{49}$

7. Solve for x: $x - 4 = \log_2 \frac{1}{64}$

8. Given $\log_b 2 = 0.3562$ and $\log_b 5 = 0.8271$, evaluate $\log_b \sqrt[4]{8/25}$.

9. Write $5 \ln x - \frac{1}{2} \ln y + 6 \ln z$ as a single logarithm.

10. Using your calculator and the change of base formula, evaluate $\log_9 28$.

11. Use your calculator to solve for N: $\log_{10} N = 0.6646$

12. Graph $y = \log_4 x$ by hand.

13. Determine the domain of $f(x) = \log_3(x^2 - 9)$.

14. Graph $y = \ln(x - 2)$ by hand.

15. True or false: $\dfrac{\ln x}{\ln y} = \ln(x - y)$

16. Solve for x: $5^x = 41$

17. Solve for x: $x - x^2 = \log_5 \frac{1}{25}$

18. Solve for x: $\log_2 x + \log_2(x - 3) = 2$

19. Solve for x: $\dfrac{e^x + e^{-x}}{3} = 4$

20. Six thousand dollars is deposited into a fund at an annual percentage rate of 13%. Find the time required for the investment to double if the interest is compounded continuously.

21. Use a graphing utility to find the points of intersection of the graphs of $y = \ln (3x)$ and $y = e^x - 4$.

22. Use a graphing utility to find the power model $y = ax^b$ for the data $(1, 1)$, $(2, 5)$, $(3, 8)$, and $(4, 17)$.

CHAPTER 4
Trigonometric Functions

CHAPTER 4
Trigonometric Functions

Section 4.1   Radian and Degree Measure

You should know the following basic facts about angles, their measurement, and their applications.

■ Types of Angles:

  (a) Acute: Measure between 0° and 90°.

  (b) Right: Measure 90°.

  (c) Obtuse: Measure between 90° and 180°.

  (d) Straight: Measure 180°.

■ α and β are complementary if $\alpha + \beta = 90°$. They are supplementary if $\alpha + \beta = 180°$.

■ Two angles in standard position that have the same terminal side are called coterminal angles.

■ To convert degrees to radians, use $1° = \pi/180$ radians.

■ To convert radians to degrees, use 1 radian $= (180/\pi)°$.

■ $1' =$ one minute $= 1/60$ of $1°$

■ $1'' =$ one second $= 1/60$ of $1' = 1/3600$ of $1°$

■ The length of a circular arc is $s = r\theta$ where θ is measured in radians.

■ Speed $=$ distance/time

■ Angular speed $= \theta/t = s/rt$

Solutions to Odd-Numbered Exercises

1.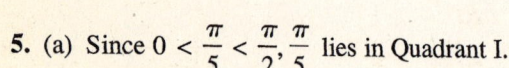

The angle shown is approximately 2 radians.

3.

The angle shown is approximately -3 radians.

5. (a) Since $0 < \dfrac{\pi}{5} < \dfrac{\pi}{2}, \dfrac{\pi}{5}$ lies in Quadrant I.

   (b) Since $\pi < \dfrac{7\pi}{5} < \dfrac{3\pi}{2}, \dfrac{7\pi}{5}$ lies in Quadrant III.

7. (a) Since $-\dfrac{\pi}{2} < -1 < 0$; -1 lies in Quadrant IV.

   (b) Since $-\pi < -2 < -\dfrac{\pi}{2}$; -2 lies in Quadrant III.

9. (a)

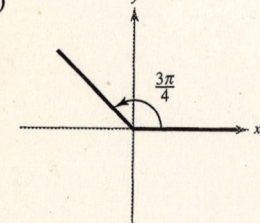

(b)

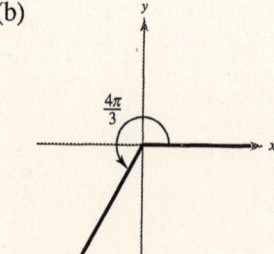

11. (a)

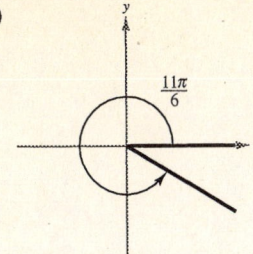

(b)

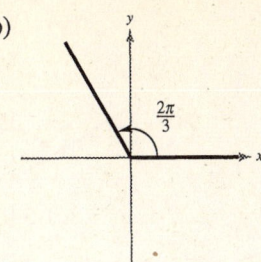

13. (a) Coterminal angles for $\dfrac{\pi}{6}$

$$\dfrac{\pi}{6} + 2\pi = \dfrac{13\pi}{6}$$

$$\dfrac{\pi}{6} - 2\pi = -\dfrac{11\pi}{6}$$

(b) Coterminal angles for $\dfrac{2\pi}{3}$

$$\dfrac{2\pi}{3} + 2\pi = \dfrac{8\pi}{3}$$

$$\dfrac{2\pi}{3} - 2\pi = -\dfrac{4\pi}{3}$$

15. (a) Coterminal angles for $-\dfrac{7\pi}{4}$

$$-\dfrac{7\pi}{4} + 2\pi = \dfrac{\pi}{4}$$

$$-\dfrac{7\pi}{4} - 2\pi = \dfrac{-15\pi}{4}$$

(b) Coterminal angles for $-\dfrac{11\pi}{6}$

$$-\dfrac{11\pi}{6} + 2\pi = \dfrac{\pi}{6}$$

$$-\dfrac{11\pi}{6} - 2\pi = \dfrac{-23\pi}{6}$$

17. (a) Complement: $\dfrac{\pi}{2} - \dfrac{\pi}{3} = \dfrac{\pi}{6}$

Supplement: $\pi - \dfrac{\pi}{3} = \dfrac{2\pi}{3}$

19. (a) Complement: $\dfrac{\pi}{2} - \dfrac{\pi}{12} = \dfrac{5\pi}{12}$

Supplement: $\pi - \dfrac{\pi}{12} = \dfrac{11\pi}{12}$

21. Complement: $\dfrac{\pi}{2} - 1 \approx 0.57$

Supplement: $\pi - 1 \approx 2.14$

23.

The angle shown is approximately $210°$.

25.

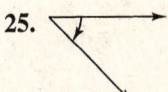

The angle shown is approximately $-45°$.

27. (a) Since $90° < 150° < 180°$, $150°$ lies in Quadrant II.

(b) Since $270° < 282° < 360°$, $282°$ lies in Quadrant IV.

29. (a) Since $-180° < -132° \, 50' < -90°$, $-132° \, 50'$ lies in Quadrant III.

(b) Since $-360° < -336° \, 30' < -270°$, $-336° \, 30'$ lies in Quadrant I.

31. (a)

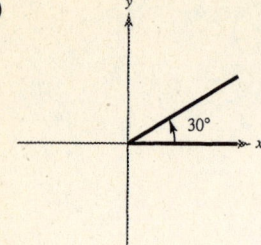

(b)

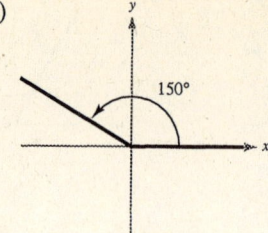

33. (a)

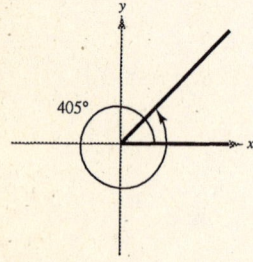

(b)

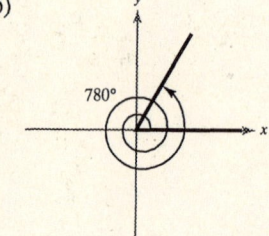

35. (a) Coterminal angles for 52°

$52° + 360° = 412°$

$52° - 360° = -308°$

(b) Coterminal angles for $-36°$

$-36° + 360° = 324°$

$-36° - 360° = -396°$

37. (a) Coterminal angles for 300°

$300° + 360° = 660°$

$300° - 360° = -60°$

(b) Coterminal angles for 230°

$230° + 360° = 590°$

$230° - 360° = -130°$

39. Complement: $90° - 36° = 54°$

Supplement: $180° - 36° = 144°$

41. Complement: Not possible (158° > 90°)

Supplement: $180° - 158° = 22°$

43. Complement: Not possible (99° > 90°)

Supplement: $180° - 99° = 81°$

45. (a) $30° = 30\left(\dfrac{\pi}{180}\right) = \dfrac{\pi}{6}$

(b) $150° = 150\left(\dfrac{\pi}{180}\right) = \dfrac{5\pi}{6}$

47. (a) $-20° = -20\left(\dfrac{\pi}{180}\right) = -\dfrac{\pi}{9}$

(b) $-240° = -240\left(\dfrac{\pi}{180}\right) = -\dfrac{4\pi}{3}$

49. (a) $\dfrac{3\pi}{2} = \dfrac{3\pi}{2}\left(\dfrac{180}{\pi}\right)° = 270°$

(b) $-\dfrac{7\pi}{6} = -\dfrac{7\pi}{6}\left(\dfrac{180}{\pi}\right)° = -210°$

51. (a) $\dfrac{7\pi}{3} = \dfrac{7\pi}{3}\left(\dfrac{180°}{\pi}\right) = 420°$

(b) $-\dfrac{13\pi}{60} = -\dfrac{13\pi}{60}\left(\dfrac{180°}{\pi}\right) = -39°$

53. $115° = 115\left(\dfrac{\pi}{180}\right) \approx 2.007$ radians

55. $-216.35° = -216.35\left(\frac{\pi}{180}\right) \approx -3.776$ radians

57. $-0.78° = -0.78\left(\frac{\pi}{180}\right) \approx -0.014$ radians

59. $\frac{\pi}{7} = \frac{\pi}{7}\left(\frac{180°}{\pi}\right) \approx 25.714°$

61. $6.5\pi = 6.5\pi\left(\frac{180°}{\pi}\right) = 1170°$

63. $-2 = -2\left(\frac{180}{\pi}\right) \approx -114.592°$

65. $64° \, 45' = 64° + \left(\frac{45}{60}\right)° = 64.75°$

67. $85° \, 18' \, 30'' = 85° + \left(\frac{18}{60}\right)° + \left(\frac{30}{3600}\right)° \approx 85.308°$

69. $-125° \, 36'' = -125° - \left(\frac{36}{3600}\right)° = -125.01°$

71. $280.6° = 280° + 0.6(60)' = 280° \, 36'$

73. $-345.12° = -345° \, 7' \, 12''$

75. $-0.355 = -0.355\left(\frac{180°}{\pi}\right)$

$\approx -20.34° = -20° \, 20' \, 24''$

77. $s = r\theta$

$6 = 5\theta$

$\theta = \frac{6}{5}$ radians

79. $s = r\theta$

$32 = 7\theta$

$\theta = \frac{32}{7} = 4\frac{4}{7}$ radians

81. $s = r\theta$

$8 = 15\theta$

$\theta = \frac{8}{15}$ radians

83. $s = r\theta$

$35 = 14.5\theta$

$\theta = \frac{70}{29} \approx 2.414$ radians

85. $s = r\theta, \; \theta$ in radians

$s = 14(180)\left(\frac{\pi}{180}\right) = 14\pi \approx 43.982$ inches

87. $s = r\theta, \; \theta$ in radians

$s = 27\left(\frac{2\pi}{3}\right) = 18\pi$ meters ≈ 56.55 meters

89. $\theta = 42° \, 7' \, 45'' - 25° \, 46' \, 26''$

$= 16° \, 21' \, 19'' \approx 0.28545$ radian

$s = r\theta = 4000(0.28545) \approx 1141.81$ miles

91. $\theta = \frac{s}{r} = \frac{450}{6378} \approx 0.07056$ radian $\approx 4.04°$

$\approx 4° \, 2' \, 33.02''$

93. $\theta = \frac{s}{r} = \frac{2.5}{6} = \frac{25}{60} = \frac{5}{12}$ radian $\approx 23.87°$

95. (a) single axel: $1\frac{1}{2}$ revolutions $= 360° + 180° = 540°$

$= 2\pi + \pi = 3\pi$ radians

(b) double axel: $2\frac{1}{2}$ revolutions $= 720° + 180° = 900°$

$= 4\pi + \pi = 5\pi$ radians

(c) triple axel: $3\frac{1}{2}$ revolutions $= 1260°$

$= 7\pi$ radians

97. (a) $\dfrac{\text{Revolutions}}{\text{Second}} = \dfrac{2400}{60} = 40 \text{ rev/sec}$

Angular speed $= (2\pi)(40) = 80\pi \text{ rad/sec}$

(b) Radius of saw blade $= \dfrac{7.5}{2} = 3.75 \text{ in.}$

Radius in feet $= \dfrac{3.75 \text{ in.}}{12 \text{ in./ft}} = 0.3125 \text{ ft}$

$\text{Speed} = \dfrac{s}{t} = \dfrac{r\theta}{t} = r\dfrac{\theta}{t}$

$= r(\text{angular speed})$

$= 0.3125(80\pi) = 78.54 \text{ ft/sec}$

99. False, 1 radian $= \left(\dfrac{180}{\pi}\right)^{\circ} \approx 57.3°$, so one radian is much larger than one degree.

101. True: $\dfrac{2\pi}{3} + \dfrac{\pi}{4} + \dfrac{\pi}{12} = \dfrac{8\pi + 3\pi + \pi}{12} = \pi = 180°$

103. If θ is constant, the length of the arc is proportional to the radius ($s = r\theta$), and hence increasing.

105. $A = \dfrac{1}{2}r^2\theta = \dfrac{1}{2}(10)^2 \cdot \dfrac{\pi}{3} = \dfrac{50}{3}\pi$ square meters

107. $A = \frac{1}{2}r^2\theta,\ s = r\theta$

(a) $\theta = 0.8 \implies A = \frac{1}{2}r^2(0.8) = 0.4r^2$ Domain: $r > 0$

$\qquad\qquad\qquad s = r\theta = r(0.8)$ Domain: $r > 0$

The area function changes more rapidly for $r > 1$ because it is quadratic and the arc length function is linear.

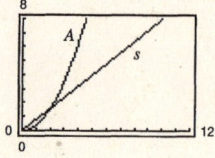

(b) $r = 10 \implies A = \frac{1}{2}(10^2)\theta = 50\theta$ Domain: $0 < \theta < 2\pi$

$\qquad\qquad\qquad s = r\theta = 10\theta$ Domain: $0 < \theta < 2\pi$

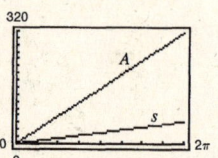

109.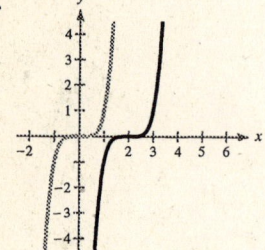

111.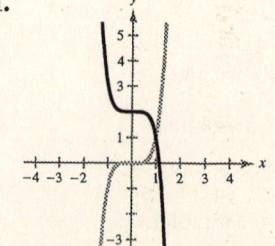

113.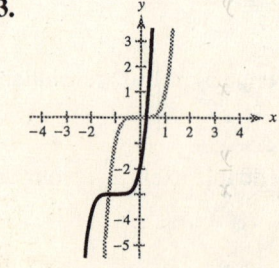

Section 4.2   Trigonometric Functions: The Unit Circle

Solutions to Odd-Numbered Exercises

1. $\sin \theta = y = \dfrac{15}{17}$

$\cos \theta = x = -\dfrac{8}{17}$

$\tan \theta = \dfrac{y}{x} = -\dfrac{15}{8}$

$\cot \theta = \dfrac{x}{y} = -\dfrac{8}{15}$

$\sec \theta = \dfrac{1}{x} = -\dfrac{17}{8}$

$\csc \theta = \dfrac{1}{y} = \dfrac{17}{15}$

3. $\sin \theta = y = -\dfrac{5}{13}$

$\cos \theta = x = \dfrac{12}{13}$

$\tan \theta = \dfrac{y}{x} = -\dfrac{5}{12}$

$\cot \theta = \dfrac{x}{y} = -\dfrac{12}{5}$

$\sec \theta = \dfrac{1}{x} = \dfrac{13}{12}$

$\csc \theta = \dfrac{1}{y} = -\dfrac{13}{5}$

5. $t = \dfrac{\pi}{4}$ corresponds to $\left(\dfrac{\sqrt{2}}{2}, \dfrac{\sqrt{2}}{2} \right)$.

7. $t = \dfrac{7\pi}{6}$ corresponds to $\left(-\dfrac{\sqrt{3}}{2}, -\dfrac{1}{2} \right)$.

9. $t = \dfrac{2\pi}{3}$ corresponds to $\left(-\dfrac{1}{2}, \dfrac{\sqrt{3}}{2} \right)$.

11. $t = \dfrac{3\pi}{2}$ corresponds to $(0, -1)$.

13. $t = \dfrac{\pi}{4}$ corresponds to $\left(\dfrac{\sqrt{2}}{2}, \dfrac{\sqrt{2}}{2} \right)$.

$\sin t = y = \dfrac{\sqrt{2}}{2}$

$\cos t = x = \dfrac{\sqrt{2}}{2}$

$\tan t = \dfrac{y}{x} = 1$

15. $t = -\dfrac{\pi}{6}$ corresponds to $\left(\dfrac{\sqrt{3}}{2}, -\dfrac{1}{2} \right)$.

$\sin t = y = -\dfrac{1}{2}$

$\cos t = x = \dfrac{\sqrt{3}}{2}$

$\tan t = \dfrac{y}{x} = -\dfrac{\sqrt{3}}{3}$

17. $t = -\dfrac{7\pi}{4}$ corresponds to $\left(\dfrac{\sqrt{2}}{2}, \dfrac{\sqrt{2}}{2} \right)$

$\sin t = y = \dfrac{\sqrt{2}}{2}$

$\cos t = x = \dfrac{\sqrt{2}}{2}$

$\tan t = \dfrac{y}{x} = 1$

19. $t = \dfrac{5\pi}{3}$ corresponds to $\left(\dfrac{1}{2}, -\dfrac{\sqrt{3}}{2} \right)$

$\sin t = y = -\dfrac{\sqrt{3}}{2}$

$\cos t = x = \dfrac{1}{2}$

$\tan t = \dfrac{y}{x} = -\sqrt{3}$

21. $t = -\dfrac{3\pi}{2}$ corresponds to $(0, 1)$.

$\sin t = y = 1$

$\cos t = x = 0$

$\tan t = \dfrac{y}{x}$ is undefined.

23. $t = \dfrac{3\pi}{4}$ corresponds to $\left(-\dfrac{\sqrt{2}}{2}, \dfrac{\sqrt{2}}{2}\right)$.

$\sin t = y = \dfrac{\sqrt{2}}{2}$ $\qquad$ $\csc t = \dfrac{1}{y} = \sqrt{2}$

$\cos t = x = -\dfrac{\sqrt{2}}{2}$ $\qquad$ $\sec t = \dfrac{1}{x} = -\sqrt{2}$

$\tan t = \dfrac{y}{x} = -1$ $\qquad$ $\cot t = \dfrac{x}{y} = -1$

25. $t = \dfrac{\pi}{2}$ corresponds to $(0, 1)$.

$\sin t = y = 1$ $\qquad$ $\csc t = \dfrac{1}{y} = 1$

$\cos t = x = 0$ $\qquad$ $\sec t = \dfrac{1}{x}$ is undefined.

$\tan t = \dfrac{y}{x}$ is undefined. $\qquad$ $\cot t = \dfrac{x}{y} = 0$

27. $t = -\dfrac{2\pi}{3}$ corresponds to $\left(-\dfrac{1}{2}, -\dfrac{\sqrt{3}}{2}\right)$

$\sin t = y = -\dfrac{\sqrt{3}}{2}$ $\qquad$ $\csc t = -\dfrac{2\sqrt{3}}{3}$

$\cos t = x = -\dfrac{1}{2}$ $\qquad$ $\sec \theta = -2$

$\tan t = \dfrac{y}{x} = \sqrt{3}$ $\qquad$ $\cot \theta = \dfrac{\sqrt{3}}{3}$

29. $\sin 5\pi = \sin \pi = 0$

31. $\cos \dfrac{8\pi}{3} = \cos \dfrac{2\pi}{3} = -\dfrac{1}{2}$

33. $\cos\left(-\dfrac{13\pi}{6}\right) = \cos\left(-\dfrac{\pi}{6}\right) = \cos\left(\dfrac{11\pi}{6}\right) = \dfrac{\sqrt{3}}{2}$

35. $\sin\left(-\dfrac{9\pi}{4}\right) = \sin\left(-\dfrac{\pi}{4}\right) = -\dfrac{\sqrt{2}}{2}$

37. $\sin t = \dfrac{1}{3}$

(a) $\sin(-t) = -\sin t = -\dfrac{1}{3}$

(b) $\csc(-t) = -\csc t = -3$

39. $\cos(-t) = -\dfrac{1}{5}$

(a) $\cos t = \cos(-t) = -\dfrac{1}{5}$

(b) $\sec(-t) = \dfrac{1}{\cos(-t)} = -5$

41. $\sin t = \dfrac{4}{5}$

(a) $\sin(\pi - t) = \sin t = \dfrac{4}{5}$

(b) $\sin(t + \pi) = -\sin t = -\dfrac{4}{5}$

43. $\sin\left(\dfrac{\pi}{6}\right) = 0.5$

45. $\csc 1.3 \approx 1.0378$

47. $\cos(-1.7) \approx -0.1288$

49. $\csc 0.8 = \dfrac{1}{\sin 0.8} \approx 1.3940$

51. $\sec 22.8 = \dfrac{1}{\cos 22.8} \approx -1.4486$

53. (a) $\sin 5 \approx -1$ $\qquad$ (b) $\cos 2 \approx -0.4$

55. (a) $\sin t = 0.25$ $\qquad$ (b) $\cos t = -0.25$

$\qquad$ $t \approx 0.25$ or 2.89 $\qquad$ $t \approx 1.82$ or 4.46

57. $\quad I = 5e^{-2t} \sin t$

$\quad I(0.7) = 5e^{-1.4} \sin 0.7 \approx 0.79$ amperes

59. $y(t) = \frac{1}{4}\cos 6t$ (a) $y(0) = \frac{1}{4}\cos 0 = 0.2500$ ft   (b) $y\left(\frac{1}{4}\right) = \frac{1}{4}\cos\frac{3}{2} \approx 0.0177$ ft   (c) $y\left(\frac{1}{2}\right) = \frac{1}{4}\cos 3 \approx -0.2475$ ft

61. False. $\sin\left(\dfrac{-4\pi}{3}\right) = \dfrac{\sqrt{3}}{2} > 0$

63. False. 0 corresponds to $(1, 0)$.

65. (a) The points have y-axis symmetry.

(b) $\sin t_1 = \sin(\pi - t_1)$ since they have the same y-value.

(c) $-\cos t_1 = \cos(\pi - t_1)$ since the x-values have opposite signs.

67. $\cos 1.5 \approx 0.0707$, $2\cos 0.75 \approx 1.4634$

Thus, $\cos 2t \neq 2\cos t$.

69. $\cos\theta = x = \cos(-\theta)$

$\sec\theta = \dfrac{1}{\cos\theta} = \dfrac{1}{\cos(-\theta)} = \sec(-\theta)$

71. $h(t) = f(t)g(t)$ is odd:

$h(-t) = f(-t)g(-t) = -f(t)g(t) = -h(t)$

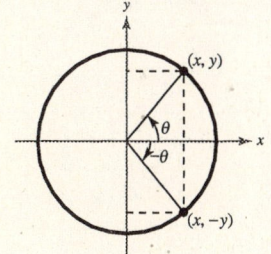

73.
$$f(x) = \tfrac{1}{2}(3x - 2)$$
$$y = \tfrac{1}{2}(3x - 2)$$
$$x = \tfrac{1}{2}(3y - 2)$$
$$2x = 3y - 2$$
$$2x + 2 = 3y$$
$$\tfrac{2}{3}(x + 1) = y$$
$$f^{-1}(x) = \tfrac{2}{3}(x + 1)$$

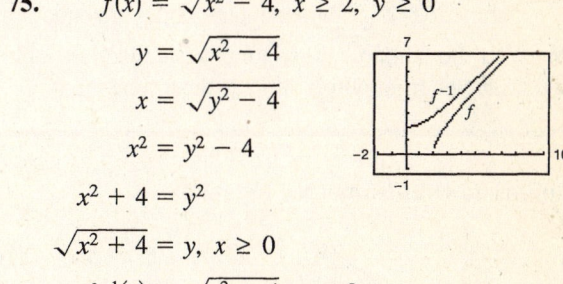

75.
$$f(x) = \sqrt{x^2 - 4},\ x \geq 2,\ y \geq 0$$
$$y = \sqrt{x^2 - 4}$$
$$x = \sqrt{y^2 - 4}$$
$$x^2 = y^2 - 4$$
$$x^2 + 4 = y^2$$
$$\sqrt{x^2 + 4} = y,\ x \geq 0$$
$$f^{-1}(x) = \sqrt{x^2 + 4},\ x \geq 0$$

77. $f(x) = \dfrac{2x}{x - 3}$

Asymptotes: $x = 3$, $y = 1$

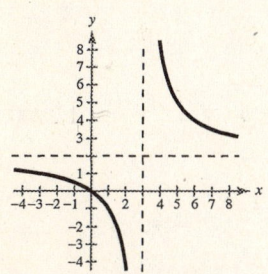

79. $f(x) = \dfrac{x^2 + 3x - 10}{2x^2 - 8} = \dfrac{(x - 2)(x + 5)}{2(x - 2)(x + 2)}$

$\quad = \dfrac{x + 5}{2(x + 2)}, \quad x \neq 2$

Asymptotes: $x = -2$, $y = \dfrac{1}{2}$

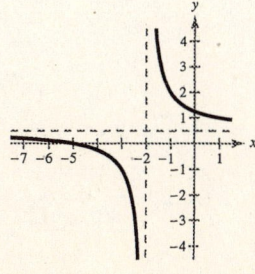

81. $C(10,000) = 50$ dollars/pound

$C(100,000) = 9.5$ dollars/pound

$C(1,000,000) = 5.45$ dollars/pound

As $x \to \infty$, $C \to 5$ dollars/pound.

83. $f(t) = 95 - 12 \log_{10}(t + 1), \quad 0 \le t \le 12$

$f(3) = 95 - 12 \log_{10}(3 + 1) \approx 87.8$

$f(6) = 95 - 12 \log_{10}(6 + 1) \approx 84.9$

Section 4.3   Right Triangle Trigonometry

■ You should know the right triangle definition of trigonometric functions.

(a) $\sin \theta = \dfrac{\text{opp}}{\text{hyp}}$

(b) $\cos \theta = \dfrac{\text{adj}}{\text{hyp}}$

(c) $\tan \theta = \dfrac{\text{opp}}{\text{adj}}$

(d) $\csc \theta = \dfrac{\text{hyp}}{\text{opp}}$

(e) $\sec \theta = \dfrac{\text{hyp}}{\text{adj}}$

(f) $\cot \theta = \dfrac{\text{adj}}{\text{opp}}$

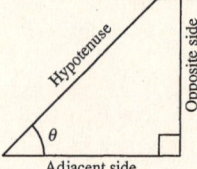

■ You should know the following identities.

(a) $\sin \theta = \dfrac{1}{\csc \theta}$

(b) $\csc \theta = \dfrac{1}{\sin \theta}$

(c) $\cos \theta = \dfrac{1}{\sec \theta}$

(d) $\sec \theta = \dfrac{1}{\cos \theta}$

(e) $\tan \theta = \dfrac{1}{\cot \theta}$

(f) $\cot \theta = \dfrac{1}{\tan \theta}$

(g) $\tan \theta = \dfrac{\sin \theta}{\cos \theta}$

(h) $\cot \theta = \dfrac{\cos \theta}{\sin \theta}$

(i) $\sin^2 \theta + \cos^2 \theta = 1$

(j) $1 + \tan^2 \theta = \sec^2 \theta$

(k) $1 + \cot^2 \theta = \csc^2 \theta$

■ You should know that two acute angles α and β are complementary if $\alpha + \beta = 90°$, and cofunctions of complementary angles are equal.

■ You should know the trigonometric function values of 30°, 45°, and 60°, or be able to construct triangles from which you can determine them.

Solutions to Odd-Numbered Exercises

1.

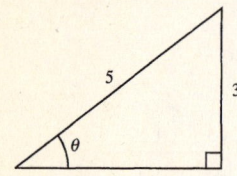

$\text{adj} = \sqrt{5^2 - 3^2} = \sqrt{16} = 4$

$\sin \theta = \dfrac{\text{opp}}{\text{hyp}} = \dfrac{3}{5}$ $\csc \theta = \dfrac{\text{hyp}}{\text{opp}} = \dfrac{5}{3}$

$\cos \theta = \dfrac{\text{adj}}{\text{hyp}} = \dfrac{4}{5}$ $\sec \theta = \dfrac{\text{hyp}}{\text{adj}} = \dfrac{5}{4}$

$\tan \theta = \dfrac{\text{opp}}{\text{adj}} = \dfrac{3}{4}$ $\cot \theta = \dfrac{\text{adj}}{\text{opp}} = \dfrac{4}{3}$

3.

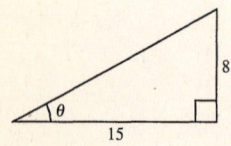

$\text{hyp} = \sqrt{8^2 + 15^2} = 17$

$\sin \theta = \dfrac{\text{opp}}{\text{hyp}} = \dfrac{8}{17}$ $\csc \theta = \dfrac{\text{hyp}}{\text{opp}} = \dfrac{17}{8}$

$\cos \theta = \dfrac{\text{adj}}{\text{hyp}} = \dfrac{15}{17}$ $\sec \theta = \dfrac{\text{hyp}}{\text{adj}} = \dfrac{17}{15}$

$\tan \theta = \dfrac{\text{opp}}{\text{adj}} = \dfrac{8}{15}$ $\cot \theta = \dfrac{\text{adj}}{\text{opp}} = \dfrac{15}{8}$

5.

adj $= \sqrt{3^2 - 1^2} = \sqrt{8} = 2\sqrt{2}$

$\sin \theta = \dfrac{\text{opp}}{\text{hyp}} = \dfrac{1}{3}$ $\csc \theta = \dfrac{\text{hyp}}{\text{opp}} = 3$

$\cos \theta = \dfrac{\text{adj}}{\text{hyp}} = \dfrac{2\sqrt{2}}{3}$ $\sec \theta = \dfrac{\text{hyp}}{\text{adj}} = \dfrac{3}{2\sqrt{2}} = \dfrac{3\sqrt{2}}{4}$

$\tan \theta = \dfrac{\text{opp}}{\text{adj}} = \dfrac{1}{2\sqrt{2}} = \dfrac{\sqrt{2}}{4}$ $\cot \theta = \dfrac{\text{adj}}{\text{opp}} = 2\sqrt{2}$

adj $= \sqrt{6^2 - 2^2} = \sqrt{32} = 4\sqrt{2}$

$\sin \theta = \dfrac{\text{opp}}{\text{hyp}} = \dfrac{2}{6} = \dfrac{1}{3}$ $\csc \theta = \dfrac{\text{hyp}}{\text{opp}} = \dfrac{6}{2} = 3$

$\cos \theta = \dfrac{\text{adj}}{\text{hyp}} = \dfrac{4\sqrt{2}}{6} = \dfrac{2\sqrt{2}}{3}$ $\sec \theta = \dfrac{\text{hyp}}{\text{adj}} = \dfrac{6}{4\sqrt{2}} = \dfrac{3}{2\sqrt{2}} = \dfrac{3\sqrt{2}}{4}$

$\tan \theta = \dfrac{\text{opp}}{\text{adj}} = \dfrac{2}{4\sqrt{2}} = \dfrac{1}{2\sqrt{2}} = \dfrac{\sqrt{2}}{4}$ $\cot \theta = \dfrac{\text{adj}}{\text{opp}} = \dfrac{4\sqrt{2}}{2} = 2\sqrt{2}$

The function values are the same since the triangles are similar and the corresponding sides are proportional.

7.

opp $= \sqrt{10^2 - 8^2} = 6$

$\sin \theta = \dfrac{\text{opp}}{\text{hyp}} = \dfrac{6}{10} = \dfrac{3}{5}$ $\csc \theta = \dfrac{\text{hyp}}{\text{opp}} = \dfrac{10}{6} = \dfrac{5}{3}$

$\cos \theta = \dfrac{\text{adj}}{\text{hyp}} = \dfrac{8}{10} = \dfrac{4}{5}$ $\sec \theta = \dfrac{\text{hyp}}{\text{adj}} = \dfrac{10}{8} = \dfrac{5}{4}$

$\tan \theta = \dfrac{\text{opp}}{\text{adj}} = \dfrac{6}{8} = \dfrac{3}{4}$ $\cot \theta = \dfrac{\text{adj}}{\text{opp}} = \dfrac{8}{6} = \dfrac{4}{3}$

opp $= \sqrt{2.5^2 - 2^2} = 1.5$

$\sin \theta = \dfrac{\text{opp}}{\text{hyp}} = \dfrac{1.5}{2.5} = \dfrac{3}{5}$ $\csc \theta = \dfrac{\text{hyp}}{\text{opp}} = \dfrac{2.5}{1.5} = \dfrac{5}{3}$

$\cos \theta = \dfrac{\text{adj}}{\text{hyp}} = \dfrac{2}{2.5} = \dfrac{4}{5}$ $\sec \theta = \dfrac{\text{hyp}}{\text{adj}} = \dfrac{2.5}{2} = \dfrac{5}{4}$

$\tan \theta = \dfrac{\text{opp}}{\text{adj}} = \dfrac{1.5}{2} = \dfrac{3}{4}$ $\cot \theta = \dfrac{\text{adj}}{\text{opp}} = \dfrac{2}{1.5} = \dfrac{4}{3}$

The function values are the same since the triangles are similar and the corresponding sides are proportional.

9. Given: $\sin \theta = \dfrac{5}{6} = \dfrac{\text{opp}}{\text{hyp}}$

$$5^2 + (\text{adj})^2 = 6^2$$

$$\text{adj} = \sqrt{11}$$

$$\cos \theta = \frac{\text{adj}}{\text{hyp}} = \frac{\sqrt{11}}{6}$$

$$\tan \theta = \frac{\text{opp}}{\text{adj}} = \frac{5}{\sqrt{11}} = \frac{5\sqrt{11}}{11}$$

$$\cot \theta = \frac{\text{adj}}{\text{opp}} = \frac{\sqrt{11}}{5}$$

$$\sec \theta = \frac{\text{hyp}}{\text{adj}} = \frac{6}{\sqrt{11}} = \frac{6\sqrt{11}}{11}$$

$$\csc \theta = \frac{\text{hyp}}{\text{opp}} = \frac{6}{5}$$

11. Given: $\sec \theta = 4 = \dfrac{4}{1} = \dfrac{\text{hyp}}{\text{adj}}$

$$(\text{opp})^2 + 1^2 = 4^2$$

$$\text{opp} = \sqrt{15}$$

$$\sin \theta = \frac{\sqrt{15}}{4}$$

$$\cos \theta = \frac{1}{4}$$

$$\tan \theta = \sqrt{15}$$

$$\cot \theta = \frac{1}{\sqrt{15}} = \frac{\sqrt{15}}{15}$$

$$\csc \theta = \frac{4}{\sqrt{15}} = \frac{4\sqrt{15}}{15}$$

13. Given: $\tan \theta = 3 = \dfrac{3}{1} = \dfrac{\text{opp}}{\text{adj}}$

$$3^2 + 1^2 = (\text{hyp})^2$$

$$\text{hyp} = \sqrt{10}$$

$$\sin \theta = \frac{3\sqrt{10}}{10}$$

$$\cos \theta = \frac{\sqrt{10}}{10} \qquad \sec \theta = \sqrt{10}$$

$$\cot \theta = \frac{1}{3} \qquad \csc \theta = \frac{\sqrt{10}}{3}$$

15. Given: $\cot \theta = \dfrac{9}{4} = \dfrac{\text{adj}}{\text{opp}}$

$$4^2 + 9^2 = (\text{hyp})^2$$

$$\text{hyp} = \sqrt{97}$$

$$\sin \theta = \frac{4}{\sqrt{97}} = \frac{4\sqrt{97}}{97}$$

$$\cos \theta = \frac{9}{\sqrt{97}} = \frac{9\sqrt{97}}{97} \qquad \sec \theta = \frac{\sqrt{97}}{9}$$

$$\tan \theta = \frac{4}{9} \qquad \csc \theta = \frac{\sqrt{97}}{4}$$

Function	θ (deg)	θ (rad)	Function Value
17. sin	30°	$\dfrac{\pi}{6}$	$\dfrac{1}{2}$
19. tan	60°	$\dfrac{\pi}{3}$	$\sqrt{3}$
21. cot	60°	$\dfrac{\pi}{3}$	$\dfrac{\sqrt{3}}{3}$
23. cos	30°	$\dfrac{\pi}{6}$	$\dfrac{\sqrt{3}}{2}$
25. cot	45°	$\dfrac{\pi}{4}$	1

27. $\sin 60° = \dfrac{\sqrt{3}}{2}$, $\cos 60° = \dfrac{1}{2}$

(a) $\tan 60° = \dfrac{\sin 60°}{\cos 60°} = \sqrt{3}$

(b) $\sin 30° = \cos 60° = \dfrac{1}{2}$

(c) $\cos 30° = \sin 60° = \dfrac{\sqrt{3}}{2}$

(d) $\cot 60° = \dfrac{\cos 60°}{\sin 60°} = \dfrac{1}{\sqrt{3}} = \dfrac{\sqrt{3}}{3}$

29. $\csc \theta = 3$, $\sec \theta = \dfrac{3\sqrt{2}}{4}$

(a) $\sin \theta = \dfrac{1}{\csc \theta} = \dfrac{1}{3}$

(b) $\cos \theta = \dfrac{1}{\sec \theta} = \dfrac{2\sqrt{2}}{3}$

(c) $\tan \theta = \dfrac{\sin \theta}{\cos \theta} = \dfrac{1/3}{(2\sqrt{2})/3} = \dfrac{\sqrt{2}}{4}$

(d) $\sec(90° - \theta) = \csc \theta = 3$

31. $\cos \alpha = \dfrac{1}{4}$

(a) $\sec \alpha = \dfrac{1}{\cos \alpha} = 4$

(b) $\sin^2 \alpha + \cos^2 \alpha = 1$

$$\sin^2 \alpha + \left(\dfrac{1}{4}\right)^2 = 1$$

$$\sin^2 \alpha = \dfrac{15}{16}$$

$$\sin \alpha = \pm \dfrac{\sqrt{15}}{4}$$

(c) $\cot \alpha = \dfrac{\cos \alpha}{\sin \alpha}$

$$= \pm \dfrac{1/4}{\sqrt{15}/4}$$

$$= \pm \dfrac{1}{\sqrt{15}}$$

$$= \pm \dfrac{\sqrt{15}}{15}$$

(d) $\sin(90° - \alpha) = \cos \alpha = \dfrac{1}{4}$

33. $\tan \theta \cot \theta = \tan \theta \left(\dfrac{1}{\tan \theta}\right) = 1$

35. $\tan \theta \cos \theta = \left(\dfrac{\sin \theta}{\cos \theta}\right)\cos \theta = \sin \theta$

37. $(1 + \cos \theta)(1 - \cos \theta) = 1 - \cos^2 \theta$

$$= (\sin^2 \theta + \cos^2 \theta) - \cos^2 \theta$$

$$= \sin^2 \theta$$

39. $\dfrac{\sin \theta}{\cos \theta} + \dfrac{\cos \theta}{\sin \theta} = \dfrac{\sin^2 \theta + \cos^2 \theta}{\sin \theta \cos \theta}$

$$= \dfrac{1}{\sin \theta \cos \theta}$$

$$= \dfrac{1}{\sin \theta} \cdot \dfrac{1}{\cos \theta}$$

$$= \csc \theta \sec \theta$$

41. (a) $\sin 41° \approx 0.6561$

(b) $\cos 87° \approx 0.0523$

43. (a) $\sec 42° \, 12' = \sec 42.2° = \dfrac{1}{\cos 42.2°} \approx 1.3499$

(b) $\csc 48° \, 7' = \dfrac{1}{\sin\left(48 + \frac{7}{60}\right)°} \approx 1.3432$

45. Make sure that your calculator is in radian mode.

(a) $\cot \dfrac{\pi}{16} = \dfrac{1}{\tan(\pi/16)} \approx 5.0273$

(b) $\tan \dfrac{\pi}{8} \approx 0.4142$

47. (a) $\sin \theta = \dfrac{1}{2} \implies \theta = 30° = \dfrac{\pi}{6}$

(b) $\csc \theta = 2 \implies \theta = 30° = \dfrac{\pi}{6}$

49. (a) $\sec \theta = 2 \implies \theta = 60° = \dfrac{\pi}{3}$

(b) $\cot \theta = 1 \implies \theta = 45° = \dfrac{\pi}{4}$

51. (a) $\csc \theta = \dfrac{2\sqrt{3}}{3} \implies \theta = 60° = \dfrac{\pi}{3}$

(b) $\sin \theta = \dfrac{\sqrt{2}}{2} \implies \theta = 45° = \dfrac{\pi}{4}$

53. $\cot 60° = \dfrac{x}{38}$

$\dfrac{\sqrt{3}}{3} = \dfrac{x}{38}$

$\dfrac{38\sqrt{3}}{3} = x$

55. $\sin 50° = \dfrac{y}{15}$

$y = 15 \cdot \sin 50° \approx 11.4907 \approx 11.5$

57. (a)

(b) $\tan \theta = \dfrac{6}{5}$ and $\tan \theta = \dfrac{h}{21}$ Thus, $\dfrac{6}{5} = \dfrac{h}{21}$.

(c) $h = \dfrac{6(21)}{5} = 25.2$ feet

59. $\tan \theta = \dfrac{\text{opp}}{\text{adj}}$

$\tan 58° = \dfrac{w}{100}$

$w = 100 \tan 58° \approx 160.03$ feet

61. (a)

(b) $\sin \theta = \dfrac{\text{opp}}{\text{hyp}}$

$\sin \theta = \dfrac{10/3}{20} = \dfrac{1}{6}$

(c) Using the table feature, you obtain $\theta \approx 9.59°$.

63. (a) $\sin(58°) = \dfrac{200}{x}$

$x = \dfrac{200}{\sin(58°)}$

≈ 235.84 ft

(b) $\tan(58°) = \dfrac{200}{y}$

$y = \dfrac{200}{\tan(58°)} \approx 124.97$ ft

65.

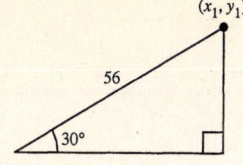

$$\sin 30° = \frac{y_1}{56}$$

$$y_1 = (\sin 30°)(56) = \left(\frac{1}{2}\right)(56) = 28$$

$$\cos 30° = \frac{x_1}{56}$$

$$x_1 = \cos 30°(56) = \frac{\sqrt{3}}{2}(56) = 28\sqrt{3}$$

$$(x_1, y_1) = \left(28\sqrt{3}, 28\right)$$

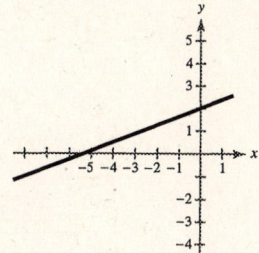

$$\sin 60° = \frac{y_2}{56}$$

$$y_2 = \sin 60°(56) = \left(\frac{\sqrt{3}}{2}\right)(56) = 28\sqrt{3}$$

$$\cos 60° = \frac{x_2}{56}$$

$$x_2 = (\cos 60°)(56) = \left(\frac{1}{2}\right)(56) = 28$$

$$(x_2, y_2) = \left(28, 28\sqrt{3}\right)$$

67. True.

$$\sin 60° \csc 60° = \sin 60° \frac{1}{\sin 60°}$$

$$= 1$$

69. True. $1 + \cot^2 \theta = \csc^2 \theta$ for all θ

71. (a)

θ	0°	20°	40°	60°	80°
$\sin \theta$	0	0.3420	0.6428	0.8660	0.9848
$\cos \theta$	1	0.9397	0.7660	0.5000	0.1736
$\tan \theta$	0	0.3640	0.8391	1.7321	5.6713

(b) Sine and tangent are increasing, cosine is decreasing.

(c) In each case, $\tan \theta = \dfrac{\sin \theta}{\cos \theta}$.

73. $y = -x - 9$

Intercepts: $(0, -9), (-9, 0)$

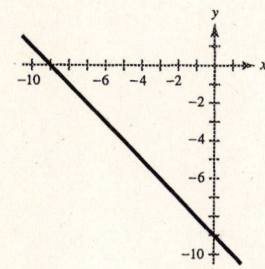

75. $-3x + 8y = 16$

Intercepts: $(0, 2), \left(-\frac{16}{3}, 0\right)$

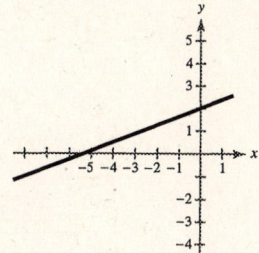

77.

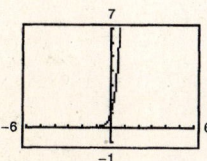

79.

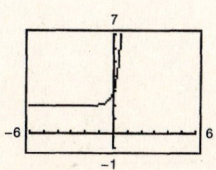

81.

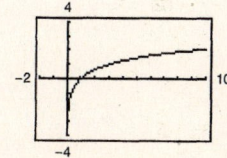

83.

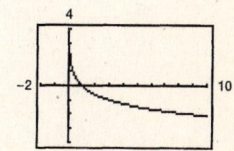

Section 4.4   Trigonometric Functions of Any Angle

■ Know the Definitions of Trigonometric Functions of Any Angle.

If θ is in standard position, (x, y) a point on the terminal side and $r = \sqrt{x^2 + y^2} \neq 0$, then:

$$\sin \theta = \frac{y}{r} \qquad\qquad \csc \theta = \frac{r}{y}, \ y \neq 0$$

$$\cos \theta = \frac{x}{r} \qquad\qquad \sec \theta = \frac{r}{x}, \ x \neq 0$$

$$\tan \theta = \frac{y}{x}, \ x \neq 0 \qquad \cot \theta = \frac{x}{y}, \ y \neq 0$$

■ You should know the signs of the trigonometric functions in each quadrant.

■ You should know the trigonometric function values of the quadrant angles 0, $\dfrac{\pi}{2}$, π, and $\dfrac{3\pi}{2}$.

■ You should be able to find reference angles.

■ You should be able to evaluate trigonometric functions of any angle. (Use reference angles.)

■ You should know that the period of sine and cosine is 2π.

■ You should know which trigonometric functions are odd and even.

Even: $\cos x$ and $\sec x$

Odd: $\sin x$, $\tan x$, $\cot x$, $\csc x$

Solutions to Odd-Numbered Exercises

1. (a) $(x, y) = (4, 3)$

$\qquad r = \sqrt{16 + 9} = 5$

$\qquad \sin \theta = \dfrac{y}{r} = \dfrac{3}{5} \qquad \csc \theta = \dfrac{r}{y} = \dfrac{5}{3}$

$\qquad \cos \theta = \dfrac{x}{r} = \dfrac{4}{5} \qquad \sec \theta = \dfrac{r}{x} = \dfrac{5}{4}$

$\qquad \tan \theta = \dfrac{y}{x} = \dfrac{3}{4} \qquad \cot \theta = \dfrac{x}{y} = \dfrac{4}{3}$

(b) $(x, y) = (-8, -15)$

$\qquad r = \sqrt{64 + 225} = 17$

$\qquad \sin \theta = \dfrac{y}{r} = -\dfrac{15}{17} \qquad \csc \theta = \dfrac{r}{y} = -\dfrac{17}{15}$

$\qquad \cos \theta = \dfrac{x}{r} = -\dfrac{8}{17} \qquad \sec \theta = \dfrac{r}{x} = -\dfrac{17}{8}$

$\qquad \tan \theta = \dfrac{y}{x} = \dfrac{15}{8} \qquad \cot \theta = \dfrac{x}{y} = \dfrac{8}{15}$

3. (a) $(x, y) = \left(-\sqrt{3}, -1\right)$

$\qquad r = \sqrt{3 + 1} = 2$

$\qquad \sin \theta = \dfrac{y}{r} = -\dfrac{1}{2} \qquad \csc \theta = \dfrac{r}{y} = -2$

$\qquad \cos \theta = \dfrac{x}{r} = \dfrac{-\sqrt{3}}{2} \qquad \sec \theta = \dfrac{r}{x} = \dfrac{-2\sqrt{3}}{3}$

$\qquad \tan \theta = \dfrac{y}{x} = \dfrac{\sqrt{3}}{3} \qquad \cot \theta = \dfrac{x}{y} = \sqrt{3}$

(b) $(x, y) = (-2, 2)$

$\qquad r = \sqrt{4 + 4} = 2\sqrt{2}$

$\qquad \sin \theta = \dfrac{y}{r} = \dfrac{\sqrt{2}}{2} \qquad \csc \theta = \dfrac{r}{y} = \sqrt{2}$

$\qquad \cos \theta = \dfrac{x}{r} = -\dfrac{\sqrt{2}}{2} \qquad \sec \theta = \dfrac{r}{x} = -\sqrt{2}$

$\qquad \tan \theta = \dfrac{y}{x} = -1 \qquad \cot \theta = \dfrac{x}{y} = -1$

5. $(x, y) = (7, 24)$

$r = \sqrt{49 + 576} = 25$

$\sin \theta = \dfrac{y}{r} = \dfrac{24}{25}$ $\csc \theta = \dfrac{r}{y} = \dfrac{25}{24}$

$\cos \theta = \dfrac{x}{r} = \dfrac{7}{25}$ $\sec \theta = \dfrac{r}{x} = \dfrac{25}{7}$

$\tan \theta = \dfrac{y}{x} = \dfrac{24}{7}$ $\cot \theta = \dfrac{x}{y} = \dfrac{7}{24}$

7. $(x, y) = (5, -12)$

$r = \sqrt{5^2 + (-12)^2} = \sqrt{25 + 144} = \sqrt{169} = 13$

$\sin \theta = \dfrac{y}{r} = -\dfrac{12}{13}$ $\csc \theta = \dfrac{r}{y} = -\dfrac{13}{12}$

$\cos \theta = \dfrac{x}{r} = \dfrac{5}{13}$ $\sec \theta = \dfrac{r}{x} = \dfrac{13}{5}$

$\tan \theta = \dfrac{y}{x} = -\dfrac{12}{5}$ $\cot \theta = \dfrac{x}{y} = -\dfrac{5}{12}$

9. $(x, y) = (-4, 10)$

$r = \sqrt{16 + 100} = 2\sqrt{29}$

$\sin \theta = \dfrac{y}{r} = \dfrac{5\sqrt{29}}{29}$ $\csc \theta = \dfrac{r}{y} = \dfrac{\sqrt{29}}{5}$

$\cos \theta = \dfrac{x}{r} = -\dfrac{2\sqrt{29}}{29}$ $\sec \theta = \dfrac{r}{x} = -\dfrac{\sqrt{29}}{2}$

$\tan \theta = \dfrac{y}{x} = -\dfrac{5}{2}$ $\cot \theta = \dfrac{x}{y} = -\dfrac{2}{5}$

11. $(x, y) = (-10, 8)$

$r = \sqrt{(-10)^2 + 8^2} = \sqrt{164} = 2\sqrt{41}$

$\sin \theta = \dfrac{y}{r} = \dfrac{8}{2\sqrt{41}} = \dfrac{4\sqrt{41}}{41}$ $\csc \theta = \dfrac{r}{y} = \dfrac{\sqrt{41}}{4}$

$\cos \theta = \dfrac{x}{r} = \dfrac{-10}{2\sqrt{41}} = -\dfrac{5\sqrt{41}}{41}$ $\sec \theta = \dfrac{r}{x} = -\dfrac{\sqrt{41}}{5}$

$\tan \theta = \dfrac{y}{x} = \dfrac{8}{-10} = -\dfrac{4}{5}$ $\cot \theta = \dfrac{x}{y} = -\dfrac{5}{4}$

13. $\sin \theta < 0 \implies \theta$ lies in Quadrant III or Quadrant IV.

$\cos \theta < 0 \implies \theta$ lies in Quadrant II or Quadrant III.

$\sin \theta < 0$ *and* $\cos \theta < 0 \implies \theta$ lies in Quadrant III.

15. $\cot \theta > 0 \implies \theta$ lies in Quadrant I or Quadrant III.

$\cot \theta > 0 \implies \theta$ lies in Quadrant I or Quadrant IV.

$\cot \theta > 0$ and $\cot \theta > 0 \implies \theta$ lies in Quadrant I.

17. $\sin \theta = \dfrac{y}{r} = \dfrac{3}{5} \implies x^2 = 25 - 9 = 16$

θ in Quadrant II $\implies x = -4$

$\sin \theta = \dfrac{y}{r} = \dfrac{3}{5}$ $\csc \theta = \dfrac{r}{y} = \dfrac{5}{3}$

$\cos \theta = \dfrac{x}{r} = -\dfrac{4}{5}$ $\sec \theta = \dfrac{r}{x} = -\dfrac{5}{4}$

$\tan \theta = \dfrac{y}{x} = -\dfrac{3}{4}$ $\cot \theta = \dfrac{x}{y} = -\dfrac{4}{3}$

19. $\sin \theta < 0 \implies y < 0$

$\tan \theta = \dfrac{y}{x} = \dfrac{-15}{8} \implies r = 17$

$\sin \theta = \dfrac{y}{r} = -\dfrac{15}{17}$ $\csc \theta = \dfrac{r}{y} = -\dfrac{17}{15}$

$\cos \theta = \dfrac{x}{r} = \dfrac{8}{17}$ $\sec \theta = \dfrac{r}{x} = \dfrac{17}{8}$

$\cot \theta = \dfrac{x}{y} = -\dfrac{8}{15}$

21. $\sec \theta = \dfrac{r}{x} = \dfrac{2}{-1} \implies y^2 = 4 - 1 = 3$

$\sin \theta \geq 0 \implies y = \sqrt{3}$

$\sin \theta = \dfrac{y}{r} = \dfrac{\sqrt{3}}{2}$ $\csc \theta = \dfrac{r}{y} = \dfrac{2\sqrt{3}}{3}$

$\cos \theta = \dfrac{x}{r} = -\dfrac{1}{2}$ $\sec \theta = \dfrac{r}{x} = -2$

$\tan \theta = \dfrac{y}{x} = -\sqrt{3}$ $\cot \theta = \dfrac{x}{y} = -\dfrac{\sqrt{3}}{3}$

23. $\cot \theta$ is undefined $\implies \theta = n\pi$.

$\dfrac{\pi}{2} \leq \theta \leq \dfrac{3\pi}{2} \implies \theta = \pi, y = 0, x = -r$

$\sin \theta = \dfrac{y}{r} = 0$ $\csc \theta = \dfrac{r}{y}$ is undefined.

$\cos \theta = \dfrac{x}{r} = \dfrac{-r}{r} = -1$ $\sec \theta = \dfrac{r}{x} = -1$

$\tan \theta = \dfrac{y}{x} = \dfrac{0}{x} = 0$ $\cot \theta$ is undefined.

25. To find a point on the terminal side of θ, use any point on the line $y = -x$ that lies in Quadrant II. $(-1, 1)$ is one such point.

$x = -1, y = 1, r = \sqrt{2}$

$\sin \theta = \dfrac{1}{\sqrt{2}} = \dfrac{\sqrt{2}}{2}$ $\csc \theta = \sqrt{2}$

$\cos \theta = -\dfrac{1}{\sqrt{2}} = -\dfrac{\sqrt{2}}{2}$ $\sec \theta = -\sqrt{2}$

$\tan \theta = -1$ $\cot \theta = -1$

27. To find a point on the terminal side of θ, use any point on the line $y = 2x$ that lies in Quadrant III. $(-1, -2)$ is one such point.

$x = -1, y = -2, r = \sqrt{5}$

$\sin \theta = -\dfrac{2}{\sqrt{5}} = -\dfrac{2\sqrt{5}}{5}$ $\csc \theta = \dfrac{\sqrt{5}}{-2} = -\dfrac{\sqrt{5}}{2}$

$\cos \theta = -\dfrac{1}{\sqrt{5}} = -\dfrac{\sqrt{5}}{5}$ $\sec \theta = \dfrac{\sqrt{5}}{-1} = -\sqrt{5}$

$\tan \theta = \dfrac{-2}{-1} = 2$ $\cot \theta = \dfrac{-1}{-2} = \dfrac{1}{2}$

29. $(x, y) = (-1, 0)$

$\sec \pi = \dfrac{r}{x}$

$= \dfrac{1}{-1} = -1$

31. $(x, y) = (0, -1)$

$\cot\left(\dfrac{3\pi}{2}\right) = \dfrac{x}{y}$

$= \dfrac{0}{-1} = 0$

33. $(x, y) = (1, 0)$

$\sec 0 = \dfrac{r}{x} = \dfrac{1}{1} = 1$

35. $(x, y) = (-1, 0)$

$\cot \pi = \dfrac{x}{y} = -\dfrac{1}{0}$

undefined

37. $\theta = 120°$

$\theta' = 180° - 120° = 60°$

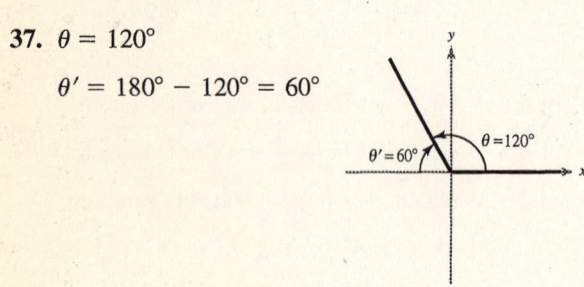

39. $\theta = -135°$ coterminal with $225°$.

$\theta' = 225° - 180° = 45°$

41. $\theta = \dfrac{5\pi}{3}, \theta' = 2\pi - \dfrac{5\pi}{3} = \dfrac{\pi}{3}$

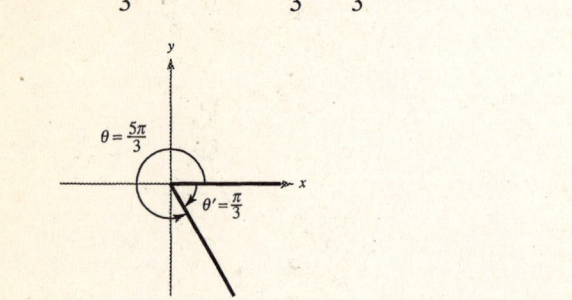

43. $\theta = -\dfrac{5\pi}{6}$ coterminal with $\dfrac{7\pi}{6}$.

$\theta' = \dfrac{7\pi}{6} - \pi = \dfrac{\pi}{6}$

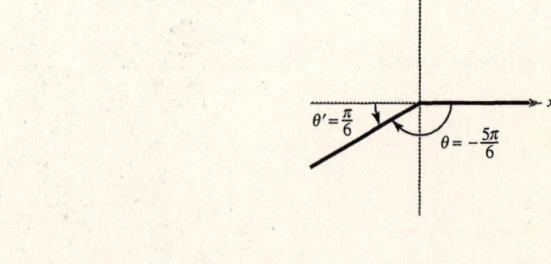

45. $\theta = 208°$

$\theta' = 208° - 180° = 28°$

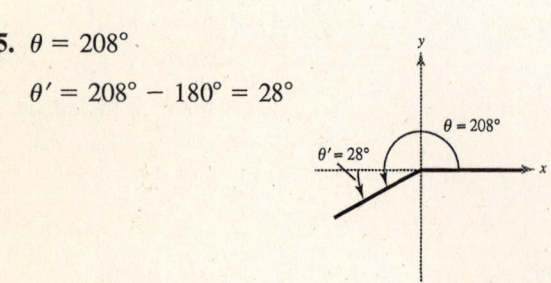

47. $\theta = -292°$

$\theta' = 360° - 292° = 68°$

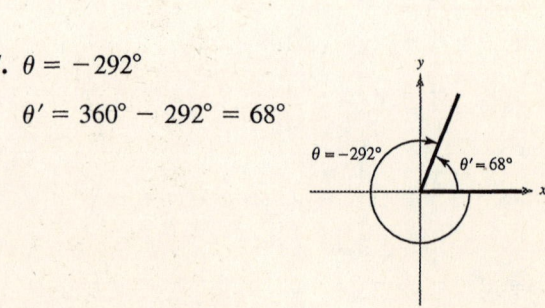

49. $\theta = \dfrac{11\pi}{5}$ coterminal with $\dfrac{\pi}{5}$.

$\theta' = \dfrac{\pi}{5}$

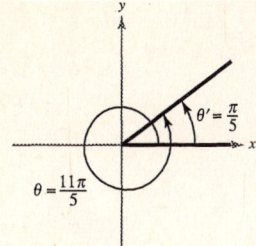

51. $\theta = -3.68$ coterminal with $2\pi - 3.68 \approx 2.6032$

$\theta' = \pi - (2\pi - 3.68) \approx 0.5384$

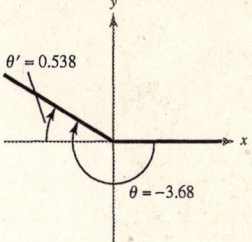

53. $\theta' = 45°$, Quadrant III

$\sin 225° = -\sin 45° = -\dfrac{\sqrt{2}}{2}$

$\cos 225° = -\cos 45° = -\dfrac{\sqrt{2}}{2}$

$\tan 225° = \tan 45° = 1$

55. $\theta = -750°$ coterminal with $330°$, Quadrant IV

$\theta' = 360° - 330° = 30°$

$\sin(-750°) = -\sin 30° = -\dfrac{1}{2}$

$\cos(-750°) = \cos 30° = \dfrac{\sqrt{3}}{2}$

$\tan(-750°) = -\tan 30° = -\dfrac{\sqrt{3}}{3}$

57. $\theta = -240°$ coterminal with $120°$, Quadrant II

$\theta' = 180° - 120° = 60°$

$\sin(-240°) = \sin 60° = \dfrac{\sqrt{3}}{2}$

$\cos(-240°) = -\cos 60° = -\dfrac{1}{2}$

$\tan(-240°) = -\tan 60° = -\sqrt{3}$

59. $\theta = \dfrac{5\pi}{3}$, Quadrant IV

$\theta' = 2\pi - \dfrac{5\pi}{3} = \dfrac{\pi}{3}$

$\sin\left(\dfrac{5\pi}{3}\right) = -\sin\left(\dfrac{\pi}{3}\right) = -\dfrac{\sqrt{3}}{2}$

$\cos\left(\dfrac{5\pi}{3}\right) = \cos\left(\dfrac{\pi}{3}\right) = \dfrac{1}{2}$

$\tan\left(\dfrac{5\pi}{3}\right) = -\tan\left(\dfrac{\pi}{3}\right) = -\sqrt{3}$

61. $\theta' = \dfrac{\pi}{6}$, Quadrant IV

$\sin\left(-\dfrac{\pi}{6}\right) = -\sin\dfrac{\pi}{6} = -\dfrac{1}{2}$

$\cos\left(-\dfrac{\pi}{6}\right) = \cos\dfrac{\pi}{6} = \dfrac{\sqrt{3}}{2}$

$\tan\left(-\dfrac{\pi}{6}\right) = -\tan\dfrac{\pi}{6} = -\dfrac{\sqrt{3}}{3}$

63. $\theta' = \dfrac{\pi}{4}$, Quadrant II

$\sin\dfrac{11\pi}{4} = \sin\dfrac{\pi}{4} = \dfrac{\sqrt{2}}{2}$

$\cos\dfrac{11\pi}{4} = -\cos\dfrac{\pi}{4} = -\dfrac{\sqrt{2}}{2}$

$\tan\dfrac{11\pi}{4} = -\tan\dfrac{\pi}{4} = -1$

65. $\theta = -\dfrac{17\pi}{6}$ coterminal with $\dfrac{7\pi}{6}$.

$\theta' = \dfrac{7\pi}{6} - \pi = \dfrac{\pi}{6}$

Quadrant III

$\sin\left(-\dfrac{17\pi}{6}\right) = -\sin\left(\dfrac{\pi}{6}\right) = -\dfrac{1}{2}$

$\cos\left(-\dfrac{17\pi}{6}\right) = -\cos\left(\dfrac{\pi}{6}\right) = -\dfrac{\sqrt{3}}{2}$

$\tan\left(-\dfrac{17\pi}{6}\right) = \tan\left(\dfrac{\pi}{6}\right) = \dfrac{\sqrt{3}}{3}$

67. $\qquad \sin\theta = -\dfrac{3}{5}$

$\sin^2\theta + \cos^2\theta = 1$

$\cos^2\theta = 1 - \sin^2\theta$

$\cos^2\theta = 1 - \left(-\dfrac{3}{5}\right)^2$

$\cos^2\theta = 1 - \dfrac{9}{25}$

$\cos^2\theta = \dfrac{16}{25}$

$\cos\theta > 0$ in Quadrant IV.

$\cos\theta = \dfrac{4}{5}$

69. $\tan\theta = \dfrac{3}{2}$

$\sec^2\theta = 1 + \tan^2\theta$

$\sec^2\theta = 1 + \left(\dfrac{3}{2}\right)^2$

$\sec^2\theta = 1 + \dfrac{9}{4}$

$\sec^2\theta = \dfrac{13}{4}$

$\sec\theta < 0$ in Quadrant III.

$\sec\theta = -\dfrac{\sqrt{13}}{2}$

71. $\cos\theta = \dfrac{5}{8}$

$\cos\theta = \dfrac{1}{\sec\theta} \Rightarrow \sec\theta = \dfrac{1}{\cos\theta}$

$\sec\theta = \dfrac{1}{5/8} = \dfrac{8}{5}$

73. $\sin(\theta) = \dfrac{1}{3}$

$\sin^2\theta + \cos^2\theta = 1$

$\cos^2\theta = 1 - \left(\dfrac{1}{3}\right)^2 = \dfrac{8}{9}$. Quadrant II

$\cos\theta = -\dfrac{2\sqrt{2}}{3}$

75. $\sin 10° \approx 0.1736$

77. $\tan 245° \approx 2.1445$

79. $\cos(-110°) \approx -0.3420$

81. $\sec(-280°) = \dfrac{1}{\cos(-280°)}$

≈ 5.7588

83. $\sin 0.65 \approx 0.6052$

85. $\cos(-1.81) \approx -0.2369$

87. $\tan\left(\dfrac{2\pi}{9}\right) \approx 0.8391$

89. $\csc\left(-\dfrac{8\pi}{9}\right) = \dfrac{1}{\sin\left(-\dfrac{8\pi}{9}\right)} \approx -2.9238$

91. (a) $\sin \theta = \dfrac{1}{2} \implies$ reference angle is $30°$ or $\dfrac{\pi}{6}$ and θ is in Quadrant I or Quadrant II.

Values in degrees: $30°, 150°$

Values in radian: $\dfrac{\pi}{6}, \dfrac{5\pi}{6}$

(b) $\sin \theta = -\dfrac{1}{2} \implies$ reference angle is $30°$ or $\dfrac{\pi}{6}$ and θ is in Quadrant III or Quadrant IV.

Values in degrees: $210°, 330°$

Values in radians: $\dfrac{7\pi}{6}, \dfrac{11\pi}{6}$

93. (a) $\csc \theta = \dfrac{2\sqrt{3}}{3} \implies$ reference angle is $60°$ or $\dfrac{\pi}{3}$ and θ is in Quadrant I or Quadrant II.

Values in degrees: $60°, 120°$

Values in radians: $\dfrac{\pi}{3}, \dfrac{2\pi}{3}$

(b) $\cot \theta = -1 \implies$ reference angle is $45°$ or $\dfrac{\pi}{4}$ and θ is in Quadrant II or Quadrant IV.

Values in degrees: $135°, 315°$

Values in radians: $\dfrac{3\pi}{4}, \dfrac{7\pi}{4}$

95. (a) $\sec \theta = -\dfrac{2\sqrt{3}}{3} \implies$ reference angle is $\dfrac{\pi}{6}$ or $30°$, and θ is in Quadrant II or Quadrant III.

Value in degrees: $150°, 210°$

Value in radians: $\dfrac{5\pi}{6}, \dfrac{7\pi}{6}$

(b) $\cos \theta = -\dfrac{1}{2} \implies$ reference angle is $\dfrac{\pi}{3}$ or $60°$, and θ is in Quadrant II or Quadrant III.

Values in degrees: $120°, 240°$

Values in radians: $\dfrac{2\pi}{3}, \dfrac{4\pi}{3}$

97. $T = 49.5 + 20.5 \cos\left(\dfrac{\pi t}{6} - \dfrac{7\pi}{6}\right)$

(a) January: $t = 1 \implies T = 49.5 + 20.5 \cos\left(\dfrac{\pi(1)}{6} - \dfrac{7\pi}{6}\right) = 29°$

(b) July: $t = 7 \implies T = 70°$

(c) December: $t = 12 \implies T \approx 31.75°$

99. $\sin \theta = \dfrac{6}{d} \implies d = \dfrac{6}{\sin \theta}$

(a) $\theta = 30°$

$d = \dfrac{6}{\sin 30°} = \dfrac{6}{(1/2)} = 12$ miles

(b) $\theta = 90°$

$d = \dfrac{6}{\sin 90°} = \dfrac{6}{1} = 6$ miles

(c) $\theta = 120°$

$d = \dfrac{6}{\sin 120°} \approx 6.9$ miles

101. True. The reference angle for $\theta = 151°$ is $\theta' = 180° - 151° = 29°$, and sine is positive in Quadrants I and II.

103. (a)

θ	0°	20°	40°	60°	80°
$\sin \theta$	0	0.3420	0.6428	0.8660	0.9848
$\sin(180° - \theta)$	0	0.3420	0.6428	0.8660	0.9848

(b) It appears that $\sin \theta = \sin(180° - \theta)$.

105. $3x - 7 = 14$

$3x = 21$

$x = 7$

107. $x^2 - 2x - 5 = 0$

$x = \dfrac{2 \pm \sqrt{4 + 20}}{2} = 1 \pm \sqrt{6}$ $(3.449, -1.449)$

109. $\dfrac{3}{x - 1} = \dfrac{x + 2}{9}$

$27 = (x - 1)(x + 2)$

$x^2 + x - 29 = 0$

$x = \dfrac{-1 \pm \sqrt{1 + 4(29)}}{2} = \dfrac{-1 \pm \sqrt{117}}{2}$

$x \approx -5.9083, \quad 4.9083$

111. $4^{3-x} = 726$

$3 - x = \log_4 726$

$x = 3 - \log_4 726 = 3 - \dfrac{\ln 726}{\ln 4} \approx -1.752$

113. $\ln x = -6$

$x = e^{-6} \approx 0.002479 \approx 0.002$

Section 4.5   Graphs of Sine and Cosine Functions

■   You should be able to graph $y = a \sin(bx - c)$ and $y = a \cos(bx - c)$.

■   Amplitude: $|a|$

■   Period: $\dfrac{2\pi}{|b|}$

■   Shift: Solve $bx - c = 0$ and $bx - c = 2\pi$.

■   Key increments: $\dfrac{1}{4}$ (period)

Solutions to Odd-Numbered Exercises

1. $y = 3 \sin 2x$

Period: $\dfrac{2\pi}{2} = \pi$

Amplitude: $|3| = 3$

```
Xmin = -2π
Xmax = 2π
Xscl = π/2
Ymin = -4
Ymax = 4
Yscl = 1
```

3. $y = \dfrac{5}{2} \cos \dfrac{x}{2}$

Period: $\dfrac{2\pi}{1/2} = 4\pi$

Amplitude: $\left|\dfrac{5}{2}\right| = \dfrac{5}{2}$

```
Xmin = -4π
Xmax = 4π
Xscl = π
Ymin = -3
Ymax = 3
Yscl = 1
```

5. $y = \dfrac{2}{3} \sin \pi x$

Period: $\dfrac{2\pi}{\pi} = 2$

Amplitude: $\left|\dfrac{2}{3}\right| = \dfrac{2}{3}$

> Xmin $= -\pi$
> Xmax $= \pi$
> Xscl $= \pi/2$
> Ymin $= -1$
> Ymax $= 1$
> Yscl $= .5$

7. $y = -2 \sin x$

Period: $\dfrac{2\pi}{1} = 2\pi$

Amplitude: $|-2| = 2$

9. $y = 3 \sin 10x$

Period: $\dfrac{2\pi}{10} = \dfrac{\pi}{5}$

Amplitude: $|3| = 3$

11. $y = \dfrac{1}{4} \cos \dfrac{2x}{3}$

Period: $\dfrac{2\pi}{2/3} = 3\pi$

Amplitude: $\left|\dfrac{1}{4}\right| = \dfrac{1}{4}$

13. $y = \dfrac{1}{3} \sin 4\pi x$

Period: $\dfrac{2\pi}{4\pi} = \dfrac{1}{2}$

Amplitude: $\left|\dfrac{1}{3}\right| = \dfrac{1}{3}$

15. $f(x) = \sin x$

$g(x) = \sin(x - \pi)$

The graph of g is a horizontal shift to the right π units of the graph of f (a phase shift).

17. $f(x) = \cos 2x$

$g(x) = -\cos 2x$

The graph of g is a reflection in the x-axis of the graph of f.

19. $f(x) = \cos x$

$g(x) = -5 \cos x$

The graph of g has five times the amplitude of f, and reflected in the x-axis.

21. $f(x) = \sin 2x$

$g(x) = 5 + \sin 2x$

The graph of g is a vertical shift upward of five units of the graph of f.

23. The graph of g has twice the amplitude as the graph of f. The period is the same.

25. The graph of g is a horizontal shift π units to the right of the graph of f.

27. $f(x) = \sin x$

Period: 2π

Amplitude: 1

$g(x) = -4 \sin x$

Period: 2π

Amplitude: $|-4| = 4$

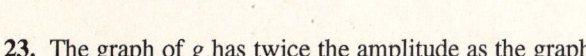

29. $f(x) = \cos x$

Period: 2π

Amplitude: 1

$g(x) = 4 + \cos x$
is a vertical shift of the graph of $f(x)$ four units upward.

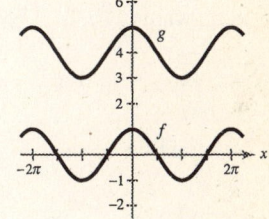

31. $f(x) = -\dfrac{1}{2} \sin \dfrac{x}{2}$

Period: 4π

Amplitude: $\dfrac{1}{2}$

$g(x) = 3 - \dfrac{1}{2} \sin \dfrac{x}{2}$ is

the graph of $f(x)$ shifted vertically three units upward.

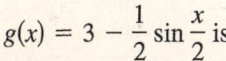

33. $f(x) = 2 \cos x$

Period: 2π

Amplitude: 2

$g(x) = 2 \cos(x + \pi)$ is the graph of $f(x)$ shifted π units to the left.

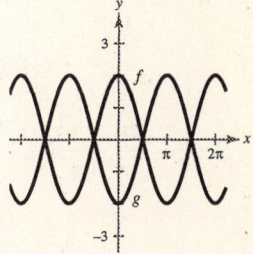

35. $f(x) = \sin x$, $g(x) = \cos\left(x - \dfrac{\pi}{2}\right)$

$\sin x = \cos\left(x - \dfrac{\pi}{2}\right)$.

Period: 2π
Amplitude: 1

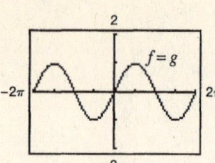

37. $f(x) = \cos x$

$g(x) = -\sin\left(x - \dfrac{\pi}{2}\right) = \sin\left(\dfrac{\pi}{2} - x\right) = \cos x$

Thus, $f(x) = g(x)$.

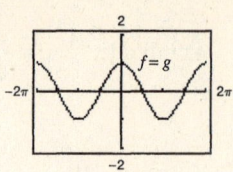

39. $y = 3 \sin x$

Period: 2π
Amplitude: 3
Key points: $(0, 0), \left(\dfrac{\pi}{2}, 3\right), (\pi, 0), \left(\dfrac{3\pi}{2}, -3\right), (2\pi, 0)$

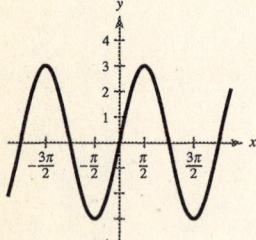

41. $y = \cos \dfrac{x}{2}$

Period: 4π
Amplitude: 1
Key points: $(0, 1), (\pi, 0), (2\pi, -1), (3\pi, 0), (4\pi, 1)$

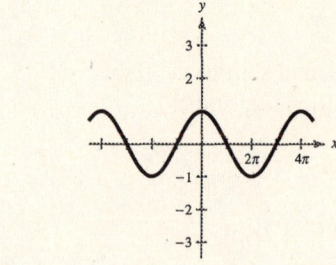

43. $y = -2 \sin \dfrac{2\pi x}{3}$

Period: $\dfrac{2\pi}{2\pi/3} = 3$
Amplitude: 2
Key points: $(0, 0), \left(\dfrac{3}{4}, -2\right), \left(\dfrac{3}{2}, 0\right), \left(\dfrac{9}{4}, 2\right), (3, 0)$

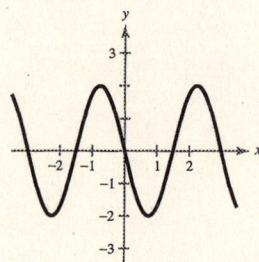

45. $y = \sin\left(x - \dfrac{\pi}{4}\right)$; $a = 1$, $b = 1$, $c = \dfrac{\pi}{4}$

Period: 2π
Amplitude: 1

Shift: Set $x - \dfrac{\pi}{4} = 0$ and $x - \dfrac{\pi}{4} = 2\pi$

$\qquad x = \dfrac{\pi}{4} \qquad\qquad x = \dfrac{9\pi}{4}$

Key points: $\left(\dfrac{\pi}{4}, 0\right), \left(\dfrac{3\pi}{4}, 1\right), \left(\dfrac{5\pi}{4}, 0\right), \left(\dfrac{7\pi}{4}, -1\right), \left(\dfrac{9\pi}{4}, 0\right)$

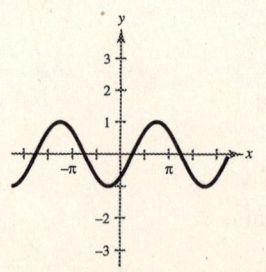

47. $y = -8 \cos(x + \pi)$

Period: 2π

Amplitude: 8

Key points: $(-\pi, -8), \left(-\frac{\pi}{2}, 0\right), (0, 8), \left(\frac{\pi}{2}, 0\right), (\pi, -8)$

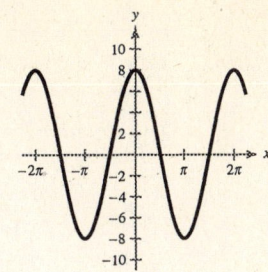

49. $y = 1 + \frac{1}{2} \cos 4\pi t$

Period: $\frac{1}{2}$

Amplitude: $\frac{1}{2}$

Vertical shift one unit upward

Key points: $\left(0, \frac{3}{2}\right), \left(\frac{1}{8}, 1\right), \left(\frac{1}{4}, \frac{1}{2}\right), \left(\frac{3}{8}, 1\right), \left(\frac{1}{2}, \frac{3}{2}\right)$

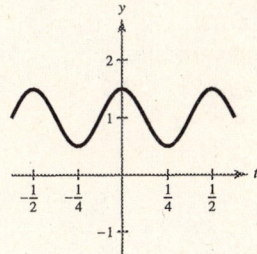

51. $y = 2 - 2 \sin \dfrac{2\pi x}{3}$

Vertical shift two units upward of the graph in Exercise #43

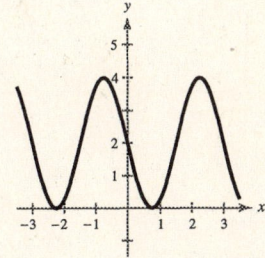

53. $y = \frac{2}{3} \cos\left(\frac{x}{2} - \frac{\pi}{4}\right);\ a = \frac{2}{3},\ b = \frac{1}{2},\ c = \frac{\pi}{4}$

Period: 4π

Amplitude: $\dfrac{2}{3}$

Shift: Set $\dfrac{x}{2} - \dfrac{\pi}{4} = 0$ and $\dfrac{x}{2} - \dfrac{\pi}{4} = 2\pi$

$x = \dfrac{\pi}{2} \qquad\qquad x = \dfrac{9\pi}{2}$

Key points: $\left(\frac{\pi}{2}, \frac{2}{3}\right), \left(\frac{3\pi}{2}, 0\right), \left(\frac{5\pi}{2}, \frac{-2}{3}\right), \left(\frac{7\pi}{2}, 0\right), \left(\frac{9\pi}{2}, \frac{2}{3}\right)$

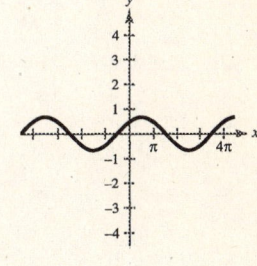

55. $y = -2 \sin(4x + \pi)$

Amplitude: 2

Period: $\dfrac{\pi}{2}$

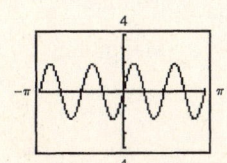

57. $y = \cos\left(2\pi x - \dfrac{\pi}{2}\right) + 1$

Amplitude: 1

Period: 1

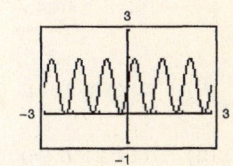

59. $y = 5 \sin(\pi - 2x) + 10$

Amplitude: 5

Period: π

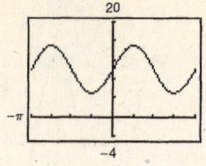

61. $y = \frac{1}{100} \sin 120\pi t$

Amplitude: $\frac{1}{100}$

Period: $\frac{1}{60}$

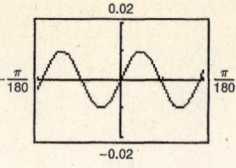

63. $f(x) = a \cos x + d$

Amplitude: $\frac{1}{2}[8 - 0] = 4$

Since $f(x)$ is the graph of $g(x) = 4 \cos x$ reflected about the x-axis and shifted vertically four units upward, we have $a = -4$ and $d = 4$. Thus, $f(x) = -4 \cos x + 4 = 4 - 4 \cos x$.

65. $f(x) = a \cos x + d$

Amplitude: $\frac{1}{2}[7 - (-5)] = 6$

Graph of f is the graph of $g(x) = 6 \cos x$ reflected about the x-axis and shifted vertically one unit upward. Thus $f(x) = -6 \cos x + 1$.

67. $f(x) = a \sin(bx - c)$

Amplitude: $|a| = 3$

Since the graph is reflected about the x-axis, we have $a = -3$.

Period: $\dfrac{2\pi}{b} = \pi \implies b = 2$

Phase shift: $c = 0$

Thus, $f(x) = -3 \sin 2x$.

69. $f(x) = a \sin(bx - c)$

Amplitude: $a = 1$

Period: $2\pi \implies b = 1$

Phase shift: $bx - c = 0$ when $x = \dfrac{\pi}{4}$.

$$(1)\left(\frac{\pi}{4}\right) - c = 0 \implies c = \frac{\pi}{4}$$

Thus, $f(x) = \sin\left(x - \dfrac{\pi}{4}\right)$.

71. $y_1 = \sin x$

$y_2 = -\dfrac{1}{2}$

In the interval $[-2\pi, 2\pi]$, $\sin x = -\frac{1}{2}$ when

$x = -\dfrac{5\pi}{6}, -\dfrac{\pi}{6}, \dfrac{7\pi}{6}, \dfrac{11\pi}{6}$.

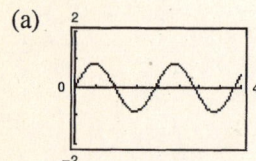

73. $v = 0.85 \sin \dfrac{\pi t}{3}$

(a)

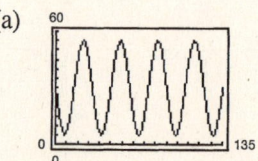

(b) Time for one cycle = one period = $\dfrac{2\pi}{\pi/3} = 6$ sec

(c) Cycles per min = $\dfrac{60}{6} = 10$ cycles per min

(d) The period would change.

75. $h = 25 \sin \dfrac{\pi}{15}(t - 75) + 30$

(a)

(b) Minimum: $30 - 25 = 5$ feet

Maximum: $30 + 25 = 55$ feet

77. $C = 30.3 + 21.6 \sin\left(\dfrac{2\pi t}{365} + 10.9\right)$

(a) Period $= \dfrac{2\pi}{b} = \dfrac{2\pi}{(2\pi/365)} = 365$ days

This is to be expected: 365 days = 1 year

(b) The constant 30.3 gallons is the average daily fuel consumption.

(c)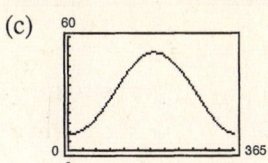

Consumption exceeds 40 gallons/day when $124 \le x \le 252$. (Graph C together with $y = 40$.) (Beginning of May through part of September)

79. (a)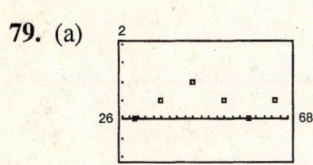

(b) Using a graphing utility, you obtain

$y = 0.508 \sin(0.216x - 1.616) + 0.539$.

(c)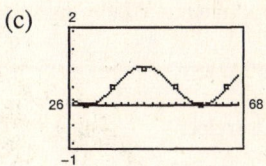

The model is a good fit.

(d) Period: $\dfrac{2\pi}{b} = \dfrac{2\pi}{0.216} \approx 29.09$ days

(e) For June 22, 2007, $x = 365 + 173 = 538$.

$y \approx 1.05$ or 105%. Hence, 100% illumination.

81. True. The period is $\dfrac{2\pi}{3/10} = \dfrac{20\pi}{3}$.

83. True

85. (a) $h(x) = \cos^2 x$ is even.

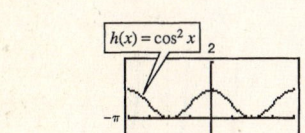

(b) $h(x) = \sin^2 x$ is even.

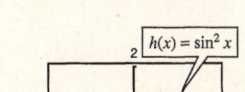

(c) $h(x) = \sin x \cos x$ is odd.

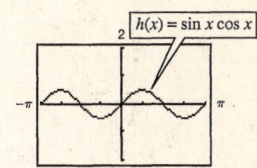

87.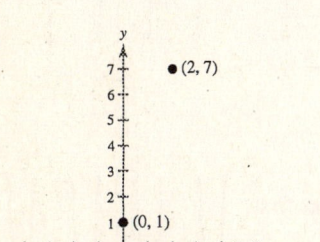

Slope $= \dfrac{7 - 1}{2 - 0} = 3$

89. $8.5 = 8.5\left(\dfrac{180°}{\pi}\right) \approx 487.014°$

Section 4.6   Graphs of Other Trigonometric Functions

- ■ You should be able to graph:

  $y = a \tan(bx - c)$ $y = a \cot(bx - c)$

  $y = a \sec(bx - c)$ $y = a \csc(bx - c)$

- ■ When graphing $y = a \sec(bx - c)$ or $y = a \csc(bx - c)$ you should know to first graph $y = a \cos(bx - c)$ or $y = a \sin(bx - c)$ since

  (a) The intercepts of sine and cosine are vertical asymptotes of cosecant and secant.

  (b) The maximums of sine and cosine are local minimums of cosecant and secant.

  (c) The minimums of sine and cosine are local maximums of cosecant and secant.

- ■ You should be able to graph using a damping factor.

Solutions to Odd-Numbered Exercises

1. $y = \sec \dfrac{x}{2}$

Period: $\dfrac{2\pi}{1/2} = 4\pi$

Matches graph (c).

3. $y = \tan 2x$

Period: $\dfrac{\pi}{2}$

Matches graph (e).

5. $y = \cot \dfrac{\pi x}{2}$

Period: $\dfrac{\pi}{\pi/2} = 2$

Matches graph (a).

7. $y = \dfrac{1}{2} \tan x$

Period: π

Two consecutive asymptotes: $x = -\dfrac{\pi}{2}$ and $x = \dfrac{\pi}{2}$

x	$-\dfrac{\pi}{4}$	0	$\dfrac{\pi}{4}$
y	$-\dfrac{1}{2}$	0	$\dfrac{1}{2}$

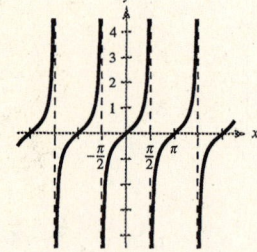

9. $y = -2 \tan 2x$

Period: $\dfrac{\pi}{2}$

Two consecutive asymptotes:

$2x = -\dfrac{\pi}{2} \implies x = -\dfrac{\pi}{4}$

$2x = \dfrac{\pi}{2} \implies x = \dfrac{\pi}{4}$

x	$-\dfrac{\pi}{8}$	0	$\dfrac{\pi}{8}$
y	2	0	-2

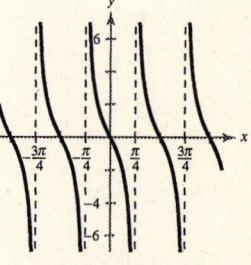

11. $y = -\dfrac{1}{2} \sec x$

Graph $y = -\dfrac{1}{2} \cos x$ first.

Period: 2π

One cycle: 0 to 2π

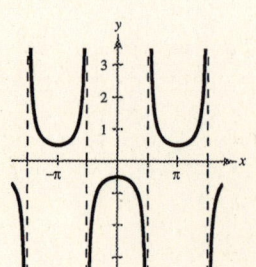

13. $y = -\sec \pi x$

Graph $y = -\cos \pi x$ first.

Period: $\dfrac{2\pi}{\pi} = 2$

One cycle: 0 to 2

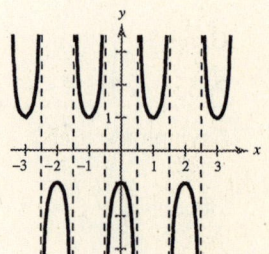

15. $y = \sec \pi x - 3$

Reflect the graph in Exercise #13 about the x-axis and then shift it vertically down three units.

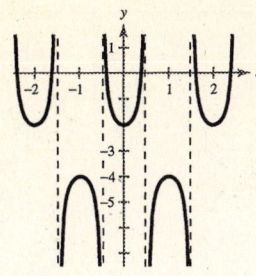

17. $y = 3 \csc \dfrac{x}{2}$

Graph $y = 3 \sin \dfrac{x}{2}$ first.

Period: $\dfrac{2\pi}{1/2} = 4\pi$

One cycle: 0 to 4π

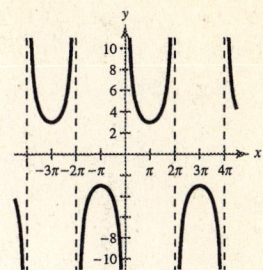

19. $y = \dfrac{1}{2} \cot \dfrac{x}{2}$

Period: $\dfrac{\pi}{1/2} = 2\pi$

Two consecutive asymptotes: $\dfrac{x}{2} = 0 \implies x = 0$

$\dfrac{x}{2} = \pi \implies x = 2\pi$

x	$\dfrac{\pi}{2}$	π	$\dfrac{3\pi}{2}$
y	$\dfrac{1}{2}$	0	$-\dfrac{1}{2}$

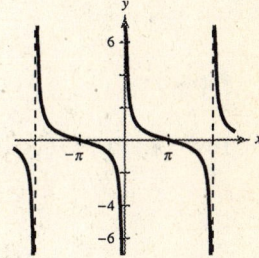

21. $y = 2 \tan \dfrac{\pi x}{4}$

Period: $\dfrac{\pi}{\pi/4} = 4$

Two consecutive asymptotes: $\dfrac{\pi x}{4} = -\dfrac{\pi}{2} \implies x = -2$

$\dfrac{\pi x}{4} = \dfrac{\pi}{2} \implies x = 2$

x	-1	0	1
y	-2	0	2

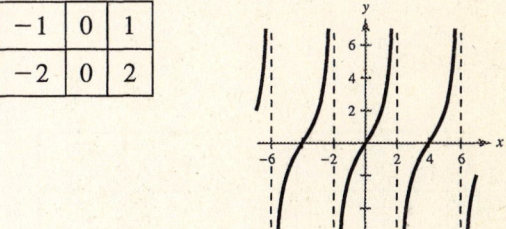

23. $y = \dfrac{1}{2} \sec 2x$

Graph $y = \dfrac{1}{2} \cos 2x$ first.

Period: $\dfrac{2\pi}{2} = \pi$

One cycle: 0 to π

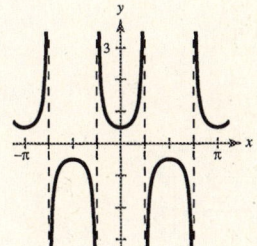

25. $y = \csc(\pi - x)$

Graph $y = \sin(\pi - x)$ first.

Period: 2π

Asymptotes: Set $\pi - x = 0$ and $\pi - x = 2\pi$

$\qquad\qquad\qquad x = \pi \qquad\qquad x = -\pi$

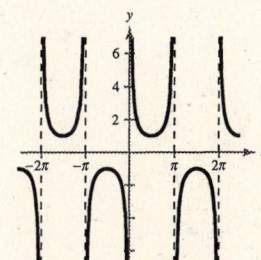

27. $y = 2 \cot\left(x - \dfrac{\pi}{2}\right)$

Period: π

Two consecutive asymptotes: $x - \dfrac{\pi}{2} = 0 \implies x = \dfrac{\pi}{2}$

$$x - \dfrac{\pi}{2} = \pi \implies x = \dfrac{3\pi}{2}$$

x	$\dfrac{3\pi}{4}$	π	$\dfrac{5\pi}{4}$
y	2	0	-2

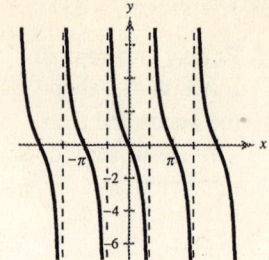

29. $y = \tan\left(x - \dfrac{\pi}{4}\right)$

Period: π

Two consecutive asymptotes: $x - \dfrac{\pi}{4} = -\dfrac{\pi}{2} \implies x = -\dfrac{\pi}{4}$

$$x - \dfrac{\pi}{4} = \dfrac{\pi}{2} \implies x = \dfrac{3\pi}{4}$$

x	0	$\dfrac{\pi}{4}$	$\dfrac{\pi}{2}$
y	-1	0	1

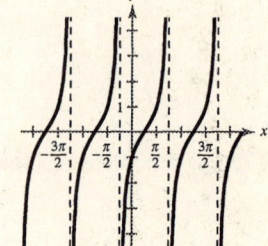

31. $y = 2 \csc 3x = \dfrac{2}{\sin(3x)}$

Period: $\dfrac{2\pi}{3}$

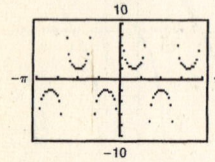 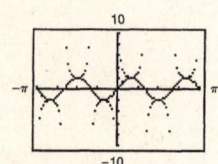

33. $y = -2 \sec 4x$

$$= \dfrac{-2}{\cos 4x}$$

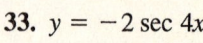

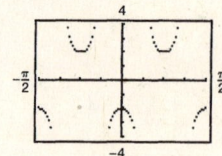 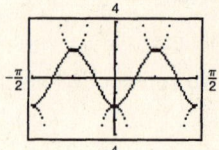

35. $y = \dfrac{1}{3} \sec\left(\dfrac{\pi x}{2} + \dfrac{\pi}{2}\right) = \dfrac{1}{3 \cos\left(\dfrac{\pi x}{2} + \dfrac{\pi}{2}\right)}$

Period: 4

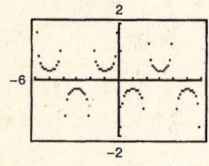 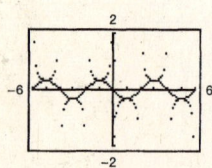

37. $\tan x = 1$

$$x = -\dfrac{7\pi}{4}, -\dfrac{3\pi}{4}, \dfrac{\pi}{4}, \dfrac{5\pi}{4}$$

39. $\sec x = -2$

$$x = \pm\dfrac{2\pi}{3}, \pm\dfrac{4\pi}{3}$$

41. The graph of $f(x) = \sec x$ has y-axis symmetry. Thus, the function is even.

43. The function

$$f(x) = \csc 2x = \dfrac{1}{\sin 2x}$$

has origin symmetry. Thus, the function is odd.

45. $y_1 = \sin x \csc x$ and $y_2 = 1$

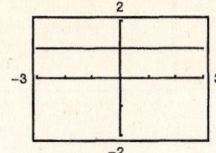

Not equivalent because y_1 is not defined at 0.

$$\sin x \csc x = \sin x\left(\frac{1}{\sin x}\right) = 1, \quad \sin x \neq 0$$

47. $y_1 = \dfrac{\cos x}{\sin x}$ and $y_2 = \cot x = \dfrac{1}{\tan x}$

Equivalent

$$\cot x = \frac{\cos x}{\sin x}$$

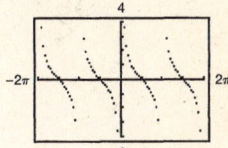

49. $f(x) = x \cos x$

As $x \to 0, f(x) \to 0$.

Odd function

$$f\left(\frac{3\pi}{2}\right) = 0$$

Matches graph (d).

51. $g(x) = |x| \sin x$

As $x \to 0, g(x) \to 0$.

Odd function

$$g(2\pi) = 0$$

Matches graph (b).

53. $f(x) = \sin x + \cos\left(x + \dfrac{\pi}{2}\right), g(x) = 0$

$f(x) = g(x)$

The graph is the line $y = 0$.

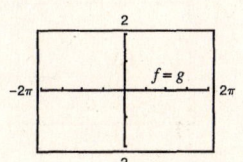

55. $f(x) = \sin^2 x, g(x) = \frac{1}{2}(1 - \cos 2x)$

$f(x) = g(x)$

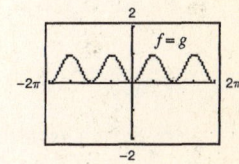

57. $f(x) = e^{-x} \cos x$

Damping factor: e^{-x}

As $x \to \infty, f(x) \to 0$.

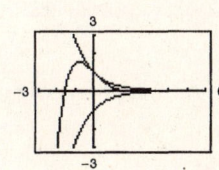

59. $g(x) = e^{-x^2/2} \sin x$

Damping factor: $y = e^{-x^2/2}$

$-e^{-x^2/2} \leq g(x) \leq e^{-x^2/2}$

As $x \to \pm\infty, g(x) \to 0$.

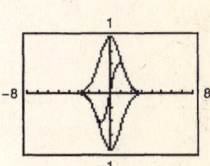

61. $y = \dfrac{6}{x} + \cos x$

As $x \to 0$ from the right, $y \to \infty$.

As $x \to 0$ from the left, $y \to -\infty$.

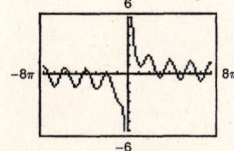

63.

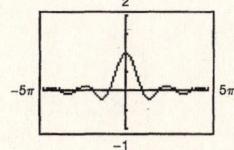

As $x \to 0, \dfrac{\sin x}{x} \to 1$.

65. $f(x) = \dfrac{\tan x}{x}$

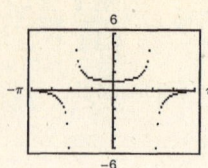

As $x \to 0, f(x) \to 1$.

67. $\tan x = \dfrac{5}{d}$

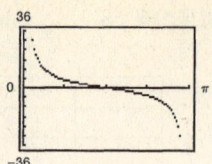

$d = \dfrac{5}{\tan x} = 5 \cot x$

69. As the predator population increases, the number of prey decrease. When the number of prey is small, the number of predators decreases.

71. (a)

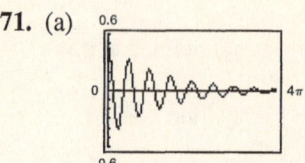

(b) The displacement function is approximately periodic, but damped. It approaches 0 as t increases.

73. (a) Yes. For each t there corresponds one and only one value of y.

(c) $y = 12(0.221)^t \cos(8.2t)$

To obtain this model, first fit an exponential model $y = ab^t$ to the data points $(0, 12)$, $(0.7622, 3.76)$, and $(1.5476, 1.16)$. This yields $y = 12(0.2210)^t$. Using

$$\dfrac{2\pi}{0.7622} \approx 8.2$$

for the cosine term, you obtain the model above.

(b) One way to determine the frequency is to note that the time between the first and second maximum points is $t = 0.7622 - 0 = 0.7622$. Thus, the frequency is approximately $(0.7622)^{-1} = 1.3$ oscillations per second.

(d) $\ln 0.221 \approx -1.51 \implies y = 12e^{-1.5t} \cos(8.2t)$

(e)

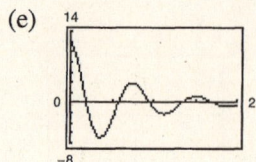

75. True. $-\dfrac{3\pi}{2} + \pi = -\dfrac{\pi}{2}$ and $x = -\dfrac{\pi}{2}$ is a vertical asymptote for the tangent function.

77. As x approaches $\dfrac{\pi}{2}$ from the left, $\tan x \to \infty$.

As x approaches $\dfrac{\pi}{2}$ from the right, $\tan x \to -\infty$.

79. $f(x) = 2 \sin x$, $g(x) = \dfrac{1}{2} \csc x$

(a)

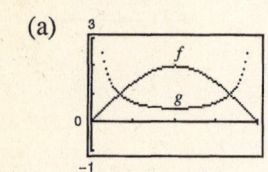

(b) $f(x) > g(x)$ for $\dfrac{\pi}{6} < x < \dfrac{5\pi}{6}$.

(c) As $x \to \pi$, $2 \sin x \to 0$ and $\dfrac{1}{2} \csc x \to \infty$, since $g(x)$ is the reciprocal of $f(x)$.

81. Distributive Property

83. Additive Identity Property

85. Not one-to-one

87. $y = \sqrt{3x - 14}$, $x \geq \dfrac{14}{3}$, $y \geq 0$

$x = \sqrt{3y - 14}$, $y \geq \dfrac{14}{3}$, $x \geq 0$

$x^2 = 3y - 14$

$y = \dfrac{1}{3}(x^2 + 14)$

$f^{-1}(x) = \dfrac{1}{3}(x^2 + 14)$, $x \geq 0$

Section 4.7   Inverse Trigonometric Functions

■ You should know the definitions, domains, and ranges of $y = \arcsin x$, $y = \arccos x$, and $y = \arctan x$.

Function	Domain	Range
$y = \arcsin x \implies x = \sin y$	$-1 \le x \le 1$	$-\dfrac{\pi}{2} \le y \le \dfrac{\pi}{2}$
$y = \arccos x \implies x = \cos y$	$-1 \le x \le 1$	$0 \le y \le \pi$
$y = \arctan x \implies x = \tan y$	$-\infty < x < \infty$	$-\dfrac{\pi}{2} < y < \dfrac{\pi}{2}$

■ You should know the inverse properties of the inverse trigonometric functions.

$$\sin(\arcsin x) = x, \ -1 \le x \le 1 \ \text{and} \ \arcsin(\sin y) = y, \ -\frac{\pi}{2} \le y \le \frac{\pi}{2}$$

$$\cos(\arccos x) = x, \ -1 \le x \le 1 \ \text{and} \ \arccos(\cos y) = y, \ 0 \le y \le \pi$$

$$\tan(\arctan x) = x \ \text{and} \ \arctan(\tan y) = y, \ -\frac{\pi}{2} < y < \frac{\pi}{2}$$

■ You should be able to use the triangle technique to convert trigonometric functions or inverse trigonometric functions into algebraic expressions.

Solutions to Odd-Numbered Exercises

1. (a) $\arcsin \dfrac{1}{2} = \dfrac{\pi}{6}$ (b) $\arcsin 0 = 0$

3. (a) $\arctan \dfrac{\sqrt{3}}{3} = \dfrac{\pi}{6}$ (b) $\arctan(-1) = -\dfrac{\pi}{4}$

5. (a) $y = \arctan\left(-\sqrt{3}\right) \implies \tan y = -\sqrt{3}$ for

$$-\frac{\pi}{2} < y < \frac{\pi}{2} \implies y = -\frac{\pi}{3}$$

(b) $y = \arctan \sqrt{3} \implies \tan y = \sqrt{3} \implies y = \dfrac{\pi}{3}$

7. (a) $y = \sin^{-1} \dfrac{\sqrt{3}}{2} \implies \sin y = \dfrac{\sqrt{3}}{2}$ for

$$-\frac{\pi}{2} \le y \le \frac{\pi}{2} \implies y = \frac{\pi}{3}$$

(b) $y = \tan^{-1}\left(\dfrac{-\sqrt{3}}{3}\right) \implies \tan y = \dfrac{-\sqrt{3}}{3}$

$$\implies y = -\frac{\pi}{6}$$

9. $y = \arccos x$

(a)

x	-1	-0.8	-0.6	-0.4	-0.2
y	3.1416	2.4981	2.2143	1.9823	1.7722

x	0	0.2	0.4	0.6	0.8	1.0
y	1.5708	1.3694	1.1593	0.9273	0.6435	0

(b)

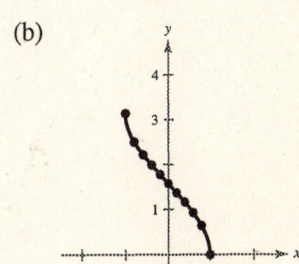

—CONTINUED—

9. —CONTINUED—

(c)

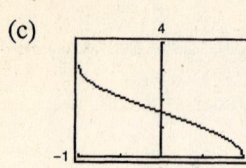

(d) Intercepts are $\left(0, \dfrac{\pi}{2}\right)$ and $(1, 0)$.

No symmetry

11. $y = \arctan x \leftrightarrow \tan y = x$

$$\left(-\sqrt{3}, -\dfrac{\pi}{3}\right), \left(-\dfrac{\sqrt{3}}{3}, -\dfrac{\pi}{6}\right), \left(1, \dfrac{\pi}{4}\right)$$

13. $\cos^{-1}(0.75) \approx 0.72$

15. $\arcsin(-0.75) \approx -0.85$

17. $\arctan(-6) \approx -1.41$

19. $\sin^{-1}(0.19) \approx 0.19$

21. $\arccos(-0.51) \approx 2.11$

23. $\tan^{-1} 1.32 \approx 0.92$

25. $f(x) = \tan x$ and $g(x) = \arctan x$

Graph: $y_1 = \tan x$

$y_2 = \tan^{-1} x$

$y_3 = x$

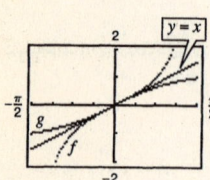

27. $\tan \theta = \dfrac{x}{8}$

$\theta = \arctan \dfrac{x}{8}$

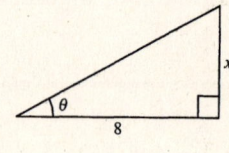

29. $\sin \theta = \dfrac{x + 2}{5}$

$\theta = \arcsin\left(\dfrac{x + 2}{5}\right)$

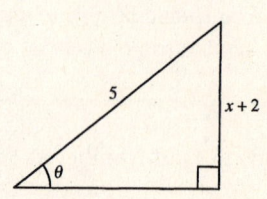

31. $\sin(\arcsin 0.7) = 0.7$

33. $\cos[\arccos(-0.3)] = -0.3$

35. $\arcsin(\sin 3\pi) = \arcsin(0) = 0$

Note: 3π is not in the range of the arcsine function.

37. $\arctan\left(\tan \dfrac{11\pi}{6}\right) = \arctan\left(-\dfrac{\sqrt{3}}{3}\right) = -\dfrac{\pi}{6}$

39. Let $y = \arctan \dfrac{4}{3}$. Then

$\tan y = \dfrac{4}{3}, 0 < y < \dfrac{\pi}{2}$, and

$\sin y = \dfrac{4}{5}$.

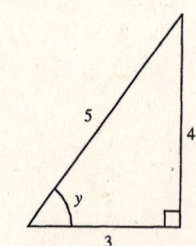

41. Let $y = \arcsin \dfrac{24}{25}$. Then

$\sin y = \dfrac{24}{25}$, and $\cos y = \dfrac{7}{25}$.

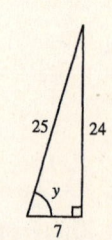

43. Let $y = \arctan\left(-\dfrac{3}{5}\right)$. Then,

$\tan y = -\dfrac{3}{5}$, $-\dfrac{\pi}{2} < y < 0$ and $\sec y = \dfrac{\sqrt{34}}{5}$.

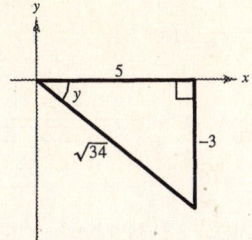

45. Let $y = \arccos\left(-\dfrac{2}{3}\right)$. Then,

$\cos y = -\dfrac{2}{3}$, $\dfrac{\pi}{2} < y < \pi$ and $\sin y = \dfrac{\sqrt{5}}{3}$.

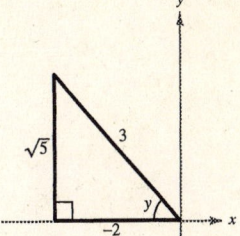

47. Let $y = \arctan x$. Then,

$\tan y = x$ and $\cot y = \dfrac{1}{x}$.

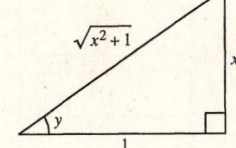

49. Let $y = \arccos(x + 2)$, $\cos y = x + 2$.

Opposite side: $\sqrt{1 - (x + 2)^2}$

$\sin y = \dfrac{\sqrt{1 - (x + 2)^2}}{1} = \sqrt{-x^2 - 4x - 3}$

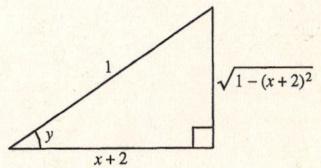

51. Let $y = \arccos\dfrac{x}{5}$. Then $\cos y = \dfrac{x}{5}$, and

$\tan y = \dfrac{\sqrt{25 - x^2}}{x}$.

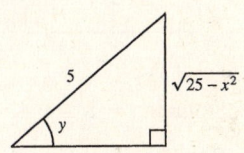

53. Let $y = \arctan\dfrac{x}{\sqrt{7}}$. Then $\tan y = \dfrac{x}{\sqrt{7}}$ and

$\csc y = \dfrac{\sqrt{7 + x^2}}{x}$.

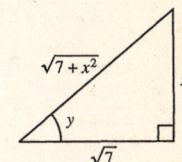

55. $f(x) = \sin(\arctan 2x)$, $g(x) = \dfrac{2x}{\sqrt{1 + 4x^2}}$

Let $y = \arctan 2x$. Then, $\tan y = 2x = \dfrac{2x}{1}$ and $\sin y = \dfrac{2x}{\sqrt{1 + 4x^2}}$.

$g(x) = \dfrac{2x}{\sqrt{1 + 4x^2}} = f(x)$

The graph has horizontal asymptotes at $y = \pm 1$.

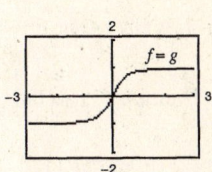

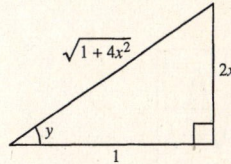

57. Let $y = \arctan \dfrac{14}{x}$. Then $\tan y = \dfrac{14}{x}$ and

$\sin y = \dfrac{14}{\sqrt{196 + x^2}}$. Thus, $y = \arcsin\left(\dfrac{14}{\sqrt{196 + x^2}}\right)$.

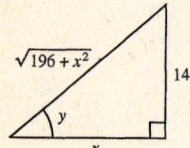

61. $y = 2 \arccos x$

Domain: $-1 \le x \le 1$

Range: $0 \le y \le 2\pi$

Vertical stretch of $f(x) = \arccos x$

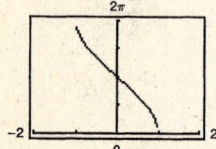

65. $f(x) = \arctan 2x$

Domain: all real numbers

Range: $-\dfrac{\pi}{2} < y < \dfrac{\pi}{2}$

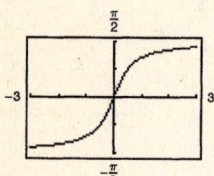

69. $f(t) = 3 \cos 2t + 3 \sin 2t$

$= \sqrt{3^2 + 3^2} \sin\left(2t + \arctan\dfrac{3}{3}\right)$

$= 3\sqrt{2} \sin(2t + \arctan 1)$

$= 3\sqrt{2} \sin\left(2t + \dfrac{\pi}{4}\right)$

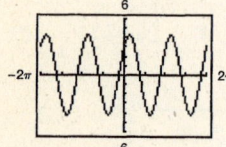

The graphs are the same.

59. Let $y = \arccos \dfrac{3}{\sqrt{x^2 - 2x + 10}}$. Then,

$\cos y = \dfrac{3}{\sqrt{x^2 - 2x + 10}} = \dfrac{3}{\sqrt{(x-1)^2 + 9}}$

and $\sin y = \dfrac{|x-1|}{\sqrt{(x-1)^2 + 9}}$. Thus,

$y = \arcsin \dfrac{|x-1|}{\sqrt{(x-1)^2 + 9}} = \arcsin \dfrac{|x-1|}{\sqrt{x^2 - 2x + 10}}$.

63. The graph of $f(x) = \arcsin(x - 2)$ is a horizontal translation of the graph of $y = \arcsin x$ by two units.

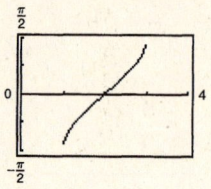

67. $h(v) = \tan(\arccos v) = \dfrac{\sqrt{1 - v^2}}{v}$

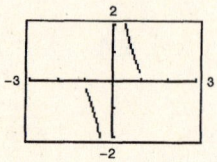 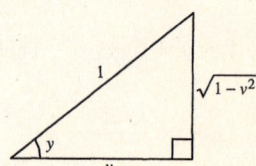

Domain: $-1 \le v \le 1, v \ne 0$

Range: all real numbers

71. (a) $\sin \theta = \dfrac{10}{s} \implies \theta = \arcsin\left(\dfrac{10}{s}\right)$

(b) $s = 52$: $\theta = \arcsin\left(\dfrac{10}{52}\right) \approx 0.1935, \; (\approx 11.1°)$

$s = 26$: $\theta = \arcsin\left(\dfrac{10}{26}\right) \approx 0.3948, \; (\approx 22.6°)$

73. (a) $\tan \theta = \dfrac{s}{750}$

   $\theta = \arctan\left(\dfrac{s}{750}\right)$

(b) When $s = 400$, $\theta = \arctan\left(\dfrac{400}{750}\right) \approx 0.49$ radian, $(\approx 28°)$.

   When $s = 1600$, $\theta \approx 1.13$ radians, $(\approx 65°)$.

75. (a) $\tan \theta = \dfrac{6}{x}$

   $\theta = \arctan\left(\dfrac{6}{x}\right)$

(b) When $x = 10$, $\theta = \arctan\left(\dfrac{6}{10}\right) \approx 0.54$ radian, $(\approx 31°)$.

   When $x = 3$, $\theta = \arctan\left(\dfrac{6}{3}\right) \approx 1.11$ radians, $(\approx 63°)$.

77. False. $\arcsin \dfrac{1}{2} = \dfrac{\pi}{6}$

79. $y = \operatorname{arccot} x$ if and only if $\cot y = x$.

Domain: $-\infty < x < \infty$

Range: $0 < x < \pi$

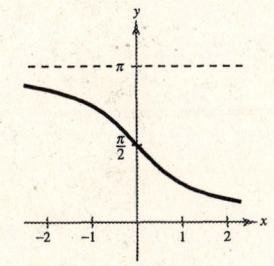

81. $y = \operatorname{arccsc} x$ if and only if $\csc y = x$.

Domain: $(-\infty, -1] \cup [1, \infty)$

Range: $\left[-\dfrac{\pi}{2}, 0\right) \cup \left(0, \dfrac{\pi}{2}\right]$

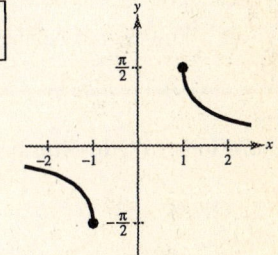

83. Let $y = \arcsin(-x)$. Then,

   $\sin y = -x$

   $-\sin y = x$

   $\sin(-y) = x$

   $-y = \arcsin x$

   $y = -\arcsin x.$

Therefore, $\arcsin(-x) = -\arcsin x.$

85. $\arcsin x + \arccos x = \dfrac{\pi}{2}$

Let $\alpha = \arcsin x \implies \sin \alpha = x$

Let $\beta = \arccos x \implies \cos \beta = x$

Hence, $\sin \alpha = \cos \beta \implies \alpha$ and β are

complementary angles $\implies \alpha + \beta = \dfrac{\pi}{2}$

$\implies \arcsin x + \arccos x = \dfrac{\pi}{2}$

87. $\dfrac{4}{4\sqrt{2}} = \dfrac{1}{\sqrt{2}} = \dfrac{\sqrt{2}}{2}$

89. $\dfrac{2\sqrt{3}}{6} = \dfrac{\sqrt{3}}{3}$

91. $\sin \theta = \dfrac{5}{6}$

Adjacent side: $\sqrt{6^2 - 5^2} = \sqrt{11}$

$\cos \theta = \dfrac{\sqrt{11}}{6} \qquad \tan \theta = \dfrac{5}{\sqrt{11}} = \dfrac{5\sqrt{11}}{11}$

$\csc \theta = \dfrac{6}{5} \qquad \sec \theta = \dfrac{6}{\sqrt{11}} = \dfrac{6\sqrt{11}}{11}$

$\cot \theta = \dfrac{\sqrt{11}}{5}$

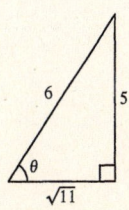

93. $\sin \theta = \dfrac{3}{4}$

Adjacent side: $\sqrt{16 - 9} = \sqrt{7}$

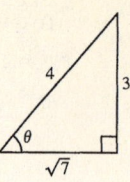

$\cos \theta = \dfrac{\sqrt{7}}{4}$ $\qquad$ $\tan \theta = \dfrac{3}{\sqrt{7}} = \dfrac{3\sqrt{7}}{7}$

$\csc \theta = \dfrac{4}{3}$ $\qquad$ $\sec \theta = \dfrac{4}{\sqrt{7}} = \dfrac{4\sqrt{7}}{7}$

$\cot \theta = \dfrac{\sqrt{7}}{3}$

Section 4.8   Applications and Models

- You should be able to solve right triangles.
- You should be able to solve right triangle applications.
- You should be able to solve applications of simple harmonic motion: $d = a \sin wt$ or $d = a \cos wt$.

Solutions to Odd-Numbered Exercises

1. Given: $A = 20°$, $b = 10$

$B = 90° - 20° = 70°$

$\tan A = \dfrac{a}{b} \implies a = b \tan A = 10 \tan 20° \approx 3.64$

$\cos A = \dfrac{b}{c} \implies c = \dfrac{b}{\cos A} = \dfrac{10}{\cos 20°} \approx 10.64$

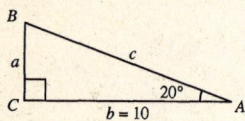

3. Given: $B = 71°$, $b = 24$

$\tan B = \dfrac{b}{a} \implies a = \dfrac{b}{\tan B} = \dfrac{24}{\tan 71°} \approx 8.26$

$\sin B = \dfrac{b}{c} \implies c = \dfrac{b}{\sin B} = \dfrac{24}{\sin 71°} \approx 25.38$

$A = 90° - 71° = 19°$

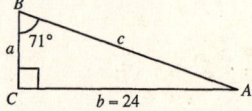

5. Given: $a = 6$, $b = 16$

$c^2 = a^2 + b^2 \implies c = \sqrt{292} \approx 17.09$

$\tan A = \dfrac{a}{b} = \dfrac{6}{16} \implies A = \arctan\left(\dfrac{3}{8}\right) \approx 20.56°$

$B = 90° - 20.56° = 69.44°$

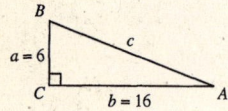

7. Given: $b = 16$, $c = 48$

$a = \sqrt{c^2 - b^2} = \sqrt{2048} \approx 45.25$

$\cos A = \dfrac{16}{48} = \dfrac{1}{3} \implies A = \arccos\left(\dfrac{1}{3}\right) \approx 70.53°$

$B \approx 90° - 70.53° = 19.47°$

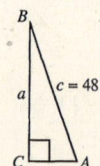

9. $A = 12° 15'$, $c = 430.5$

$B = 90° - 12° 15' = 77° 45'$

$\sin 12° 15' = \dfrac{a}{430.5}$

$a = 430.5 \sin 12° 15' \approx 91.34$

$\cos 12° 15' = \dfrac{b}{430.5}$

$b = 430.5 \cos 12° 15' \approx 420.70$

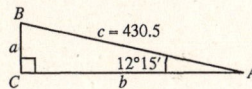

11. $\tan \theta = \dfrac{h}{(1/2)b}$

$h = \dfrac{1}{2}b \tan \theta$

$h = \dfrac{1}{2}(4) \tan 52° \approx 2.56$ in.

13. $\tan \theta = \dfrac{h}{(1/2)b} \implies h = \dfrac{1}{2}b \tan \theta = \dfrac{1}{2}(14.2) \tan(41.6°) \approx 6.30$ feet

15. (a)

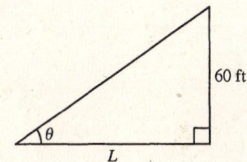

(b) $\tan \theta = \dfrac{60}{L}$

$L = \dfrac{60}{\tan \theta}$

$= 60 \cot \theta$

17. $\sin 80° = \dfrac{h}{20}$

$h = 20 \sin 80°$

≈ 19.70 feet

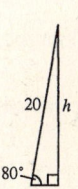

(c)

θ	10°	20°	30°	40°	50°
L	340	165	104	72	50

(d) No, the shadow lengths do not increase in equal increments. The cotangent function is not linear.

19. (a)

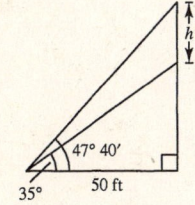

(c) $h \approx 19.9$ feet

(b) Let the height of the church $= x$ and the height of the church and steeple $= y$. Then:

$\tan 35° = \dfrac{x}{50}$ and $\tan 47° 40' = \dfrac{y}{50}$

$x = 50 \tan 35°$ and $y = 50 \tan 47° 40'$

$h = y - x = 50(\tan 47° 40' - \tan 35°)$

21. $\sin 31.5° = \dfrac{x}{4000}$

$x = 4000 \sin 31.5°$

≈ 2089.99 feet

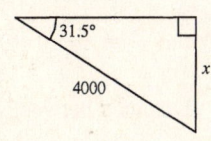

23. $\tan \theta = \dfrac{75}{95}$

$\theta = \arctan \dfrac{15}{19} \approx 38.29°$

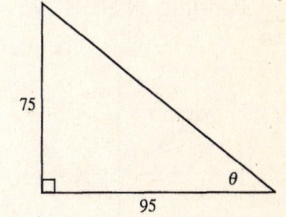

25. $\sin \alpha = \dfrac{4000}{16,500} \Rightarrow \alpha \approx 14.03°$

Hence, $\theta \approx 90° - \alpha \approx 75.97°$.

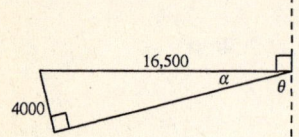

27. Since the airplane speed is

$$\left(275\, \frac{\text{ft}}{\text{sec}}\right)\left(60\, \frac{\text{sec}}{\text{min}}\right) = 16,500\, \frac{\text{ft}}{\text{min}},$$

after one minute its distance travelled is 16,500 feet.

$$\sin 18° = \frac{a}{16,500}$$

$$a = 16,500 \sin 18°$$

$$\approx 5099 \text{ ft}$$

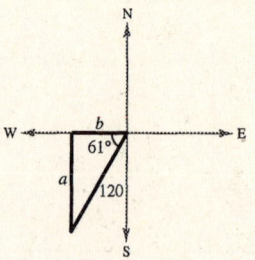

29. $\sin 9.5° = \dfrac{x}{4}$

$x = 4 \sin 9.5°$

≈ 0.66 mile

31. $90° - 29° = 61°$

$(20)(6) = 120$ nautical miles

$\sin 61° = \dfrac{a}{120} \Rightarrow a = 120 \sin 61° \approx 104.95$ nautical miles

$\cos 61° = \dfrac{b}{120} \Rightarrow b = 120 \cos 61° \approx 58.18$ nautical miles

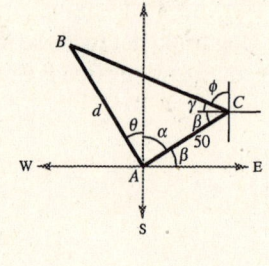

33. $\theta = 32°, \phi = 68°$ *Note: ABC form a right triangle.*

(a) $\alpha = 90° - 32° = 58°$

Bearing from A to C: N 58° E

(b) $\beta = \theta = 32°$

$\gamma = 90° - \phi = 22°$

$C = \beta + \gamma = 54°$

$\tan C = \dfrac{d}{50} \Rightarrow \tan 54° = \dfrac{d}{50} \Rightarrow d \approx 68.82$ m

35. $\tan \theta = \dfrac{45}{30} = \dfrac{3}{2} \Rightarrow \theta \approx 56.31°$

Bearing: N 56.31° W

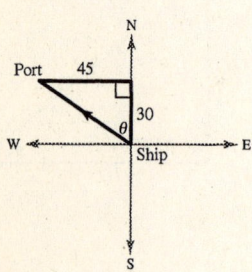

37. $\tan 6.5° = \dfrac{350}{d} \Rightarrow d \approx 3071.91$ ft

$\tan 4° = \dfrac{350}{D} \Rightarrow D \approx 5005.23$ ft

Distance between ships: $D - d \approx 1933.3$ ft

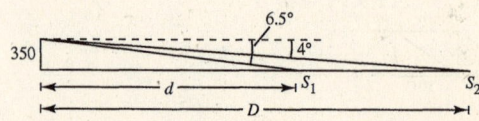

39. $\tan 57° = \dfrac{a}{x} \implies x = a \cot 57°$

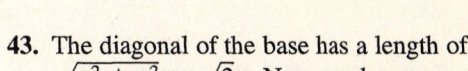

$\tan 16° = \dfrac{a}{x + (55/6)}$

$\tan 16° = \dfrac{a}{a \cot 57° + (55/6)}$

$\cot 16° = \dfrac{a \cot 57° + (55/6)}{a}$

$a \cot 16° - a \cot 57° = \dfrac{55}{6} \implies a \approx 3.23$ miles $\approx 17{,}054$ feet

41. $L_1:\ 3x - 2y = 5 \implies y = \dfrac{3}{2}x - \dfrac{5}{2} \implies m_1 = \dfrac{3}{2}$

$L_2:\ x + y = 1 \implies y = -x + 1 \implies m_2 = -1$

$\tan \alpha = \left| \dfrac{-1 - (3/2)}{1 + (-1)(3/2)} \right| = \left| \dfrac{-5/2}{-1/2} \right| = 5$

$\alpha = \arctan 5 \approx 78.7°$

43. The diagonal of the base has a length of $\sqrt{a^2 + a^2} = \sqrt{2}\,a$. Now, we have:

$\tan \theta = \dfrac{a}{\sqrt{2}\,a} = \dfrac{1}{\sqrt{2}}$

$\theta = \arctan \dfrac{1}{\sqrt{2}}$

$\theta \approx 35.3°$

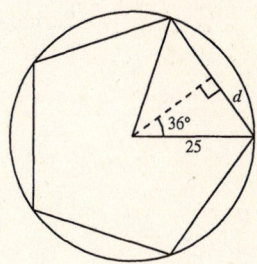

45. $\cos 30° = \dfrac{b}{r}$

$b = \cos 30° r$

$b = \dfrac{\sqrt{3}\,r}{2}$

$y = 2b = 2\left(\dfrac{\sqrt{3}\,r}{2} \right) = \sqrt{3}\,r$

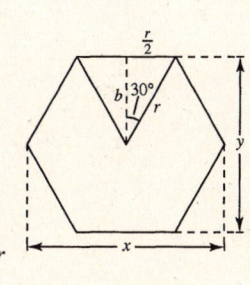

47. $\sin 36° = \dfrac{d}{25} \implies d \approx 14.695$

Length of side: $2d \approx 29.39$ inches

49. $\tan 35° = \dfrac{b}{10}$

$b = 10 \tan 35° \approx 7$

$\cos 35° = \dfrac{10}{a}$

$a = \dfrac{10}{\cos 35°} \approx 12.2$

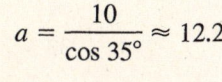
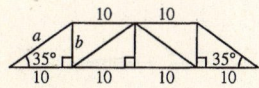

51. $d = 0$ when $t = 0$, $a = 8$, period $= 2$

Use $d = a \sin wt$ since $d = 0$ when $t = 0$.

$\dfrac{2\pi}{w} = 2 \implies w = \pi$

Thus, $d = 8 \sin \pi t$.

53. $d = 3$ when $t = 0$, $a = 3$, period $= 1.5$

Use $d = a \cos wt$ since $d = 3$ when $t = 0$.

$$\frac{2\pi}{w} = 1.5 \implies w = \frac{4}{3}\pi$$

Thus, $d = 3 \cos\left(\frac{4}{3}\pi t\right)$.

55. $d = 4 \cos 8\pi t$

(a) Maximum displacement = amplitude = 4

(b) Frequency $= \dfrac{\omega}{2\pi} = \dfrac{8\pi}{2\pi}$

$\qquad\qquad = 4$ cycles per unit of time

(c) $d = 4 \cos(8\pi(5)) = 4$

(d) $8\pi t = \dfrac{\pi}{2} \implies t = \dfrac{1}{16}$

57. $d = \dfrac{1}{16} \sin 140\pi t$

(a) Maximum displacement = amplitude $= \dfrac{1}{16}$

(b) Frequency $= \dfrac{\omega}{2\pi} = \dfrac{140\pi}{2\pi}$

$\qquad\qquad = 70$ cycles per unit of time

(c) $d = 0$

(d) $140\pi t = \pi \implies t = \dfrac{1}{140}$

59. $d = a \sin \omega t$

$$\text{Period} = \frac{2\pi}{\omega} = \frac{1}{\text{frequency}}$$

$$\frac{2\pi}{\omega} = \frac{1}{264}$$

$$\omega = 2\pi(264) = 528\pi$$

61. $y = \dfrac{1}{4} \cos 16t, \; t > 0$

(a)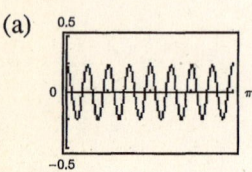

(b) Period: $\dfrac{2\pi}{16} = \dfrac{\pi}{8}$ seconds

(c) $\dfrac{1}{4} \cos 16t = 0$ when

$$16t = \frac{\pi}{2} \implies t = \frac{\pi}{32} \text{ seconds.}$$

63. (a), (b)

Base 1	Base 2	Altitude	Area
8	$8 + 16 \cos 10°$	$8 \sin 10°$	22.1
8	$8 + 16 \cos 20°$	$8 \sin 20°$	42.5
8	$8 + 16 \cos 30°$	$8 \sin 30°$	59.7
8	$8 + 16 \cos 40°$	$8 \sin 40°$	72.7
8	$8 + 16 \cos 50°$	$8 \sin 50°$	80.5
8	$8 + 16 \cos 60°$	$8 \sin 60°$	83.1
8	$8 + 16 \cos 70°$	$8 \sin 70°$	80.7

Maximum ≈ 83.1 square feet

(c) $A = \frac{1}{2}(b_1 + b_2)h$

$\quad = \frac{1}{2}[8 + 8 + 16 \cos \theta] 8 \sin \theta$

$\quad = 64(1 + \cos \theta)\sin \theta$

(d) Maximum area is approximately 83.1 square feet for $\theta = 60°$.

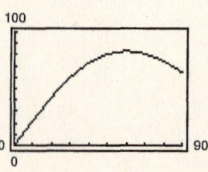

65. (a)

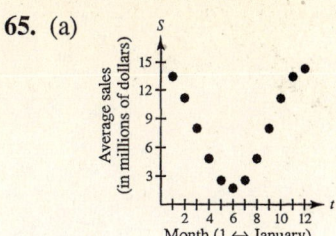

(b) $a = \frac{1}{2}(14.30 - 1.70) = 6.3$

$$\frac{2\pi}{b} = 12 \implies b = \frac{\pi}{6}$$

Shift: $d = 14.3 - 6.3 = 8$

$S = d + a \cos bt$

$$S = 8 + 6.3 \cos\left(\frac{\pi t}{6}\right)$$

The model is a good fit.

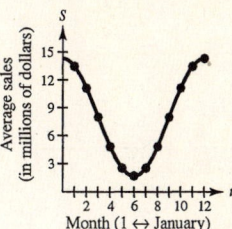

(c) Period: $\dfrac{2\pi}{\pi/6} = 12$

This corresponds to the 12 months in a year. Since the sales of outerwear is seasonal, this is reasonable.

(d) The amplitude represents the maximum displacement from the average sale of 8 million dollars. Sales are greatest in December (cold weather + holidays) and least in June.

67. False. The other acute angle is $90° - 48.1° = 41.9°$. Then

$$\tan(41.9°) = \frac{\text{opp}}{\text{adj}} = \frac{a}{22.56} \implies a = 22.56 \cdot \tan(41.9°).$$

69. $y - 2 = 4(x + 1)$

$4x - y + 6 = 0$

71.   Slope $= \dfrac{6 - 2}{-2 - 3} = -\dfrac{4}{5}$

$$y - 2 = -\frac{4}{5}(x - 3)$$

$$5y - 10 = -4x + 12$$

$$4x + 5y - 22 = 0$$

73. Domain: $(-\infty, \infty)$ **75.** Domain: $(-\infty, \infty)$ **77.** Domain: all $x \neq 0, 2$ **79.** Domain: $(-\infty, \infty)$

81. $e^{2x} = 54$

$2x = \ln 54$

$x = \frac{1}{2}\ln 54 \approx 1.994$

83. $\ln(x^2 + 1) = 3.2$

$x^2 + 1 = e^{3.2}$

$x = \pm\sqrt{e^{3.2} - 1} \approx \pm 4.851$

Review Exercises for Chapter 4

Solutions to Odd-Numbered Exercises

1. 40° or 0.7 radians

3. (a)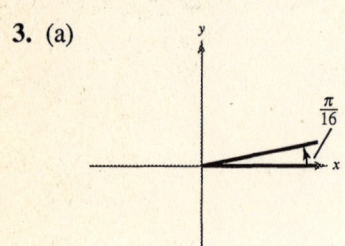

(b) Quadrant I (c) $\dfrac{\pi}{16} + 2\pi = \dfrac{33\pi}{16}$

$\dfrac{\pi}{16} - 2\pi = -\dfrac{31\pi}{16}$

5. (a)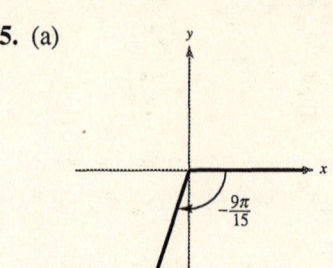

(b) Quadrant III (c) $-\dfrac{9\pi}{15} + 2\pi = \dfrac{21\pi}{15}$

$-\dfrac{9\pi}{15} - 2\pi = -\dfrac{39\pi}{15}$

7. Complement: $\dfrac{\pi}{2} - \dfrac{\pi}{8} = \dfrac{3\pi}{8}$

Supplement: $\pi - \dfrac{\pi}{8} = \dfrac{7\pi}{8}$

9. Complement: $\dfrac{\pi}{2} - \dfrac{3\pi}{10} = \dfrac{\pi}{5}$

Supplement: $\pi - \dfrac{3\pi}{10} = \dfrac{7\pi}{10}$

11. (a)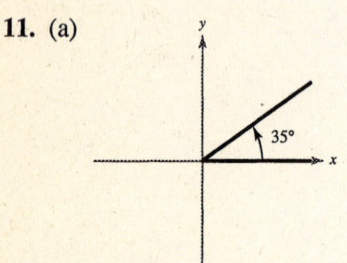

(b) Quadrant I (c) $35° + 360° = 395°$

$35° - 360° = -325°$

13. (a)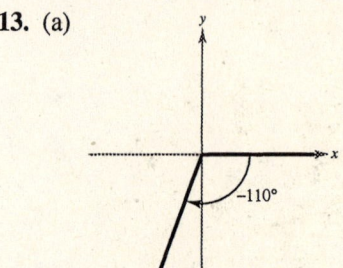

(b) Quadrant III (c) $-110° + 360° = 250°$

$-110° - 360° = -470°$

15. Complement: $90° - 8° = 82°$

Supplement: $180° - 8° = 172°$

17. Complement: not possible

Supplement: $180° - 171° = 9°$

19. $135° \, 16' \, 45'' = \left(135 + \frac{16}{60} + \frac{45}{3600}\right)^{\circ} \approx 135.28°$

21. $5° \, 22' \, 53'' = \left(5 + \frac{22}{60} + \frac{53}{3600}\right)^{\circ} \approx 5.38°$

23. $135.29° = 135° + (0.29)(60)' = 135° \, 17' \, 24''$

25. $-85.36° = -[85 + 0.36(60')] = -85° \, 21' \, 36''$

27. $480° = 480° \cdot \dfrac{\pi \, \text{rad}}{180°} = \dfrac{8\pi}{3} \, \text{rad} \approx 8.3776 \, \text{rad}$

29. $-33° = -33°\left(\dfrac{\pi \, \text{rad}}{180°}\right) \approx -0.5760 \, \text{rad}$

31. $\dfrac{5\pi}{7} = \dfrac{5\pi}{7}\left(\dfrac{180°}{\pi}\right) \approx 128.57°$

33. $-3.5 = -3.5\left(\dfrac{180°}{\pi}\right) \approx -200.54°$

35. $s = r\theta$

$25 = 12\theta$

$\theta = \dfrac{25}{12} \approx 2.083$

37. $s = r\theta$

$s = 20(138°)\dfrac{\pi}{180°}$

$s \approx 48.171 \text{ m}$

39. In one revolution, the arc length traveled is $s = 2\pi r = 2\pi(6) = 12\pi$ cm. The time required for one revolution is

$t = \dfrac{1}{500} \text{ minutes} = \dfrac{1}{500}(60) = \dfrac{3}{25} \text{ seconds.}$

Linear speed $= \dfrac{s}{t} = \dfrac{12\pi}{3/25} = 100\pi \text{ cm/sec}$

41. $t = \dfrac{7\pi}{4}$ corresponds to $\left(\dfrac{\sqrt{2}}{2}, -\dfrac{\sqrt{2}}{2} \right)$.

43. $\cos \dfrac{5\pi}{6} = -\dfrac{\sqrt{3}}{2}$

$\sin \dfrac{5\pi}{6} = \dfrac{1}{2}$

$(x, y) = \left(-\dfrac{\sqrt{3}}{2}, \dfrac{1}{2} \right)$

45. $\sin \dfrac{7\pi}{6} = -\dfrac{1}{2}$

$\cos \dfrac{7\pi}{6} = -\dfrac{\sqrt{3}}{2}$

$\tan \dfrac{7\pi}{6} = \dfrac{1}{\sqrt{3}} = \dfrac{\sqrt{3}}{3}$

$\cot \dfrac{7\pi}{6} = \sqrt{3}$

$\sec \dfrac{7\pi}{6} = -\dfrac{2}{\sqrt{3}} = -\dfrac{2\sqrt{3}}{3}$

$\csc \dfrac{7\pi}{6} = -2$

47. $\sin\left(-\dfrac{2\pi}{3} \right) = \sin\left(\dfrac{4\pi}{3} \right) = -\dfrac{\sqrt{3}}{2}$

$\cos\left(-\dfrac{2\pi}{3} \right) = -\dfrac{1}{2}$

$\tan\left(-\dfrac{2\pi}{3} \right) = \sqrt{3}$

$\cot\left(-\dfrac{2\pi}{3} \right) = \dfrac{1}{\sqrt{3}} = \dfrac{\sqrt{3}}{3}$

$\sec\left(-\dfrac{2\pi}{3} \right) = -2$

$\csc\left(-\dfrac{2\pi}{3} \right) = -\dfrac{2}{\sqrt{3}} = -\dfrac{2\sqrt{3}}{3}$

49. $\sin\left(\dfrac{11\pi}{4} \right) = \sin\left(\dfrac{3\pi}{4} \right) = \dfrac{\sqrt{2}}{2}$

51. $\sin\left(-\dfrac{17\pi}{6} \right) = \sin\left(\dfrac{7\pi}{6} \right) = -\dfrac{1}{2}$

53. $\cot 2.3 = \dfrac{1}{\tan 2.3} \approx -0.8935$

55. $\cos \dfrac{5\pi}{3} = \dfrac{1}{2}$

57. The hypotenuse is $\sqrt{12^2 + 10^2} = \sqrt{244} = 2\sqrt{61}$.

$$\sin \theta = \frac{\text{opp}}{\text{hyp}} = \frac{10}{2\sqrt{61}} = \frac{5}{\sqrt{61}} = \frac{5\sqrt{61}}{61} \qquad \cos \theta = \frac{\text{adj}}{\text{hyp}} = \frac{12}{2\sqrt{61}} = \frac{6}{\sqrt{61}} = \frac{6\sqrt{61}}{61}$$

$$\csc \theta = \frac{1}{\sin \theta} = \frac{\sqrt{61}}{5} \qquad\qquad\qquad \sec \theta = \frac{1}{\cos \theta} = \frac{\sqrt{61}}{6}$$

$$\tan \theta = \frac{\text{opp}}{\text{adj}} = \frac{10}{12} = \frac{5}{6} \qquad\qquad\quad \cot \theta = \frac{1}{\tan \theta} = \frac{6}{5}$$

59. The opposite side is $\sqrt{9^2 - 4^2} = \sqrt{81 - 16} = \sqrt{65}$

$$\sin \theta = \frac{\text{opp}}{\text{hyp}} = \frac{\sqrt{65}}{9} \qquad\qquad\qquad \cos \theta = \frac{\text{adj}}{\text{hyp}} = \frac{4}{9}$$

$$\csc \theta = \frac{1}{\sin \theta} = \frac{9}{\sqrt{65}} = \frac{9\sqrt{65}}{65} \qquad \sec \theta = \frac{1}{\cos \theta} = \frac{9}{4}$$

$$\tan \theta = \frac{\text{opp}}{\text{adj}} = \frac{\sqrt{65}}{4} \qquad\qquad\quad \cot \theta = \frac{1}{\tan \theta} = \frac{4}{\sqrt{65}} = \frac{4\sqrt{65}}{65}$$

61. $\csc \theta \tan \theta = \dfrac{1}{\sin \theta} \cdot \dfrac{\sin \theta}{\cos \theta} = \dfrac{1}{\cos \theta} = \sec \theta$

63. (a) $\cos 84° \approx 0.1045$

(b) $\sin 6° \approx 0.1045$

65. (a) $\cos \dfrac{\pi}{4} \approx 0.7071$

(b) $\sec \dfrac{\pi}{4} \approx 1.4142$

67. $\tan 62° = \dfrac{x}{125}$

$x = 125 \tan 62° \approx 235.09$ feet

69. $x = 12, y = 16, r = \sqrt{144 + 256} = \sqrt{400} = 20$

$$\sin \theta = \frac{y}{r} = \frac{4}{5} \qquad \csc \theta = \frac{r}{y} = \frac{5}{4}$$

$$\cos \theta = \frac{x}{r} = \frac{3}{5} \qquad \sec \theta = \frac{r}{x} = \frac{5}{3}$$

$$\tan \theta = \frac{y}{x} = \frac{4}{3} \qquad \cot \theta = \frac{x}{y} = \frac{3}{4}$$

71. $x = -7, y = 2, r = \sqrt{49 + 4} = \sqrt{53}$

$$\sin \theta = \frac{y}{r} = \frac{2}{\sqrt{53}} = \frac{2\sqrt{53}}{53} \qquad \csc \theta = \frac{\sqrt{53}}{2}$$

$$\cos \theta = \frac{x}{r} = -\frac{7}{\sqrt{53}} = -\frac{7\sqrt{53}}{53} \qquad \sec \theta = -\frac{\sqrt{53}}{7}$$

$$\tan \theta = \frac{y}{x} = -\frac{2}{7} \qquad\qquad\quad \cot \theta = -\frac{7}{2}$$

73. $x = 2, y = 5, r = \sqrt{2^2 + 5^2} = \sqrt{29}$

$$\sin \theta = \frac{y}{r} = \frac{5}{\sqrt{29}} = \frac{5\sqrt{29}}{29} \qquad \csc \theta = \frac{\sqrt{29}}{5}$$

$$\cos \theta = \frac{x}{r} = \frac{2}{\sqrt{29}} = \frac{2\sqrt{29}}{29} \qquad \sec \theta = \frac{\sqrt{29}}{2}$$

$$\tan \theta = \frac{y}{x} = \frac{5}{2} \qquad\qquad\quad \cot \theta = \frac{2}{5}$$

75. $\sec \theta = \dfrac{6}{5}$, $\tan \theta < 0 \implies \theta$ is in Quadrant IV.

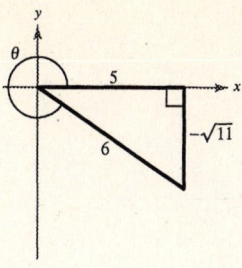

$r = 6, x = 5, y = -\sqrt{36 - 25} = -\sqrt{11}$

$$\sin \theta = \frac{y}{r} = -\frac{\sqrt{11}}{6} \qquad\qquad \csc \theta = -\frac{6\sqrt{11}}{11}$$

$$\cos \theta = \frac{x}{r} = \frac{5}{6} \qquad\qquad \sec \theta = \frac{6}{5}$$

$$\tan \theta = \frac{y}{x} = -\frac{\sqrt{11}}{5} \qquad\qquad \cot \theta = -\frac{5\sqrt{11}}{11}$$

77. $\sin \theta = \dfrac{3}{8}$, $\cos \theta < 0 \implies \theta$ is in Quadrant II.

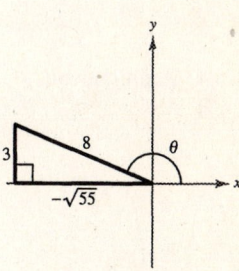

$y = 3, r = 8, x = -\sqrt{55}$

$$\sin \theta = \frac{y}{r} = \frac{3}{8} \qquad\qquad \csc \theta = \frac{8}{3}$$

$$\cos \theta = \frac{x}{r} = -\frac{\sqrt{55}}{8} \qquad\qquad \sec \theta = -\frac{8}{\sqrt{55}} = -\frac{8\sqrt{55}}{55}$$

$$\tan \theta = \frac{y}{x} = -\frac{3}{\sqrt{55}} = -\frac{3\sqrt{55}}{55} \qquad\qquad \cot \theta = -\frac{\sqrt{55}}{3}$$

79. Reference angle: $264° - 180° = 84°$

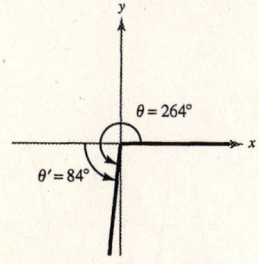

81. Coterminal angle: $2\pi - \dfrac{6\pi}{5} = \dfrac{4\pi}{5}$

Reference angle: $\pi - \dfrac{4\pi}{5} = \dfrac{\pi}{5}$

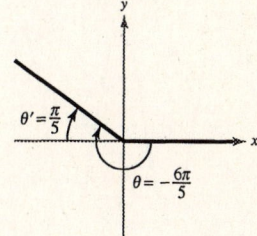

83. $240°$ is in Quadrant III with reference angle $60°$.

$$\sin 240° = -\sin 60° = -\frac{\sqrt{3}}{2}$$

$$\cos 240° = -\cos 60° = -\frac{1}{2}$$

$$\tan 240° = \frac{-\sqrt{3}/2}{-1/2} = \sqrt{3}$$

85. $-210°$ is coterminal with $150°$ in Quadrant II with reference angle $30°$.

$$\sin(-210°) = \sin(30°) = \frac{1}{2}$$

$$\cos(-210°) = -\cos(30°) = -\frac{\sqrt{3}}{2}$$

$$\tan(-210°) = \frac{1/2}{-\sqrt{3}/2} = -\frac{1}{\sqrt{3}} = -\frac{\sqrt{3}}{3}$$

87. $-9\pi/4$ is coterminal with $7\pi/4$ in Quadrant IV with reference angle $\pi/4$.

$$\sin\left(-\frac{9\pi}{4}\right) = -\sin\left(\frac{\pi}{4}\right) = -\frac{\sqrt{2}}{2}$$

$$\cos\left(-\frac{9\pi}{4}\right) = \cos\left(\frac{\pi}{4}\right) = \frac{\sqrt{2}}{2}$$

$$\tan\left(-\frac{9\pi}{4}\right) = \frac{-\sqrt{2}/2}{\sqrt{2}/2} = -1$$

89. $\sin\left(\frac{\pi}{2}\right) = 1$

$$\cos\left(\frac{\pi}{2}\right) = 0$$

$\tan\left(\frac{\pi}{2}\right)$ is undefined.

91. $\tan 33° \approx 0.6494$

93. $\sec\frac{12\pi}{5} \approx \frac{1}{\cos(12\pi/5)} \approx 3.2361$

95. $f(x) = 3 \sin x$, Amplitude: 3

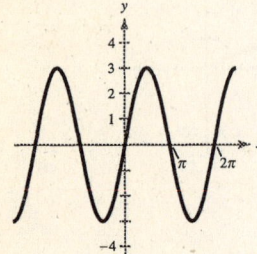

97. $f(x) = \frac{1}{4} \cos x$, Amplitude: $\frac{1}{4}$

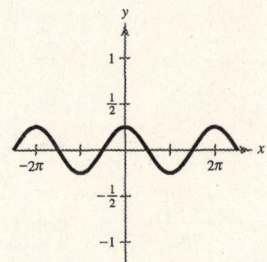

99. Period: $\frac{2\pi}{\pi} = 2$; Amplitude: 5

101. Period: $\frac{2\pi}{2} = \pi$; Amplitude: 3.4

103. $y = 3 \cos 2\pi x$

Amplitude: 3

Period: $\frac{2\pi}{2\pi} = 1$

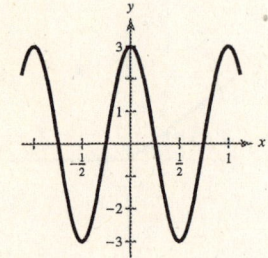

105. $f(x) = 5 \sin\frac{2x}{5}$

Amplitude: 5

Period: $\frac{2\pi}{2/5} = 5\pi$

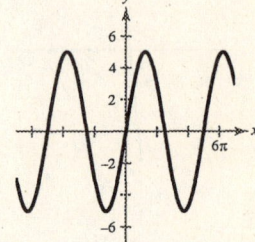

107. $f(x) = -\frac{5}{2} \cos\left(\frac{x}{4}\right)$

Amplitude: $\frac{5}{2}$

Period: $\frac{2\pi}{1/4} = 8\pi$

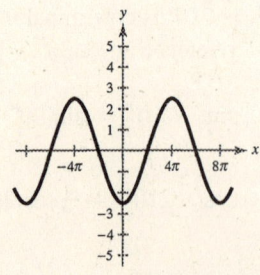

109. $f(x) = \frac{5}{2} \sin(x - \pi)$

Amplitude: $\frac{5}{2}$

Period: 2π

Shift:

$x - \pi = 0$ and $x - \pi = 2\pi$

$x = \pi \qquad\qquad x = 3\pi$

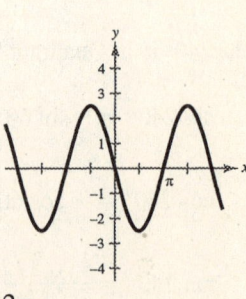

111. $f(x) = 2 - \cos \dfrac{\pi x}{2}$

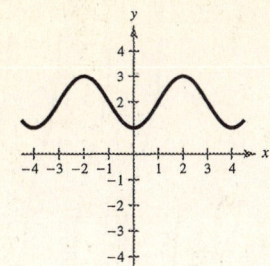

113. $f(x) = -3 \cos\left(\dfrac{x}{2} - \dfrac{\pi}{4}\right)$

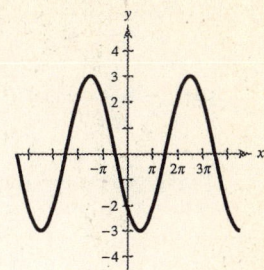

115. $f(x) = -2 \cos\left(x - \dfrac{\pi}{4}\right)$

117. $f(x) = -4 \cos\left(2x - \dfrac{\pi}{2}\right)$

119. $S = 48.4 - 6.1 \cos \dfrac{\pi t}{6}$

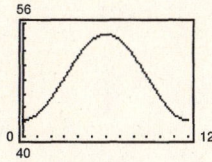

Maximum sales: $t = 6$ (June)

Minimum sales: $t = 12$ (December)

121. $f(x) = -\tan \dfrac{\pi x}{4}$

Period $= \dfrac{\pi}{(\pi/4)} = 4$

Asymptotes:
$x = -2$, $x = 2$

Reflected in x-axis

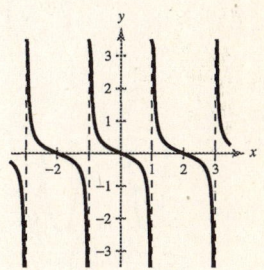

x	-1	0	1
y	1	0	-1

123. $f(x) = -\dfrac{1}{4} \tan \dfrac{\pi x}{2}$

Period: $\dfrac{\pi}{\pi/2} = 2$

Two consecutive asymptotes: $\dfrac{\pi x}{2} = -\dfrac{\pi}{2} \implies x = -1$

$\dfrac{\pi x}{2} = \dfrac{\pi}{2} \implies x = 1$

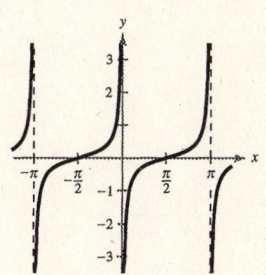

125. $f(x) = \dfrac{1}{4} \tan\left(x - \dfrac{\pi}{2}\right)$

Period: π

Two consecutive asymptotes: $x - \dfrac{\pi}{2} = -\dfrac{\pi}{2} \implies x = 0$

$x - \dfrac{\pi}{2} = \dfrac{\pi}{2} \implies x = \pi$

127. $f(x) = 3 \cot \dfrac{x}{2}$

Period: $\dfrac{\pi}{1/2} = 2\pi$

Two consecutive asymptotes: $\dfrac{x}{2} = 0 \implies x = 0$

$$\dfrac{x}{2} = \pi \implies x = 2\pi$$

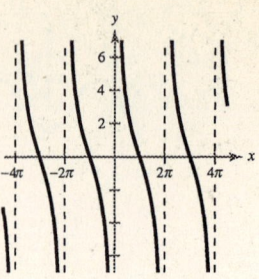

129. $f(x) = \dfrac{1}{2} \cot\left(x - \dfrac{\pi}{2}\right)$

Period: π

Two consecutive asymptotes: $x - \dfrac{\pi}{2} = 0 \implies x = \dfrac{\pi}{2}$

$$x - \dfrac{\pi}{2} = \pi \implies x = \dfrac{3\pi}{2}$$

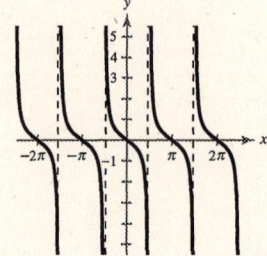

131. $f(x) = \frac{1}{4} \sec x$

Period: 2π

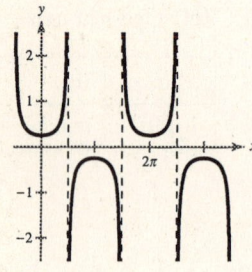

133. $f(x) = \frac{1}{4} \csc 2x$

Period: π

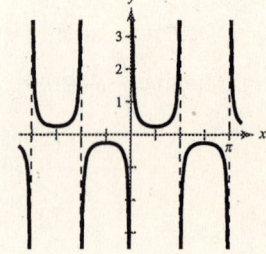

135. $f(x) = \sec\left(x - \dfrac{\pi}{4}\right)$

Secant function shifted $\dfrac{\pi}{4}$ to right

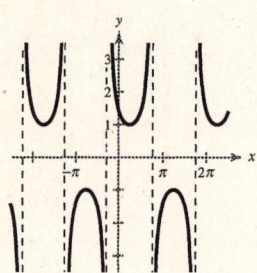

137. $f(x) = 2 \sec(x - \pi)$

$y = 2 \sec x$ shifted π to the right

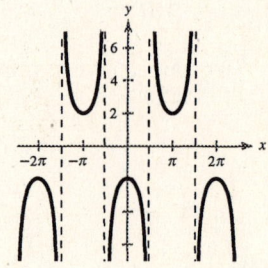

139. $f(x) = \csc\left(3x - \dfrac{\pi}{2}\right)$

Graph $y = \sin\left(3x - \dfrac{\pi}{2}\right)$ first.

Period: $\dfrac{2\pi}{3}$

Shift: $3x - \dfrac{\pi}{2} = 0$ and $3x - \dfrac{\pi}{2} = 2\pi$

$$x = \dfrac{\pi}{6} \qquad x = \dfrac{5\pi}{6}$$

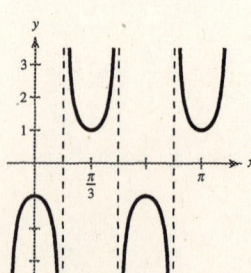

141. $f(x) = e^x \sin 2x$

Damping factor: $y = e^x$

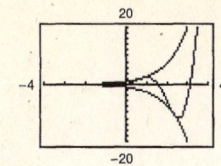

143. $f(x) = e^x \cos x$

Damping factor: e^x

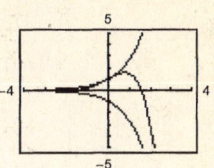

145. (a) $\arcsin 1 = \dfrac{\pi}{2}$ because $\sin \dfrac{\pi}{2} = 1$.

(b) $\arcsin 4$ is undefined.

147. (a) $\arccos\left(\dfrac{\sqrt{2}}{2}\right) = \dfrac{\pi}{4}$ because $\cos \dfrac{\pi}{4} = \dfrac{\sqrt{2}}{2}$.

(b) $\arccos\left(-\dfrac{\sqrt{3}}{2}\right) = \dfrac{5\pi}{6}$ because $\cos \dfrac{5\pi}{6} = -\dfrac{\sqrt{3}}{2}$.

149. $\arccos(0.42) \approx 1.14$

151. $\sin^{-1}(-0.94) \approx -1.22$

153. $\arctan(-12) \approx -1.49$

155. $\tan^{-1}(0.81) \approx 0.68$

157. $\sin\theta = \dfrac{x+3}{16} \implies \theta = \arcsin\left(\dfrac{x+3}{16}\right)$

159. Let $y = \arcsin(x-1)$. Then,

$\sin y = (x-1) = \dfrac{x-1}{1}$ and

$\sec y = \dfrac{1}{\sqrt{-x^2 + 2x}}$

$= \dfrac{\sqrt{-x^2 + 2x}}{-x^2 + 2x}$.

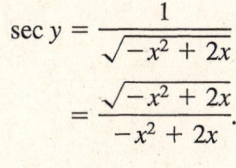

161. Let $y = \arccos \dfrac{x^2}{4 - x^2}$. Then $\cos y = \dfrac{x^2}{4 - x^2}$ and

$\sin y = \dfrac{\sqrt{(4 - x^2)^2 - (x^2)^2}}{4 - x^2}$.

$= \dfrac{\sqrt{16 - 8x^2}}{4 - x^2}$

$= \dfrac{2\sqrt{4 - 2x^2}}{4 - x^2}$.

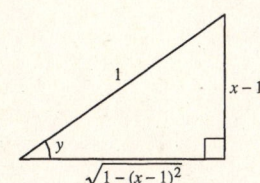

163. $\sin(1° \, 10') = \dfrac{a}{3.5}$

$a = 3.5 \sin(1° \, 10') = 3.5 \sin\left(\dfrac{7°}{6}\right) \approx 0.0713$ or 71 meters

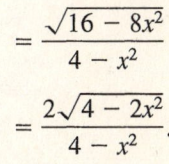

not drawn to scale

165. $\tan 14° = \dfrac{y}{37,000} \implies y = 37,000 \tan 14° \approx 9225.1$ feet

$\tan 58° = \dfrac{x+y}{37,000} \implies x + y = 37,000 \tan 58° \approx 59,212.4$ feet

$x = 59,212.4 - 9225.1 \approx 49,987.2$ feet

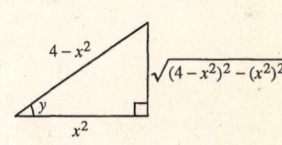

The towns are approximately 50,000 feet apart or 9.47 miles.

167. Use cosine model with amplitude 3 feet.

Period = 15 seconds

$y = 3 \cos\left(\dfrac{2\pi}{15} t\right)$

169. False. $y = \sin\theta$ is a function, but it is not one-to-one.

171. $\tan \theta = \dfrac{0.672s^2}{3000}$

(a)

s	10	20	30	40	50	60
θ	1.28°	5.12°	11.40°	19.72°	29.25°	38.88°

(b) θ increases at an increasing rate. The function is not linear.

Chapter 4   Practice Test

1. Express 350° in radian measure.

2. Express $(5\pi)/9$ in degree measure.

3. Convert $135°\,14'\,12''$ to decimal form.

4. Convert $-22.569°$ to D° M′ S″ form.

5. If $\cos\theta = \frac{2}{3}$, use the trigonometric identities to find $\tan\theta$.

6. Find θ given $\sin\theta = 0.9063$.

7. Solve for x in the figure below.

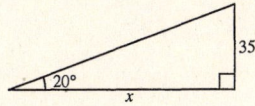

8. Find the magnitude of the reference angle for $\theta = (6\pi)/5$.

9. Evaluate csc 3.92.

10. Find $\sec\theta$ given that θ lies in Quadrant III and $\tan\theta = 6$.

11. Graph $y = 3\sin\dfrac{x}{2}$.

12. Graph $y = -2\cos(x - \pi)$.

13. Graph $y = \tan 2x$.

14. Graph $y = -\csc\left(x + \dfrac{\pi}{4}\right)$.

15. Graph $y = 2x + \sin x$, using a graphing calculator.

16. Graph $y = 3x\cos x$, using a graphing calculator.

17. Evaluate arcsin 1.

18. Evaluate $\arctan(-3)$.

19. Evaluate $\sin\left(\arccos\dfrac{4}{\sqrt{35}}\right)$.

20. Write an algebraic expression for $\cos\left(\arcsin\dfrac{x}{4}\right)$.

For Exercises 21–23, solve the right triangle.

21. $A = 40°$, $c = 12$

22. $B = 6.84°$, $a = 21.3$

23. $a = 5$, $b = 9$

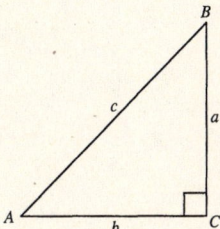

24. A 20-foot ladder leans against the side of a barn. Find the height of the top of the ladder if the angle of elevation of the ladder is 67°.

25. An observer in a lighthouse 250 feet above sea level spots a ship off the shore. If the angle of depression to the ship is 5°, how far out is the ship?

C H A P T E R   5
Analytic Trigonometry

C H A P T E R 5
Analytic Trigonometry

Section 5.1   Using Fundamental Identities

■ You should know the fundamental trigonometric identities.

(a) Reciprocal Identities

$$\sin u = \frac{1}{\csc u} \qquad\qquad \csc u = \frac{1}{\sin u}$$

$$\cos u = \frac{1}{\sec u} \qquad\qquad \sec u = \frac{1}{\cos u}$$

$$\tan u = \frac{1}{\cot u} = \frac{\sin u}{\cos u} \qquad\qquad \cot u = \frac{1}{\tan u} = \frac{\cos u}{\sin u}$$

(b) Pythagorean Identities

$$\sin^2 u + \cos^2 u = 1$$
$$1 + \tan^2 u = \sec^2 u$$
$$1 + \cot^2 u = \csc^2 u$$

(c) Cofunction Identities

$$\sin\left(\frac{\pi}{2} - u\right) = \cos u \qquad\qquad \cos\left(\frac{\pi}{2} - u\right) = \sin u$$

$$\tan\left(\frac{\pi}{2} - u\right) = \cot u \qquad\qquad \cot\left(\frac{\pi}{2} - u\right) = \tan u$$

$$\sec\left(\frac{\pi}{2} - u\right) = \csc u \qquad\qquad \csc\left(\frac{\pi}{2} - u\right) = \sec u$$

(d) Negative Angle Identities

$$\sin(-x) = -\sin x \qquad\qquad \csc(-x) = -\csc x$$
$$\cos(-x) = \cos x \qquad\qquad \sec(-x) = \sec x$$
$$\tan(-x) = -\tan x \qquad\qquad \cot(-x) = -\cot x$$

■ You should be able to use these fundamental identities to find function values.

■ You should be able to convert trigonometric expressions to equivalent forms by using the fundamental identities.

■ You should be able to check your answers with a graphing utility.

Solutions to Odd-Numbered Exercises

1. $\sin x = \dfrac{\sqrt{3}}{2}$, $\cos x = \dfrac{1}{2}$ $\implies$ x is in Quadrant I.

$$\tan x = \frac{\sin x}{\cos x} = \frac{\sqrt{3}/2}{1/2} = \sqrt{3} \qquad\qquad \cot x = \frac{1}{\tan x} = \frac{1}{\sqrt{3}} = \frac{\sqrt{3}}{3}$$

$$\sec x = \frac{1}{\cos x} = 2 \qquad\qquad\qquad \csc x = \frac{1}{\sin x} = \frac{2}{\sqrt{3}} = \frac{2\sqrt{3}}{3}$$

3. $\sec \theta = \sqrt{2}$, $\sin \theta = -\dfrac{\sqrt{2}}{2} \implies \theta$ is in Quadrant IV.

$$\cos \theta = \frac{1}{\sec \theta} = \frac{1}{\sqrt{2}} = \frac{\sqrt{2}}{2}$$

$$\tan \theta = \frac{\sin \theta}{\cos \theta} = \frac{-\sqrt{2}/2}{\sqrt{2}/2} = -1$$

$$\cot \theta = \frac{1}{\tan \theta} = -1$$

$$\csc \theta = -\sqrt{2}$$

5. $\tan x = \dfrac{7}{24}$, $\sec x = \dfrac{-25}{24} \implies x$ is in Quadrant III.

$$\cot x = \frac{24}{7}$$

$$\cos x = -\frac{24}{25}$$

$$\sin x = -\sqrt{1 - \cos^2 x} = -\frac{7}{25}$$

$$\csc x = \frac{1}{\sin x} = -\frac{25}{7}$$

7. $\sec \phi = -\dfrac{13}{12}$, $\sin \phi = \dfrac{5}{13}$

$$\cos \phi = \frac{1}{\sec \phi} = -\frac{12}{13}$$

$$\tan \phi = \frac{\sin \phi}{\cos \phi} = -\frac{5}{12}$$

$$\cot \phi = \frac{1}{\tan \theta} = -\frac{12}{5}$$

$$\csc \phi = \frac{1}{\sin \phi} = \frac{13}{5}$$

9. $\sin(-x) = -\sin x = -\dfrac{2}{3} \implies \sin x = \dfrac{2}{3}$

$$\sin x = \frac{2}{3}, \quad \tan x = -\frac{2\sqrt{5}}{5} \implies x \text{ is in Quadrant II.}$$

$$\cos x = -\sqrt{1 - \sin^2 x} = -\sqrt{1 - \frac{4}{9}} = -\frac{\sqrt{5}}{3}$$

$$\cot x = \frac{1}{\tan x} = -\frac{\sqrt{5}}{2}$$

$$\sec x = \frac{1}{\cos x} = -\frac{3\sqrt{5}}{5}$$

$$\csc x = \frac{1}{\sin x} = \frac{3}{2}$$

11. $\tan \theta = 2$, $\sin \theta < 0 \implies \theta$ is in Quadrant III.

$$\sec \theta = -\sqrt{\tan^2 \theta + 1} = -\sqrt{5}$$

$$\cos \theta = \frac{1}{-\sqrt{5}} = -\frac{\sqrt{5}}{5}$$

$$\cot \theta = \frac{1}{2}$$

$$\sin \theta = -\sqrt{1 - \cos^2 \theta}$$

$$= -\sqrt{1 - \frac{1}{5}} = -\sqrt{\frac{4}{5}} = \frac{-2}{\sqrt{5}} = \frac{-2\sqrt{5}}{5}$$

$$\csc \theta = -\frac{\sqrt{5}}{2}$$

13. $\csc \theta$ is undefined and $\cos \theta < 0 \implies \theta = \pi$.

$$\sin \theta = 0$$

$$\cos \theta = -1$$

$$\tan \theta = 0$$

$$\cot \theta \text{ undefined}$$

$$\sec \theta = -1$$

15. $\sec x \cos x = \dfrac{1}{\cos x} \cdot \cos x = 1$

Matches (d)

17. $\cot^2 x - \csc^2 x = \cot^2 x - (1 + \cot^2 x) = -1$

Matches (b)

19. $\dfrac{\sin(-x)}{\cos(-x)} = \dfrac{-\sin x}{\cos x} = -\tan x$

The expression is matched with (e).

21. $\sin x \sec x = \sin x \left(\dfrac{1}{\cos x} \right) = \tan x$

Matches (b)

23. $\sec^4 x - \tan^4 x = (\sec^2 x + \tan^2 x)(\sec^2 x - \tan^2 x)$

$= (\sec^2 x + \tan^2 x)(1)$

$= \sec^2 x + \tan^2 x$

The expression is matched with (f).

25. $\dfrac{\sec^2 x - 1}{\sin^2 x} = \dfrac{\tan^2 x}{\sin^2 x} = \dfrac{\sin^2 x}{\cos^2 x} \cdot \dfrac{1}{\sin^2 x} = \sec^2 x$

The expression is matched with (e).

27. $\cot x \sin x = \dfrac{\cos x}{\sin x} \sin x = \cos x$

29. $\sin \phi(\csc \phi - \sin \phi) = \sin \phi \csc \phi - \sin^2 \phi$

$= \sin \phi \cdot \dfrac{1}{\sin \phi} - \sin^2 \phi$

$= 1 - \sin^2 \phi$

$= \cos^2 \phi$

31. $\dfrac{\cot x}{\csc x} = \dfrac{\cos x / \sin x}{1 / \sin x}$

$= \dfrac{\cos x}{\sin x} \cdot \dfrac{\sin x}{1} = \cos x$

33. $\sec \alpha \dfrac{\sin \alpha}{\tan \alpha} = \dfrac{1}{\cos \alpha}(\sin \alpha) \cot \alpha$

$= \dfrac{1}{\cos \alpha}(\sin \alpha)\left(\dfrac{\cos \alpha}{\sin \alpha}\right) = 1$

35. $\sin\left(\dfrac{\pi}{2} - x\right) \csc x = \cos x \cdot \dfrac{1}{\sin x} = \cot x$

37. $\dfrac{\cos^2 y}{1 - \sin y} = \dfrac{1 - \sin^2 y}{1 - \sin y}$

$= \dfrac{(1 + \sin y)(1 - \sin y)}{1 - \sin y}$

$= 1 + \sin y$

39. $\sin \theta + \cos \theta \cot \theta = \sin \theta + \cos \theta \dfrac{\cos \theta}{\sin \theta}$

$= \dfrac{\sin^2 \theta + \cos^2 \theta}{\sin \theta}$

$= \dfrac{1}{\sin \theta}$

$= \csc \theta$

41. $\dfrac{\cos \theta}{1 - \sin \theta} = \dfrac{\cos \theta}{1 - \sin \theta} \cdot \dfrac{1 + \sin \theta}{1 + \sin \theta}$

$= \dfrac{\cos \theta(1 + \sin \theta)}{1 - \sin^2 \theta}$

$= \dfrac{\cos \theta(1 + \sin \theta)}{\cos^2 \theta}$

$= \dfrac{1 + \sin \theta}{\cos \theta}$

$= \sec \theta + \tan \theta$

43. $\dfrac{1 + \cos \theta}{\sin \theta} + \dfrac{\sin \theta}{1 + \cos \theta} = \dfrac{1 + 2 \cos \theta + \cos^2 \theta + \sin^2 \theta}{\sin \theta(1 + \cos \theta)}$

$= \dfrac{2 + 2 \cos \theta}{\sin \theta(1 + \cos \theta)}$

$= \dfrac{2(1 + \cos \theta)}{\sin \theta(1 + \cos \theta)}$

$= \dfrac{2}{\sin \theta}$

$= 2 \csc \theta$

45. $\csc\theta\tan\theta = \dfrac{1}{\sin\theta}\cdot\dfrac{\sin\theta}{\cos\theta}$

$ = \dfrac{1}{\cos\theta}$

$ = \sec\theta$

47. $1 - \dfrac{\sin^2\theta}{1-\cos\theta} = \dfrac{1-\cos\theta-\sin^2\theta}{1-\cos\theta}$

$\phantom{1 - \dfrac{\sin^2\theta}{1-\cos\theta}} = \dfrac{\cos^2\theta-\cos\theta}{1-\cos\theta}$

$\phantom{1 - \dfrac{\sin^2\theta}{1-\cos\theta}} = \dfrac{\cos\theta(\cos\theta-1)}{1-\cos\theta}$

$\phantom{1 - \dfrac{\sin^2\theta}{1-\cos\theta}} = -\cos\theta$

49. $\dfrac{\cot(-\theta)}{\csc\theta} = \dfrac{\cos(-\theta)}{\sin(-\theta)}\sin\theta$

$\phantom{\dfrac{\cot(-\theta)}{\csc\theta}} = \dfrac{\cos\theta}{-\sin\theta}\sin\theta$

$\phantom{\dfrac{\cot(-\theta)}{\csc\theta}} = -\cos\theta$

51. $\cot^2 x - \cot^2 x\cos^2 x = \cot^2 x(1-\cos^2 x)$

$ = \dfrac{\cos^2 x}{\sin^2 x}\sin^2 x$

$ = \cos^2 x$

53. $\dfrac{\cos^2 x - 4}{\cos x - 2} = \dfrac{(\cos x + 2)(\cos x - 2)}{\cos x - 2} = \cos x + 2$

55. $\tan^4 x + 2\tan^2 x + 1 = (\tan^2 x + 1)^2$

$ = (\sec^2 x)^2 = \sec^4 x$

57. $\sin^4 x - \cos^4 x = (\sin^2 x + \cos^2 x)(\sin^2 x - \cos^2 x)$

$ = (1)(\sin^2 x - \cos^2 x)$

$ = \sin^2 x - \cos^2 x$

59. $\csc^3 x - \csc^2 x - \csc x + 1 = \csc^2 x(\csc x - 1) - (\csc x - 1)$

$ = (\csc^2 x - 1)(\csc x - 1)$

$ = \cot^2 x(\csc x - 1)$

61. $(\sin x + \cos x)^2 = \sin^2 x + 2\sin x\cos x + \cos^2 x$

$ = (\sin^2 x + \cos^2 x) + 2\sin x\cos x$

$ = 1 + 2\sin x\cos x$

63. $(\sec x + 1)(\sec x - 1) = \sec^2 x - 1 = \tan^2 x$

65. $\dfrac{1}{1+\cos x} + \dfrac{1}{1-\cos x} = \dfrac{1-\cos x + 1 + \cos x}{(1+\cos x)(1-\cos x)}$

$\phantom{\dfrac{1}{1+\cos x} + \dfrac{1}{1-\cos x}} = \dfrac{2}{1-\cos^2 x}$

$\phantom{\dfrac{1}{1+\cos x} + \dfrac{1}{1-\cos x}} = \dfrac{2}{\sin^2 x}$

$\phantom{\dfrac{1}{1+\cos x} + \dfrac{1}{1-\cos x}} = 2\csc^2 x$

67. $\tan x - \dfrac{\sec^2 x}{\tan x} = \dfrac{\tan^2 x - \sec^2 x}{\tan x}$

$\phantom{\tan x - \dfrac{\sec^2 x}{\tan x}} = \dfrac{-1}{\tan x}$

$\phantom{\tan x - \dfrac{\sec^2 x}{\tan x}} = -\cot x$

69. $\dfrac{\sin^2 y}{1-\cos y} = \dfrac{1-\cos^2 y}{1-\cos y}$

$\phantom{\dfrac{\sin^2 y}{1-\cos y}} = \dfrac{(1+\cos y)(1-\cos y)}{1-\cos y}$

$\phantom{\dfrac{\sin^2 y}{1-\cos y}} = 1 + \cos y$

71. $\dfrac{3}{\sec x - \tan x}\cdot\dfrac{\sec x + \tan x}{\sec x + \tan x} = \dfrac{3(\sec x + \tan x)}{\sec^2 x - \tan^2 x}$

$\phantom{\dfrac{3}{\sec x - \tan x}\cdot\dfrac{\sec x + \tan x}{\sec x + \tan x}} = \dfrac{3(\sec x + \tan x)}{1}$

$\phantom{\dfrac{3}{\sec x - \tan x}\cdot\dfrac{\sec x + \tan x}{\sec x + \tan x}} = 3(\sec x + \tan x)$

73. $y_1 = \cos\left(\dfrac{\pi}{2} - x\right)$, $y_2 = \sin x$

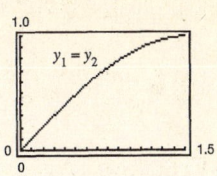

x	0.2	0.4	0.6	0.8	1.0	1.2	1.4
y_1	0.1987	0.3894	0.5646	0.7174	0.8415	0.9320	0.9854
y_2	0.1987	0.3894	0.5646	0.7174	0.8415	0.9320	0.9854

Conjecture: $y_1 = y_2$

75. $y_1 = \dfrac{\cos x}{1 - \sin x}$, $y_2 = \dfrac{1 + \sin x}{\cos x}$

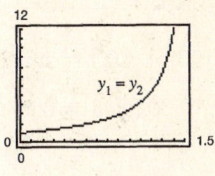

x	0.2	0.4	0.6	0.8	1.0	1.2	1.4
y_1	1.2230	1.5085	1.8958	2.4650	3.4082	5.3319	11.6814
y_2	1.2230	1.5085	1.8958	2.4650	3.4082	5.3319	11.6814

Conjecture: $y_1 = y_2$

77. $y_1 = \cos x \cot x + \sin x = \csc x$

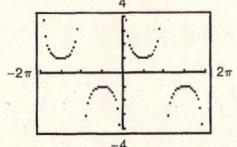

79. $y_1 = \sec x - \dfrac{\cos x}{1 + \sin x} = \tan x$

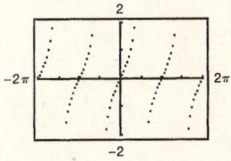

81. $\sqrt{25 - x^2} = \sqrt{25 - (5\sin\theta)^2}$, $x = 5\sin\theta$

$\qquad = \sqrt{25 - 25\sin^2\theta}$

$\qquad = \sqrt{25(1 - \sin^2\theta)}$

$\qquad = \sqrt{25\cos^2\theta}$

$\qquad = 5\cos\theta$

83. $\sqrt{x^2 - 9} = \sqrt{(3\sec\theta)^2 - 9}$, $x = 3\sec\theta$

$\qquad = \sqrt{9\sec^2\theta - 9}$

$\qquad = \sqrt{9(\sec^2\theta - 1)}$

$\qquad = \sqrt{9\tan^2\theta}$

$\qquad = 3\tan\theta$

85. $\sin\theta = \sqrt{1 - \cos^2\theta}$

Let $y_1 = \sin x$ and $y_2 = \sqrt{1 - \cos^2 x}$, $0 \le x < 2\pi$.

$y_1 = y_2$ for $0 \le x \le \pi$, so we have

$\sin\theta = \sqrt{1 - \cos^2\theta}$ for $0 \le \theta \le \pi$.

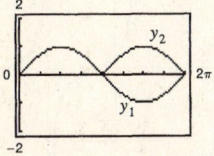

87. $\sec\theta = \sqrt{1 + \tan^2\theta}$

Let $y_1 = \dfrac{1}{\cos x}$ and $y_2 = \sqrt{1 + \tan^2 x}$, $0 \le x < 2\pi$.

$y_1 = y_2$ for $0 \le x < \dfrac{\pi}{2}$ and $\dfrac{3\pi}{2} < x < 2\pi$, so we have

$\sec\theta = \sqrt{1 + \tan^2\theta}$ for $0 \le \theta < \dfrac{\pi}{2}$ and

$\dfrac{3\pi}{2} < \theta < 2\pi$.

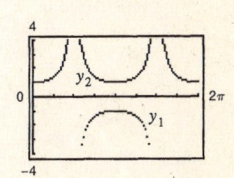

89. $\ln|\cos\theta| - \ln|\sin\theta| = \ln\dfrac{|\cos\theta|}{|\sin\theta|} = \ln|\cot\theta|$

91. $\ln(1 + \sin x) - \ln|\sec x| = \ln\left|\dfrac{1 + \sin x}{\sec x}\right|$

$= \ln|\cos x(1 + \sin x)|$

93. (a) $\csc^2 132° - \cot^2 132° \approx 1.8107 - 0.8107 = 1$

(b) $\csc^2\dfrac{2\pi}{7} - \cot^2\dfrac{2\pi}{7} \approx 1.6360 - 0.6360 = 1$

95. $\cos\left(\dfrac{\pi}{2} - \theta\right) = \sin\theta$

(a) $\theta = 80°$

$\cos(90° - 80°) = \sin 80°$

$0.9848 = 0.9848$

(b) $\theta = 0.8$

$\cos\left(\dfrac{\pi}{2} - 0.8\right) = \sin 0.8$

$0.7174 = 0.7174$

97. $\csc x \cot x - \cos x = \dfrac{1}{\sin x} \cdot \dfrac{\cos x}{\sin x} - \cos x$

$= \cos x(\csc^2 x - 1)$

$= \cos x \cdot \cot^2 x$

99. True for all $\theta \neq n\pi$

$\sin\theta \cdot \csc\theta = \sin\theta\left(\dfrac{1}{\sin\theta}\right) = 1$

101. As $x \to \dfrac{\pi^-}{2}$, $\sin x \to 1$ and $\csc x \to 1$.

103. As $x \to \dfrac{\pi^-}{2}$, $\tan x \to \infty$ and $\cot x \to 0$.

105. $\sin\theta$

$\cos\theta = \pm\sqrt{1 - \sin^2\theta}$

$\tan\theta = \dfrac{\sin\theta}{\cos\theta} = \pm\dfrac{\sin\theta}{\sqrt{1 - \sin^2\theta}}$

$\csc\theta = \dfrac{1}{\sin\theta}$

$\sec\theta = \dfrac{\pm 1}{\sqrt{1 - \sin^2\theta}}$

$\cot\theta = \pm\dfrac{\sqrt{1 - \sin^2\theta}}{\sin\theta}$

107. $\sin\theta = \dfrac{\text{opp}}{\text{hyp}}$, $\cos\theta = \dfrac{\text{adj}}{\text{hyp}}$

From the Pythagorean Theorem,

$(\text{opp})^2 + (\text{adj})^2 = (\text{hyp})^2$

$\sin^2\theta + \cos^2\theta = 1.$

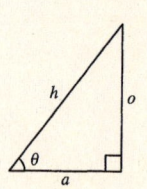

109. $f(x) = \dfrac{1}{2}\sin\pi x$

Period: $\dfrac{2\pi}{\pi} = 2$

Amplitude: $\dfrac{1}{2}$

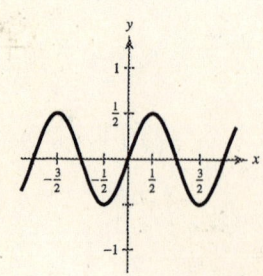

111. $f(x) = \dfrac{1}{2}\cot\left(x + \dfrac{\pi}{4}\right)$

Period: π

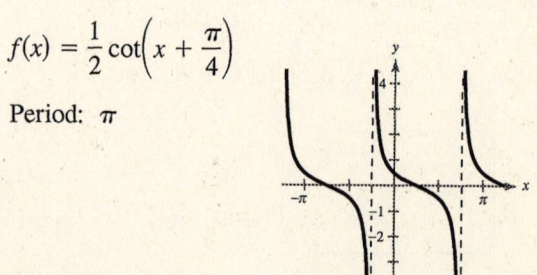

Section 5.2   Verifying Trigonometric Identities

- ■ You should know the difference between an expression, a conditional equation, and an identity.
- ■ You should be able to solve trigonometric identities, using the following techniques.
   - (a) Work with *one* side at a time. Do not "cross" the equal sign.
   - (b) Use algebraic techniques such as combining fractions, factoring expressions, rationalizing denominators, and squaring binomials.
   - (c) Use the fundamental identities.
   - (d) Convert all the terms into sines and cosines.

Solutions to Odd-Numbered Exercises

1. $\sin t \csc t = \sin t \left(\dfrac{1}{\sin t} \right) = 1$

3. $\dfrac{\csc^2 x}{\cot x} = \dfrac{1}{\sin^2 x} \cdot \dfrac{\sin x}{\cos x} = \dfrac{1}{\sin x \cdot \cos x}$

$$= \csc x \cdot \sec x$$

5. $\cos^2 \beta - \sin^2 \beta = (1 - \sin^2 \beta) - \sin^2 \beta$

$$= 1 - 2 \sin^2 \beta$$

7. $\tan^2 \theta + 5 = (\sec^2 \theta - 1) + 5$

$$= \sec^2 \theta + 4$$

9. $(1 + \sin x)(1 - \sin x) = 1 - \sin^2 x = \cos^2 x$

11. $\dfrac{1}{\sec x \tan x} = \cos x \cdot \dfrac{\cos x}{\sin x}$

$$= \dfrac{\cos^2 x}{\sin x}$$

$$= \dfrac{1 - \sin^2 x}{\sin x}$$

$$= \dfrac{1}{\sin x} - \sin x$$

$$= \csc x - \sin x$$

x	0.2	0.4	0.6	0.8	1.0	1.2	1.4
y_1	4.8348	2.1785	1.2064	0.6767	0.3469	0.1409	0.0293
y_2	4.8348	2.1785	1.2064	0.6767	0.3469	0.1409	0.0293

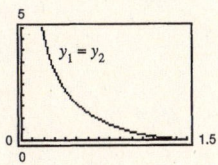

13. $\csc x - \sin x = \dfrac{1}{\sin x} - \sin x$

$$= \dfrac{1 - \sin^2 x}{\sin x}$$

$$= \dfrac{\cos^2 x}{\sin x}$$

$$= \cos x \cdot \dfrac{\cos x}{\sin x}$$

$$= \cos x \cdot \cot x$$

x	0.2	0.4	0.6	0.8	1.0	1.2	1.4
y_1	4.8348	2.1785	1.2064	0.6767	0.3469	0.1409	0.0293
y_2	4.8348	2.1785	1.2064	0.6767	0.3469	0.1409	0.0293

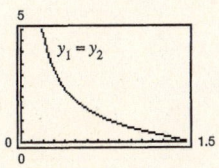

15. $\sin x + \cos x \cot x = \sin x + \cos x \dfrac{\cos x}{\sin x}$

$$= \frac{\sin^2 x + \cos^2 x}{\sin x}$$

$$= \frac{1}{\sin x}$$

$$= \csc x$$

x	0.2	0.4	0.6	0.8	1.0	1.2	1.4
y_1	5.0335	2.5679	1.7710	1.3940	1.1884	1.0729	1.0148
y_2	5.0335	2.5679	1.7710	1.3940	1.1884	1.0729	1.0148

17. $\dfrac{1}{\tan x} + \dfrac{1}{\cot x} = \dfrac{\cot x + \tan x}{\tan x \cdot \cot x}$

$$= \cot x + \tan x$$

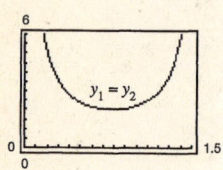

x	0.2	0.4	0.6	0.8	1.0	1.2	1.4
y_1	5.1359	2.7880	2.1458	2.0009	2.1995	2.9609	5.9704
y_2	5.1359	2.7880	2.1458	2.0009	2.1995	2.9609	5.9704

19. The error is in line 1: $\cot(-x) \neq \cot x$.

21. $\sin^{1/2} x \cos x - \sin^{5/2} x \cos x = \sin^{1/2} x \cos x (1 - \sin^2 x) = \sin^{1/2} x \cos x \cdot \cos^2 x = \cos^3 x \sqrt{\sin x}$

23. $\tan\left(\dfrac{\pi}{2} - x\right) \sec x = \cot x \cdot \sec x$

$$= \frac{\cos x}{\sin x} \cdot \frac{1}{\cos x}$$

$$= \frac{1}{\sin x}$$

$$= \csc x$$

25. $\dfrac{\csc(-x)}{\sec(-x)} = \dfrac{1/\sin(-x)}{1/\cos(-x)}$

$$= \frac{\cos(-x)}{\sin(-x)}$$

$$= \frac{\cos x}{-\sin x}$$

$$= -\cot x$$

27. $\dfrac{\cos(-\theta)}{1 + \sin(-\theta)} = \dfrac{\cos \theta}{1 - \sin \theta} \cdot \dfrac{1 + \sin \theta}{1 + \sin \theta}$

$$= \frac{\cos \theta(1 + \sin \theta)}{1 - \sin^2 \theta}$$

$$= \frac{\cos \theta(1 + \sin \theta)}{\cos^2 \theta}$$

$$= \frac{1 + \sin \theta}{\cos \theta}$$

$$= \frac{1}{\cos \theta} + \frac{\sin \theta}{\cos \theta}$$

$$= \sec \theta + \tan \theta$$

29. $\dfrac{\sin x \cos y + \cos x \sin y}{\cos x \cos y - \sin x \sin y} = \dfrac{\dfrac{\sin x \cos y}{\cos x \cos y} + \dfrac{\cos x \sin y}{\cos x \cos y}}{\dfrac{\cos x \cos y}{\cos x \cos y} - \dfrac{\sin x \sin y}{\cos x \cos y}} = \dfrac{\tan x + \tan y}{1 - \tan x \tan y}$

31. $\dfrac{\cos x - \cos y}{\sin x + \sin y} + \dfrac{\sin x - \sin y}{\cos x + \cos y} = \dfrac{(\cos x + \cos y)(\cos x - \cos y) + (\sin x + \sin y)(\sin x - \sin y)}{(\sin x + \sin y)(\cos x + \cos y)}$

$$= \dfrac{\cos^2 x - \cos^2 y + \sin^2 x - \sin^2 y}{(\sin x + \sin y)(\cos x + \cos y)}$$

$$= \dfrac{1 - 1}{(\sin x + \sin y)(\cos x + \cos y)}$$

$$= 0$$

33. $\sqrt{\dfrac{1 + \sin \theta}{1 - \sin \theta}} = \sqrt{\dfrac{1 + \sin \theta}{1 - \sin \theta} \cdot \dfrac{1 + \sin \theta}{1 + \sin \theta}}$

$$= \sqrt{\dfrac{(1 + \sin \theta)^2}{1 - \sin^2 \theta}}$$

$$= \sqrt{\dfrac{(1 + \sin \theta)^2}{\cos^2 \theta}}$$

$$= \dfrac{1 + \sin \theta}{|\cos \theta|}$$

Note: Check your answer with a graphing utility. What happens if you leave off the absolute value?

35. $\cos^2 x + \cos^2\left(\dfrac{\pi}{2} - x\right) = \cos^2 x + \sin^2 x = 1$

37. $\sin x \csc\left(\dfrac{\pi}{2} - x\right) = \sin x \sec x$

$$= \sin x\left(\dfrac{1}{\cos x}\right)$$

$$= \tan x$$

39. $2 \sec^2 x - 2 \sec^2 x \sin^2 x - \sin^2 x - \cos^2 x = 2 \sec^2 x(1 - \sin^2 x) - (\sin^2 x + \cos^2 x)$

$$= 2 \sec^2 x(\cos^2 x) - 1$$

$$= 2 \cdot \dfrac{1}{\cos^2 x} \cdot \cos^2 x - 1$$

$$= 2 - 1 = 1$$

41. $\dfrac{\tan x \cot x}{\cos x} = \dfrac{1}{\cos x} = \sec x$

43. $\csc^4 x - 2 \csc^2 x + 1 = (\csc^2 x - 1)^2$

$$= (\cot^2 x)^2 = \cot^4 x$$

45. $\sec^4 \theta - \tan^4 \theta = (\sec^2 \theta + \tan^2 \theta)(\sec^2 \theta - \tan^2 \theta)$

$$= (1 + \tan^2 \theta + \tan^2 \theta)(1)$$

$$= 1 + 2 \tan^2 \theta$$

47. $\dfrac{\sin \beta}{1 - \cos \beta} \cdot \dfrac{1 + \cos \beta}{1 + \cos \beta} = \dfrac{\sin \beta(1 + \cos \beta)}{1 - \cos^2 \beta}$

$$= \dfrac{\sin \beta(1 + \cos \beta)}{\sin^2 \beta}$$

$$= \dfrac{1 + \cos \beta}{\sin \beta}$$

49. $\dfrac{\tan^3 \alpha - 1}{\tan \alpha - 1} = \dfrac{(\tan \alpha - 1)(\tan^2 \alpha + \tan \alpha + 1)}{\tan \alpha - 1} = \tan^2 \alpha + \tan \alpha + 1$

51. It appears that $y_1 = 1$. Analytically,

$$\frac{1}{\cot x + 1} + \frac{1}{\tan x + 1} = \frac{\tan x + 1 + \cot x + 1}{(\cot x + 1)(\tan x + 1)}$$

$$= \frac{\tan x + \cot x + 2}{\cot x \tan x + \cot x + \tan x + 1}$$

$$= \frac{\tan x + \cot x + 2}{\tan x + \cot x + 2}$$

$$= 1.$$

53. It appears that $y_1 = \sin x$. Analytically,

$$\frac{1}{\sin x} - \frac{\cos^2 x}{\sin x} = \frac{1 - \cos^2 x}{\sin x} = \frac{\sin^2 x}{\sin x} = \sin x.$$

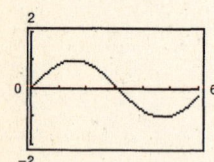

55. $\ln|\cot \theta| = \ln\left|\dfrac{\cos \theta}{\sin \theta}\right|$

$$= \ln \frac{|\cos \theta|}{|\sin \theta|}$$

$$= \ln|\cos \theta| - \ln|\sin \theta|$$

57. $-\ln(1 + \cos \theta) = \ln(1 + \cos \theta)^{-1}$

$$= \ln\left[\frac{1}{1 + \cos \theta} \cdot \frac{1 - \cos \theta}{1 - \cos \theta}\right]$$

$$= \ln \frac{1 - \cos \theta}{1 - \cos^2 \theta}$$

$$= \ln \frac{1 - \cos \theta}{\sin^2 \theta}$$

$$= \ln(1 - \cos \theta) - \ln \sin^2 \theta$$

$$= \ln(1 - \cos \theta) - 2 \ln|\sin \theta|$$

59. $\sin^2 25° + \sin^2 65° = \sin^2 25° + \cos^2 25° = 1$

61. $\cos^2 20° + \cos^2 52° + \cos^2 38° + \cos^2 70° = \cos^2 20° + \cos^2 52^2 + \sin^2(90° - 38°) + \sin^2(90° - 70°)$

$$= \cos^2 20° + \cos^2 52^2 + \sin^2 52° + \sin^2 20°$$

$$= (\cos^2 20° + \sin^2 20°) + (\cos^2 52° + \sin^2 52°)$$

$$= 1 + 1$$

$$= 2$$

63. $\tan^5 x = \tan^3 x \cdot \tan^2 x$

$$= \tan^3 x(\sec^2 x - 1)$$

$$= \tan^3 x \sec^2 x - \tan^3 x$$

65. $(\sin^2 x - \sin^4 x) \cos x = \sin^2 x(1 - \sin^2 x) \cos x$

$$= \sin^2 x \cdot \cos^2 x \cdot \cos x$$

$$= \cos^3 x \sin^2 x$$

67. $\mu W \cos \theta = W \sin \theta$

$$\mu = \frac{W \sin \theta}{W \cos \theta} = \frac{\sin \theta}{\cos \theta} = \tan \theta, \; W \neq 0$$

69. True

71. False. Just because the equation is true for one value of θ, you cannot conclude that the equation is an identity. For example,

$$\sin^2 \frac{\pi}{4} + \cos^2 \frac{\pi}{4} = 1 \neq 1 + \tan^2 \frac{\pi}{4}.$$

73. $\sqrt{\tan^2 x} = |\tan x|$

Let $x = 3\pi/4$. Then,

$$\sqrt{\tan^2 x} = \sqrt{(-1)^2} = 1 \neq \tan\left(\frac{3\pi}{4}\right) = -1.$$

75. When n is even, $\cos\left[\dfrac{(2n+1)\pi}{2}\right] = \cos \dfrac{\pi}{2} = 0.$

When n is odd, $\cos\left[\dfrac{(2n+1)\pi}{2}\right] = \cos \dfrac{3\pi}{2} = 0.$

Thus, $\cos\left[\dfrac{(2n+1)\pi}{2}\right] = 0$ for all n.

77. $(x - 1)(x - 8i)(x + 8i) = (x - 1)(x^2 + 64)$
$$= x^3 - x^2 + 64x - 64$$

Answers will vary.

79. $(x - 4)(x - 6 - i)(x - 6 + i) = (x - 4)((x - 6)^2 + 1)$
$$= (x - 4)(x^2 - 12x + 37)$$
$$= x^3 - 16x^2 + 85x - 148$$

Answers will vary.

81. $f(x) = 2^x + 3$

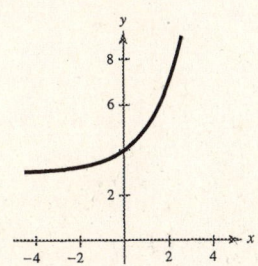

83. $f(x) = 2^{-x} + 1$

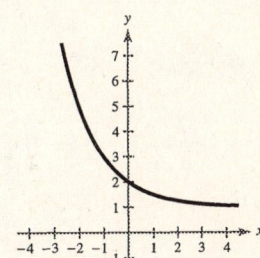

85. $\csc \theta > 0$ and $\tan \theta < 0 \implies$ Quadrant II

87. $\sec \theta > 0$ and $\sin \theta < 0 \implies$ Quadrant IV

89. $B = 80°, A = 90° - 80° = 10°$

$$\sin A = \frac{a}{c} \implies c = \frac{a}{\sin A} \approx 92.14$$

$$\tan A = \frac{a}{b} \implies b = \frac{a}{\tan A} \approx 90.74$$

91. $c = \sqrt{a^2 + b^2} \approx 16.12$

$$\tan A = \frac{a}{b} \implies A = \arctan\left(\frac{a}{b}\right)$$
$$= \arctan\left(\frac{14}{8}\right) \approx 60.26°$$

$$\tan B = \frac{b}{a} \implies B = \arctan\left(\frac{8}{14}\right) \approx 29.74°$$

Section 5.3   Solving Trigonometric Equations

- ■ You should be able to identify and solve trigonometric equations.
- ■ A trigonometric equation is a conditional equation. It is true for a specific set of values.
- ■ To solve trigonometric equations, use algebraic techniques such as collecting like terms, taking square roots, factoring, squaring, converting to quadratic form, using formulas, and using inverse functions. Study the examples in this section.
- ■ Use your graphing utility to calculate solutions and verify results.

Solutions to Odd-Numbered Exercises

1. $2 \cos x - 1 = 0$

  (a) $2 \cos \dfrac{\pi}{3} - 1 = 2\left(\dfrac{1}{2}\right) - 1 = 0$

  (b) $2 \cos \dfrac{5\pi}{3} - 1 = 2\left(\dfrac{1}{2}\right) - 1 = 0$

3. $3 \tan^2 2x - 1 = 0$

  (a) $3\left[\tan\left(\dfrac{2\pi}{12}\right)\right]^2 - 1 = 3 \tan^2 \dfrac{\pi}{6} - 1$

                $= 3\left(\dfrac{1}{\sqrt{3}}\right)^2 - 1 = 0$

  (b) $3\left[\tan\left(\dfrac{10\pi}{12}\right)\right]^2 - 1 = 3 \tan^2 \dfrac{5\pi}{6} - 1$

                $= 3\left(-\dfrac{1}{\sqrt{3}}\right)^2 - 1 = 0$

5. $2 \sin^2 x - \sin x - 1 = 0$

  (a) $x = \dfrac{\pi}{2}$: $2 \sin^2\left(\dfrac{\pi}{2}\right) - \sin\left(\dfrac{\pi}{2}\right) - 1 = 2 - 1 - 1 = 0$

  (b) $x = \dfrac{7\pi}{6}$: $2 \sin^2\left(\dfrac{7\pi}{6}\right) - \sin\left(\dfrac{7\pi}{6}\right) - 1 = 2\left(\dfrac{1}{4}\right) - \left(-\dfrac{1}{2}\right) - 1 = 0$

7. $2 \cos x + 1 = 0$

    $2 \cos x = -1$

     $\cos x = -\dfrac{1}{2}$

        $x = \dfrac{2\pi}{3} + 2n\pi$

    or $x = \dfrac{4\pi}{3} + 2n\pi$

9. $\sqrt{3} \sec x - 2 = 0$

    $\sqrt{3} \sec x = 2$

     $\sec x = \dfrac{2}{\sqrt{3}}$

     $\cos x = \dfrac{\sqrt{3}}{2}$

        $x = \dfrac{\pi}{6} + 2n\pi$

    or $x = \dfrac{11\pi}{6} + 2n\pi$

11. $3 \csc^2 x - 4 = 0$

     $\csc^2 x = \dfrac{4}{3}$

      $\csc x = \pm\dfrac{2}{\sqrt{3}}$

      $\sin x = \pm\dfrac{\sqrt{3}}{2}$

        $x = \dfrac{\pi}{3} + n\pi$

    or $x = \dfrac{2\pi}{3} + n\pi$

13. $4\cos^2 x - 1 = 0$

$$\cos^2 x = \frac{1}{4}$$

$$\cos x = \pm\frac{1}{2}$$

$$x = \frac{\pi}{3} + n\pi$$

$$\text{or } x = \frac{2\pi}{3} + n\pi$$

15. $\sin^2 x = 3\cos^2 x$

$$\sin^2 x - 3(1 - \sin^2 x) = 0$$

$$4\sin^2 x = 3$$

$$\sin x = \pm\frac{\sqrt{3}}{2}$$

$$x = \frac{\pi}{3} + n\pi$$

$$\text{or } x = \frac{2\pi}{3} + n\pi$$

17. $\tan x + \sqrt{3} = 0$

$$\tan x = -\sqrt{3}$$

$$x = \frac{2\pi}{3}, \frac{5\pi}{3}$$

19. $\csc^2 x - 2 = 0$

$$\csc^2 x = 2$$

$$\csc x = \pm\sqrt{2}$$

$$\sin x = \pm\frac{1}{\sqrt{2}}$$

$$x = \frac{\pi}{4}, \frac{3\pi}{4}, \frac{5\pi}{4}, \frac{7\pi}{4}$$

21. $3\tan^3 x - \tan x = 0$

$$\tan x(3\tan^2 x - 1) = 0$$

$$\tan x = 0 \quad \text{or} \quad 3\tan^2 x - 1 = 0$$

$$x = 0, \pi \qquad \tan x = \pm\frac{\sqrt{3}}{3}$$

$$x = \frac{\pi}{6}, \frac{5\pi}{6}, \frac{7\pi}{6}, \frac{11\pi}{6}$$

23. $\sec^2 x - \sec x - 2 = 0$

$$(\sec x - 2)(\sec x + 1) = 0$$

$$\sec x - 2 = 0 \quad \text{or} \quad \sec x + 1 = 0$$

$$\sec x = 2 \qquad\qquad \sec x = -1$$

$$x = \frac{\pi}{3}, \frac{5\pi}{3} \qquad\qquad x = \pi$$

25. $2\sin x + \csc x = 0$

$$2\sin x + \frac{1}{\sin x} = 0$$

$$2\sin^2 x + 1 = 0$$

Since $2\sin^2 x + 1 > 0$, there are no solutions.

27. $\cos x + \sin x \tan x = 2$

$$\cos x + \frac{\sin^2 x}{\cos x} = 2$$

$$\frac{\cos^2 x + \sin^2 x}{\cos x} = 2$$

$$\frac{1}{\cos x} = 2$$

$$\cos x = \frac{1}{2}$$

$$x = \frac{\pi}{3}, \frac{5\pi}{3}$$

29.

$$\sec^2 x + \tan x = 3$$

$$(1 + \tan^2 x) + \tan x = 3$$

$$\tan^2 x + \tan x - 2 = 0$$

$$(\tan x + 2)(\tan x - 1) = 0$$

$$\tan x = -2 \qquad \text{or} \qquad \tan x = 1$$

$$x \approx 2.0344, 5.1760 \qquad x = \frac{\pi}{4}, \frac{5\pi}{4}$$

31. $2 \sin^2 x + 3 \sin x + 1 = 0$

$$y = 2 \sin^2 x + 3 \sin x + 1$$

$$x \approx 3.6652, 5.7596, 4.7124$$

33. $y = 4 \sin^2 x - 2 \cos x - 1$

$$x \approx 0.8614, 5.4218$$

35. $y = \csc x + \cot x - 1 = \dfrac{1}{\sin x} + \dfrac{\cos x}{\sin x} - 1$

$$x \approx 1.5708 \qquad \left(\frac{\pi}{2}\right)$$

37. $\dfrac{\cos x \cot x}{1 - \sin x} = 3$

Graph $y = \dfrac{\cos x}{(1 - \sin x) \tan x} - 3$.

The solutions are approximately
$x \approx 0.5236, x \approx 2.6180$.

39. $\sin 2x = -\dfrac{\sqrt{3}}{2}$

$$2x = \frac{4\pi}{3} + 2n\pi \quad \text{or} \quad 2x = \frac{5\pi}{3} + 2n\pi$$

$$x = \frac{2\pi}{3} + n\pi \qquad x = \frac{5\pi}{6} + n\pi$$

41. $2 \sin^2 2x = 1$

$$\sin^2 2x = \frac{1}{2}$$

$$\sin 2x = \pm\frac{\sqrt{2}}{2}$$

$$2x = \frac{\pi}{4} + \frac{n\pi}{2}$$

$$x = \frac{\pi}{8} + \frac{n\pi}{4}$$

43. $\tan 3x(\tan x - 1) = 0$

$$3x = n\pi \quad \text{or} \quad x = \frac{\pi}{4} + n\pi$$

$$x = \frac{n\pi}{3} \qquad x = \frac{\pi}{4} + n\pi$$

45. $\cos \dfrac{x}{2} = \dfrac{\sqrt{2}}{2}$

$$\frac{x}{2} = \frac{\pi}{4} + 2n\pi \quad \text{or} \quad \frac{x}{2} = \frac{7\pi}{4} + 2n\pi$$

$$x = \frac{\pi}{2} + 4n\pi \qquad x = \frac{7\pi}{2} + 4n\pi$$

47. $y = \sin \dfrac{\pi x}{2} + 1$

From the graph in the textbook we see that the curve has x-intercepts at $x = -1$ and at $x = 3$.

49. $y = \tan^2\left(\dfrac{\pi x}{6}\right) - 3$

From the graph in the textbook
we see that the curve has
x-intercepts at $x = \pm 2$.

51. $2 \cos x - \sin x = 0$

Graph $y_1 = 2 \cos x - \sin x$
and estimate the zeros.

$x \approx 1.1071, \, 4.2487$

53. $x \tan x - 1 = 0$

Graph $y_1 = x \tan x - 1$ and
estimate the zeros.

$x \approx 0.8603, \, 3.4256$

55. $\sec^2 x + 0.5 \tan x - 1 = 0$

Graph $y_1 = \dfrac{1}{(\cos x)^2} + 0.5 \tan x - 1$.

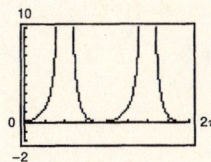

The x-intercepts occur at $x = 0$, $x \approx 2.6779$,
$x \approx 3.1416$ and $x \approx 5.8195$.

57. $12 \sin^2 x - 13 \sin x + 3 = 0$

Graph $y_1 = 12 \sin^2 x - 13 \sin x + 3$.

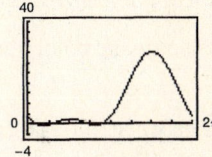

The x-intercepts occur at $x \approx 0.3398$, $x \approx 0.8481$,
$x \approx 2.2935$, and $x \approx 2.8018$.

59. $3 \tan^2 x + 5 \tan x - 4 = 0, \ \left[-\dfrac{\pi}{2}, \dfrac{\pi}{2} \right]$

$$\tan x = \frac{-5 \pm \sqrt{25 - 4(-4)(3)}}{2(3)}$$

$$= \frac{-5 \pm \sqrt{73}}{2(3)}$$

$$x \approx -1.154, \, 0.5354$$

61. $4 \cos^2 x - 2 \sin x + 1 = 0, \ \left[-\dfrac{\pi}{2}, \dfrac{\pi}{2} \right]$

$$4(1 - \sin^2 x) - 2 \sin x + 1 = 0$$

$$-4 \sin^2 x - 2 \sin x + 5 = 0$$

$$\sin x = \frac{2 \pm \sqrt{4 - 4(-4)(5)}}{2(-4)}$$

$$= \frac{2 \pm \sqrt{84}}{-8} = \frac{1 \pm \sqrt{21}}{-4}$$

$$x \approx 1.110$$

63. (a) $f(x) = \sin x + \cos x$

Maximum: $(0.7854, 1.4142)$

Minimum: $(3.9270, -1.4142)$

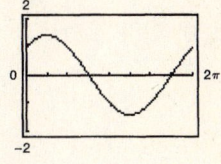

(b) $\cos x - \sin x = 0$

$$\cos x = \sin x$$

$$1 = \frac{\sin x}{\cos x}$$

$$\tan x = 1$$

$$x = \frac{\pi}{4}, \frac{5\pi}{4}$$

$$f\left(\frac{\pi}{4}\right) = \sin \frac{\pi}{4} + \cos \frac{\pi}{4} = \frac{\sqrt{2}}{2} + \frac{\sqrt{2}}{2} = \sqrt{2}$$

$$f\left(\frac{5\pi}{4}\right) = \sin \frac{5\pi}{4} + \cos \frac{5\pi}{4} = -\sin \frac{\pi}{4} + \left(-\cos \frac{\pi}{4} \right)$$

$$= -\frac{\sqrt{2}}{2} - \frac{\sqrt{2}}{2} = -\sqrt{2}$$

Therefore, the maximum point in the interval $[0, 2\pi)$
is $\left(\pi/4, \sqrt{2} \right)$ and the minimum point is $\left(5\pi/4, -\sqrt{2} \right)$.

65. $f(x) = \tan \dfrac{\pi x}{4}$

$\tan 0 = 0$, but 0 is not positive. By graphing $y = \tan \dfrac{\pi x}{4} - x$, you see that

the smallest positive fixed point is $x = 1$.

67. $f(x) = \cos \dfrac{1}{x}$

(a) The domain of $f(x)$ is all real numbers except 0.

(b) The graph has y-axis symmetry and a horizontal asymptote at $y = 1$.

(c) As $x \to 0$, $f(x)$ oscillates between -1 and 1.

(d) There are an infinite number of solutions in the interval $[-1, 1]$.

$$\dfrac{1}{x} = \dfrac{\pi}{2} + n\pi = \dfrac{\pi + 2n\pi}{2} \implies x = \dfrac{2}{\pi(2n + 1)}$$

(e) The greatest solution appears to occur at $x \approx 0.6366$.

69.
$$y = \dfrac{1}{12}(\cos 8t - 3 \sin 8t)$$

$$\dfrac{1}{12}(\cos 8t - 3 \sin 8t) = 0$$

$$\cos 8t = 3 \sin 8t$$

$$\dfrac{1}{3} = \tan 8t$$

$$8t = 0.32175 + n\pi$$

$$t = 0.04 + \dfrac{n\pi}{8}$$

In the interval $0 \le t \le 1$, $t = 0.04, 0.43$, and 0.83 second.

71. $S = 74.50 - 43.75 \cos \dfrac{\pi t}{6}$

t	1	2	3	4	5	6	7	8	9	10	11	12
S	36.6	52.6	74.5	96.4	112.4	118.3	112.4	96.4	74.5	52.6	36.6	30.8

$S > 100$ for $t = 5, 6, 7$ (May, June, July).

73. $r = \frac{1}{32}v_0{}^2 \sin 2\theta$

$300 = \frac{1}{32}(100)^2 \sin 2\theta$

$\sin 2\theta = 0.96$

$2\theta \approx 1.287$ or $2\theta \approx \pi - 1.287 \approx 1.855$

$\theta \approx 0.6435 \approx 37°$ or $\theta \approx 0.9275 \approx 53°$

75. (a)

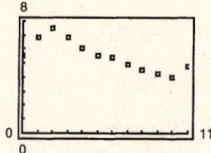

(b) $r = 1.39 \sin(0.48t + 0.42) + 5.51$

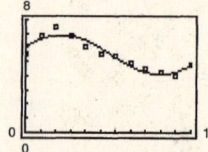

(c) The constant term 5.51 gives the average unemployment rate.

(d) The business cycle is the period

$$\dfrac{2\pi}{0.48} \approx 13 \text{ years.}$$

(e) 2007

77. False. $\sin x - x = 0$ has one solution, $x = 0$.

79. False. The equation has no solution because $-1 \le \sin x \le 1$.

81. $124° = 124° \left(\dfrac{\pi}{180°} \right) \approx 2.164$ radians

83. $-0.41° = -0.41° \left(\dfrac{\pi}{180°} \right) \approx -0.007$ radian

85. $\tan 30° = \dfrac{14}{x} \implies x = \dfrac{14}{\tan 30°} = \dfrac{14}{\sqrt{3}/3} \approx 24.249$

87. $\tan 87.5° = \dfrac{x}{100}$

$\qquad x = 100 \tan 87.5°$

$\qquad\qquad \approx 2290.4 \text{ feet} \approx 0.43 \text{ mile}$

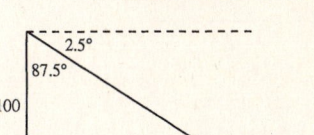

Section 5.4   Sum and Difference Formulas

- You should memorize the sum and difference formulas.

  $\sin(u \pm v) = \sin u \cos v \pm \cos u \sin v$

  $\cos(u \pm v) = \cos u \cos v \mp \sin u \sin v$

  $\tan(u \pm v) = \dfrac{\tan u \pm \tan v}{1 \mp \tan u \tan v}$

- You should be able to use these formulas to find the values of the trigonometric functions of angles whose sums or differences are special angles.

- You should be able to use these formulas to solve trigonometric equations.

Solutions to Odd-Numbered Exercises

1. (a) $\cos(240° - 0°) = \cos(240°) = -\dfrac{1}{2}$

  (b) $\cos(240°) - \cos 0° = -\dfrac{1}{2} - 1 = -\dfrac{3}{2}$

3. (a) $\cos\left(\dfrac{\pi}{6} + \dfrac{\pi}{3} \right) = \cos\dfrac{\pi}{6} \cos\dfrac{\pi}{3} - \sin\dfrac{\pi}{6} \sin\dfrac{\pi}{3}$

  $\qquad = \dfrac{\sqrt{3}}{2} \cdot \dfrac{1}{2} - \dfrac{1}{2} \cdot \dfrac{\sqrt{3}}{2} = 0$

  (b) $\cos\dfrac{\pi}{6} + \cos\dfrac{\pi}{3} = \dfrac{\sqrt{3}}{2} + \dfrac{1}{2} = \dfrac{\sqrt{3} + 1}{2}$

5. (a) $\sin(315° - 60°) = \sin 315° \cos 60° - \cos 315° \sin 60°$

  $\qquad = -\dfrac{\sqrt{2}}{2} \cdot \dfrac{1}{2} - \dfrac{\sqrt{2}}{2} \cdot \dfrac{\sqrt{3}}{2} = \dfrac{-\sqrt{2} - \sqrt{6}}{4}$

  (b) $\sin 315° - \sin 60° = -\dfrac{\sqrt{2}}{2} - \dfrac{\sqrt{3}}{2} = -\dfrac{\sqrt{2} + \sqrt{3}}{2}$

7. $\sin 105° = \sin(60° + 45°)$

$\qquad = \sin 60° \cos 45° + \sin 45° \cos 60°$

$\qquad = \dfrac{\sqrt{3}}{2} \cdot \dfrac{\sqrt{2}}{2} + \dfrac{\sqrt{2}}{2} \cdot \dfrac{1}{2}$

$\qquad = \dfrac{\sqrt{2}}{4}(\sqrt{3} + 1)$

$\cos 105° = \cos(60° + 45°)$

$\qquad = \cos 60° \cos 45° - \sin 60° \sin 45°$

$\qquad = \dfrac{1}{2} \cdot \dfrac{\sqrt{2}}{2} - \dfrac{\sqrt{3}}{2} \cdot \dfrac{\sqrt{2}}{2}$

$\qquad = \dfrac{\sqrt{2}}{4}(1 - \sqrt{3})$

$\tan 105° = \tan(60° + 45°)$

$\qquad = \dfrac{\tan 60° + \tan 45°}{1 - \tan 60° \tan 45°}$

$\qquad = \dfrac{\sqrt{3} + 1}{1 - \sqrt{3}} = \dfrac{\sqrt{3} + 1}{1 - \sqrt{3}} \cdot \dfrac{1 + \sqrt{3}}{1 + \sqrt{3}}$

$\qquad = \dfrac{4 + 2\sqrt{3}}{-2} = -2 - \sqrt{3}$

9. $\sin 195° = \sin(225° - 30°)$

$\qquad = \sin 225° \cos 30° - \sin 30° \cos 225°$

$\qquad = -\sin 45° \cos 30° + \sin 30° \cos 45°$

$\qquad = -\dfrac{\sqrt{2}}{2} \cdot \dfrac{\sqrt{3}}{2} + \dfrac{1}{2} \cdot \dfrac{\sqrt{2}}{2} = \dfrac{\sqrt{2}}{4}(1 - \sqrt{3})$

$\cos 195° = \cos(225° - 30°)$

$\qquad = \cos 225° \cos 30° + \sin 225° \sin 30°$

$\qquad = -\cos 45° \cos 30° - \sin 45° \sin 30°$

$\qquad = -\dfrac{\sqrt{2}}{2} \cdot \dfrac{\sqrt{3}}{2} - \dfrac{\sqrt{2}}{2} \cdot \dfrac{1}{2}$

$\qquad = -\dfrac{\sqrt{2}}{4}(\sqrt{3} + 1)$

$\tan 195° = \tan(225° - 30°)$

$\qquad = \dfrac{\tan 225° - \tan 30°}{1 + \tan 225° \tan 30°}$

$\qquad = \dfrac{\tan 45° - \tan 30°}{1 + \tan 45° \tan 30°}$

$\qquad = \dfrac{1 - (\sqrt{3}/3)}{1 + (\sqrt{3}/3)} = \dfrac{3 - \sqrt{3}}{3 + \sqrt{3}} \cdot \dfrac{3 - \sqrt{3}}{3 - \sqrt{3}}$

$\qquad = \dfrac{12 - 6\sqrt{3}}{6} = 2 - \sqrt{3}$

11. $\sin \dfrac{11\pi}{12} = \sin\left(\dfrac{3\pi}{4} + \dfrac{\pi}{6}\right)$

$\qquad = \sin \dfrac{3\pi}{4} \cos \dfrac{\pi}{6} + \sin \dfrac{\pi}{6} \cos \dfrac{3\pi}{4}$

$\qquad = \dfrac{\sqrt{2}}{2} \cdot \dfrac{\sqrt{3}}{2} + \dfrac{1}{2}\left(-\dfrac{\sqrt{2}}{2}\right) = \dfrac{\sqrt{2}}{4}(\sqrt{3} - 1)$

$\cos \dfrac{11\pi}{12} = \cos\left(\dfrac{3\pi}{4} + \dfrac{\pi}{6}\right)$

$\qquad = \cos \dfrac{3\pi}{4} \cos \dfrac{\pi}{6} - \sin \dfrac{3\pi}{4} \sin \dfrac{\pi}{6}$

$\qquad = -\dfrac{\sqrt{2}}{2} \cdot \dfrac{\sqrt{3}}{2} - \dfrac{\sqrt{2}}{2} \cdot \dfrac{1}{2} = -\dfrac{\sqrt{2}}{4}(\sqrt{3} + 1)$

$\tan \dfrac{11\pi}{12} = \tan\left(\dfrac{3\pi}{4} + \dfrac{\pi}{6}\right)$

$\qquad = \dfrac{\tan(3\pi/4) + \tan(\pi/6)}{1 - \tan(3\pi/4)\tan(\pi/6)}$

$\qquad = \dfrac{-1 + (\sqrt{3}/3)}{1 - (-1)(\sqrt{3}/3)} = \dfrac{-3 + \sqrt{3}}{3 + \sqrt{3}} \cdot \dfrac{3 - \sqrt{3}}{3 - \sqrt{3}} = \dfrac{-12 + 6\sqrt{3}}{6} = -2 + \sqrt{3}$

13. $-\dfrac{\pi}{12} = \dfrac{\pi}{6} - \dfrac{\pi}{4}$

$$\sin\left(-\dfrac{\pi}{12}\right) = \sin\left(\dfrac{\pi}{6} - \dfrac{\pi}{4}\right)$$

$$= \sin\dfrac{\pi}{6}\cos\dfrac{\pi}{4} - \sin\dfrac{\pi}{4}\cos\dfrac{\pi}{6}$$

$$= \dfrac{1}{2}\cdot\dfrac{\sqrt{2}}{2} - \dfrac{\sqrt{2}}{2}\cdot\dfrac{\sqrt{3}}{2} = \dfrac{\sqrt{2}}{4}\left(1 - \sqrt{3}\right)$$

$$\cos\left(-\dfrac{\pi}{12}\right) = \cos\left(\dfrac{\pi}{6} - \dfrac{\pi}{4}\right)$$

$$= \cos\dfrac{\pi}{6}\cos\dfrac{\pi}{4} + \sin\dfrac{\pi}{6}\sin\dfrac{\pi}{4}$$

$$= \dfrac{\sqrt{3}}{2}\cdot\dfrac{\sqrt{2}}{2} + \dfrac{1}{2}\cdot\dfrac{\sqrt{2}}{2} = \dfrac{\sqrt{2}}{4}\left(\sqrt{3} + 1\right)$$

$$\tan\left(-\dfrac{\pi}{12}\right) = \tan\left(\dfrac{\pi}{6} - \dfrac{\pi}{4}\right)$$

$$= \dfrac{\tan(\pi/6) - \tan(\pi/4)}{1 + \tan(\pi/6)\tan(\pi/4)}$$

$$= \dfrac{\left(\sqrt{3}/3\right) - 1}{1 + \left(\sqrt{3}/3\right)} = \dfrac{\sqrt{3} - 3}{\sqrt{3} + 3}\cdot\dfrac{\sqrt{3} - 3}{\sqrt{3} - 3}$$

$$= \dfrac{12 - 6\sqrt{3}}{-6} = -2 + \sqrt{3}$$

15. $\sin 75° = \sin(30° + 45°)$

$$= \sin 30°\cos 45° + \sin 45°\cos 30°$$

$$= \dfrac{1}{2}\cdot\dfrac{\sqrt{2}}{2} + \dfrac{\sqrt{2}}{2}\cdot\dfrac{\sqrt{3}}{2}$$

$$= \dfrac{\sqrt{2}}{4}\left(1 + \sqrt{3}\right)$$

$\cos 75° = \cos(30° + 45°)$

$$= \cos 30°\cos 45° - \sin 30°\sin 45°$$

$$= \dfrac{\sqrt{3}}{2}\cdot\dfrac{\sqrt{2}}{2} - \dfrac{1}{2}\cdot\dfrac{\sqrt{2}}{2}$$

$$= \dfrac{\sqrt{2}}{4}\left(\sqrt{3} - 1\right)$$

$\tan 75° = \tan(30° + 45°)$

$$= \dfrac{\tan 30° + \tan 45°}{1 - \tan 30°\tan 45°}$$

$$= \dfrac{\left(\sqrt{3}/3\right) + 1}{1 - \left(\sqrt{3}/3\right)} = \dfrac{\sqrt{3} + 3}{3 - \sqrt{3}}\cdot\dfrac{3 + \sqrt{3}}{3 + \sqrt{3}}$$

$$= \dfrac{6\sqrt{3} + 12}{6} = \sqrt{3} + 2$$

17. $-225°$ is coterminal with $135°$, and lies in Quadrant II.

$$\sin(-225°) = \dfrac{\sqrt{2}}{2}$$

$$\cos(-225°) = -\dfrac{\sqrt{2}}{2}$$

$$\tan(-225°) = -1$$

19. $\dfrac{13\pi}{12} = \dfrac{3\pi}{4} + \dfrac{\pi}{3}$

$$\sin\dfrac{13\pi}{12} = \sin\left(\dfrac{3\pi}{4} + \dfrac{\pi}{3}\right) = \sin\dfrac{3\pi}{4}\cos\dfrac{\pi}{3} + \sin\dfrac{\pi}{3}\cos\dfrac{3\pi}{4}$$

$$= \dfrac{\sqrt{2}}{2}\cdot\dfrac{1}{2} + \dfrac{\sqrt{3}}{2}\left(-\dfrac{\sqrt{2}}{2}\right) = \dfrac{\sqrt{2} - \sqrt{6}}{4}$$

$$\cos\dfrac{13\pi}{12} = \cos\left(\dfrac{3\pi}{4} + \dfrac{\pi}{3}\right) = \cos\dfrac{3\pi}{4}\cos\dfrac{\pi}{3} - \sin\dfrac{3\pi}{4}\sin\dfrac{\pi}{3}$$

$$= \left(-\dfrac{\sqrt{2}}{2}\right)\left(\dfrac{1}{2}\right) - \left(\dfrac{\sqrt{2}}{2}\right)\left(\dfrac{\sqrt{3}}{2}\right) = -\dfrac{\sqrt{6} + \sqrt{2}}{4}$$

$$\tan\dfrac{13\pi}{12} = \tan\left(\dfrac{3\pi}{4} + \dfrac{\pi}{3}\right) = \dfrac{\tan(3\pi/4) + \tan(\pi/3)}{1 - \tan(3\pi/4)\tan(\pi/3)}$$

$$= \dfrac{(-1) + \sqrt{3}}{1 - (-1)\sqrt{3}} = \dfrac{\sqrt{3} - 1}{\sqrt{3} + 1} = 2 - \sqrt{3}$$

21. $-\dfrac{7\pi}{12} = \dfrac{\pi}{6} - \dfrac{3\pi}{4}$

$$\sin\left(-\dfrac{7\pi}{12}\right) = \sin\left(\dfrac{\pi}{6} - \dfrac{3\pi}{4}\right) = \sin\dfrac{\pi}{6}\cos\dfrac{3\pi}{4} - \sin\dfrac{3\pi}{4}\cos\dfrac{\pi}{6}$$

$$= \dfrac{1}{2}\left(-\dfrac{\sqrt{2}}{2}\right) - \dfrac{\sqrt{2}}{2}\left(\dfrac{\sqrt{3}}{2}\right) = -\dfrac{\sqrt{2}+\sqrt{6}}{4}$$

$$\cos\left(-\dfrac{7\pi}{12}\right) = \cos\left(\dfrac{\pi}{6} - \dfrac{3\pi}{4}\right) = \cos\dfrac{\pi}{6}\cos\dfrac{3\pi}{4} + \sin\dfrac{\pi}{6}\sin\dfrac{3\pi}{4}$$

$$= \dfrac{\sqrt{3}}{2}\left(-\dfrac{\sqrt{2}}{2}\right) + \dfrac{1}{2}\cdot\dfrac{\sqrt{2}}{2} = \dfrac{\sqrt{2}-\sqrt{6}}{4}$$

$$\tan\left(-\dfrac{7\pi}{12}\right) = \tan\left(\dfrac{\pi}{6} - \dfrac{3\pi}{4}\right) = \dfrac{\tan(\pi/6) - \tan(3\pi/4)}{1 + \tan(\pi/6)\tan(3\pi/4)}$$

$$= \dfrac{\left(\sqrt{3}/3\right) - (-1)}{1 + \left(\sqrt{3}/3\right)(-1)} = \dfrac{3 + \sqrt{3}}{3 - \sqrt{3}} = 2 + \sqrt{3}$$

23. $\cos 60° \cos 10° - \sin 60° \sin 10° = \cos(60° + 10°)$
$$= \cos 70°$$

25. $\dfrac{\tan 325° - \tan 86°}{1 + \tan 325° \tan 86°} = \tan(325° - 86°) = \tan 239°$

27. $\sin 3.5 \cos 1.2 - \cos 3.5 \sin 1.2 = \sin(3.5 - 1.2)$
$$= \sin 2.3$$

29. $\cos\dfrac{\pi}{7}\cos\dfrac{\pi}{5} - \sin\dfrac{\pi}{7}\sin\dfrac{\pi}{5} = \cos\left(\dfrac{\pi}{7} + \dfrac{\pi}{5}\right)$
$$= \cos\dfrac{12\pi}{35}$$

31. $y_1 = \sin\left(\dfrac{\pi}{6} + x\right)$

$= \sin\dfrac{\pi}{6}\cos x + \sin x \cdot \cos\dfrac{\pi}{6}$

$= \dfrac{1}{2}\cos x + \dfrac{\sqrt{3}}{2}\sin x$

$= \dfrac{1}{2}\left(\cos x + \sqrt{3}\sin x\right)$

$= y_2$

x	0.2	0.4	0.6	0.8	1.0	1.2	1.4
y_1	0.6621	0.7978	0.9017	0.9696	0.9989	0.9883	0.9384
y_2	0.6621	0.7978	0.9017	0.9696	0.9989	0.9883	0.9384

33. $y_1 = \cos(x + \pi)\cos(x - \pi)$

$= (\cos x \cdot \cos \pi - \sin x \cdot \sin \pi)$

$\quad [\cos x \cos \pi + \sin x \sin \pi]$

$= [-\cos x][-\cos x]$

$= \cos^2 x$

$= y_2$

x	0.2	0.4	0.6	0.8	1.0	1.2	1.4
y_1	0.9605	0.8484	0.6812	0.4854	0.2919	0.1313	0.0289
y_2	0.9605	0.8484	0.6812	0.4854	0.2919	0.1313	0.0289

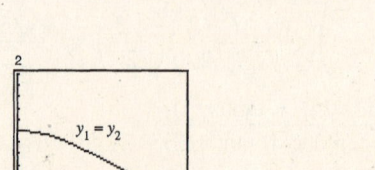

For Exercises 35 and 37, we have:

$\sin u = \frac{5}{13}$ and *u* in Quadrant II $\Rightarrow$ $\cos u = -\frac{12}{13}$

$\cos v = -\frac{3}{5}$ and *v* in Quadrant II $\Rightarrow$ $\sin v = \frac{4}{5}$

$\tan u = -\frac{5}{12}$ and $\tan v = -\frac{4}{3}$.

35. $\sin(u + v) = \sin u \cos v + \sin v \cos u$

$$= \frac{5}{13}\left(-\frac{3}{5}\right) + \frac{4}{5}\left(-\frac{12}{13}\right)$$

$$= -\frac{63}{65}$$

37. $\tan(u + v) = \dfrac{\tan u + \tan v}{1 - \tan u \tan v}$

$$= \frac{(-5/12) - (4/3)}{1 - (-5/12)(-4/3)}$$

$$= \frac{-63/36}{16/36}$$

$$= -\frac{63}{16}$$

For Exercises 39 and 41, we have:

$\sin u = -\frac{7}{25}$ and *u* in Quadrant III $\Rightarrow$ $\cos u = -\frac{24}{25}$

$\cos v = -\frac{4}{5}$ and *v* in Quadrant III $\Rightarrow$ $\sin v = -\frac{3}{5}$.

(*Note:* $u \approx 196.26°$ and $v \approx 216.87°$)

39. $\cos(u + v) = \cos u \cos v - \sin u \sin v$

$$= \left(-\frac{24}{25}\right)\left(-\frac{4}{5}\right) - \left(-\frac{7}{25}\right)\left(-\frac{3}{5}\right)$$

$$= \frac{75}{125} = \frac{3}{5}$$

41. $\sin(v - u) = \sin v \cos u - \sin u \cos v$

$$= \left(-\frac{3}{5}\right)\left(-\frac{24}{25}\right) - \left(-\frac{7}{25}\right)\left(-\frac{4}{5}\right)$$

$$= \frac{44}{125}$$

43. $\sin(\arcsin x + \arccos x) = \sin(\arcsin x)\cos(\arccos x) + \sin(\arccos x)\cos(\arcsin x)$

$$= x \cdot x + \sqrt{1 - x^2} \cdot \sqrt{1 - x^2}$$

$$= x^2 + 1 - x^2$$

$$= 1$$

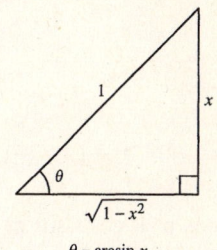

$\theta = \arcsin x$

$\alpha = \arccos x$

45. Let:

$u = \arctan 2x$ and $v = \arccos x$

$\tan u = 2x$ $\cos v = x$

$\sin(\arctan 2x - \arccos x) = \sin(u - v)$

$$= \sin u \cos v - \cos u \sin v$$

$$= \frac{2x}{\sqrt{4x^2 + 1}}(x) - \frac{1}{\sqrt{4x^2 + 1}}\left(\sqrt{1 - x^2}\right)$$

$$= \frac{2x^2 - \sqrt{1 - x^2}}{\sqrt{4x^2 + 1}}$$

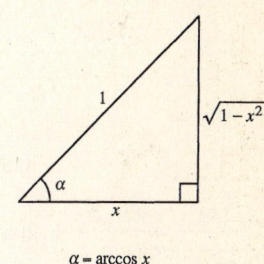

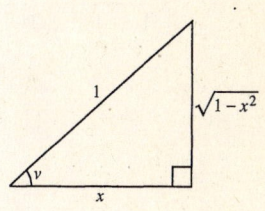

47. $\sin\left(\dfrac{\pi}{2} + x\right) = \sin\dfrac{\pi}{2}\cos x + \sin x \cos\dfrac{\pi}{2} = (1)\cos x + 0 = \cos x$

49. $\tan(x + \pi) - \tan(\pi - x) = \dfrac{\tan x + \tan \pi}{1 - \tan x \cdot \tan \pi} - \dfrac{\tan \pi - \tan x}{1 + \tan \pi \tan x}$

$$= \dfrac{\tan x}{1} - \left(-\dfrac{\tan x}{1}\right)$$

$$= 2\tan x$$

51. $\sin(x + y) + \sin(x - y) = \sin x \cos y + \sin y \cos x + \sin x \cos y - \sin y \cos x$

$$= 2\sin x \cos y$$

53. $\cos(x + y)\cos(x - y) = [\cos x \cos y - \sin x \sin y][\cos x \cos y + \sin x \sin y]$

$$= \cos^2 x \cos^2 y - \sin^2 x \sin^2 y$$

$$= \cos^2 x(1 - \sin^2 y) - \sin^2 x \sin^2 y$$

$$= \cos^2 x - \sin^2 y(\cos^2 x + \sin^2 x)$$

$$= \cos^2 x - \sin^2 y$$

55.
$$\sin\left(x + \dfrac{\pi}{3}\right) + \sin\left(x - \dfrac{\pi}{3}\right) = 1$$

$$\sin x \cos\dfrac{\pi}{3} + \cos x \sin\dfrac{\pi}{3} + \sin x \cos\dfrac{\pi}{3} - \cos x \sin\dfrac{\pi}{3} = 1$$

$$2\sin x(0.5) = 1$$

$$\sin x = 1$$

$$x = \dfrac{\pi}{2}$$

57.
$$\tan(x + \pi) + 2\sin(x + \pi) = 0$$

$$\dfrac{\tan x + \tan \pi}{1 - \tan x \tan \pi} + 2(\sin x \cos \pi + \cos x \sin \pi) = 0$$

$$\dfrac{\tan x + 0}{1 - \tan x(0)} + 2[\sin x(-1) + \cos x(0)] = 0$$

$$\dfrac{\tan x}{1} - 2\sin x = 0$$

$$\dfrac{\sin x}{\cos x} = 2\sin x$$

$$\sin x = 2\sin x \cos x$$

$$\sin x(1 - 2\cos x) = 0$$

$$\sin x = 0 \quad \text{or} \quad \cos x = \dfrac{1}{2}$$

$$x = 0, \pi \qquad\qquad x = \dfrac{\pi}{3}, \dfrac{5\pi}{3}$$

59. Graph $y_1 = \cos\left(x + \dfrac{\pi}{4}\right) + \cos\left(x - \dfrac{\pi}{4}\right)$ and $y_2 = 1$.

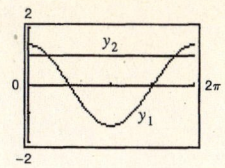

The points of intersection occur at $x \approx 0.7854$ and $x \approx 5.4978$.

61. $\tan(x + \pi) - \cos\left(x + \dfrac{\pi}{2}\right) = 0$

Answers: $0.0,\ 3.1416\ (x = 0,\ \pi)$

63. $y_1 + y_2 = A \cos 2\pi\left(\dfrac{t}{T} - \dfrac{x}{\lambda}\right) + A \cos 2\pi\left(\dfrac{t}{T} + \dfrac{x}{\lambda}\right)$

$$= A\left[\cos\left(\dfrac{2\pi t}{T}\right)\cos\left(\dfrac{2\pi x}{\lambda}\right) + \sin\left(\dfrac{2\pi t}{T}\right)\sin\left(\dfrac{2\pi x}{\lambda}\right)\right] + A\left[\cos\left(\dfrac{2\pi t}{T}\right)\cos\left(\dfrac{2\pi x}{\lambda}\right) - \sin\left(\dfrac{2\pi t}{T}\right)\sin\left(\dfrac{2\pi x}{\lambda}\right)\right]$$

$$= 2A \cos\left(\dfrac{2\pi t}{T}\right)\cos\left(\dfrac{2\pi x}{\lambda}\right)$$

65. False. See page 508.

67. $\cos(n\pi + \theta) = \cos n\pi \cos \theta - \sin n\pi \sin \theta$

$$= (-1)^n(\cos \theta) - (0)(\sin \theta)$$

$$= (-1)^n(\cos \theta), \text{ where } n \text{ is an integer.}$$

69. $C = \arctan \dfrac{b}{a} \implies \tan C = \dfrac{b}{a} \implies \sin C = \dfrac{b}{\sqrt{a^2 + b^2}}, \cos C = \dfrac{a}{\sqrt{a^2 + b^2}}$

$$\sqrt{a^2 + b^2}\, \sin(B\theta + C) = \sqrt{a^2 + b^2}\left(\sin B\theta \cdot \dfrac{a}{\sqrt{a^2 + b^2}} + \dfrac{b}{\sqrt{a^2 + b^2}} \cdot \cos B\theta\right) = a \sin B\theta + b \cos B\theta$$

71. $\sin \theta + \cos \theta$

$a = 1,\ b = 1,\ B = 1$

(a) $C = \arctan \dfrac{b}{a} = \arctan 1 = \dfrac{\pi}{4}$

$\sin \theta + \cos \theta = \sqrt{a^2 + b^2}\, \sin(B\theta + C)$

$$= \sqrt{2}\, \sin\left(\theta + \dfrac{\pi}{4}\right)$$

(b) $C = \arctan \dfrac{a}{b} = \arctan 1 = \dfrac{\pi}{4}$

$\sin \theta + \cos \theta = \sqrt{a^2 + b^2}\, \cos(B\theta - C)$

$$= \sqrt{2}\, \cos\left(\theta - \dfrac{\pi}{4}\right)$$

73. $12 \sin 3\theta + 5 \cos 3\theta$

$a = 12,\ b = 5,\ B = 3$

(a) $C = \arctan \dfrac{b}{a} = \arctan \dfrac{5}{12} \approx 0.3948$

$12 \sin 3\theta + 5 \cos 3\theta = \sqrt{a^2 + b^2}\, \sin(B\theta + C)$

$$\approx 13 \sin(3\theta + 0.3948)$$

(b) $C = \arctan \dfrac{a}{b} = \arctan \dfrac{12}{5} \approx 1.1760$

$12 \sin 3\theta + 5 \cos 3\theta = \sqrt{a^2 + b^2}\, \cos(B\theta - C)$

$$\approx 13 \cos(3\theta - 1.1760)$$

75. $C = \arctan \dfrac{b}{a} = \dfrac{\pi}{2} \implies a = 0$

$\sqrt{a^2 + b^2} = 2 \implies b = 2$

$B = 1$

$2 \sin\left(\theta + \dfrac{\pi}{2}\right) = (0)(\sin\theta) + (2)(\cos\theta) = 2\cos\theta$

77. $\dfrac{\cos(x+h) - \cos x}{h} = \dfrac{\cos x \cos h - \sin x \sin h - \cos x}{h}$

$\qquad = \dfrac{\cos x(\cos h - 1)}{h} - \dfrac{\sin x \sin h}{h}$

79. From the figure, it appears that $u + v = w$. Assume that u, v, and w are all in Quadrant I. From the figure:

$\tan u = \dfrac{s}{3s} = \dfrac{1}{3}$

$\tan v = \dfrac{s}{2s} = \dfrac{1}{2}$

$\tan w = \dfrac{s}{s} = 1$

$\tan(u + v) = \dfrac{\tan u + \tan v}{1 - \tan u \tan v}$

$\qquad = \dfrac{(1/3) + (1/2)}{1 - (1/3)(1/2)}$

$\qquad = \dfrac{5/6}{1 - (1/6)}$

$\qquad = 1 = \tan w.$

Thus, $\tan(u + v) = \tan w$. Because u, v, and w are all in Quadrant I, we have

$\arctan[\tan(u + v)] = \arctan[\tan w]$

$\qquad u + v = w.$

81. $x = 0$: $y = -\frac{1}{2}(0 - 10) + 14 = 5 + 14 = 19$

y-intercept: $(0, 19)$

$y = 0$: $0 = -\frac{1}{2}(x - 10) + 14$

$\qquad = -\frac{1}{2}x + 19 \implies x = 38$

x-intercept: $(38, 0)$

83. $x = 0$: $|2(0) - 9| - 5 = 9 - 5 = 4.$

y-intercept: $(0, 4)$

$y = 0$: $|2x - 9| = 5 \implies x = 7, 2.$

x-intercepts: $(2, 0), (7, 0)$

85. $\arccos\left(\dfrac{\sqrt{3}}{2}\right) = \dfrac{\pi}{6}$ because $\cos\dfrac{\pi}{6} = \dfrac{\sqrt{3}}{2}.$

87. $\arcsin 1 = \dfrac{\pi}{2}$ because $\sin\dfrac{\pi}{2} = 1.$

Section 5.5   Multiple-Angle and Product-to-Sum Formulas

■ You should know the following double-angle formulas.

(a) $\sin 2u = 2 \sin u \cos u$ (b) $\cos 2u = \cos^2 u - \sin^2 u$ (c) $\tan 2u = \dfrac{2 \tan u}{1 - \tan^2 u}$

$= 2 \cos^2 u - 1$

$= 1 - 2 \sin^2 u$

■ You should be able to reduce the power of a trigonometric function.

(a) $\sin^2 u = \dfrac{1 - \cos 2u}{2}$ (b) $\cos^2 u = \dfrac{1 + \cos 2u}{2}$ (c) $\tan^2 u = \dfrac{1 - \cos 2u}{1 + \cos 2u}$

■ You should be able to use the half-angle formulas.

(a) $\sin \dfrac{u}{2} = \pm \sqrt{\dfrac{1 - \cos u}{2}}$ (b) $\cos \dfrac{u}{2} = \pm \sqrt{\dfrac{1 + \cos u}{2}}$ (c) $\tan \dfrac{u}{2} = \dfrac{1 - \cos u}{\sin u} = \dfrac{\sin u}{1 + \cos u}$

■ You should be able to use the product-sum formulas.

(a) $\sin u \sin v = \dfrac{1}{2}[\cos(u - v) - \cos(u + v)]$ (b) $\cos u \cos v = \dfrac{1}{2}[\cos(u - v) + \cos(u + v)]$

(c) $\sin u \cos v = \dfrac{1}{2}[\sin(u + v) + \sin(u - v)]$ (d) $\cos u \sin v = \dfrac{1}{2}[\sin(u + v) - \sin(u - v)]$

■ You should be able to use the sum-product formulas.

(a) $\sin x + \sin y = 2 \sin\left(\dfrac{x + y}{2}\right) \cos\left(\dfrac{x - y}{2}\right)$ (b) $\sin x - \sin y = 2 \cos\left(\dfrac{x + y}{2}\right) \sin\left(\dfrac{x - y}{2}\right)$

(c) $\cos x + \cos y = 2 \cos\left(\dfrac{x + y}{2}\right) \cos\left(\dfrac{x - y}{2}\right)$ (d) $\cos x - \cos y = -2 \sin\left(\dfrac{x + y}{2}\right) \sin\left(\dfrac{x - y}{2}\right)$

Solutions to Odd-Numbered Exercises

Figure for Exercises 1–7:

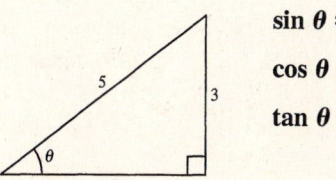

$\sin \theta = \dfrac{3}{5}$

$\cos \theta = \dfrac{4}{5}$

$\tan \theta = \dfrac{3}{4}$

1. $\sin \theta = \dfrac{3}{5}$

3. $\cos 2\theta = 2 \cos^2 \theta - 1$

$= 2\left(\dfrac{4}{5}\right)^2 - 1$

$= \dfrac{32}{25} - \dfrac{25}{25}$

$= \dfrac{7}{25}$

5. $\tan 2\theta = \dfrac{2 \tan \theta}{1 - \tan^2 \theta}$

$= \dfrac{2(3/4)}{1 - (3/4)^2}$

$= \dfrac{3/2}{1 - (9/16)}$

$= \dfrac{3}{2} \cdot \dfrac{16}{7} = \dfrac{24}{7}$

7. $\csc 2\theta = \dfrac{1}{\sin 2\theta}$

$= \dfrac{1}{2 \sin \theta \cos \theta}$

$= \dfrac{1}{2(3/5)(4/5)}$

$= \dfrac{25}{24}$

9. $\sin 2x - \sin x = 0$

Solutions: 0, 1.047, 3.142, 5.236

Analytically:

$$\sin 2x - \sin x = 0$$

$$2 \sin x \cos x - \sin x = 0$$

$$\sin x (2 \cos x - 1) = 0$$

$\sin x = 0$ or $2 \cos x - 1 = 0$

$x = 0, \pi$ $\qquad \cos x = \dfrac{1}{2}$

$x = 0, \dfrac{\pi}{3}, \pi, \dfrac{5\pi}{3}$ $\qquad x = \dfrac{\pi}{3}, \dfrac{5\pi}{3}$

11. $4 \sin x \cos x = 1$

$x \approx 0.2618, 1.3090, 3.4034, 4.4506$

Analytically:

$$4 \sin x \cos x = 1$$

$$2 \sin(2x) = 1$$

$$\sin(2x) = \dfrac{1}{2}$$

$$2x = \dfrac{\pi}{6}, \dfrac{5\pi}{6}, \dfrac{13\pi}{6}, \dfrac{17\pi}{6}$$

$$x = \dfrac{\pi}{12}, \dfrac{5\pi}{12}, \dfrac{13\pi}{12}, \dfrac{17\pi}{12}$$

13. $\cos 2x - \cos x = 0$

$x \approx 0, 2.094, 4.189, (6.283 \text{ not in interval})$

Analytically:

$$\cos 2x - \cos x = 0$$

$$2 \cos^2 x - 1 - \cos x = 0$$

$$(2 \cos x + 1)(\cos x - 1) = 0$$

$$\cos x = -\dfrac{1}{2}, \qquad \cos x = 1$$

$$x = \dfrac{2\pi}{3}, \dfrac{4\pi}{3}, 0, (2\pi \text{ not in interval})$$

15. Solutions: 0, 1.571, 3.142, 4.712

$$\sin 4x = -2 \sin 2x$$

$$\sin 4x + 2 \sin 2x = 0$$

$$2 \sin 2x \cos 2x + 2 \sin 2x = 0$$

$$2 \sin 2x (\cos 2x + 1) = 0$$

$2 \sin 2x = 0$ or $\cos 2x + 1 = 0$

$\sin 2x = 0$ $\qquad \cos 2x = -1$

$2x = n\pi$ $\qquad 2x = \pi + 2n\pi$

$x = \dfrac{n}{2}\pi$ $\qquad x = \dfrac{\pi}{2} + n\pi$

$x = 0, \dfrac{\pi}{2}, \pi, \dfrac{3\pi}{2}$ $\qquad x = \dfrac{\pi}{2}, \dfrac{3\pi}{2}$

17. $\sin u = \dfrac{3}{5}, \; 0 < u < \dfrac{\pi}{2} \implies \cos u = \dfrac{4}{5}$

$$\sin 2u = 2 \sin u \cos u = 2 \cdot \dfrac{3}{5} \cdot \dfrac{4}{5} = \dfrac{24}{25}$$

$$\cos 2u = \cos^2 u - \sin^2 u = \dfrac{16}{25} - \dfrac{9}{25} = \dfrac{7}{25}$$

$$\tan 2u = \dfrac{2 \tan u}{1 - \tan^2 u} = \dfrac{2(3/4)}{1 - (9/16)} = \dfrac{24}{7}$$

19. $\tan u = \dfrac{1}{2}, \; \pi < u < \dfrac{3\pi}{2} \implies \sin u = -\dfrac{1}{\sqrt{5}}$ and

$$\cos u = -\dfrac{2}{\sqrt{5}}$$

$$\sin 2u = 2 \sin u \cos u = 2\left(-\dfrac{1}{\sqrt{5}}\right)\left(-\dfrac{2}{\sqrt{5}}\right) = \dfrac{4}{5}$$

$$\cos 2u = \cos^2 u - \sin^2 u = \left(-\dfrac{2}{\sqrt{5}}\right)^2 - \left(-\dfrac{1}{\sqrt{5}}\right)^2 = \dfrac{3}{5}$$

$$\tan 2u = \dfrac{2 \tan u}{1 - \tan^2 u} = \dfrac{2(1/2)}{1 - (1/4)} = \dfrac{4}{3}$$

21. $\sec u = -\dfrac{5}{2}, \ \dfrac{\pi}{2} < u < \pi$

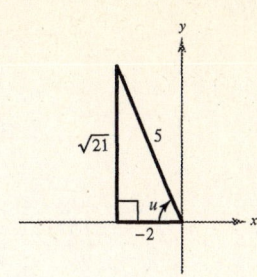

$\cos u = -\dfrac{2}{5} \implies \sin u = \dfrac{\sqrt{21}}{5}$

$\sin 2u = 2 \sin u \cos u = 2\left(\dfrac{\sqrt{21}}{5}\right)\left(-\dfrac{2}{5}\right) = \dfrac{-4\sqrt{21}}{25}$

$\cos 2u = \cos^2 u - \sin^2 u = \dfrac{4}{25} - \dfrac{21}{25} = -\dfrac{17}{25}$

$\tan 2u = \dfrac{2 \tan u}{1 - \tan^2 u} = \dfrac{2\left(\sqrt{21}/-2\right)}{1 - (21/4)} = \dfrac{-\sqrt{21}}{-17/4} = \dfrac{4\sqrt{21}}{17}$

23. $8 \sin x \cos x = 4(2 \sin x \cos x) = 4 \sin 2x$ **25.** $6 - 12 \sin^2 x = 6(1 - 2 \sin^2 x) = 6 \cos 2x$

27. $\cos^4 x = (\cos^2 x)(\cos^2 x) = \left(\dfrac{1 + \cos 2x}{2}\right)\left(\dfrac{1 + \cos 2x}{2}\right) = \dfrac{1 + 2 \cos 2x + \cos^2 2x}{4}$

$\qquad = \dfrac{1 + 2 \cos 2x + (1 + \cos 4x)/2}{4} = \dfrac{2 + 4 \cos 2x + 1 + \cos 4x}{8}$

$\qquad = \dfrac{3 + 4 \cos 2x + \cos 4x}{8} = \dfrac{1}{8}(3 + 4 \cos 2x + \cos 4x)$

29. $(\sin^2 x)(\cos^2 x) = \left(\dfrac{1 - \cos 2x}{2}\right)\left(\dfrac{1 + \cos 2x}{2}\right)$

$\qquad = \dfrac{1 - \cos^2 2x}{4} = \dfrac{1}{4}\left(1 - \dfrac{1 + \cos 4x}{2}\right) = \dfrac{1}{8}(2 - 1 - \cos 4x) = \dfrac{1}{8}(1 - \cos 4x)$

31. $\sin^2 x \cos^4 x = \sin^2 x \cos^2 x \cos^2 x = \left(\dfrac{1 - \cos 2x}{2}\right)\left(\dfrac{1 + \cos 2x}{2}\right)\left(\dfrac{1 + \cos 2x}{2}\right)$

$\qquad = \dfrac{1}{8}(1 - \cos 2x)(1 + \cos 2x)(1 + \cos 2x)$

$\qquad = \dfrac{1}{8}(1 - \cos^2 2x)(1 + \cos 2x)$

$\qquad = \dfrac{1}{8}(1 + \cos 2x - \cos^2 2x - \cos^3 2x)$

$\qquad = \dfrac{1}{8}\left[1 + \cos 2x - \left(\dfrac{1 + \cos 4x}{2}\right) - \cos 2x\left(\dfrac{1 + \cos 4x}{2}\right)\right]$

$\qquad = \dfrac{1}{16}[2 + 2 \cos 2x - 1 - \cos 4x - \cos 2x - \cos 2x \cos 4x]$

$\qquad = \dfrac{1}{16}\left[1 + \cos 2x - \cos 4x - \left(\dfrac{1}{2} \cos 2x + \dfrac{1}{2} \cos 6x\right)\right]$

$\qquad = \dfrac{1}{32}(2 + 2 \cos 2x - 2 \cos 4x - \cos 2x - \cos 6x)$

$\qquad = \dfrac{1}{32}(2 + \cos 2x - 2 \cos 4x - \cos 6x)$

Figure for Exercises 33–39:

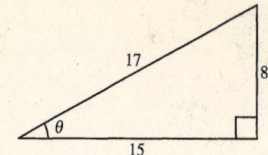

$$\sin \theta = \tfrac{8}{17}$$

$$\cos \theta = \tfrac{15}{17}$$

33. $\cos \dfrac{\theta}{2} = \sqrt{\dfrac{1 + \cos \theta}{2}} = \sqrt{\dfrac{1 + (15/17)}{2}} = \sqrt{\dfrac{16}{17}} = \dfrac{4}{\sqrt{17}} = \dfrac{4\sqrt{17}}{17}$

35. $\tan \dfrac{\theta}{2} = \dfrac{\sin \theta}{1 + \cos \theta} = \dfrac{8/17}{1 + (15/17)} = \dfrac{8}{32} = \dfrac{1}{4}$

37. $\csc \dfrac{\theta}{2} = \dfrac{1}{\sin(\theta/2)} = \dfrac{1}{\sqrt{(1 - \cos\theta)/2}} = \dfrac{1}{\sqrt{[1 - (15/17)]/2}} = \dfrac{1}{1/\sqrt{17}} = \sqrt{17}$

39. $2 \sin \dfrac{\theta}{2} \cos \dfrac{\theta}{2} = 2\left(\dfrac{1}{\sqrt{17}}\right)\left(\dfrac{4\sqrt{17}}{17}\right) = \dfrac{8}{17}$ $(= \sin \theta)$

41. $\sin 15° = \sin\left(\dfrac{1}{2} \cdot 30°\right) = \sqrt{\dfrac{1 - \cos 30°}{2}} = \sqrt{\dfrac{1 - (\sqrt{3}/2)}{2}} = \dfrac{1}{2}\sqrt{2 - \sqrt{3}}$

$\cos 15° = \cos\left(\dfrac{1}{2} \cdot 30°\right) = \sqrt{\dfrac{1 + \cos 30°}{2}} = \sqrt{\dfrac{1 + (\sqrt{3}/2)}{2}} = \dfrac{1}{2}\sqrt{2 + \sqrt{3}}$

$\tan 15° = \tan\left(\dfrac{1}{2} \cdot 30°\right) = \dfrac{\sin 30°}{1 + \cos 30°} = \dfrac{1/2}{1 + (\sqrt{3}/2)} = \dfrac{1}{2 + \sqrt{3}} = 2 - \sqrt{3}$

43. $\sin 112° \, 30' = \sin\left(\dfrac{1}{2} \cdot 225°\right) = \sqrt{\dfrac{1 - \cos 225°}{2}} = \sqrt{\dfrac{1 + (\sqrt{2}/2)}{2}} = \dfrac{1}{2}\sqrt{2 + \sqrt{2}}$

$\cos 112° \, 30' = \cos\left(\dfrac{1}{2} \cdot 225°\right) = -\sqrt{\dfrac{1 + \cos 225°}{2}} = -\sqrt{\dfrac{1 - (\sqrt{2}/2)}{2}} = -\dfrac{1}{2}\sqrt{2 - \sqrt{2}}$

$\tan 112° \, 30' = \tan\left(\dfrac{1}{2} \cdot 225°\right) = \dfrac{\sin 225°}{1 + \cos 225°} = \dfrac{-\sqrt{2}/2}{1 - (\sqrt{2}/2)} = -1 - \sqrt{2}$

45. $\sin \dfrac{\pi}{8} = \sin\left[\dfrac{1}{2}\left(\dfrac{\pi}{4}\right)\right] = \sqrt{\dfrac{1 - \cos(\pi/4)}{2}} = \dfrac{1}{2}\sqrt{2 - \sqrt{2}}$

$\cos \dfrac{\pi}{8} = \cos\left[\dfrac{1}{2}\left(\dfrac{\pi}{4}\right)\right] = \sqrt{\dfrac{1 + \cos(\pi/4)}{2}} = \dfrac{1}{2}\sqrt{2 + \sqrt{2}}$

$\tan \dfrac{\pi}{8} = \tan\left[\dfrac{1}{2}\left(\dfrac{\pi}{4}\right)\right] = \dfrac{\sin(\pi/4)}{1 + \cos(\pi/4)} = \dfrac{\sqrt{2}/2}{1 + (\sqrt{2}/2)} = \sqrt{2} - 1$

47. $\sin \dfrac{3\pi}{8} = \sin\left(\dfrac{1}{2} \cdot \dfrac{3\pi}{4}\right) = \sqrt{\dfrac{1 - \cos(3\pi/4)}{2}} = \sqrt{\dfrac{1 + (\sqrt{2}/2)}{2}} = \dfrac{1}{2}\sqrt{2 + \sqrt{2}}$

$\cos \dfrac{3\pi}{8} = \cos\left(\dfrac{1}{2} \cdot \dfrac{3\pi}{4}\right) = \sqrt{\dfrac{1 + \cos(3\pi/4)}{2}} = \sqrt{\dfrac{1 - (\sqrt{2}/2)}{2}} = \dfrac{1}{2}\sqrt{2 - \sqrt{2}}$

$\tan \dfrac{3\pi}{8} = \tan\left(\dfrac{1}{2} \cdot \dfrac{3\pi}{4}\right) = \dfrac{\sin(3\pi/4)}{1 + \cos(3\pi/4)} = \dfrac{\sqrt{2}/2}{1 - (\sqrt{2}/2)} = \dfrac{\sqrt{2}}{2 - \sqrt{2}} = \sqrt{2} + 1$

49. $\sin u = \dfrac{5}{13}, \dfrac{\pi}{2} < u < \pi \Rightarrow \cos u = -\dfrac{12}{13}$

$$\sin\left(\frac{u}{2}\right) = \sqrt{\frac{1 - \cos u}{2}} = \sqrt{\frac{1 + (12/13)}{2}} = \frac{5\sqrt{26}}{26}$$

$$\cos\left(\frac{u}{2}\right) = \sqrt{\frac{1 + \cos u}{2}} = \sqrt{\frac{1 - (12/13)}{2}} = \frac{\sqrt{26}}{26}$$

$$\tan\left(\frac{u}{2}\right) = \frac{\sin u}{1 + \cos u} = \frac{5/13}{1 - (12/13)} = \frac{5}{1} = 5$$

51. $\tan u = -\dfrac{8}{5}, \dfrac{3\pi}{2} < u < 2\pi$, Quadrant IV

$$\sin u = -\frac{8}{\sqrt{89}}, \cos u = \frac{5}{\sqrt{89}}$$

$$\sin\left(\frac{u}{2}\right) = \sqrt{\frac{1 - \cos u}{2}}$$

$$= \sqrt{\frac{1 - (5/\sqrt{89})}{2}} = \sqrt{\frac{\sqrt{89} - 5}{2\sqrt{89}}} = \sqrt{\frac{89 - 5\sqrt{89}}{178}}$$

$$\cos\left(\frac{u}{2}\right) = -\sqrt{\frac{1 + \cos u}{2}} = -\sqrt{\frac{1 + (5/\sqrt{89})}{2}} = -\sqrt{\frac{\sqrt{89} + 5}{2\sqrt{89}}} = -\sqrt{\frac{89 + 5\sqrt{89}}{178}}$$

$$\tan\left(\frac{u}{2}\right) = \frac{1 - \cos u}{\sin u} = \frac{1 - (5/\sqrt{89})}{-8/\sqrt{89}} = \frac{5 - \sqrt{89}}{8}$$

53. $\csc u = -\dfrac{5}{3}, \pi < u < \dfrac{3\pi}{2}$, Quadrant III

$$\sin u = -\frac{3}{5}, \cos u = -\frac{4}{5}$$

$$\sin\left(\frac{u}{2}\right) = \sqrt{\frac{1 - \cos u}{2}} = \sqrt{\frac{1 + (4/5)}{2}} = \frac{3}{\sqrt{10}} = \frac{3\sqrt{10}}{10}$$

$$\cos\left(\frac{u}{2}\right) = -\sqrt{\frac{1 + \cos u}{2}} = -\sqrt{\frac{1 - (4/5)}{2}} = \frac{-1}{\sqrt{10}} = -\frac{\sqrt{10}}{10}$$

$$\tan\left(\frac{u}{2}\right) = \frac{1 - \cos u}{\sin u} = \frac{1 + (4/5)}{-3/5} = -3$$

55. $\sqrt{\dfrac{1 - \cos 6x}{2}} = |\sin 3x|$

57. $-\sqrt{\dfrac{1 - \cos 8x}{1 + \cos 8x}} = -\dfrac{\sqrt{(1 - \cos 8x)/2}}{\sqrt{(1 + \cos 8x)/2}}$

$$= -\left|\frac{\sin 4x}{\cos 4x}\right|$$

$$= -|\tan 4x|$$

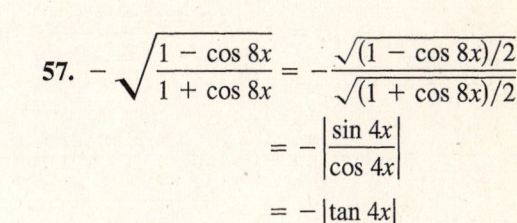

59. $\sin \dfrac{x}{2} - \cos x = 0$

$$\pm \sqrt{\dfrac{1 - \cos x}{2}} = \cos x$$

$$\dfrac{1 - \cos x}{2} = \cos^2 x$$

$$0 = 2\cos^2 x + \cos x - 1$$

$$= (2\cos x - 1)(\cos x + 1)$$

$$\cos x = \dfrac{1}{2} \quad \text{or} \quad \cos x = -1$$

$$x = \dfrac{\pi}{3}, \dfrac{5\pi}{3} \qquad x = \pi$$

By checking these values in the original equations, we see that $x = \pi/3$ and $x = 5\pi/3$ are the only solutions. $x = \pi$ is extraneous.

61. $\cos \dfrac{x}{2} - \sin x = 0$

$$\pm \sqrt{\dfrac{1 + \cos x}{2}} = \sin x$$

$$\dfrac{1 + \cos x}{2} = \sin^2 x$$

$$1 + \cos x = 2\sin^2 x$$

$$1 + \cos x = 2 - 2\cos^2 x$$

$$2\cos^2 x + \cos x - 1 = 0$$

$$(2\cos x - 1)(\cos x + 1) = 0$$

$$2\cos x - 1 = 0 \quad \text{or} \quad \cos x + 1 = 0$$

$$\cos x = \dfrac{1}{2} \qquad\qquad \cos x = -1$$

$$x = \dfrac{\pi}{3}, \dfrac{5\pi}{3} \qquad\qquad x = \pi$$

$$x = \dfrac{\pi}{3}, \pi, \dfrac{5\pi}{3}$$

$\pi/3$, π, and $5\pi/3$ are all solutions to the equation.

63. $6 \sin \dfrac{\pi}{3} \cos \dfrac{\pi}{3} = 6 \cdot \dfrac{1}{2}\left[\sin\left(\dfrac{\pi}{3} + \dfrac{\pi}{3}\right) + \sin\left(\dfrac{\pi}{3} - \dfrac{\pi}{3}\right) \right]$

$$= 3\left[\sin \dfrac{2\pi}{3} + \sin 0 \right] = 3 \sin \dfrac{2\pi}{3}$$

65. $\sin 5\theta \cos 3\theta = \dfrac{1}{2}[\sin(5\theta + 3\theta) + \sin(5\theta - 3\theta)]$

$$= \dfrac{1}{2}(\sin 8\theta + \sin 2\theta)$$

67. $5 \cos(-5\beta) \cos 3\beta = 5 \cdot \frac{1}{2}[\cos(-5\beta - 3\beta) + \cos(-5\beta + 3\beta)]$

$$= \tfrac{5}{2}[\cos(-8\beta) + \cos(-2\beta)]$$

$$= \tfrac{5}{2}(\cos 8\beta + \cos 2\beta)$$

69. $\sin 5\theta - \sin \theta = 2 \cos\left(\dfrac{5\theta + \theta}{2}\right) \sin\left(\dfrac{5\theta - \theta}{2}\right) = 2 \cos 3\theta \cdot \sin 2\theta$

71. $\sin(\alpha + \beta) - \sin(\alpha - \beta) = 2 \cos\left(\dfrac{\alpha + \beta + \alpha - \beta}{2}\right) \sin\left(\dfrac{\alpha + \beta - \alpha + \beta}{2}\right) = 2 \cos \alpha \sin \beta$

73. $\cos\left(\theta + \dfrac{\pi}{2}\right) - \cos\left(\theta - \dfrac{\pi}{2}\right) = -2 \sin\left(\dfrac{\theta + (\pi/2) + \theta - (\pi/2)}{2}\right) \sin\left(\dfrac{\theta + (\pi/2) - \theta + (\pi/2)}{2}\right)$

$$= -2 \sin \theta \sin \dfrac{\pi}{2} = -2\sin \theta$$

75. $\sin 195° + \sin 105° = 2 \sin\left(\dfrac{195° + 105°}{2}\right) \cos\left(\dfrac{195° - 105°}{2}\right)$

$$= 2 \sin(150°) \cos(45°) = 2\left(\dfrac{1}{2}\right)\left(\dfrac{\sqrt{2}}{2}\right) = \dfrac{\sqrt{2}}{2}$$

77. $\cos\dfrac{5\pi}{12} + \cos\dfrac{\pi}{12} = 2\cos\left(\dfrac{(5\pi/12) + (\pi/12)}{2}\right)\cos\left(\dfrac{(5\pi/12) - (\pi/12)}{2}\right)$

$$= 2\cos\left(\dfrac{\pi}{4}\right)\cos\left(\dfrac{\pi}{6}\right) = 2\left(\dfrac{\sqrt{2}}{2}\right)\left(\dfrac{\sqrt{3}}{2}\right) = \dfrac{2\sqrt{6}}{4} = \dfrac{\sqrt{6}}{2}$$

79. $\qquad\qquad \sin 6x + \sin 2x = 0$

$$2\sin\left(\dfrac{6x + 2x}{2}\right)\cos\left(\dfrac{6x - 2x}{2}\right) = 0$$

$$\sin 4x \cos 2x = 0$$

$\sin 4x = 0 \quad\text{or}\quad \cos 2x = 0$

$\qquad 4x = n\pi \qquad\qquad 2x = \dfrac{\pi}{2} + n\pi$

$\qquad x = \dfrac{n\pi}{4} \qquad\qquad x = \dfrac{\pi}{4} + \dfrac{n\pi}{2}$

In the interval we have

$$x = 0, \dfrac{\pi}{4}, \dfrac{\pi}{2}, \dfrac{3\pi}{4}, \pi, \dfrac{5\pi}{4}, \dfrac{3\pi}{2}, \dfrac{7\pi}{4}.$$

81. $\dfrac{\cos 2x}{\sin 3x - \sin x} - 1 = 0$

$$\dfrac{\cos 2x}{\sin 3x - \sin x} = 1$$

$$\dfrac{\cos 2x}{2\cos 2x \sin x} = 1$$

$$2\sin x = 1$$

$$\sin x = \dfrac{1}{2}$$

$$x = \dfrac{\pi}{6}, \dfrac{5\pi}{6}$$

83. $\sin^2 \alpha = \left(\dfrac{5}{13}\right)^2 = \dfrac{25}{169}$

$\sin^2 \alpha = 1 - \cos^2 \alpha$

$$= 1 - \left(\dfrac{12}{13}\right)^2 = 1 - \dfrac{144}{169} = \dfrac{25}{169}$$

85. $\sin \alpha \cos \beta = \left(\dfrac{5}{13}\right)\left(\dfrac{4}{5}\right) = \dfrac{4}{13}$

$$\sin \alpha \cos \beta = \cos\left(\dfrac{\pi}{2} - \alpha\right)\sin\left(\dfrac{\pi}{2} - \beta\right)$$

$$= \left(\dfrac{5}{13}\right)\left(\dfrac{4}{5}\right) = \dfrac{4}{13}$$

87. $\csc 2\theta = \dfrac{1}{\sin 2\theta}$

$$= \dfrac{1}{2\sin\theta\cos\theta}$$

$$= \dfrac{1}{\sin\theta} \cdot \dfrac{1}{2\cos\theta}$$

$$= \dfrac{\csc\theta}{2\cos\theta}$$

89. $\cos^2 2\alpha - \sin^2 2\alpha = \cos[2(2\alpha)]$

$$= \cos 4\alpha$$

91. $(\sin x + \cos x)^2 = \sin^2 x + 2\sin x \cos x + \cos^2 x$

$$= (\sin^2 x + \cos^2 x) + 2\sin x \cos x$$

$$= 1 + \sin 2x$$

93. $\sec \dfrac{u}{2} = \dfrac{1}{\cos(u/2)}$

$= \pm \sqrt{\dfrac{2}{1 + \cos u}} = \pm \sqrt{\dfrac{2 \sin u}{\sin u (1 + \cos u)}} = \pm \sqrt{\dfrac{2 \sin u}{\sin u + \sin u \cos u}}$

$= \pm \sqrt{\dfrac{(2 \sin u)/(\cos u)}{(\sin u)/(\cos u) + (\sin u \cos u)/(\cos u)}} = \pm \sqrt{\dfrac{2 \tan u}{\tan u + \sin u}}$

95. $\cos 3\beta = \cos(2\beta + \beta)$

$= \cos 2\beta \cos \beta - \sin 2\beta \sin \beta$

$= (\cos^2 \beta - \sin^2 \beta) \cos \beta - 2 \sin \beta \cos \beta \sin \beta$

$= \cos^3 \beta - \sin^2 \beta \cos \beta - 2 \sin^2 \beta \cos \beta$

$= \cos^3 \beta - 3 \sin^2 \beta \cos \beta$

97. $\dfrac{\cos 4x - \cos 2x}{2 \sin 3x} = \dfrac{-2 \sin\left(\dfrac{4x + 2x}{2}\right) \sin\left(\dfrac{4x - 2x}{2}\right)}{2 \sin 3x}$

$= \dfrac{-2 \sin 3x \sin x}{2 \sin 3x}$

$= -\sin x$

99. $\sin^2 x = \dfrac{1 - \cos 2x}{2} = \dfrac{1}{2} - \dfrac{\cos 2x}{2}$

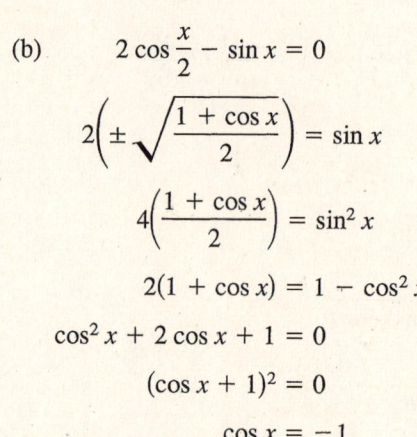

101. $f(x) = \cos^4 x = \dfrac{1}{8}(3 + 4 \cos 2x + \cos 4x)$

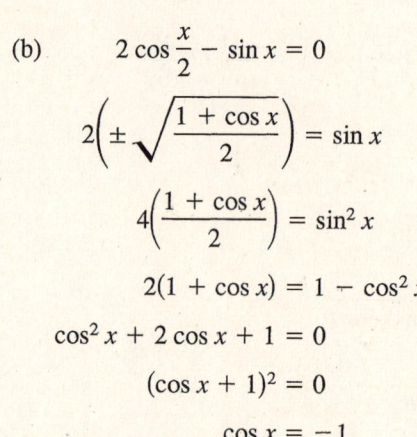

103. (a) $y = 4 \sin \dfrac{x}{2} + \cos x$

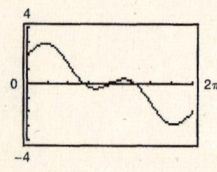

Maximum: $(\pi, 3)$

(b) $\quad 2 \cos \dfrac{x}{2} - \sin x = 0$

$2\left(\pm \sqrt{\dfrac{1 + \cos x}{2}}\right) = \sin x$

$4\left(\dfrac{1 + \cos x}{2}\right) = \sin^2 x$

$2(1 + \cos x) = 1 - \cos^2 x$

$\cos^2 x + 2 \cos x + 1 = 0$

$(\cos x + 1)^2 = 0$

$\cos x = -1$

$x = \pi$

105. (a) $y = 2 \cos \dfrac{x}{2} + \sin 2x$

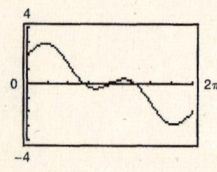

Maximum: $(0.699, 2.864)$

Minimum: $(5.584, -2.864)$

(b) $2 \cos 2x - \sin \dfrac{x}{2} = 0$ has four zeros on $[0, 2\pi)$.

Two of the zeros are $x = 0.699$ and $x = 5.584$.

(The other two are $2.608, 3.675$)

107. $\sin(2 \arcsin x) = 2 \sin(\arcsin x) \cos(\arcsin x)$

$$= 2x\sqrt{1 - x^2}$$

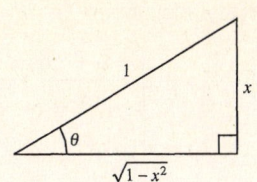

109. $\cos(2 \arcsin x) = 1 - 2 \sin^2(\arcsin x)$

$$= 1 - 2x^2$$

111. $\cos(2 \arctan x) = 1 - 2 \sin^2(\arctan x)$

$$= 1 - 2\left(\frac{x}{\sqrt{1 + x^2}}\right)^2$$

$$= 1 - \frac{2x^2}{1 + x^2}$$

$$= \frac{1 - x^2}{1 + x^2}$$

113. $r = \dfrac{1}{32} v_0{}^2 \sin 2\theta$

$$= \frac{1}{32} v_0{}^2 (2 \sin \theta \cos \theta)$$

$$= \frac{1}{16} v_0{}^2 \sin \theta \cos \theta$$

115. $\sin \dfrac{\theta}{2} = \dfrac{1}{M}$

(a) $\sin \dfrac{\theta}{2} = \dfrac{1}{1} = 1 \implies \dfrac{\theta}{2} = \dfrac{\pi}{2} \implies \theta = \pi = 180°$

(b) $\sin \dfrac{\theta}{2} = \dfrac{1}{4.5} = \dfrac{2}{9}$

$$\frac{\theta}{2} = \arcsin\left(\frac{2}{9}\right) \approx 0.2241$$

$$\theta \approx 0.4482 \approx 25.7°$$

(c) $M = 1 \implies$ Speed $= 760$ mph

$$M = 4.5 \implies \frac{\text{Speed}}{760} = 4.5 \implies \text{Speed} = 3420 \text{ mph}$$

(d) $\sin \dfrac{\theta}{2} = \sqrt{\dfrac{1 - \cos \theta}{2}} = \dfrac{1}{M}$

$$\frac{1 - \cos \theta}{2} = \frac{1}{M^2}$$

$$1 - \cos \theta = \frac{2}{M^2}$$

$$\cos \theta = 1 - \frac{2}{M^2}$$

117. False. If $x = \pi$, $\sin \dfrac{x}{2} = \sin \dfrac{\pi}{2} = 1$, whereas $-\sqrt{\dfrac{1 - \cos \pi}{2}} = -1$.

119. $f(x) = 2 \sin x \left[2 \cos^2\left(\dfrac{x}{2}\right) - 1 \right]$

(a)

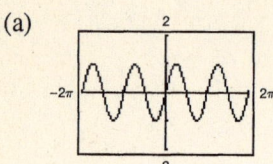

(b) The graph appears to be that of $y = \sin 2x$.

(c) $2 \sin x \left[2 \cos^2\left(\dfrac{x}{2}\right) - 1 \right] = 2 \sin x \left[2\dfrac{1 + \cos x}{2} - 1 \right]$

$\qquad\qquad = 2 \sin x [\cos x]$

$\qquad\qquad = \sin 2x$

121. Answers will vary.

123.

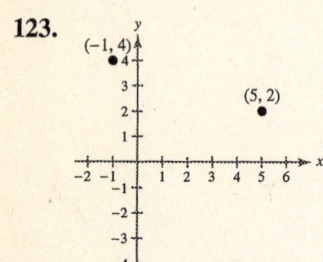

Distance: $\sqrt{(5 + 1)^2 + (2 - 4)^2} = \sqrt{40} = 2\sqrt{10}$

Midpoint: $\left(\dfrac{-1 + 5}{2}, \dfrac{4 + 2}{2} \right) = (2, 3)$

125.

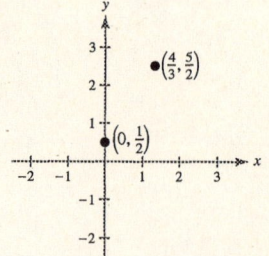

Distance: $\sqrt{\left(\dfrac{4}{3}\right)^2 + \left(\dfrac{5}{2} - \dfrac{1}{2}\right)^2} = \sqrt{\dfrac{16}{9} + 4}$

$\qquad\qquad = \dfrac{\sqrt{52}}{3} = \dfrac{2\sqrt{13}}{3}$

Midpoint: $\left(\dfrac{0 + (4/3)}{2}, \dfrac{(1/2) + (5/2)}{2} \right) = \left(\dfrac{2}{3}, \dfrac{3}{2} \right)$

127. (a) Complement: $90° - 55° = 35°$

Supplement: $180° - 55° = 125°$

(b) Complement: none

Supplement: $180° - 162° = 18°$

129. (a) Complement: $\dfrac{\pi}{2} - \dfrac{\pi}{18} = \dfrac{8\pi}{18} = \dfrac{4\pi}{9}$

Supplement: $\pi - \dfrac{\pi}{18} = \dfrac{17\pi}{18}$

(b) Complement: $\dfrac{\pi}{2} - \dfrac{9\pi}{20} = \dfrac{\pi}{20}$

Supplement: $\pi - \dfrac{9\pi}{20} = \dfrac{11\pi}{20}$

131. $s = r\theta \implies \theta = \dfrac{s}{r} = \dfrac{7}{15} \approx 0.467 \text{ rad}$

133. $f(x) = \dfrac{3}{2} \cos(2x)$

Period: $\dfrac{2\pi}{2} = \pi$

Amplitude: $\dfrac{3}{2}$

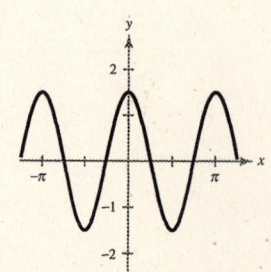

135. $f(x) = \dfrac{1}{2} \tan(2\pi x)$

Period: $\dfrac{\pi}{2\pi} = \dfrac{1}{2}$

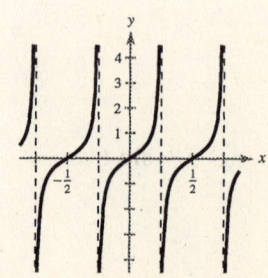

Review Exercises for Chapter 5

Solutions to Odd-Numbered Exercises

1. $\dfrac{1}{\cos x} = \sec x$ **3.** $\dfrac{1}{\sec x} = \cos x$ **5.** $\sqrt{1 - \cos^2 x} = \pm \sin x$ **7.** $\csc\left(\dfrac{\pi}{2} - x\right) = \sec x$

9. $\sec(-x) = \sec x$

11. $\sin x = \dfrac{4}{5}, \cos x = \dfrac{3}{5}$, Quadrant I

$$\tan x = \frac{\sin x}{\cos x} = \frac{4}{3}$$

$$\cot x = \frac{3}{4}$$

$$\sec x = \frac{5}{3}$$

$$\csc x = \frac{5}{4}$$

13. $\sin\left(\dfrac{\pi}{2} - x\right) = \cos x$

$$= \frac{1}{\sqrt{2}} = \frac{\sqrt{2}}{2},$$

$$\sin x = -\frac{1}{\sqrt{2}} = -\frac{\sqrt{2}}{2},$$

Quadrant IV

$$\tan x = -1$$

$$\cot x = -1$$

$$\sec x = \sqrt{2}$$

$$\csc x = -\sqrt{2}$$

15. $\dfrac{1}{\cot^2 x + 1} = \dfrac{1}{\csc^2 x} = \sin^2 x$

17. $\dfrac{\sin^2 \alpha - \cos^2 \alpha}{\sin^2 \alpha - \sin \alpha \cos \alpha} = \dfrac{(\sin \alpha + \cos \alpha)(\sin \alpha - \cos \alpha)}{\sin \alpha(\sin \alpha - \cos \alpha)}$

$$= \frac{\sin \alpha + \cos \alpha}{\sin \alpha}$$

$$= 1 + \cot \alpha$$

19. $\tan^2 \theta \left(\csc^2 \theta - 1\right) = \tan^2 \theta(\cot^2 \theta)$

$$= \tan^2 \theta\left(\frac{1}{\tan^2 \theta}\right)$$

$$= 1$$

21. $\tan\left(\dfrac{\pi}{2} - x\right) \sec x = \cot x \sec x$

$$= \frac{\cos x}{\sin x} \cdot \frac{1}{\cos x}$$

$$= \frac{1}{\sin x} = \csc x$$

23. $\sin^{-1/2} x \cos x = \dfrac{\cos x}{\sin^{1/2} x}$

$$= \frac{\cos x}{\sqrt{\sin x}} \cdot \frac{\sqrt{\sin x}}{\sqrt{\sin x}}$$

$$= \frac{\cos x}{\sin x}\sqrt{\sin x}$$

$$= \cot x \sqrt{\sin x}$$

25. $\cos x(\tan^2 x + 1) = \cos x \sec^2 x$

$$= \frac{1}{\sec x} \sec^2 x$$

$$= \sec x$$

27. $\sin^3 \theta + \sin \theta \cos^2 \theta = \sin \theta(\sin^2 \theta + \cos^2 \theta)$

$$= \sin \theta$$

29. $\sin^5 x \cos^2 x = \sin^4 x \cos^2 x \sin x$

$$= (1 - \cos^2 x)^2 \cos^2 x \sin x$$

$$= (1 - 2\cos^2 x + \cos^4 x)\cos^2 x \sin x$$

$$= (\cos^2 x - 2\cos^4 x + \cos^6 x)\sin x$$

31. $\sqrt{\dfrac{1-\sin\theta}{1+\sin\theta}} = \sqrt{\dfrac{1-\sin\theta}{1+\sin\theta}\cdot\dfrac{1-\sin\theta}{1-\sin\theta}} = \sqrt{\dfrac{(1-\sin\theta)^2}{1-\sin^2\theta}} = \sqrt{\dfrac{(1-\sin\theta)^2}{\cos^2\theta}} = \dfrac{|1-\sin\theta|}{|\cos\theta|} = \dfrac{1-\sin\theta}{|\cos\theta|}$

Note: We can drop the absolute value on $1-\sin\theta$ since it is always nonnegative.

33. $\dfrac{\csc(-x)}{\sec(-x)} = -\dfrac{\csc x}{\sec x} = -\dfrac{\cos x}{\sin x} = -\cot x$

35. $\sin^2 x + \sin^2\left(\dfrac{\pi}{2}-x\right) = \sin^2 x + \cos^2 x = 1$

37. $2\sin x - 1 = 0$

$\sin x = \dfrac{1}{2}$

$x = \dfrac{\pi}{6} + 2n\pi$

$x = \dfrac{5\pi}{6} + 2n\pi$

39. $\sin x = \sqrt{3} - \sin x$

$2\sin x = \sqrt{3}$

$\sin x = \dfrac{\sqrt{3}}{2}$

$x = \dfrac{\pi}{3} + 2n\pi$

$x = \dfrac{2\pi}{3} + 2n\pi$

41. $3\sqrt{3}\tan x = 3$

$\tan x = \dfrac{1}{\sqrt{3}}$

$x = \dfrac{\pi}{6} + n\pi$

43. $3\csc^2 x = 4$

$\csc^2 x = \dfrac{4}{3}$

$\sin^2 x = \dfrac{3}{4}$

$\sin x = \pm\dfrac{\sqrt{3}}{2}$

$x = \dfrac{\pi}{3} + n\pi$

$x = \dfrac{2\pi}{3} + n\pi$

45. $4\cos^2 x - 3 = 0$

$\cos^2 x = \dfrac{3}{4}$

$\cos x = \pm\dfrac{\sqrt{3}}{2}$

$x = \dfrac{\pi}{6} + n\pi$

$x = \dfrac{5\pi}{6} + n\pi$

47. $\sin x - \tan x = 0$

$\sin x - \dfrac{\sin x}{\cos x} = 0$

$\sin x\cos x - \sin x = 0$

$\sin x(\cos x - 1) = 0$

$\sin x = 0 \quad \text{or} \quad \cos x - 1 = 0$

$x = n\pi \qquad\qquad \cos x = 1$

49. $2\cos^2 x - \cos x - 1 = 0$

$(2\cos x + 1)(\cos x - 1) = 0$

$2\cos x + 1 = 0 \qquad \text{or} \quad \cos x - 1 = 0$

$\cos x = -\dfrac{1}{2} \qquad\qquad \cos x = 1$

$x = \dfrac{2\pi}{3}, \dfrac{4\pi}{3} \qquad\qquad x = 0$

51. $\cos^2 x + \sin x = 1$

$1 - \sin^2 x + \sin x = 1$

$\sin x(\sin x - 1) = 0$

$\sin x = 0 \qquad \text{or} \quad \sin x = 1$

$x = 0, \pi \qquad\qquad x = \dfrac{\pi}{2}$

53. $2\sin 2x = \sqrt{2}$

$\sin 2x = \dfrac{\sqrt{2}}{2}$

$2x = \dfrac{\pi}{4}, \dfrac{3\pi}{4}, \dfrac{9\pi}{4}, \dfrac{11\pi}{4}$

$x = \dfrac{\pi}{8}, \dfrac{3\pi}{8}, \dfrac{9\pi}{8}, \dfrac{11\pi}{8}$

55. $\cos 4x(\cos x - 1) = 0$

$\cos 4x = 0 \text{ or } \cos x - 1 = 0$

$4x = \dfrac{\pi}{2}, \dfrac{3\pi}{2}, \dfrac{5\pi}{2}, \dfrac{7\pi}{2}, \dfrac{9\pi}{2}, \dfrac{11\pi}{2}, \dfrac{13\pi}{2}, \dfrac{15\pi}{2}$

or $\cos x = 1$

$x = \dfrac{\pi}{8}, \dfrac{3\pi}{8}, \dfrac{5\pi}{8}, \dfrac{7\pi}{8}, \dfrac{9\pi}{8}, \dfrac{11\pi}{8}, \dfrac{13\pi}{8}, \dfrac{15\pi}{8}, 0$

57.
$$\cos 4x - 7 \cos 2x = 8$$
$$2 \cos^2 2x - 1 - 7 \cos 2x = 8$$
$$2 \cos^2 2x - 7 \cos 2x - 9 = 0$$
$$(2 \cos 2x - 9)(\cos 2x + 1) = 0$$
$$2 \cos 2x - 9 = 0 \quad \text{or} \quad \cos 2x + 1 = 0$$

$$\cos 2x = \frac{9}{2} \qquad \qquad \cos 2x = -1$$

No solution
$$2x = \pi + 2n\pi$$
$$x = \frac{\pi}{2} + n\pi$$
$$x = \frac{\pi}{2}, \frac{3\pi}{2}$$

59. $\sin^2 x - 2 \sin x = 0$
$$\sin x(\sin x - 2) = 0$$
$$\sin x = 0 \quad \text{or} \quad \sin x = 2 \text{ (impossible)}$$
$$x = 0, \pi$$

61. $\tan^2 \theta + \tan \theta - 12 = 0$
$$(\tan \theta + 4)(\tan \theta - 3) = 0$$
$$\tan \theta = -4 \quad \text{or} \quad \tan \theta = 3$$
$$\theta = 1.8158, \, 4.9574, \, 1.2490, \, 4.3906$$

63. $\sin 285° = \sin(315° - 30°)$
$$= \sin 315° \cos 30° - \cos 315° \sin 30°$$
$$= \left(-\frac{\sqrt{2}}{2}\right)\left(\frac{\sqrt{3}}{2}\right) - \left(\frac{\sqrt{2}}{2}\right)\left(\frac{1}{2}\right) = -\frac{\sqrt{6} + \sqrt{2}}{4}$$
$$\cos 285° = \cos(315° - 30°)$$
$$= \cos 315° \cos 30° + \sin 315° \sin 30°$$
$$= \left(\frac{\sqrt{2}}{2}\right)\left(\frac{\sqrt{3}}{2}\right) + \left(-\frac{\sqrt{2}}{2}\right)\left(\frac{1}{2}\right) = \frac{\sqrt{6} - \sqrt{2}}{4}$$
$$\tan 285° = -\frac{\sqrt{6} + \sqrt{2}}{\sqrt{6} - \sqrt{2}} = -2 - \sqrt{3}$$

65. $\sin \dfrac{25\pi}{12} = \sin\left(\dfrac{11\pi}{6} + \dfrac{\pi}{4}\right) = \sin \dfrac{11\pi}{6} \cos \dfrac{\pi}{4} + \sin \dfrac{\pi}{4} \cos \dfrac{11\pi}{6}$
$$= \left(-\frac{1}{2}\right)\frac{\sqrt{2}}{2} + \frac{\sqrt{2}}{2} \cdot \frac{\sqrt{3}}{2} = \frac{\sqrt{6} - \sqrt{2}}{4}$$
$$\cos \frac{25\pi}{12} = \cos\left(\frac{11\pi}{6} + \frac{\pi}{4}\right) = \cos \frac{11\pi}{6} \cos \frac{\pi}{4} - \sin \frac{11\pi}{6} \sin \frac{\pi}{4}$$
$$= \frac{\sqrt{3}}{2} \cdot \frac{\sqrt{2}}{2} - \left(-\frac{1}{2}\right)\frac{\sqrt{2}}{2} = \frac{\sqrt{6} + \sqrt{2}}{4}$$
$$\tan \frac{25\pi}{12} = \frac{\sin(25\pi/12)}{\cos(25\pi/12)} = \frac{\sqrt{6} - \sqrt{2}}{\sqrt{6} + \sqrt{2}} = 2 - \sqrt{3}$$

67. $\sin 140° \cos 50° + \cos 140° \sin 50° = \sin(140° + 50°) = \sin 190°$

69. $\dfrac{\tan 25° + \tan 10°}{1 - \tan 25° \tan 10°} = \tan(25° + 10°) = \tan 35°$

For Exercises 71–75:

$\sin u = \dfrac{3}{4}$, u **in Quadrant II** $\Rightarrow \cos u = -\dfrac{\sqrt{7}}{4}$, $\tan u = -\dfrac{3}{\sqrt{7}} = \dfrac{-3\sqrt{7}}{7}$

$\cos v = -\dfrac{5}{13}$, v **in Quadrant II** $\Rightarrow \sin v = \dfrac{12}{13}$, $\tan v = -\dfrac{12}{5}$

71. $\sin(u + v) = \sin u \cos v + \cos u \sin v$

$= \left(\dfrac{3}{4}\right)\left(-\dfrac{5}{13}\right) + \left(-\dfrac{\sqrt{7}}{4}\right)\left(\dfrac{12}{13}\right)$

$= -\dfrac{15}{52} - \dfrac{12\sqrt{7}}{52}$

$= \dfrac{-3\left(5 + 4\sqrt{7}\right)}{52}$

73. $\tan(u - v) = \dfrac{\tan u - \tan v}{1 + \tan u \tan v}$

$= \dfrac{-\dfrac{3\sqrt{7}}{7} + \dfrac{12}{5}}{1 + \left(\dfrac{-3\sqrt{7}}{7}\right)\left(-\dfrac{12}{5}\right)}$

$= \dfrac{84 - 15\sqrt{7}}{35 + 36\sqrt{7}}$

$= \dfrac{507\sqrt{7} - 960}{1121}$

75. $\cos(u + v) = \cos u \cos v - \sin u \sin u$

$= \left(-\dfrac{\sqrt{7}}{4}\right)\left(-\dfrac{5}{13}\right) - \left(\dfrac{3}{4}\right)\left(\dfrac{12}{13}\right)$

$= \dfrac{5\sqrt{7} - 36}{52}$

77. $\cos\left(x + \dfrac{\pi}{2}\right) = \cos x \cos \dfrac{\pi}{2} - \sin x \sin \dfrac{\pi}{2}$

$= (\cos x)(0) - (\sin x)(1)$

$= -\sin x$

79. $\cot\left(\dfrac{\pi}{2} - x\right) = \dfrac{\cos[(\pi/2) - x]}{\sin[(\pi/2) - x]}$

$= \dfrac{\cos(\pi/2) \cos x + \sin(\pi/2) \sin x}{\sin(\pi/2) \cos x - \sin x \cos(\pi/2)}$

$= \dfrac{\sin x}{\cos x}$

$= \tan x$

81. $\cos 3x = \cos(2x + x)$

$= \cos 2x \cos x - \sin 2x \sin x$

$= (\cos^2 x - \sin^2 x) \cos x - 2 \sin x \cos x \sin x$

$= \cos^3 x - 3 \sin^2 x \cos x$

$= \cos^3 x - 3 \cos x(1 - \cos^2 x)$

$= \cos^3 x - 3 \cos x + 3 \cos^3 x$

$= 4 \cos^3 x - 3 \cos x$

83. $\sin\left(x + \dfrac{\pi}{2}\right) - \sin\left(x - \dfrac{\pi}{2}\right) = \sqrt{3}$

$$\cos x + \cos x = \sqrt{3}$$

$$\cos x = \frac{\sqrt{3}}{2}$$

$$x = \frac{\pi}{6}, \frac{11\pi}{6}$$

85. $\sin u = -\dfrac{5}{7}, \ \pi < u < \dfrac{3\pi}{2}$, Quadrant III

$$\cos^2 u = 1 - \left(-\frac{5}{7}\right)^2 = \frac{24}{49} \implies \cos u = -\frac{2\sqrt{6}}{7}$$

$$\sin 2u = 2 \sin u \cos u = 2\left(-\frac{5}{7}\right)\left(-\frac{2\sqrt{6}}{7}\right) = \frac{20\sqrt{6}}{49}$$

$$\cos 2u = 1 - 2 \sin^2 u$$

$$= 1 - 2\left(-\frac{5}{7}\right)^2 = 1 - \frac{50}{49} = -\frac{1}{49}$$

$$\tan 2u = \frac{\sin 2u}{\cos 2u} = \frac{20\sqrt{6}}{-1} = -20\sqrt{6}$$

87. $\tan u = -\dfrac{2}{9}, \ \dfrac{\pi}{2} < u < \pi$, Quadrant II

$$\sec^2 u = \tan^2 u + 1 = \frac{4}{81} + 1 = \frac{85}{81} \implies \sec u = -\frac{\sqrt{85}}{9}$$

$$\cos u = \frac{-9\sqrt{85}}{85}, \ \sin u = (\tan u)(\cos u) = \frac{2\sqrt{85}}{85}$$

$$\sin 2u = 2 \sin u \cos u = 2\left(\frac{2\sqrt{85}}{85}\right)\left(\frac{-9\sqrt{85}}{85}\right) = -\frac{36}{85}$$

$$\cos 2u = 1 - 2 \sin^2 u = 1 - 2\left(\frac{4}{85}\right) = \frac{77}{85}$$

$$\tan 2u = \frac{\sin 2u}{\cos 2u} = -\frac{36}{77}$$

89. $6 \sin x \cos x = 3[2 \sin x \cos x] = 3 \sin 2x$

91. $1 - 4 \sin^2 x \cos^2 x = 1 - (2 \sin x \cos x)^2$

$$= 1 - \sin^2 2x = \cos^2 2x$$

93. $\quad r = \frac{1}{32} v_0{}^2 \sin 2\theta$

$$100 = \tfrac{1}{32}(80)^2 \sin 2\theta$$

$$\sin 2\theta = 0.5$$

$$2\theta = 30° \quad \text{or} \quad 2\theta = 180° - 30° = 150°$$

$$\theta = 15° \qquad \theta = 75°$$

95. $\sin^6 x = \left(\dfrac{1 - \cos 2x}{2}\right)^3 = \dfrac{1}{8}(1 - 3 \cos 2x + 3 \cos^2 2x - \cos^3 2x)$

$$= \frac{1}{8}\left[1 - 3 \cos 2x + 3\left(\frac{1 + \cos 4x}{2}\right) - \cos 2x\left(\frac{1 + \cos 4x}{2}\right)\right]$$

$$= \frac{1}{8}\left(1 - 3 \cos 2x + \frac{3}{2} + \frac{3}{2} \cos 4x - \frac{1}{2} \cos 2x - \frac{1}{2} \cos 2x \cos 4x\right)$$

$$= \frac{1}{16}\left(5 - 7 \cos 2x + 3 \cos 4x - \frac{1}{2}[\cos 2x + \cos 6x]\right)$$

$$= \frac{1}{32}(10 - 15 \cos 2x + 6 \cos 4x - \cos 6x)$$

97. $\cos^4 2x = \left(\dfrac{1 + \cos 4x}{2}\right)^2 = \dfrac{1}{4}(1 + 2\cos 4x + \cos^2 4x)$

$$= \dfrac{1}{4}\left(1 + 2\cos 4x + \dfrac{1 + \cos 8x}{2}\right)$$

$$= \dfrac{1}{8}(2 + 4\cos 4x + 1 + \cos 8x)$$

$$= \dfrac{1}{8}(3 + 4\cos 4x + \cos 8x)$$

99. $\sin 105° = \sin\left(\dfrac{1}{2} \cdot 210°\right) = \sqrt{\dfrac{1 - \cos 210°}{2}} = \sqrt{\dfrac{1 + \sqrt{3}/2}{2}} = \dfrac{\sqrt{2 + \sqrt{3}}}{2}$

$\cos 105° = \cos\left(\dfrac{1}{2} \cdot 210°\right) = -\sqrt{\dfrac{1 + \cos 210°}{2}} = -\sqrt{\dfrac{1 - \sqrt{3}/2}{2}} = \dfrac{-\sqrt{2 - \sqrt{3}}}{2}$

$\tan 105° = \tan\left(\dfrac{1}{2} \cdot 210°\right) = \dfrac{\sin 210°}{1 + \cos 210°} = \dfrac{-\dfrac{1}{2}}{1 - \dfrac{\sqrt{3}}{2}} = \dfrac{1}{\sqrt{3} - 2} = -2 - \sqrt{3}$

101. $\sin\left(\dfrac{7\pi}{8}\right) = \sin\left(\dfrac{1}{2} \cdot \dfrac{7\pi}{4}\right) = \sqrt{\dfrac{1 - \cos 7\pi/4}{2}} = \sqrt{\dfrac{1 - \sqrt{2}/2}{2}} = \dfrac{\sqrt{2 - \sqrt{2}}}{2}$

$\cos\left(\dfrac{7\pi}{8}\right) = \cos\left(\dfrac{1}{2} \cdot \dfrac{7\pi}{4}\right) = -\sqrt{\dfrac{1 + \cos 7\pi/4}{2}} = -\sqrt{\dfrac{1 + \sqrt{2}/2}{2}} = \dfrac{-\sqrt{2 + \sqrt{2}}}{2}$

$\tan\left(\dfrac{7\pi}{8}\right) = \tan\left(\dfrac{1}{2} \cdot \dfrac{7\pi}{4}\right) = \dfrac{\sin\dfrac{7\pi}{4}}{1 + \cos\dfrac{7\pi}{4}} = \dfrac{-\dfrac{\sqrt{2}}{2}}{1 + \dfrac{\sqrt{2}}{2}} = \dfrac{-\sqrt{2}}{2 + \sqrt{2}} = 1 - \sqrt{2}$

103. $\sin u = \dfrac{3}{5}, 0 < u < \dfrac{\pi}{2} \implies \cos u = \dfrac{4}{5}$

$\sin\left(\dfrac{u}{2}\right) = \sqrt{\dfrac{1 - \cos u}{2}} = \sqrt{\dfrac{1 - (4/5)}{2}} = \dfrac{1}{\sqrt{10}} = \dfrac{\sqrt{10}}{10}$

$\cos\left(\dfrac{u}{2}\right) = \sqrt{\dfrac{1 + \cos u}{2}} = \sqrt{\dfrac{1 + (4/5)}{2}} = \dfrac{3}{\sqrt{10}} = \dfrac{3\sqrt{10}}{10}$

$\tan\left(\dfrac{u}{2}\right) = \dfrac{1 - \cos u}{\sin u} = \dfrac{1 - (4/5)}{3/5} = \dfrac{1}{3}$

105. $\cos u = -\dfrac{2}{7}, \dfrac{\pi}{2} < u < \pi \implies \sin u = \sqrt{1 - \dfrac{4}{49}} = \dfrac{\sqrt{45}}{7}$

$\sin\left(\dfrac{u}{2}\right) = \sqrt{\dfrac{1 - \cos u}{2}} = \sqrt{\dfrac{1 + (2/7)}{2}} = \sqrt{\dfrac{9}{14}} = \dfrac{3\sqrt{14}}{14}$

$\cos\left(\dfrac{u}{2}\right) = \sqrt{\dfrac{1 + \cos u}{2}} = \sqrt{\dfrac{1 - (2/7)}{2}} = \sqrt{\dfrac{5}{14}} = \dfrac{\sqrt{70}}{14}$

$\tan\left(\dfrac{u}{2}\right) = \dfrac{1 - \cos u}{\sin u} = \dfrac{1 + (2/7)}{\sqrt{45}/7} = \dfrac{9}{\sqrt{45}} = \dfrac{9\sqrt{45}}{45} = \dfrac{\sqrt{45}}{5} = \dfrac{3\sqrt{5}}{5}$

107. $-\sqrt{\dfrac{1 + \cos 10x}{2}} = -|\cos(5x)|$

109. Volume V of the trough will be the area A of the isosceles triangle times the length l of the trough.

not drawn to scale

$V = A \cdot l$

$A = \dfrac{1}{2}bh$

$\cos\dfrac{\theta}{2} = \dfrac{h}{0.5} \implies h = 0.5\cos\dfrac{\theta}{2}$

$\sin\dfrac{\theta}{2} = \dfrac{b/2}{0.5} \implies \dfrac{b}{2} = 0.5\sin\dfrac{\theta}{2}$

$A = 0.5\sin\dfrac{\theta}{2}\,0.5\cos\dfrac{\theta}{2} = (0.5)^2\sin\dfrac{\theta}{2}\cos\dfrac{\theta}{2} = 0.25\sin\dfrac{\theta}{2}\cos\dfrac{\theta}{2}$ square meters

$V = (0.25)(4)\sin\dfrac{\theta}{2}\cos\dfrac{\theta}{2}$ cubic meters $= \sin\dfrac{\theta}{2}\cos\dfrac{\theta}{2}$ cubic meters

111. $6\sin\dfrac{\pi}{4}\cos\dfrac{\pi}{4} = 6\left[\dfrac{1}{2}\sin\left(\dfrac{\pi}{4} + \dfrac{\pi}{4}\right) + \sin\left(\dfrac{\pi}{4} - \dfrac{\pi}{4}\right)\right]$

$= 3\left(\sin\dfrac{\pi}{2} + \sin\theta\right) = 3$

113. $\sin 3\alpha \sin 2\alpha = \dfrac{1}{2}[\cos(3\alpha - 2\alpha) - \cos(3\alpha + 2\alpha)$

$= \dfrac{1}{2}(\cos\alpha - \cos 5\alpha)$

115. $\cos 3\theta + \cos 2\theta = 2\cos\left(\dfrac{3\theta + 2\theta}{2}\right)\cos\left(\dfrac{3\theta - 2\theta}{2}\right)$

$= 2\cos\dfrac{5\theta}{2}\cos\dfrac{\theta}{2}$

117. $\sin\left(x + \dfrac{\pi}{4}\right) - \sin\left(x - \dfrac{\pi}{4}\right) = 2\cos x \sin\dfrac{\pi}{4}$

$= \sqrt{2}\cos x$

119. $y = 1.5\sin 8t - 0.5\cos 8t$

$a = \dfrac{3}{2},\ b = -\dfrac{1}{2},\ B = 8,\ C = \arctan\left(-\dfrac{1/2}{3/2}\right)$

$y = \sqrt{\left(\dfrac{3}{2}\right)^2 + \left(\dfrac{1}{2}\right)^2}\sin\left(8t + \arctan\left(-\dfrac{1}{3}\right)\right)$

$y = \dfrac{1}{2}\sqrt{10}\sin\left(8t - \arctan\dfrac{1}{3}\right)$

121. The amplitude is $\sqrt{10}/2$.

123. If $\dfrac{\pi}{2} < \theta < \pi$, then $\cos\dfrac{\theta}{2} < 0$. False, if

$\dfrac{\pi}{2} < \theta < \pi \implies \dfrac{\pi}{4} < \dfrac{\theta}{2} < \dfrac{\pi}{2}$,

which is in Quadrant I $\implies \cos\left(\dfrac{\theta}{2}\right) > 0$.

125. $4\sin(-x)\cos(-x) = -2\sin 2x$. True.

$4\sin(-x)\cos(-x) = 4(-\sin x)(\cos x)$

$= -4\sin x\cos x$

$= -2(2\sin x\cos x)$

$= -2\sin 2x$

127. Answers will vary. See page 480.

129. $y_1 = \sec^2\left(\dfrac{\pi}{2} - x\right) = \csc^2 x$

$y_2 = \cot^2 x$

$\csc^2 x = \cot^2 x + 1$

Let $y_3 = y_2 + 1 = \cot^2 x + 1 = y_1$.

Chapter 5   Practice Test

1. Find the value of the other five trigonometric functions, given $\tan x = \frac{4}{11}$, $\sec x < 0$.

2. Simplify $\dfrac{\sec^2 x + \csc^2 x}{\csc^2 x(1 + \tan^2 x)}$.

3. Rewrite as a single logarithm and simplify $\ln|\tan \theta| - \ln|\cot \theta|$.

4. True or false: $\cos\left(\dfrac{\pi}{2} - x\right) = \dfrac{1}{\csc x}$

5. Factor and simplify: $\sin^4 x + (\sin^2 x)\cos^2 x$

6. Multiply and simplify: $(\csc x + 1)(\csc x - 1)$

7. Rationalize the denominator and simplify:

$$\frac{\cos^2 x}{1 - \sin x}$$

8. Verify:

$$\frac{1 + \cos \theta}{\sin \theta} + \frac{\sin \theta}{1 + \cos \theta} = 2 \csc \theta$$

9. Verify:

$$\tan^4 x + 2 \tan^2 x + 1 = \sec^4 x$$

10. Use the sum or difference formulas to determine:

(a) $\sin 105°$ (b) $\tan 15°$

11. Simplify: $(\sin 42°)\cos 38° - (\cos 42°)\sin 38°$

12. Verify: $\tan\left(\theta + \dfrac{\pi}{4}\right) = \dfrac{1 + \tan \theta}{1 - \tan \theta}$

13. Write $\sin(\arcsin x - \arccos x)$ as an algebraic expression in x.

14. Use the double-angle formulas to determine:

(a) $\cos 120°$ (b) $\tan 300°$

15. Use the half-angle formulas to determine:

(a) $\sin 22.5°$ (b) $\tan \dfrac{\pi}{12}$

16. Given $\sin \theta = 4/5$, θ lies in Quadrant II, find $\cos \theta/2$.

17. Use the power-reducing identities to write $(\sin^2 x)\cos^2 x$ in terms of the first power of cosine.

18. Rewrite as a sum: $6(\sin 5\theta)\cos 2\theta$

19. Rewrite as a product: $\sin(x + \pi) + \sin(x - \pi)$

20. Verify: $\dfrac{\sin 9x + \sin 5x}{\cos 9x - \cos 5x} = -\cot 2x$

21. Verify: $(\cos u)\sin v = \frac{1}{2}[\sin(u + v) - \sin(u - v)]$

22. Find all solutions in the interval $[0, 2\pi)$:

$4 \sin^2 x = 1$

23. Find all solutions in the interval $[0, 2\pi)$:

$\tan^2 \theta + \left(\sqrt{3} - 1\right)\tan \theta - \sqrt{3} = 0$

24. Find all solutions in the interval $[0, 2\pi)$:

$\sin 2x = \cos x$

25. Use the Quadratic Formula to find all solutions in the interval $[0, 2\pi)$:

$\tan^2 x - 6 \tan x + 4 = 0$

CHAPTER 6
Additional Topics in Trigonometry

CHAPTER 6
Additional Topics in Trigonometry

Section 6.1   Law of Sines

■ If ABC is any oblique triangle with sides a, b, and c, then the Law of Sines says

$$\frac{a}{\sin A} = \frac{b}{\sin B} = \frac{c}{\sin C}.$$

■ You should be able to use the Law of Sines to solve an oblique triangle for the remaining three parts, given:

(a) Two angles and any side (AAS or ASA)

(b) Two sides and an angle opposite one of them (SSA)

   1. If A is acute and $h = b \sin A$:

     (a) $a < h$, no triangle is possible.

     (b) $a = h$ or $a > b$, one triangle is possible.

     (c) $h < a < b$, two triangles are possible.

   2. If A is obtuse and $h = b \sin A$:

     (a) $a \leq b$, no triangle is possible.

     (b) $a > b$, one triangle is possible.

■ The area of any triangle equals one-half the product of the lengths of two sides times the sine of their included angle.

$$A = \tfrac{1}{2}ab \sin C = \tfrac{1}{2}ac \sin B = \tfrac{1}{2}bc \sin A$$

Solutions to Odd-Numbered Exercises

1. Given: $A = 30°$, $B = 45°$, $a = 12$

$C = 180° - 30° - 45° = 105°$

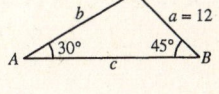

$b = \dfrac{a}{\sin A}(\sin B) = \dfrac{12}{\sin 30°}(\sin 45°) = 12\sqrt{2} \approx 16.97$

$c = \dfrac{a}{\sin A}(\sin C) = \dfrac{12}{\sin 30°}(\sin 105°) \approx 23.18$

3. Given: $A = 10°$, $B = 60°$, $a = 4.5$

$C = 180° - 10° - 60° = 110°$

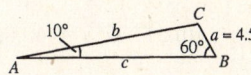

$b = \dfrac{a}{\sin A}(\sin B) = \dfrac{4.5}{\sin 10°}(\sin 60°) \approx 22.44$

$c = \dfrac{a}{\sin A}(\sin C) = \dfrac{4.5}{\sin 10°}(\sin 110°) \approx 24.35$

5. Given: $A = 36°$, $a = 8$, $b = 5$

$$\sin B = \frac{b \sin A}{a} = \frac{5 \sin(36°)}{8} \approx 0.3674 \implies B \approx 21.6°$$

$$C = 180° - A - B \approx 180° - 36° - 21.6° = 122.4°$$

$$c = \frac{a}{\sin A}(\sin C) = \frac{8}{\sin(36°)} \sin(122.4°) \approx 11.49$$

7. Given: $A = 102.4°$, $C = 16.7°$, $a = 21.6$

$$B = 180° - A - C = 180° - 102.4° - 16.7° = 60.9°$$

$$b = \frac{a}{\sin A}(\sin B) = \frac{21.6}{\sin 102.4°}(\sin 60.9°) \approx 19.32$$

$$c = \frac{a}{\sin A}(\sin C) = \frac{21.6}{\sin 102.4°}(\sin 16.7°) \approx 6.36$$

9. Given: $A = 83° \, 20'$, $C = 54.6°$, $c = 18.1$

$$B = 180° - A - C = 180° - 83° \, 20' - 54° \, 36' = 42° \, 4'$$

$$a = \frac{c}{\sin C}(\sin A) = \frac{18.1}{\sin 54.6°}(\sin 83° \, 20') \approx 22.05$$

$$b = \frac{c}{\sin C}(\sin B) = \frac{18.1}{\sin 54.6°}(\sin 42° \, 4') \approx 14.88$$

11. Given: $B = 15° \, 30'$, $a = 4.5$, $b = 6.8$

$$\sin A = \frac{a \sin B}{b} = \frac{4.5 \sin 15° \, 30'}{6.8} \approx 0.17685 \implies A \approx 10° \, 11'$$

$$C = 180° - A - B \approx 180° - 10° \, 11' - 15° \, 30' = 154° \, 19'$$

$$c = \frac{b}{\sin B}(\sin C) = \frac{6.8}{\sin 15° \, 30'}(\sin 154° \, 19') \approx 11.03$$

13. Given: $A = 110° \, 15'$, $a = 48$, $b = 16$

$$\sin B = \frac{b \sin A}{a} = \frac{16 \sin 110° \, 15'}{48} \approx 0.31273 \implies B \approx 18° \, 13'$$

$$C = 180° - A - B \approx 180° - 110° \, 15' - 18° \, 13' = 51° \, 32'$$

$$c = \frac{a}{\sin A}(\sin C) = \frac{48}{\sin 110° \, 15'}(\sin 51° \, 32') \approx 40.06$$

15. Given: $A = 110°$, $a = 125$, $b = 100$

$$\sin B = \frac{b \sin A}{a} = \frac{100 \sin 110°}{125} \approx 0.75175 \implies B \approx 48.74°$$

$$C = 180 - A - B \approx 21.26°$$

$$c = \frac{a}{\sin A}(\sin C) = \frac{125 \sin 21.26°}{\sin 110°} \approx 48.23$$

17. Given: $A = 76°, a = 18, b = 20$

$$\sin B = \frac{b \sin A}{a} = \frac{20 \sin 76°}{18} \approx 1.078$$

No solution

19. Given: $A = 58°, a = 11.4, b = 12.8$

$$\sin B = \frac{b \sin A}{a} = \frac{12.8 \sin 58°}{11.4} \approx 0.9522 \Rightarrow B \approx 72.21° \text{ or } 107.79°$$

Case 1

$B \approx 72.21°$

$C = 180° - 58° - 72.21° = 49.79°$

$c = \dfrac{a}{\sin A}(\sin C) = \dfrac{11.4}{\sin 58°}(\sin 49.79°) \approx 10.27$

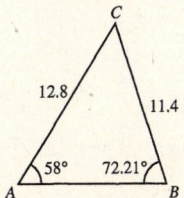

Case 2

$B \approx 107.79$

$C \approx 180° - 58° - 107.79° = 14.21°$

$c = \dfrac{a}{\sin A}(\sin C) = \dfrac{11.4}{\sin 58°}(\sin 14.21°) \approx 3.30$

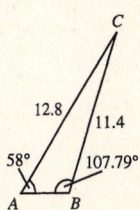

21. Area $= \frac{1}{2}ab \sin C$

$= \frac{1}{2}(6)(10)\sin(110°)$

≈ 28.2 square units

23. Area $= \frac{1}{2}bc \sin A$

$= \frac{1}{2}(67)(85)\sin(38° \, 45')$

≈ 1782.3 square units

25. Area $= \frac{1}{2}ac \sin B$

$= \frac{1}{2}(103)(58)\sin(74° \, 30')$

≈ 2878.4 square units

27. Angle $CAB = 70°$

Angle $B = 20° + 14° = 34°$

(a)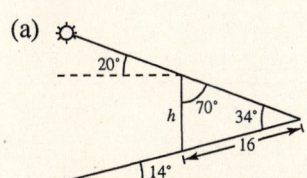

(b) $\dfrac{16}{\sin 70°} = \dfrac{h}{\sin 34°}$

(c) $h = \dfrac{16 \sin 34°}{\sin 70°} \approx 9.52$ meters

29. $\sin A = \dfrac{a \sin B}{b} = \dfrac{500 \sin(46°)}{720} \approx 0.4995$

$A \approx 29.97°$

$\angle ACD = 90° - 29.97° \approx 60°$

Bearing: S 60° W or (240° in plane navigation)

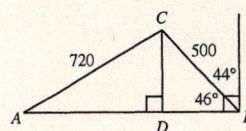

31. (a)

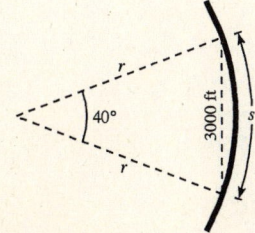

(b) $r = \dfrac{3000 \sin[1/2(180° - 40°)]}{\sin 40°} \approx 4385.71$ feet

(c) $s \approx 40°\left(\dfrac{\pi}{180°}\right)4385.71 \approx 3061.80$ feet

33. $\angle ACD = 65°$

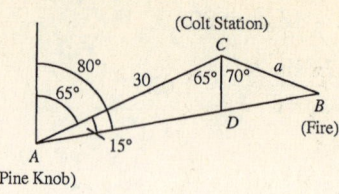

$\angle ADC = 180° - 65° - 15° = 100°$

$\angle CDB = 180° - 100° = 80°$

$\angle B = 180° - 80° - 70° = 30°$

$a = \dfrac{b}{\sin B}(\sin A) = \dfrac{30}{\sin 30°}(\sin 15°) \approx 15.53 \text{ km}$

$c = \dfrac{b}{\sin B}(\sin C) = \dfrac{30}{\sin 30°}(\sin 135°) \approx 42.43 \text{ km}$

35. $\dfrac{\sin(42° - \theta)}{10} = \dfrac{\sin 48°}{17}$

$\sin(42° - \theta) = \dfrac{10}{17}\sin 48° \approx 0.4371$

$42° - \theta \approx 25.919$

$\theta \approx 16.1°$

37. (a) $\sin \alpha = \dfrac{5.45}{58.36} \approx 0.0934$

$\alpha \approx 5.36°$

(b) $\dfrac{d}{\sin \beta} = \dfrac{58.36}{\sin \theta} \implies \sin \beta = \dfrac{d \sin \theta}{58.36}$

$\beta = \sin^{-1}\left[\dfrac{d \sin \theta}{58.36}\right]$

(c) $\theta + \beta + 90° + 5.36° = 180° \implies \beta = 84.64° - \theta$

$d = \sin \beta\left(\dfrac{58.36}{\sin \theta}\right) = \sin(84.64° - \theta)\dfrac{58.36}{\sin \theta}$

(d)

θ	10°	20°	30°	40°	50°	60°
d	324.1	154.2	95.2	63.8	43.3	28.1

39. False. If just the three angles are known, the triangle cannot be solved.

41. Yes, the Law of Sines can be used to solve a right triangle if you are given at least one side.

43. $\tan \theta = \dfrac{\sin \theta}{\cos \theta} = -\dfrac{12}{5},$ $\sec \theta = \dfrac{13}{5}$

$\cot \theta = -\dfrac{5}{12},$ $\csc \theta = -\dfrac{13}{12}$

45. $\tan \theta = -\dfrac{1}{11}; \cos \theta = -\dfrac{11}{\sqrt{122}} = -\dfrac{11\sqrt{122}}{122}$

$\sin \theta = \left(-\dfrac{1}{11}\right)\left(-\dfrac{11}{\sqrt{122}}\right) = \dfrac{1}{\sqrt{122}} = \dfrac{\sqrt{122}}{122};$

$\csc \theta = \sqrt{122}$

47. $6 \sin 8\theta \cos 3\theta = 6\left(\tfrac{1}{2}\right)[\sin(8\theta + 3\theta) + \sin(8\theta - 3\theta)]$

$= 3(\sin 11\theta + \sin 5\theta)$

49. $3 \cos \dfrac{\pi}{6} \sin \dfrac{5\pi}{3} = 3\left(\dfrac{1}{2}\right)\left[\sin\left(\dfrac{\pi}{6} + \dfrac{5\pi}{3}\right) - \sin\left(\dfrac{\pi}{6} - \dfrac{5\pi}{3}\right)\right]$

$= \dfrac{3}{2}\left[\sin\left(\dfrac{11\pi}{6}\right) - \sin\left(-\dfrac{3\pi}{2}\right)\right]$

$= \dfrac{3}{2}\left[-\dfrac{1}{2} - 1\right] = -\dfrac{9}{4}$

Section 6.2   Law of Cosines

■ If ABC is any oblique triangle with sides a, b, and c, then the Law of Cosines says:

(a) $a^2 = b^2 + c^2 - 2bc \cos A$ or $\cos A = \dfrac{b^2 + c^2 - a^2}{2bc}$

(b) $b^2 = a^2 + c^2 - 2ac \cos B$ or $\cos B = \dfrac{a^2 + c^2 - b^2}{2ac}$

(c) $c^2 = a^2 + b^2 - 2ab \cos C$ or $\cos C = \dfrac{a^2 + b^2 - c^2}{2ab}$

■ You should be able to use the Law of Cosines to solve an oblique triangle for the remaining three parts, given:

(a) Three sides (SSS)

(b) Two sides and their included angle (SAS)

■ Given any triangle with sides of lengths a, b, and c, then the area of the triangle is

$$\text{Area} = \sqrt{s(s-a)(s-b)(s-c)}, \text{ where } s = \frac{a+b+c}{2}. \qquad \text{(Heron's Formula)}$$

Solutions to Odd-Numbered Exercises

1. Given: $a = 6$, $b = 8$, $c = 12$

$$\cos A = \frac{b^2 + c^2 - a^2}{2bc} = \frac{64 + 144 - 36}{2(8)(12)} \approx 0.8958 \implies A \approx 26.4°$$

$$\sin B = \frac{b \sin A}{a} \approx \frac{8 \sin 26.4°}{6} \approx 0.5928 \implies B \approx 36.3°$$

$$C \approx 180° - 26.4° - 36.3° = 117.3°$$

3. Given: $A = 50°, b = 15, c = 30$

$$a^2 = b^2 + c^2 - 2bc \cos A = 225 + 900 - 2(15)(30) \cos 50° \approx 546.49 \implies a \approx 23.38$$

$$\cos B = \frac{a^2 + c^2 - b^2}{2ac} \approx \frac{547.56 + 900 - 225}{2(23.4)(30)} \approx 0.8708 \implies B \approx 29.4°$$

$$C = 180° - A - B \approx 180° - 50° - 29.5° = 100.6°$$

5. Given: $a = 9$, $b = 12$, $c = 15$

$$\cos C = \frac{a^2 + b^2 - c^2}{2ab} = \frac{81 + 144 - 225}{2(9)(12)} = 0 \implies C = 90°$$

$$\sin A = \frac{9}{15} = \frac{3}{5} \implies A \approx 36.9°$$

$$B \approx 180° - 90° - 36.9° = 53.1°$$

7. Given: $a = 75.4, b = 48, c = 48$

$$\cos A = \frac{b^2 + c^2 - a^2}{2bc} = \frac{48^2 + 48^2 - 75.4^2}{2(48)(48)} \approx -0.2338 \implies A \approx 103.5°$$

$$\sin B = \frac{b \sin A}{a} \approx \frac{48 \sin (103.5°)}{75.4} \approx 0.6190 \implies B \approx 38.2°$$

$C = B \approx 38.2°$ (Because of roundoff error, $A + B + C \neq 180°$.)

9. Given: $B = 8° \, 15' = 8.25°, a = 26, c = 18$

$$b^2 = a^2 + c^2 - 2ac \cos B = 26^2 + 18^2 - 2(26)(18) \cos(8.25°) \approx 73.6863 \implies b \approx 8.58$$

$$\sin C = \frac{c \sin B}{b} \approx \frac{18 \sin(8.25°)}{8.58} \approx 0.3 \implies C \approx 17.5° \approx 17° \, 31'$$

$$A = 180° - B - C \approx 180° - 8.25° - 17.5° = 154.2° \approx 154° \, 14'$$

11. Given: $B = 75° \, 20' \approx 75.33°, a = 6.2, c = 9.5$

$$b^2 = a^2 + c^2 - 2ac \cos B$$

$$= 6.2^2 + 9.5^2 - 2(6.2)(9.5) \cos(75.33°)$$

$$= 98.86$$

$$b \approx 9.94$$

$$\sin C = \frac{c \sin B}{b} = \frac{9.5 \sin(75.33°)}{9.94}$$

$$\approx 0.9246 \implies C \approx 67.6°$$

$$A = 180° - B - C \approx 37.1°$$

13. $d^2 = 4^2 + 8^2 - 2(4)(8) \cos 30°$

$$\approx 24.57 \implies d \approx 4.96$$

$$2\phi = 360° - 2\theta \implies \phi = 150°$$

$$c^2 = 4^2 + 8^2 - 2(4)(8) \cos 150° \approx 135.43$$

$$c \approx 11.64$$

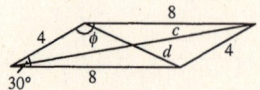

15. $\cos \phi = \dfrac{10^2 + 14^2 - 20^2}{2(10)(14)}$

$$\phi \approx 111.8°$$

$$2\theta \approx 360° - 2(111.80°)$$

$$\theta = 68.2°$$

$$d^2 = 10^2 + 14^2 - 2(10)(14) \cos 68.2°$$

$$d \approx 13.86$$

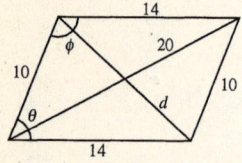

17. $\cos \alpha = \dfrac{15^2 + 12.5^2 - 10^2}{2(15)(12.5)} = 0.75 \implies \alpha \approx 41.41°$

$$\cos \beta = \frac{15^2 + 10^2 - 12.5^2}{2(15)(10)} = 0.5625 \implies \beta \approx 55.77°$$

$$\delta = 180° - 41.41° - 55.77° \approx 82.82°$$

$$\mu = 180° - \delta \approx 97.18°$$

$$b^2 = 12.5^2 + 10^2 - 2(12.5)(10) \cos(97.18°) \approx 287.50$$

$$b \approx 16.96$$

$$\sin \omega = \frac{10}{16.96} \sin \mu \approx 0.585 \implies \omega \approx 35.8°$$

$$\sin \epsilon = \frac{12.5}{16.96} \sin \mu \approx 0.731 \implies \epsilon \approx 47°$$

$$\theta = \alpha + \omega \approx 77.2°$$

$$\phi = \beta + \epsilon \approx 102.8°$$

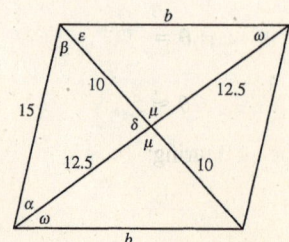

19. Given: $a = 5, b = 8, c = 10$

$$s = \frac{a + b + c}{2} = \frac{23}{2} = 11.5$$

$$\text{Area} = \sqrt{s(s - a)(s - b)(s - c)}$$

$$= \sqrt{11.5(6.5)(3.5)(1.5)}$$

$$\approx 19.81 \text{ square units}$$

21. Given: $a = 3.5, b = 10.2, c = 9$

$$s = \frac{a + b + c}{2} = 11.35$$

$$\text{Area} = \sqrt{s(s - a)(s - b)(s - c)}$$

$$= \sqrt{11.35(7.85)(1.15)(2.35)}$$

$$\approx 15.52 \text{ square units}$$

23. Given: $a = 10.59, b = 6.65, c = 12.31$

$$s = \frac{a + b + c}{2} = 14.775$$

$$\text{Area} = \sqrt{s(s - a)(s - b)(s - c)}$$

$$= \sqrt{14.775(4.185)(8.125)(2.465)}$$

$$\approx 35.19 \text{ square units}$$

25. $B = 105° + 32° = 137°$

$$b^2 = a^2 + c^2 - 2ac \cdot \cos B$$

$$= 648^2 + 810^2 - 2(648)(810) \cos(137°)$$

$$= 1,843,749.862$$

$$b = 1357.8 \text{ miles}$$

From the Law of Sines:

$$\frac{a}{\sin A} = \frac{b}{\sin B} \implies \sin A = \frac{a}{b} \sin B = \frac{648}{1357.8} \sin(137°) \approx 0.32548$$

$$\implies A \approx 19° \implies \text{Bearing S 56° W (or 236° for airplane navigation)}$$

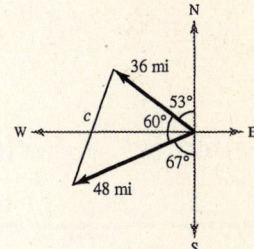

27. Angle at $B = 180° - 80° = 100°$

$$b^2 = 240^2 + 380^2 - 2(240)(380) \cos 100°$$

$$\approx 233,673.4 \implies b \approx 483.4 \text{ meters}$$

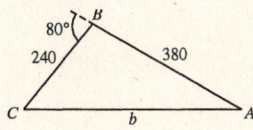

29. $C = 180° - 53° - 67° = 60°$

$$c^2 = a^2 + b^2 - 2ab \cos C$$

$$= 36^2 + 48^2 - 2(36)(48)(0.5) = 1872$$

$$c \approx 43.3 \text{ mi}$$

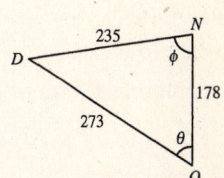

31. (a) $\cos \theta = \dfrac{273^2 + 178^2 - 235^2}{2(273)(178)}$

   $\theta \approx 58.4°$

   Bearing: N 58.4° W

(b) $\cos \phi = \dfrac{235^2 + 178^2 - 273^2}{2(235)(178)}$

   $\phi \approx 81.5°$

   Bearing: S 81.5° W

33. $\overline{RS} = \sqrt{8^2 + 10^2} = \sqrt{164} = 2\sqrt{41} \approx 12.8$ ft

$\overline{PQ} = \dfrac{1}{2}\sqrt{16^2 + 10^2} = \dfrac{1}{2}\sqrt{356} = \sqrt{89} \approx 9.4$ ft

$\tan P = \dfrac{10}{16}$

$P = \arctan\dfrac{5}{8} \approx 32.0°$

$\overline{QS} = \sqrt{8^2 + 9.4^2 - 2(8)(9.4)\cos 32°}$

$\approx \sqrt{24.81} \approx 5.0$ ft

35. $s = \dfrac{a + b + c}{2} = \dfrac{145 + 257 + 290}{2} = 346$

Area $= \sqrt{s(s-a)(s-b)(s-c)}$

$= \sqrt{346(201)(89)(56)}$

$\approx 18{,}617.7$ square feet

37. (a) $7^2 = 1.5^2 + x^2 - 2(1.5)(x)\cos\theta$

$49 = 2.25 + x^2 - 3x\cos\theta$

(b) $x^2 - 3x\cos\theta = 46.75$

$x^2 - 3x\cos\theta + \left(\dfrac{3\cos\theta}{2}\right)^2 = 46.75 + \left(\dfrac{3\cos\theta}{2}\right)^2$

$\left[x - \dfrac{3\cos\theta}{2}\right]^2 = \dfrac{187}{4} + \dfrac{9\cos^2\theta}{4}$

$x - \dfrac{3\cos\theta}{2} = \pm\sqrt{\dfrac{187 + 9\cos^2\theta}{4}}$

Choosing the positive values of x, we have
$x = \frac{1}{2}\left(3\cos\theta + \sqrt{9\cos^2\theta + 187}\right)$.

(c)
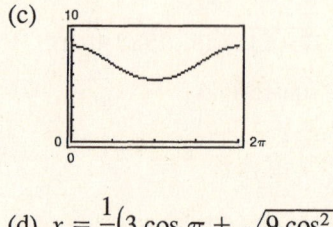

(d) $x = \dfrac{1}{2}\left(3\cos\pi + \sqrt{9\cos^2\pi + 187}\right)$

$= 5.5$

≈ 6 inches

39. False. This is not a triangle! $5 + 10 < 16$

41. False. $s = \dfrac{a+b+c}{2}$, not $\dfrac{a+b+c}{3}$.

43. $\arcsin(-1) = -\dfrac{\pi}{2}$ because $\sin\left(-\dfrac{\pi}{2}\right) = -1$.

45. $\tan^{-1}\left(\sqrt{3}\right) = \dfrac{\pi}{3}$ because $\tan\left(\dfrac{\pi}{3}\right) = \sqrt{3}$.

47. $\cos\dfrac{5\pi}{6} - \cos\dfrac{\pi}{3} = -2\sin\left(\dfrac{\dfrac{5\pi}{6} + \dfrac{\pi}{3}}{2}\right)\sin\left(\dfrac{\dfrac{5\pi}{6} - \dfrac{\pi}{3}}{2}\right) = -2\sin\dfrac{7\pi}{12}\sin\dfrac{\pi}{4}$

Section 6.3   Vectors in the Plane

■ A vector **v** is the collection of all directed line segments that are equivalent to a given directed line segment $\overrightarrow{PQ}$.

■ You should be able to *geometrically* perform the operations of vector addition and scalar multiplication.

■ The component form of the vector with initial point $P = (p_1, p_2)$ and terminal point $Q = (q_1, q_2)$ is

$\overrightarrow{PQ} = \langle q_1 - p_1, q_2 - p_2 \rangle = \langle v_1, v_2 \rangle = \mathbf{v}.$

■ The magnitude of $\mathbf{v} = \langle v_1, v_2 \rangle$ is given by $\|\mathbf{v}\| = \sqrt{{v_1}^2 + {v_2}^2}$.

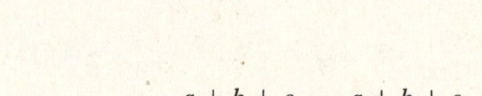

—CONTINUED—

—CONTINUED—

■ You should be able to perform the operations of scalar multiplication and vector addition in component form.

■ You should know the following properties of vector addition and scalar multiplication.

    (a) $\mathbf{u} + \mathbf{v} = \mathbf{v} + \mathbf{u}$

    (b) $(\mathbf{u} + \mathbf{v}) + \mathbf{w} = \mathbf{u} + (\mathbf{v} + \mathbf{w})$

    (c) $\mathbf{u} + \mathbf{0} = \mathbf{u}$

    (d) $\mathbf{u} + (-\mathbf{u}) = \mathbf{0}$

    (e) $c(d\mathbf{u}) = (cd)\mathbf{u}$

    (f) $(c + d)\mathbf{u} = c\mathbf{u} + d\mathbf{u}$

    (g) $c(\mathbf{u} + \mathbf{v}) = c\mathbf{u} + c\mathbf{v}$

    (h) $1(\mathbf{u}) = \mathbf{u}, 0\mathbf{u} = \mathbf{0}$

    (i) $\|c\mathbf{v}\| = |c|\,\|\mathbf{v}\|$

■ A unit vector in the direction of $\mathbf{v}$ is given by $\mathbf{u} = \dfrac{\mathbf{v}}{\|\mathbf{v}\|}$.

■ The standard unit vectors are $\mathbf{i} = \langle 1, 0 \rangle$ and $\mathbf{j} = \langle 0, 1 \rangle$. $\mathbf{v} = \langle v_1, v_2 \rangle$ can be written as $\mathbf{v} = v_1\mathbf{i} + v_2\mathbf{j}$.

■ A vector $\mathbf{v}$ with magnitude $\|\mathbf{v}\|$ and direction θ can be written as $\mathbf{v} = a\mathbf{i} + b\mathbf{j} = \mathbf{v}(\cos\theta)\mathbf{i} + \mathbf{v}(\sin\theta)\mathbf{j}$ where $\tan\theta = b/a$.

Solutions to Odd-Numbered Exercises

1. $\mathbf{u} = \langle 6 - 2, 5 - 4 \rangle = \langle 4, 1 \rangle = \mathbf{v}$

3. Initial point: $(0, 0)$

    Terminal point: $(4, 3)$

    $\mathbf{v} = \langle 4 - 0, 3 - 0 \rangle = \langle 4, 3 \rangle$

    $\|\mathbf{v}\| = \sqrt{4^2 + 3^2} = 5$

5. Initial point: $(2, 2)$

    Terminal point: $(-1, 4)$

    $\mathbf{v} = \langle -1 - 2, 4 - 2 \rangle = \langle -3, 2 \rangle$

    $\|\mathbf{v}\| = \sqrt{(-3)^2 + 2^2} = \sqrt{13} \approx 3.61$

7. Initial point: $(3, -2)$

    Terminal point: $(3, 3)$

    $\mathbf{v} = \langle 3 - 3, 3 - (-2) \rangle = \langle 0, 5 \rangle$

    $\|\mathbf{v}\| = 5$

9. Initial point: $\left(\dfrac{5}{2}, 1\right)$

    Terminal point: $\left(-2, -\dfrac{3}{2}\right)$

    $\mathbf{v} = \left\langle -2 - \dfrac{5}{2}, -\dfrac{3}{2} - 1 \right\rangle = \left\langle -\dfrac{9}{2}, -\dfrac{5}{2} \right\rangle$

    $\|\mathbf{v}\| = \sqrt{\left(-\dfrac{9}{2}\right)^2 + \left(-\dfrac{5}{2}\right)^2}$

    $= \sqrt{\dfrac{81 + 25}{4}} = \dfrac{1}{2}\sqrt{106} \approx 5.15$

11. Initial point: $(-3, -5)$

    Terminal point: $(5, 1)$

    $\mathbf{v} = \langle 5 - (-3), 1 - (-5) \rangle = \langle 8, 6 \rangle$

    $\|\mathbf{v}\| = \sqrt{8^2 + 6^2} = \sqrt{100} = 10$

13. −**v**

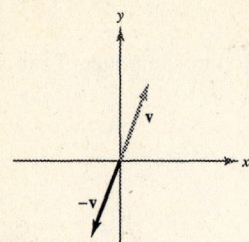

15. **u** + **v**

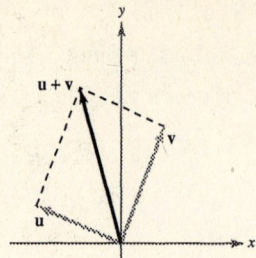

17. **u** + 2**v**

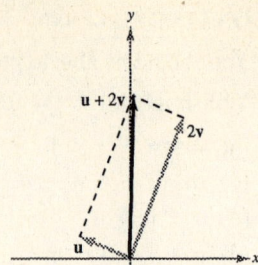

19. **u** = ⟨4, 2⟩, **v** = ⟨7, 1⟩

(a) **u** + **v** = ⟨11, 3⟩

(b) **u** − **v** = ⟨−3, 1⟩

(c) 2**u** − 3**v** = ⟨8, 4⟩ − ⟨21, 3⟩ = ⟨−13, 1⟩

(d) **v** + 4**u** = ⟨7, 1⟩ + ⟨16, 8⟩ = ⟨23, 9⟩

21. **u** = ⟨−6, −8⟩, **v** = ⟨2, 4⟩

(a) **u** + **v** = ⟨−4, −4⟩

(b) **u** − **v** = ⟨−8, −12⟩

(c) 2**u** − 3**v** = ⟨−12, −16⟩ − ⟨6, 12⟩ = ⟨−18, −28⟩

(d) **v** + 4**u** = ⟨2, 4⟩ + 4⟨−6, −8⟩ = ⟨−22, −28⟩

23. **u** = **i** + **j**, **v** = 2**i** − 3**j**

(a) **u** + **v** = 3**i** − 2**j**

(b) **u** − **v** = −**i** + 4**j**

(c) 2**u** − 3**v** = (2**i** + 2**j**) − (6**i** − 9**j**) = −4**i** + 11**j**

(d) **v** + 4**u** = (2**i** − 3**j**) + (4**i** + 4**j**) = 6**i** + **j**

25. $\|⟨6, 0⟩\| = 6$

Unit vector = $\frac{1}{6}⟨6, 0⟩ = ⟨1, 0⟩$

27. **v** = ⟨−1, 1⟩

$\|\mathbf{v}\| = \sqrt{2}$

Unit vector = $\frac{1}{\|\mathbf{v}\|}\mathbf{v}$

$= \left⟨-\frac{1}{\sqrt{2}}, \frac{1}{\sqrt{2}}\right⟩ = \left⟨-\frac{\sqrt{2}}{2}, \frac{\sqrt{2}}{2}\right⟩$

29. $\|\mathbf{v}\| = \|⟨-24, -7⟩\| = \sqrt{(-24)^2 + (-7)^2} = 25$

Unit vector = $\frac{1}{25}⟨-24, -7⟩ = \left⟨-\frac{24}{25}, -\frac{7}{25}\right⟩$

31. $\mathbf{u} = \frac{1}{\|\mathbf{v}\|}\mathbf{v}$

$= \frac{1}{\sqrt{16 + 9}}(4\mathbf{i} - 3\mathbf{j}) = \frac{1}{5}(4\mathbf{i} - 3\mathbf{j}) = \frac{4}{5}\mathbf{i} - \frac{3}{5}\mathbf{j}$

33. $\mathbf{u} = \frac{1}{2}(2\mathbf{j}) = \mathbf{j}$

35. $8\left(\frac{1}{\|\mathbf{u}\|}\mathbf{u}\right) = 8\left(\frac{1}{\sqrt{5^2 + 6^2}}⟨5, 6⟩\right)$

$= \frac{8}{\sqrt{61}}⟨5, 6⟩$

$= \left⟨\frac{40\sqrt{61}}{61}, \frac{48\sqrt{61}}{61}\right⟩$

37. $7\left(\frac{1}{\|\mathbf{u}\|}\mathbf{u}\right) = 7\left(\frac{1}{\sqrt{3^2 + 4^2}}⟨3, 4⟩\right)$

$= \frac{7}{5}⟨3, 4⟩$

$= \left⟨\frac{21}{5}, \frac{28}{5}\right⟩$

$= \frac{21}{5}\mathbf{i} + \frac{28}{5}\mathbf{j}$

39. $8\left(\frac{1}{\|\mathbf{u}\|}\mathbf{u}\right) = 8\left(\frac{1}{2}⟨-2, 0⟩\right)$

$= 4⟨-2, 0⟩$

$= ⟨-8, 0⟩$

$= -8\mathbf{i}$

41. $\mathbf{v} = \langle 4 - (-3), 5 - 1 \rangle = \langle 7, 4 \rangle = 7\mathbf{i} + 4\mathbf{j}$

43. $\mathbf{v} = \langle 2 - (-1), 3 - (-5) \rangle = \langle 3, 8 \rangle = 3\mathbf{i} + 8\mathbf{j}$

45. $\mathbf{v} = \frac{3}{2}\mathbf{u}$

$\quad = \frac{3}{2}(2\mathbf{i} - \mathbf{j})$

$\quad = 3\mathbf{i} - \frac{3}{2}\mathbf{j}$

$\quad = \left\langle 3, -\frac{3}{2} \right\rangle$

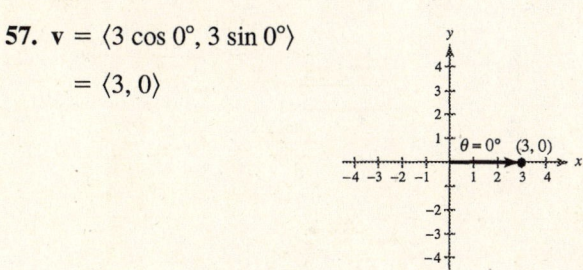

47. $\mathbf{v} = \mathbf{u} + 2\mathbf{w}$

$\quad = (2\mathbf{i} - \mathbf{j}) + 2(\mathbf{i} + 2\mathbf{j})$

$\quad = 4\mathbf{i} + 3\mathbf{j}$

$\quad = \langle 4, 3 \rangle$

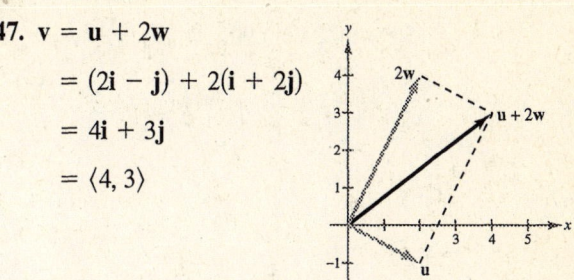

49. $\mathbf{v} = \frac{1}{2}(3\mathbf{u} + \mathbf{w})$

$\quad = \frac{1}{2}(3\langle 2, -1 \rangle + \langle 1, 2 \rangle)$

$\quad = \left\langle \frac{7}{2}, -\frac{1}{2} \right\rangle$

51. $\mathbf{v} = 5(\cos 30°\mathbf{i} + \sin 30°\mathbf{j})$

$\quad \|\mathbf{v}\| = 5, \ \theta = 30°$

53. $\mathbf{v} = 6\mathbf{i} - 6\mathbf{j}$

$\quad \|\mathbf{v}\| = \sqrt{6^2 + (-6)^2} = \sqrt{72} = 6\sqrt{2}$

$\quad \tan \theta = -\frac{6}{6} = -1$

Since $\mathbf{v}$ lies in Quadrant IV, $\theta = 315°$.

55. $\mathbf{v} = -2\mathbf{i} + 5\mathbf{j}$

$\quad \|\mathbf{v}\| = \sqrt{(-2)^2 + 5^2} = \sqrt{29}$

$\quad \tan \theta = -\frac{5}{2}$

Since $\mathbf{v}$ lies in Quadrant II, $\theta \approx 111.8°$.

57. $\mathbf{v} = \langle 3 \cos 0°, 3 \sin 0° \rangle$

$\quad = \langle 3, 0 \rangle$

59. $\mathbf{v} = \left\langle 3\sqrt{2} \cos 150°, 3\sqrt{2} \sin 150° \right\rangle$

$\quad = \left\langle -\frac{3\sqrt{6}}{2}, \frac{3\sqrt{2}}{2} \right\rangle$

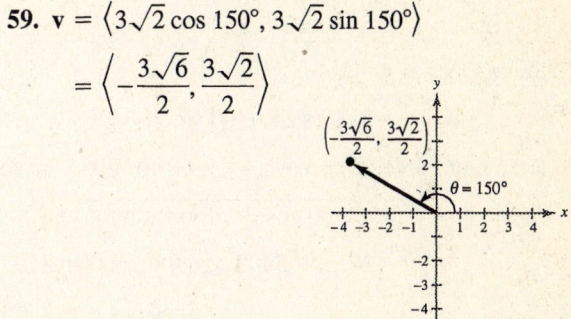

61. $\mathbf{v} = 2\left(\dfrac{1}{\sqrt{3^2 + 1^2}} \right)(\mathbf{i} + 3\mathbf{j})$

$\quad = \dfrac{2}{\sqrt{10}}(\mathbf{i} + 3\mathbf{j})$

$\quad = \dfrac{\sqrt{10}}{5}\mathbf{i} + \dfrac{3\sqrt{10}}{5}\mathbf{j}$

$\quad = \left\langle \dfrac{\sqrt{10}}{5}, \dfrac{3\sqrt{10}}{5} \right\rangle$

$\tan \theta = \dfrac{3}{1} \implies \theta \approx 71.57$

63. $\mathbf{u} = \langle 5 \cos 60°, 5 \sin 60° \rangle = \left\langle \dfrac{5}{2}, \dfrac{5\sqrt{3}}{2} \right\rangle$

$\mathbf{v} = \langle 5 \cos 90°, 5 \sin 90° \rangle = \langle 0, 5 \rangle$

$\mathbf{u} + \mathbf{v} = \left\langle \dfrac{5}{2}, \dfrac{5\sqrt{3}}{2} \right\rangle + \langle 0, 5 \rangle = \left\langle \dfrac{5}{2}, 5 + \dfrac{5}{2}\sqrt{3} \right\rangle$

65. $\mathbf{u} = \langle 20 \cos 45°, 20 \sin 45° \rangle = \left\langle 10\sqrt{2}, 10\sqrt{2} \right\rangle$

$\mathbf{v} = \langle 50 \cos 150°, 50 \sin 150° \rangle = \left\langle -25\sqrt{3}, 25 \right\rangle$

$\mathbf{u} + \mathbf{v} = \left\langle 10\sqrt{2} - 25\sqrt{3}, 10\sqrt{2} + 25 \right\rangle$

67. $\mathbf{v} = \mathbf{i} + \mathbf{j}$

$\mathbf{w} = 2(\mathbf{i} - \mathbf{j})$

$\mathbf{u} = \mathbf{v} - \mathbf{w} = -\mathbf{i} + 3\mathbf{j}$

$\|\mathbf{v}\| = \sqrt{2}$

$\|\mathbf{w}\| = 2\sqrt{2}$

$\|\mathbf{v} - \mathbf{w}\| = \sqrt{10}$

$\cos \alpha = \dfrac{\|\mathbf{v}\|^2 + \|\mathbf{w}\|^2 - \|\mathbf{v} - \mathbf{w}\|^2}{2\|\mathbf{v}\| \, \|\mathbf{w}\|} = \dfrac{2 + 8 - 10}{2\sqrt{2} \cdot 2\sqrt{2}} = 0$

$\alpha = 90°$

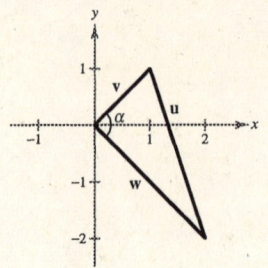

69. $\mathbf{u} = 400 \cos 25°\mathbf{i} + 400 \sin 25°\mathbf{j}$

$\mathbf{v} = 300 \cos 70°\mathbf{i} + 300 \sin 70°\mathbf{j}$

$\mathbf{u} + \mathbf{v} \approx 465.13\mathbf{i} + 450.96\mathbf{j}$

$\|\mathbf{u} + \mathbf{v}\| \approx \sqrt{(465.13)^2 + (450.96)^2} \approx 647.9$

$\alpha = \arctan\!\left(\dfrac{450.96}{465.13}\right) \approx 44.1°$

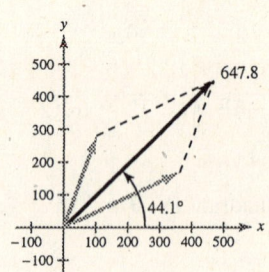

71. Force One: $\mathbf{u} = 45\mathbf{i}$

Force Two: $\mathbf{v} = 60 \cos \theta\mathbf{i} + 60 \sin \theta\mathbf{j}$

Resultant Force: $\mathbf{u} + \mathbf{v} = (45 + 60 \cos \theta)\mathbf{i} + 60 \sin \theta\mathbf{j}$

$\|\mathbf{u} + \mathbf{v}\| = \sqrt{(45 + 60 \cos \theta)^2 + (60 \sin \theta)^2} = 90$

$2025 + 5400 \cos \theta + 3600 = 8100$

$5400 \cos \theta = 2475$

$\cos \theta = \dfrac{2475}{5400} \approx 0.4583$

$\theta \approx 62.7°$

73. $\mathbf{u} = (2000 \cos 30°)\,\mathbf{i} + (2000 \sin 30°)\mathbf{j}$

$\approx 1732.05\mathbf{i} + 1000\mathbf{j}$

$\mathbf{v} = (900 \cos(-45°))\mathbf{i} + (900 \sin(-45°))\mathbf{j}$

$\approx 636.4\mathbf{i} + -636.4\mathbf{j}$

$\mathbf{u} + \mathbf{v} \approx 2368.4\mathbf{i} + 363.6\mathbf{j}$

$\|\mathbf{u} + \mathbf{v}\| \approx \sqrt{(2368.4)^2 + (363.6)^2} \approx 2396.15$

$\tan \theta \approx \dfrac{363.6}{2368.4} \approx 0.1535 \implies \theta \approx 8.7°$

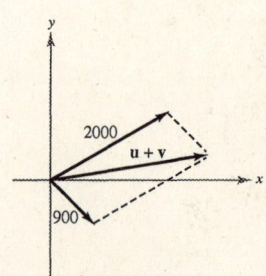

75. Horizontal component of velocity: $70 \cos 40° \approx 53.62$ ft/sec

Vertical component of velocity: $70 \sin 40° \approx 45.0$ ft/sec

77. Rope $\overrightarrow{AC}$: $\mathbf{u} = 10\mathbf{i} - 24\mathbf{j}$

The vector lies in Quadrant IV and its reference angle is $\arctan\left(\frac{12}{5}\right)$.

$$\mathbf{u} = \|\mathbf{u}\| \left[\cos\left(\arctan \tfrac{12}{5}\right)\mathbf{i} - \sin\left(\arctan \tfrac{12}{5}\right)\mathbf{j} \right]$$

Rope $\overrightarrow{BC}$: $\mathbf{v} = -20\mathbf{i} - 24\mathbf{j}$

The vector lies in Quadrant III and its reference angle is $\arctan\left(\frac{6}{5}\right)$.

$$\mathbf{v} = \|\mathbf{v}\| \left[-\cos\left(\arctan \tfrac{6}{5}\right)\mathbf{i} - \sin\left(\arctan \tfrac{6}{5}\right)\mathbf{j} \right]$$

Resultant: $\mathbf{u} + \mathbf{v} = -5000\mathbf{j}$

$$\|\mathbf{u}\| \cos\left(\arctan \tfrac{12}{5}\right) - \|\mathbf{v}\| \cos\left(\arctan \tfrac{6}{5}\right) = 0$$

$$-\|\mathbf{u}\| \sin\left(\arctan \tfrac{12}{5}\right) - \|\mathbf{v}\| \sin\left(\arctan \tfrac{6}{5}\right) = -5000$$

Solving this system of equations yields:

$T_{AC} = \|\mathbf{u}\| \approx 3611.1$ pounds

$T_{BC} = \|\mathbf{v}\| \approx 2169.5$ pounds

79. (a) Tow line 1: $\mathbf{u} = \|\mathbf{u}\| (\cos \theta \mathbf{i} + \sin \theta \mathbf{j})$

Tow line 2: $\mathbf{v} = \|\mathbf{u}\| (\cos(-\theta)\mathbf{i} + \sin(-\theta)\mathbf{j})$

Resultant: $\mathbf{u} + \mathbf{v} = 6000\mathbf{i} = [\|\mathbf{u}\| \cos \theta + \|\mathbf{u}\| \cos(-\theta)]\mathbf{i}$

$$\implies 6000 = 2\|\mathbf{u}\| \cos \theta \implies \|\mathbf{u}\| \approx 3000 \text{ sec.}$$

$$T = \|\mathbf{u}\| = 3000 \sec \theta$$

Domain: $0° \le \theta < 90°$

(b)

θ	10°	20°	30°	40°	50°	60°
T	3046.3	3192.5	3464.1	3916.2	4667.2	6000.0

(c)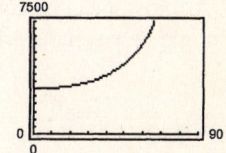

(d) The tension increases because the component in the direction of the motion of the barge decreases.

81. Airspeed: $\mathbf{v} = 860(\cos 302°\mathbf{i} + \sin 302°\mathbf{j})$

Groundspeed: $\mathbf{u} = 800(\cos 310°\mathbf{i} + \sin 310°\mathbf{j})$

$\mathbf{w} + \mathbf{v} = \mathbf{u}$

$\mathbf{w} = \mathbf{u} - \mathbf{v} = 800(\cos 310°\mathbf{i} + \sin 310°\mathbf{j}) - 860(\cos 302°\mathbf{i} + \sin 302°\mathbf{j}) \approx 58.50\mathbf{i} + 116.49\mathbf{j}$

$\|\mathbf{w}\| = \sqrt{58.50^2 + 116.49^2} \approx 130.35$ km/hr

$$\theta = \arctan\left(\frac{116.49}{58.50}\right) \approx 63.3°$$

Direction: N 26.7° E

83. (a) $\mathbf{u} = 220\mathbf{i}$, $\mathbf{v} = 150\cos 30°\mathbf{i} + 150\sin 30°\mathbf{j}$

$\mathbf{u} + \mathbf{v} = \left(220 + 75\sqrt{3}\right)\mathbf{i} + 75\mathbf{j}$

$\|\mathbf{u} + \mathbf{v}\| = \sqrt{\left(220 + 75\sqrt{3}\right)^2 + 75^2} \approx 357.85$ newtons

$\tan\theta = \dfrac{75}{220 + 75\sqrt{3}} \implies \theta \approx 12.1°$

(b) $\mathbf{u} + \mathbf{v} = 220\mathbf{i} + (150\cos\theta\mathbf{i} + 150\sin\theta\mathbf{j})$

$M = \|\mathbf{u} + \mathbf{v}\| = \sqrt{220^2 + 150^2(\cos^2\theta + \sin^2\theta) + 2(220)(150)\cos\theta}$

$= \sqrt{70{,}900 + 66{,}000\cos\theta} = 10\sqrt{709 + 660\cos\theta}$

$\alpha = \arctan\left(\dfrac{15\sin\theta}{22 + 15\cos\theta}\right)$

(c)

θ	0°	30°	60°	90°	120°	150°	180°
M	370.0	357.9	322.3	266.3	194.7	117.2	70.0
α	0°	12.1°	23.8°	34.3°	41.9°	39.8°	0°

(d)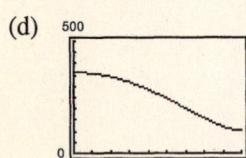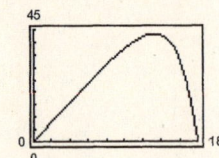

(e) For increasing θ, the two vectors tend to work against each other resulting in a decrease in the magnitude of the resultant.

85. True. See page 548.

87. True. In fact, $a = b = 0$.

89. (a) The angle between them is 0°.

(b) The angle between them is 180°.

(c) No. At most it can be equal to the sum when the angle between them is 0°.

91. Let $\mathbf{v} = (\cos\theta)\mathbf{i} + (\sin\theta)\mathbf{j}$.

$\|\mathbf{v}\| = \sqrt{\cos^2\theta + \sin^2\theta} = \sqrt{1} = 1$

Therefore, $\mathbf{v}$ is a unit vector for any value of θ.

93. $\mathbf{u} = \langle 5 - 1, 2 - 6\rangle = \langle 4, -4\rangle$

$\mathbf{v} = \langle 9 - 4, 4 - 5\rangle = \langle 5, -1\rangle$

$\mathbf{u} - \mathbf{v} = \langle -1, -3\rangle$

$\mathbf{v} - \mathbf{u} = \langle 1, 3\rangle$

95. $\left(\dfrac{6x^4}{7y^{-2}}\right)(14x^{-1}y^5) = \dfrac{12x^4y^5y^2}{x}$

$= 12x^3y^7,\ x \neq 0, y \neq 0$

97. $(18x)^0(4xy)^2(3x^{-1}) = \dfrac{16x^2y^2(3)}{x}$

$= 48xy^2,\ x \neq 0$

99. $(2.1 \times 10^9)(3.4 \times 10^{-4}) = 7.14 \times 10^5$

101. $\sin\theta = \dfrac{x}{7} \implies \sqrt{49 - x^2} = 7\cos\theta$

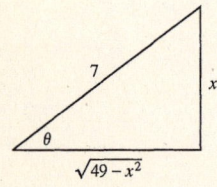

103. $\cot\theta = \dfrac{x}{10} \implies \sqrt{x^2 + 100} = 10 \cdot \csc\theta$

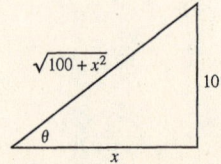

105. $\cos x(\cos x + 1) = 0$

$$\cos x = 0 \implies x = \frac{\pi}{2} + n\pi$$

$$\cos x = -1 \implies x = \pi + 2n\pi$$

107. $3 \sec x + 4 = 10$

$$\sec x = 2$$

$$\cos x = \frac{1}{2}$$

$$x = \frac{\pi}{3} + 2n\pi, \frac{5\pi}{3} + 2n\pi$$

Section 6.4   Vectors and Dot Products

- Know the definition of the dot product of $\mathbf{u} = \langle u_1, u_2 \rangle$ and $\mathbf{v} = \langle v_1, v_2 \rangle$.

  $\mathbf{u} \cdot \mathbf{v} = u_1 v_1 + u_2 v_2$

- Know the following properties of the dot product:

  1. $\mathbf{u} \cdot \mathbf{v} = \mathbf{v} \cdot \mathbf{u}$
  2. $\mathbf{0} \cdot \mathbf{v} = 0$
  3. $\mathbf{u} \cdot (\mathbf{v} + \mathbf{w}) = \mathbf{u} \cdot \mathbf{v} + \mathbf{u} \cdot \mathbf{w}$
  4. $\mathbf{v} \cdot \mathbf{v} = \|\mathbf{v}\|^2$
  5. $c(\mathbf{u} \cdot \mathbf{v}) = c\mathbf{u} \cdot \mathbf{v} = \mathbf{u} \cdot c\mathbf{v}$

- If θ is the angle between two nonzero vectors $\mathbf{u}$ and $\mathbf{v}$, then

  $$\cos \theta = \frac{\mathbf{u} \cdot \mathbf{v}}{\|\mathbf{u}\| \, \|\mathbf{v}\|}.$$

- The vectors $\mathbf{u}$ and $\mathbf{v}$ are orthogonal if $\mathbf{u} \cdot \mathbf{v} = 0$.

- Know the definition of vector components. $\mathbf{u} = \mathbf{w}_1 + \mathbf{w}_2$ where $\mathbf{w}_1$ and $\mathbf{w}_2$ are orthogonal, and $\mathbf{w}_1$ is parallel to $\mathbf{v}$. $\mathbf{w}_1$ is called the projection of $\mathbf{u}$ onto $\mathbf{v}$ and is denoted by

  $$\mathbf{w}_1 = \text{proj}_\mathbf{v}\mathbf{u} = \left(\frac{\mathbf{u} \cdot \mathbf{v}}{\|\mathbf{v}\|^2}\right)\mathbf{v}.$$

  Then we have $\mathbf{w}_2 = \mathbf{u} - \mathbf{w}_1$.

- Know the definition of work.

  1. Projection form: $W = \|\text{proj}_{\overrightarrow{PQ}}\mathbf{F}\| \, \|\overrightarrow{PQ}\|$
  2. Dot product form: $W = \mathbf{F} \cdot \overrightarrow{PQ}$

Solutions to Odd-Numbered Exercises

1. $\mathbf{u} \cdot \mathbf{v} = \langle 6, 3 \rangle \cdot \langle 2, -4 \rangle = 6(2) + 3(-4) = 0$

3. $\mathbf{u} \cdot \mathbf{v} = \langle 5, 1 \rangle \cdot \langle 3, -1 \rangle = 5(3) + 1(-1) = 14$

5. $\mathbf{u} = \langle 2, 2 \rangle$

$\mathbf{u} \cdot \mathbf{u} = 2(2) + 2(2) = 8$, scalar

7. $\mathbf{u} = \langle 2, 2 \rangle, \mathbf{v} = \langle -3, 4 \rangle, \mathbf{w} = \langle 1, -4 \rangle$

$\mathbf{u} \cdot \mathbf{v} = 2(-3) + 2(4) = 2$

$(\mathbf{u} \cdot \mathbf{v})\mathbf{w} = 2\langle 1, -4 \rangle = \langle 2, -8 \rangle$, vector

9. $\mathbf{u} = \langle 2, 2 \rangle, \mathbf{v} = \langle -3, 4 \rangle$

$\mathbf{u} \cdot 2\mathbf{v} = 2\mathbf{u} \cdot \mathbf{v} = 4(-3) + 4(4) = 4$, scalar

11. $\mathbf{u} = \langle -5, 12 \rangle$

$\|\mathbf{u}\| = \sqrt{\mathbf{u} \cdot \mathbf{u}} = \sqrt{(-5)^2 + 12^2} = 13$

13. $\mathbf{u} = 20\mathbf{i} + 25\mathbf{j}$

$\|\mathbf{u}\|\sqrt{\mathbf{u} \cdot \mathbf{u}} = \sqrt{(20)^2 + (25)^2} = \sqrt{1025} = 5\sqrt{41}$

15. $\mathbf{u} = -4\mathbf{j}$

$\|\mathbf{u}\| = \sqrt{\mathbf{u} \cdot \mathbf{u}} = \sqrt{(-4)(-4)} = 4$

17. $\mathbf{u} = \langle -1, 0 \rangle$, $\mathbf{v} = \langle 0, 2 \rangle$

$$\cos \theta = \frac{\mathbf{u} \cdot \mathbf{v}}{\|\mathbf{u}\| \, \|\mathbf{v}\|} = \frac{0}{(1)(2)} = 0 \implies \theta = 90°$$

19. $\mathbf{u} = 3\mathbf{i} + 4\mathbf{j}$, $\mathbf{v} = -2\mathbf{i} + 3\mathbf{j}$

$$\cos \theta = \frac{\mathbf{u} \cdot \mathbf{v}}{\|\mathbf{u}\| \, \|\mathbf{v}\|} = \frac{-6 + 12}{(5)(\sqrt{13})} = \frac{6}{5\sqrt{13}}$$

$$\theta = \arccos\left(\frac{6}{5\sqrt{13}}\right) \approx 70.56°$$

21. $\mathbf{u} = 2\mathbf{i}$, $\mathbf{v} = -3\mathbf{j}$

$$\cos \theta = \frac{\mathbf{u} \cdot \mathbf{v}}{\|\mathbf{u}\| \, \|\mathbf{v}\|} = \frac{0}{(2)(3)} = 0 \implies \theta = 90°$$

23. $\quad \mathbf{u} = \left(\cos \frac{\pi}{3}\right)\mathbf{i} + \left(\sin \frac{\pi}{3}\right)\mathbf{j} = \frac{1}{2}\mathbf{i} + \frac{\sqrt{3}}{2}\mathbf{j}$

$$\mathbf{v} = \left(\cos \frac{3\pi}{4}\right)\mathbf{i} + \left(\sin \frac{3\pi}{4}\right)\mathbf{j} = -\frac{\sqrt{2}}{2}\mathbf{i} + \frac{\sqrt{2}}{2}\mathbf{j}$$

$$\|\mathbf{u}\| = \|\mathbf{v}\| = 1$$

$$\cos \theta = \frac{\mathbf{u} \cdot \mathbf{v}}{\|\mathbf{u}\| \, \|\mathbf{v}\|} = \mathbf{u} \cdot \mathbf{v} = \left(\frac{1}{2}\right)\left(-\frac{\sqrt{2}}{2}\right) + \left(\frac{\sqrt{3}}{2}\right)\left(\frac{\sqrt{2}}{2}\right) = \frac{-\sqrt{2} + \sqrt{6}}{4}$$

$$\theta = \arccos\left(\frac{-\sqrt{2} + \sqrt{6}}{4}\right) = 75° = \frac{5\pi}{12}$$

25. $\mathbf{u} = 3\mathbf{i} + 4\mathbf{j}$, $\mathbf{v} = -7\mathbf{i} + 5\mathbf{j}$

$$\cos \theta = \frac{\mathbf{u} \cdot \mathbf{v}}{\|\mathbf{u}\| \, \|\mathbf{v}\|} = -\frac{1}{(5)(\sqrt{74})} \implies \theta \approx 91.33°$$

27. $\mathbf{u} = 5\mathbf{i} + 5\mathbf{j}$, $\mathbf{v} = -8\mathbf{i} + 8\mathbf{j}$

$$\cos \theta = \frac{\mathbf{u} \cdot \mathbf{v}}{\|\mathbf{u}\| \, \|\mathbf{v}\|} = 0 \implies \theta = 90°$$

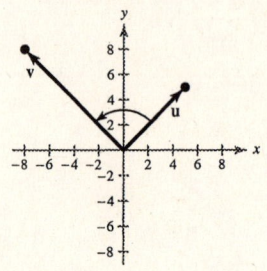

29. $P = (1, 2)$, $Q = (3, 4)$, $R = (2, 5)$

$\overrightarrow{PQ} = \langle 2, 2 \rangle$, $\overrightarrow{PR} = \langle 1, 3 \rangle$, $\overrightarrow{QR} = \langle -1, 1 \rangle$

$$\cos \alpha = \frac{\overrightarrow{PQ} \cdot \overrightarrow{PR}}{\|\overrightarrow{PQ}\| \, \|\overrightarrow{PR}\|} = \frac{8}{(2\sqrt{2})(\sqrt{10})} \implies \alpha = \arccos \frac{2}{\sqrt{5}} \approx 26.6°$$

$$\cos \beta = \frac{\overrightarrow{PQ} \cdot \overrightarrow{QR}}{\|\overrightarrow{PQ}\| \, \|\overrightarrow{QR}\|} = 0 \implies \beta = 90°$$

Thus, $\gamma \approx 180° - 26.6° - 90° = 63.4°$.

31. $\mathbf{u} \cdot \mathbf{v} = \|\mathbf{u}\| \, \|\mathbf{v}\| \cos \theta$

$$= (4)(10) \cos \frac{2\pi}{3} = 40\left(-\frac{1}{2}\right) = -20$$

33. $\mathbf{u} = \langle -12, 30 \rangle$, $\mathbf{v} = \left\langle \frac{1}{2}, -\frac{5}{4} \right\rangle$

$\mathbf{u} = -24\mathbf{v} \implies \mathbf{u}$ and $\mathbf{v}$ are parallel.

35. $\mathbf{u} = \frac{1}{4}(3\mathbf{i} - \mathbf{j})$, $\mathbf{v} = 5\mathbf{i} + 6\mathbf{j}$

$\mathbf{u} \neq k\mathbf{v} \implies$ Not parallel

$\mathbf{u} \cdot \mathbf{v} \neq 0 \implies$ Not orthogonal

Neither

37. $\mathbf{u} = 2\mathbf{i} - 2\mathbf{j}$, $\mathbf{v} = -\mathbf{i} - \mathbf{j}$

$\mathbf{u} \cdot \mathbf{v} = 0 \implies \mathbf{u}$ and $\mathbf{v}$ are orthogonal.

39. $\mathbf{u} = \langle 3, 4 \rangle$, $\mathbf{v} = \langle 8, 2 \rangle$

$$\mathbf{w}_1 = \text{proj}_\mathbf{v}\mathbf{u} = \left(\frac{\mathbf{u} \cdot \mathbf{v}}{\|\mathbf{v}\|^2}\right)\mathbf{v}$$

$$= \left(\frac{32}{68}\right)\mathbf{v} = \frac{8}{17}\langle 8, 2 \rangle = \frac{16}{17}\langle 4, 1 \rangle$$

$$\mathbf{w}_2 = \mathbf{u} - \mathbf{w}_1 = \langle 3, 4 \rangle - \frac{16}{17}\langle 4, 1 \rangle = \frac{13}{17}\langle -1, 4 \rangle$$

$$\mathbf{u} = \mathbf{w}_1 + \mathbf{w}_2 = \frac{16}{17}\langle 4, 1 \rangle + \frac{13}{17}\langle -1, 4 \rangle$$

41. $\mathbf{u} = \langle 0, 3 \rangle$, $\mathbf{v} = \langle 2, 15 \rangle$

$$\mathbf{w}_1 = \text{proj}_\mathbf{v}\mathbf{u} = \left(\frac{\mathbf{u} \cdot \mathbf{v}}{\|\mathbf{v}\|^2}\right)\mathbf{v} = \frac{45}{229}\langle 2, 15 \rangle$$

$$\mathbf{w}_2 = \mathbf{u} - \mathbf{w}_1 = \langle 0, 3 \rangle - \frac{45}{229}\langle 2, 15 \rangle$$

$$= \left\langle -\frac{90}{229}, \frac{12}{229} \right\rangle = \frac{6}{229}\langle -15, 2 \rangle$$

$$\mathbf{u} = \mathbf{w}_1 + \mathbf{w}_2 = \frac{45}{229}\langle 2, 15 \rangle + \frac{6}{229}\langle -15, 2 \rangle$$

43. $\text{proj}_\mathbf{v}\mathbf{u} = \mathbf{u}$ since they are parallel.

$$\text{proj}_\mathbf{v}\mathbf{u} = \frac{\mathbf{u} \cdot \mathbf{v}}{\|\mathbf{v}\|^2}\mathbf{v}$$

$$= \frac{18 + 8}{36 + 16}\mathbf{v} = \frac{26}{52}\langle 6, 4 \rangle = \langle 3, 2 \rangle = \mathbf{u}$$

45. $\text{proj}_\mathbf{v}\mathbf{u} = \mathbf{0}$ since they are perpendicular.

$$\text{proj}_\mathbf{v}\mathbf{u} = \frac{\mathbf{u} \cdot \mathbf{v}}{\|\mathbf{v}\|^2}\mathbf{v} = \mathbf{0}, \text{ since } \mathbf{u} \cdot \mathbf{v} = 0.$$

47. $\mathbf{u} = \langle 2, 6 \rangle$

For $\mathbf{v}$ to be orthogonal to $\mathbf{u}$, $\mathbf{u} \cdot \mathbf{v}$ must equal 0. Two possibilities: $\langle 6, -2 \rangle$ and $\langle -6, 2 \rangle$

49. $\mathbf{u} = \frac{1}{2}\mathbf{i} - \frac{3}{4}\mathbf{j}$

For $\mathbf{v}$ to be orthogonal to $\mathbf{u}$, $\mathbf{u} \cdot \mathbf{v}$ must equal 0. Two possibilities: $\left\langle \frac{3}{4}, \frac{1}{2} \right\rangle$ and $\left\langle -\frac{3}{4}, -\frac{1}{2} \right\rangle$

51. $W = \|\text{proj}_{\overrightarrow{PQ}}\mathbf{v}\| \|\overrightarrow{PQ}\|$ where $\overrightarrow{PQ} = \langle 4, 7 \rangle$ and $\mathbf{v} = \langle 1, 4 \rangle$.

$$\text{proj}_{\overrightarrow{PQ}}\mathbf{v} = \left(\frac{\mathbf{v} \cdot \overrightarrow{PQ}}{\|\overrightarrow{PQ}\|^2}\right)\overrightarrow{PQ} = \left(\frac{32}{65}\right)\langle 4, 7 \rangle$$

$$W = \|\text{proj}_{\overrightarrow{PQ}}\mathbf{v}\| \|\overrightarrow{PQ}\| = \left(\frac{32\sqrt{65}}{65}\right)\left(\sqrt{65}\right) = 32$$

53. $\mathbf{u} = \langle 1245, 2600 \rangle$, $\mathbf{v} = \langle 12.20, 8.50 \rangle$

$\mathbf{u} \cdot \mathbf{v} = 1245(12.20) + 2600(8.50) = \$37,289$

This gives the total revenue that can be earned by selling all of the units.

55. (a) $\mathbf{F} = -30,000\overrightarrow{\mathbf{j}}$, Gravitational force

$\mathbf{v} = \langle \cos(d°), \sin(d°) \rangle$

$$\mathbf{w}_1 = \text{proj}_\mathbf{v}\mathbf{F} = \left(\frac{\mathbf{F} \cdot \mathbf{v}}{\|\mathbf{v}\|^2}\right)\mathbf{v} = (\mathbf{F} \cdot \mathbf{v})\mathbf{v} = -30,000 \sin(d°)\langle \cos d°, \sin d° \rangle$$

$$= \langle -30,000 \sin d° \cos d°, -30,000 \sin^2 d° \rangle$$

Force needed: $30,000 \sin(d°)$

(b)

d	0°	1°	2°	3°	4°	5°	6°	7°	8°	9°	10°
Force	0	523.6	1047.0	1570.1	2092.7	2614.7	3135.9	3656.1	4175.2	4693.0	5209.4

—CONTINUED—

55. **—CONTINUED—**

(c) $\mathbf{w}_2 = \mathbf{F} - \mathbf{w}_1 = -30{,}000\mathbf{j} + 2614.7(\cos(5°)\mathbf{i} + \sin(5°)\mathbf{j}) \approx 2604.75\mathbf{i} - 29{,}772.11\mathbf{j}$

$\|\mathbf{w}_2\| \approx 29{,}885.8$ pounds

57. (a) $\mathbf{F} = 15{,}691\langle \cos 30°, \sin 30°\rangle$

$\overrightarrow{PQ} = d\langle 1, 0\rangle$

$W = \mathbf{F} \cdot \overrightarrow{PQ} = 15{,}691\dfrac{\sqrt{3}}{2}d \approx 13{,}588.8d$

(b)

d	0	200	400	800
Work	0	2,717,761	5,435,522	10,871,044

59. $W = (\cos 25°)(20)(40) \approx 725.05$ foot-pounds

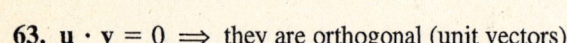

61. True. $\mathbf{u} \cdot \mathbf{v} = 0$

63. $\mathbf{u} \cdot \mathbf{v} = 0 \implies$ they are orthogonal (unit vectors).

65. (a) $\text{proj}_\mathbf{v}\,\mathbf{u} = \mathbf{u} \implies \mathbf{u}$ and $\mathbf{v}$ are parallel.

(b) $\text{proj}_\mathbf{v}\,\mathbf{u} = 0 \implies \mathbf{u}$ and $\mathbf{v}$ are orthogonal.

67. Use the Law of Cosines on the triangle:

$\|\mathbf{u} - \mathbf{v}\|^2 = \|\mathbf{u}\|^2 + \|\mathbf{v}\|^2 - 2\|\mathbf{u}\|\,\|\mathbf{v}\|$

$= \|\mathbf{u}\|^2 + \|\mathbf{v}\|^2 - 2\mathbf{u} \cdot \mathbf{v}$

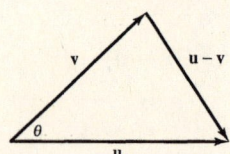

69. $g(x) = f(x - 4)$ is a horizontal shift of f four units to the right.

71. $g(x) = f(x) + 6$ is a vertical shift of f six units upward.

73. $\sqrt{-4} - 1 = 2i - 1$

$= -1 + 2i$

75. $3i(4 - 5i) = 12i + 15$

$= 15 + 12i$

77. $(1 + 3i)(1 - 3i) = 1 - (3i)^2$

$= 1 + 9$

$= 10$

79. $\dfrac{3}{1 + i} + \dfrac{2}{2 - 3i} = \dfrac{3}{1 + i} \cdot \dfrac{1 - i}{1 - i} + \dfrac{2}{2 - 3i} \cdot \dfrac{2 + 3i}{2 + 3i}$

$= \dfrac{3 - 3i}{2} + \dfrac{4 + 6i}{13}$

$= \dfrac{39 - 39i + 8 + 12i}{26}$

$= \dfrac{47}{26} - \dfrac{27}{26}i$

81. $-2i$

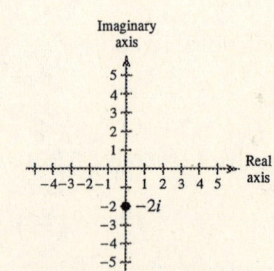

83. $1 + 8i$

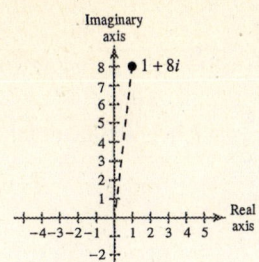

85. Let x be the number of people presently in the group.

Each share is $\dfrac{250{,}000}{x}$.

Also, $\dfrac{250{,}000}{x} - 6250 = \dfrac{250{,}000}{x + 2}$.

Solving this equation, $x = 8$ people.

Section 6.5   Trigonometric Form of a Complex Number

- You should be able to graphically represent complex numbers.
- The absolute value of the complex numbers $z = a + bi$ is $|z| = \sqrt{a^2 + b^2}$.
- The trigonometric form of the complex number $z = a + bi$ is $z = r(\cos\theta + i\sin\theta)$ where
  - (a) $a = r\cos\theta$
  - (b) $b = r\sin\theta$
  - (c) $r = \sqrt{a^2 + b^2}$; r is called the modulus of z.
  - (d) $\tan\theta = b/a$; θ is called the argument of z.
- Given $z_1 = r_1(\cos\theta_1 + i\sin\theta_1)$ and $z_2 = r_2(\cos\theta_2 + i\sin\theta_2)$:
  - (a) $z_1 z_2 = r_1 r_2[\cos(\theta_1 + \theta_2) + i\sin(\theta_1 + \theta_2)]$
  - (b) $\dfrac{z_1}{z_2} = \dfrac{r_1}{r_2}[\cos(\theta_1 - \theta_2) + i\sin(\theta_1 - \theta_2)]$, $z_2 \neq 0$
- You should know DeMoivre's Theorem: If $z = r(\cos\theta + i\sin\theta)$, then for any positive integer n,
  $$z^n = r^n(\cos n\theta + i\sin n\theta).$$
- You should know that for any positive integer n, $z = r(\cos\theta + i\sin\theta)$ has n distinct nth roots given by
  $$\sqrt[n]{r}\left[\cos\left(\frac{\theta + 2\pi k}{n}\right) + i\sin\left(\frac{\theta + 2\pi k}{n}\right)\right]$$
  where $k = 0, 1, 2, \ldots, n - 1$.

Solutions to Odd-Numbered Exercises

1. $|6i| = 6$

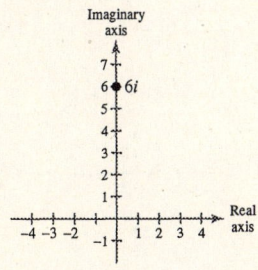

3. $|-4| = \sqrt{(-4)^2 + 0^2}$
       $= \sqrt{16} = 4$

5. $|-4 + 4i| = \sqrt{(-4)^2 + (4)^2}$
       $= \sqrt{32} = 4\sqrt{2}$

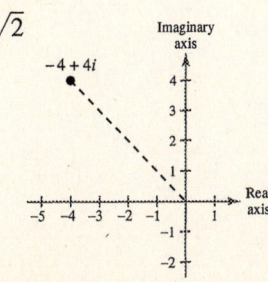

7. $|3 + 6i| = \sqrt{9 + 36}$
       $= \sqrt{45} = 3\sqrt{5}$

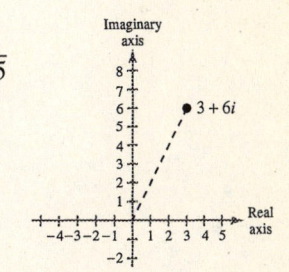

9. $z = 3i$

$r = \sqrt{0^2 + 3^2} = \sqrt{9} = 3$

$\tan \theta = \dfrac{3}{0}$, undefined $\Rightarrow \theta = \dfrac{\pi}{2}$

$z = 3\left(\cos \dfrac{\pi}{2} + i \sin \dfrac{\pi}{2}\right)$

11. $z = -2 - 2i$

$r = \sqrt{(-2)^2 + (-2)^2} = \sqrt{8} = 2\sqrt{2}$

$\tan \theta = \dfrac{-2}{-2} = 1$, θ is in Quadrant III.

$\theta = \dfrac{5\pi}{4}$

$z = 2\sqrt{2}\left(\cos \dfrac{5\pi}{4} + i \sin \dfrac{5\pi}{4}\right)$

13. $z = 5 - 5i$

$r = \sqrt{5^2 + (-5)^2} = \sqrt{50} = 5\sqrt{2}$

$\tan \theta = -\dfrac{5}{5} = -1 \Rightarrow \theta = \dfrac{7\pi}{4}$

$z = 5\sqrt{2}\left(\cos \dfrac{7\pi}{4} + i \sin \dfrac{7\pi}{4}\right)$

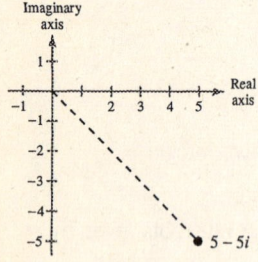

15. $z = \sqrt{3} + i$

$r = \sqrt{\left(\sqrt{3}\right)^2 + 1^2} = \sqrt{4} = 2$

$\tan \theta = \dfrac{1}{\sqrt{3}} = \dfrac{\sqrt{3}}{3} \Rightarrow \theta = \dfrac{\pi}{6}$

$z = 2\left(\cos \dfrac{\pi}{6} + i \sin \dfrac{\pi}{6}\right)$

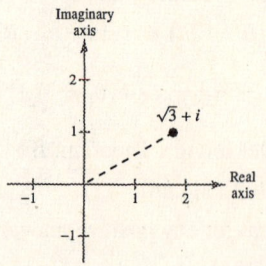

17. $z = -2(1 + \sqrt{3}i)$

$r = \sqrt{(-2)^2 + \left(-2\sqrt{3}\right)^2} = \sqrt{16} = 4$

$\tan \theta = \dfrac{\sqrt{3}}{1} = \sqrt{3} \Rightarrow \theta = \dfrac{4\pi}{3}$

$z = 4\left(\cos \dfrac{4\pi}{3} + i \sin \dfrac{4\pi}{3}\right)$

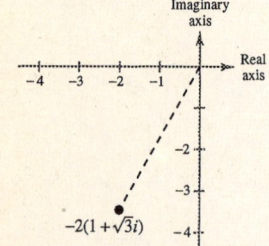

19. $z = -8i$

$r = \sqrt{0 + (-8)^2} = \sqrt{64} = 8$

$\tan \theta = -\dfrac{8}{0}$, undefined $\Rightarrow \theta = \dfrac{3\pi}{2}$

$z = 8\left(\cos \dfrac{3\pi}{2} + i \sin \dfrac{3\pi}{2}\right)$

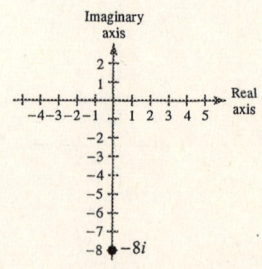

21. $z = -7 + 4i$

$r = \sqrt{49 + 16} = \sqrt{65}$

$\tan \theta = \dfrac{4}{-7} \implies \theta \approx 2.62$ radians or $150.26°$

$z = \sqrt{65}(\cos 150.26° + i \sin 150.26°)$

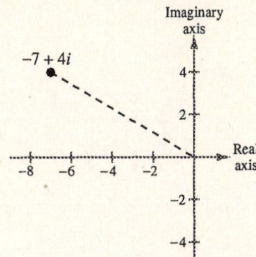

23. $z = 3$

$r = \sqrt{3^2 + 0^2} = 3$

$\tan \theta = \dfrac{0}{7} = 0 \implies \theta = 0°$

$z = 3(\cos 0° + i \sin 0°)$

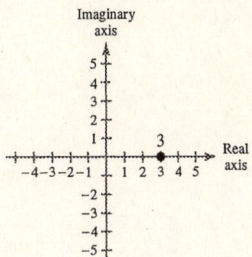

25. $z = 3 + \sqrt{3}i$

$r = \sqrt{9 + 3} = \sqrt{12} = 2\sqrt{3}$

$\tan \theta = \dfrac{\sqrt{3}}{3} \implies \theta = \dfrac{\pi}{6}$ or $30°$

$z = 2\sqrt{3}\left(\cos \dfrac{\pi}{6} + i \sin \dfrac{\pi}{6}\right)$

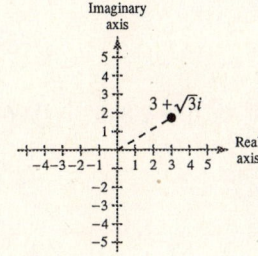

27. $z = -1 - 2i$

$r = \sqrt{1^2 + 2^2} = \sqrt{5}$

$\tan \theta = \dfrac{-2}{-1} = 2 \implies \theta \approx 243.4°$

$z = \sqrt{5}(\cos 243.4° + i \sin 243.4°)$

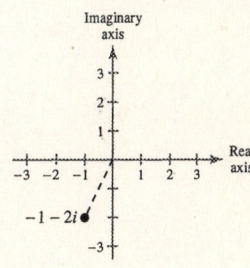

29. $z = 5 + 2i$

$r = \sqrt{25 + 4} = \sqrt{29} \approx 5.385$

$\tan \theta = \dfrac{2}{5} \implies \theta \approx 21.80°$

$z = \sqrt{29}(\cos 21.80° + i \sin 21.80°)$

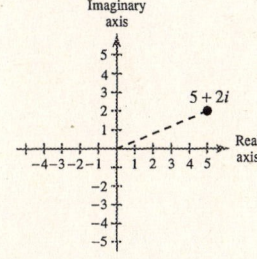

31. $z = 3\sqrt{2} - 7i$

$r = \sqrt{18 + 49} = \sqrt{67} \approx 8.185$

$\tan \theta = \dfrac{-7}{3\sqrt{2}} \approx -1.6499 \implies \theta \approx 301.22°$

$z = \sqrt{67}(\cos 301.22° + i \sin 301.22°)$

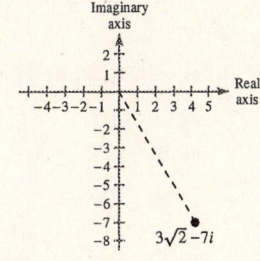

33. $2(\cos 120° + i \sin 120°) = 2\left(-\frac{1}{2} + \frac{\sqrt{3}}{2}i\right)$
$$= -1 + \sqrt{3}i$$

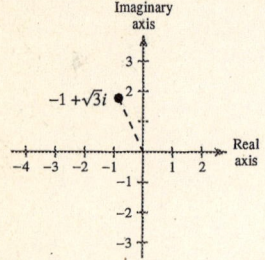

35. $\frac{3}{2}(\cos 330° + i \sin 330°) = \frac{3}{2}\left(\frac{\sqrt{3}}{2} - \frac{1}{2}i\right)$
$$= \frac{3\sqrt{3}}{4} - \frac{3}{4}i$$

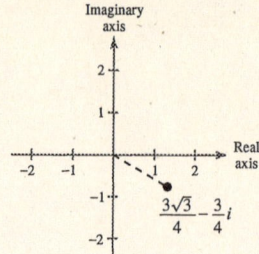

37. $3.75\left(\cos \frac{3\pi}{4} + i \sin \frac{3\pi}{4}\right) = -\frac{15\sqrt{2}}{8} + \frac{15\sqrt{2}}{8}i$

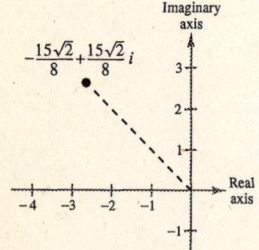

39. $6\left(\cos \frac{\pi}{3} + i \sin \frac{\pi}{3}\right) = 6\left(\frac{1}{2} + \frac{\sqrt{3}}{2}i\right) = 3 + 3\sqrt{3}i$

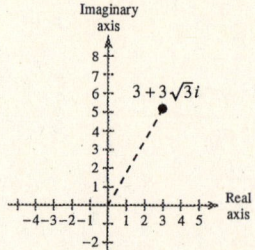

41. $4\left(\cos \frac{3\pi}{2} + i \sin \frac{3\pi}{2}\right) = 4(0 - i) = -4i$

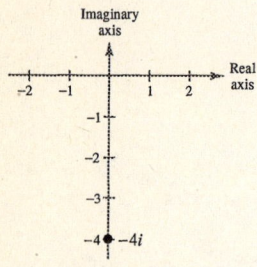

43. $3[\cos(18° \, 45') + i \sin(18° \, 45')] \approx 2.8408 + 0.9643i$

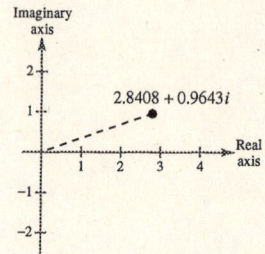

45. $5\left(\cos \frac{\pi}{9} + i \sin \frac{\pi}{9}\right) \approx 4.6985 + 1.7101i$

47. $9(\cos 58° + i \sin 58°) \approx 4.7693 + 7.6324i$

49.

The absolute value of each power is 1.

$z^3 = \frac{\sqrt{2}}{2}(-1 + i)$ $z^2 = i$

$z^4 = -1$ $z = \frac{\sqrt{2}}{2}(1 + i)$

51. $\left[3\left(\cos\dfrac{\pi}{3}+i\sin\dfrac{\pi}{3}\right)\right]\left[4\left(\cos\dfrac{\pi}{6}+i\sin\dfrac{\pi}{6}\right)\right]=(3)(4)\left[\cos\left(\dfrac{\pi}{3}+\dfrac{\pi}{6}\right)+i\sin\left(\dfrac{\pi}{6}+\dfrac{\pi}{3}\right)\right]=12\left(\cos\dfrac{\pi}{2}+i\sin\dfrac{\pi}{2}\right)$

53. $\left[\dfrac{5}{3}(\cos 140°+i\sin 140°)\right]\left[\dfrac{2}{3}(\cos 60°+i\sin 60°)\right]=\left(\dfrac{5}{3}\right)\left(\dfrac{2}{3}\right)[\cos(140°+60°)+i\sin(140°+60°)]$

$$=\dfrac{10}{9}(\cos 200°+i\sin 200°)$$

55. $\left[\dfrac{11}{20}(\cos 290°+i\sin 290°)\right]\left[\dfrac{2}{5}(\cos 200°+i\sin 200°)\right]=\left(\dfrac{11}{20}\right)\left(\dfrac{2}{5}\right)[\cos(290°+200°)+i\sin(290°+200°)]$

$$=\dfrac{11}{50}(\cos 490°+i\sin 490°)$$

$$=\dfrac{11}{50}(\cos 130°+i\sin 130°)$$

57. $\dfrac{\cos 50°+i\sin 50°}{\cos 20°+i\sin 20°}=\cos(50°-20°)+i\sin(50°-20°)=\cos 30°+i\sin 30°$

59. $\dfrac{2(\cos 120°+i\sin 120°)}{4(\cos 40°+i\sin 40°)}=\dfrac{1}{2}[\cos(120°-40°)+i\sin(120°-40°)]=\dfrac{1}{2}(\cos 80°+ +i\sin 80°)$

61. $\dfrac{18(\cos 54°+i\sin 54°)}{3(\cos 102°+i\sin 102°)}=6(\cos(54°-102°)+i\sin(54°-102°))=6(\cos(-48°)+i\sin(-48°))$

63. (a) $2+2i=2\sqrt{2}(\cos 45°+i\sin 45°)$

$1-i=\sqrt{2}[\cos(-45°)+i\sin(-45°)]$

(b) $(2+2i)(1-i)=\left[2\sqrt{2}(\cos 45°+i\sin 45°)\right]\left[\sqrt{2}(\cos(-45°)+i\sin(-45°))\right]=4(\cos 0°+i\sin 0°)=4$

(c) $(2+2i)(1-i)=2-2i+2i-2i^2=2+2=4$

65. (a) $-2i=2[\cos(-90°)+i\sin(-90°)]$

$1+i=\sqrt{2}(\cos 45°+i\sin 45°)$

(b) $-2i(1+i)=2[\cos(-90°)+i\sin(-90°)]\left[\sqrt{2}(\cos 45°+i\sin 45°)\right]$

$$=2\sqrt{2}[\cos(-45°)+i\sin(-45°)]$$

$$=2\sqrt{2}\left[\dfrac{1}{\sqrt{2}}-\dfrac{1}{\sqrt{2}}i\right]=2-2i$$

(c) $-2i(1+i)=-2i-2i^2=-2i+2=2-2i$

67. (a) $5=5(\cos 0°+i\sin 0°)$

$2+3i\approx\sqrt{13}(\cos 56.31°+i\sin 56.31°)$

(b) $\dfrac{5}{2+3i}\approx\dfrac{5(\cos 0°+i\sin 0°)}{\sqrt{13}(\cos 56.31°+i\sin 56.31°)}=\dfrac{5\sqrt{13}}{13}[\cos(-56.31°)+i\sin(-56.31°)]\approx 0.7692-1.154i$

(c) $\dfrac{5}{2+3i}=\dfrac{5}{2+3i}\cdot\dfrac{2-3i}{2-3i}=\dfrac{10-15i}{13}=\dfrac{10}{13}-\dfrac{15}{13}i\approx 0.7692-1.154i$

69. Let $z = x + iy$ such that:

$$|z| = 2 \implies 2 = \sqrt{x^2 + y^2} \implies 4 = x^2 + y^2$$

Circle with radius of 2

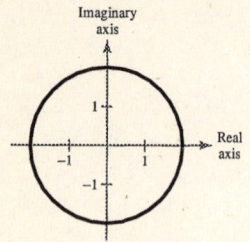

71. $\theta = \dfrac{\pi}{6}$

Let $z = x + iy$ such that:

$$\tan \frac{\pi}{6} = \frac{y}{x} \implies \frac{y}{x} = \frac{1}{\sqrt{3}} \implies y = \frac{1}{\sqrt{3}}x$$

Line

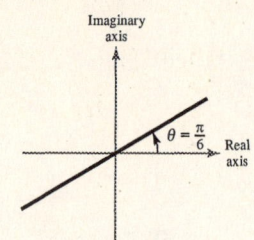

73. $(1 + i)^3 = \left[\sqrt{2}\left(\cos\dfrac{\pi}{4} + i \sin \dfrac{\pi}{4} \right) \right]^3$

$$= (\sqrt{2})^3 \left(\cos \frac{3\pi}{4} + i \sin \frac{3\pi}{4} \right)$$

$$= 2\sqrt{2}\left(-\frac{\sqrt{2}}{2} + \frac{\sqrt{2}}{2}i \right)$$

$$= -2 + 2i$$

75. $(-1 + i)^{10} = \left[\sqrt{2}\left(\cos\dfrac{3\pi}{4} + i \sin \dfrac{3\pi}{4} \right) \right]^{10}$

$$= (\sqrt{2})^{10}\left(\cos \frac{30\pi}{4} + i \sin \frac{30\pi}{4} \right)$$

$$= 32\left[\cos\left(\frac{3\pi}{2} + 6\pi \right) + i \sin\left(\frac{3\pi}{2} + 6\pi \right) \right]$$

$$= 32\left(\cos \frac{3\pi}{2} + i \sin \frac{3\pi}{2} \right)$$

$$= 32[0 + i(-1)]$$

$$= -32i$$

77. $2(\sqrt{3} + i)^5 = 2\left[2\left(\cos\dfrac{\pi}{6} + i \sin \dfrac{\pi}{6} \right) \right]^5$

$$= 2\left[2^5\left(\cos \frac{5\pi}{6} + i \sin \frac{5\pi}{6} \right) \right]$$

$$= 64\left(-\frac{\sqrt{3}}{2} + \frac{1}{2}i \right)$$

$$= -32\sqrt{3} + 32i$$

79. $[5(\cos 20° + i \sin 20°)]^3 = 5^3(\cos 60° + i \sin 60°)$

$$= \frac{125}{2} + \frac{125\sqrt{3}}{2}i$$

81. $\left(\cos\dfrac{5\pi}{4} + i \sin \dfrac{5\pi}{4} \right)^{10} = \cos \dfrac{25\pi}{2} + i \sin \dfrac{25\pi}{2}$

$$= \cos\left(12\pi + \frac{\pi}{2} \right) + i \sin\left(12\pi + \frac{\pi}{2} \right) = \cos \frac{\pi}{2} + i \sin \frac{\pi}{2} = i$$

83. $[4(\cos 2.8 + i \sin 2.8)]^5 = 4^5(\cos 14 + i \sin 14)$

$$\approx 140.02 + 1014.38i$$

85. $(3 - 2i)^5 = -597 - 122i$

87. $[3(\cos 15° + i \sin 15°)]^4 = 81(\cos 60° + i \sin 60°)$

$$= \frac{81}{2} + \frac{81\sqrt{3}}{2}i$$

89. $\left[-\dfrac{1}{2}(1 + \sqrt{3}i) \right]^6 = \left[\cos\dfrac{4\pi}{3} + i \sin \dfrac{4\pi}{3} \right]^6$

$$= \cos 8\pi + i \sin 8\pi$$

$$= 1$$

91. (a) In trigonometric form we have:

$2(\cos 30° + i \sin 30°)$

$2(\cos 150° + i \sin 150°)$

$2(\cos 270° + i \sin 270°)$

(c) $[2(\cos 30° + i \sin 30°)]^3 = 8i$

$[2(\cos 150° + i \sin 150°)]^3 = 8i$

$[2(\cos 270° + i \sin 270°)]^3 = 8i$

(b) There are three roots evenly spaced around a circle of radius 2. Therefore, they represent the cube roots of some number of modulus 8. Cubing them shows that they are all cube roots of $8i$.

93. (a) Square roots of $5(\cos 120° + i \sin 120°)$:

$$\sqrt{5}\left[\cos\left(\frac{120° + 360°k}{2}\right) + i \sin\left(\frac{120° + 360°k}{2}\right)\right],\ k = 0, 1$$

$\sqrt{5}(\cos 60° + i \sin 60°)$

$\sqrt{5}(\cos 240° + i \sin 240°)$

(c) $\dfrac{\sqrt{5}}{2} + \dfrac{\sqrt{15}}{2}i,\ -\dfrac{\sqrt{5}}{2} - \dfrac{\sqrt{15}}{2}i$

(b)

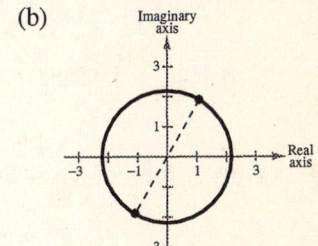

95. (a) Fourth roots of $16\left(\cos\dfrac{4\pi}{3} + i \sin\dfrac{4\pi}{3}\right)$:

$$\sqrt[4]{16}\left[\cos\left(\frac{(4\pi/3) + 2k\pi}{4}\right) + i \sin\left(\frac{(4\pi/3) + 2k\pi}{4}\right)\right],\ k = 0, 1, 2, 3$$

$2\left(\cos\dfrac{\pi}{3} + i \sin\dfrac{\pi}{3}\right)$

$2\left(\cos\dfrac{5\pi}{6} + i \sin\dfrac{5\pi}{6}\right)$

$2\left(\cos\dfrac{4\pi}{3} + i \sin\dfrac{4\pi}{3}\right)$

$2\left(\cos\dfrac{11\pi}{6} + i \sin\dfrac{11\pi}{6}\right)$

(c) $1 + \sqrt{3}i,\ -\sqrt{3} + i,\ -1 - \sqrt{3}i,\ \sqrt{3} - i$

(b)

97. (a) Cube roots of $-27i = 27\left(\cos\dfrac{3\pi}{2} + i \sin\dfrac{3\pi}{2}\right)$:

$$(27)^{1/3}\left[\cos\left(\frac{(3\pi/2) + 2k\pi}{3}\right) + i \sin\left(\frac{(3\pi/2) + 2k\pi}{3}\right)\right],\ k = 0, 1, 2$$

$3\left(\cos\dfrac{\pi}{2} + i \sin\dfrac{\pi}{2}\right)$

$3\left(\cos\dfrac{7\pi}{6} + i \sin\dfrac{7\pi}{6}\right)$

$3\left(\cos\dfrac{11\pi}{6} + i \sin\dfrac{11\pi}{6}\right)$

(c) $3i,\ -\dfrac{3\sqrt{3}}{2} - \dfrac{3}{2}i,\ \dfrac{3\sqrt{3}}{2} - \dfrac{3}{2}i$

(b)

99. (a) Cube roots of $-\dfrac{125}{2}(1 + \sqrt{3}i) = 125\left(\cos\dfrac{4\pi}{3} + i\sin\dfrac{4\pi}{3}\right)$:

(b)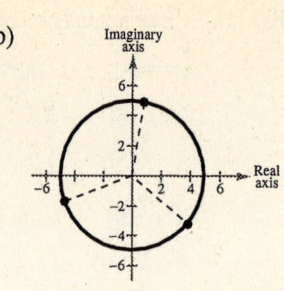

$$\sqrt[3]{125}\left[\cos\left(\dfrac{(4\pi/3) + 2k\pi}{3}\right) + i\sin\left(\dfrac{(4\pi/3) + 2k\pi}{3}\right)\right], \quad k = 0, 1, 2$$

$$5\left(\cos\dfrac{4\pi}{9} + i\sin\dfrac{4\pi}{9}\right)$$

$$5\left(\cos\dfrac{10\pi}{9} + i\sin\dfrac{10\pi}{9}\right)$$

$$5\left(\cos\dfrac{16\pi}{9} + i\sin\dfrac{16\pi}{9}\right)$$

(c) $0.8682 + 4.9240i, \ -4.6985 - 1.7101i, \ 3.8302 - 3.2139i$

101. (a) Cube roots of $64i = 64\left(\cos\dfrac{\pi}{2} + i\sin\dfrac{\pi}{2}\right)$:

(b)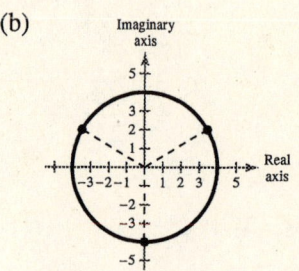

$$(64)^{1/3}\left[\cos\left(\dfrac{(\pi/2) + 2k\pi}{3}\right) + i\sin\left(\dfrac{(\pi/2) + 2k\pi}{3}\right)\right], \quad k = 0, 1, 2$$

$$4\left(\cos\dfrac{\pi}{6} + i\sin\dfrac{\pi}{6}\right)$$

$$4\left(\cos\dfrac{5\pi}{6} + i\sin\dfrac{5\pi}{6}\right)$$

$$4\left(\cos\dfrac{9\pi}{6} + i\sin\dfrac{9\pi}{6}\right) = 4\left(\cos\dfrac{3\pi}{2} + i\sin\dfrac{3\pi}{2}\right)$$

(c) $2\sqrt{3} + 2i, \ -2\sqrt{3} + 2i, \ -4i$

103. (a) Fifth roots of $1 = \cos 0 + i\sin 0$:

(b)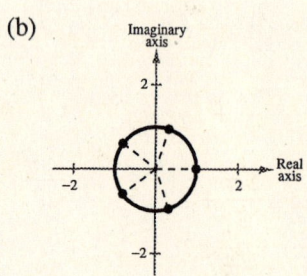

$$\cos\dfrac{2k\pi}{5} + i\sin\dfrac{2k\pi}{5}, \quad k = 0, 1, 2, 3, 4$$

$$\cos 0 + i\sin 0$$

$$\cos\dfrac{2\pi}{5} + i\sin\dfrac{2\pi}{5}$$

$$\cos\dfrac{4\pi}{5} + i\sin\dfrac{4\pi}{5}$$

$$\cos\dfrac{6\pi}{5} + i\sin\dfrac{6\pi}{5}$$

$$\cos\dfrac{8\pi}{5} + i\sin\dfrac{8\pi}{5}$$

(c) $1, 0.3090 + 0.9511i, \ -0.8090 + 0.5878i, \ -0.8090 - 0.5878i, \ 0.3090 - 0.9511i$

105. (a) Cube roots of $-125 = 125(\cos 180° + i \sin 180°)$ are:

$$\sqrt[3]{125}\left[\cos\left(\frac{180 + 360k}{3}\right) + i \sin\left(\frac{180 + 360k}{3}\right)\right], \quad k = 0, 1, 2$$

$5(\cos 60° + i \sin 60°)$

$5(\cos 180° + i \sin 180°)$

$5(\cos 300° + i \sin 300°)$

(b)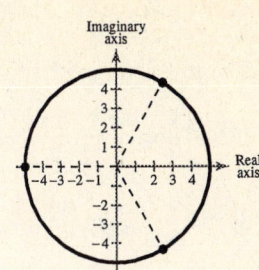

(c) $\dfrac{5}{2} + \dfrac{5\sqrt{3}}{2}i, -5, \dfrac{5}{2} - \dfrac{5\sqrt{3}}{2}i$

107. (a) Fifth roots of $128(-1 + i) = 128\sqrt{2}(\cos 135° + i \sin 135°)$ are:

$$2\sqrt{2}(\cos 27° + i \sin 27°) = 2\sqrt{2}\left(\cos\frac{3\pi}{20} + i \sin\frac{3\pi}{20}\right)$$

$2\sqrt{2}(\cos 99° + i \sin 99°)$

$2\sqrt{2}(\cos 171° + i \sin 171°)$

$2\sqrt{2}(\cos 243° + i \sin 243°)$

$2\sqrt{2}(\cos 315° + i \sin 315°)$

(b)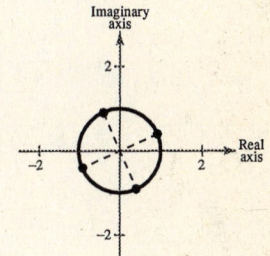

(c) $2.52 + 1.28i, -0.44 + 2.79i, -2.79 + 0.44i, -1.28 - 2.52i, 2 - 2i$

109. $x^4 - i = 0$

$x^4 = i$

The solutions are the fourth roots of $i = \cos\dfrac{\pi}{2} + i \sin\dfrac{\pi}{2}$:

$$\sqrt[4]{1}\left[\cos\left(\frac{(\pi/2) + 2k\pi}{4}\right) + i \sin\left(\frac{(\pi/2) + 2k\pi}{4}\right)\right], \quad k = 0, 1, 2, 3$$

$\cos\dfrac{\pi}{8} + i \sin\dfrac{\pi}{8}$

$\cos\dfrac{5\pi}{8} + i \sin\dfrac{5\pi}{8}$

$\cos\dfrac{9\pi}{8} + i \sin\dfrac{9\pi}{8}$

$\cos\dfrac{13\pi}{8} + i \sin\dfrac{13\pi}{8}$

111. $x^5 = -243$

The solutions are the fifth roots of $-243 = 243[\cos \pi + i \sin \pi]$:

$$\sqrt[5]{243}\left[\cos\left(\frac{\pi + 2k\pi}{5}\right) + i \sin\left(\frac{\pi + 2k\pi}{5}\right)\right], \quad k = 0, 1, 2, 3, 4$$

$$3\left(\cos \frac{\pi}{5} + i \sin \frac{\pi}{5}\right)$$

$$3\left(\cos \frac{3\pi}{5} + i \sin \frac{3\pi}{5}\right)$$

$$3\left(\cos \frac{5\pi}{5} + i \sin \frac{5\pi}{5}\right) = -3$$

$$3\left(\cos \frac{7\pi}{5} + i \sin \frac{7\pi}{5}\right)$$

$$3\left(\cos \frac{9\pi}{5} + i \sin \frac{9\pi}{5}\right)$$

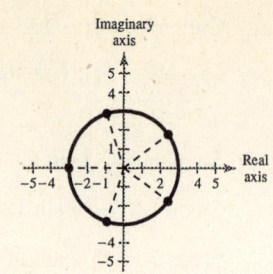

113. $x^4 = -16i$

The solutions are the fourth roots of $-16i = 16\left[\cos \frac{3\pi}{2} + i \sin \frac{3\pi}{2}\right]$:

$$\sqrt[4]{16}\left[\cos\left(\frac{(3\pi/2) + 2k\pi}{4}\right) + i \sin\left(\frac{(3\pi/2) + 2k\pi}{4}\right)\right], \quad k = 0, 1, 2, 3$$

$$2\left[\cos\left(\frac{3\pi}{8}\right) + i \sin\left(\frac{3\pi}{8}\right)\right]$$

$$2\left[\cos\left(\frac{7\pi}{8}\right) + i \sin\left(\frac{7\pi}{8}\right)\right]$$

$$2\left[\cos\left(\frac{11\pi}{8}\right) + i \sin\left(\frac{11\pi}{8}\right)\right]$$

$$2\left[\cos\left(\frac{15\pi}{8}\right) + i \sin\left(\frac{15\pi}{8}\right)\right]$$

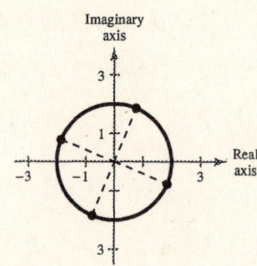

115. $x^3 - (1 - i) = 0$

$$x^3 = 1 - i = \sqrt{2}(\cos 315° + i \sin 315°)$$

The solutions are the cube roots of $1 - i$:

$$\sqrt[3]{\sqrt{2}}\left[\cos\left(\frac{315° + 360°k}{3}\right) + i \sin\left(\frac{315° + 360°k}{3}\right)\right], \quad k = 0, 1, 2$$

$$\sqrt[6]{2}(\cos 105° + i \sin 105°)$$

$$\sqrt[6]{2}(\cos 225° + i \sin 225°)$$

$$\sqrt[6]{2}(\cos 345° + i \sin 345°)$$

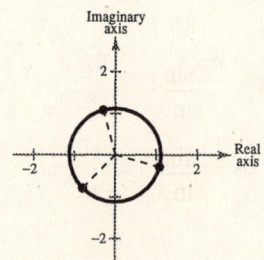

117. True. $\left[\frac{1}{2}\left(1 - \sqrt{3}i\right)\right]^9 = \left[\frac{1}{2} - \frac{\sqrt{3}}{2}i\right]^9 = -1$ **119.** True

121. $z = r(\cos \theta + i \sin \theta)$

$\bar{z} = r(\cos \theta - i \sin \theta)$

$\quad = r(\cos(-\theta) + i \sin(-\theta))$

123. $z = r(\cos \theta + i \sin \theta)$

$-z = -r(\cos \theta + i \sin \theta)$

$\quad = r(-\cos \theta - i \sin \theta)$

$\quad = r(\cos(\theta + \pi) + i \sin(\theta + \pi))$

125. $d = 16 \cos\left(\dfrac{\pi}{4}t\right)$

Maximum displacement: 16

Lowest possible t-value: $\dfrac{\pi}{4}t = \dfrac{\pi}{2} \implies t = 2$

127. $d = \dfrac{1}{8} \cos(12\pi t)$

Maximum displacement: $\dfrac{1}{8}$

Lowest possible t-value: $12\pi t = \dfrac{\pi}{2} \implies t = \dfrac{1}{24}$

129. $2 \cos(x + \pi) + 2 \cos(x - \pi) = 0$

$4 \cos x \cos \pi = 0$

$\cos x = 0$

$x = \dfrac{\pi}{2}, \dfrac{3\pi}{2}$

131. $\sin\left(x - \dfrac{\pi}{3}\right) - \sin\left(x + \dfrac{\pi}{3}\right) = \dfrac{3}{2}$

$\dfrac{1}{2}\left[\sin\left(x + \dfrac{\pi}{3}\right) - \sin\left(x - \dfrac{\pi}{3}\right)\right] = \left(-\dfrac{3}{2}\right)\dfrac{1}{2}$

$\cos x \sin \dfrac{\pi}{3} = -\dfrac{3}{4}$

$\cos x\left(\dfrac{\sqrt{3}}{2}\right) = -\dfrac{3}{4}$

$\cos x = -\dfrac{\sqrt{3}}{2}$

$x = \dfrac{5\pi}{6}, \dfrac{7\pi}{6}$

Review Exercises for Chapter 6

Solutions to Odd-Numbered Exercises

1. Given: $A = 21°, B = 42°, a = 6$

$C = 180° - 21° - 42° = 117°$

$b = \dfrac{a \sin B}{\sin A} = \dfrac{6 \sin(42°)}{\sin(21°)} \approx 11.20$

$c = \dfrac{a \sin C}{\sin A} = \dfrac{6 \sin 117°}{\sin 21°} \approx 14.92$

3. Given: $A = 75°, a = 2.5, b = 16.5$

$\sin B = \dfrac{b \sin A}{a} = \dfrac{16.5 \sin 75°}{2.5}$

$\approx 6.375 \implies$ no triangle formed

No solution

5. Given: $B = 115°, a = 9, b = 14.5$

$\sin A = \dfrac{a \sin B}{b} = \dfrac{9 \sin 115°}{14.5} \approx 0.5625 \implies A \approx 34.2°$

$C \approx 180° - 115° - 34.2° = 30.8°$

$c = \dfrac{b}{\sin B}(\sin C) \approx \dfrac{14.5}{\sin 115°}(\sin 30.8°) \approx 8.18$

7. Given: $A = 15°$, $a = 5$, $b = 10$

$$\sin B = \frac{b \sin A}{a} = \frac{10 \sin 15°}{5} \approx 0.5176 \implies B \approx 31.2° \text{ or } 148.8°$$

Case 1: $B \approx 31.2°$ | Case 2: $B \approx 148.8°$

$C \approx 180° - 15° - 31.2° = 133.8°$ | $C \approx 180° - 15° - 148.8° = 16.2°$

$$c = \frac{a \sin C}{\sin A} \approx \frac{5 \sin 133.8°}{\sin 15°} \approx 13.94 \qquad c = \frac{a \sin C}{\sin A} \approx \frac{5 \sin 16.2°}{\sin 15°} \approx 5.4$$

9. Given: $B = 25°$, $a = 6.2$, $b = 4$

$$\sin A = \frac{a \sin B}{b} \approx \frac{6.2 \sin 25°}{4} \approx 0.6551 \implies A \approx 40.9° \text{ or } 139.1°$$

Case 1: $A \approx 40.9°$ | Case 2: $A \approx 139.1°$

$C \approx 180° - 25° - 40.9° = 114.1°$ | $C \approx 180° - 25° - 139.1° = 15.9°$

$$c = \frac{b \sin C}{\sin B} \approx \frac{4 \sin 114.1°}{\sin 25°} \approx 8.64 \qquad c = \frac{b \sin C}{\sin B} \approx \frac{4 \sin 15.9°}{\sin 25°} \approx 2.60$$

11. $A = 27°$, $b = 5$, $c = 8$

Area $= \frac{1}{2}bc \sin A$

$\quad = \frac{1}{2}(5)(8)(\sin 27°)$

$\quad \approx 9.08$ square units

13. $C = 122°$, $b = 18$, $a = 29$

Area $= \frac{1}{2}ab \sin C$

$\quad = \frac{1}{2}(29)(18) \sin 122°$

$\quad \approx 221.34$ square units

15. $h = 50 \tan 17° \approx 15.3$ meters

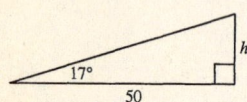

17. $\sin 28° = \dfrac{h}{75}$

$\quad h = 75 \sin 28° \approx 35.21$ feet

$\cos 28° = \dfrac{x}{75}$

$\quad x = 75 \cos 28° \approx 66.22$ feet

$\tan 45° = \dfrac{H}{x}$

$\quad H = x \tan 45° \approx 66.22$ feet

Height of tree:

$H - h \approx 31$ feet

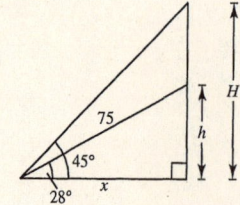

19. Given: $a = 9$, $b = 12$, $c = 20$

$$\cos C = \frac{a^2 + b^2 - c^2}{2ab}$$

$$= \frac{81 + 144 - 400}{2(9)(12)}$$

$$\approx -0.8102 \implies C \approx 144.1°$$

$$\sin A = \frac{a \sin C}{c}$$

$$= \frac{9 \sin(144.1°)}{20}$$

$$\approx 0.264 \implies A \approx 15.3$$

$$B = 180° - 144.1° - 15.3° = 20.6°$$

21. Given: $a = 6, b = 9, C = 45°$

$c^2 = a^2 + b^2 - 2ab \cos C = 36 + 81 - 2(6)(9)(\cos 45°) \approx 40.63 \implies c \approx 6.37$

$\cos B = \dfrac{a^2 + c^2 - b^2}{2ac} \approx \dfrac{36 + 40.63 - 81}{2(6)(6.37)} \approx -0.0572 \implies B \approx 93.3°$

$A \approx 180° - 45° - 93.3° = 41.7°$

23. Given: $B = 110°, a = 4, c = 4$

$b^2 = a^2 + c^2 - 2ac \cos B = 16 + 16 - 2(4)(4)(\cos 110) \approx 42.94 \implies b \approx 6.55$

$\sin A = \dfrac{a \sin B}{b} \approx \dfrac{4 \sin 110°}{6.6} \approx 0.5695 \implies A \approx 35°$

$c = a \implies C = A \approx 35°$

25. Given: $B = 150°, a = 10, c = 20$

$b^2 = a^2 + c^2 - 2ac \cos B \approx 100 + 400 - 400(-0.8660) \approx 846.4 \implies b \approx 29.09$

$\sin C = \dfrac{c \sin B}{b} \approx \dfrac{20(0.5)}{29.09} \approx 0.3437 \implies C \approx 20.1°$

$\sin A = \dfrac{a \sin B}{b} \approx \dfrac{10(0.5)}{29.09} \approx 0.1719 \implies A \approx 9.9°$

27. Given: $a = 8.9, b = 6.1, c = 10.5$

$\cos C = \dfrac{a^2 + b^2 - c^2}{2ab} = \dfrac{8.9^2 + 6.1^2 - 10.5^2}{2(8.9)(6.1)} \approx 0.0568 \implies C \approx 86.7°$

$\sin A = \dfrac{a \sin C}{c} = \dfrac{8.9 \sin(86.7°)}{10.5} \approx 0.846 \implies A \approx 57.8°$

$B \approx 180° - 86.7° - 57.8° = 35.5°$

29. $a^2 = 5^2 + 8^2 - 2(5)(8) \cos 152° \approx 159.6 \implies a \approx 12.63$ ft

$b^2 = 5^2 + 8^2 - 2(5)(8) \cos 28° \approx 18.36 \implies b \approx 4.285$ ft

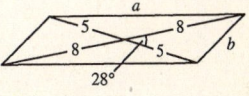

31. Angle between planes is $5° + 67° = 72°$.

In two hours, distances from airport are 850 miles and 1060 miles.

By the Law of Cosines,

$d^2 = 850^2 + 1060^2 - 2(850)(1060) \cos(72°) \approx 1,289,251.376$

$d \approx 1135.5$ miles.

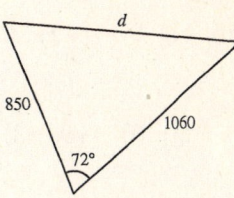

33. $a = 4, b = 5, c = 7$

$s = \dfrac{a + b + c}{2} = \dfrac{4 + 5 + 7}{2} = 8$

Area $= \sqrt{s(s - a)(s - b)(s - c)}$

$= \sqrt{8(4)(3)(1)}$

≈ 9.798 square units

35. $a = 64.8, b = 49.2, c = 24.1$

$s = \dfrac{a + b + c}{2} = \dfrac{64.8 + 49.2 + 24.1}{2} = 69.05$

Area $= \sqrt{s(s - a)(s - b)(s - c)}$

$= \sqrt{69.05(4.25)(19.85)(44.95)}$

≈ 511.7 square units

37. $\mathbf{u} = \langle 4 - (-2), 6 - 1 \rangle = \langle 6, 5 \rangle$

$\mathbf{v} = \langle 6 - 0, 3 - (-2) \rangle = \langle 6, 5 \rangle$

$\mathbf{u} = \mathbf{v}$

39. Initial point: $(-5, 4)$

Terminal point: $(2, -1)$

$\mathbf{v} = \langle 2 - (-5), -1 - 4 \rangle = \langle 7, -5 \rangle$

41. Initial point: $(0, 10)$

Terminal point: $(7, 3)$

$\mathbf{v} = \langle 7 - 0, 3 - 10 \rangle = \langle 7, -7 \rangle$

43. $8 \cos 120°\mathbf{i} + 8 \sin 120°\mathbf{j} = \langle -4, 4\sqrt{3} \rangle$

45. (a) $\mathbf{u} + \mathbf{v} = \langle -1, -3 \rangle + \langle -3, 6 \rangle = \langle -4, 3 \rangle$

(b) $\mathbf{u} - \mathbf{v} = \langle 2, -9 \rangle$

(c) $3\mathbf{u} = \langle -3, -9 \rangle$

(d) $2\mathbf{v} + 5\mathbf{u} = \langle -6, 12 \rangle + \langle -5, -15 \rangle = \langle -11, -3 \rangle$

47. (a) $\mathbf{u} + \mathbf{v} = \langle -5, 2 \rangle + \langle 4, 4 \rangle = \langle -1, 6 \rangle$

(b) $\mathbf{u} - \mathbf{v} = \langle -9, -2 \rangle$

(c) $3\mathbf{u} = \langle -15, 6 \rangle$

(d) $2\mathbf{v} + 5\mathbf{u} = \langle 8, 8 \rangle + \langle -25, 10 \rangle = \langle -17, 18 \rangle$

49. (a) $\mathbf{u} + \mathbf{v} = \langle 2, -1 \rangle + \langle 5, 3 \rangle = \langle 7, 2 \rangle$

(b) $\mathbf{u} - \mathbf{v} = \langle -3, -4 \rangle$

(c) $3\mathbf{u} = \langle 6, -3 \rangle$

(d) $2\mathbf{v} + 5\mathbf{u} = \langle 10, 6 \rangle + \langle 10, -5 \rangle = \langle 20, 1 \rangle$

51. (a) $\mathbf{u} + \mathbf{v} = \langle 4, 0 \rangle + \langle -1, 6 \rangle = \langle 3, 6 \rangle$

(b) $\mathbf{u} - \mathbf{v} = \langle 5, -6 \rangle$

(c) $3\mathbf{u} = \langle 12, 0 \rangle$

(d) $2\mathbf{v} + 5\mathbf{u} = \langle -2, 12 \rangle + \langle 20, 0 \rangle = \langle 18, 12 \rangle$

53. $3\mathbf{v} = 3(10\mathbf{i} + 3\mathbf{j}) = 30\mathbf{i} + 9\mathbf{j} = \langle 30, 9 \rangle$

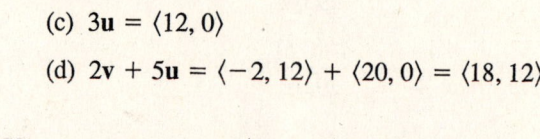

55. $\mathbf{w} = 4\mathbf{u} + 5\mathbf{v} = 4(6\mathbf{i} - 5\mathbf{j}) + 5(10\mathbf{i} + 3\mathbf{j})$

$= 74\mathbf{i} - 5\mathbf{j}$

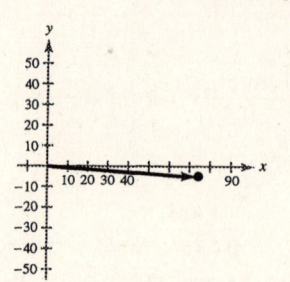

57. $\|\mathbf{u}\| = 6$

Unit vector: $\dfrac{1}{6}\langle 0, -6 \rangle = \langle 0, -1 \rangle$

59. $\|\mathbf{v}\| = \sqrt{5^2 + (-2)^2} = \sqrt{29}$

Unit vector: $\dfrac{1}{\sqrt{29}}\langle 5, -2 \rangle = \left\langle \dfrac{5}{\sqrt{29}}, \dfrac{-2}{\sqrt{29}} \right\rangle$

61. $\mathbf{u} = \langle 1 - (-8), -5 - 3 \rangle = \langle 9, -8 \rangle = 9\mathbf{i} - 8\mathbf{j}$

63. $\mathbf{v} = -10\mathbf{i} + 10\mathbf{j}$

$\|\mathbf{v}\| = \sqrt{(-10)^2 + (10)^2} = \sqrt{200} = 10\sqrt{2}$

$\tan \theta = \dfrac{10}{-10} = -1 \implies \theta = 135°$ since $\mathbf{v}$ is in Quadrant II.

$\mathbf{v} = 10\sqrt{2}(\cos 135°\mathbf{i} + \sin 135°\mathbf{j})$

65. $\mathbf{u} = 15[(\cos 20°)\mathbf{i} + (\sin 20°)\mathbf{j}]$

$\mathbf{v} = 20[(\cos 63°)\mathbf{i} + (\sin 63°)\mathbf{j}]$

$\mathbf{u} + \mathbf{v} \approx 23.1752\mathbf{i} + 22.9504\mathbf{j}$

$\|\mathbf{u} + \mathbf{v}\| \approx 32.62$

$\tan \theta = \dfrac{22.9504}{23.1752} \Rightarrow \theta \approx 44.72°$

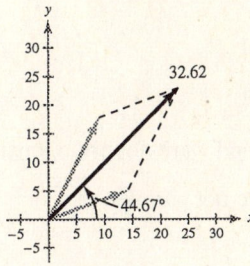

67. $\mathbf{F}_1 = 250\langle\cos 60°, \sin 60°\rangle$

$\mathbf{F}_2 = 100\langle\cos 150°, \sin 150°\rangle$

$\mathbf{F}_3 = 200\langle\cos(-90°), \sin(-90°)\rangle$

$\mathbf{F} = \mathbf{F}_1 + \mathbf{F}_2 + \mathbf{F}_3 \approx \langle 38.39746, 66.50635\rangle$

$\tan \theta \approx \dfrac{66.50635}{38.39746} \Rightarrow \theta \approx 60°$

$\|\mathbf{F}\| = \sqrt{38.39746^2 + 66.50635^2} \approx 76.8$ pounds

69. Rope One:

$\mathbf{u} = \|\mathbf{u}\|(\cos 30°\mathbf{i} - \sin 30°\mathbf{j}) = \|\mathbf{u}\|\left(\dfrac{\sqrt{3}}{2}\mathbf{i} - \dfrac{1}{2}\mathbf{j}\right)$

Rope Two:

$\mathbf{v} = \|\mathbf{u}\|(-\cos 30°\mathbf{i} - \sin 30°\mathbf{j}) = \|\mathbf{u}\|\left(-\dfrac{\sqrt{3}}{2}\mathbf{i} - \dfrac{1}{2}\mathbf{j}\right)$

Resultant: $\mathbf{u} + \mathbf{v} = -\|\mathbf{u}\|\mathbf{j} = -180\mathbf{j}$

$\|\mathbf{u}\| = 180$

Therefore, the tension on each rope is $\|\mathbf{u}\| = 180$ lb.

71. Airplane velocity: $\mathbf{u} = 430\langle\cos 315°, \sin 315°\rangle$

Wind velocity: $\mathbf{w} = 35\langle\cos 60°, \sin 60°\rangle$

$\mathbf{u} + \mathbf{w} \approx \langle 321.5559, -273.7450\rangle$

$\|\mathbf{u} + \mathbf{w}\| \approx 422.3$ mph

$\tan(\mathbf{u} + \mathbf{w}) = \dfrac{-273.7450}{321.5559}$

$\approx -0.8513 \Rightarrow \theta \approx -40.4°$

The bearing from the north is $90° + 40.4° = 130.4°$.

73. $\mathbf{u} \cdot \mathbf{v} = \langle 0, -2\rangle \cdot \langle 1, 10\rangle = 0 - 20 = -20$

75. $\mathbf{u} \cdot \mathbf{v} = \langle 6, -1\rangle \cdot \langle 2, 5\rangle = 6(2) + (-1)(5) = 7$

77. $\mathbf{u} \cdot \mathbf{u} = \langle -3, -4\rangle \cdot \langle -3, -4\rangle$

$= 9 + 16 = 25 = \|\mathbf{u}\|^2$

79. $4\mathbf{u} \cdot \mathbf{v} = 4\langle -3, -4\rangle \cdot \langle 2, 1\rangle$

$= 4(-6 - 4) = -40$

81. $\mathbf{u} = \langle 2\sqrt{2}, -4\rangle, \mathbf{v} = \langle -\sqrt{2}, 1\rangle$

$\cos \theta = \dfrac{\mathbf{u} \cdot \mathbf{v}}{\|\mathbf{u}\|\,\|\mathbf{v}\|} = \dfrac{-8}{(\sqrt{24})(\sqrt{3})} \Rightarrow \theta \approx 160.5°$

83. $\mathbf{u} = \cos\dfrac{7\pi}{4}\mathbf{i} + \sin\dfrac{7\pi}{4}\mathbf{j} = \left\langle \dfrac{1}{\sqrt{2}}, -\dfrac{1}{\sqrt{2}}\right\rangle$

$\mathbf{v} = \cos\dfrac{5\pi}{6}\mathbf{i} + \sin\dfrac{5\pi}{6}\mathbf{j} = \left\langle -\dfrac{\sqrt{3}}{2}, \dfrac{1}{2}\right\rangle$

$\cos \theta = \dfrac{\mathbf{u} \cdot \mathbf{v}}{\|\mathbf{u}\|\,\|\mathbf{v}\|} = \dfrac{\left(-\sqrt{3}/2\sqrt{2}\right) - \left(1/2\sqrt{2}\right)}{(1)(1)}$

$\approx -0.966 \Rightarrow \theta \approx 165°$ or $\dfrac{11\pi}{12}$

85. $\cos \theta = \dfrac{\mathbf{u} \cdot \mathbf{v}}{\|\mathbf{u}\|\,\|\mathbf{v}\|}$

$= 0 \Rightarrow \theta = 90°$

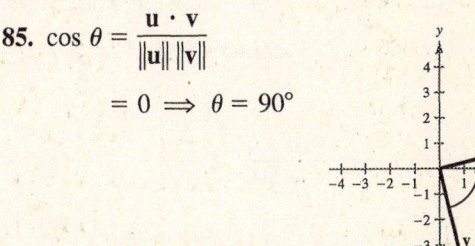

87. $\cos \theta = \dfrac{\mathbf{u} \cdot \mathbf{v}}{\|\mathbf{u}\| \|\mathbf{v}\|} = \dfrac{70 - 15}{\sqrt{74}\sqrt{109}}$

$\qquad\qquad \approx 0.612$

$\qquad\qquad \Rightarrow \theta \approx 52.2°$

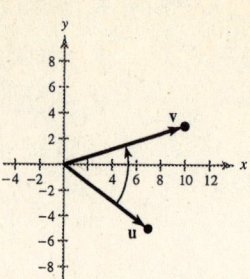

89. $\mathbf{u} = \langle 39, -12 \rangle, \mathbf{v} = \langle -26, 8 \rangle$

$\quad \mathbf{u} \cdot \mathbf{v} = 39(-26) + (-12)(8)$

$\qquad = -1110 \neq 0 \Rightarrow \mathbf{u}$ and $\mathbf{v}$ are not orthogonal.

$\quad \mathbf{v} = -\frac{2}{3}\mathbf{u} \Rightarrow \mathbf{u}$ and $\mathbf{v}$ are parallel.

91. $\mathbf{u} = \langle 8, 5 \rangle, \mathbf{v} = \langle -2, 4 \rangle$

$\quad \mathbf{u} \cdot \mathbf{v} = 8(-2) + (5)(4)$

$\qquad = 4 \neq 0 \Rightarrow \mathbf{u}$ and $\mathbf{v}$ are not orthogonal.

$\quad \mathbf{u} \neq k\mathbf{v} \Rightarrow \mathbf{u}$ and $\mathbf{v}$ are not parallel.

$\quad$ Neither

93. $\mathbf{u} = \langle -4, 3 \rangle, \ \mathbf{v} = \langle -8, -2 \rangle$

$\quad \text{proj}_{\mathbf{v}}\mathbf{u} = \left(\dfrac{\mathbf{u} \cdot \mathbf{v}}{\|\mathbf{v}\|^2}\right)\mathbf{v} = \left(\dfrac{26}{68}\right)\langle -8, -2 \rangle = -\dfrac{13}{17}\langle 4, 1 \rangle$

$\quad \mathbf{u} - \text{proj}_{\mathbf{v}}\mathbf{u} = \langle -4, 3 \rangle - \left\langle \dfrac{-52}{17}, \dfrac{-13}{17} \right\rangle$

$\qquad\qquad = \left\langle -\dfrac{16}{17}, \dfrac{64}{17} \right\rangle$

$\quad \mathbf{u} = \left\langle -\dfrac{52}{17}, -\dfrac{13}{17} \right\rangle + \left\langle -\dfrac{16}{17}, \dfrac{64}{17} \right\rangle$

95. $\mathbf{u} = \langle 2, 7 \rangle, \mathbf{v} = \langle 1, -1 \rangle$

$\quad \text{proj}_{\mathbf{v}}\mathbf{u} = \left(\dfrac{\mathbf{u} \cdot \mathbf{v}}{\|\mathbf{v}\|^2}\right)\mathbf{v} = \dfrac{-5}{2}\langle 1, -1 \rangle = \left\langle -\dfrac{5}{2}, \dfrac{5}{2} \right\rangle$

$\quad \mathbf{u} - \text{proj}_{\mathbf{v}}\mathbf{u} = \langle 2, 7 \rangle - \left\langle -\dfrac{5}{2}, \dfrac{5}{2} \right\rangle = \left\langle \dfrac{9}{2}, \dfrac{9}{2} \right\rangle$

$\quad \mathbf{u} = \left\langle -\dfrac{5}{2}, \dfrac{5}{2} \right\rangle + \left\langle \dfrac{9}{2}, \dfrac{9}{2} \right\rangle$

97. 48 inches = 4 feet

$\quad$ Work $= 18,000(4) = 72,000 \ \text{ft} \cdot \text{lb}$

99. $|-i| = 1$

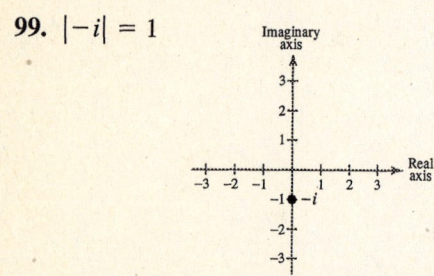

101. $|7 - 5i| = \sqrt{7^2 + (-5)^2}$

$\qquad\qquad = \sqrt{74}$

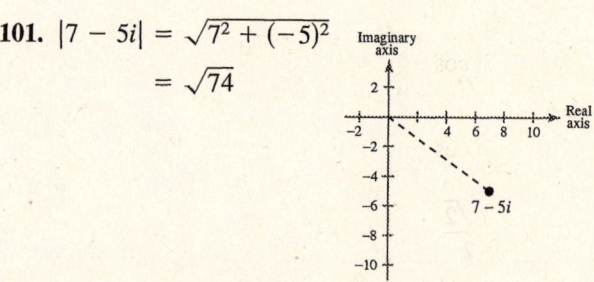

103. $z = 5 - 2i, \ r = \sqrt{25 + 4} = \sqrt{29}, \ \theta = \arctan\left(-\frac{2}{5}\right) \approx 338.2°$

$\quad z = \sqrt{29}(\cos 338.2° + i \sin 338.2°)$

105. $5 - 5i$

$\quad r = \sqrt{5^2 + (-5)^2} = \sqrt{50} = 5\sqrt{2}$

$\quad \tan \theta = \dfrac{-5}{5} = -1 \Rightarrow \theta \approx 315°$ since the

$\quad$ complex number is in Quadrant IV.

$\quad 5 - 5i = 5\sqrt{2}(\cos 315° + i \sin 315°)$

107. $5 + 12i$

$\quad r = \sqrt{5^2 + 12^2} = \sqrt{169} = 13$

$\quad \tan \theta = \dfrac{12}{5} \Rightarrow \theta \approx 67.38°$ since the number is

$\quad$ in Quadrant I.

$\quad 5 + 12i \approx 13(\cos 67.38° + i \sin 67.38°)$

109. $\left[\dfrac{5}{2}\left(\cos\dfrac{\pi}{2}+i\sin\dfrac{\pi}{2}\right)\right]\left[4\left(\cos\dfrac{\pi}{4}+i\sin\dfrac{\pi}{4}\right)\right]=10\left[\cos\dfrac{3\pi}{4}+i\sin\dfrac{3\pi}{4}\right]$

111. $\dfrac{20(\cos 320°+i\sin 320°)}{5(\cos 80°+i\sin 80°)}=4[\cos 240°+i\sin 240°]$

113. $\left[5\left(\cos\dfrac{\pi}{12}+i\sin\dfrac{\pi}{12}\right)\right]^4=5^4\left(\dfrac{4\pi}{12}+i\sin\dfrac{4\pi}{12}\right)$

$=625\left(\cos\dfrac{\pi}{3}+i\sin\dfrac{\pi}{3}\right)$

$=625\left(\dfrac{1}{2}+\dfrac{\sqrt{3}}{2}i\right)$

$=\dfrac{625}{2}+\dfrac{625\sqrt{3}}{2}i$

115. $(2+3i)^6\approx\left[\sqrt{13}(\cos 56.3°+i\sin 56.3°)\right]^6$

$=13^3(\cos 337.9°+i\sin 337.9°)$

$\approx 13^3(0.9263-0.3769i)$

$\approx 2035-828i$

117. (a) Sixth roots of $-729i=729\left(\cos\dfrac{3\pi}{2}+i\sin\dfrac{3\pi}{2}\right)$:

(b)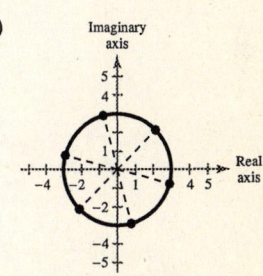

$\sqrt[6]{729}\left(\cos\dfrac{(3\pi/2)+2k\pi}{6}+i\sin\dfrac{(3\pi/2)+2k\pi}{6}\right),\ k=0,1,2,3,4,5$

$3\left(\cos\dfrac{\pi}{4}+i\sin\dfrac{\pi}{4}\right)$

$3\left(\cos\dfrac{7\pi}{12}+i\sin\dfrac{7\pi}{12}\right)$

$3\left(\cos\dfrac{11\pi}{12}+i\sin\dfrac{11\pi}{12}\right)$

$3\left(\cos\dfrac{5\pi}{4}+i\sin\dfrac{5\pi}{4}\right)$

$3\left(\cos\dfrac{19\pi}{12}+i\sin\dfrac{19\pi}{12}\right)$

$3\left(\cos\dfrac{23\pi}{12}+i\sin\dfrac{23\pi}{12}\right)$

(c) $\dfrac{3\sqrt{2}}{2}+\dfrac{3\sqrt{2}}{2}i,\ -0.7765+2.898i,\ -2.898+0.7765i,\ -\dfrac{3\sqrt{2}}{2}-\dfrac{3\sqrt{2}}{2}i,\ 0.7765-2.898i,\ 2.898-0.7765i$

119. (a) Cube roots of $8=8(\cos 0+i\sin 0)$:

(b)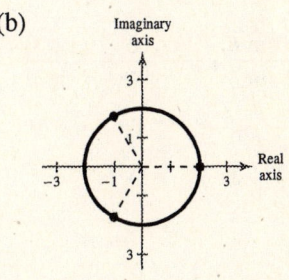

$\sqrt[3]{8}\left(\cos\left(\dfrac{2k\pi}{3}\right)+i\sin\left(\dfrac{2k\pi}{3}\right)\right),\ k=0,1,2$

$2(\cos 0+i\sin 0)$

$2\left(\cos\dfrac{2\pi}{3}+i\sin\dfrac{2\pi}{3}\right)$

$2\left(\cos\dfrac{4\pi}{3}+i\sin\dfrac{4\pi}{3}\right)$

(c) $2,\ -1+\sqrt{3}i,\ -1-\sqrt{3}i$

121. $x^4 + 256 = 0$

$$x^4 = -256 = 256(\cos \pi + i \sin \pi)$$

$$\sqrt[4]{-256} = 4\left[\cos\left(\frac{\pi + 2\pi k}{4}\right) + i \sin\left(\frac{\pi + 2\pi k}{4}\right)\right], \; k = 0, 1, 2, 3$$

$$4\left(\cos \frac{\pi}{4} + i \sin \frac{\pi}{4}\right) = \frac{4\sqrt{2}}{2} + \frac{4\sqrt{2}}{2}i = 2\sqrt{2} + 2\sqrt{2}i$$

$$4\left(\cos \frac{3\pi}{4} + i \sin \frac{3\pi}{4}\right) = -\frac{4\sqrt{2}}{2} + \frac{4\sqrt{2}}{2}i = -2\sqrt{2} + 2\sqrt{2}i$$

$$4\left(\cos \frac{5\pi}{4} + i \sin \frac{5\pi}{4}\right) = -\frac{4\sqrt{2}}{2} - \frac{4\sqrt{2}}{2}i = -2\sqrt{2} - 2\sqrt{2}i$$

$$4\left(\cos \frac{7\pi}{4} + i \sin \frac{7\pi}{4}\right) = \frac{4\sqrt{2}}{2} - \frac{4\sqrt{2}}{2}i = 2\sqrt{2} - 2\sqrt{2}i$$

123. $x^3 + 8i = 0$

$$x^3 = -8i$$

$$-8i = 8\left(\cos \frac{3\pi}{2} + i \sin \frac{3\pi}{2}\right)$$

$$\sqrt[3]{-8i} = \sqrt[3]{8}\left[\cos \frac{(3\pi/2) + 2\pi k}{3} + i \sin \frac{(3\pi/2) + 2\pi k}{3}\right], \; k = 0, 1, 2$$

$$2\left(\cos \frac{\pi}{2} + i \sin \frac{\pi}{2}\right) = 2i$$

$$2\left(\cos \frac{7\pi}{6} + i \sin \frac{7\pi}{6}\right) = -\sqrt{3} - i$$

$$2\left(\cos \frac{11\pi}{6} + i \sin \frac{11\pi}{6}\right) = \sqrt{3} - i$$

125. True

127. Length and direction characterize vectors in plane.

129. $z_1 z_2 = 2(\cos \theta + i \sin \theta)2(\cos(\pi - \theta) + i \sin(\pi - \theta))$

$$= 4(\cos \theta + i \sin \theta)(-\cos \theta + i \sin \theta)$$

$$= 4(-\cos^2 \theta - \sin^2 \theta) = -4$$

$$\frac{z_1}{z_2} = \frac{2(\cos \theta + i \sin \theta)}{2(\cos(\pi - \theta) + i \sin(\pi - \theta))}$$

$$= \frac{\cos \theta + i \sin \theta}{\cos \theta + i \sin \theta} \cdot \frac{-\cos \theta - i \sin \theta}{-\cos \theta - i \sin \theta}$$

$$= -\cos^2 \theta + \sin^2 \theta - 2 \sin \theta \cos \theta i$$

$$= -(\cos \theta + i \sin \theta)^2$$

$$= -\left(\frac{z_1}{2}\right)^2$$

$$= -\frac{z_1{}^2}{4}$$

Chapter 6   Practice Test

For Exercises 1 and 2, use the Law of Sines to find the remaining sides and angles of the triangle.

1. $A = 40°$, $B = 12°$, $b = 100$

2. $C = 150°$, $a = 5$, $c = 20$

3. Find the area of the triangle: $a = 3$, $b = 6$, $C = 130°$

4. Determine the number of solutions to the triangle: $a = 10$, $b = 35$, $A = 22.5°$

For Exercises 5 and 6, use the Law of Cosines to find the remaining sides and angles of the triangle.

5. $a = 49$, $b = 53$, $c = 38$

6. $C = 29°$, $a = 100$, $b = 300$

7. Use Heron's Formula to find the area of the triangle: $a = 4.1$, $b = 6.8$, $c = 5.5$.

8. A ship travels 40 miles due east, then adjusts its course 12° southward. After traveling 70 miles in that direction, how far is the ship from its point of departure?

9. $\mathbf{w} = 4\mathbf{u} - 7\mathbf{v}$ where $\mathbf{u} = 3\mathbf{i} + \mathbf{j}$ and $\mathbf{v} = -\mathbf{i} + 2\mathbf{j}$. Find $\mathbf{w}$.

10. Find a unit vector in the direction of $\mathbf{v} = 5\mathbf{i} - 3\mathbf{j}$.

11. Find the dot product and the angle between $\mathbf{u} = 6\mathbf{i} + 5\mathbf{j}$ and $\mathbf{v} = 2\mathbf{i} - 3\mathbf{j}$.

12. $\mathbf{v}$ is a vector of magnitude 4 making an angle of 30° with the positive x-axis. Find $\mathbf{v}$ in component form.

13. Find the projection of $\mathbf{u}$ onto $\mathbf{v}$ given $\mathbf{u} = \langle 3, -1 \rangle$ and $\mathbf{v} = \langle -2, 4 \rangle$.

14. Give the trigonometric form of $z = 5 - 5i$.

15. Give the standard form of $z = 6(\cos 225° + i \sin 225°)$.

16. Multiply $[7 (\cos 23° + i \sin 23°)][4(\cos 7° + i \sin 7°)]$.

17. Divide $\dfrac{9\left(\cos \dfrac{5\pi}{4} + i \sin \dfrac{5\pi}{4}\right)}{3(\cos \pi + i \sin \pi)}$.

18. Find $(2 + 2i)^8$.

19. Find the cube roots of $8\left(\cos \dfrac{\pi}{3} + i \sin \dfrac{\pi}{3}\right)$.

20. Find all the solutions to $x^4 + i = 0$.

CHAPTER 7
Linear Systems and Matrices

CHAPTER 7
Linear Systems and Matrices

Section 7.1   Solving Systems of Equations

■ You should be able to solve systems of equations by the method of substitution.

1. Solve one of the equations for one of the variables.

2. Substitute this expression into the other equation and solve.

3. Back-substitute into the first equation to find the value of the other variable.

4. Check your answer in each of the original equations.

■ You should be able to find solutions graphically. (See Example 5 in textbook.)

Solutions to Odd-Numbered Exercises

1. (a) $4(0) - (-3) \overset{?}{=} 1$

$6(0) + (-3) \overset{?}{=} -6$

$3 \neq 1$

$-3 \neq -6$

No, $(0, -3)$ is not a solution.

(b) $4(-1) - (-5) \overset{?}{=} 1$

$6(-1) + (-5) \overset{?}{=} -6$

$1 = 1$

$-11 \neq -6$

No, $(-1, -5)$ is not a solution.

(c) $4\left(-\frac{3}{2}\right) - (3) \overset{?}{=} 1$

$6\left(-\frac{3}{2}\right) + (3) \overset{?}{=} -6$

$-9 \neq 1$

$-6 = -6$

No, $\left(-\frac{3}{2}, 3\right)$ is not a solution.

(d) $4\left(-\frac{1}{2}\right) - (-3) \overset{?}{=} 1$

$6\left(-\frac{1}{2}\right) + (-3) \overset{?}{=} -6$

$1 = 1$

$-6 = -6$

Yes, $\left(-\frac{1}{2}, -3\right)$ is a solution.

3. (a) $0 \overset{?}{=} -2e^{-2}$

$3(-2) - 0 \overset{?}{=} 2$

$0 \neq -2e^{-2}$

$-6 \neq 2$

No, $(-2, 0)$ is not a solution.

(b) $-2 \overset{?}{=} -2e^0$

$3(0) - (-2) \overset{?}{=} 2$

$-2 = -2$

$2 = 2$

Yes, $(0, -2)$ is a solution.

(c) $-3 \overset{?}{=} -2e^0$

$3(0) - (-3) \overset{?}{=} 2$

$-3 \neq -2$

$3 \neq 2$

No, $(0, -3)$ is not a solution.

(d) $-5 \overset{?}{=} -2e^{-1}$

$3(-1) - (-5) \overset{?}{=} 2$

$-5 \neq -2e^{-1}$

$2 = 2$

No, $(-1, -5)$ is not a solution.

5. $\begin{cases} 2x + y = 6 & \text{Equation 1} \\ -x + y = 0 & \text{Equation 2} \end{cases}$

Solve for y in Equation 1: $y = 6 - 2x$

Substitute for y in Equation 2: $-x + (6 - 2x) = 0$

Solve for x: $-3x + 6 = 0 \implies x = 2$

Back-substitute $x = 2$: $y = 6 - 2(2) = 2$

Answer: $(2, 2)$

7. $\begin{cases} x - y = -4 & \text{Equation 1} \\ x^2 - y = -2 & \text{Equation 2} \end{cases}$

Solve for y in Equation 1: $y = x + 4$

Substitute for y in Equation 2: $x^2 - (x + 4) = -2$

Solve for x: $x^2 - x - 2 = 0 \implies (x + 1)(x - 2) = 0 \implies x = -1, 2$

Back-substitute $x = -1$: $y = -1 + 4 = 3$

Back-substitute $x = 2$: $y = 2 + 4 = 6$

Answers: $(-1, 3), (2, 6)$

9. $\begin{cases} 3x + y = 2 & \text{Equation 1} \\ x^3 - 2 + y = 0 & \text{Equation 2} \end{cases}$

Solve for y in Equation 1: $y = 2 - 3x$

Substitute for y in Equation 2: $x^3 - 2 + (2 - 3x) = 0$

Solve for x: $x^3 - 3x = 0 \implies x(x^2 - 3) = 0 \implies x = 0, \pm\sqrt{3}$

Back-substitute: $x = 0$: $y = 2$

$$x = \sqrt{3}: \ y = 2 - 3\sqrt{3}$$
$$x = -\sqrt{3}: \ y = 2 + 3\sqrt{3}$$

Solutions: $(0, 2), \left(\sqrt{3}, 2 - 3\sqrt{3}\right), \left(-\sqrt{3}, 2 + 3\sqrt{3}\right)$

11. $\begin{cases} x^2 + y = 0 & \text{Equation 1} \\ x^2 - 4x - y = 0 & \text{Equation 2} \end{cases}$

Solve for y in Equation 1: $y = -x^2$

Substitute for y in Equation 2: $x^2 - 4x - (-x^2) = 0$

Solve for x: $2x^2 - 4x = 0 \implies 2x(x - 2) = 0 \implies x = 0, 2$

Back-substitute $x = 0$: $y = -0^2 = 0$

Back-substitute $x = 2$: $y = -2^2 = -4$

Answers: $(0, 0), (2, -4)$

13. $\begin{cases} -\frac{7}{2}x - y = -18 & \text{Equation 1} \\ 8x^2 - 2y^3 = 0 & \text{Equation 2} \end{cases}$

Solve for x in Equation 1: $-\frac{7}{2}x = y - 18 \implies x = -\frac{2}{7}y + \frac{36}{7}$

Substitute for x in Equation 2: $8\left(-\frac{2}{7}y + \frac{36}{7}\right)^2 - 2y^3 = 0$

Solve for x: $-2y^3 + 8\left(\frac{4}{49}y^2 - \frac{144}{49}y + \frac{36^2}{49}\right) = 0$

$$49y^3 - 16y^2 + 576y - 5184 = 0$$

$$(y - 4)(49y^2 + 180y + 1296) = 0$$

Hence, $y = 4$ and $x = -\frac{2}{7}(4) + \frac{36}{7} = 4.$

Solution: $(4, 4)$

15. $\begin{cases} x - y = 0 & \text{Equation 1} \\ 5x - 3y = 10 & \text{Equation 2} \end{cases}$

Solve for y in Equation 1: $y = x$

Substitute for y in Equation 2: $5x - 3x = 10$

Solve for x: $2x = 10 \implies x = 5$

Back-substitute in Equation 1: $y = x = 5$

Answer: $(5, 5)$

17. $\begin{cases} 2x - y + 2 = 0 & \text{Equation 1} \\ 4x + y - 5 = 0 & \text{Equation 2} \end{cases}$

Solve for y in Equation 1: $y = 2x + 2$

Substitute for y in Equation 2: $4x + (2x + 2) - 5 = 0$

Solve for x: $4x + (2x + 2) - 5 = 0 \implies 6x - 3 = 0 \implies x = \frac{1}{2}$

Back-substitute $x = \frac{1}{2}$: $y = 2x + 2 = 2\left(\frac{1}{2}\right) + 2 = 3$

Answer: $\left(\frac{1}{2}, 3\right)$

19. $\begin{cases} 1.5x + 0.8y = 2.3 \implies 15x + 8y = 23 \\ 0.3x - 0.2y = 0.1 \implies 3x - 2y = 1 \end{cases}$

Solve for y in Equation 2: $-2y = 1 - 3x$

$$y = \frac{3x - 1}{2}$$

Substitute for y in Equation 1: $15x + 8\left(\frac{3x - 1}{2}\right) = 23$

$$15x + 12x - 4 = 23$$

$$27x = 27$$

$$x = 1$$

Then, $y = \dfrac{3x - 1}{2} = \dfrac{3(1) - 1}{2} = 1.$

Solution: $(1, 1)$

21. $\begin{cases} \frac{1}{5}x + \frac{1}{2}y = 8 & \text{Equation 1} \\ x + y = 20 & \text{Equation 2} \end{cases}$

Solve for x in Equation 2: $x = 20 - y$

Substitute for x in Equation 1: $\frac{1}{5}(20 - y) + \frac{1}{2}y = 8$

Solve for y: $4 + \dfrac{3}{10}y = 8 \implies y = \dfrac{40}{3}$

Back-substitute $y = \dfrac{40}{3}$: $x = 20 - y$

$$= 20 - \frac{40}{3} = \frac{20}{3}$$

Answer: $\left(\dfrac{20}{3}, \dfrac{40}{3}\right)$

23. $\begin{cases} 6x + 5y = -3 & \text{Equation 1} \\ -x - \frac{5}{6}y = -7 & \text{Equation 2} \end{cases}$

Solve for x in Equation 2: $x = 7 - \frac{5}{6}y$

Substitute for x in Equation 1: $6\left(7 - \frac{5}{6}y\right) + 5y = -3$

Solve for y: $\qquad\qquad 42 - 5y + 5y = -3$

$\qquad\qquad\qquad\qquad\qquad 42 = -3$ Inconsistent

No solution.

25. $\begin{cases} -\frac{5}{3}x + y = 5 & \text{Equation 1} \\ -5x + 3y = 6 & \text{Equation 2} \end{cases}$

Solve for y in Equation 1: $y = 5 + \frac{5}{3}x$

Substitute for y in Equation 2: $-5x + 3\left(5 + \frac{5}{3}x\right) = 6$

Solve for x: $-5x + 15 + 5x = 6$

$\qquad\qquad\qquad 15 \neq 6$ Inconsistent

No solution

27. $\begin{cases} x^3 - y = 0 & \text{Equation 1} \\ x - y = 0 & \text{Equation 2} \end{cases}$

Solve for y in Equation 2: $y = x$

Substitute for y in Equation 1: $x^3 - x = 0$

Solve for x: $x(x - 1)(x + 1) = 0 \implies x = 0, 1, -1$

Back-substitute: $x = 0 \implies y = 0$

$\qquad\qquad\qquad x = 1 \implies y = 1$

$\qquad\qquad\qquad x = -1 \implies y = -1$

Answer: $(0, 0), (1, 1), (-1, -1)$

29. $\begin{cases} -x + 2y = 2 \\ 3x + y = 15 \end{cases}$

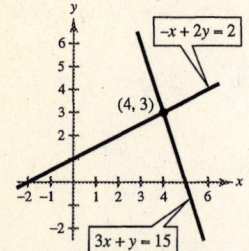

Point of intersection: $(4, 3)$

31. $\begin{cases} x - 3y = -2 \\ 5x + 3y = 17 \end{cases}$

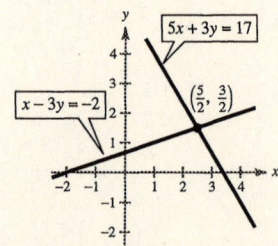

Point of intersection: $\left(\frac{5}{2}, \frac{3}{2}\right)$

33. $\begin{cases} x + y = 4 \\ x^2 + y^2 - 4x = 0 \end{cases}$

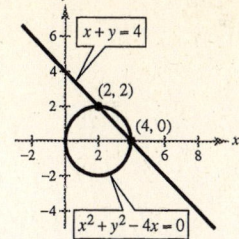

Points of intersection: $(2, 2), (4, 0)$

35. $\begin{cases} x - y + 3 = 0 \\ x^2 - 4x + 7 = y \end{cases}$

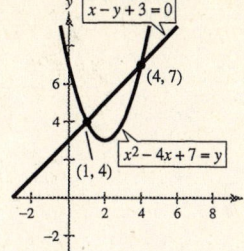

Points of intersection: $(1, 4), (4, 7)$

37. $\begin{cases} 7x + 8y = 24 \implies y_1 = -\frac{7}{8}x + 3 \\ x - 8y = 8 \implies y_2 = \frac{1}{8}x - 1 \end{cases}$

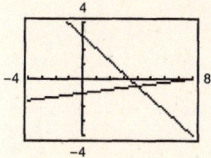

Point of intersection: $\left(4, -\frac{1}{2}\right)$

39. $\begin{cases} 2x - y + 3 = 0 \implies y_1 = 2x + 3 \\ x^2 + y^2 - 4x = 0 \implies y_2 = \sqrt{4x - x^2}, y_3 = -\sqrt{4x - x^2} \end{cases}$

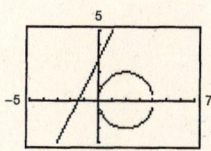

No points of intersection

41. $\begin{cases} x^2 + y^2 = 8 \implies y_1 = \sqrt{8 - x^2} \text{ and } y_2 = -\sqrt{8 - x^2} \\ y = x^2 \implies y_3 = x^2 \end{cases}$

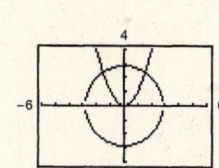

Points of intersection: $\left(\pm\sqrt{\dfrac{-1 + \sqrt{33}}{2}}, \dfrac{-1 + \sqrt{33}}{2} \right) \approx (\pm 1.54, 2.37)$

43. $\begin{cases} y = e^x \\ x - y + 1 = 0 \implies y = x + 1 \end{cases}$

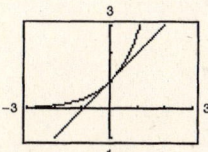

Point of intersection: $(0, 1)$

45. $\begin{cases} x + 2y = 8 \implies y_1 = 4 - x/2 \\ y = 2 + \ln x \implies y_2 = 2 + \ln x \end{cases}$

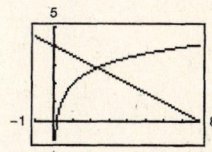

Point of intersection: Approximately $(2.318, 2.841)$

47. $\begin{cases} y = \sqrt{x} + 4 \\ y = 2x + 1 \end{cases}$

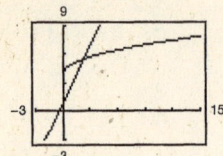

Points of intersection: $\left(\frac{9}{4}, \frac{11}{2}\right)$

49. $\begin{cases} x^2 + y^2 = 169 \implies y_1 = \sqrt{169 - x^2} \text{ and} \\ \qquad\qquad\qquad y_2 = -\sqrt{169 - x^2} \\ x^2 - 8y = 104 \implies y_3 = \frac{1}{8}x^2 - 13 \end{cases}$

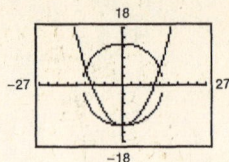

Points of intersection: $(0, -13), (\pm 12, 5)$

51. $\begin{cases} y = 2x & \text{Equation 1} \\ y = x^2 + 1 & \text{Equation 2} \end{cases}$

Substitute for y in Equation 2: $2x = x^2 + 1$

Solve for x: $x^2 - 2x + 1 = (x - 1)^2 = 0 \implies x = 1$

Back-substitute $x = 1$ in Equation 1: $y = 2x = 2$

Answer: $(1, 2)$

53. $\begin{cases} 3x - 7y + 6 = 0 & \text{Equation 1} \\ \quad x^2 - y^2 = 4 & \text{Equation 2} \end{cases}$

Solve for y in Equation 1: $y = \dfrac{3x + 6}{7}$

Substitute for y in Equation 2: $x^2 - \left(\dfrac{3x + 6}{7}\right)^2 = 4$

Solve for x: $x^2 - \left(\dfrac{9x^2 + 36x + 36}{49}\right) = 4$

$$49x^2 - (9x^2 + 36x + 36) = 196$$

$$40x^2 - 36x - 232 = 0$$

$10x^2 - 9x - 58 = 0 \implies x = \dfrac{9 \pm \sqrt{81 + 40(58)}}{20} \implies x = \dfrac{29}{10}, -2$

Back-substitute $x = \dfrac{29}{10}$: $y = \dfrac{3x + 6}{7} = \dfrac{3(29/10) + 6}{7} = \dfrac{21}{10}$

Back-substitute $x = -2$: $y = \dfrac{3x + 6}{7} = 0$

Answers: $\left(\dfrac{29}{10}, \dfrac{21}{10}\right), (-2, 0)$

55. $\begin{cases} y = 2x + 1 \\ y = \sqrt{x + 2} \end{cases}$

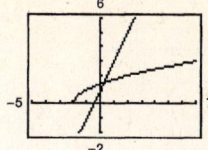

Point of intersection: $\left(\frac{1}{4}, \frac{3}{2}\right)$

57. $\begin{cases} y - e^{-x} = 1 \implies y = e^{-x} + 1 \\ y - \ln x = 3 \implies y = \ln x + 3 \end{cases}$

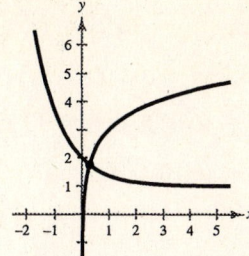

Point of intersection: Approximately $(0.287, 1.751)$

59. $\begin{cases} y = x^3 - 2x^2 + 1 & \text{Equation 1} \\ y = 1 - x^2 & \text{Equation 2} \end{cases}$

Substitute for y in Equation 2: $x^3 - 2x^2 + 1 = 1 - x^2$

Solve for x: $x^3 - x^2 = 0$

$$x^2(x - 1) = 0 \implies x = 0, 1$$

Back-substitute: $x = 0 \implies y = 1$

$$x = 1 \implies y = 0$$

Answers: $(0, 1), (1, 0)$

61. $\begin{cases} xy - 1 = 0 & \text{Equation 1} \\ 2x - 4y + 7 = 0 & \text{Equation 2} \end{cases}$

Solve for y in Equation 1: $y = \dfrac{1}{x}$

Substitute for y in Equation 2: $2x - 4\left(\dfrac{1}{x}\right) + 7 = 0$

Solve for x: $2x^2 - 4 + 7x = 0 \implies (2x - 1)(x + 4) = 0 \implies x = \dfrac{1}{2}, -4$

Back-substitute $x = \dfrac{1}{2}$: $y = \dfrac{1}{1/2} = 2$

Back-substitute $x = -4$: $y = \dfrac{1}{-4} = -\dfrac{1}{4}$

Answers: $\left(\dfrac{1}{2}, 2\right), \left(-4, -\dfrac{1}{4}\right)$

63. $C = 8650x + 250,000,\ R = 9950x$

$R = C$

$9950x = 8650x + 250,000$

$1300x = 250,000$

$x \approx 192$ units

$R \approx \$1,910,400$

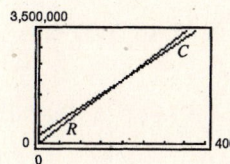

65. $C = 5.5\sqrt{x} + 10,000, \; R = 3.29x$

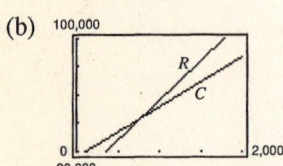

$$R = C$$
$$3.29x = 5.5\sqrt{x} + 10,000$$
$$3.29x - 10,000 = 5.5\sqrt{x}$$
$$10.8241x^2 - 65,800x + 100,000,000 = 30.25x$$
$$10.8241x^2 - 65,830.25x + 100,000,000 = 0$$
$$x \approx 3133 \text{ units}$$

In order for the revenue to break even with the cost, 3133 units must be sold, $R \approx \$10,308$.

67. (a) $C = 35.45x + 16,000$

$R = 55.95x$

(b)

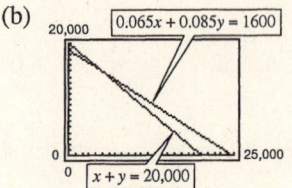

$C = R$ for $x \approx 780$ units

(c) $35.45x + 16,000 = 55.95x$

$$16,000 = 20.5x$$
$$x = \frac{16,000}{20.5} \approx 780 \text{ units}$$

69. $0.06x = 0.03x + 350$

$0.03x = 350$

$x \approx \$11,666.67$

To make the straight commission offer better, you would have to sell more than $\$11,666.67$ per week.

71. (a) $\begin{cases} x + y = 20,000 \\ 0.065x + 0.085y = 1600 \end{cases}$

(b)

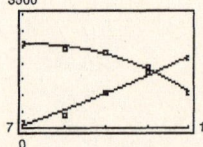

(c) The curves intersect at $x = 5000$. Thus, $\$5000$ should be invested at 6.5%.

As x increases, y decreases and the amount of interest decreases.

73. (a) $F_{\text{VCR}} = -107.86x^2 + 1583.6x - 3235$

$F_{\text{DVD}} = 21.14x^2 + 143.8x - 1939$

(b)

(c) The curves intersect at $x = 10.17$, so DVD sales exceed VCR sales in 2000.

(d) $21.14x^2 + 143.8x - 1939 > 107.86x^2 + 1583.6x - 3235$

$129x^2 - 1439.8x + 1296 > 0$

Solving this inequality for x, you obtain $x > 10.17$.

(e) The answers are the same.

75. $2l + 2w = 30 \Rightarrow l + w = 15$

$\quad l = w + 3 \Rightarrow (w + 3) + w = 15$

$\qquad\qquad\qquad\qquad\qquad 2w = 12$

$\qquad\qquad\qquad\qquad\qquad w = 6$

$l = w + 3 = 9$

Dimensions: 6 meters $\times$ 9 meters

77. $2l + 2w = 40 \Rightarrow l + w = 20 \Rightarrow w = 20 - l$

$\quad lw = 96 \Rightarrow l(20 - l) = 96$

$\qquad\qquad\qquad 20l - l^2 = 96$

$\qquad\qquad\qquad 0 = l^2 - 20l + 96$

$\qquad\qquad\qquad 0 = (l - 8)(l - 12)$

$\qquad\qquad\qquad l = 8 \text{ or } l = 12$

$l = 12, w = 8$

Since the length is supposed to be greater than the width, we have $l = 12$ miles and $w = 8$ miles.

79. False. You could solve for x first.

81. The system has no solution if you arrive at a false statement, ie. $4 = 8$, or you have a quadratic equation with a negative discriminant, which would yield imaginary roots.

83. (a) The line $y = 2x$ intersects the parabola $y = x^2$ at two points, $(0, 0)$ and $(2, 4)$.

(b) The line $y = 0$ intersects $y = x^2$ at $(0, 0)$ only.

(c) The line $y = x - 2$ does not intersect $y = x^2$.

(Other answers possible.)

85. $(-2, 7), (5, 5)$

$$m = \frac{5 - 7}{5 - (-2)} = -\frac{2}{7}$$

$$y - 7 = -\frac{2}{7}(x - (-2))$$

$$7y - 49 = -2x - 4$$

$$2x + 7y - 45 = 0$$

87. $(6, 3), (10, 3)$

$$m = \frac{3 - 3}{10 - 6} = 0 \Rightarrow \text{The line is horizontal.}$$

$$y = 3 \Rightarrow y - 3 = 0$$

89. $\left(\dfrac{3}{5}, 0\right), (4, 6)$

$$m = \frac{6 - 0}{4 - \frac{3}{5}} = \frac{6}{\frac{17}{5}} = \frac{30}{17}$$

$$y - 6 = \frac{30}{17}(x - 4)$$

$$17y - 102 = 30x - 120$$

$$0 = 30x - 17y - 18$$

91. Domain: all $x \neq 6$

Vertical asymptotes: $x = 6$

Horizontal asymptote: $y = 0$

93. Domain: all $x \neq \pm 4$

Vertical asymptotes: $x = \pm 4$

Horizontal asymptote: $y = 1$

Section 7.2   Systems of Linear Equations in Two Variables

■ You should be able to solve a linear system by the method of elimination.

1. Obtain coefficients for either x or y that differ only in sign. This is done by multiplying all the terms of one or both equations by appropriate constants.

2. Add the equations to eliminate one of the variables and then solve for the remaining variable.

3. Use back-substitution into either original equation and solve for the other variable.

4. Check your answer.

■ You should know that for a system of two linear equations, one of the following is true.

(a) There are infinitely many solutions; the lines are identical. The system is consistent.

(b) There is no solution; the lines are parallel. The system is inconsistent.

(c) There is one solution; the lines intersect at one point. The system is consistent.

Solutions to Odd-Numbered Exercises

1. $\begin{cases} 2x + y = 5 & \text{Equation 1} \\ x - y = 1 & \text{Equation 2} \end{cases}$

Add to eliminate y: $3x = 6 \implies x = 2$

Substitute $x = 2$ in Equation 2:
$2 - y = 1 \implies y = 1$

Answer: $(2, 1)$

3. $\begin{cases} x + y = 0 & \text{Equation 1} \\ 3x + 2y = 1 & \text{Equation 2} \end{cases}$

Multiply Equation 1 by -2: $-2x - 2y = 0$

Add this to Equation 2 to eliminate y: $x = 1$

Substitute $x = 1$ in Equation 1:
$1 + y = 0 \implies y = -1$

Answer: $(1, -1)$

5. $\begin{cases} x - y = 2 & \text{Equation 1} \\ -2x + 2y = 5 & \text{Equation 2} \end{cases}$

Multiply Equation 1 by 2: $2x - 2y = 4$

Add this to Equation 2: $0 = 9$

There are no solutions.

7. $\begin{cases} x + 2y = 4 & \text{Equation 1} \\ x - 2y = 1 & \text{Equation 2} \end{cases}$

Add to eliminate y:

$$2x = 5$$

$$x = \frac{5}{2}$$

Substitute $x = \frac{5}{2}$ in Equation 1:

$$\frac{5}{2} + 2y = 4 \implies y = \frac{3}{4}$$

Answer: $\left(\frac{5}{2}, \frac{3}{4}\right)$

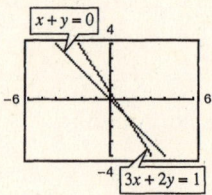

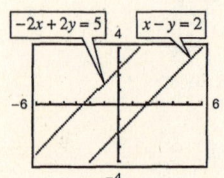

9. $\begin{cases} 2x + 3y = 18 & \text{Equation 1} \\ 5x - y = 11 & \text{Equation 2} \end{cases}$

Multiply Equation 2 by 3: $15x - 3y = 33$

Add this to Equation 1 to eliminate y:

$17x = 51 \implies x = 3$

Substitute $x = 3$ in Equation 1:

$6 + 3y = 18 \implies y = 4$

Answer: $(3, 4)$

11. $\begin{cases} 3r + 2s = 10 & \text{Equation 1} \\ 2r + 5s = 3 & \text{Equation 2} \end{cases}$

Multiply Equation 1 by 2 and Equation 2 by -3:

$6r + 4s = 20$

$-6r - 15s = -9$

Add to eliminate r:

$-11s = 11 \implies s = -1$

Substitute $s = -1$ in Equation 1:

$3r + 2(-1) = 10 \implies r = 4$

Answer: $(4, -1)$

13. $\begin{cases} 5u + 6v = 24 & \text{Equation 1} \\ 3u + 5v = 18 & \text{Equation 2} \end{cases}$

Multiply Equation 1 by 3 and Equation 2 by (-5): $15u + 18v = 72$

$-15u - 25v = -90$

Add to eliminate u: $-7v = -18 \implies v = \frac{18}{7}$

Substitute $v = \frac{18}{7}$ in Equation 2: $3u + 5\left(\frac{18}{7}\right) = 18 \implies u = \frac{12}{7}$

Answer: $\left(\frac{12}{7}, \frac{18}{7}\right)$

15. $\begin{cases} 1.8x + 1.2y = 4 & \text{Equation 1} \\ 9x + 6y = 3 & \text{Equation 2} \end{cases}$

Multiply Equation 1 by (-5): $-9x - 6y = -20$

Add this to Equation 2: $0 = -17$

Inconsistent; no solution

17. $\begin{cases} 2x - 5y = 0 \implies y = \frac{2}{5}x \\ x - y = 3 \implies y = x - 3 \end{cases}$ passes through $(0, 0)$
    y-intercept $(0, -3)$

Matches (b)

19. $\begin{cases} 2x - 5y = 0 \implies y = \frac{2}{5}x \\ 2x - 3y = -4 \implies y = \frac{2}{3}x + \frac{4}{3} \end{cases}$ passes through $(0, 0)$
    y-intercept $\left(0, \frac{4}{3}\right)$

Matches (c)

21. $\begin{cases} 4x + 3y = 3 & \text{Equation 1} \\ 3x + 11y = 13 & \text{Equation 2} \end{cases}$

Multiply Equation 1 by 3 and Equation 2 by -4:

$\begin{cases} 12x + 9y = 9 \\ -12x - 44y = -52 \end{cases}$

Add to eliminate x: $-35y = -43 \implies y = \frac{43}{35}$

Substitute $y = \frac{43}{35}$ into Equation 1:

$4x + 3\left(\frac{43}{35}\right) = 3 \implies x = -\frac{6}{35}$

Answer: $\left(-\frac{6}{35}, \frac{43}{35}\right)$

23. $\begin{cases} \dfrac{x}{4} + \dfrac{y}{6} = 1 & \text{Equation 1} \\ -3x - 2y = 0 & \text{Equation 2} \end{cases}$

Multiply Equation 1 by 12 and add to Equation 2:

$$0 = 12$$

Inconsistent. No solution

25. $\begin{cases} \frac{3}{4}x + y = \frac{1}{8} & \text{Equation 1} \\ \frac{9}{4}x + 3y = \frac{3}{8} & \text{Equation 2} \end{cases}$

Multiply Equation 1 by -3: $-\frac{9}{4}x - 3y = -\frac{3}{8}$

Add this to Equation 2: $\qquad\qquad 0 = 0$

There are an infinite number of solutions.

The solutions consist of all (x, y) satisfying

$\frac{3}{4}x + y = \frac{1}{8}$, or $6x + 8y = 1$.

27. $\begin{cases} \dfrac{x+3}{4} + \dfrac{y-1}{3} = 1 & \text{Equation 1} \\ 2x - y = 12 & \text{Equation 2} \end{cases}$

Multiply Equation 1 by 12 and Equation 2 by 4

$$\begin{cases} 3x + 4y = 7 \\ 8x - 4y = 48 \end{cases}$$

Add to eliminate y: $11x = 55 \Rightarrow x = 5$

Substitute $x = 5$ into Equation 2:

$$2(5) - y = 12 \Rightarrow y = -2$$

Answer: $(5, -2)$

29. $\begin{cases} 2.5x - 3y = 1.5 & \text{Equation 1} \\ 10x - 12y = 6 & \text{Equation 2 multiplied by 5} \end{cases}$

Multiply Equation 1 by (-4):

$$-10x + 12y = -6$$

Add this to Equation 2 to eliminate x:

$$0 = 0$$

The solution set consists of all points lying on the line

$$10x - 12y = 6.$$

All points on the line

$$5x - 6y = 3$$

Let $x = a$, then $y = \frac{5}{6}a - \frac{1}{2}$.

Answer: $\left(a, \frac{5}{6}a - \frac{1}{2}\right)$, where a is any real number.

31. $\begin{cases} 0.2x - 0.5y = -27.8 & \text{Equation 1} \\ 0.3x + 0.4y = 68.7 & \text{Equation 2} \end{cases}$

Multiply Equation 1 by 40 and Equation 2 by 50:

$$\begin{cases} 8x - 20y = -1112 \\ 15x + 20y = 3435 \end{cases}$$

Adding the equation eliminates y:

$$23x = 2323 \Rightarrow x = 101$$

Substitute $x = 101$ into Equation 1:

$$8(101) - 20y = -1112 \Rightarrow y = 96$$

Solution: $(101, 96)$

33. $\begin{cases} 2x - 5y = 0 \Rightarrow y = \frac{2}{5}x \\ x - y = 3 \Rightarrow y = x - 3 \end{cases}$

The system is consistent. There is one solution, $(5, 2)$.

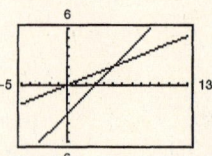

35. $\begin{cases} \frac{3}{5}x - y = 3 \Rightarrow y = \frac{3}{5}x - 3 \\ -3x + 5y = 9 \Rightarrow y = \frac{1}{5}(3x + 9) = \frac{3}{5}x + \frac{9}{5} \end{cases}$

The lines are parallel. The system is inconsistent.

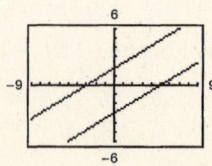

37. $\begin{cases} 8x - 14y = 5 \implies y = (8x - 5)/14 = \frac{4}{7}x - \frac{5}{14} \\ 2x - 3.5y = 1.25 \implies y = (2x - 1.25)/3.5 = \frac{4}{7}x - \frac{5}{14} \end{cases}$

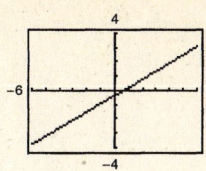

The system is consistent. The solution set consists of all points on the line $y = \frac{4}{7}x - \frac{5}{14}$, or $8x - 14y = 5$.

39. $\begin{cases} 6y = 42 \implies y = 7 \\ 6x - y = 16 \implies y = 6x - 16 \end{cases}$

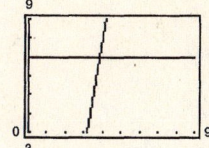

Solution: $\left(\frac{23}{6}, 7\right) \approx (3.833, 7)$

41. $\begin{cases} \frac{3}{2}x - \frac{1}{5}y = 8 \\ -2x + 3y = 3 \end{cases}$

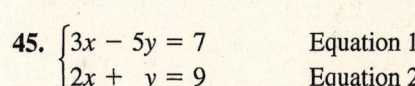
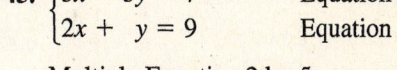

Solution: $(6, 5)$

43. $\begin{cases} 0.5x + 2.2y = 9 \\ 6x + 0.4y = -22 \end{cases}$

Solution: $(-4, 5)$

45. $\begin{cases} 3x - 5y = 7 \qquad \text{Equation 1} \\ 2x + y = 9 \qquad \text{Equation 2} \end{cases}$

Multiply Equation 2 by 5:

$$10x + 5y = 45$$

Add this to Equation 1:

$$13x = 52 \implies x = 4$$

Back-substitute $x = 4$ into Equation 2:

$$2(4) + y = 9 \implies y = 1$$

Solution: $(4, 1)$

47. $\begin{cases} y = 4x + 3 \\ y = -5x - 12 \end{cases}$

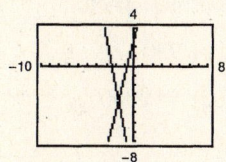

The lines intersect at

$\left(-\frac{5}{3}, -\frac{11}{3}\right) \approx (-1.667, -3.667)$

49. $\begin{cases} x - 5y = 21 \\ 6x + 5y = 21 \end{cases}$

Adding the equations, $7x = 42 \implies x = 6$.

Back-substituting, $x - 5y = 6 - 5y = 21 \implies$

$$-5y = 15 \implies y = -3$$

Solution: $(6, -3)$

51. $\begin{cases} -2x + 8y = 19 \qquad \text{Equation 1} \\ y = x - 3 \qquad \text{Equation 2} \end{cases}$

Substituting into Equation 1,

$$-2x + 8(x - 3) = 19 \implies 6x = 43$$

$$\implies x = \frac{43}{6}$$

Back-substituting, $y = x - 3 = \frac{43}{6} - 3 = \frac{25}{6}$

Solution: $\left(\frac{43}{6}, \frac{25}{6}\right)$

53. There are infinitely many systems that have the solution $(0, 8)$. One possible system is

$$\begin{cases} x + y = 8 \\ -x + y = 8 \end{cases}$$

55. There are infinitely many systems that have the solution $\left(3, \frac{5}{2}\right)$. One possible system is:

$$2(3) + 2\left(\tfrac{5}{2}\right) = 11 \implies 2x + 2y = 11$$
$$3 - 4\left(\tfrac{5}{2}\right) = -7 \implies x - 4y = -7$$

57.

$$\text{Demand} = \text{Supply}$$
$$50 - 0.5x = 0.125x$$
$$50 = 0.625x$$
$$x = 80 \text{ units}$$
$$p = \$10$$

Answer: $(80, 10)$

59.

$$\text{Demand} = \text{Supply}$$
$$140 - 0.00002x = 80 + 0.00001x$$
$$60 = 0.00003x$$
$$x = 2{,}000{,}000 \text{ units}$$
$$p = \$100.00$$

Answer: $(2{,}000{,}000, 100)$

61. Let x = the ground speed and y = the wind speed.

$$\begin{cases} 3.6(x - y) = 1800 \quad \text{Equation 1} \\ 3(x + y) = 1800 \quad \text{Equation 2} \end{cases}$$

$$\begin{aligned} x - y &= 500 \\ \underline{x + y} &= \underline{600} \\ 2x \quad\;\; &= 1110 \\ x \quad &= 550 \end{aligned}$$

Substituting $x = 550$ in Equation 2: $550 + y = 600$
$$y = 50$$

Answer: $x = 550$ mph, $y = 50$ mph

63. (a) Let x = the number of liters of 40%

$\quad\quad y$ = the number of liters of 65%

$$\begin{cases} x + y = 20 \\ 0.4x + 0.65y = 0.5(20) = 10 \end{cases}$$

(b) $y_1 = 20 - x$

$$y_2 = \frac{10 - 0.4x}{0.65}$$

As x increases, y decreases

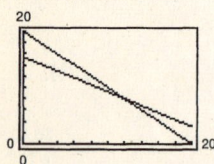

(c) The lines intersect at $(12, 8)$.

12 liters of 40% solution, 8 liters of 65% solution

65. Let x = amount invested at 7.5%

Let y = amount invested at 6%

$$\begin{cases} x + y = 15{,}000 \\ 0.075x + 0.06y = 990 \end{cases}$$

From Equation 1, $y = 15{,}000 - x$. Then,

$$0.075x + 0.06(15{,}000 - x) = 990$$
$$0.015x = 90 \implies x = 6000, y = 9000$$

$9000 in the 6% fund.

67. Let x = number of adult tickets sold, y = number of child tickets sold.

$$\begin{cases} x + y = 500 & \text{Equation 1} \\ 7.5x + 4y = \$3312.50 & \text{Equation 2} \end{cases}$$

$$\begin{aligned} -4x - 4y &= -2000.00 \\ 7.5x + 4y &= 3312.50 \\ \hline 3.5x &= 1312.50 \\ x &= 375 \\ 375 + y &= 500 \\ y &= 125 \end{aligned}$$

Answer: $x = 375$ adult tickets, $y = 125$ child tickets

69.
$$\begin{cases} 5b + 10a = 20.2 \implies -10b - 20a = -40.4 \\ 10b + 30a = 50.1 \implies 10b + 30a = 50.1 \end{cases}$$

$$\begin{aligned} 10a &= 9.7 \\ a &= 0.97 \\ b &= 2.10 \end{aligned}$$

Least squares regression line: $y = 0.97x + 2.10$

71. (a) $\begin{cases} 4b + 7a = 174 \implies 28b + 49a = 1218 \\ 7b + 13.5a = 322 \implies -28b - 54a = -1288 \end{cases}$

Adding, $-5a = -70 \implies a = 14$, $b = 19$.

Thus, $y = 14x + 19$

(c)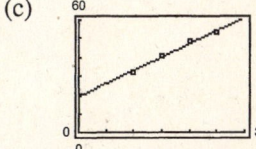

(b) Using a graphing utility, you obtain $y = 14x + 19$.

(d) If $x = 1.6$, (160 pounds/acre),
$y = 14(1.6) + 19 = 41.4$ bushels per acre.

73. True. A consistent linear system has either one solution or an infinite number of solutions.

75. $\begin{cases} 100y - x = 200 & \text{Equation 1} \\ 99y - x = -198 & \text{Equation 2} \end{cases}$

Subtract Equation 2 from Equation 1 to eliminate

x: $y = 398$

Substitute $y = 398$ into Equation 1:

$100(398) - x = 200 \implies x = 39,600$

Solution: $(39,600, 398)$

The lines are not parallel. The scale on the axes must be changed to see the point of intersection.

77. No, it is not possible for a consistent system of linear equations to have exactly two solutions. Either the lines will intersect once or they will coincide and then the system would have infinite solutions.

79. $\begin{cases} 4x - 8y = -3 & \text{Equation 1} \\ 2x + ky = 16 & \text{Equation 2} \end{cases}$

Multiply Equation 2 by -2: $-4x - 2ky = -32$

Add this to Equation 1: $-8y - 2ky = -35$

The system is inconsistent if $-8y - 2ky = 0$.

This occurs when $k = -4$. Note that for $k = -4$, the two original equations represent parallel lines.

81. Subtracting the two equations:

$$vxe^x - v(x + 1)e^x = -e^x \ln x$$

$$vxe^x - vxe^x - ve^x = -e^x \ln x$$

$$ve^x = e^x \ln x$$

$$v = \ln x$$

Finally, $ue^x + vxe^x = ue^x + \ln x \cdot x \cdot e^x = 0$

$$\implies ue^x = -x \ln x \cdot e^x$$

$$\implies u = -x \ln x$$

83. $-11 - 6x \geq 33$

$-6x \geq 44$

$x \leq -\frac{44}{6} = -\frac{22}{3}$

85. $|x - 8| < 10$

$-10 < x - 8 < 10$

$-2 < x < 18$

87. $2x^2 + 3x - 35 < 0$

$(2x - 7)(x + 5) < 0$

Critical numbers: $\frac{7}{2}, -5$.
Testing the three

intervals, $-5 < x < \frac{7}{2}$.

89. $\ln x + \ln 6 = \ln 6x$

91. $\log_9 12 - \log_9 x = \log_9 \frac{12}{x}$

Section 7.3 Multivariable Linear Systems

- ■ You should know the operations that lead to equivalent systems of linear equations:
  - (a) Interchange any two equations.
  - (b) Multiply all terms of an equation by a nonzero constant.
  - (c) Replace an equation by the sum of itself and a constant multiple of any other equation in the system.
- ■ You should be able to use the method of elimination.

Solutions to Odd-Numbered Exercises

1. (a) $3(2) - 5 + 0 \overset{?}{=} 1$ Yes

     $2(2) - 3(0) \overset{?}{=} -14$ No

     $5(5) + 2(0) \overset{?}{=} 8$ No

     No, $(2, 5, 0)$ is not a solution.

     (b) $3(-2) - 0 + 4 \overset{?}{=} 1$ No

      $2(-2) - 3(4) \overset{?}{=} -14$ No

      $5(0) + 2(4) \overset{?}{=} 8$ Yes

      No, $(-2, 0, 4)$ is not a solution.

—CONTINUED—

1. —CONTINUED—

(c) $3(0) - (-1) + 3 \stackrel{?}{=} 1$ No

   $2(0) - 3(3) \stackrel{?}{=} -14$ No

   $5(-1) + 2(3) \stackrel{?}{=} 8$ No

   No, $(0, -1, 3)$ is not a solution.

(d) $3(-1) - 0 + 4 \stackrel{?}{=} 1$ Yes

   $2(-1) - 3(4) \stackrel{?}{=} -14$ Yes

   $5(0) + 2(4) \stackrel{?}{=} 8$ Yes

   Yes, $(-1, 0, 4)$ is a solution.

3. (a) $4(0) + 1 - 1 \stackrel{?}{=} 0$ Yes

   $-8(0) - 6(1) + 1 \stackrel{?}{=} -\frac{7}{4}$ No

   $3(0) - 1 \stackrel{?}{=} -\frac{9}{4}$ No

   No, $(0, 1, 1)$ is not a solution.

(b) $4\left(-\frac{3}{2}\right) + \frac{5}{4} - \left(-\frac{5}{4}\right) \stackrel{?}{=} 0$ No

   $-8\left(-\frac{3}{2}\right) - 6\left(\frac{5}{4}\right) + \left(-\frac{5}{4}\right) \stackrel{?}{=} -\frac{7}{4}$ No

   $3\left(-\frac{3}{2}\right) - \left(\frac{5}{4}\right) \stackrel{?}{=} -\frac{9}{4}$ No

   No, $\left(-\frac{3}{2}, \frac{5}{4}, -\frac{5}{4}\right)$ is not a solution.

(c) $4\left(-\frac{1}{2}\right) + \frac{3}{4} - \left(-\frac{5}{4}\right) \stackrel{?}{=} 0$ Yes

   $-8\left(-\frac{1}{2}\right) - 6\left(\frac{3}{4}\right) - \frac{5}{4} \stackrel{?}{=} -\frac{7}{4}$ Yes

   $3\left(-\frac{1}{2}\right) - \frac{3}{4} \stackrel{?}{=} -\frac{9}{4}$ Yes

   Yes, $\left(-\frac{1}{2}, \frac{3}{4}, -\frac{5}{4}\right)$ is a solution.

(d) $4\left(-\frac{1}{2}\right) + 2 - 0 \stackrel{?}{=} 0$ Yes

   $-8\left(-\frac{1}{2}\right) - 6(2) + 0 \stackrel{?}{=} -\frac{7}{2}$ No

   $3\left(-\frac{1}{2}\right) - 2 \stackrel{?}{=} -\frac{9}{4}$ No

   No, $\left(-\frac{1}{2}, 2, 0\right)$ is not a solution.

5. $\begin{cases} 2x - y + 5z = 24 \\ \quad\ y + 2z = \ 4 \\ \qquad\quad z = \ 6 \end{cases}$ Equation 1
Equation 2
Equation 3

Back-substitute $z = 6$ into Equation 2

   $y + 2(6) = 4$

   $y = -8$

Back-substitute $y = -8$ and $z = 6$ into Equation 1

   $2x - (-8) + 5(6) = 24$

   $2x = -14$

   $x = -7$

Answer: $(-7, -8, 6)$

7. $\begin{cases} 2x + y - 3z = 10 \\ \quad\ y + \ z = 12 \\ \qquad\quad z = \ 2 \end{cases}$ Equation 1
Equation 2
Equation 3

Back-substitute $z = 2$ into Equation 2

   $y + 2 = 12$

   $y = 10$

Back-substitute $y = 10$ and $z = 2$ into Equation 1

   $2x + 10 - 3(2) = 10$

   $2x = 6$

   $x = 3$

Answer: $(3, 10, 2)$

9. $\begin{cases} 4x - 2y + z = 8 \\ \quad\ -\ y + z = 4 \\ \qquad\qquad z = 2 \end{cases}$ Equation 1
Equation 2
Equation 3

Back-substitute $z = 2$ into Equation 2

   $-y + 2 = 4$

   $y = -2$

Back-substitute $y = -2$ and $z = 2$ into Equation 1

   $4x - 2(-2) + 2 = 8$

   $4x = 2$

   $x = \frac{1}{2}$

Answer: $\left(\frac{1}{2}, -2, 2\right)$

11. $\begin{cases} x - 2y + 3z = 5 \\ -x + 3y - 5z = 4 \\ 2x \quad\ - 3z = 0 \end{cases}$ Equation 1
Equation 2
Equation 3

Add Equation 1 to Equation 2.

   $y - 2z = 9$ New Equation 2

This is the first step in putting the system in row-echelon form.

$\begin{cases} x - 2y + 3z = 5 \\ \quad\ y - 2z = 9 \\ 2x \quad\ - 3z = 0 \end{cases}$

13. $\begin{cases} x + y + z = 6 \\ 2x - y + z = 3 \\ 3x \quad\;\; - z = 0 \end{cases}$ Equation 1   Equation 2   Equation 3

$\begin{cases} x + y + z = 6 \\ \quad -3y - z = -9 \\ \quad -3y - 4z = -18 \end{cases}$ (-2) Eq. 1 + Eq. 2 (-3) Eq. 1 + Eq. 3

$\begin{cases} x + y + z = 6 \\ \quad -3y - z = -9 \\ \quad\quad -3z = -9 \end{cases}$ (-1) Eq. 2 + Eq. 3

$-3z = -9 \implies z = 3$

$-3y - 3 = -9 \implies y = 2$

$x + 2 + 3 = 6 \implies x = 1$

Answer: $(1, 2, 3)$

15. $\begin{cases} 2x \quad\quad + 2z = 2 \\ 5x + 3y \quad\quad = 4 \\ \quad\;\; 3y - 4z = 4 \end{cases}$ Equation 1   Equation 2   Equation 3

$\begin{cases} x \quad\quad + z = 1 \\ 5x + 3y \quad\quad = 4 \\ \quad\; 3y - 4z = 4 \end{cases}$ $\left(\frac{1}{2}\right)$ Eq. 1

$\begin{cases} x \quad\quad + z = 1 \\ \quad\; 3y - 5z = -1 \\ \quad\; 3y - 4z = 4 \end{cases}$ (-5) Eq. 1 + Eq. 2

$\begin{cases} x \quad\quad + z = 1 \\ \quad\; 3y - 5z = -1 \\ \quad\quad\;\; z = 5 \end{cases}$ (-1) Eq. 2 + Eq. 3

$3y - 5(5) = -1 \implies y = 8$

$x + 5 = 1 \implies x = -4$

Answer: $(-4, 8, 5)$

17. $\begin{cases} \quad\quad 6y + 4z = -18 \\ 3x + 3y \quad\quad = 9 \\ 2x \quad\quad - 3z = 12 \end{cases}$ Equation 1   Equation 2   Equation 3

$\begin{cases} 3x + 3y \quad\quad = 9 \\ \quad\quad 6y + 4z = -18 \\ 2x \quad\quad - 3z = 12 \end{cases}$ Interchange equations 1 and 2

$\begin{cases} x + y \quad\quad = 3 \\ \quad\quad 6y + 4z = -18 \\ 2x \quad\quad - 3z = 12 \end{cases}$ $\frac{1}{3}$ (New Eq. 1)

$\begin{cases} x + y \quad\quad = 3 \\ \quad\quad 6y + 4z = -18 \\ \quad\; -2y - 3z = 6 \end{cases}$ (-2) Eq. 1 + Eq. 3

$\begin{cases} x + y \quad\quad = 3 \\ \quad\; -2y - 3z = 6 \\ \quad\quad 6y + 4z = -18 \end{cases}$ Interchange the equations

$\begin{cases} x + y \quad\quad = 3 \\ \quad\; -2y - 3z = 6 \\ \quad\quad\;\; -5z = 0 \end{cases}$ 3 Eq. 2 + Eq. 3

$-5z = 0 \implies z = 0$

$-2y - 3(0) = 6 \implies y = -3$

$x + (-3) = 3 \implies x = 6$

Answer: $(6, -3, 0)$

19. $\begin{cases} x + y - 2z = 3 \\ 3x - 2y + 4z = 1 \\ 2x - 3y + 6z = 8 \end{cases}$ Interchange the equations.

$\begin{cases} x + y - 2z = 3 \\ \quad -5y + 10z = -8 \\ \quad -5y + 10z = 10 \end{cases}$ -3 Eq. 1 + Eq. 2 -2 Eq. 1 + Eq. 3

$\begin{cases} x + y - 2z = 3 \\ \quad -5y + 10z = -8 \\ \quad\quad\quad 0 = 18 \end{cases}$ $-$Eq. 2 + Eq. 3

No solution, inconsistent

21. $\begin{cases} 3x + 3y + 5z = 1 \\ 3x + 5y + 9z = 0 \\ 5x + 9y + 17z = 0 \end{cases}$

$\begin{cases} 6x + 6y + 10z = 2 \\ 3x + 5y + 9z = 0 \\ 5x + 9y + 17z = 0 \end{cases}$ 2 Eq. 1

$\begin{cases} x - 3y - 7z = 2 \\ 3x + 5y + 9z = 0 \\ 5x + 9y + 17z = 0 \end{cases}$ $-$Eq. 3 + Eq. 1

$\begin{cases} x - 3y - 7z = 2 \\ 14y + 30z = -6 \\ 24y + 52z = -10 \end{cases}$ -3 Eq. 1 + Eq. 2 -5 Eq. 1 + Eq. 3

$\begin{cases} x - 3y - 7z = 2 \\ 84y + 180z = -36 \\ 84y + 182z = -35 \end{cases}$ 6 Eq. 2   3.5 Eq. 3

$\begin{cases} x - 3y - 7z = 2 \\ 84y + 180z = -36 \\ 2z = 1 \end{cases}$ $-$Eq. 2 + Eq. 3

$2z = 1 \implies z = \frac{1}{2}$

$84y + 180\left(\frac{1}{2}\right) = -36 \implies y = -\frac{3}{2}$

$x - 3\left(-\frac{3}{2}\right) - 7\left(\frac{1}{2}\right) = 2 \implies x = 1$

Answer: $\left(1, -\frac{3}{2}, \frac{1}{2}\right)$

23. $\begin{cases} x + 2y - 7z = -4 \\ 2x + y + z = 13 \\ 3x + 9y - 36z = -33 \end{cases}$

$\begin{cases} x + 2y - 7z = -4 \\ -3y + 15z = 21 \\ 3y - 15z = -21 \end{cases}$ -2 Eq. 1 + Eq. 2 -3 Eq. 1 + Eq. 3

$\begin{cases} x + 2y - 7z = -4 \\ -3y + 15z = 21 \\ 0 = 0 \end{cases}$ Eq. 2 + Eq. 3

$\begin{cases} x + 2y - 7z = -4 \\ y - 5z = -7 \end{cases}$ $-\frac{1}{3}$ Eq. 2

$\begin{cases} x + 3z = 10 \\ y - 5z = -7 \end{cases}$ -2 Eq. 2 + Eq. 1

Let $z = a$, then:

$y = 5a - 7$

$x = -3a + 10$

Answer: $(-3a + 10, 5a - 7, a)$

25. $\begin{cases} 3x - 3y + 6z = 6 \\ x + 2y - z = 5 \\ 5x - 8y + 13z = 7 \end{cases}$

$\begin{cases} x - y + 2z = 2 \\ x + 2y - z = 5 \\ 5x - 8y + 13z = 7 \end{cases}$

$\begin{cases} x - y + 2z = 2 \\ 3y - 3z = 3 \\ -3y + 3z = -3 \end{cases}$

$\begin{cases} x - y + 2z = 2 \\ y - z = 1 \\ 0 = 0 \end{cases}$

$\begin{cases} x + z = 3 \\ y - z = 1 \end{cases}$

Let $z = a$, then:

$y = a + 1$

$x = -a + 3$

Answer: $(-a + 3, a + 1, a)$

27. $\begin{cases} x - 2y + 3z = 4 \\ 3x - y + 2z = 0 \\ x + 3y - 4z = -2 \end{cases}$ Equation 1   Equation 2   Equation 3

$\begin{cases} x - 2y + 3z = 4 \\ 5y - 7z = -12 \\ 5y - 7z = -6 \end{cases}$ -3 Eq. 1 + Eq. 2 -1 Eq. 1 + Eq. 3

$\begin{cases} x - 2y + 3z = 4 \\ 5y - 7z = -12 \\ 0 = 6 \end{cases}$ $-$Eq. 2 + Eq. 3

No solution, inconsistent.

29. $\begin{cases} x + 4z = 1 \\ x + y + 10z = 10 \\ 2x - y + 2z = -5 \end{cases}$

$\begin{cases} x + 4z = 1 \\ y + 6z = 9 \\ -y - 6z = -7 \end{cases}$ $-$Eq. 1 + Eq. 2 -2 Eq. + Eq. 3

$\begin{cases} x + 4z = 1 \\ y + 6z = 9 \\ 0 = 2 \end{cases}$ Eq. 2 + Eq. 3

No solution, inconsistent.

31. $\begin{cases} x - 2y + 5z = 2 \\ 4x \qquad - z = 0 \end{cases}$

$\begin{cases} x - 2y + 5z = 2 \\ 8y - 21z = -8 \end{cases}$ $\qquad -4\,\text{Eq. 1} + \text{Eq. 2}$

$\begin{cases} x - 2y + 5z = 2 \\ y - \frac{21}{8}z = -1 \end{cases}$ $\qquad \frac{1}{8}\,\text{Eq. 3}$

$\begin{cases} x \qquad - \frac{1}{4}z = 0 \\ y - \frac{21}{8}z = -1 \end{cases}$ $\qquad 2\,\text{Eq. 2} + \text{Eq. 1}$

Let $z = a$. Then $y = \frac{21}{8}a - 1$ and $x = \frac{1}{4}a$

Solution: $\left(\frac{1}{4}a, \frac{21}{8}a - 1, a\right)$

33. $\begin{cases} 2x - 3y + z = -2 \\ -4x + 9y \qquad = 7 \end{cases}$

$\begin{cases} 2x - 3y + z = -2 \\ 3y + 2z = 3 \end{cases}$ $\qquad 2\,\text{Eq. 1} + \text{Eq. 2}$

$\begin{cases} 2x \qquad + 3z = 1 \\ 3y + 2z = 3 \end{cases}$ $\qquad \text{Eq. 2} + \text{Eq. 1}$

Let $z = a$, then:

$y = -\frac{2}{3}a + 1$

$x = -\frac{3}{2}a + \frac{1}{2}$

Answer: $\left(-\frac{3}{2}a + \frac{1}{2}, -\frac{2}{3}a + 1, a\right)$

35. $\begin{cases} x - 3y + 2z = 18 \\ 5x - 13y + 12z = 80 \end{cases}$ $\qquad$ Equation 1 $\qquad$ Equation 2

$\begin{cases} x - 3y + 2z = 18 \\ 2y + 2z = -10 \end{cases}$ $\qquad -5\,\text{Eq. 1} + \text{Eq. 2}$

$\begin{cases} x - 3y + 2z = 18 \\ y + z = -5 \end{cases}$ $\qquad \frac{1}{2}\,\text{Eq. 2}$

$\begin{cases} x \qquad + 5z = 3 \\ y + z = -5 \end{cases}$ $\qquad 3\,\text{Eq. 2} + \text{Eq. 1}$

Let $z = a$, then $y = -a - 5$, and $x = -5a + 3$.

Answer: $(-5a + 3, -a - 5, a)$

37. $\begin{cases} x \qquad\qquad + 3w = 4 \\ 2y - z - w = 0 \\ 3y \qquad - 2w = 1 \\ 2x - y + 4z \qquad = 5 \end{cases}$

$\begin{cases} x \qquad\qquad + 3w = 4 \\ 2y - z - w = 0 \\ 3y \qquad - 2w = 1 \\ - y + 4z - 6w = -3 \end{cases}$ $\qquad -2\,\text{Eq. 1} + \text{Eq. 4}$

$\begin{cases} x \qquad\qquad + 3w = 4 \\ y - 4z + 6w = 3 \\ 2y - z - w = 0 \\ 3y \qquad - 2w = 1 \end{cases}$ $\qquad \begin{array}{l} -\text{Eq. 4 and} \\ \text{interchange} \\ \text{the equations} \end{array}$

$\begin{cases} x \qquad\qquad + 3w = 4 \\ y - 4z + 6w = 3 \\ 7z - 13w = -6 \\ 12z - 20w = -8 \end{cases}$ $\qquad \begin{array}{l} -2\,\text{Eq. 2} + \text{Eq. 3} \\ \\ -3\,\text{Eq. 2} + \text{Eq. 4} \end{array}$

$\begin{cases} x \qquad\qquad + 3w = 4 \\ y - 4z + 6w = 3 \\ z - 3w = -2 \\ 12z - 20w = -8 \end{cases}$ $\qquad -\frac{1}{2}\,\text{Eq. 4} + \text{Eq. 3}$

$\begin{cases} x \qquad\qquad + 3w = 4 \\ y - 4z + 6w = 3 \\ z - 3w = -2 \\ 16w = 16 \end{cases}$ $\qquad -12\,\text{Eq. 3} + \text{Eq. 4}$

$16w = 16 \implies w = 1$

$z - 3(1) = -2 \implies z = 1$

$y - 4(1) + 6(1) = 3 \implies y = 1$

$x + 3(1) = 4 \implies x = 1$

Answer: $(1, 1, 1, 1)$

39. There are an infinite number of linear systems that has $(4, -1, 2)$ as their solution. One such system is as follows:

$$3(4) + (-1) - (2) = 9 \implies 3x + y - z = 9$$
$$(4) + 2(-1) - (2) = 0 \implies x + 2y - z = 0$$
$$-(4) + (-1) + 3(2) = 1 \implies -x + y + 3z = 1$$

41. There are infinite numbers of linear systems that have $\left(3, -\frac{1}{2}, \frac{7}{4}\right)$ as their solution. One such system is as follows:

$$1(3) + 2\left(-\frac{1}{2}\right) + 4\left(\frac{7}{4}\right) = 9 \implies x + 2y + 4z = 9$$
$$4\left(-\frac{1}{2}\right) + 8\left(\frac{7}{4}\right) = 12 \implies 4y + 8z = 12$$
$$4\left(\frac{7}{4}\right) = 7 \implies 4z = 7$$

43. Plane: $2x + 3y + 4z = 12$

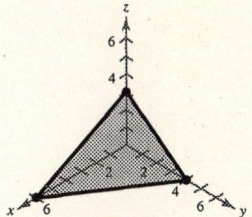

Four points are:

$$(6, 0, 0), \ (0, 4, 0), \ (0, 0, 3), \ (4, 0, 1)$$

45. Plane: $2x + y + z = 4$

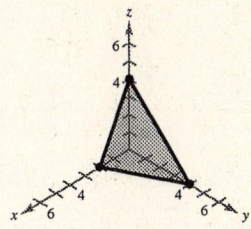

Four points are:

$$(2, 0, 0), \ (0, 4, 0), \ (0, 0, 4), \ (0, 2, 2)$$

47. $\dfrac{7}{x^2 - 14x} = \dfrac{7}{x(x - 14)} = \dfrac{A}{x} + \dfrac{B}{x - 14}$

49. $\dfrac{12}{x^3 - 10x^2} = \dfrac{12}{x^2(x - 10)} = \dfrac{A}{x} + \dfrac{B}{x^2} + \dfrac{C}{x - 10}$

51. $\dfrac{4x^2 + 3}{(x - 5)^3} = \dfrac{A}{(x - 5)} + \dfrac{B}{(x - 5)^2} + \dfrac{C}{(x - 5)^3}$

53. (a) $\dfrac{1}{x^2 - 1} = \dfrac{A}{x + 1} + \dfrac{B}{x - 1}$

$$1 = A(x - 1) + B(x + 1) = (A + B)x + (B - A)$$

$$\begin{cases} A + B = 0 \\ -A + B = 1 \end{cases}$$

$$2B = \implies B = \tfrac{1}{2} \implies A = -\tfrac{1}{2}$$

$$\dfrac{1}{x^2 - 1} = \dfrac{-\dfrac{1}{2}}{x + 1} + \dfrac{\dfrac{1}{2}}{x - 1} = \dfrac{1}{2}\left[\dfrac{1}{x - 1} - \dfrac{1}{x + 1}\right]$$

(c)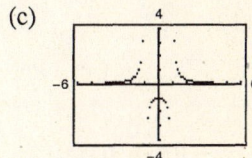

(b) Checking, $\dfrac{1}{2}\left[\dfrac{1}{x - 1} - \dfrac{1}{x + 1}\right] = \dfrac{1}{2}\left[\dfrac{x + 1 - (x - 1)}{x^2 - 1}\right] = \dfrac{1}{x^2 - 1}$

55. (a) $\dfrac{1}{x^2 + x} = \dfrac{1}{x(x + 1)} = \dfrac{A}{x} + \dfrac{B}{x + 1}$

$$1 = A(x + 1) + Bx = (A + B)x + A$$

$$\begin{cases} A + B = 0 \\ A = 1 \implies B = -1 \end{cases}$$

$$\dfrac{1}{x^2 + x} = \dfrac{1}{x} + \dfrac{-1}{x + 1} = \dfrac{1}{x} - \dfrac{1}{x + 1}$$

(c)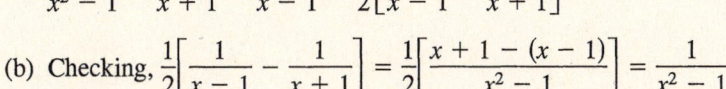

(b) Checking, $\dfrac{1}{x} - \dfrac{1}{x + 1} = \dfrac{(x + 1) - x}{x(x + 1)} = \dfrac{1}{x^2 + x}$

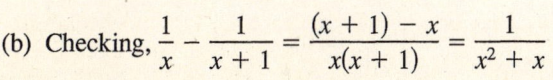

57. (a) $\dfrac{1}{2x^2 + x} = \dfrac{1}{x(2x + 1)} = \dfrac{A}{2x + 1} + \dfrac{B}{x}$

(c)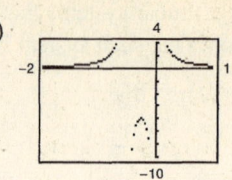

$1 = Ax + B(2x + 1) = (A + 2B)x + B$

$\begin{cases} A + 2B = 0 \\ B = 1 \end{cases} \Rightarrow A = -2$

$\dfrac{1}{2x^2 + x} = \dfrac{-2}{2x + 1} + \dfrac{1}{x} = \dfrac{1}{x} - \dfrac{2}{2x + 1}$

(b) Checking, $\dfrac{1}{x} - \dfrac{2}{2x + 1} = \dfrac{2x + 1 - 2x}{x(2x + 1)} = \dfrac{1}{2x^2 + x}$

59. (a) $\dfrac{5 - x}{2x^2 + x - 1} = \dfrac{5 - x}{(2x - 1)(x + 1)} = \dfrac{A}{2x - 1} + \dfrac{B}{x + 1}$

(c)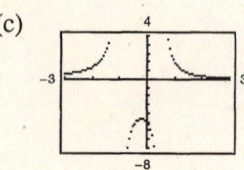

$5 - x = A(x + 1) + B(2x - 1) = (A + 2B)x + (A - B)$

$\begin{cases} A + 2B = -1 \Rightarrow A = -1 - 2B \\ A - B = 5 \end{cases}$

$(-1 - 2B) - B = 5 \Rightarrow B = -2 \quad \text{and} \quad A = 3$

$\dfrac{5 - x}{2x^2 + x - 1} = \dfrac{3}{2x - 1} + \dfrac{-2}{x + 1}$

(b) Checking, $\dfrac{3}{2x - 1} + \dfrac{-2}{x + 1} = \dfrac{3(x + 1) - 2(2x - 1)}{(2x - 1)(x + 1)} = \dfrac{-x + 5}{2x^2 + x - 1}$

61. (a) $\dfrac{x^2 + 12x + 12}{x^3 - 4x} = \dfrac{x^2 + 12x + 12}{x(x - 2)(x + 2)} = \dfrac{A}{x} + \dfrac{B}{x + 2} + \dfrac{C}{x - 2}$

(c)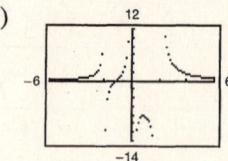

$x^2 + 12x + 12 = A(x + 2)(x - 2) + Bx(x - 2) + Cx(x + 2)$

$\qquad\qquad\qquad = (A + B + C)x^2 + (-2B + 2C)x + (-4A)$

$\begin{cases} A + B + C = 1 \\ -2B + 2C = 12 \\ -4A = 12 \end{cases} \Rightarrow A = -3$

$\begin{cases} B + C = 4 \\ -B + C = 6 \end{cases}$

$2C = 10 \Rightarrow C = 5 \Rightarrow B = -1$

$\dfrac{x^2 + 12x + 12}{x^3 - 4x} = \dfrac{-3}{x} + \dfrac{-1}{x + 2} + \dfrac{5}{x - 2}$

(b) Checking, $\dfrac{-3}{x} + \dfrac{-1}{x + 2} + \dfrac{5}{x - 2} = \dfrac{-3(x^2 - 4) - 1(x)(x - 2) + 5x(x + 2)}{x(x^2 - 4)}$

$\qquad\qquad\qquad = \dfrac{x^2 + 12x + 12}{x^3 - 4x}$

63. (a) $\dfrac{4x^2 + 2x - 1}{x^2(x + 1)} = \dfrac{A}{x} + \dfrac{B}{x^2} + \dfrac{C}{x + 1}$

$4x^2 + 2x - 1 = Ax(x + 1) + B(x + 1) + Cx^2$

$\qquad\qquad\qquad = (A + C)x^2 + (A + B)x + B$

$\begin{cases} A \qquad\; + C = \;\;\;4 \\ A + B \qquad\; = \;\;\;2 \\ \qquad B \qquad\;\; = -1 \end{cases}$

$B = -1 \implies A = 3 \implies C = 1$

$\dfrac{4x^2 + 2x - 1}{x^2(x + 1)} = \dfrac{3}{x} + \dfrac{-1}{x^2} + \dfrac{1}{x + 1}$

(b) Checking, $\dfrac{3}{x} + \dfrac{-1}{x^2} + \dfrac{1}{x + 1} = \dfrac{3x(x + 1) - (x + 1) + x^2}{x^2(x + 1)} = \dfrac{4x^2 + 2x - 1}{x^2(x + 1)}$

(c)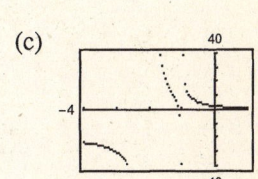

65. (a) $\dfrac{27 - 7x}{x(x - 3)^2} = \dfrac{A}{x} + \dfrac{B}{x - 3} + \dfrac{C}{(x - 3)^2}$

$27 - 7x = A(x - 3)^2 + Bx(x - 3) + Cx$

$\qquad\qquad = (A + B)x^2 + (-6A - 3B + C)x + 9a$

$\begin{cases} \;\;\;A + \;\;B \qquad\quad = \;\;\;0 \\ -6A - 3B + C = -7 \\ \;\;\;9A \qquad\qquad\;\; = \;\;27 \end{cases}$

$A = 3 \implies B = -3 \implies C = -7 + 18 - 9 = 2$

$\dfrac{27 - 7x}{x(x - 3)^2} = \dfrac{3}{x} + \dfrac{-3}{x - 3} + \dfrac{2}{(x - 3)^2}$

(b) Checking, $\dfrac{3}{x} + \dfrac{-3}{x - 3} + \dfrac{2}{(x - 3)^2} = \dfrac{3(x - 3)^2 - 3(x)(x - 3) + 2x}{x(x - 3)^2} = \dfrac{-7x + 27}{x(x - 3)^2}$

(c)

67. (a) $\dfrac{2x^3 - x^2 + x + 5}{x^2 + 3x + 2} = 2x - 7 + \dfrac{18x + 19}{(x + 1)(x + 2)}$

$\dfrac{18x + 19}{(x + 1)(x + 2)} = \dfrac{A}{x + 1} + \dfrac{B}{x + 2}$

$18x + 19 = A(x + 2) + B(x + 1) = (A + B)x + (2A + B)$

$\begin{cases} \;A + B = 18 \\ 2A + B = 19 \end{cases}$

$A = 1 \implies B = 17$

$\dfrac{2x^3 - x^2 + x + 5}{x^2 + 3x + 2} = 2x - 7 + \dfrac{1}{x + 1} + \dfrac{17}{x + 2}$

(c)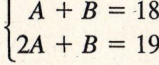

(b) Checking,

$2x - 7 + \dfrac{1}{x + 1} + \dfrac{17}{x + 2} = \dfrac{(2x - 7)(x^2 + 3x + 2) + (x + 2) + 17(x + 1)}{x^2 + 3x + 2} = \dfrac{2x^3 - x^2 + x + 5}{x^2 + 3x + 2}$

69. (a) $\dfrac{x^4}{(x-1)^3} = x + 3 + \dfrac{6x^2 - 8x + 3}{(x-1)^3}$

$\dfrac{6x^2 - 8x + 3}{(x-1)^3} = \dfrac{A}{x-1} + \dfrac{B}{(x-1)^2} + \dfrac{C}{(x-1)^3}$

$6x^2 - 8x + 3 = A(x-1)^2 + B(x-1) + C$

$\qquad\qquad = Ax^2 + (-2A + B)x + (A - B + C)$

$\begin{cases} A & = & 6 \\ -2A + B & = & -8 \\ A - B + C & = & 3 \end{cases}$

$A = 6 \implies B = -8 + 2(6) = 4 \implies C = 3 - 6 + 4 = 1$

$\dfrac{x^4}{(x-1)^3} = \dfrac{6}{x-1} + \dfrac{4}{(x-1)^2} + \dfrac{1}{(x-1)^3} + x + 3$

(b) Checking,

$\dfrac{6}{x-1} + \dfrac{4}{(x-1)^2} + \dfrac{1}{(x-1)^3} + x + 3 = \dfrac{6(x-1)^2 + 4(x-1) + 1 + (x-1)^3(x+3)}{(x-1)^3}$

$\qquad\qquad = \dfrac{6x^2 - 12x + 6 + 4x - 3 + (x^4 - 6x^2 + 8x - 3)}{(x-1)^3}$

$\qquad\qquad = \dfrac{x^4}{(x-1)^3}$

(c)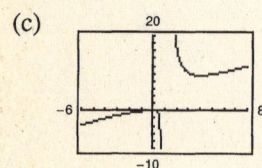

71. $\dfrac{x - 12}{x(x-4)} = \dfrac{A}{x} + \dfrac{B}{x-4}$

$x - 12 = A(x-4) + Bx$

$\begin{cases} A + B & = & 1 \\ -4A & = & -12 \end{cases} \implies A = 3, B = -2$

$\dfrac{x - 12}{x(x-4)} = \dfrac{3}{x} - \dfrac{2}{x-4}$

$y = \dfrac{x - 12}{x(x-4)}$ 　　　 $y = \dfrac{3}{x}, y = -\dfrac{2}{x-4}$

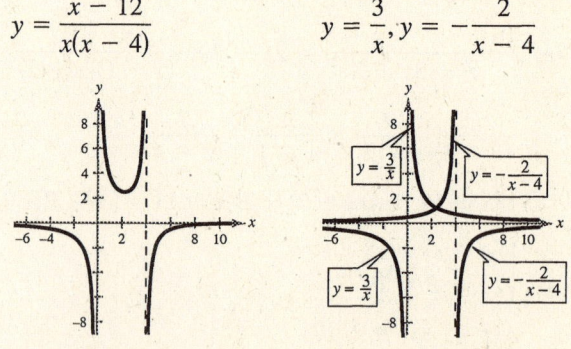

Vertical asymptotes: 　　 Vertical asymptotes:
$x = 0$ and $x = 4$ 　　　 $x = 0$ and $x = 4$

The combination of the vertical asymptotes of the terms of the decompositions are the same as the vertical asymptotes of the rational function.

73. $s = \frac{1}{2}at^2 + v_0t + s_0$

$(1, 128), (2, 80), (3, 0)$

$$\begin{cases} 128 = \frac{1}{2}a + v_0 + s_0 \implies a + 2v_0 + 2s_0 = 256 \\ 80 = 2a + 2v_0 + s_0 \implies 2a + 2v_0 + s_0 = 80 \\ 0 = \frac{9}{2}a + 3v_0 + s_0 \implies 9a + 6v_0 + 2s_0 = 0 \end{cases}$$

Solving this system yields $a = -32, v_0 = 0, s_0 = 144$.

Thus, $s = \frac{1}{2}(-32)t^2 + (0)t + 144$

$ = -16t^2 + 144.$

75. $s = \frac{1}{2}at^2 + v_0t + a_0$

$(1, 452), (2, 372), (3, 260)$

$$\begin{cases} 452 = \frac{1}{2}a + v_0 + s_0 \implies a + 2v_0 + 2s_0 = 904 \\ 372 = 2a + 2v_0 + s_0 \implies 2a + 2v_0 + s_0 = 372 \\ 260 = \frac{9}{2}a + 3v_0 + s_0 \implies 9a + 6v_0 + 2s_0 = 520 \end{cases}$$

Solving this system yields $a = -32, v_0 = -32, s_0 = 500$

Thus, $s = \frac{1}{2}(-32)t^2 - 32t + 500$

$ = -16t^2 - 32t + 500$

77. $y = ax^2 + bx + c$ passing through $(0, 0), (2, -2), (4, 0)$

$(0, 0)$: $0 = c$

$(2, -2)$: $-2 = 4a + 2b + c \implies -1 = 2a + b$

$(4, 0)$: $0 = 16a + 4b + c \implies 0 = 4a + b$

Answer: $a = \frac{1}{2}, b = -2, c = 0$

The equation of the parabola is $y = \frac{1}{2}x^2 - 2x.$

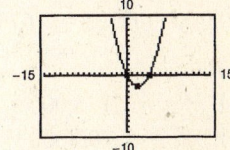

79. $y = ax^2 + bx + c$ passing through $(2, 0), (3, -1), (4, 0)$

$(2, 0)$: $0 = 4a + 2b + c$

$(3, -1)$: $-1 = 9a + 3b + c \implies -1 = 5a + b$

$(4, 0)$: $0 = 16a + 4b + c \implies 0 = 12a + 2b$

Answer: $a = 1, b = -6, c = 8$

The equation of the parabola is $y = x^2 - 6x + 8.$

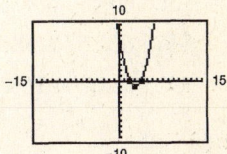

81. $x^2 + y^2 + Dx + Ey + F = 0$ passing through $(0, 0), (2, 2), (4, 0)$

$(0, 0)$: $F = 0$

$(2, 2)$: $8 + 2D + 2E + F = 0 \implies D + E = -4$

$(4, 0)$: $16 + 4D + F = 0 \implies D = -4$ and $E = 0$

The equation of the circle is $x^2 + y^2 - 4x = 0.$

To graph, let $y_1 = \sqrt{4x - x^2}$ and $y_2 = -\sqrt{4x - x^2}.$

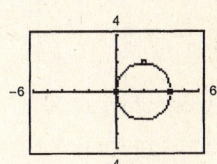

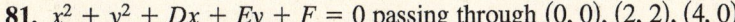

83. $x^2 + y^2 + Dx + Ey + F = 0$ passing through $(-3, -1), (2, 4), (-6, 8)$

$(-3, -1)$: $\quad 10 - 3D - E + F = 0 \implies \quad 10 = \quad 3D + \quad E - F$

$(\quad 2, \quad 4)$: $\quad 20 + 2D + 4E + F = 0 \implies \quad 20 = -2D - 4E - F$

$(-6, \quad 8)$: $100 - 6D + 8E + F = 0 \implies 100 = \quad 6D - 8E - F$

Answer: $D = 6, E = -8, F = 0$

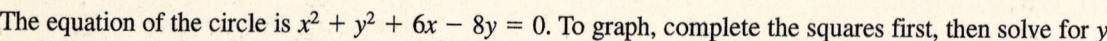

The equation of the circle is $x^2 + y^2 + 6x - 8y = 0$. To graph, complete the squares first, then solve for y.

$(x^2 + 6x + 9) + (y^2 - 8y + 16) = 0 + 9 + 16$

$$(x + 3)^2 + (y - 4)^2 = 25$$

$$(y - 4)^2 = 25 - (x + 3)^2$$

$$y - 4 = \pm\sqrt{25 - (x + 3)^2}$$

$$y = 4 \pm \sqrt{25 - (x + 3)^2}$$

Let $y_1 = 4 + \sqrt{25 - (x + 3)^2}$ and $y_2 = 4 - \sqrt{25 - (x + 3)^2}$.

85. Let $x =$ amount at 8%.

Let $y =$ amount at 9%.

Let $z =$ amount at 10%.

$$\begin{cases} x + \quad y + \quad z = 775{,}000 \\ 0.08x + 0.09y + 0.10z = \quad 67{,}000 \\ x \qquad\qquad = \qquad 4z \end{cases}$$

$$\begin{cases} x + y + \quad z = \quad 775{,}000 \\ 8x + 9y + 10z = 6{,}700{,}000 \\ x \qquad - 4z = \qquad\qquad 0 \end{cases}$$

Solving this system, you obtain

$x \approx 366{,}666.67, y \approx 316{,}666.67, z \approx 91{,}666.67$

Answer: $x =$ \$366,666.67 at 8%

$\qquad\quad y =$ \$316,666.67 at 9%

$\qquad\quad z =$ \$91,666.67 at 10%

87. Let $C =$ amount in certificates of deposit

Let $M =$ amount in municipal bonds

Let $B =$ amount in blue chip stocks

Let $G =$ amount in growth or speculative stocks

$$\begin{cases} C + M + B + G = 500{,}000 \\ 0.08C + 0.09M + 0.12B + 0.15G = 0.10(500{,}000) \\ M = \frac{1}{4}(500{,}000) \end{cases}$$

Solving this system, you obtain $M = 125{,}000$ and

$C = 156{,}250 + 0.75s$

$M = 125{,}000$

$B = 218750 - 1.75s$

$G = s$

89. Let $x =$ gallons of spray X.

Let $y =$ gallons of spray Y.

Let $z =$ gallons of spray Z.

Chemical A: $\frac{1}{5}x + \frac{1}{2}z = 12$

Chemical B: $\frac{2}{5}x + \frac{1}{2}z = 16 \implies x = 20, z = 16$

Chemical C: $\frac{2}{5}x + \quad y = 26 \implies y = 18$

Answer: 20 liters of spray X

$\qquad\quad$ 18 liters of spray Y

$\qquad\quad$ 16 liters of spray Z

91.

	Product	
Truck	A	B
Large	6	3
Medium	4	4
Small	0	3

Possible solutions:

(1) 4 medium trucks

(2) 2 large trucks, 1 medium truck, 2 small trucks

(3) 3 large trucks, 1 medium truck, 1 small truck

(4) 3 large trucks, 3 small trucks

93.
$$\begin{cases} I_1 - I_2 + I_3 = 0 \\ 3I_1 + 2I_2 \quad\;\; = 7 \\ \quad\;\; 2I_2 + 4I_3 = 8 \end{cases}$$
Equation 1
Equation 2
Equation 3

$$\begin{cases} I_1 - I_2 + I_3 = 0 \\ \quad 5I_2 - 3I_3 = 7 \\ \quad 2I_2 + 4I_3 = 8 \end{cases}$$ -3 Eq. 1 + Eq. 2

$$\begin{cases} I_1 - I_2 + I_3 = 0 \\ \quad 10I_2 - 6I_3 = 14 \\ \quad 10I_2 + 20I_3 = 40 \end{cases}$$ 2 Eq. 2
5 Eq. 3

$$\begin{cases} I_1 - I_2 + I_3 = 0 \\ \quad 10I_2 - 6I_3 = 14 \\ \qquad\qquad 26I_3 = 26 \end{cases}$$ $-$Eq. 2 + Eq. 3

$26I_3 = 26 \implies I_3 = 1$

$10I_2 - 6(1) = 14 \implies I_2 = 2$

$I_1 - 2 + 1 = 0 \implies I_1 = 1$

Answer: $I_1 = 1$ ampere, $I_2 = 2$ amperes,
$\qquad\qquad I_3 = 1$ ampere

95. Least squares regression parabola through
$(-4, 5), (-2, 6), (2, 6), (4, 2)$

$$\begin{cases} 4c \quad\;\; + 40a = \;\; 19 \\ \quad\; 40b \qquad\quad = -12 \\ 40c \quad\; + 544a = \; 160 \end{cases}$$

Solving this system yields

$a = -\frac{5}{24}, b = -\frac{3}{10},$ and $c = \frac{41}{6}.$

Thus, $y = -\frac{5}{24}x^2 - \frac{3}{10}x + \frac{41}{6}.$

97. Least squares regression parabola through
$(0, 0), (2, 2), (3, 6), (4, 12)$

$$\begin{cases} 4c + \;\; 9b + \;\; 29a = \;\; 20 \\ 9c + 29b + \;\; 99a = \;\; 70 \\ 29c + 99b + 353a = 254 \end{cases}$$

Solving this system yields

$a = 1, b = -1,$ and $c = 0.$ Thus, $y = x^2 - x.$

99. (a) $\begin{cases} a(30)^2 + b(30) + c = \;\; 55 \\ a(40)^2 + b(40) + c = 105 \\ a(50)^2 + b(50) + c = 188 \end{cases}$

Solving the system, $a = 0.165, b = -6.55$ and
$c = 103$

$y = 0.165x^2 - 6.55x + 103$

(b)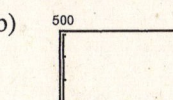

(c) For $x = 70, y = 453$ feet.

101. (a) $\dfrac{2000(4 - 3x)}{(11 - 7x)(7 - 4x)} = \dfrac{A}{11 - 7x} + \dfrac{B}{7 - 4x}$, $0 \le x \le 1$

$2000(4 - 3x) = A(7 - 4x) + B(11 - 7x)$

$\begin{cases} -6000 = -4A - 7B \\ 8000 = 7A + 11B \end{cases} \Rightarrow \begin{aligned} A &= -2000 \\ B &= 2000 \end{aligned}$

$\dfrac{2000(4 - 3x)}{(11 - 7x)(7 - 4x)} = \dfrac{-2000}{11 - 7x} + \dfrac{2000}{7 - 4x}$

$= \dfrac{2000}{7 - 4x} - \dfrac{2000}{11 - 7x}$

(b) $y_1 = \dfrac{2000}{7 - 4x}$

$y_2 = \dfrac{2000}{11 - 7x}$

103. False. The coefficient of y in the second equation is not 1.

105. False. The correct form is

$$\frac{A}{x + 10} + \frac{B}{x - 10} + \frac{C}{(x - 10)^2}.$$

107. $\dfrac{1}{a^2 - x^2} = \dfrac{1}{(a + x)(a - x)} = \dfrac{A}{a + x} + \dfrac{B}{a - x}$

$1 = A(a - x) + B(a + x) = (-A + B)x + (Aa + Ba)$

$\begin{cases} -A + B = 0 \\ Aa + Ba = 1 \end{cases} \Rightarrow A = \dfrac{1}{2a}, B = \dfrac{1}{2a}$

$\dfrac{1}{a^2 - x^2} = \dfrac{\frac{1}{2a}}{a + x} + \dfrac{\frac{1}{2a}}{a - x} = \dfrac{1}{2a}\left[\dfrac{1}{a + x} + \dfrac{1}{a - x}\right]$

109. $\dfrac{1}{y(a - y)} = \dfrac{A}{y} + \dfrac{B}{a - y}$

$1 = A(a - y) + By = (-A + B)y + aA$

$A = \dfrac{1}{a}, B = \dfrac{1}{a}$

$\dfrac{1}{y(a - y)} = \dfrac{1}{a}\left(\dfrac{1}{y} + \dfrac{1}{a - y}\right)$

111. No, they are not equivalent. In the second system, the constant in the second equation should be -11 and the coefficient of z in the third equation should be 2.

113. $\begin{cases} y + \quad \lambda = \\ x + \quad \lambda = \\ x + y - 10 = \end{cases} \Rightarrow \begin{aligned} x = y = -\lambda \\ \Rightarrow 2x - 10 = 0 \end{aligned}$

$x = 5$

$y = 5$

$\lambda = -5$

115. $\begin{cases} 2x - 2x\lambda = 0 \Rightarrow x = x\lambda \\ -2y + \lambda = 0 \Rightarrow 2y = \lambda \\ y - x^2 = 0 \Rightarrow y = x^2 \end{cases}$

From the first equation, $x = 0$ or $\lambda = 1$.

If $x = 0$, then $y = 0^2 = 0$ and $\lambda = 0$.

If $x \ne 0$, then $\lambda = 1 \Rightarrow y = \frac{1}{2}$ and $x = \pm\sqrt{\frac{1}{2}}$.

Thus, the solutions are:

(1) $x = y = \lambda = 0$

(2) $x = \dfrac{\sqrt{2}}{2}$, $y = \dfrac{1}{2}$, $\lambda = 1$

(3) $x = -\dfrac{\sqrt{2}}{2}$, $y = \dfrac{1}{2}$, $\lambda = 1$

117. $y = -3x + 7$

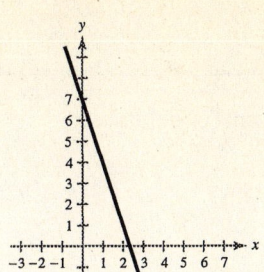

119. $y = -2x^2$

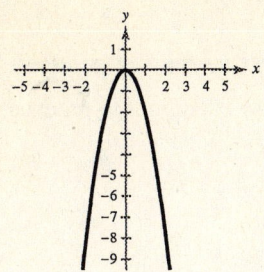

121. $y = -x^2(x - 3)$

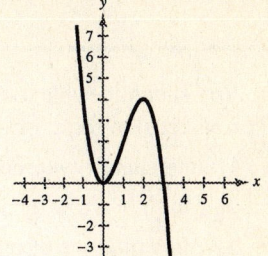

123. $f(x) = x^3 + x^2 - 12x = x(x^2 + x - 12) = x(x + 4)(x - 3) \Rightarrow x = 0, -4, 3$

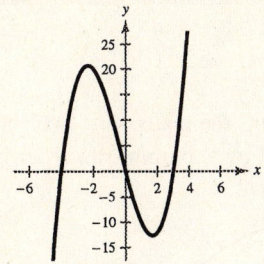

125. $f(x) = 2x^3 + 5x^2 - 21x - 36 = (2x + 3)(x + 4)(x - 3) \Rightarrow x = -\frac{3}{2}, -4, 3$

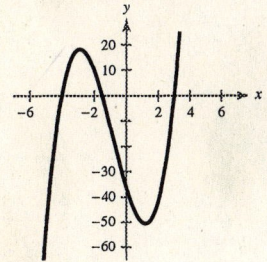

127. $y = 4^{-x-4} - 5$

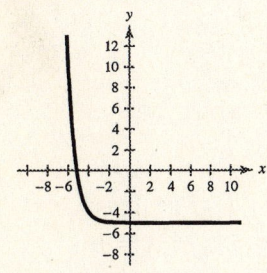

x	-7	-6	-5	-4	-3	-1	0
y	59	11	-1	-4	-4.75	-4.98	-4.996

129. $y = 2.9^{0.8x} - 3$

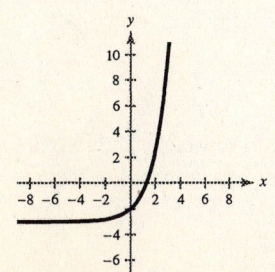

x	-3	-2	-1	0	1	2	3
y	-2.9	-2.8	-2.6	-2	$-.66$	2.5	9.9

Section 7.4   Matrices and Systems of Equations

■ You should be able to use elementary row operations to produce a row-echelon form (or reduced row-echelon form) of a matrix.

1. Interchange two rows

2. Multiply a row by a nonzero constant

3. Add a multiple of one row to another row

■ You should be able to use either Gaussian elimination with back-substitution or Gauss-Jordan elimination to solve a system of linear equations.

Solutions to Odd-Numbered Exercises

1. Since the matrix has one row and two columns, its order is 1×2.

3. Since the matrix has three rows and one column, its order is 3×1.

5. Since the matrix has two rows and two columns, its order is 2×2.

7. $\begin{cases} 4x - 5y = 33 \\ -x + 5y = -27 \end{cases}$

$\begin{bmatrix} 4 & -5 & \vdots & 33 \\ -1 & 5 & \vdots & -27 \end{bmatrix}$

9. $\begin{cases} x + 10y - 2z = 2 \\ 5x - 3y + 4z = 0 \\ 2x + y = 6 \end{cases}$

$\begin{bmatrix} 1 & 10 & -2 & \vdots & 2 \\ 5 & -3 & 4 & \vdots & 0 \\ 2 & 1 & 0 & \vdots & 6 \end{bmatrix}$

11. $\begin{bmatrix} 1 & 2 & \vdots & 7 \\ 2 & -3 & \vdots & 4 \end{bmatrix}$

$\begin{cases} x + 2y = 7 \\ 2x - 3y = 4 \end{cases}$

13. $\begin{bmatrix} 9 & 12 & 3 & \vdots & 0 \\ -2 & 18 & 5 & \vdots & -10 \\ 1 & 7 & -8 & \vdots & -4 \end{bmatrix}$

$\begin{cases} 9x + 12y + 3z = 0 \\ -2x + 18y + 5z = 10 \\ x + 7y - 8z = -4 \end{cases}$

15. $\begin{bmatrix} 1 & 4 & 3 \\ 2 & 10 & 5 \end{bmatrix}$

$-2R_1 + R_2 \rightarrow \begin{bmatrix} 1 & 4 & 3 \\ 0 & \boxed{2} & -1 \end{bmatrix}$

17. $\begin{bmatrix} 1 & 1 & 4 & -1 \\ 3 & 8 & 10 & 3 \\ -2 & 1 & 12 & 6 \end{bmatrix} \begin{matrix} \\ -3R_1 + R_2 \rightarrow \\ 2R_1 + R_3 \rightarrow \end{matrix} \begin{bmatrix} 1 & 1 & 4 & -1 \\ 0 & 5 & \boxed{-2} & \boxed{6} \\ 0 & 3 & \boxed{20} & \boxed{4} \end{bmatrix} \begin{matrix} \\ \frac{1}{5}R_2 \rightarrow \\ \\ \end{matrix} \begin{bmatrix} 1 & 1 & 4 & -1 \\ 0 & 1 & -\frac{2}{5} & \frac{6}{5} \\ 0 & 3 & 20 & 4 \end{bmatrix}$

19. Add 5 times Row 2 to Row 1.

21. Interchange Rows 1 and 2.

23. $\begin{bmatrix} 1 & 0 & 0 & 0 \\ 0 & 1 & 1 & 5 \\ 0 & 0 & 0 & 0 \end{bmatrix}$

This matrix is in reduced row-echelon form.

25. $\begin{bmatrix} 2 & 0 & 4 & 0 \\ 0 & -1 & 3 & 6 \\ 0 & 0 & 1 & 5 \end{bmatrix}$

The first nonzero entries in rows one and two are not one. The matrix is not in row-echelon form.

27.
$$\begin{bmatrix} 1 & 2 & 3 \\ 2 & -1 & -4 \\ 3 & 1 & -1 \end{bmatrix}$$

(a) $\begin{bmatrix} 1 & 2 & 3 \\ 0 & -5 & -10 \\ 3 & 1 & -1 \end{bmatrix}$ (b) $\begin{bmatrix} 1 & 2 & 3 \\ 0 & -5 & -10 \\ 0 & -5 & -10 \end{bmatrix}$

(c) $\begin{bmatrix} 1 & 2 & 3 \\ 0 & -5 & -10 \\ 0 & 0 & 0 \end{bmatrix}$ (d) $\begin{bmatrix} 1 & 2 & 3 \\ 0 & 1 & 2 \\ 0 & 0 & 0 \end{bmatrix}$

(e) $\begin{bmatrix} 1 & 0 & -1 \\ 0 & 1 & 2 \\ 0 & 0 & 0 \end{bmatrix}$ This matrix is in reduced row-echelon form.

29. (See Exercise 27.) (Answer is series of screens.)

(a)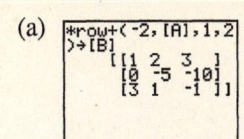
(b)
(c)
(d)
(e)

31.
$$\begin{bmatrix} 1 & 1 & 0 & 5 \\ -2 & -1 & 2 & -10 \\ 3 & 6 & 7 & 14 \end{bmatrix}$$

$\begin{matrix} 2R_1 + R_2 \rightarrow \\ -3R_1 + R_3 \rightarrow \end{matrix} \begin{bmatrix} 1 & 1 & 0 & 5 \\ 0 & 1 & 2 & 0 \\ 0 & 3 & 7 & -1 \end{bmatrix}$

$\begin{matrix} \\ \\ -3R_2 + R_3 \rightarrow \end{matrix} \begin{bmatrix} 1 & 1 & 0 & 5 \\ 0 & 1 & 2 & 0 \\ 0 & 0 & 1 & -1 \end{bmatrix}$

33.
$$\begin{bmatrix} 1 & -1 & -1 & 1 \\ 5 & -4 & 1 & 8 \\ -6 & 8 & 18 & 0 \end{bmatrix}$$

$\begin{matrix} -5R_1 + R_2 \rightarrow \\ 6R_1 + R_3 \rightarrow \end{matrix} \begin{bmatrix} 1 & -1 & -1 & 1 \\ 0 & 1 & 6 & 3 \\ 0 & 2 & 12 & 6 \end{bmatrix}$

$\begin{matrix} \\ \\ -2R_2 + R_3 \rightarrow \end{matrix} \begin{bmatrix} 1 & -1 & -1 & 1 \\ 0 & 1 & 6 & 3 \\ 0 & 0 & 0 & 0 \end{bmatrix}$

35.
$$\begin{bmatrix} 3 & 3 & 3 \\ -1 & 0 & -4 \\ 2 & 4 & -2 \end{bmatrix}$$

$\frac{1}{3}R_1 \rightarrow \begin{bmatrix} 1 & 1 & 1 \\ -1 & 0 & -4 \\ 2 & 4 & -2 \end{bmatrix}$

$\begin{matrix} R_1 + R_2 \rightarrow \\ -2R_1 + R_3 \rightarrow \end{matrix} \begin{bmatrix} 1 & 1 & 1 \\ 0 & 1 & -3 \\ 0 & 2 & -4 \end{bmatrix}$

$\begin{matrix} -R_2 + R_1 \rightarrow \\ \\ -2R_2 + R_3 \rightarrow \end{matrix} \begin{bmatrix} 1 & 0 & 4 \\ 0 & 1 & -3 \\ 0 & 0 & 2 \end{bmatrix}$

$\frac{1}{2}R_3 \rightarrow \begin{bmatrix} 1 & 0 & 4 \\ 0 & 1 & -3 \\ 0 & 0 & 1 \end{bmatrix}$

$\begin{matrix} -4R_3 + R_1 \rightarrow \\ 3R_3 + R_2 \rightarrow \end{matrix} \begin{bmatrix} 1 & 0 & 0 \\ 0 & 1 & 0 \\ 0 & 0 & 1 \end{bmatrix}$

37.
$$\begin{bmatrix} -3 & 5 & 1 & 12 \\ 1 & -1 & 1 & 4 \end{bmatrix}$$

$\begin{matrix} R_1 \rightarrow \\ R_2 \rightarrow \end{matrix} \begin{bmatrix} 1 & -1 & 1 & 4 \\ -3 & 5 & 1 & 12 \end{bmatrix}$

$3R_1 + R_2 \rightarrow \begin{bmatrix} 1 & -1 & 1 & 4 \\ 0 & 2 & 4 & 24 \end{bmatrix}$

$\frac{1}{2}R_2 \begin{bmatrix} 1 & -1 & 1 & 4 \\ 0 & 1 & 2 & 12 \end{bmatrix}$

$R_2 + R_1 \rightarrow \begin{bmatrix} 1 & 0 & 3 & 16 \\ 0 & 1 & 2 & 12 \end{bmatrix}$

39. $\begin{cases} x - 2y = \ \ 4 \\ \ \ \ \ \ \ \ \ y = -3 \end{cases}$

$x - 2(-3) = \ \ 4$

$x = -2$

Answer: $(-2, -3)$

41. $\begin{cases} x - y + 2z = \ \ 4 \\ \ \ \ \ \ y - \ z = \ \ 2 \\ \ \ \ \ \ \ \ \ \ \ \ \ \ z = -2 \end{cases}$

$y - (-2) = \ \ 2$

$y = \ \ 0$

$x - 0 + 2(-2) = \ \ 4$

$x = \ \ 8$

Answer: $(8, 0, -2)$

43. $\begin{bmatrix} 1 & 0 & \vdots & 7 \\ 0 & 1 & \vdots & -5 \end{bmatrix}$

$x = \ \ 7$

$y = -5$

Answer: $(7, -5)$

45. $\begin{bmatrix} 1 & 0 & 0 & \vdots & -4 \\ 0 & 1 & 0 & \vdots & -8 \\ 0 & 0 & 1 & \vdots & 2 \end{bmatrix}$

$x = -4$

$y = -8$

$z = \ \ 2$

Answer: $(-4, -8, 2)$

47. $\begin{cases} x + 2y = 7 \\ 2x + \ y = 8 \end{cases}$

$\begin{bmatrix} 1 & 2 & \vdots & 7 \\ 2 & 1 & \vdots & 8 \end{bmatrix} -2R_1 + R_2 \rightarrow \begin{bmatrix} 1 & 2 & \vdots & 7 \\ 0 & -3 & \vdots & -6 \end{bmatrix} -\frac{1}{3}R_2 \rightarrow \begin{bmatrix} 1 & 2 & \vdots & 7 \\ 0 & 1 & \vdots & 2 \end{bmatrix}$

$y = 2$

$x + 2(2) = 7 \implies x = 3$

Answer: $(3, 2)$

49. $\begin{cases} -x + \ y = -22 \\ 3x + 4y = \ \ \ 4 \\ 4x - 8y = \ \ 32 \end{cases}$

$\begin{bmatrix} -1 & 1 & \vdots & -22 \\ 3 & 4 & \vdots & 4 \\ 4 & -8 & \vdots & 32 \end{bmatrix}$

$\begin{matrix} \\ 3R_1 + R_2 \\ 4R_1 + R_3 \end{matrix} \begin{bmatrix} -1 & 1 & \vdots & -22 \\ 0 & 7 & \vdots & -62 \\ 0 & -4 & \vdots & -56 \end{bmatrix}$

$\begin{matrix} \\ R_2 \\ R_3 \end{matrix} \begin{bmatrix} -1 & 1 & & -22 \\ 0 & -4 & & -56 \\ 0 & 7 & & -62 \end{bmatrix}$

$\begin{matrix} \\ -\frac{1}{4}R_2 \\ -7R_2 + R_3 \end{matrix} \begin{bmatrix} -1 & 1 & & -22 \\ 0 & 1 & & 14 \\ 0 & 0 & & -160 \end{bmatrix}$

No solution, Inconsistent

51. $\begin{cases} 8x - 4y = 13 \\ 5x + 2y = \ \ 7 \end{cases}$

$\begin{bmatrix} 8 & -4 & 13 \\ 5 & 2 & 7 \end{bmatrix} \begin{matrix} 3R_1 \rightarrow \\ 5R_2 \rightarrow \end{matrix} \begin{bmatrix} 24 & -12 & 39 \\ 25 & 10 & 35 \end{bmatrix}$

$\begin{matrix} -R_2 + R_1 \rightarrow \\ \frac{1}{5}R_2 \rightarrow \end{matrix} \begin{bmatrix} -1 & -22 & 4 \\ 5 & 2 & 7 \end{bmatrix}$

$5R_1 + R_2 \rightarrow \begin{bmatrix} -1 & -22 & 4 \\ 0 & -108 & 27 \end{bmatrix}$

$\begin{matrix} -1R_1 \rightarrow \\ -\frac{1}{108}R_2 \rightarrow \end{matrix} \begin{bmatrix} 1 & 22 & -4 \\ 0 & 1 & -\frac{1}{4} \end{bmatrix}$

$y = -\frac{1}{4}$

$x + 22\left(-\frac{1}{4}\right) = -4 \implies x = \frac{3}{2}$

Answer: $\left(\frac{3}{2}, -\frac{1}{4}\right)$

53. $\begin{cases} -x + 2y = 1.5 \\ 2x - 4y = 3.0 \end{cases}$

$$\begin{bmatrix} -1 & 2 & \vdots & 1.5 \\ 2 & -4 & \vdots & 3.0 \end{bmatrix}$$

$2R_1 + R_2 \rightarrow \begin{bmatrix} -1 & 2 & \vdots & 1.5 \\ 0 & 0 & \vdots & 6.0 \end{bmatrix}$

The system is inconsistent and there is no solution.

55. $\begin{cases} x & - 3z = -2 \\ 3x + y - 2z = 5 \\ 2x + 2y + z = 4 \end{cases}$

$$\begin{bmatrix} 1 & 0 & -3 & \vdots & -2 \\ 3 & 1 & -2 & \vdots & 5 \\ 2 & 2 & 1 & \vdots & 4 \end{bmatrix}$$

$\begin{matrix} \\ -3R_1 + R_2 \rightarrow \\ -2R_1 + R_3 \rightarrow \end{matrix} \begin{bmatrix} 1 & 0 & -3 & \vdots & -2 \\ 0 & 1 & 7 & \vdots & 11 \\ 0 & 2 & 7 & \vdots & 8 \end{bmatrix}$

$\begin{matrix} \\ \\ -2R_2 + R_3 \rightarrow \end{matrix} \begin{bmatrix} 1 & 0 & -3 & \vdots & -2 \\ 0 & 1 & 7 & \vdots & 11 \\ 0 & 0 & -7 & \vdots & -14 \end{bmatrix}$

$\begin{matrix} \\ \\ -\frac{1}{7}R_3 \rightarrow \end{matrix} \begin{bmatrix} 1 & 0 & -3 & \vdots & -2 \\ 0 & 1 & 7 & \vdots & 11 \\ 0 & 0 & 1 & \vdots & 2 \end{bmatrix}$

$$z = 2$$

$$y + 7(2) = 11 \implies y = -3$$

$$x - 3(2) = -2 \implies x = 4$$

Answer: $(4, -3, 2)$

57. $\begin{cases} x + y - 5z = 3 \\ x & - 2z = 1 \\ 2x - y - z = 0 \end{cases}$

$$\begin{bmatrix} 1 & 1 & -5 & \vdots & 3 \\ 1 & 0 & -2 & \vdots & 1 \\ 2 & -1 & -1 & \vdots & 0 \end{bmatrix}$$

$\begin{matrix} \\ -R_1 + R_2 \rightarrow \\ -2R_1 + R_3 \rightarrow \end{matrix} \begin{bmatrix} 1 & 1 & -5 & \vdots & 3 \\ 0 & -1 & 3 & \vdots & -2 \\ 0 & -3 & 9 & \vdots & -6 \end{bmatrix}$

$\begin{matrix} \\ \\ -3R_2 + R_3 \rightarrow \end{matrix} \begin{bmatrix} 1 & 1 & -5 & \vdots & 3 \\ 0 & -1 & 3 & \vdots & -2 \\ 0 & 0 & 0 & \vdots & 0 \end{bmatrix}$

$\begin{matrix} R_2 + R_1 \rightarrow \\ -R_2 \rightarrow \\ \\ \end{matrix} \begin{bmatrix} 1 & 0 & -2 & \vdots & 1 \\ 0 & 1 & -3 & \vdots & 2 \\ 0 & 0 & 0 & \vdots & 0 \end{bmatrix}$

Let $z = a$, any real number

$$y - 3a = 2 \implies y = 3a + 2$$

$$x - 2a = 1 \implies x = 2a + 1$$

Answer: $(2a + 1, 3a + 2, a)$

59. $\begin{cases} -x + y - z = -14 \\ 2x - y + z = 21 \\ 3x + 2y + z = 19 \end{cases}$

$$\begin{bmatrix} -1 & 1 & -1 & \vdots & -14 \\ 2 & -1 & 1 & \vdots & 21 \\ 3 & 2 & 1 & \vdots & 19 \end{bmatrix}$$

$\begin{matrix} \\ 2R_1 + R_2 \\ 3R_1 + R_3 \end{matrix} \begin{bmatrix} -1 & 1 & -1 & \vdots & -14 \\ 0 & 1 & -1 & \vdots & -7 \\ 0 & 5 & -2 & \vdots & -23 \end{bmatrix}$

$\begin{matrix} -R_1 \\ \\ -5R_2 + R_3 \end{matrix} \begin{bmatrix} 1 & -1 & 1 & \vdots & 14 \\ 0 & 1 & -1 & \vdots & -7 \\ 0 & 0 & 3 & \vdots & 12 \end{bmatrix}$

$$3z = 12 \implies z = 4$$

$$y - 4 = 7 \implies y = -3$$

$$x - (-3) + 4 = 14 \implies x = 7$$

Answer: $(7, -3, 4)$

61. $\begin{cases} 3x + 3y + 12z = 6 \\ x + y + 4z = 2 \\ 2x + 5y + 20z = 10 \\ -x + 2y + 8z = 4 \end{cases}$ $\begin{bmatrix} 3 & 3 & 12 & \vdots & 6 \\ 1 & 1 & 4 & \vdots & 2 \\ 2 & 5 & 20 & \vdots & 10 \\ -1 & 2 & 8 & \vdots & 4 \end{bmatrix} \Rightarrow \begin{bmatrix} 1 & 0 & 0 & \vdots & 0 \\ 0 & 0 & 0 & \vdots & 0 \\ 0 & 1 & 4 & \vdots & 2 \\ 0 & 0 & 0 & \vdots & 0 \end{bmatrix}$

Let $z = a$, any real number

$y = -4a + 2$

$x = 0$

Answer: $(0, -4a + 2, a)$

63. $\begin{bmatrix} 2 & 1 & -1 & 2 & \vdots & -6 \\ 3 & 4 & 0 & 1 & \vdots & 1 \\ 1 & 5 & 2 & 6 & \vdots & -3 \\ 5 & 2 & -1 & -1 & \vdots & 3 \end{bmatrix}$ row reduces to $\begin{bmatrix} 1 & 0 & 0 & 0 & \vdots & 1 \\ 0 & 1 & 0 & 0 & \vdots & 0 \\ 0 & 0 & 1 & 0 & \vdots & 4 \\ 0 & 0 & 0 & 1 & \vdots & -2 \end{bmatrix}$

Answer: $(1, 0, 4, -2)$

65. $\begin{cases} x + y + z = 0 \\ 2x + 3y + z = 0 \\ 3x + 5y + z = 0 \end{cases}$ $\begin{bmatrix} 1 & 1 & 1 & \vdots & 0 \\ 2 & 3 & 1 & \vdots & 0 \\ 3 & 5 & 1 & \vdots & 0 \end{bmatrix} \Rightarrow \begin{bmatrix} 1 & 0 & 2 & \vdots & 0 \\ 0 & 1 & -1 & \vdots & 0 \\ 0 & 0 & 0 & \vdots & 0 \end{bmatrix}$

Let $z = a$, any real number

$y = a$

$x = -2a$

Answer: $(-2a, a, a)$

67. Yes, the systems yield the same solutions.

(a) $z = -3;\ y = 5(-3) + 16 = 1;$
$\quad x = 2(1) - (-3) - 6 = -1$

Answer: $(-1, 1, -3)$

(b) $z = -3,\ y = -3(-3) - 8 = 1,$
$\quad x = -1 + 2(-3) + 6 = -1$

Answer: $(-1, 1, -3)$

69. No, solutions are different.

(a) $z = 8,\ y = 7(8) - 54 = 2,$
$\quad x = 4(2) - 5(8) + 27 = -5$

Answer: $(-5, 2, 8)$

(b) $z = 8,\ y = -5(8) + 42 = 2,$
$\quad x = 6(2) - 8 + 15 = 19$

Answer: $(19, 2, 8)$

71. $f(x) = ax^2 + bx + c$

$\begin{cases} f(1) = a + b + c = 8 \\ f(2) = 4a + 2b + c = 13 \\ f(3) = 9a + 3b + c = 20 \end{cases}$

$$\begin{bmatrix} 1 & 1 & 1 & \vdots & 8 \\ 4 & 2 & 1 & \vdots & 13 \\ 9 & 3 & 1 & \vdots & 20 \end{bmatrix}$$

$\begin{matrix} \\ -4R_1 + R_2 \rightarrow \\ -9R_1 + R_3 \rightarrow \end{matrix} \begin{bmatrix} 1 & 1 & 1 & \vdots & 8 \\ 0 & -2 & -3 & \vdots & -19 \\ 0 & -6 & -8 & \vdots & -52 \end{bmatrix}$

$\begin{matrix} \\ -\frac{1}{2}R_2 \rightarrow \\ -3R_2 + R_3 \rightarrow \end{matrix} \begin{bmatrix} 1 & 1 & 1 & \vdots & 8 \\ 0 & 1 & \frac{3}{2} & \vdots & \frac{19}{2} \\ 0 & 0 & 1 & \vdots & 5 \end{bmatrix}$

$$c = 5$$

$$b + \tfrac{3}{2}(5) = \tfrac{19}{2} \implies b = 2$$

$$a + 2 + 5 = 8 \implies a = 1$$

Answer: $y = x^2 + 2x + 5$

73. x = amount at 7%

y = amount at 8%

z = amount at 10%

$\begin{cases} x + \quad y + \quad z = 1{,}500{,}000 \\ 0.07x + 0.08y + 0.1z = \quad 130{,}500 \\ 4x \quad - \quad z = \qquad 0 \end{cases}$

$$\begin{bmatrix} 1 & 1 & 1 & \vdots & 1{,}500{,}000 \\ 0.07 & 0.08 & 0.1 & \vdots & 130{,}500 \\ 4 & 0 & -1 & \vdots & 0 \end{bmatrix}$$

$\begin{matrix} \\ -0.07R_1 + R_2 \\ -4R_1 + R_2 \end{matrix} \begin{bmatrix} 1 & 1 & 1 & \vdots & 1{,}500{,}000 \\ 0 & 0.01 & 0.03 & \vdots & 25{,}500 \\ 0 & -4 & -5 & \vdots & -6{,}000{,}000 \end{bmatrix}$

$\begin{matrix} \\ 100R_2 \\ 4R_2 + R_3 \end{matrix} \begin{bmatrix} 1 & 1 & 1 & \vdots & 1{,}500{,}000 \\ 0 & 1 & 3 & \vdots & 2{,}550{,}000 \\ 0 & 0 & 7 & \vdots & 4{,}200{,}000 \end{bmatrix}$

$7z = 4{,}200{,}000 \implies z = 600{,}00$

$y + 3(600{,}000) = 2{,}550{,}000 \implies y = 750{,}000$

$x + 750{,}000 + 600{,}000 = 1{,}500{,}000 \implies x = 150{,}000$

Answers: $150{,}000 at 7%

$\qquad\qquad$ $750{,}000 at 8%

$\qquad\qquad$ $600{,}000 at 10%

75. $\begin{cases} I_1 - I_2 + I_3 = 0 \\ 2I_1 + 2I_2 \qquad = 7 \\ \quad 2I_2 + 4I_3 = 8 \end{cases}$

$$\begin{bmatrix} 1 & -1 & 1 & \vdots & 0 \\ 2 & 2 & 0 & \vdots & 7 \\ 0 & 2 & 4 & \vdots & 8 \end{bmatrix}$$

$-2R_1 + R_2 \rightarrow \begin{bmatrix} 1 & -1 & 1 & \vdots & 0 \\ 0 & 4 & -2 & \vdots & 7 \\ 0 & 2 & 4 & \vdots & 8 \end{bmatrix}$

$\begin{matrix} R_3 \rightarrow \\ R_2 \rightarrow \end{matrix} \begin{bmatrix} 1 & -1 & 1 & \vdots & 0 \\ 0 & 2 & 4 & \vdots & 8 \\ 0 & 4 & -2 & \vdots & 7 \end{bmatrix}$

$\tfrac{1}{2}R_2 \rightarrow \begin{bmatrix} 1 & -1 & 1 & \vdots & 0 \\ 0 & 1 & 2 & \vdots & 4 \\ 0 & 4 & -2 & \vdots & 7 \end{bmatrix}$

$-4R_2 + R_3 \rightarrow \begin{bmatrix} 1 & -1 & 1 & \vdots & 0 \\ 0 & 1 & 2 & \vdots & 4 \\ 0 & 0 & -10 & \vdots & -9 \end{bmatrix}$

$-\tfrac{1}{10}R_3 \rightarrow \begin{bmatrix} 1 & -1 & 1 & \vdots & 0 \\ 0 & 1 & 2 & \vdots & 4 \\ 0 & 0 & 1 & \vdots & \frac{9}{10} \end{bmatrix}$

$I_3 = \tfrac{9}{10}$ amperes

$I_2 + 2\left(\tfrac{9}{10}\right) = 4 \implies I_2 = \tfrac{11}{5}$ amperes

$I_1 - \tfrac{11}{5} + \tfrac{9}{10} = 0 \implies I_1 = \tfrac{13}{10}$ amperes

77. (a)
$(9, 43.90)$
$(10, 42.10)$
$(11, 34.10)$

$$\begin{cases} 81a + 9b + c = 43.9 \\ 100a + 10b + c = 42.1 \\ 121a + 11b + c = 34.10 \end{cases}$$

$$\begin{bmatrix} 81 & 9 & 1 & \vdots & 43.9 \\ 100 & 10 & 1 & \vdots & 42.1 \\ 121 & 11 & 1 & \vdots & 34.1 \end{bmatrix} \Rightarrow \begin{bmatrix} 1 & 0 & 0 & \vdots & -3.1 \\ 0 & 1 & 0 & \vdots & 57.1 \\ 0 & 0 & 1 & \vdots & -218.9 \end{bmatrix}$$

$a = 3.1, b = 57.1, c = -218.9$

$y = -3.1t^2 + 57.1t - 218.9$

(b)

(c) For 2002, $t = 12$ and $y = -3.1(12)^2 + 57.1(12) - 218.9 = 19.9$

(d) For 2005, $t = 15$ and $y = -59.9$, which is unreasonable.

79. (a)
$$\begin{cases} x_1 + x_3 = 600 \\ x_1 = x_2 + x_4 \Rightarrow x_1 - x_2 - x_4 = 0 \\ x_2 + x_5 = 500 \\ x_3 + x_6 = 600 \\ x_4 + x_7 = x_6 \Rightarrow x_4 - x_6 + x_7 = 0 \\ x_5 + x_7 = 500 \end{cases}$$

$$\begin{bmatrix} 1 & 0 & 1 & 0 & 0 & 0 & 0 & \vdots & 600 \\ 1 & -1 & 0 & -1 & 0 & 0 & 0 & \vdots & 0 \\ 0 & 1 & 0 & 0 & 1 & 0 & 0 & \vdots & 500 \\ 0 & 0 & 1 & 0 & 0 & 1 & 0 & \vdots & 600 \\ 0 & 0 & 0 & 1 & 0 & -1 & 1 & \vdots & 0 \\ 0 & 0 & 0 & 0 & 1 & 0 & 1 & \vdots & 500 \end{bmatrix}$$

$$\begin{matrix} \\ -R_1 + R_2 \rightarrow \\ R_2 + R_3 \rightarrow \\ R_3 + R_4 \rightarrow \\ R_4 + R_5 \rightarrow \\ -R_5 + R_6 \rightarrow \end{matrix} \begin{bmatrix} 1 & 0 & 1 & 0 & 0 & 0 & 0 & \vdots & 600 \\ 0 & -1 & -1 & -1 & 0 & 0 & 0 & \vdots & -600 \\ 0 & 0 & -1 & -1 & 1 & 0 & 0 & \vdots & -100 \\ 0 & 0 & 0 & -1 & 1 & 1 & 0 & \vdots & 500 \\ 0 & 0 & 0 & 0 & 1 & 0 & 1 & \vdots & 500 \\ 0 & 0 & 0 & 0 & 0 & 0 & 0 & \vdots & 0 \end{bmatrix}$$

$$\begin{matrix} \\ -R_3 + R_2 \rightarrow \\ -R_4 + R_3 \rightarrow \\ -R_4 \rightarrow \\ \\ \end{matrix} \begin{bmatrix} 1 & 0 & 1 & 0 & 0 & 0 & 0 & \vdots & 600 \\ 0 & -1 & 0 & 0 & -1 & 0 & 0 & \vdots & -500 \\ 0 & 0 & -1 & 0 & 0 & -1 & 0 & \vdots & -600 \\ 0 & 0 & 0 & 1 & -1 & -1 & 0 & \vdots & -500 \\ 0 & 0 & 0 & 0 & 1 & 0 & 1 & \vdots & 500 \\ 0 & 0 & 0 & 0 & 0 & 0 & 0 & \vdots & 0 \end{bmatrix}$$

Let $x_7 = t$ and $x_6 = s$, then:

$x_5 = 500 - t$

$x_4 = -500 + s + (500 - t) = s - t$

$x_3 = 600 - s$

$x_2 = 500 - (500 - t) = t$

$x_1 = 600 - (600 - s) = s$

(b) If $x_6 = x_7 = 0$, then $s = t = 0$, and

$x_1 = 0$

$x_2 = 0$

$x_3 = 600$

$x_4 = 0$

$x_5 = 500$

$x_6 = x_7 = 0$

(c) If $x_5 = 1000$ and $x_6 = 0$, then $s = 0$ and $t = -500$.

Thus, $x_1 = 0$

$x_2 = -500$

$x_3 = 600$

$x_4 = 500$

$x_5 = 1000$

$x_6 = 0$

$x_7 = -500$

81. False. It is a 2×4 matrix.

83. $\begin{cases} x + 3z = -2 & \text{Equation 1} \\ y + 4z = 1 & \text{Equation 2} \end{cases}$

(Equation 1) + (Equation 2) $\rightarrow$ new Equation 1

(Equation 1) + 2(Equation 2) $\rightarrow$ new Equation 2

2(Equation 1) + (Equation 2) $\rightarrow$ new Equation 3

$\begin{cases} x + y + 7z = -1 \\ x + 2y + 11z = 0 \\ 2x + y + 10z = -3 \end{cases}$

85. The row operation $-2R_1 + R_2$ was not performed on the last column. Nor was $-R_2 + R_1$.

87. $f(x) = \dfrac{7}{-x - 1}$

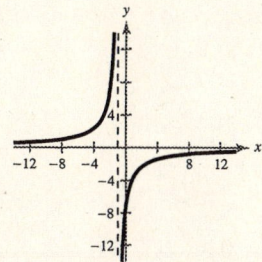

Asymptotes: $x = -1, y = 0$

89. $f(x) = \dfrac{x^2 - 2x - 3}{x - 4} = x + 2 + \dfrac{5}{x - 4}$

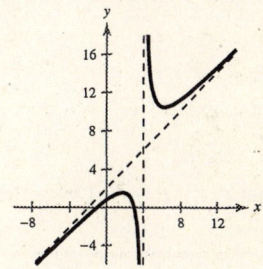

Asymptotes: $x = 4, y = x + 2$

91. $f(x) = \dfrac{2x^2 - 4x}{3x - x^2} = \dfrac{2x(x - 2)}{x(3 - x)} = \dfrac{2(x - 2)}{3 - x}, \quad x \neq 0$

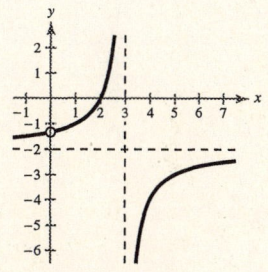

Asymptotes: $x = 3, y = -2$

Hole at $x = 0$

Section 7.5   Operations with Matrices

- $A = B$ if and only if they have the same order and $a_{ij} = b_{ij}$.
- You should be able to perform the operations of matrix addition, scalar multiplication, and matrix multiplication.
- Some properties of matrix addition and scalar multiplication are:
  - (a) $A + B = B + A$
  - (b) $A + (B + C) = (A + B) + C$
  - (c) $(cd)A = c(dA)$
  - (d) $1A = A$
  - (e) $c(A + B) = cA + cB$
  - (f) $(c + d)A = cA + dA$
- Some properties of matrix multiplication are:
  - (a) $A(BC) = (AB)C$
  - (b) $A(B + C) = AB + AC$
  - (c) $(A + B)C = AC + BC$
  - (d) $c(AB) = (cA)B = A(cB)$
- You should remember that $AB \neq BA$ in general.

Solutions to Odd-Numbered Exercises

1. $x = -4, \ y = 22$

3. $2x + 7 = 5 \implies x = -1$

$\qquad 3y = 12 \implies y = 4$

$\qquad 3z - 14 = 4 \implies z = 6$

5. (a) $A + B = \begin{bmatrix} 1 & -1 \\ 2 & -1 \end{bmatrix} + \begin{bmatrix} 2 & -1 \\ -1 & 8 \end{bmatrix} = \begin{bmatrix} 1+2 & -1-1 \\ 2-1 & -1+8 \end{bmatrix} = \begin{bmatrix} 3 & -2 \\ 1 & 7 \end{bmatrix}$

(b) $A - B = \begin{bmatrix} 1 & -1 \\ 2 & -1 \end{bmatrix} - \begin{bmatrix} 2 & -1 \\ -1 & 8 \end{bmatrix} = \begin{bmatrix} 1-2 & -1+1 \\ 2+1 & -1-8 \end{bmatrix} = \begin{bmatrix} -1 & 0 \\ 3 & -9 \end{bmatrix}$

(c) $3A = 3\begin{bmatrix} 1 & -1 \\ 2 & -1 \end{bmatrix} = \begin{bmatrix} 3(1) & 3(-1) \\ 3(2) & 3(-1) \end{bmatrix} = \begin{bmatrix} 3 & -3 \\ 6 & -3 \end{bmatrix}$

(d) $3A - 2B = \begin{bmatrix} 3 & -3 \\ 6 & -3 \end{bmatrix} - 2\begin{bmatrix} 2 & -1 \\ -1 & 8 \end{bmatrix} = \begin{bmatrix} 3 & -3 \\ 6 & -3 \end{bmatrix} + \begin{bmatrix} -4 & 2 \\ 2 & -16 \end{bmatrix} = \begin{bmatrix} -1 & -1 \\ 8 & -19 \end{bmatrix}$

7. $A = \begin{bmatrix} 8 & -1 \\ 2 & 3 \\ -4 & 5 \end{bmatrix}, B = \begin{bmatrix} 1 & 6 \\ -1 & -5 \\ 1 & 10 \end{bmatrix}$

(a) $A + B = \begin{bmatrix} 9 & 5 \\ 1 & -2 \\ -3 & 15 \end{bmatrix}$
(b) $A - B = \begin{bmatrix} 7 & -7 \\ 3 & 8 \\ -5 & -5 \end{bmatrix}$
(c) $3A = \begin{bmatrix} 24 & -3 \\ 6 & 9 \\ -12 & 15 \end{bmatrix}$

(d) $3A - 2B = \begin{bmatrix} 24 & -3 \\ 6 & 9 \\ -12 & 15 \end{bmatrix} - \begin{bmatrix} 2 & 12 \\ -2 & -10 \\ 2 & 20 \end{bmatrix} = \begin{bmatrix} 22 & -15 \\ 8 & 19 \\ -14 & -5 \end{bmatrix}$

9. $A = \begin{bmatrix} 4 & 5 & -1 & 3 & 4 \\ 1 & 2 & -2 & -1 & 0 \end{bmatrix}$

$B = \begin{bmatrix} 1 & 0 & -1 & 1 & 0 \\ -6 & 8 & 2 & -3 & -7 \end{bmatrix}$

(a) $A + B = \begin{bmatrix} 5 & 5 & -2 & 4 & 4 \\ -5 & 10 & 0 & -4 & -7 \end{bmatrix}$

(b) $A - B = \begin{bmatrix} 3 & 5 & 0 & 2 & 4 \\ 7 & -6 & -4 & 2 & 7 \end{bmatrix}$

(c) $3A = \begin{bmatrix} 12 & 15 & -3 & 9 & 12 \\ 3 & 6 & -6 & -3 & 0 \end{bmatrix}$

(d) $3A - 2B = \begin{bmatrix} 12 & 15 & -3 & 9 & 12 \\ 3 & 6 & -6 & -3 & 0 \end{bmatrix} - \begin{bmatrix} 2 & 0 & -2 & 2 & 0 \\ -12 & 16 & 4 & -6 & -14 \end{bmatrix}$

$\qquad\qquad = \begin{bmatrix} 10 & 15 & -1 & 7 & 12 \\ 15 & -10 & -10 & 3 & 14 \end{bmatrix}$

11. $A = \begin{bmatrix} 6 & 0 & 3 \\ -1 & -4 & 0 \end{bmatrix}$, $B = \begin{bmatrix} 8 & -1 \\ 4 & -3 \end{bmatrix}$

(a) $A + B$ not possible   (b) $A - B$ not possible   (c) $3A = \begin{bmatrix} 18 & 0 & 9 \\ -3 & -12 & 0 \end{bmatrix}$ (d) $3A - 2B$ not possible

13. $\begin{bmatrix} -5 & 0 \\ 3 & -6 \end{bmatrix} + \begin{bmatrix} 7 & 1 \\ -2 & -1 \end{bmatrix} + \begin{bmatrix} -10 & -8 \\ 14 & 6 \end{bmatrix}$

$\qquad = \begin{bmatrix} -5 & 0 \\ 3 & -6 \end{bmatrix} + \begin{bmatrix} -3 & -7 \\ 12 & 5 \end{bmatrix} = \begin{bmatrix} -8 & -7 \\ 15 & -1 \end{bmatrix}$

15. $4\left(\begin{bmatrix} -4 & 0 & 1 \\ 0 & 2 & 3 \end{bmatrix} - \begin{bmatrix} 2 & 1 & -2 \\ 3 & -6 & 0 \end{bmatrix} \right) = 4 \begin{bmatrix} -6 & -1 & 3 \\ -3 & 8 & 3 \end{bmatrix} = \begin{bmatrix} -24 & -4 & 12 \\ -12 & 32 & 12 \end{bmatrix}$

17. $\begin{bmatrix} 2 & 5 \\ -1 & -4 \end{bmatrix} + \begin{bmatrix} -3 & 0 \\ 2 & 2 \end{bmatrix} = \begin{bmatrix} -1 & 5 \\ 1 & -2 \end{bmatrix}$

19. $-\tfrac{1}{2} \begin{bmatrix} 3.211 & 6.829 \\ -1.004 & 4.914 \\ 0.055 & -3.889 \end{bmatrix} - 8 \begin{bmatrix} 1.630 & -3.090 \\ 5.256 & 8.335 \\ -9.768 & 4.251 \end{bmatrix} = \begin{bmatrix} -14.645 & 21.305 \\ -41.546 & -69.137 \\ 78.117 & -32.064 \end{bmatrix}$

21. $-3\left(\begin{bmatrix} 0 & -3 \\ 7 & 2 \end{bmatrix} + \begin{bmatrix} -6 & 3 \\ 8 & 1 \end{bmatrix} \right) - \begin{bmatrix} 4 & -4 \\ 7 & -9 \end{bmatrix} = -3 \begin{bmatrix} -6 & 0 \\ 15 & 3 \end{bmatrix} - \begin{bmatrix} 4 & -4 \\ 7 & -9 \end{bmatrix} = \begin{bmatrix} 18 & 0 \\ -45 & -9 \end{bmatrix} - \begin{bmatrix} 4 & -4 \\ 7 & -9 \end{bmatrix}$

$\qquad\qquad\qquad\qquad\qquad\qquad = \begin{bmatrix} 14 & 4 \\ -52 & 0 \end{bmatrix}$

23. $X = 3 \begin{bmatrix} -2 & -1 \\ 1 & 0 \\ 3 & -4 \end{bmatrix} - 2 \begin{bmatrix} 0 & 3 \\ 2 & 0 \\ -4 & -1 \end{bmatrix} = \begin{bmatrix} -6 & -3 \\ 3 & 0 \\ 9 & -12 \end{bmatrix} - \begin{bmatrix} 0 & 6 \\ 4 & 0 \\ -8 & -2 \end{bmatrix} = \begin{bmatrix} -6 & -9 \\ -1 & 0 \\ 17 & -10 \end{bmatrix}$

25. $X = -\frac{3}{2}A + \frac{1}{2}B = -\frac{3}{2}\begin{bmatrix} -2 & -1 \\ 1 & 0 \\ 3 & -4 \end{bmatrix} + \frac{1}{2}\begin{bmatrix} 0 & 3 \\ 2 & 0 \\ -4 & -1 \end{bmatrix} = \begin{bmatrix} 3 & 3 \\ -\frac{1}{2} & 0 \\ -\frac{13}{2} & \frac{11}{2} \end{bmatrix}$

27. *A* is 3×2 and *B* is 3×3 $\implies$ *AB* is not defined.

29. $AB = \begin{bmatrix} -1 & 6 \\ -4 & 5 \\ 0 & 3 \end{bmatrix}\begin{bmatrix} 2 & 3 \\ 0 & 9 \end{bmatrix} = \begin{bmatrix} -2 & 51 \\ -8 & 33 \\ 0 & 27 \end{bmatrix}$

31. *A* is 3×3, *B* is 3×3 $\implies$ *AB* is 3×3.

$AB = \begin{bmatrix} 5 & 0 & 0 \\ 0 & -8 & 0 \\ 0 & 0 & 7 \end{bmatrix}\begin{bmatrix} \frac{1}{5} & 0 & 0 \\ 0 & -\frac{1}{8} & 0 \\ 0 & 0 & \frac{1}{2} \end{bmatrix} = \begin{bmatrix} 1 & 0 & 0 \\ 0 & 1 & 0 \\ 0 & 0 & \frac{7}{2} \end{bmatrix}$

33. $AB = \begin{bmatrix} 5 \\ 6 \end{bmatrix}\begin{bmatrix} -3 & -1 & -5 & -9 \end{bmatrix} = \begin{bmatrix} -15 & -5 & -25 & -45 \\ -18 & -6 & -30 & -54 \end{bmatrix}$

35. (a) $AB = \begin{bmatrix} 1 & 2 \\ 5 & 2 \end{bmatrix}\begin{bmatrix} 2 & -1 \\ -1 & 8 \end{bmatrix} = \begin{bmatrix} 2-2 & -1+16 \\ 10-2 & -5+16 \end{bmatrix} = \begin{bmatrix} 0 & 15 \\ 8 & 11 \end{bmatrix}$

(b) $BA = \begin{bmatrix} 2 & -1 \\ -1 & 8 \end{bmatrix}\begin{bmatrix} 1 & 2 \\ 5 & 2 \end{bmatrix} = \begin{bmatrix} 2-5 & 4-2 \\ -1+40 & -2+16 \end{bmatrix} = \begin{bmatrix} -3 & 2 \\ 39 & 14 \end{bmatrix}$

(c) $A^2 = \begin{bmatrix} 1 & 2 \\ 5 & 2 \end{bmatrix}\begin{bmatrix} 1 & 2 \\ 5 & 2 \end{bmatrix} = \begin{bmatrix} 1+10 & 2+4 \\ 5+10 & 10+4 \end{bmatrix} = \begin{bmatrix} 11 & 6 \\ 15 & 14 \end{bmatrix}$

37. (a) $AB = \begin{bmatrix} 3 & -1 \\ 1 & 3 \end{bmatrix}\begin{bmatrix} 1 & -3 \\ 3 & 1 \end{bmatrix} = \begin{bmatrix} 3-3 & -9-1 \\ 1+9 & -3+3 \end{bmatrix} = \begin{bmatrix} 0 & -10 \\ 10 & 0 \end{bmatrix}$

(b) $BA = \begin{bmatrix} 1 & -3 \\ 3 & 1 \end{bmatrix}\begin{bmatrix} 3 & -1 \\ 1 & 3 \end{bmatrix} = \begin{bmatrix} 3-3 & -1-9 \\ 9+1 & -3+3 \end{bmatrix} = \begin{bmatrix} 0 & -10 \\ 10 & 0 \end{bmatrix}$

(c) $A^2 = \begin{bmatrix} 3 & -1 \\ 1 & 3 \end{bmatrix}\begin{bmatrix} 3 & -1 \\ 1 & 3 \end{bmatrix} = \begin{bmatrix} 9-1 & -3-3 \\ 3+3 & -1+9 \end{bmatrix} = \begin{bmatrix} 8 & -6 \\ 6 & 8 \end{bmatrix}$

39. (a) $AB = \begin{bmatrix} 7 \\ 8 \\ -1 \end{bmatrix}\begin{bmatrix} 1 & 1 & 2 \end{bmatrix} = \begin{bmatrix} 7 & 7 & 14 \\ 8 & 8 & 16 \\ -1 & -1 & -2 \end{bmatrix}$

(b) $BA = \begin{bmatrix} 1 & 1 & 2 \end{bmatrix}\begin{bmatrix} 7 \\ 8 \\ -1 \end{bmatrix} = [7 + 8 - 2] = [13]$ (c) A^2 is not defined.

41. $AB = \begin{bmatrix} 70 & -17 & 73 \\ 32 & 11 & 6 \\ 16 & -38 & 70 \end{bmatrix}$

43. $\begin{bmatrix} -3 & 8 & -6 & 8 \\ -12 & 15 & 9 & 6 \\ 5 & -1 & 1 & 5 \end{bmatrix}\begin{bmatrix} 3 & 1 & 6 \\ 24 & 15 & 14 \\ 16 & 10 & 21 \\ 8 & -4 & 10 \end{bmatrix} = \begin{bmatrix} 151 & 25 & 48 \\ 516 & 279 & 387 \\ 47 & -20 & 87 \end{bmatrix}$

45. A is 2×4 and B is $2 \times 4 \implies AB$ is not defined.

47. $\left(\begin{bmatrix} 3 & 1 \\ 0 & -2 \end{bmatrix} \begin{bmatrix} 1 & 0 \\ -2 & 2 \end{bmatrix} \right) \begin{bmatrix} 1 & 0 \\ 2 & 4 \end{bmatrix} = \begin{bmatrix} 1 & 2 \\ 4 & -4 \end{bmatrix} \begin{bmatrix} 1 & 0 \\ 2 & 4 \end{bmatrix} = \begin{bmatrix} 5 & 8 \\ -4 & -16 \end{bmatrix}$

49. $\begin{bmatrix} 0 & 2 & -2 \\ 4 & 1 & 2 \end{bmatrix} \left(\begin{bmatrix} 4 & 0 \\ 0 & -1 \\ -1 & 2 \end{bmatrix} + \begin{bmatrix} -2 & 3 \\ -3 & 5 \\ 0 & -3 \end{bmatrix} \right) = \begin{bmatrix} 0 & 2 & -2 \\ 4 & 1 & 2 \end{bmatrix} \begin{bmatrix} 2 & 3 \\ -3 & 4 \\ -1 & -1 \end{bmatrix} = \begin{bmatrix} -4 & 10 \\ 3 & 14 \end{bmatrix}$

51. $\begin{bmatrix} 1 & 2 & \vdots & 4 \\ 3 & 2 & \vdots & 0 \end{bmatrix}$

(a) $\begin{bmatrix} 1 & 2 \\ 3 & 2 \end{bmatrix} \begin{bmatrix} 2 \\ 1 \end{bmatrix} = \begin{bmatrix} 4 \\ 8 \end{bmatrix} \implies \begin{bmatrix} 2 \\ 1 \end{bmatrix}$ is not a solution.　　(b) $\begin{bmatrix} 1 & 2 \\ 3 & 2 \end{bmatrix} \begin{bmatrix} -2 \\ 3 \end{bmatrix} = \begin{bmatrix} 4 \\ 0 \end{bmatrix} \implies \begin{bmatrix} -2 \\ 3 \end{bmatrix}$ is a solution.

(c) $\begin{bmatrix} 1 & 2 \\ 3 & 2 \end{bmatrix} \begin{bmatrix} -4 \\ 4 \end{bmatrix} = \begin{bmatrix} 4 \\ -4 \end{bmatrix} \implies \begin{bmatrix} -4 \\ 4 \end{bmatrix}$ is not a　　(d) $\begin{bmatrix} 1 & 2 \\ 3 & 2 \end{bmatrix} \begin{bmatrix} 2 \\ -3 \end{bmatrix} = \begin{bmatrix} -4 \\ 0 \end{bmatrix} \implies \begin{bmatrix} 2 \\ -3 \end{bmatrix}$ is not a

　　　solution.　　　　　　　　　　　　　　　　　　　　　　solution.

53. $\begin{bmatrix} -2 & -3 & \vdots & -6 \\ 4 & 2 & \vdots & 20 \end{bmatrix}$

(a) $\begin{bmatrix} -2 & -3 \\ 4 & 2 \end{bmatrix} \begin{bmatrix} 3 \\ 0 \end{bmatrix} = \begin{bmatrix} -6 \\ 12 \end{bmatrix} \implies \begin{bmatrix} 3 \\ 0 \end{bmatrix}$ is not a　　(b) $\begin{bmatrix} -2 & -3 \\ 4 & 2 \end{bmatrix} \begin{bmatrix} 6 \\ -2 \end{bmatrix} = \begin{bmatrix} -6 \\ 20 \end{bmatrix} \implies \begin{bmatrix} 6 \\ -2 \end{bmatrix}$ is a

　　　solution.　　　　　　　　　　　　　　　　　　　　　　solution.

(c) $\begin{bmatrix} -2 & -3 \\ 4 & 2 \end{bmatrix} \begin{bmatrix} -6 \\ 6 \end{bmatrix} = \begin{bmatrix} -6 \\ -12 \end{bmatrix} \implies \begin{bmatrix} -6 \\ 6 \end{bmatrix}$ is not　　(d) $\begin{bmatrix} -2 & -3 \\ 4 & 2 \end{bmatrix} \begin{bmatrix} 4 \\ 2 \end{bmatrix} = \begin{bmatrix} -14 \\ 20 \end{bmatrix} \implies \begin{bmatrix} 4 \\ 2 \end{bmatrix}$ is not

　　　a solution.　　　　　　　　　　　　　　　　　　　　　a solution.

55. (a) $A = \begin{bmatrix} -1 & 1 \\ -2 & 1 \end{bmatrix}$, $X = \begin{bmatrix} x_1 \\ x_2 \end{bmatrix}$, $B = \begin{bmatrix} 4 \\ 0 \end{bmatrix}$

(b) By Gauss-Jordan elimination on

$$\begin{bmatrix} -1 & 1 & \vdots & 4 \\ -2 & 1 & \vdots & 0 \end{bmatrix} \begin{matrix} -R_1 \to \\ 2R_1 + R_2 \to \end{matrix} \begin{bmatrix} 1 & -1 & \vdots & -4 \\ 0 & -1 & \vdots & -8 \end{bmatrix} \begin{matrix} R_2 + R_1 \to \\ -R_2 \to \end{matrix} \begin{bmatrix} 1 & 0 & \vdots & 4 \\ 0 & 1 & \vdots & 8 \end{bmatrix},$$

we have $x_1 = 4$ and $x_2 = 8$. Thus, $X = \begin{bmatrix} 4 \\ 8 \end{bmatrix}$.

57. (a) $A = \begin{bmatrix} -2 & -3 \\ 6 & 1 \end{bmatrix}$, $X = \begin{bmatrix} x_1 \\ x_2 \end{bmatrix}$, $B = \begin{bmatrix} -4 \\ -36 \end{bmatrix}$

(b) $\begin{bmatrix} -2 & -3 & \vdots & -4 \\ 6 & 1 & \vdots & -36 \end{bmatrix}$

$3R_1 + R_2 \begin{bmatrix} -2 & -3 & \vdots & -4 \\ 0 & -8 & \vdots & -48 \end{bmatrix}$

$(-\frac{1}{8})R_2 \begin{bmatrix} -2 & -3 & \vdots & -4 \\ 0 & 1 & \vdots & 6 \end{bmatrix}$

$3R_2 + R_1 \begin{bmatrix} -2 & 0 & \vdots & 14 \\ 0 & 1 & \vdots & 6 \end{bmatrix}$

$-\frac{1}{2}R_1 \begin{bmatrix} 1 & 0 & \vdots & -7 \\ 0 & 1 & \vdots & 6 \end{bmatrix}$

$x_1 = -7, x_2 = 6$. Thus, $X = \begin{bmatrix} -7 \\ 6 \end{bmatrix}$.

59. (a) $A = \begin{bmatrix} 1 & -2 & 3 \\ -1 & 3 & -1 \\ 2 & -5 & 5 \end{bmatrix}$, $X = \begin{bmatrix} x_1 \\ x_2 \\ x_3 \end{bmatrix}$, $B = \begin{bmatrix} 9 \\ -6 \\ 17 \end{bmatrix}$

(b) $\begin{bmatrix} 1 & -2 & 3 & \vdots & 9 \\ -1 & 3 & -1 & \vdots & -6 \\ 2 & -5 & 5 & \vdots & 17 \end{bmatrix}$

$\begin{matrix} R_1 + R_2 \rightarrow \\ -2R_1 + R_3 \rightarrow \end{matrix} \begin{bmatrix} 1 & -2 & 3 & \vdots & 9 \\ 0 & 1 & 2 & \vdots & 3 \\ 0 & -1 & -1 & \vdots & -1 \end{bmatrix}$

$\begin{matrix} 2R_2 + R_1 \rightarrow \\ \\ R_2 + R_3 \rightarrow \end{matrix} \begin{bmatrix} 1 & 0 & 7 & \vdots & 15 \\ 0 & 1 & 2 & \vdots & 3 \\ 0 & 0 & 1 & \vdots & 2 \end{bmatrix}$

$\begin{matrix} -7R_3 + R_1 \rightarrow \\ -2R_3 + R_2 \rightarrow \end{matrix} \begin{bmatrix} 1 & 0 & 0 & \vdots & 1 \\ 0 & 1 & 0 & \vdots & -1 \\ 0 & 0 & 1 & \vdots & 2 \end{bmatrix}$

$x_1 = 1, x_2 = -1, x_3 = 2$. Thus, $X = \begin{bmatrix} 1 \\ -1 \\ 2 \end{bmatrix}$.

61. (a) $A = \begin{bmatrix} 1 & -5 & 2 \\ -3 & 1 & -1 \\ 0 & -2 & 5 \end{bmatrix}$, $X = \begin{bmatrix} x_1 \\ x_2 \\ x_3 \end{bmatrix}$, $B = \begin{bmatrix} -20 \\ 8 \\ -16 \end{bmatrix}$

(b) $\begin{bmatrix} 1 & -5 & 2 & \vdots & -20 \\ -3 & 1 & -1 & \vdots & 8 \\ 0 & -2 & 5 & \vdots & -16 \end{bmatrix}$ $3R_1 + R_2 \begin{bmatrix} 1 & -5 & 2 & \vdots & -20 \\ 0 & -14 & 5 & \vdots & -52 \\ 0 & -2 & 5 & \vdots & -16 \end{bmatrix}$

$\begin{matrix} R_2 \\ R_3 \end{matrix} \begin{bmatrix} 1 & -5 & 2 & \vdots & -20 \\ 0 & -2 & 5 & \vdots & -16 \\ 0 & -14 & 5 & \vdots & -52 \end{bmatrix}$

$-7R_2 + R_3 \begin{bmatrix} 1 & -5 & 2 & \vdots & -20 \\ 0 & -2 & 5 & \vdots & -16 \\ 0 & 0 & -30 & \vdots & 60 \end{bmatrix}$

$-\frac{1}{30}R_3 \begin{bmatrix} 1 & -5 & 2 & \vdots & -20 \\ 0 & -2 & 5 & \vdots & -16 \\ 0 & 0 & 1 & \vdots & -2 \end{bmatrix}$

$\begin{matrix} -2R_3 + R_1 \\ -5R_3 + R_2 \end{matrix} \begin{bmatrix} 1 & -5 & 0 & \vdots & -16 \\ 0 & -2 & 0 & \vdots & -6 \\ 0 & 0 & 1 & \vdots & -2 \end{bmatrix}$

$(-\frac{1}{2})R_2 \begin{bmatrix} 1 & -5 & 0 & \vdots & -16 \\ 0 & 1 & 0 & \vdots & 3 \\ 0 & 0 & 1 & \vdots & -2 \end{bmatrix}$

$5R_2 + R_1 \begin{bmatrix} 1 & 0 & 0 & \vdots & -1 \\ 0 & 1 & 0 & \vdots & 3 \\ 0 & 0 & 1 & \vdots & -2 \end{bmatrix}$

$x_1 = -1, x_2 = 3, x_3 = -2$. Thus, $X = \begin{bmatrix} -1 \\ 3 \\ -2 \end{bmatrix}$.

63. $A = \begin{bmatrix} 2 & 0 \\ 4 & 5 \end{bmatrix}$

$f(A) = A^2 - 5A + 2I = \begin{bmatrix} 2 & 0 \\ 4 & 5 \end{bmatrix}\begin{bmatrix} 2 & 0 \\ 4 & 5 \end{bmatrix} - 5\begin{bmatrix} 2 & 0 \\ 4 & 5 \end{bmatrix} + 2\begin{bmatrix} 1 & 0 \\ 0 & 1 \end{bmatrix} = \begin{bmatrix} -4 & 0 \\ 8 & 2 \end{bmatrix}$

65. $1.20\begin{bmatrix} 70 & 50 & 25 \\ 35 & 100 & 70 \end{bmatrix} = \begin{bmatrix} 84 & 60 & 30 \\ 42 & 120 & 84 \end{bmatrix}$

67. $BA = [3.50 \quad 6.00]\begin{bmatrix} 125 & 100 & 75 \\ 100 & 175 & 125 \end{bmatrix}$

$= [1037.50 \quad 1400 \quad 1012.50]$

The entries in the last matrix BA represent the profit for both crops at each of the three outlets.

69. $ST = \begin{bmatrix} 3 & 2 & 2 & 3 & 0 \\ 0 & 2 & 3 & 4 & 3 \\ 4 & 2 & 1 & 3 & 2 \end{bmatrix}\begin{bmatrix} 840 & 1100 \\ 1200 & 1350 \\ 1450 & 1650 \\ 2650 & 3000 \\ 3050 & 3200 \end{bmatrix} = \begin{bmatrix} \$15,770 & \$18,300 \\ \$26,500 & \$29,250 \\ \$21,260 & \$24,150 \end{bmatrix}$

The entries represent the wholesale and retail prices of the inventory at each outlet.

71. $P^2 = \begin{bmatrix} 0.6 & 0.1 & 0.1 \\ 0.2 & 0.7 & 0.1 \\ 0.2 & 0.2 & 0.8 \end{bmatrix}\begin{bmatrix} 0.6 & 0.1 & 0.1 \\ 0.2 & 0.7 & 0.1 \\ 0.2 & 0.2 & 0.8 \end{bmatrix} = \begin{bmatrix} 0.40 & 0.15 & 0.15 \\ 0.28 & 0.53 & 0.17 \\ 0.32 & 0.32 & 0.68 \end{bmatrix}$

This product represents the changes in party affiliation after *two* elections.

73. True

For 75–81, A is of order 2×3, B is of order 2×3, C is of order 3×2 and D is of order 2×2.

75. $A + 2C$ is not possible. A and C are not of the same order.

77. AB is not possible. The number of columns of A does not equal the number of rows of B.

79. $BC - D$ is possible. The resulting order is 2×2.

81. $D(A - 3B)$ is possible. The resulting order is 2×3.

83. $AC = \begin{bmatrix} 0 & 1 \\ 0 & 1 \end{bmatrix}\begin{bmatrix} 2 & 3 \\ 2 & 3 \end{bmatrix} = \begin{bmatrix} 2 & 3 \\ 2 & 3 \end{bmatrix}$

$BC = \begin{bmatrix} 1 & 0 \\ 1 & 0 \end{bmatrix}\begin{bmatrix} 2 & 3 \\ 2 & 3 \end{bmatrix} = \begin{bmatrix} 2 & 3 \\ 2 & 3 \end{bmatrix}$

$AC = BC$, but $A \neq B$.

85. (a) $A^2 = \begin{bmatrix} i & 0 \\ 0 & i \end{bmatrix} \begin{bmatrix} i & 0 \\ 0 & i \end{bmatrix} = \begin{bmatrix} -1 & 0 \\ 0 & -1 \end{bmatrix}$ and $i^2 = -1$

$A^3 = A^2A = \begin{bmatrix} -1 & 0 \\ 0 & -1 \end{bmatrix} \begin{bmatrix} i & 0 \\ 0 & i \end{bmatrix} = \begin{bmatrix} -i & 0 \\ 0 & -i \end{bmatrix}$ and $i^3 = -i$

$A^4 = A^3A = \begin{bmatrix} -i & 0 \\ 0 & -i \end{bmatrix} \begin{bmatrix} i & 0 \\ 0 & i \end{bmatrix} = \begin{bmatrix} 1 & 0 \\ 0 & 1 \end{bmatrix}$ and $i^4 = 1$

(b) $B^2 = \begin{bmatrix} 0 & -i \\ i & 0 \end{bmatrix} \begin{bmatrix} 0 & -i \\ i & 0 \end{bmatrix} = \begin{bmatrix} 1 & 0 \\ 0 & 1 \end{bmatrix}$,

The identity matrix

87. (a) $A = \begin{bmatrix} 0 & 2 \\ 0 & 0 \end{bmatrix}, B = \begin{bmatrix} 0 & 2 & 3 \\ 0 & 0 & 4 \\ 0 & 0 & 0 \end{bmatrix}$

(b) A^2 and B^3 are both zero matrices.

(c) If A is 4×4, then A^4 will be the zero matrix.

(d) If A is $n \times n$, then A^n is the zero matrix.

89. $3 \ln 4 - \frac{1}{3}\ln(x^2 + 3) = \ln 4^3 - \ln(x^2 + 3)^{1/3}$

$= \ln\left[\dfrac{64}{(x^2 + 3)^{1/3}} \right]$

91. $\frac{1}{2}[2 \ln(x + 5) + \ln x - \ln(x - 8)] = \ln(x + 5) + \ln x^{1/2} - \ln(x - 8)^{1/2}$

$= \ln\left[\dfrac{(x + 5)\sqrt{x}}{\sqrt{x - 8}} \right]$

Section 7.6   The Inverse of a Square Matrix

■ You should be able to find the inverse, if it exists, of a square matrix.

    (a) Write the $n \times 2n$ matrix that consists of the given matrix A on the left and the $n \times n$ identity matrix I on the right to obtain $[A \;\vdots\; I]$. Note that we separate the matrices A and I by a dotted line. We call this process **adjoining** the matrices A and I.

    (b) If possible, row reduce A to I using elementary row operations on the *entire* matrix $[A \;\vdots\; I]$. The result will be the matrix $[I \;\vdots\; A^{-1}]$. If this is not possible, then A is not invertible.

    (c) Check your work by multiplying to see that $AA^{-1} = I = A^{-1}A$.

■ You should be able to use inverse matrices to solve systems of equation.

■ You should be able to find inverses using a graphing utility.

Solutions to Odd-Numbered Exercises

1. $AB = \begin{bmatrix} 2 & 1 \\ 5 & 3 \end{bmatrix} \begin{bmatrix} 3 & -1 \\ -5 & 2 \end{bmatrix} = \begin{bmatrix} 2(3) + 1(-5) & 2(-1) + 1(2) \\ 5(3) + 3(-5) & 5(-1) + 3(2) \end{bmatrix} = \begin{bmatrix} 1 & 0 \\ 0 & 1 \end{bmatrix}$

$BA = \begin{bmatrix} 3 & -1 \\ -5 & 2 \end{bmatrix} \begin{bmatrix} 2 & 1 \\ 5 & 3 \end{bmatrix} = \begin{bmatrix} 3(2) + (-1)(5) & 3(1) + (-1)(3) \\ -5(2) + 2(5) & -5(1) + 2(3) \end{bmatrix} = \begin{bmatrix} 1 & 0 \\ 0 & 1 \end{bmatrix}$

3. $AB = \begin{bmatrix} 1 & 2 \\ 3 & 4 \end{bmatrix} \begin{bmatrix} -2 & 1 \\ \frac{3}{2} & -\frac{1}{2} \end{bmatrix} = \begin{bmatrix} -2+3 & 1-1 \\ -6+6 & 3-2 \end{bmatrix} = \begin{bmatrix} 1 & 0 \\ 0 & 1 \end{bmatrix}$

$BA = \begin{bmatrix} -2 & 1 \\ \frac{3}{2} & -\frac{1}{2} \end{bmatrix} \begin{bmatrix} 1 & 2 \\ 3 & 4 \end{bmatrix} = \begin{bmatrix} -2+3 & -4+4 \\ \frac{3}{2}-\frac{3}{2} & 3-2 \end{bmatrix} = \begin{bmatrix} 1 & 0 \\ 0 & 1 \end{bmatrix}$

5. $AB = \begin{bmatrix} 2 & -17 & 11 \\ -1 & 11 & -7 \\ 0 & 3 & -2 \end{bmatrix} \begin{bmatrix} 1 & 1 & 2 \\ 2 & 4 & -3 \\ 3 & 6 & -5 \end{bmatrix}$

$= \begin{bmatrix} 2-34+33 & 2-68+66 & 4+51-55 \\ -1+22-21 & -1+44-42 & -2-33+35 \\ 6-6 & 12-12 & -9+10 \end{bmatrix} = \begin{bmatrix} 1 & 0 & 0 \\ 0 & 1 & 0 \\ 0 & 0 & 1 \end{bmatrix}$

$BA = \begin{bmatrix} 1 & 1 & 2 \\ 2 & 4 & -3 \\ 3 & 6 & -5 \end{bmatrix} \begin{bmatrix} 2 & -17 & 11 \\ -1 & 11 & -7 \\ 0 & 3 & -2 \end{bmatrix} = \begin{bmatrix} 2-1 & -17+11+6 & 11-7-4 \\ 4-4 & -34+44-9 & 22-28+6 \\ 6-6 & -51+66-15 & 33-42+10 \end{bmatrix} = \begin{bmatrix} 1 & 0 & 0 \\ 0 & 1 & 0 \\ 0 & 0 & 1 \end{bmatrix}$

7. $AB = \frac{1}{3} \begin{bmatrix} -2 & 2 & 3 \\ 1 & -1 & 0 \\ 0 & 1 & 4 \end{bmatrix} \begin{bmatrix} -4 & -5 & 3 \\ -4 & -8 & 3 \\ 1 & 2 & 0 \end{bmatrix} = \frac{1}{3} \begin{bmatrix} 8-8+3 & 10-16+6 & -6+6 \\ -4+4 & -5+8 & 3-3 \\ -4+4 & -8+8 & 3 \end{bmatrix}$

$= \frac{1}{3} \begin{bmatrix} 3 & 0 & 0 \\ 0 & 3 & 0 \\ 0 & 0 & 3 \end{bmatrix} = \begin{bmatrix} 1 & 0 & 0 \\ 0 & 1 & 0 \\ 0 & 0 & 1 \end{bmatrix}$

$BA = \frac{1}{3} \begin{bmatrix} -4 & -5 & 3 \\ -4 & -8 & 3 \\ 1 & 2 & 0 \end{bmatrix} \begin{bmatrix} -2 & 2 & 3 \\ 1 & -1 & 0 \\ 0 & 1 & 4 \end{bmatrix} = \frac{1}{3} \begin{bmatrix} 8-5 & -8+5+3 & -12+12 \\ 8-8 & -8+8+3 & -12+12 \\ -2+2 & 2-2 & 3 \end{bmatrix} = \begin{bmatrix} 1 & 0 & 0 \\ 0 & 1 & 0 \\ 0 & 0 & 1 \end{bmatrix}$

9. $AB = \begin{bmatrix} -1 & -4 \\ 1 & 2 \end{bmatrix} \begin{bmatrix} 1 & 2 \\ -\frac{1}{2} & -\frac{1}{2} \end{bmatrix} = \begin{bmatrix} 1 & 0 \\ 0 & 1 \end{bmatrix}; BA = \begin{bmatrix} 1 & 0 \\ 0 & 1 \end{bmatrix}$

11. $AB = \begin{bmatrix} 1.6 & 2 \\ -3.5 & -4.5 \end{bmatrix} \begin{bmatrix} 22.5 & 10 \\ -17.5 & -8 \end{bmatrix} = \begin{bmatrix} 1 & 0 \\ 0 & 1 \end{bmatrix}; BA = \begin{bmatrix} 1 & 0 \\ 0 & 1 \end{bmatrix}$

13. $[A \;\vdots\; I] = \begin{bmatrix} 2 & 0 & \vdots & 1 & 0 \\ 0 & 3 & \vdots & 0 & 1 \end{bmatrix}$

$\begin{matrix} \frac{1}{2}R_1 \to \\ \frac{1}{3}R_2 \to \end{matrix} \begin{bmatrix} 1 & 0 & \vdots & \frac{1}{2} & 0 \\ 0 & 1 & \vdots & 0 & \frac{1}{3} \end{bmatrix} = [I \;\vdots\; A^{-1}]$

$A^{-1} = \begin{bmatrix} \frac{1}{2} & 0 \\ 0 & \frac{1}{3} \end{bmatrix} = \frac{1}{6} \begin{bmatrix} 3 & 0 \\ 0 & 2 \end{bmatrix}$

15. $[A \;\vdots\; I] = \begin{bmatrix} 1 & -2 & \vdots & 1 & 0 \\ 2 & -3 & \vdots & 0 & 1 \end{bmatrix}$

$-2R_1 + R_2 \to \begin{bmatrix} 1 & -2 & \vdots & 1 & 0 \\ 0 & 1 & \vdots & -2 & 1 \end{bmatrix}$

$2R_2 + R_1 \to \begin{bmatrix} 1 & 0 & \vdots & -3 & 2 \\ 0 & 1 & \vdots & -2 & 1 \end{bmatrix} = [I \;\vdots\; A^{-1}]$

$A^{-1} = \begin{bmatrix} -3 & 2 \\ -2 & 1 \end{bmatrix}$

17. $[A \ \vdots \ I] = \begin{bmatrix} -1 & 1 & \vdots & 1 & 0 \\ -2 & 1 & \vdots & 0 & 1 \end{bmatrix}$

$-2R_1 + R_2 \rightarrow \begin{bmatrix} -1 & 1 & \vdots & 1 & 0 \\ 0 & -1 & \vdots & -2 & 1 \end{bmatrix}$

$R_2 + R_1 \rightarrow \begin{bmatrix} -1 & 0 & \vdots & -1 & 1 \\ 0 & -1 & \vdots & -2 & 1 \end{bmatrix}$

$\begin{matrix} -R_1 \rightarrow \\ -R_2 \rightarrow \end{matrix} \begin{bmatrix} 1 & 0 & \vdots & 1 & -1 \\ 0 & 1 & \vdots & 2 & -1 \end{bmatrix} = [I \ \vdots \ A^{-1}]$

$A^{-1} = \begin{bmatrix} 1 & -1 \\ 2 & -1 \end{bmatrix}$

19. $A = \begin{bmatrix} 2 & 7 & 1 \\ -3 & -9 & 2 \end{bmatrix}$

A has no inverse because it is not square.

21. $[A \ \vdots \ I] = \begin{bmatrix} 1 & 1 & 1 & \vdots & 1 & 0 & 0 \\ 3 & 5 & 4 & \vdots & 0 & 1 & 0 \\ 3 & 6 & 5 & \vdots & 0 & 0 & 1 \end{bmatrix}$

$\begin{matrix} -3R_1 + R_2 \rightarrow \\ -3R_1 + R_3 \rightarrow \end{matrix} \begin{bmatrix} 1 & 1 & 1 & \vdots & 1 & 0 & 0 \\ 0 & 2 & 1 & \vdots & -3 & 1 & 0 \\ 0 & 3 & 2 & \vdots & -3 & 0 & 1 \end{bmatrix}$

$\begin{matrix} -R_2 + R_1 \rightarrow \\ \frac{1}{2}R_2 \rightarrow \\ -3R_2 + R_3 \rightarrow \end{matrix} \begin{bmatrix} 1 & 0 & \frac{1}{2} & \vdots & \frac{5}{2} & -\frac{1}{2} & 0 \\ 0 & 1 & \frac{1}{2} & \vdots & -\frac{3}{2} & \frac{1}{2} & 0 \\ 0 & 0 & \frac{1}{2} & \vdots & \frac{3}{2} & -\frac{3}{2} & 1 \end{bmatrix}$

$\begin{matrix} -R_3 + R_1 \rightarrow \\ -R_3 + R_2 \rightarrow \\ 2R_3 \rightarrow \end{matrix} \begin{bmatrix} 1 & 0 & 0 & \vdots & 1 & 1 & -1 \\ 0 & 1 & 0 & \vdots & -3 & 2 & -1 \\ 0 & 0 & 1 & \vdots & 3 & -3 & 2 \end{bmatrix}$

$= [I \ \vdots \ A^{-1}]$

$A^{-1} = \begin{bmatrix} 1 & 1 & -1 \\ -3 & 2 & -1 \\ 3 & -3 & 2 \end{bmatrix}$

23. $[A \ \vdots \ I] = \begin{bmatrix} -5 & 0 & 0 & \vdots & 1 & 0 & 0 \\ 2 & 0 & 0 & \vdots & 0 & 1 & 0 \\ -1 & 5 & 7 & \vdots & 0 & 0 & 1 \end{bmatrix}$

$\begin{matrix} (-\frac{1}{5})R_1 \\ (-2)R_1 + R_2 \end{matrix} \begin{bmatrix} 1 & 0 & 0 & \vdots & -\frac{1}{5} & 0 & 0 \\ 0 & 0 & 0 & \vdots & \frac{2}{5} & 1 & 0 \\ -1 & 5 & 7 & \vdots & 0 & 0 & 1 \end{bmatrix}$

Not invertible   (row of zeros)

A^1 does not exist

25. Not invertible. A^{-1} does not exist.

27. $A = \begin{bmatrix} -\frac{1}{2} & \frac{3}{4} & \frac{1}{4} \\ 1 & 0 & -\frac{3}{2} \\ 0 & -1 & \frac{1}{2} \end{bmatrix}$

$A^{-1} = \begin{bmatrix} -12 & -5 & -9 \\ -4 & -2 & -4 \\ -8 & -4 & -6 \end{bmatrix}$

29. $A = \begin{bmatrix} 0.1 & 0.2 & 0.3 \\ -0.3 & 0.2 & 0.2 \\ 0.5 & 0.4 & 0.4 \end{bmatrix}$

$A^{-1} = \frac{5}{11} \begin{bmatrix} 0 & -4 & 2 \\ -22 & 11 & 11 \\ 22 & -6 & -8 \end{bmatrix}$

31. $A = \begin{bmatrix} -1 & 0 & 1 & 0 \\ 0 & 2 & 0 & -1 \\ 2 & 0 & -1 & 0 \\ 0 & -1 & 0 & 1 \end{bmatrix}$

$A^{-1} = \begin{bmatrix} 1 & 0 & 1 & 0 \\ 0 & 1 & 0 & 1 \\ 2 & 0 & 1 & 0 \\ 0 & 1 & 0 & 2 \end{bmatrix}$

33. $\begin{bmatrix} 5 & 1 \\ -2 & -2 \end{bmatrix}^{-1} = \dfrac{1}{5(-2) - (-2)(1)}\begin{bmatrix} -2 & -1 \\ 2 & 5 \end{bmatrix} = \dfrac{1}{-8}\begin{bmatrix} -2 & -1 \\ 2 & 5 \end{bmatrix} = \begin{bmatrix} \frac{1}{4} & \frac{1}{8} \\ -\frac{1}{4} & -\frac{5}{8} \end{bmatrix}$

35. $\begin{bmatrix} \frac{7}{2} & -\frac{3}{4} \\ \frac{1}{5} & \frac{4}{5} \end{bmatrix}^{-1} = \dfrac{1}{\left(\frac{7}{2}\right)\left(\frac{4}{5}\right) - \left(-\frac{3}{4}\right)\left(\frac{1}{5}\right)}\begin{bmatrix} \frac{4}{5} & \frac{3}{4} \\ -\frac{1}{5} & \frac{7}{2} \end{bmatrix}$

$= \dfrac{20}{59}\begin{bmatrix} \frac{4}{5} & \frac{3}{4} \\ -\frac{1}{5} & \frac{7}{2} \end{bmatrix} = \dfrac{1}{59}\begin{bmatrix} 16 & 15 \\ -4 & 70 \end{bmatrix}$

37. $\begin{bmatrix} x \\ y \end{bmatrix} = \begin{bmatrix} -3 & 2 \\ -2 & 1 \end{bmatrix}\begin{bmatrix} 5 \\ 10 \end{bmatrix} = \begin{bmatrix} 5 \\ 0 \end{bmatrix}$

Answer: $(5, 0)$

39. $\begin{bmatrix} x \\ y \end{bmatrix} = \begin{bmatrix} -3 & 2 \\ -2 & 1 \end{bmatrix}\begin{bmatrix} 4 \\ 2 \end{bmatrix}\begin{bmatrix} -8 \\ -6 \end{bmatrix}$

Answer: $(-8, -6)$

41. $\begin{bmatrix} x \\ y \\ z \end{bmatrix} = \begin{bmatrix} 1 & 1 & -1 \\ -3 & 2 & -1 \\ 3 & -3 & 2 \end{bmatrix}\begin{bmatrix} 0 \\ 5 \\ 2 \end{bmatrix} = \begin{bmatrix} 3 \\ 8 \\ -11 \end{bmatrix}$

Answer: $(3, 8, -11)$

43. $\begin{bmatrix} x_1 \\ x_2 \\ x_3 \\ x_4 \end{bmatrix} = \begin{bmatrix} -24 & 7 & 1 & -2 \\ -10 & 3 & 0 & -1 \\ -29 & 7 & 3 & -2 \\ 12 & -3 & -1 & 1 \end{bmatrix}\begin{bmatrix} 0 \\ 1 \\ -1 \\ 2 \end{bmatrix} = \begin{bmatrix} 2 \\ 1 \\ 0 \\ 0 \end{bmatrix}$

Answer: $(2, 1, 0, 0)$

45. $A = \begin{bmatrix} 3 & 4 \\ 5 & 3 \end{bmatrix}$

$A^{-1} = \dfrac{1}{9 - 20}\begin{bmatrix} 3 & -4 \\ -5 & 3 \end{bmatrix}$

$\begin{bmatrix} x \\ y \end{bmatrix} = -\dfrac{1}{11}\begin{bmatrix} 3 & -4 \\ -5 & 3 \end{bmatrix}\begin{bmatrix} -2 \\ 4 \end{bmatrix} = -\dfrac{1}{11}\begin{bmatrix} -22 \\ 22 \end{bmatrix}$

$= \begin{bmatrix} 2 \\ -2 \end{bmatrix}$

Answer: $(2, -2)$

47. $A = \begin{bmatrix} -0.4 & 0.8 \\ 2 & -4 \end{bmatrix}$

$A^{-1} = \dfrac{1}{1.6 - 1.6}\begin{bmatrix} -4 & -0.8 \\ -2 & -0.4 \end{bmatrix} \Rightarrow A^{-1}$

does not exist.

[The system actually has no solution.]

49. $A = \begin{bmatrix} -\frac{1}{4} & \frac{3}{8} \\ \frac{3}{2} & \frac{3}{4} \end{bmatrix}$

$A^{-1} = \begin{bmatrix} -1 & \frac{1}{2} \\ 2 & \frac{1}{3} \end{bmatrix}$

$\begin{bmatrix} x \\ y \end{bmatrix} = A^{-1}b = \begin{bmatrix} -1 & \frac{1}{2} \\ 2 & \frac{1}{3} \end{bmatrix}\begin{bmatrix} -2 \\ -12 \end{bmatrix} = \begin{bmatrix} -4 \\ -8 \end{bmatrix}$

Answer: $(-4, -8)$

51. $A = \begin{bmatrix} 4 & -1 & 1 \\ 2 & 2 & 3 \\ 5 & -2 & 6 \end{bmatrix}$

$A^{-1} = \frac{1}{55}\begin{bmatrix} 18 & 4 & -5 \\ 3 & 19 & -10 \\ -14 & 3 & 10 \end{bmatrix}$

$\begin{bmatrix} x \\ y \\ z \end{bmatrix} = \frac{1}{55}\begin{bmatrix} 18 & 4 & -5 \\ 3 & 19 & -10 \\ -14 & 3 & 10 \end{bmatrix}\begin{bmatrix} -5 \\ 10 \\ 1 \end{bmatrix} = \frac{1}{55}\begin{bmatrix} -55 \\ 165 \\ 110 \end{bmatrix}$

$= \begin{bmatrix} -1 \\ 3 \\ 2 \end{bmatrix}$

Answer: $(-1, 3, 2)$

53. $A = \begin{bmatrix} 5 & -3 & 2 \\ 2 & 2 & -3 \\ -1 & 7 & -8 \end{bmatrix}$

A^{-1} does not exist. [The system actually has an infinite number of solutions of the form

$x = 0.3125t + 0.8125$

$y = 1.1875t + 0.6875$

$z = t$

where t is any real number.]

55. $\begin{bmatrix} 7 & -3 & 0 & 2 & \vdots & 41 \\ -2 & 1 & 0 & -1 & \vdots & -13 \\ 4 & 0 & 1 & -2 & \vdots & 12 \\ -1 & 1 & 0 & -1 & \vdots & -8 \end{bmatrix}$ row reduces to $\begin{bmatrix} 1 & 0 & 0 & 0 & \vdots & 5 \\ 0 & 1 & 0 & 0 & \vdots & 0 \\ 0 & 0 & 1 & 0 & \vdots & -2 \\ 0 & 0 & 0 & 1 & \vdots & 3 \end{bmatrix}$

Solution: $(5, 0, -2, 3)$

For 57 and 59 use $A = \begin{bmatrix} 1 & 1 & 1 \\ 0.065 & 0.07 & 0.09 \\ 0 & 2 & -1 \end{bmatrix}$. Using the methods of this section, we have $A^{-1} = \frac{1}{11}\begin{bmatrix} 50 & -600 & -4 \\ -13 & 200 & 5 \\ -26 & 400 & -1 \end{bmatrix}$.

57. $X = A^{-1}B = \frac{1}{11}\begin{bmatrix} 50 & -600 & -4 \\ -13 & 200 & 5 \\ -26 & 400 & -1 \end{bmatrix}\begin{bmatrix} 25{,}000 \\ 1900 \\ 0 \end{bmatrix} = \begin{bmatrix} 10{,}000 \\ 5000 \\ 10{,}000 \end{bmatrix}$

Answer: $10,000 in AAA bonds, $5000 in A-bonds and $10,000 in B-bonds.

59. $X = A^{-1}B = \frac{1}{11}\begin{bmatrix} 50 & -600 & -4 \\ -13 & 200 & 5 \\ -26 & 400 & -1 \end{bmatrix}\begin{bmatrix} 65{,}000 \\ 5050 \\ 0 \end{bmatrix} = \begin{bmatrix} 20{,}000 \\ 15{,}000 \\ 30{,}000 \end{bmatrix}$

Answer: $20,000 in AAA bonds, $15,000 in A-bonds and $30,000 in B-bonds.

61. (a) $A = \begin{bmatrix} 2 & 0 & 4 \\ 0 & 1 & 4 \\ 1 & 1 & -1 \end{bmatrix}$ $A^{-1} = \frac{1}{14}\begin{bmatrix} 5 & -4 & 4 \\ -4 & 6 & 8 \\ 1 & 2 & -2 \end{bmatrix}$

$\begin{bmatrix} I_1 \\ I_2 \\ I_3 \end{bmatrix} = \frac{1}{14}\begin{bmatrix} 5 & -4 & 4 \\ -4 & 6 & 8 \\ 1 & 2 & -2 \end{bmatrix}\begin{bmatrix} 14 \\ 28 \\ 0 \end{bmatrix} = \begin{bmatrix} -3 \\ 8 \\ 5 \end{bmatrix}$

Answer: $I_1 = -3$ amps, $I_2 = 8$ amps, $I_3 = 5$ amps

—CONTINUED—

61.   —CONTINUED—

(b) $A = \begin{bmatrix} 2 & 0 & 4 \\ 0 & 1 & 4 \\ 1 & 1 & -1 \end{bmatrix}$ $A^{-1} = \frac{1}{14} \begin{bmatrix} 5 & -4 & 4 \\ -4 & 6 & 8 \\ 1 & 2 & -2 \end{bmatrix}$

$\begin{bmatrix} I_1 \\ I_2 \\ I_3 \end{bmatrix} = \frac{1}{14} \begin{bmatrix} 5 & -4 & 4 \\ -4 & 6 & 8 \\ 1 & 2 & -2 \end{bmatrix} \begin{bmatrix} 10 \\ 10 \\ 0 \end{bmatrix} = \begin{bmatrix} 5/7 \\ 10/7 \\ 15/7 \end{bmatrix}$

Answer: $I_1 = 5/7$ amps, $I_2 = 10/7$ amps, $I_3 = 15/7$ amps

63. True. $AA^{-1} = A^{-1}A = I$

65. $AA^{-1} = \begin{bmatrix} a & b \\ c & d \end{bmatrix} \left(\frac{1}{ad - bc} \right) \begin{bmatrix} d & -b \\ -c & a \end{bmatrix} = \frac{1}{ad - bc} \begin{bmatrix} a & b \\ c & d \end{bmatrix} \begin{bmatrix} d & -b \\ -c & a \end{bmatrix}$

$= \frac{1}{ad - bc} \begin{bmatrix} ad - bc & 0 \\ 0 & ad - bc \end{bmatrix} = \begin{bmatrix} 1 & 0 \\ 0 & 1 \end{bmatrix}$

$A^{-1}A = \frac{1}{ad - bc} \begin{bmatrix} d & -b \\ -c & a \end{bmatrix} \begin{bmatrix} a & b \\ c & d \end{bmatrix} = \frac{1}{ad - bc} \begin{bmatrix} ad - bc & 0 \\ 0 & ad - bc \end{bmatrix} = \begin{bmatrix} 1 & 0 \\ 0 & 1 \end{bmatrix}$

67. $\dfrac{\left(\dfrac{9}{x} \right)}{\left(\dfrac{6}{x} + 2 \right)} = \dfrac{\left(\dfrac{9}{x} \right)}{\left(\dfrac{6 + 2x}{x} \right)} = \dfrac{9}{x} \cdot \dfrac{x}{6 + 2x} = \dfrac{9}{6 + 2x}, \quad x \neq 0$

69. $\dfrac{\dfrac{4}{x^2 - 9} + \dfrac{2}{x - 2}}{\dfrac{1}{x + 3} + \dfrac{1}{x - 3}} \cdot \dfrac{(x^2 - 9)(x - 2)}{(x^2 - 9)(x - 2)}$

$= \dfrac{4(x - 2) + 2(x^2 - 9)}{(x - 3)(x - 2) + (x + 3)(x - 2)}$

$= \dfrac{2x^2 + 4x - 26}{2x^2 - 4x}$

$= \dfrac{x^2 + 2x - 13}{x(x - 2)}, \quad x \neq \pm 3$

71. $e^{2x} + 2e^x - 15 = (e^x + 5)(e^x - 3) = 0 \implies e^x = 3 \implies x = \ln 3 \approx 1.099$

73. $7 \ln 3x = 12$

$\ln 3x = \frac{12}{7}$

$3x = e^{12/7}$

$x = \frac{1}{3} e^{12/7} \approx 1.851$

Section 7.7   The Determinant of a Square Matrix

■ You should be able to determine the determinant of a matrix of order 2×2 by using the products of the diagonals.

■ You should be able to use expansion by cofactors to find the determinant of a matrix of order 3 or greater.

■ The determinant of a triangular matrix equals the product of the entries on the main diagonal.

■ You should be able to calculate determinants using a graphing utility.

Solutions to Odd-Numbered Exercises

1. $|4| = 4$

3. $\begin{vmatrix} 8 & 4 \\ 2 & 3 \end{vmatrix} = 8(3) - 4(2) = 24 - 8 = 16$

5. $\begin{vmatrix} 6 & 2 \\ -5 & 3 \end{vmatrix} = 6(3) - (2)(-5) = 18 + 10 = 28$

7. $\begin{vmatrix} -7 & 6 \\ \frac{1}{2} & 3 \end{vmatrix} = -7(3) - 6\left(\frac{1}{2}\right) = -21 - 3 = -24$

9. $\begin{vmatrix} 2 & -1 & 0 \\ 4 & 2 & 1 \\ 4 & 2 & 1 \end{vmatrix} = 2\begin{vmatrix} 2 & 1 \\ 2 & 1 \end{vmatrix} - 4\begin{vmatrix} -1 & 0 \\ 2 & 1 \end{vmatrix} + 4\begin{vmatrix} -1 & 0 \\ 2 & 1 \end{vmatrix} = 2(0) - 4(-1) + 4(-1) = 0$

11. $\begin{vmatrix} -1 & 2 & -5 \\ 0 & 3 & 4 \\ 0 & 0 & 3 \end{vmatrix} = (-1)(3)(3) = -9$ (Upper Triangular)

13. $\begin{vmatrix} 0.3 & 0.2 & 0.2 \\ 0.2 & 0.2 & 0.2 \\ -0.4 & 0.4 & 0.3 \end{vmatrix} = -0.002$

15. $\begin{bmatrix} 3 & 4 \\ 2 & -5 \end{bmatrix}$

(a) $M_{11} = -5$

$M_{12} = 2$

$M_{21} = 4$

$M_{22} = 3$

(b) $C_{11} = M_{11} = -5$

$C_{12} = -M_{12} = -2$

$C_{21} = -M_{21} = -4$

$C_{22} = M_{22} = 3$

17. $\begin{bmatrix} 3 & -2 & 8 \\ 3 & 2 & -6 \\ -1 & 3 & 6 \end{bmatrix}$

(a) $M_{11} = \begin{vmatrix} 2 & -6 \\ 3 & 6 \end{vmatrix} = 12 + 18 = 30$

$M_{12} = \begin{vmatrix} 3 & -6 \\ -1 & 6 \end{vmatrix} = 18 - 6 = 12$

$M_{13} = \begin{vmatrix} 3 & 2 \\ -1 & 3 \end{vmatrix} = 9 + 2 = 11$

$M_{21} = \begin{vmatrix} -2 & 8 \\ 3 & 6 \end{vmatrix} = -12 - 24 = -36$

$M_{22} = \begin{vmatrix} 3 & 8 \\ -1 & 6 \end{vmatrix} = 18 + 8 = 26$

$M_{23} = \begin{vmatrix} 3 & -2 \\ -1 & 3 \end{vmatrix} = 9 - 2 = 7$

$M_{31} = \begin{vmatrix} -2 & 8 \\ 2 & -6 \end{vmatrix} = 12 - 16 = -4$

$M_{32} = \begin{vmatrix} 3 & 8 \\ 3 & -6 \end{vmatrix} = -18 - 24 = -42$

$M_{33} = \begin{vmatrix} 3 & -2 \\ 3 & 2 \end{vmatrix} = 6 + 6 = 12$

(b) $C_{11} = (-1)^2 M_{11} = \quad 30$

$C_{12} = (-1)^3 M_{12} = -12$

$C_{13} = (-1)^4 M_{13} = \quad 11$

$C_{21} = (-1)^3 M_{21} = \quad 36$

$C_{22} = (-1)^4 M_{22} = \quad 26$

$C_{23} = (-1)^5 M_{23} = \quad -7$

$C_{31} = (-1)^4 M_{31} = \quad -4$

$C_{32} = (-1)^5 M_{32} = \quad 42$

$C_{33} = (-1)^6 M_{33} = \quad 12$

19. (a) $\begin{vmatrix} -3 & 2 & 1 \\ 4 & 5 & 6 \\ 2 & -3 & 1 \end{vmatrix} = -3 \begin{vmatrix} 5 & 6 \\ -3 & 1 \end{vmatrix} - 2 \begin{vmatrix} 4 & 6 \\ 2 & 1 \end{vmatrix} + \begin{vmatrix} 4 & 5 \\ 2 & -3 \end{vmatrix} = -3(23) - 2(-8) - 22 = -75$

(b) $\begin{vmatrix} -3 & 2 & 1 \\ 4 & 5 & 6 \\ 2 & -3 & 1 \end{vmatrix} = -2 \begin{vmatrix} 4 & 6 \\ 2 & 1 \end{vmatrix} + 5 \begin{vmatrix} -3 & 1 \\ 2 & 1 \end{vmatrix} + 3 \begin{vmatrix} -3 & 1 \\ 4 & 6 \end{vmatrix} = -2(-8) + 5(-5) + 3(-22) = -75$

21. (a) $\begin{vmatrix} 6 & 0 & -3 & 5 \\ 4 & 13 & 6 & -8 \\ -1 & 0 & 7 & 4 \\ 8 & 6 & 0 & 2 \end{vmatrix} = -4 \begin{vmatrix} 0 & -3 & 5 \\ 0 & 7 & 4 \\ 6 & 0 & 2 \end{vmatrix} + 13 \begin{vmatrix} 6 & -3 & 5 \\ -1 & 7 & 4 \\ 8 & 0 & 2 \end{vmatrix} - 6 \begin{vmatrix} 6 & 0 & 5 \\ -1 & 0 & 4 \\ 8 & 6 & 2 \end{vmatrix} - 8 \begin{vmatrix} 6 & 0 & -3 \\ -1 & 0 & 7 \\ 8 & 6 & 0 \end{vmatrix}$

$= -4(-282) + 13(-298) - 6(-174) - 8(-234) = 170$

(b) $\begin{vmatrix} 6 & 0 & -3 & 5 \\ 4 & 13 & 6 & -8 \\ -1 & 0 & 7 & 4 \\ 8 & 6 & 0 & 2 \end{vmatrix} = 0 \begin{vmatrix} 4 & 6 & -8 \\ -1 & 7 & 4 \\ 8 & 0 & 2 \end{vmatrix} + 13 \begin{vmatrix} 6 & -3 & 5 \\ -1 & 7 & 4 \\ 8 & 0 & 2 \end{vmatrix} + 0 \begin{vmatrix} 6 & -3 & 5 \\ 4 & 6 & -8 \\ 8 & 0 & 2 \end{vmatrix} + 6 \begin{vmatrix} 6 & -3 & 5 \\ 4 & 6 & -8 \\ -1 & 7 & 4 \end{vmatrix}$

$= 0 + 13(-298) + 0 + 6(674) = 170$

23. Expand by Column 3.

$$\begin{vmatrix} 1 & 4 & -2 \\ 3 & 2 & 0 \\ -1 & 4 & 3 \end{vmatrix} = -2 \begin{vmatrix} 3 & 2 \\ -1 & 4 \end{vmatrix} + 3 \begin{vmatrix} 1 & 4 \\ 3 & 2 \end{vmatrix} = -2(14) + 3(-10) = -58$$

25. $\begin{vmatrix} 2 & 4 & 6 \\ 0 & 3 & 1 \\ 0 & 0 & -5 \end{vmatrix} = (2)(3)(-5) = -30$ (Upper Triangular)

27. Expand by Column 3.

$$\begin{vmatrix} 2 & 6 & 6 & 2 \\ 2 & 7 & 3 & 6 \\ 1 & 5 & 0 & 1 \\ 3 & 7 & 0 & 7 \end{vmatrix} = 6 \begin{vmatrix} 2 & 7 & 6 \\ 1 & 5 & 1 \\ 3 & 7 & 7 \end{vmatrix} - 3 \begin{vmatrix} 2 & 6 & 2 \\ 1 & 5 & 1 \\ 3 & 7 & 7 \end{vmatrix} = 6(-20) - 3(16) = -168$$

29. Expand by Column 2.

$$\begin{vmatrix} 3 & 2 & 4 & -1 & 5 \\ -2 & 0 & 1 & 3 & 2 \\ 1 & 0 & 0 & 4 & 0 \\ 6 & 0 & 2 & -1 & 0 \\ 3 & 0 & 5 & 1 & 0 \end{vmatrix} = -2 \begin{vmatrix} -2 & 1 & 3 & 2 \\ 1 & 0 & 4 & 0 \\ 6 & 2 & -1 & 0 \\ 3 & 5 & 1 & 0 \end{vmatrix} = (-2)(-2) \begin{vmatrix} 1 & 0 & 4 \\ 6 & 2 & -1 \\ 3 & 5 & 1 \end{vmatrix} = 4(103) = 412$$

31. $\begin{vmatrix} 4 & 0 & 0 & 0 \\ 6 & -5 & 0 & 0 \\ 1 & 3 & 1 & 0 \\ 1 & -2 & 7 & 3 \end{vmatrix} = (4)(-5)(1)(3) = -60$ (Lower Triangular)

33. $\det(A) = (-6)(-1)(-7)(-2)(-2) = -168$ (Upper Triangular)

35. $\begin{vmatrix} 1 & -1 & 8 & 4 \\ 2 & 6 & 0 & -4 \\ 2 & 0 & 2 & 6 \\ 0 & 2 & 8 & 0 \end{vmatrix} = -336$

37. $\begin{vmatrix} 3 & -2 & 4 & 3 & 1 \\ -1 & 0 & 2 & 1 & 0 \\ 5 & -1 & 0 & 3 & 2 \\ 4 & 7 & -8 & 0 & 0 \\ 1 & 2 & 3 & 0 & 2 \end{vmatrix} = 410$

39. (a) $\begin{vmatrix} -1 & 0 \\ 0 & 3 \end{vmatrix} = -3$

(b) $\begin{vmatrix} 2 & 0 \\ 0 & -1 \end{vmatrix} = -2$

(c) $\begin{bmatrix} -1 & 0 \\ 0 & 3 \end{bmatrix} \begin{bmatrix} 2 & 0 \\ 0 & -1 \end{bmatrix} = \begin{bmatrix} -2 & 0 \\ 0 & -3 \end{bmatrix}$

(d) $\begin{vmatrix} -2 & 0 \\ 0 & -3 \end{vmatrix} = 6$ [Note: $|AB| = |A| \, |B|$]

41. (a) $\begin{vmatrix} -1 & 2 & 1 \\ 1 & 0 & 1 \\ 0 & 1 & 0 \end{vmatrix} = 2$

(b) $\begin{vmatrix} -1 & 0 & 0 \\ 0 & 2 & 0 \\ 0 & 0 & 3 \end{vmatrix} = -6$

(c) $\begin{bmatrix} -1 & 2 & 1 \\ 1 & 0 & 1 \\ 0 & 1 & 0 \end{bmatrix} \begin{bmatrix} -1 & 0 & 0 \\ 0 & 2 & 0 \\ 0 & 0 & 3 \end{bmatrix} = \begin{bmatrix} 1 & 4 & 3 \\ -1 & 0 & 3 \\ 0 & 2 & 0 \end{bmatrix}$

(d) $\begin{vmatrix} 1 & 4 & 3 \\ -1 & 0 & 3 \\ 0 & 2 & 0 \end{vmatrix} = -12$ [Note: $|AB| = |A|\,|B|$]

43. (a) $|A| = -25$

(b) $|B| = -220$

(c) $AB = \begin{bmatrix} -7 & -16 & -1 & -28 \\ -4 & -14 & -11 & 8 \\ 13 & 4 & 4 & -4 \\ -2 & 3 & 2 & 2 \end{bmatrix}$

(d) $|AB| = 5500$ [Note: $|AB| = |A|\,|B|$]

45. $\begin{vmatrix} w & x \\ y & z \end{vmatrix} = wz - xy$

$-\begin{vmatrix} y & z \\ w & x \end{vmatrix} = -(xy - wz) = wz - xy$

Thus, $\begin{vmatrix} w & x \\ y & z \end{vmatrix} = -\begin{vmatrix} y & z \\ w & x \end{vmatrix}$.

47. $\begin{vmatrix} w & x \\ y & z \end{vmatrix} = wz - xy$

$\begin{vmatrix} w & x + cw \\ y & z + cy \end{vmatrix} = w(z + cy) - y(x + cw) = wz - xy$

Thus, $\begin{vmatrix} w & x \\ y & z \end{vmatrix} = \begin{vmatrix} w & x + cw \\ y & z + cy \end{vmatrix}$.

49. $\begin{vmatrix} 1 & x & x^2 \\ 1 & y & y^2 \\ 1 & z & z^2 \end{vmatrix} = \begin{vmatrix} y & y^2 \\ z & z^2 \end{vmatrix} - \begin{vmatrix} x & x^2 \\ z & z^2 \end{vmatrix} + \begin{vmatrix} x & x^2 \\ y & y^2 \end{vmatrix}$

$= (yz^2 - y^2z) - (xz^2 - x^2z) + (xy^2 - x^2y)$

$= yz^2 - xz^2 - y^2z + x^2z + xy(y - x)$

$= z^2(y - x) - z(y^2 - x^2) + xy(y - x)$

$= z^2(y - x) - z(y - x)(y + x) + xy(y - x)$

$= (y - x)[z^2 - z(y + x) + xy]$

$= (y - x)[z^2 - zy - zx + xy]$

$= (y - x)[z^2 - zx - zy + xy]$

$= (y - x)[z(z - x) - y(z - x)]$

$= (y - x)(z - x)(z - y)$

51. $\begin{vmatrix} x+3 & 2 \\ 1 & x+2 \end{vmatrix} = 0$

$(x+3)(x+2) - 2 = 0$

$x^2 + 5x + 4 = 0$

$(x+4)(x+1) = 0$

$x = -4, -1$

53. $\begin{vmatrix} 4u & -1 \\ -1 & 2v \end{vmatrix} = 8uv - 1$

55. $\begin{vmatrix} e^{2x} & e^{3x} \\ 2e^{2x} & 3e^{3x} \end{vmatrix} = 3e^{5x} - 2e^{5x} = e^{5x}$

57. $\begin{vmatrix} x & \ln x \\ 1 & \dfrac{1}{x} \end{vmatrix} = 1 - \ln x$

59. True. Expand along the row of zeros.

61. Let $A = \begin{bmatrix} 1 & 3 \\ -2 & 4 \end{bmatrix}$ and $B = \begin{bmatrix} -4 & 0 \\ 3 & 5 \end{bmatrix}$.

$|A| = \begin{vmatrix} 1 & 3 \\ -2 & 4 \end{vmatrix} = 10, \quad |B| = \begin{vmatrix} -4 & 0 \\ 3 & 5 \end{vmatrix} = -20$

$A + B = \begin{bmatrix} -3 & 3 \\ 1 & 9 \end{bmatrix}, \quad |A + B| = \begin{vmatrix} -3 & 3 \\ 1 & 9 \end{vmatrix} = -30$

Thus, $|A + B| \neq |A| + |B|$. Your answer may differ, depending on how you choose A and B.

63. (a) Columns 2 and 3 are interchanged.

(b) Rows 1 and 3 are interchanged.

65. (a) 5 is factored out of the first row of A.

(b) 4 and 3 are factored out of columns 2 and 3.

67. $x^2 - 3x + 2 = (x - 2)(x - 1)$

69. $4y^2 - 12y + 9 = (2y - 3)^2$

71. $3x - 10y = 46$

$x + y = -2$

$y = -x - 2$

$3x - 10(-x - 2) = 46$

$13x = 26$

$x = 2$

$y = -2 - 2 = -4$

Solution: $(2, -4)$

Section 7.8   Applications of Matrices and Determinants

■ You should be able to find the area of a triangle with vertices (x_1, y_1), (x_2, y_2), and (x_3, y_3).

$$\text{Area} = \pm\frac{1}{2}\begin{vmatrix} x_1 & y_1 & 1 \\ x_2 & y_2 & 1 \\ x_3 & y_3 & 1 \end{vmatrix}$$

The $\pm$ symbol indicates that the appropriate sign should be chosen so that the area is positive.

■ You should be able to test to see if three points, (x_1, y_1), (x_2, y_2), and (x_3, y_3), are collinear.

$$\begin{vmatrix} x_1 & y_1 & 1 \\ x_2 & y_2 & 1 \\ x_3 & y_3 & 1 \end{vmatrix} = 0, \text{ if and only if they are collinear.}$$

■ You should be able to use Cramer's Rule to solve a system of linear equations.

■ Now you should be able to solve a system of linear equations by substitution, elimination, elementary row operations on an augmented matrix, using the inverse matrix, or Cramer's Rule.

■ You should be able to encode and decode messages by using an invertible $n \times n$ matrix.

Solutions to Odd-Numbered Exercises

1. Vertices: $(-2, -3)$, $(2, -3)$, $(0, 4)$

$$\frac{1}{2}\begin{vmatrix} -2 & -3 & 1 \\ 2 & -3 & 1 \\ 0 & 4 & 1 \end{vmatrix} = \frac{1}{2}\left(-2\begin{vmatrix} -3 & 1 \\ 4 & 1 \end{vmatrix} - 2\begin{vmatrix} -3 & 1 \\ 4 & 1 \end{vmatrix}\right) = \frac{1}{2}(14 + 14). \text{ Area} = 14 \text{ square units}$$

3. Vertices: $(-2, 4)$, $(2, 3)$, $(-1, 5)$

$$\frac{1}{2}\begin{vmatrix} -2 & 4 & 1 \\ 2 & 3 & 1 \\ -1 & 5 & 1 \end{vmatrix} = \frac{1}{2}\left[-2\begin{vmatrix} 3 & 1 \\ 5 & 1 \end{vmatrix} - 4\begin{vmatrix} 2 & 1 \\ -1 & 1 \end{vmatrix} + \begin{vmatrix} 2 & 3 \\ -1 & 5 \end{vmatrix}\right]$$

$$= \frac{1}{2}\left[-2(-2) - 4(3) + 13\right] = \frac{1}{2}(5) = \frac{5}{2}$$

Area $= \frac{5}{2}$ square units

5. Vertices: $\left(0, \frac{1}{2}\right)$, $\left(\frac{5}{2}, 0\right)$, $(4, 3)$

$$\frac{1}{2}\begin{vmatrix} 0 & \frac{1}{2} & 1 \\ \frac{5}{2} & 0 & 1 \\ 4 & 3 & 1 \end{vmatrix} = \frac{1}{2}\left[-\frac{1}{2}\left(-\frac{3}{2}\right) + \frac{15}{2}\right] = \frac{1}{2}\left[\frac{33}{4}\right] = \frac{33}{8}.$$

Area $= \frac{33}{8}$ square units

7. $4 = \pm\dfrac{1}{2}\begin{vmatrix} -5 & 1 & 1 \\ 0 & 2 & 1 \\ -2 & x & 1 \end{vmatrix}$

$\pm 8 = -5\begin{vmatrix} 2 & 1 \\ x & 1 \end{vmatrix} - 2\begin{vmatrix} 1 & 1 \\ 2 & 1 \end{vmatrix}$

$\pm 8 = -5(2 - x) - 2(-1)$

$\pm 8 = 5x - 8$

$x = \dfrac{8 \pm 8}{5}$

$x = \dfrac{16}{5}$ OR $x = 0$

9. Points: $(3, -1)$, $(0, -3)$, $(12, 5)$

$\begin{vmatrix} 3 & -1 & 1 \\ 0 & -3 & 1 \\ 12 & 5 & 1 \end{vmatrix} = 3(-8) + 12(2) = 0$

The points are collinear.

11. Points: $\left(2, -\tfrac{1}{2}\right)$, $(-4, 4)$, $(6, -3)$

$\begin{vmatrix} 2 & -\tfrac{1}{2} & 1 \\ -4 & 4 & 1 \\ 6 & -3 & 1 \end{vmatrix} = 2(7) + \tfrac{1}{2}(-10) + 1(-12) = -3 \neq 0$

The points are not collinear.

13.

$\begin{vmatrix} 2 & -5 & 1 \\ 4 & x & 1 \\ 5 & -2 & 1 \end{vmatrix} = 0$

$2\begin{vmatrix} x & 1 \\ -2 & 1 \end{vmatrix} + 5\begin{vmatrix} 4 & 1 \\ 5 & 1 \end{vmatrix} + \begin{vmatrix} 4 & x \\ 5 & -2 \end{vmatrix} = 0$

$2(x + 2) + 5(-1) + (-8 - 5x) = 0$

$-3x - 9 = 0$

$x = -3$

15. $\begin{cases} -7x + 11y = -1 \\ 3x - 9y = 9 \end{cases}$

$x = \dfrac{\begin{vmatrix} -1 & 11 \\ 9 & -9 \end{vmatrix}}{\begin{vmatrix} -7 & 11 \\ 3 & -9 \end{vmatrix}} = \dfrac{-90}{30} = -3$

$y = \dfrac{\begin{vmatrix} -7 & -1 \\ 3 & 9 \end{vmatrix}}{\begin{vmatrix} -7 & 11 \\ 3 & -9 \end{vmatrix}} = \dfrac{-60}{30} = -2$

Answer: $(-3, -2)$

17. $\begin{cases} 3x + 2y = -2 \\ 6x + 4y = 4 \end{cases}$

$\begin{vmatrix} 3 & 2 \\ 6 & 4 \end{vmatrix} = 12 - 17 = 0$

Cramer's rule cannot be used.

(In fact, the system is inconsistent)

19. $\begin{cases} -0.4x + 0.8y = 1.6 \\ 0.2x + 0.3y = 2.2 \end{cases}$

$D = \begin{vmatrix} -0.4 & 0.8 \\ 0.2 & 0.3 \end{vmatrix} = -0.28$

$x = \dfrac{\begin{vmatrix} 1.6 & 0.8 \\ 2.2 & 0.3 \end{vmatrix}}{-0.28} = \dfrac{-1.28}{-0.28} = \dfrac{32}{7}$

$y = \dfrac{\begin{vmatrix} -0.4 & 1.6 \\ 0.2 & 2.2 \end{vmatrix}}{-0.28} = \dfrac{-1.20}{-0.28} = \dfrac{30}{7}$

Answer: $\left(\dfrac{32}{7}, \dfrac{30}{7}\right)$

21. $\begin{cases} 4x - y + z = -5 \\ 2x + 2y + 3z = 10 \\ 5x - 2y + 6z = 1 \end{cases}$ $D = \begin{vmatrix} 4 & -1 & 1 \\ 2 & 2 & 3 \\ 5 & -2 & 6 \end{vmatrix} = 55$

$$x = \frac{\begin{vmatrix} -5 & -1 & 1 \\ 10 & 2 & 3 \\ 1 & -2 & 6 \end{vmatrix}}{55} = \frac{-55}{55} = -1, \quad y = \frac{\begin{vmatrix} 4 & -5 & 1 \\ 2 & 10 & 3 \\ 5 & 1 & 6 \end{vmatrix}}{55} = \frac{165}{55} = 3, \quad z = \frac{\begin{vmatrix} 4 & -1 & -5 \\ 2 & 2 & 10 \\ 5 & -2 & 1 \end{vmatrix}}{55} = \frac{110}{55} = 2$$

Answer: $(-1, 3, 2)$

23. $\begin{cases} 3x + 3y + 5z = 1 \\ 3x + 5y + 9z = 2 \\ 5x + 9y + 17z = 4 \end{cases}$ $D = \begin{vmatrix} 3 & 3 & 5 \\ 3 & 5 & 9 \\ 5 & 9 & 17 \end{vmatrix} = 4$

$$x = \frac{\begin{vmatrix} 1 & 3 & 5 \\ 2 & 5 & 9 \\ 4 & 9 & 17 \end{vmatrix}}{4} = 0, \quad y = \frac{\begin{vmatrix} 3 & 1 & 5 \\ 3 & 2 & 9 \\ 5 & 4 & 17 \end{vmatrix}}{4} = -\frac{1}{2}, \quad z = \frac{\begin{vmatrix} 3 & 3 & 1 \\ 3 & 5 & 2 \\ 5 & 9 & 4 \end{vmatrix}}{4} = \frac{1}{2}$$

Answer: $\left(0, -\frac{1}{2}, \frac{1}{2}\right)$

$\qquad\qquad\quad A \qquad B \qquad C$

25. Vertices: $(0, 25), (10, 0), (28, 5)$

$\frac{1}{2}\begin{vmatrix} 0 & 25 & 1 \\ 10 & 0 & 1 \\ 28 & 5 & 1 \end{vmatrix} = 250.$ Area $= 250$ square miles

27. The uncoded row matrices are the rows of the 6×3 matrix on the left.

$\begin{matrix} C & A & L \\ L & & M \\ E & & T \\ O & M & O \\ R & R & O \\ W & & \end{matrix}$ $\begin{bmatrix} 3 & 1 & 12 \\ 12 & 0 & 13 \\ 5 & 0 & 20 \\ 15 & 13 & 15 \\ 18 & 18 & 15 \\ 23 & 0 & 0 \end{bmatrix}$ $\begin{bmatrix} 1 & -1 & 0 \\ 1 & 0 & -1 \\ -6 & 2 & 3 \end{bmatrix} = \begin{bmatrix} -68 & 21 & 35 \\ -66 & 14 & 39 \\ -115 & 35 & 60 \\ -62 & 15 & 32 \\ -54 & 12 & 27 \\ 23 & -23 & 0 \end{bmatrix}$

Answer: $[-68, 21, 35], [-66, 14, 39], [-115, 35, 60]$

$\qquad\quad [-62, 15, 32], [-54, 12, 27], [23, -23, 0]$

29. $\begin{matrix} G & O & N \\ E & & F \\ I & S & H \\ I & N & G \end{matrix}$ $\begin{bmatrix} 7 & 15 & 14 \\ 5 & 0 & 6 \\ 9 & 19 & 8 \\ 9 & 14 & 7 \end{bmatrix}$ $\begin{bmatrix} 1 & 2 & 2 \\ 3 & 7 & 9 \\ -1 & -4 & -7 \end{bmatrix} = \begin{bmatrix} 38 & 63 & 51 \\ -1 & -14 & -32 \\ 58 & 119 & 133 \\ 44 & 88 & 95 \end{bmatrix}$

Cryptogram: 38 63 51 −1 −14 −32 58 119 133 44 88 95

31. $A^{-1} = \begin{bmatrix} 1 & 2 \\ 3 & 5 \end{bmatrix}^{-1} = \begin{bmatrix} -5 & 2 \\ 3 & -1 \end{bmatrix}$

$\begin{bmatrix} 11 & 21 \\ 64 & 112 \\ 25 & 50 \\ 29 & 53 \\ 23 & 46 \\ 40 & 75 \\ 55 & 92 \end{bmatrix} \begin{bmatrix} -5 & 2 \\ 3 & -1 \end{bmatrix} = \begin{array}{cc|cc} 8 & 1 & H & A \\ 16 & 16 & P & P \\ 25 & 0 & Y & \\ 14 & 5 & N & E \\ 23 & 0 & W & \\ 25 & 5 & Y & E \\ 1 & 18 & A & R \end{array}$ Message: HAPPY NEW YEAR

33. $A^{-1} = \begin{bmatrix} 1 & 2 & 2 \\ 3 & 7 & 9 \\ -1 & -4 & -7 \end{bmatrix}^{-1} = \begin{bmatrix} -13 & 6 & 4 \\ 12 & -5 & -3 \\ -5 & 2 & 1 \end{bmatrix}$

$\begin{bmatrix} 16 & -1 & -48 \\ 5 & -20 & -65 \\ 8 & 4 & -14 \\ 41 & 83 & 89 \\ 76 & 177 & 227 \end{bmatrix} \begin{bmatrix} -13 & 6 & 4 \\ 12 & -5 & -3 \\ -5 & 2 & 1 \end{bmatrix} = \begin{array}{ccc|ccc} 20 & 5 & 19 & T & E & S \\ 20 & 0 & 15 & T & & O \\ 14 & 0 & 6 & N & & F \\ 18 & 9 & 4 & R & I & D \\ 1 & 25 & 0 & A & Y & \end{array}$ Message: TEST ON FRIDAY

35. True. Cramer's Rule requires that the determinant of the coefficient matrix be nonzero.

37. Answers will vary.

39.
$$y - 5 = \frac{5 - 3}{-1 - 7}(x + 1) = \frac{-1}{4}(x + 1)$$
$$4y - 20 = -x - 1$$
$$4y + x = 19$$
$$x + 4y - 19 = 0$$

41.
$$y + 3 = \frac{-3 + 1}{3 - 10}(x - 3) = \frac{2}{7}(x - 3)$$
$$7y + 21 = 2x - 6$$
$$7y - 2x = -27$$
$$2x - 7y - 27 = 0$$

43. $f(x) = \dfrac{2x^2}{x^2 + 4}$. Horizontal asymptote: $y = 2$

Review Exercises for Chapter 7

Solutions to Odd-Numbered Exercises

1. $\begin{cases} x + y = 2 \implies & y = 2 - x \\ x - y = 0 \implies x - (2 - x) = 0 \end{cases}$

$$2x - 2 = 0$$
$$x = 1$$
$$y = 2 - 1 = 1$$

Answer: $(1, 1)$

3. $\begin{cases} x^2 - y^2 = 9 \\ x - y = 1 \implies x = y + 1 \end{cases}$

$$(y + 1)^2 - y^2 = 9$$
$$2y + 1 = 9$$
$$y = 4$$
$$x = 5$$

Answer: $(5, 4)$

5. $\begin{cases} y = 2x^2 \\ y = x^4 - 2x^2 \implies 2x^2 = x^4 - 2x^2 \end{cases}$

$$0 = x^4 - 4x^2$$
$$0 = x^2(x^2 - 4)$$
$$0 = x^2(x + 2)(x - 2)$$
$$x = 0, x = -2, x = 2$$
$$y = 0, y = 8, y = 8$$

Answers: $(0, 0), (-2, 8), (2, 8)$

7. $\begin{cases} 5x + 6y = 7 \implies y_1 = \dfrac{1}{6}(7 - 5x) \\ -x - 4y = 0 \implies y_2 = -\dfrac{x}{4} \end{cases}$

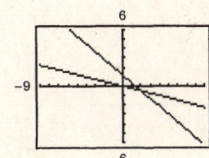

Point of intersection:

$\left(2, -\dfrac{1}{2}\right)$

9. $\begin{cases} y^2 - 2y + x = 0 \implies (y - 1)^2 = 1 - x \implies y = 1 \pm \sqrt{1 - x} \\ x + y = 0 \implies y = -x \end{cases}$

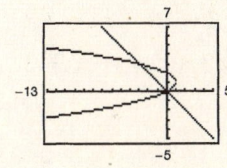

Points of intersection: $(0, 0)$ and $(-3, 3)$

11. $\begin{cases} y = 2(6 - x) \\ y = 2^{x-2} \end{cases}$

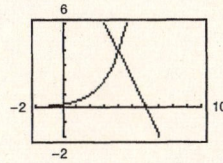

Point of intersection: $(4, 4)$

13. Revenue $= 4.95x$

Cost $= 2.85x + 10{,}000$

Break even when Revenue = Cost

$$4.95x = 2.85x + 10{,}000$$
$$2.10x = 10{,}000$$
$$x \approx 4762 \text{ units}$$

15. $\begin{cases} 2l + 2w = 480 \\ l = 1.50w \end{cases}$

$2(1.50w) + 2w = 480$

$5w = 480$

$w = 96$

$l = 144$

The dimensions are 96×144 meters.

17. $\begin{cases} 2x - y = 2 \implies 16x - 8y = 16 \\ 6x + 8y = 39 \implies 6x + 8y = 39 \end{cases}$

$\qquad\qquad\qquad 22x \qquad\quad = 55$

$\qquad\qquad\qquad\quad x = \frac{55}{22} = \frac{5}{2}$

$\qquad\qquad\qquad\quad y = 3$

Answer: $\left(\frac{5}{2}, 3\right)$

19. $\begin{cases} 1/5x + 3/10y = 7/50 \implies 20x + 30y = 14 \implies 20x + 30y = 14 \\ 2/5x + 1/2y = 1/5 \implies 4x + 5y = 2 \implies -20x - 25y = -10 \end{cases}$

$\qquad\qquad\qquad\qquad\qquad\qquad\qquad\qquad 5y = 4$

$\qquad\qquad\qquad\qquad\qquad\qquad\qquad\quad\; y = \frac{4}{5}$

$\qquad\qquad\qquad\qquad\qquad\qquad\qquad\quad\; x = -\frac{1}{2}$

Answer: $\left(-\dfrac{1}{2}, \dfrac{4}{5}\right)$ or $(-0.5, 0.8)$

21. $\begin{cases} 3x - 2y = 0 \implies 3x - 2y = 0 \\ 3x + 2(y + 5) = 10 \implies 3x + 2y = 0 \end{cases}$

$\qquad\qquad\qquad\qquad\qquad 6x \quad\;\; = 0$

$\qquad\qquad\qquad\qquad\qquad\;\; x = 0$

$\qquad\qquad\qquad\qquad\qquad\;\; y = 0$

Solution: $(0, 0)$

23. $\begin{cases} 1.25x - 2y = 3.5 \implies 5x - 8y = 14 \\ 5x - 8y = 14 \implies -5x + 8y = -14 \end{cases}$

$\qquad\qquad\qquad\qquad\qquad\qquad 0 = 0$

Infinite number of solutions

Let $y = a$, then $5x - 8a = 14 \implies x = \frac{14}{5} + \frac{8}{5}a$.

Solution: $\left(\frac{14}{5} + \frac{8}{5}a, a\right)$

25. $\begin{cases} 3x + 2y = 0 \implies y = -\frac{3}{2}x \\ x - y = 4 \implies y = x - 4 \end{cases}$

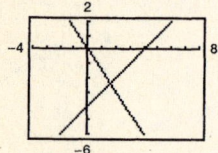

Consistent.

Answer: $(1.6, -2.4)$

27. $\begin{cases} \frac{1}{4}x - \frac{1}{5}y = 2 \implies y = \frac{5}{4}x - 10 \\ -5x + 4y = 8 \implies y = \frac{1}{4}(8 + 5x) = \frac{5}{4}x + 2 \end{cases}$

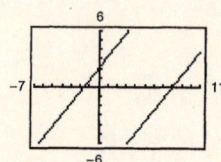

Inconsistent. Lines are parallel.

29. $\begin{cases} 2x - 2y = 8 \implies y = x - 4 \\ 4x - 1.5y = -5.5 \implies y = \frac{8}{3}x + \frac{11}{3} \end{cases}$

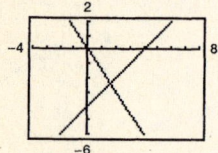

Answer: $(-4.6, -8.6)$

31. $\qquad$ Demand = Supply

$37 - 0.0002x = 22 + 0.00001x$

$15 = 0.00021x$

$x = \dfrac{500,000}{7}, p = \dfrac{159}{7}$

Point of equilibrium $\left(\dfrac{500,000}{7}, \dfrac{159}{7}\right)$

33. Let $x =$ speed of the slower plane.

Let $y =$ speed of the faster plane.

Then, distance of first plane + distance of second plane $= 275$ miles.

(rate of first plane)(time) + (rate of second plane)(time) $= 275$ miles

$$\begin{cases} x\left(\frac{40}{60}\right) + y\left(\frac{40}{60}\right) = 275 \\ \qquad\qquad\quad y = x + 25 \end{cases}$$

$$\frac{2}{3}x + \frac{2}{3}(x + 25) = 275$$

$$4x + 50 = 825$$

$$4x = 775$$

$$x = 193.75 \text{ mph}$$

$$y = x + 25 = 218.75 \text{ mph}$$

35. $\begin{cases} x - 4y + 3z = 3 & \text{Equation 1} \\ -y + z = -1 & \text{Equation 2} \\ z = -5 & \text{Equation 3} \end{cases}$

Substitute $z = -5$ into Equation 2: $-y + (-5) = -1 \implies -y = 4 \implies y = -4$

Substitute $z = -5$ and $y = -4$ into Equation 1:

$$x - 4(-4) + 3(-5) = 3$$

$$x = -16 + 15 + 3$$

$$x = 2$$

Answer: $(2, -4, -5)$

37. $\begin{cases} x + 3y - z = 13 \\ 2x - 5z = 23 \\ 4x - y - 2z = 14 \end{cases}$

$\begin{cases} x + 3y - z = 13 \\ - 6y - 3z = -3 \\ - 13y + 2z = -38 \end{cases}$

$\begin{cases} x + 3y - z = 13 \\ - 6y - 3z = -3 \\ \frac{17}{2}z = -\frac{63}{2} \end{cases}$

$$\frac{17}{2}z = -\frac{63}{2} \implies z = -\frac{63}{17}$$

$$-6y - 3\left(-\frac{63}{17}\right) = -3 \implies y = \frac{40}{17}$$

$$x + 3\left(\frac{40}{17}\right) - \left(-\frac{63}{17}\right) = 13 \implies x = \frac{38}{17}$$

Solution: $\left(\frac{38}{17}, \frac{40}{17}, -\frac{63}{17}\right)$

39. $\begin{cases} x - 2y + z = -6 \\ 2x - 3y = -7 \\ -x + 3y - 3z = 11 \end{cases}$

$\begin{cases} x - 2y + z = -6 \\   y - 2z = 5 & -2\,\text{Eq.1} + \text{Eq. 2} \\   y - 2z = 5 & \text{Eq. 1} + \text{Eq. 3} \end{cases}$

$\begin{cases} x - 2y + z = -6 \\   y - 2z = 5 \\   0 = 0 & -\text{Eq. 2} + \text{Eq. 3} \end{cases}$

Let $z = a$, then $y = 2a + 5$.

$$x - 2(2a + 5) + a = -6$$

$$x - 3a - 10 = -6$$

$$x = 3a + 4$$

Solution: $(3a + 4, 2a + 5, a)$ where a is any real number.

41. $\begin{cases} 5x - 12y + 7z = 16 \qquad \text{Equation 1} \\ 3x - 7y + 4z = 9 \qquad \text{Equation 2} \end{cases}$

3 times Eq. 1 and (-5) times Eq. 2:

$$\begin{cases} 15x - 36y + 21z = 48 \\ -15x + 35y - 20z = -45 \end{cases}$$

Adding, $-y + z = 3 \implies y = z - 3.$

$$5x - 12(z - 3) + 7z = 16$$

$$5x - 5z + 36 = 16$$

$$5x = 5z - 20$$

$$x = z - 4$$

Let $z = a$, then $x = a - 4$ and $y = a - 3$.

Solution: $(a - 4, a - 3, a)$ where a is any real number.

43. Plane: $2x - 4y + z = 8$

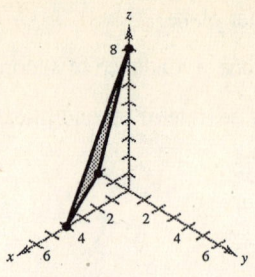

4 points on plane: $(4, 0, 0)$, $(0, -2, 0)$, $(0, 0, 8)$, $(1, 0, 6)$

45. $\dfrac{4 - x}{x^2 + 6x + 8} = \dfrac{A}{x + 2} + \dfrac{B}{x + 4}$

$4 - x = A(x + 4) + B(x + 2) = (A + B)x + (4A + 2B)$

$$\begin{cases} A + B = -1 \implies A = 3 \\ 4A + 2B = 4 \implies B = -4 \end{cases}$$

$$\dfrac{4 - x}{x^2 + 6x + 8} = \dfrac{3}{x + 2} - \dfrac{4}{x + 4}$$

47. $\dfrac{x^2 + 2x}{x^3 - x^2 + x - 1} = \dfrac{A}{x - 1} + \dfrac{Bx + C}{x^2 + 1}$

$x^2 + 2x = A(x^2 + 1) + (Bx + C)(x - 1) = (A + B)x^2 + (C - B)x + (A - C)$

$$\begin{cases} A + B = 1 \implies A = \frac{3}{2} \\ -B + C = 2 \implies B = -\frac{1}{2} \\ A - C = 0 \implies C = \frac{3}{2} \end{cases}$$

$$\dfrac{x^2 + 2x}{x^3 - x^2 + x - 1} = \dfrac{3/2}{x - 1} + \dfrac{-(1/2)x + 3/2}{x^2 + 1} = \dfrac{1}{2}\left(\dfrac{3}{x - 1} - \dfrac{x - 3}{x^2 + 1}\right)$$

49. $y = ax^2 + bx + c$ through $(0, -5)$, $(1, -2)$, and $(2, 5)$

$(0, -5)\colon -5 = c$

$$\begin{cases} (1, -2)\colon -2 = a + b + c \implies a + b = 3 \\ (2, 5)\colon 5 = 4a + 2b + c \implies 2a + b = 5 \end{cases}$$

$$\begin{cases} 2a + b = 5 \\ -a - b = -3 \end{cases}$$

$$a = 2$$

$$b = 1$$

The equation of the parabola is $y = 2x^2 + x - 5$.

51. Let x = gallons of spray X

Let y = gallons of spray Y

Let z = gallons of spray Z

$$\begin{cases} \text{Chemical A: } \frac{1}{5}x + \quad\ \frac{1}{3}z = 6 \\ \text{Chemical B: } \frac{2}{5}x + \quad\ \frac{1}{3}z = 8 \\ \text{Cehmical C: } \frac{2}{5}x + y + \frac{1}{3}z = 13 \end{cases} \qquad \begin{matrix} \text{Equation 1} \\ \text{Equation 2} \\ \text{Equation 3} \end{matrix}$$

Subtracting Eq. 2 − Eq. 1 gives $\frac{1}{5}x = 2 \implies x = 10$.

Then $z = 12$ and $y = 5$.

Answer: 10 gallons of spray X

5 gallons of spray Y

12 gallons of spray Z

53. Order 3×1

55. Order 1×1

57. $\begin{bmatrix} 3 & -10 & \vdots & 15 \\ 5 & 4 & \vdots & 22 \end{bmatrix}$

59. $\begin{bmatrix} 8 & -7 & 4 & \vdots & 12 \\ 3 & -5 & 2 & \vdots & 20 \\ 5 & 3 & -3 & \vdots & 26 \end{bmatrix}$

61. $\begin{bmatrix} 5 & 1 & 7 & \vdots & -9 \\ 4 & 2 & 0 & \vdots & 10 \\ 9 & 4 & 2 & \vdots & 3 \end{bmatrix}$ $\quad \begin{matrix} 5x + y + 7z = -9 \\ 4x + 2y \quad\quad = 10 \\ 9x + 4y + 2z = 3 \end{matrix}$

63. $\begin{bmatrix} 0 & 1 & 1 \\ 1 & 2 & 3 \\ 2 & 2 & 2 \end{bmatrix}$

$\begin{matrix} R_1 + R_2 \rightarrow \\ -R_1 + R_2 \rightarrow \\ -2R_1 + R_3 \rightarrow \end{matrix} \begin{bmatrix} 1 & 3 & 4 \\ 0 & -1 & -1 \\ 0 & -4 & -6 \end{bmatrix}$

$\begin{matrix} 3R_2 + R_1 \rightarrow \\ -R_2 \rightarrow \\ -4R_2 + R_3 \rightarrow \end{matrix} \begin{bmatrix} 1 & 0 & 1 \\ 0 & 1 & 1 \\ 0 & 0 & -2 \end{bmatrix}$

$\begin{matrix} -R_3 + R_1 \rightarrow \\ -R_3 + R_2 \rightarrow \\ -\frac{1}{2}R_3 \rightarrow \end{matrix} \begin{bmatrix} 1 & 0 & 0 \\ 0 & 1 & 0 \\ 0 & 0 & 1 \end{bmatrix}$

65. $\begin{bmatrix} 3 & -2 & 1 & 0 \\ 4 & -3 & 0 & 1 \end{bmatrix} \implies \begin{bmatrix} 1 & 0 & 3 & -2 \\ 0 & 1 & 4 & -3 \end{bmatrix}$

67. $\begin{bmatrix} 1 & 3 & 4 \\ 0 & 1 & 1 \\ 2 & 4 & 6 \end{bmatrix} \implies \begin{bmatrix} 1 & 0 & 1 \\ 0 & 1 & 1 \\ 0 & 0 & 0 \end{bmatrix}$

69. $\begin{bmatrix} 5 & 4 & \vdots & 2 \\ -1 & 1 & \vdots & -22 \end{bmatrix}$

$\begin{matrix} 4R_2 + R_1 \rightarrow \\ R_1 + R_2 \rightarrow \end{matrix} \begin{bmatrix} 1 & 8 & \vdots & -86 \\ 0 & 9 & \vdots & -108 \end{bmatrix}$

$9y = -108$

$y = -12$

$x = -8(-12) - 86 = 10$

Answer: $(10, -12)$

71.
$$\begin{bmatrix} 2 & 1 & \vdots & 0.3 \\ 3 & -1 & \vdots & -1.3 \end{bmatrix}$$

$$-R_1 + R_2 \rightarrow \begin{bmatrix} 2 & 1 & \vdots & 0.3 \\ 1 & -2 & \vdots & -1.6 \end{bmatrix}$$

$$\begin{bmatrix} 1 & -2 & \vdots & -1.6 \\ 2 & 1 & \vdots & 0.3 \end{bmatrix}$$

$$-2R_1 + R_2 \rightarrow \begin{bmatrix} 1 & -2 & \vdots & -1.6 \\ 0 & 5 & \vdots & 3.5 \end{bmatrix}$$

$5y = 3.5 \implies y = 0.7$

$x = 2(0.7) - 1.6 = -0.2$

$x = -0.2,\ y = 0.7$

73.
$$\begin{bmatrix} 2 & 3 & 3 & \vdots & 3 \\ 6 & 6 & 12 & \vdots & 13 \\ 12 & 9 & -1 & \vdots & 2 \end{bmatrix}$$

$$\begin{matrix} -3R_1 + R_2 \rightarrow \\ -6R_1 + R_3 \rightarrow \end{matrix} \begin{bmatrix} 2 & 3 & 3 & \vdots & 3 \\ 0 & -3 & 3 & \vdots & 4 \\ 0 & -9 & -19 & \vdots & -16 \end{bmatrix}$$

$$\begin{matrix} R_2 + R_1 \rightarrow \\ \\ -3R_2 + R_3 \rightarrow \end{matrix} \begin{bmatrix} 2 & 0 & 6 & \vdots & 7 \\ 0 & -3 & 3 & \vdots & 4 \\ 0 & 0 & -28 & \vdots & -28 \end{bmatrix}$$

$$\begin{matrix} \frac{1}{2}R_1 \rightarrow \\ -\frac{1}{3}R_2 \rightarrow \\ -\frac{1}{28}R_3 \rightarrow \end{matrix} \begin{bmatrix} 1 & 0 & 3 & \vdots & \frac{7}{2} \\ 0 & 1 & -1 & \vdots & -\frac{4}{3} \\ 0 & 0 & 1 & \vdots & 1 \end{bmatrix}$$

$z = 1$

$y - 1 = -\frac{4}{3} \implies y = -\frac{1}{3}$

$x + 3(1) = \frac{7}{2} \implies x = \frac{1}{2}$

Answer: $\left(\frac{1}{2}, -\frac{1}{3}, 1\right)$

75.
$$\begin{bmatrix} 3 & 21 & -29 & \vdots & -1 \\ 2 & 15 & -21 & \vdots & 0 \end{bmatrix}$$

$$-R_2 + R_1 \rightarrow \begin{bmatrix} 1 & 6 & -8 & \vdots & -1 \\ 2 & 15 & -21 & \vdots & 0 \end{bmatrix}$$

$$-2R_1 + R_2 \rightarrow \begin{bmatrix} 1 & 6 & -8 & \vdots & -1 \\ 0 & 3 & -5 & \vdots & 2 \end{bmatrix}$$

$$-2R_2 + R_1 \rightarrow \begin{bmatrix} 1 & 0 & 2 & \vdots & -5 \\ 0 & 3 & -5 & \vdots & 2 \end{bmatrix}$$

Let $z = a$, any real number.

$3y - 5a = 2 \implies y = \frac{5}{3}a + \frac{2}{3}$

$x + 2a = -5 \implies x = -2a - 5$

Answer: $\left(-2a - 5, \frac{5}{3}a + \frac{2}{3}, a\right)$

77.
$$\begin{bmatrix} -1 & 1 & 2 & \vdots & 1 \\ 2 & 3 & 1 & \vdots & -2 \\ 5 & 4 & 2 & \vdots & 4 \end{bmatrix}$$

$$\begin{matrix} -R_1 \rightarrow \\ 2R_1 + R_2 \rightarrow \\ 5R_1 + R_3 \rightarrow \end{matrix} \begin{bmatrix} 1 & -1 & -2 & \vdots & -1 \\ 0 & 5 & 5 & \vdots & 0 \\ 0 & 9 & 12 & \vdots & 9 \end{bmatrix}$$

$$\frac{1}{5}R_2 \rightarrow \begin{bmatrix} 1 & -1 & -2 & \vdots & -1 \\ 0 & 1 & 1 & \vdots & 0 \\ 0 & 9 & 12 & \vdots & 9 \end{bmatrix}$$

$$\begin{matrix} R_2 + R_1 \rightarrow \\ \\ -9R_2 + R_3 \rightarrow \end{matrix} \begin{bmatrix} 1 & 0 & -1 & \vdots & -1 \\ 0 & 1 & 1 & \vdots & 0 \\ 0 & 0 & 3 & \vdots & 9 \end{bmatrix}$$

$$\frac{1}{3}R_3 \rightarrow \begin{bmatrix} 1 & 0 & -1 & \vdots & -1 \\ 0 & 1 & 1 & \vdots & 0 \\ 0 & 0 & 1 & \vdots & 3 \end{bmatrix}$$

$$\begin{matrix} R_3 + R_1 \rightarrow \\ -R_3 + R_2 \rightarrow \end{matrix} \begin{bmatrix} 1 & 0 & 0 & \vdots & 2 \\ 0 & 1 & 0 & \vdots & -3 \\ 0 & 0 & 1 & \vdots & 3 \end{bmatrix}$$

$x = 2,\ y = -3,\ z = 3$

Answer: $(2, -3, 3)$

79.

$$\begin{bmatrix} 2 & -1 & 9 & \vdots & -8 \\ -1 & -3 & 4 & \vdots & -15 \\ 5 & 2 & -1 & \vdots & 17 \end{bmatrix}$$

$$\begin{array}{c} R_2 \\ R_1 \end{array} \begin{bmatrix} 1 & 3 & -4 & \vdots & 15 \\ 2 & -1 & 9 & \vdots & -8 \\ 5 & 2 & -1 & \vdots & 17 \end{bmatrix}$$

$$\begin{array}{c} -2R_1 + R_2 \rightarrow \\ -5R_1 + R_3 \rightarrow \end{array} \begin{bmatrix} 1 & 3 & -4 & \vdots & 15 \\ 0 & -7 & 17 & \vdots & -38 \\ 0 & -13 & 19 & \vdots & -58 \end{bmatrix}$$

$$\begin{array}{c} \\ R_3 - \frac{13}{7}R_2 \rightarrow \end{array} \begin{bmatrix} 1 & 3 & -4 & \vdots & 15 \\ 0 & -7 & 17 & \vdots & -38 \\ 0 & 0 & -\frac{88}{7} & \vdots & \frac{88}{7} \end{bmatrix} \text{ reduces to } \begin{bmatrix} 1 & 0 & 0 & \vdots & 2 \\ 0 & 1 & 0 & \vdots & 3 \\ 0 & 0 & 1 & \vdots & -1 \end{bmatrix}$$

Answer: $(2, 3, -1)$

81. $\begin{bmatrix} 1 & 2 & -1 & \vdots & 7 \\ 0 & -1 & -1 & \vdots & 4 \\ 4 & 0 & -1 & \vdots & 16 \end{bmatrix}$ reduces to $\begin{bmatrix} 1 & 0 & 0 & \vdots & 3 \\ 0 & 1 & 0 & \vdots & 0 \\ 0 & 0 & 1 & \vdots & -4 \end{bmatrix}$

Answer: $(3, 0, -4)$

83. $\begin{bmatrix} 3 & -1 & 5 & -2 & \vdots & -44 \\ 1 & 6 & 4 & -1 & \vdots & 1 \\ 5 & -1 & 1 & 3 & \vdots & -15 \\ 0 & 4 & -1 & -8 & \vdots & 58 \end{bmatrix}$ reduces to $\begin{bmatrix} 1 & 0 & 0 & 0 & \vdots & 2 \\ 0 & 1 & 0 & 0 & \vdots & 6 \\ 0 & 0 & 1 & 0 & \vdots & -10 \\ 0 & 0 & 0 & 1 & \vdots & -3 \end{bmatrix}$

Answer: $(2, 6, -10, -3)$

85. $x = 12$
  $y = -7$

87. $x + 3 = 5x - 1 \Rightarrow x = 1$
  $-4y = -44 \quad \Rightarrow y = 11$
  $y + 5 = 16 \quad\quad \Rightarrow y = 11$
  $6x = 6 \quad\quad \Rightarrow x = 1$

Answer: $x = 1, y = 11$

89. (a) $A + B = \begin{bmatrix} 7 & 3 \\ -1 & 5 \end{bmatrix} + \begin{bmatrix} 10 & -20 \\ 14 & -3 \end{bmatrix} = \begin{bmatrix} 17 & -17 \\ 13 & 2 \end{bmatrix}$

(b) $A - B = \begin{bmatrix} -3 & 23 \\ -15 & 8 \end{bmatrix}$

(c) $4A = \begin{bmatrix} 28 & 12 \\ -4 & 20 \end{bmatrix}$

(d) $A + 3B = \begin{bmatrix} 7 & 3 \\ -1 & 5 \end{bmatrix} + \begin{bmatrix} 30 & -60 \\ 42 & -9 \end{bmatrix} = \begin{bmatrix} 37 & -57 \\ 41 & -4 \end{bmatrix}$

91. (a) $A + B = \begin{bmatrix} 6 & 0 & 7 \\ 5 & -1 & 2 \\ 3 & 2 & 3 \end{bmatrix} + \begin{bmatrix} 0 & 5 & 1 \\ -4 & 8 & 6 \\ 2 & -1 & 1 \end{bmatrix} = \begin{bmatrix} 6 & 5 & 8 \\ 1 & 7 & 8 \\ 5 & 1 & 4 \end{bmatrix}$

(b) $A - B = \begin{bmatrix} 6 & -5 & 6 \\ 9 & -9 & -4 \\ 1 & 3 & 2 \end{bmatrix}$ **(c)** $4A = \begin{bmatrix} 24 & 0 & 28 \\ 20 & -4 & 8 \\ 12 & 8 & 12 \end{bmatrix}$

(d) $A + 3B = \begin{bmatrix} 6 & 0 & 7 \\ 5 & -1 & 2 \\ 3 & 2 & 3 \end{bmatrix} + \begin{bmatrix} 0 & 15 & 3 \\ -12 & 24 & 18 \\ 6 & -3 & 3 \end{bmatrix} = \begin{bmatrix} 6 & 15 & 10 \\ -7 & 23 & 20 \\ 9 & -1 & 6 \end{bmatrix}$

93. $\begin{bmatrix} 2 & 1 & 0 \\ 0 & 5 & -4 \end{bmatrix} - 3 \begin{bmatrix} 5 & 3 & -6 \\ 0 & -2 & 5 \end{bmatrix} = \begin{bmatrix} 2 & 1 & 0 \\ 0 & 5 & -4 \end{bmatrix} - \begin{bmatrix} 15 & 9 & -18 \\ 0 & -6 & 15 \end{bmatrix}$

$$= \begin{bmatrix} -13 & -8 & 18 \\ 0 & 11 & -19 \end{bmatrix}$$

95. $- \begin{bmatrix} 8 & -1 & 8 \\ -2 & 4 & 12 \\ 0 & -6 & 0 \end{bmatrix} - 5 \begin{bmatrix} -2 & 0 & -4 \\ 3 & -1 & 1 \\ 6 & 12 & -8 \end{bmatrix} = \begin{bmatrix} -8 & 1 & -8 \\ 2 & -4 & -12 \\ 0 & 6 & 0 \end{bmatrix} + \begin{bmatrix} 10 & 0 & 20 \\ -15 & 5 & -5 \\ -30 & -60 & 40 \end{bmatrix}$

$$= \begin{bmatrix} 2 & 1 & 12 \\ -13 & 1 & -17 \\ -30 & -54 & 40 \end{bmatrix}$$

97. $3 \begin{bmatrix} 8 & -2 & 5 \\ 1 & 3 & -1 \end{bmatrix} + 6 \begin{bmatrix} 4 & -2 & -3 \\ 2 & 7 & 6 \end{bmatrix} = \begin{bmatrix} 48 & -18 & -3 \\ 15 & 51 & 33 \end{bmatrix}$

99. $X = 3A - 2B = 3 \begin{bmatrix} -4 & 0 \\ 1 & -5 \\ -3 & 2 \end{bmatrix} - 2 \begin{bmatrix} 1 & 2 \\ -2 & 1 \\ 4 & 4 \end{bmatrix} = \begin{bmatrix} -14 & -4 \\ 7 & -17 \\ -17 & -2 \end{bmatrix}$

101. $X = \frac{1}{3}[B - 2A] = \frac{1}{3}\left(\begin{bmatrix} 1 & 2 \\ -2 & 1 \\ 4 & 4 \end{bmatrix} - 2 \begin{bmatrix} -4 & 0 \\ 1 & -5 \\ -3 & 2 \end{bmatrix} \right) = \frac{1}{3} \begin{bmatrix} 9 & 2 \\ -4 & 11 \\ 10 & 0 \end{bmatrix}$

103. $\begin{bmatrix} 1 & 2 \\ 5 & -4 \\ 6 & 0 \end{bmatrix} \begin{bmatrix} 6 & -2 & 8 \\ 4 & 0 & 0 \end{bmatrix} = \begin{bmatrix} 1(6) + 2(4) & 1(-2) + 2(0) & 1(8) + 2(0) \\ 5(6) + (-4)(4) & 5(-2) + (-4)(0) & 5(8) + (-4)(0) \\ 6(6) + (0)(4) & 6(-2) + (0)(0) & 6(8) + (0)(0) \end{bmatrix} = \begin{bmatrix} 14 & -2 & 8 \\ 14 & -10 & 40 \\ 36 & -12 & 48 \end{bmatrix}$

105. $AB = \begin{bmatrix} 3 & -2 & 0 \\ 1 & 4 & 9 \end{bmatrix} \begin{bmatrix} 7 & 0 \\ 5 & 3 \\ -1 & 3 \end{bmatrix} = \begin{bmatrix} 11 & -6 \\ 18 & 39 \end{bmatrix}$ **107.** $\begin{bmatrix} 4 & 1 \\ 11 & -7 \\ 12 & 3 \end{bmatrix} \begin{bmatrix} 3 & -5 & 6 \\ 2 & -2 & -2 \end{bmatrix} = \begin{bmatrix} 14 & -22 & 22 \\ 19 & -41 & 80 \\ 42 & -66 & 66 \end{bmatrix}$

109. $\begin{bmatrix} 2 & 1 \\ 6 & 0 \end{bmatrix} \left(\begin{bmatrix} 4 & 2 \\ -3 & 1 \end{bmatrix} + \begin{bmatrix} -2 & 4 \\ 0 & 4 \end{bmatrix} \right) = \begin{bmatrix} 2 & 1 \\ 6 & 0 \end{bmatrix} \begin{bmatrix} 2 & 6 \\ -3 & 5 \end{bmatrix}$

$$= \begin{bmatrix} 2(2) + 1(-3) & 2(6) + 1(5) \\ 6(2) + 0 & 6(6) + 0 \end{bmatrix}$$

$$= \begin{bmatrix} 1 & 17 \\ 12 & 36 \end{bmatrix}$$

111. Increase by 20% corresponds to multiplication by 1.2:

$$1.2\begin{bmatrix}80 & 70 & 90 & 40\\ 50 & 30 & 80 & 20\\ 90 & 60 & 100 & 50\end{bmatrix}=\begin{bmatrix}96 & 84 & 108 & 48\\ 60 & 36 & 96 & 24\\ 108 & 72 & 120 & 60\end{bmatrix}$$

113. $AB=\begin{bmatrix}-4 & -1\\ 7 & 2\end{bmatrix}\begin{bmatrix}-2 & -1\\ 7 & 4\end{bmatrix}=\begin{bmatrix}1 & 0\\ 0 & 1\end{bmatrix}; BA=I_2$

115. $\begin{bmatrix}-6 & 5 & \vdots & 1 & 0\\ -5 & 4 & \vdots & 0 & 1\end{bmatrix}$ row reduces to $\begin{bmatrix}1 & 0 & \vdots & 4 & -5\\ 0 & 1 & \vdots & 5 & -6\end{bmatrix}$

$\begin{bmatrix}-6 & 5\\ -5 & 4\end{bmatrix}^{-1}=\begin{bmatrix}4 & -5\\ 5 & -6\end{bmatrix}$

117. $\begin{bmatrix}-1 & -2 & -2 & \vdots & 1 & 0 & 0\\ 3 & 7 & 9 & \vdots & 0 & 1 & 0\\ 1 & 4 & 7 & \vdots & 0 & 0 & 1\end{bmatrix}$ row reduces to $\begin{bmatrix}1 & 0 & 0 & \vdots & 13 & 6 & -4\\ 0 & 1 & 0 & \vdots & -12 & -5 & 3\\ 0 & 0 & 1 & \vdots & 5 & 2 & -1\end{bmatrix}$

$\begin{bmatrix}-1 & -2 & -2\\ 3 & 7 & 9\\ 1 & 4 & 7\end{bmatrix}^{-1}=\begin{bmatrix}13 & 6 & -4\\ -12 & -5 & 3\\ 5 & 2 & -1\end{bmatrix}$

119. $\begin{bmatrix}2 & 6\\ 3 & -6\end{bmatrix}^{-1}=\begin{bmatrix}\frac{1}{5} & \frac{1}{5}\\ \frac{1}{10} & -\frac{1}{15}\end{bmatrix}$

121. $\begin{bmatrix}2 & 0 & 3\\ -1 & 1 & 1\\ 2 & -2 & 1\end{bmatrix}^{-1}=\begin{bmatrix}\frac{1}{2} & -1 & -\frac{1}{2}\\ \frac{1}{2} & -\frac{2}{3} & -\frac{5}{6}\\ 0 & \frac{2}{3} & \frac{1}{3}\end{bmatrix}$

123. $\begin{bmatrix}-7 & 2\\ -8 & 2\end{bmatrix}^{-1}=\frac{1}{(-7)(2)-(2)(-8)}\begin{bmatrix}2 & -2\\ 8 & -7\end{bmatrix}=\begin{bmatrix}1 & -1\\ 4 & -\frac{7}{2}\end{bmatrix}$

125. $\begin{bmatrix}-1 & 20\\ \frac{3}{10} & -6\end{bmatrix}^{-1}=\frac{1}{(-1)(-6)-(20)(\frac{3}{10})}\begin{bmatrix}-6 & -20\\ -\frac{3}{10} & -1\end{bmatrix}=\frac{1}{0}\begin{bmatrix}-6 & -20\\ -\frac{3}{10} & -1\end{bmatrix}$

Inverse does not exist.

127. $\begin{bmatrix}-1 & 4\\ 2 & -7\end{bmatrix}^{-1}=\begin{bmatrix}7 & 4\\ 2 & 1\end{bmatrix}$

$\begin{bmatrix}x\\ y\end{bmatrix}=\begin{bmatrix}7 & 4\\ 2 & 1\end{bmatrix}\begin{bmatrix}8\\ -5\end{bmatrix}=\begin{bmatrix}36\\ 11\end{bmatrix}$

Answer: (36, 11)

129. $\begin{bmatrix}3 & 2 & -1\\ 1 & -1 & 2\\ 5 & 1 & 1\end{bmatrix}^{-1}=\begin{bmatrix}-1 & -1 & 1\\ 3 & \frac{8}{3} & -\frac{7}{3}\\ 2 & \frac{7}{3} & -\frac{5}{3}\end{bmatrix}$

$\begin{bmatrix}x\\ y\\ z\end{bmatrix}=\begin{bmatrix}-1 & -1 & 1\\ 3 & \frac{8}{3} & -\frac{7}{3}\\ 2 & \frac{7}{3} & -\frac{5}{3}\end{bmatrix}\begin{bmatrix}6\\ -1\\ 7\end{bmatrix}=\begin{bmatrix}2\\ -1\\ -2\end{bmatrix}$

Answer: (2, -1, -2)

131. $\begin{bmatrix} -2 & 1 & 2 \\ -1 & -4 & 1 \\ 0 & -1 & -1 \end{bmatrix}^{-1} = \begin{bmatrix} -\frac{5}{9} & \frac{1}{9} & -1 \\ \frac{1}{9} & -\frac{2}{9} & 0 \\ -\frac{1}{9} & \frac{2}{9} & -1 \end{bmatrix}$

$\begin{bmatrix} x \\ y \\ z \end{bmatrix} = \begin{bmatrix} -\frac{5}{9} & \frac{1}{9} & -1 \\ \frac{1}{9} & -\frac{2}{9} & 0 \\ -\frac{1}{9} & \frac{2}{9} & -1 \end{bmatrix} \begin{bmatrix} -13 \\ -11 \\ 0 \end{bmatrix} = \begin{bmatrix} 6 \\ 1 \\ -1 \end{bmatrix}$

Answer: $(6, 1, -1)$

133. $\begin{cases} x + 2y = -1 \\ 3x + 4y = -5 \end{cases}$

$\begin{bmatrix} 1 & 2 \\ 3 & 4 \end{bmatrix}^{-1} = \begin{bmatrix} -2 & 1 \\ \frac{3}{2} & -\frac{1}{2} \end{bmatrix} \Rightarrow \begin{bmatrix} x \\ y \end{bmatrix} = \begin{bmatrix} -2 & 1 \\ \frac{3}{2} & -\frac{1}{2} \end{bmatrix} \begin{bmatrix} -1 \\ -5 \end{bmatrix} = \begin{bmatrix} -3 \\ 1 \end{bmatrix}$

$x = -3, y = 1$

Answer: $(-3, 1)$

135. $\begin{cases} -3x - 3y - 4z = 2 \\ \qquad\quad y + z = -1 \\ 4x + 3y + 4z = -1 \end{cases}$

$\begin{bmatrix} -3 & -3 & -4 \\ 0 & 1 & 1 \\ 4 & 3 & 4 \end{bmatrix}^{-1} = \begin{bmatrix} 1 & 0 & 1 \\ 4 & 4 & 3 \\ 4 & -3 & -3 \end{bmatrix} \Rightarrow \begin{bmatrix} x \\ y \\ z \end{bmatrix} = \begin{bmatrix} 1 & 0 & 1 \\ 4 & 4 & 3 \\ -4 & -3 & -3 \end{bmatrix} \begin{bmatrix} 2 \\ -1 \\ -1 \end{bmatrix} = \begin{bmatrix} 1 \\ 1 \\ -2 \end{bmatrix}$

$x = 1, y = 1, z = -2$

Answer: $(1, 1, -2)$

137. $\begin{vmatrix} 8 & 5 \\ 2 & -4 \end{vmatrix} = 8(-4) - 2(5) = -42$

139. $\begin{vmatrix} 50 & -30 \\ 10 & 5 \end{vmatrix} = 50(5) - (-30)(10) = 550$

141. $A = \begin{bmatrix} 2 & -1 \\ 7 & 4 \end{bmatrix}$

Minors: $M_{11} = 4$ $M_{21} = -1$

$M_{12} = 7$ $M_{22} = 2$

Cofactors: $C_{11} = 4$ $C_{21} = 1$

$C_{12} = -7$ $C_{22} = 2$

143. $A = \begin{bmatrix} 3 & 2 & -1 \\ -2 & 5 & 0 \\ 1 & 8 & 6 \end{bmatrix}$

Minors: $M_{11} = \begin{vmatrix} 5 & 0 \\ 8 & 6 \end{vmatrix} = 30, \ M_{12} = \begin{vmatrix} -2 & 0 \\ 1 & 6 \end{vmatrix} = -12, \ M_{13} = \begin{vmatrix} -2 & 5 \\ 1 & 8 \end{vmatrix} = -21$

$M_{21} = \begin{vmatrix} 2 & -1 \\ 8 & 6 \end{vmatrix} = 20, \ M_{22} = \begin{vmatrix} 3 & -1 \\ 1 & 6 \end{vmatrix} = 19, \ M_{23} = \begin{vmatrix} 3 & 2 \\ 1 & 8 \end{vmatrix} = 22$

$M_{31} = \begin{vmatrix} 2 & -1 \\ 5 & 0 \end{vmatrix} = 5, \ M_{32} = \begin{vmatrix} 3 & -1 \\ -2 & 0 \end{vmatrix} = -2, \ M_{33} = \begin{vmatrix} 3 & 2 \\ -2 & 5 \end{vmatrix} = 19$

Cofactors: $C_{11} = 30, \ C_{12} = 12, \ C_{13} = -21$

$C_{21} = -20, \ C_{22} = 19, \ C_{23} = -22$

$C_{31} = 5, \ C_{32} = 2, \ C_{33} = 19$

145. $\begin{vmatrix} -2 & 4 & 1 \\ -6 & 0 & 2 \\ 5 & 3 & 4 \end{vmatrix} = 6 \begin{vmatrix} 4 & 1 \\ 3 & 4 \end{vmatrix} - 2 \begin{vmatrix} -2 & 4 \\ 5 & 3 \end{vmatrix} = 6(13) - 2(-26) = 130$

147. $\begin{vmatrix} 1 & 0 & -2 \\ 0 & 1 & 0 \\ -2 & 0 & 1 \end{vmatrix} = 1 \begin{vmatrix} 1 & -2 \\ -2 & 1 \end{vmatrix} = 1(1) - (-2)(-2) = 1 - 4 = -3$

149. $\begin{vmatrix} 3 & 0 & -4 & 0 \\ 0 & 8 & 1 & 2 \\ 6 & 1 & 8 & 2 \\ 0 & 3 & -4 & 1 \end{vmatrix} = 3 \begin{vmatrix} 8 & 1 & 2 \\ 1 & 8 & 2 \\ 3 & -4 & 1 \end{vmatrix} + (-4) \begin{vmatrix} 0 & 8 & 2 \\ 6 & 1 & 2 \\ 0 & 3 & 1 \end{vmatrix}$ (Expansion along Row 1)

$= 3[8(8 - (-8)) - 1(1 - 6) + 2(-4 - 24)] - 4[0 - 6(8 - 6) + 0]$

$= 3[128 + 5 - 56] - 4[-12]$

$= 279$

151. $\det(A) = 8(-1)(4)(3) = -96$ (Upper Triangular)

153. $(1, 0), \ (5, 0), \ (5, 8)$

$\frac{1}{2} \begin{vmatrix} 1 & 0 & 1 \\ 5 & 0 & 1 \\ 5 & 8 & 1 \end{vmatrix} = \frac{1}{2} (32) = 16$

Area = 16 square units

155. $\frac{1}{2} \begin{vmatrix} \frac{1}{2} & 1 & 1 \\ 2 & -\frac{5}{2} & 1 \\ \frac{3}{2} & 1 & 1 \end{vmatrix} = \frac{1}{2}\left(\frac{7}{2}\right) = \frac{7}{4}$

Area = $\frac{7}{4}$ square units

157. $\begin{vmatrix} -1 & 7 & 1 \\ 2 & 5 & 1 \\ 4 & 1 & 1 \end{vmatrix} = -8$

The points are not collinear.

159. $x = \dfrac{\begin{vmatrix} 5 & 2 \\ 1 & 1 \end{vmatrix}}{\begin{vmatrix} 1 & 2 \\ -1 & 1 \end{vmatrix}} = \dfrac{3}{3} = 1$

$y = \dfrac{\begin{vmatrix} 1 & 5 \\ -1 & 1 \end{vmatrix}}{\begin{vmatrix} 1 & 2 \\ -1 & 1 \end{vmatrix}} = \dfrac{6}{3} = 2$

Answer: $(1, 2)$

161. $x = \dfrac{\begin{vmatrix} 6 & -2 \\ -23 & 3 \end{vmatrix}}{\begin{vmatrix} 5 & -2 \\ -11 & 3 \end{vmatrix}} = \dfrac{-28}{-7} = 4$

$y = \dfrac{\begin{vmatrix} 5 & 6 \\ -11 & -23 \end{vmatrix}}{\begin{vmatrix} 5 & -2 \\ -11 & 3 \end{vmatrix}} = \dfrac{-49}{-7} = 7$

Answer: $(4, 7)$

163. $x = \dfrac{\begin{vmatrix} -11 & 3 & -5 \\ -3 & -1 & 1 \\ 15 & -4 & 6 \end{vmatrix}}{\begin{vmatrix} -2 & 3 & -5 \\ 4 & -1 & 1 \\ -1 & -4 & 6 \end{vmatrix}} = \dfrac{-14}{14} = -1$

$y = \dfrac{\begin{vmatrix} -2 & -11 & -5 \\ 4 & -3 & 1 \\ -1 & 15 & 6 \end{vmatrix}}{14} = \dfrac{56}{14} = 4$

$z = \dfrac{\begin{vmatrix} -2 & 3 & -11 \\ 4 & -1 & -3 \\ -1 & -4 & 15 \end{vmatrix}}{14} = \dfrac{70}{14} = 5$

Answer: $(-1, 4, 5)$

165. $\begin{vmatrix} 1 & -3 & 2 \\ 2 & 2 & -3 \\ 1 & -7 & 8 \end{vmatrix} = 20$

$|A_1| = 0$

$|A_2| = -48$

$|A_3| = -52$

$x = 0,\ y = -\dfrac{48}{20} = -2.4,\ z = -\dfrac{52}{20} = -2.6$

Solution: $(0, -2.4, -2.6)$

167. L O O K _ O U T _ B E L O W _

$[12 \quad 15 \quad 15][11 \quad 0 \quad 15][21 \quad 20 \quad 0][2 \quad 5 \quad 12][15 \quad 23 \quad 0]$

$[12 \quad 15 \quad 15]A = [-21 \quad 6 \quad 0]$

$[11 \quad 0 \quad 15]A = [-68 \quad 8 \quad 45]$

$[21 \quad 20 \quad 0]A = [102 \quad -42 \quad -60]$

$[2 \quad 5 \quad 12]A = [-53 \quad 20 \quad 21]$

$[15 \quad 23 \quad 0]A = [99 \quad -30 \quad -69]$

Cryptogram: $-21 \quad 6 \quad 0 \quad -68 \quad 8 \quad 45 \quad 102 \quad -42 \quad -60$

$\qquad\qquad -53 \quad 20 \quad 21 \quad 99 \quad -30 \quad -69$

169. $\begin{bmatrix} -5 & 11 & -2 \\ 370 & -265 & 225 \\ -57 & 48 & -33 \\ 32 & -15 & 20 \\ 245 & -171 & 147 \end{bmatrix} \begin{bmatrix} -1 & 2 & -3 \\ 2 & 1 & 0 \\ 4 & -2 & 5 \end{bmatrix} = \begin{bmatrix} 19 & 5 & 5 \\ 0 & 25 & 15 \\ 21 & 0 & 6 \\ 18 & 9 & 4 \\ 1 & 25 & 0 \end{bmatrix}$

S	E	E
_	Y	O
U	_	F
R	I	D
A	Y	_

Message: SEE YOU FRIDAY

171. False.

173. The row operations on matrices are equivalent to the operations used in the method of elimination.

Chapter 7   Practice Test

For Exercises 1–3, solve the given system by the method of substitution.

1. $x + y = 1$

   $3x - y = 15$

2. $x - 3y = -3$

   $x^2 + 6y = 5$

3. $x + y + z = 6$

   $2x - y + 3z = 0$

   $5x + 2y - z = -3$

4. Find the two numbers whose sum is 110 and product is 2800.

5. Find the dimensions of a rectangle if its perimeter is 170 feet and its area is 2800 square feet.

For Exercises 6–7, solve the linear system by elimination.

6. $2x + 15y = 4$

   $x - 3y = 23$

7. $x + y = 2$

   $38x - 19y = 7$

8. Use a graphing utility to graph the two equations. Use the graph to approximate the solution of the system. Verify your answer analytically.

   $0.4x + 0.5y = 0.112$

   $0.3x - 0.7y = -0.131$

9. Herbert invests $17,000 in two funds that pay 11% and 13% simple interest, respectively. If he receives $2080 in yearly interest, how much is invested in each fund?

10. Find the least squares regression line for the points $(4, 3)$, $(1, 1)$, $(-1, -2)$, and $(-2, -1)$.

For Exercises 11–13, solve the system of equations.

11. $x + y = -2$

    $2x - y + z = 11$

    $4y - 3z = -20$

12. $4x - y + 5z = 4$

    $2x + y - z = 0$

    $2x + 4y + 8z = 0$

13. $3x + 2y - z = 5$

    $6x - y + 5z = 2$

14. Find the equation of the parabola $y = ax^2 + bx + c$ passing through the points $(0, -1)$, $(1, 4)$ and $(2, 13)$.

15. Find the position equation $s = \frac{1}{2}at^2 + v_0t + s_0$ given that $s = 12$ feet after 1 second, $s = 5$ feet after 2 seconds, and $s = 4$ after 3 seconds.

16. Write the matrix in reduced row-echelon form.

    $$\begin{bmatrix} 1 & -2 & 4 \\ 3 & -5 & 9 \end{bmatrix}$$

For Exercises 17–19, use matrices to solve the system of equations.

17. $3x + 5y = 3$

    $2x - y = -11$

18. $2x + 3y = -3$

    $3x + 2y = 8$

    $x + y = 1$

19. $x + 3z = -5$

    $2x + y = 0$

    $3x + y - z = 3$

20. Multiply $\begin{bmatrix} 1 & 4 & 5 \\ 2 & 0 & -3 \end{bmatrix} \begin{bmatrix} 1 & 6 \\ 0 & -7 \\ -1 & 2 \end{bmatrix}$

21. Given $A = \begin{bmatrix} 9 & 1 \\ -4 & 8 \end{bmatrix}$ and $B = \begin{bmatrix} 6 & -2 \\ 3 & 5 \end{bmatrix}$, find $3A - 5B$.

22. Find $f(A)$:

$$f(x) = x^2 - 7x + 8, \quad A = \begin{bmatrix} 3 & 0 \\ 7 & 1 \end{bmatrix}$$

23. True or false:

$(A + B)(A + 3B) = A^2 + 4AB + 3B^2$ where A and B are matrices.

(Assume that A^2, AB, and B^2 exist.)

For Exercises 24 and 25, find the inverse of the matrix, if it exists.

24. $\begin{bmatrix} 1 & 2 \\ 3 & 5 \end{bmatrix}$

25. $\begin{bmatrix} 1 & 1 & 1 \\ 3 & 6 & 5 \\ 6 & 10 & 8 \end{bmatrix}$

26. Use an inverse matrix to solve the systems.

(a) $x + 2y = 4$ (b) $x + 2y = 3$
  $3x + 5y = 1$ $3x + 5y = -2$

For Exercises 27 and 28, find the determinant of the matrix.

27. $\begin{bmatrix} 6 & -1 \\ 3 & 4 \end{bmatrix}$

28. $\begin{bmatrix} 1 & 3 & -1 \\ 5 & 9 & 0 \\ 6 & 2 & -5 \end{bmatrix}$

29. Use a graphing utility to find the determinant of the matrix.

$$\begin{bmatrix} 1 & 4 & 2 & 3 \\ 0 & 1 & -2 & 0 \\ 3 & 5 & -1 & 1 \\ 2 & 0 & 6 & 1 \end{bmatrix}$$

30. Evaluate $\begin{vmatrix} 6 & 4 & 3 & 0 & 6 \\ 0 & 5 & 1 & 4 & 8 \\ 0 & 0 & 2 & 7 & 3 \\ 0 & 0 & 0 & 9 & 2 \\ 0 & 0 & 0 & 0 & 1 \end{vmatrix}$.

31. Use a determinant to find the area of the triangle with vertices $(0, 7)$, $(5, 0)$, and $(3, 9)$.

32. Use a determinant to find the equation of the line through $(2, 7)$ and $(-1, 4)$.

For Exercises 33–35, use Cramer's Rule to find the indicated value.

33. Find x.

  $6x - 7y = 4$
  $2x + 5y = 11$

34. Find z.

  $3x \qquad + z = 1$
  $\quad y + 4z = 3$
  $x - y \qquad = 2$

35. Find y.

  $721.4x - 29.1y = 33.77$
  $45.9x + 105.6y = 19.85$

C H A P T E R 8
Sequences, Series and Probability

CHAPTER 8
Sequences, Series, and Probability

Section 8.1   Sequences and Series

■ Given the general nth term in a sequence, you should be able to find, or list, some of the terms.

■ You should be able to find an expression for the nth term of a sequence.

■ You should be able to use and evaluate factorials.

■ You should be able to use sigma notation for a sum.

Solutions to Odd-Numbered Exercises

1. $a_n = 2n + 5$

$a_1 = 2(1) + 5 = 7$

$a_2 = 2(2) + 5 = 9$

$a_3 = 2(3) + 5 = 11$

$a_4 = 2(4) + 5 = 13$

$a_5 = 2(5) + 5 = 15$

3. $a_n = 2^n$

$a_1 = 2^1 = 2$

$a_2 = 2^2 = 4$

$a_3 = 2^3 = 8$

$a_4 = 2^4 = 16$

$a_5 = 2^5 = 32$

5. $a_n = \left(-\frac{1}{2}\right)^n$

$a_1 = \left(-\frac{1}{2}\right)^1 = -\frac{1}{2}$

$a_2 = \left(-\frac{1}{2}\right)^2 = \frac{1}{4}$

$a_3 = \left(-\frac{1}{2}\right)^3 = -\frac{1}{8}$

$a_4 = \left(-\frac{1}{2}\right)^4 = \frac{1}{16}$

$a_5 = \left(-\frac{1}{2}\right)^5 = -\frac{1}{32}$

7. $a_n = \dfrac{n+1}{n}$

$a_1 = \dfrac{1+1}{1} = 2$

$a_2 = \dfrac{3}{2}$

$a_3 = \dfrac{4}{3}$

$a_4 = \dfrac{5}{4}$

$a_5 = \dfrac{6}{5}$

9. $a_n = \dfrac{n}{n^2 + 1}$

$a_1 = \dfrac{1}{1^2 + 1} = \dfrac{1}{2}$

$a_2 = \dfrac{2}{2^2 + 1} = \dfrac{2}{5}$

$a_3 = \dfrac{3}{3^2 + 1} = \dfrac{3}{10}$

$a_4 = \dfrac{4}{4^2 + 1} = \dfrac{4}{17}$

$a_5 = \dfrac{5}{5^2 + 1} = \dfrac{5}{26}$

11. $a_n = \dfrac{1 + (-1)^n}{n}$

$a_1 = 0$

$a_2 = \dfrac{2}{2} = 1$

$a_3 = 0$

$a_4 = \dfrac{2}{4} = \dfrac{1}{2}$

$a_5 = 0$

13. $a_n = 1 - \dfrac{1}{2^n}$

$a_1 = 1 - \dfrac{1}{2^1} = \dfrac{1}{2}$

$a_2 = 1 - \dfrac{1}{2^2} = 1 - \dfrac{1}{4} = \dfrac{3}{4}$

$a_3 = 1 - \dfrac{1}{2^3} = \dfrac{7}{8}$

$a_4 = 1 - \dfrac{1}{2^4} = \dfrac{15}{16}$

$a_5 = 1 - \dfrac{1}{2^5} = \dfrac{31}{32}$

15. $a_n = \dfrac{1}{n^{3/2}}$

$a_1 = \dfrac{1}{1} = 1$

$a_2 = \dfrac{1}{2^{3/2}}$

$a_3 = \dfrac{1}{3^{3/2}}$

$a_4 = \dfrac{1}{4^{3/2}} = \dfrac{1}{8}$

$a_5 = \dfrac{1}{5^{3/2}}$

17. $a_n = \dfrac{(-1)^n}{n^2}$

$a_1 = \dfrac{-1}{1} = -1$

$a_2 = \dfrac{1}{4}$

$a_3 = \dfrac{-1}{9}$

$a_4 = \dfrac{1}{16}$

$a_5 = \dfrac{-1}{25}$

19. $a_n = (2n - 1)(2n + 1)$

$a_1 = (1)(3) = 3$

$a_2 = (3)(5) = 15$

$a_3 = (5)(7) = 35$

$a_4 = (7)(9) = 63$

$a_5 = (9)(11) = 99$

21. $a_{25} = (-1)^{25}[3(25) - 2] = -73$

23. $a_{10} = \dfrac{10^2}{10^2 + 1} = \dfrac{100}{101}$

25. $a_n = \dfrac{2}{3}n$

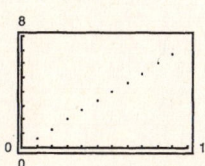

27. $a_n = 16(-0.5)^{n-1}$

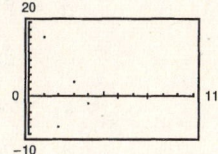

29. $a_n = \dfrac{2n}{n + 1}$

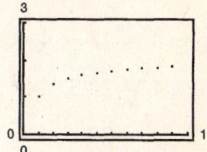

31. $a_n = 2(3n - 1) + 5$

n	1	2	3	4	5	6	7	8	9	10
a_n	9	15	21	27	33	39	45	51	57	63

33. $a_n = 1 + \dfrac{n + 1}{n}$

n	1	2	3	4	5	6	7	8	9	10
a_n	3	2.5	2.33	2.25	2.2	2.17	2.14	2.13	2.11	2.1

35. $a_n = \dfrac{8}{n+1}$

$a_n \rightarrow 0$ as $n \rightarrow \infty$

$a_1 = 4, \; a_{10} = \dfrac{8}{11}$

Matches graph (c).

37. $a_n = 4(0.5)^{n-1}$

$a_n \rightarrow 0$ as $n \rightarrow \infty$

$a_1 = 4, \; a_{10} \approx 0.008$

Matches graph (d).

39. $1, 4, 7, 10, 13, \ldots$

$a_n = 1 + (n-1)3 = 3n - 2$

41. $0, 3, 8, 15, 24, \ldots$

$a_n = n^2 - 1$

43. $\dfrac{2}{3}, \dfrac{3}{4}, \dfrac{4}{5}, \dfrac{5}{6}, \dfrac{6}{7}, \ldots$

$a_n = \dfrac{n+1}{n+2}$

45. $\dfrac{1}{2}, \dfrac{-1}{4}, \dfrac{1}{8}, \dfrac{-1}{16}, \ldots$

$a_n = \dfrac{(-1)^{n+1}}{2^n}$

47. $1 + \dfrac{1}{1}, 1 + \dfrac{1}{2}, 1 + \dfrac{1}{3}, 1 + \dfrac{1}{4}, 1 + \dfrac{1}{5}, \ldots$

$a_n = 1 + \dfrac{1}{n}$

49. $1, \dfrac{1}{2}, \dfrac{1}{6}, \dfrac{1}{24}, \dfrac{1}{120}, \ldots$

$a_n = \dfrac{1}{n!}$

51. $1, 3, 1, 3, 1, 3, \ldots$

$a_n = 2 + (-1)^n$

53. $a_1 = 28$ and $a_{k+1} = a_k - 4$

$a_1 = 28$

$a_2 = a_1 - 4 = 28 - 4 = 24$

$a_3 = a_2 - 4 = 24 - 4 = 20$

$a_4 = a_3 - 4 = 20 - 4 = 16$

$a_5 = a_4 - 4 = 16 - 4 = 12$

55. $a_1 = 3$ and $a_{k+1} = 2(a_k - 1)$

$a_1 = 3$

$a_2 = 2(a_1 - 1) = 2(3 - 1) = 4$

$a_3 = 2(a_2 - 1) = 2(4 - 1) = 6$

$a_4 = 2(a_3 - 1) = 2(6 - 1) = 10$

$a_5 = 2(a_4 - 1) = 2(10 - 1) = 18$

57. $a_1 = 6$ and $a_{k+1} = a_k + 2$

$a_1 = 6$

$a_2 = a_1 + 2 = 6 + 2 = 8$

$a_3 = a_2 + 2 = 8 + 2 = 10$

$a_4 = a_3 + 2 = 10 + 2 = 12$

$a_5 = a_4 + 2 = 12 + 2 = 14$

In general, $a_n = 2n + 4$.

59. $a_1 = 81$ and $a_{k+1} = \dfrac{1}{3}a_k$

$a_1 = 81$

$a_2 = \dfrac{1}{3}a_1 = \dfrac{1}{3}(81) = 27$

$a_3 = \dfrac{1}{3}a_2 = \dfrac{1}{3}(27) = 9$

$a_4 = \dfrac{1}{3}a_3 = \dfrac{1}{3}(9) = 3$

$a_5 = \dfrac{1}{3}a_4 = \dfrac{1}{3}(3) = 1$

In general, $a_n = 81\left(\dfrac{1}{3}\right)^{n-1} = 81(3)\left(\dfrac{1}{3}\right)^n = \dfrac{243}{3^n}$.

61. $a_n = \dfrac{1}{n!}$

$a_0 = \dfrac{1}{0!} = 1$

$a_1 = \dfrac{1}{1!} = 1$

$a_2 = \dfrac{1}{2}$

$a_3 = \dfrac{1}{3!} = \dfrac{1}{6}$

$a_4 = \dfrac{1}{4!} = \dfrac{1}{24}$

63. $a_n = \dfrac{n!}{2n + 1}$

$a_0 = \dfrac{0!}{1} = 1$

$a_1 = \dfrac{1!}{2 + 1} = \dfrac{1}{3}$

$a_2 = \dfrac{2!}{4 + 1} = \dfrac{2}{5}$

$a_3 = \dfrac{3!}{6 + 1} = \dfrac{6}{7}$

$a_4 = \dfrac{4!}{8 + 1} = \dfrac{24}{9} = \dfrac{8}{3}$

65. $a_n = \dfrac{(-1)^{2n}}{(2n)!}$

$a_0 = \dfrac{(-1)^0}{0!} = 1$

$a_1 = \dfrac{(-1)^2}{2!} = \dfrac{1}{2}$

$a_2 = \dfrac{(-1)^4}{4!} = \dfrac{1}{24}$

$a_3 = \dfrac{(-1)^6}{6!} = \dfrac{1}{720}$

$a_4 = \dfrac{(-1)^8}{8!} = \dfrac{1}{40,320}$

67. $\dfrac{2!}{4!} = \dfrac{2!}{4 \cdot 3 \cdot 2!} = \dfrac{1}{12}$

69. $\dfrac{12!}{4!8!} = \dfrac{12 \cdot 11 \cdot 10 \cdot 9 \cdot 8!}{4!8!} = \dfrac{12 \cdot 11 \cdot 10 \cdot 9}{4 \cdot 3 \cdot 2} = 495$

71. $\dfrac{(n + 1)!}{n!} = \dfrac{(n + 1)n!}{n!} = n + 1$

73. $\dfrac{(2n - 1)!}{(2n + 1)!} = \dfrac{(2n - 1)!}{(2n + 1)(2n)(2n - 1)!}$

$\qquad = \dfrac{1}{2n(2n + 1)}$

75. $\displaystyle\sum_{i=1}^{5}(2i + 1) = (2 + 1) + (4 + 1) + (6 + 1) + (8 + 1) + (10 + 1) = 35$

77. $\displaystyle\sum_{k=1}^{4}10 = 10 + 10 + 10 + 10 = 40$

79. $\displaystyle\sum_{i=0}^{4}i^2 = 0^2 + 1^2 + 2^2 + 3^2 + 4^2 = 30$

81. $\displaystyle\sum_{k=0}^{3}\dfrac{1}{k^2 + 1} = \dfrac{1}{1} + \dfrac{1}{1 + 1} + \dfrac{1}{4 + 1} + \dfrac{1}{9 + 1} = \dfrac{9}{5}$

83. $\displaystyle\sum_{i=1}^{4}[(i - 1)^2 + (i + 1)^3] = [(0)^2 + (2)^3] + [(1)^2 + (3)^3] + [(2)^2 + (4)^3] + [(3)^2 + (5)^3] = 238$

85. $\displaystyle\sum_{i=1}^{4}2^i = 2^1 + 2^2 + 2^3 + 2^4 = 30$

87. $\displaystyle\sum_{j=1}^{6}(24 - 3j) = 81$

89. $\displaystyle\sum_{k=0}^{4}\dfrac{(-1)^k}{k + 1} = \dfrac{47}{60}$

91. $\dfrac{1}{3(1)} + \dfrac{1}{3(2)} + \dfrac{1}{3(3)} + \cdots + \dfrac{1}{3(9)} = \displaystyle\sum_{i=1}^{9} \dfrac{1}{3i} \approx 0.94299$

93. $\left[2\left(\dfrac{1}{8}\right) + 3\right] + \left[2\left(\dfrac{2}{8}\right) + 3\right] + \left[2\left(\dfrac{3}{8}\right) + 3\right] + \cdots + \left[2\left(\dfrac{8}{8}\right) + 3\right] = \displaystyle\sum_{i=1}^{8} \left[2\left(\dfrac{i}{8}\right) + 3\right] = 33$

95. $3 - 9 + 27 - 81 + 243 - 729 = \displaystyle\sum_{i=1}^{6} (-1)^{i+1} 3^{i} = -546$

97. $\dfrac{1}{1^2} - \dfrac{1}{2^2} + \dfrac{1}{3^2} - \dfrac{1}{4^2} + \cdots + -\dfrac{1}{20^2} = \displaystyle\sum_{i=1}^{20} \dfrac{(-1)^{i+1}}{i^2} \approx 0.82128$

99. $\dfrac{1}{4} + \dfrac{3}{8} + \dfrac{7}{16} + \dfrac{15}{32} + \dfrac{31}{64} = \displaystyle\sum_{i=1}^{5} \dfrac{2^i - 1}{2^{i+1}} = \dfrac{129}{64} = 2.015625$

101. $\displaystyle\sum_{i=1}^{4} 5\left(\tfrac{1}{2}\right)^i = 4.6875 = \tfrac{75}{16}$

103. $\displaystyle\sum_{n=1}^{3} 4\left(-\tfrac{1}{2}\right)^n = -1.5 = -\tfrac{3}{2}$

105. $\displaystyle\sum_{i=1}^{\infty} 6\left(\tfrac{1}{10}\right)^i = 6[0.1 + 0.01 + 0.001 + \cdots]$

$= 6[0.111\ldots]$

$= 0.666\ldots$

$= \tfrac{2}{3}$

107. $\displaystyle\sum_{k=1}^{\infty} \left(\tfrac{1}{10}\right)^k = 0.1 + 0.11 + 0.111 + \cdots$

$= 0.11111$

$= \tfrac{1}{9}$

109. $A_n = 5000\left(1 + \dfrac{0.03}{4}\right)^n, \quad n = 1, 2, 3, \ldots$

(a) $A_1 = 5000\left(1 + \dfrac{0.03}{4}\right)^1 = \$5037.50$

$A_2 \approx \$5075.28$

$A_3 \approx \$5113.35$

$A_4 \approx \$5151.70$

$A_5 \approx \$5190.33$

$A_6 \approx \$5229.26$

$A_7 \approx \$5268.48$

$A_8 \approx \$5307.99$

(b) $A_{40} \approx \$6741.74$

111. $a_n = 1.37n^2 + 3.1n + 698, \quad n = 3, 4, \ldots, 11$

(a) $a_3 = 1.37(3)^2 + 3.1(3) + 698 = 719.63 \approx 720$

$a_4 \approx 732$

$a_5 \approx 748$

$a_6 \approx 766$

$a_7 \approx 787$

$a_8 \approx 810$

$a_9 \approx 837$

$a_{10} \approx 866$

$a_{11} \approx 898$

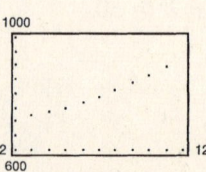

(b) The enrollment seems to be growing.

113. $a_n = 2.151n^2 + 235.9, \quad n = 4, 5, \ldots, 12$

$a_4 + a_5 + \cdots + a_{12} = 270.316 + \cdots + 545.644$

$$\approx \$3491.1 \text{ million}$$

115. True

117. $a_0 = 1, a_1 = 1, a_{k+2} = a_{k+1} + a_k$

$a_0 = 1$ $b_0 = \dfrac{1}{1} = 1$ $a_6 = 8 + 5 = 13$ $b_6 = \dfrac{21}{13}$

$a_1 = 1$ $b_1 = \dfrac{2}{1} = 2$ $a_7 = 13 + 8 = 21$ $b_7 = \dfrac{34}{21}$

$a_2 = 1 + 1 = 2$ $b_2 = \dfrac{3}{2}$ $a_8 = 21 + 13 = 34$ $b_8 = \dfrac{55}{34}$

$a_3 = 2 + 1 = 3$ $b_3 = \dfrac{5}{3}$ $a_9 = 34 + 21 = 55$ $b_9 = \dfrac{89}{55}$

$a_4 = 3 + 2 = 5$ $b_4 = \dfrac{8}{5}$ $a_{10} = 55 + 34 = 89$

$a_5 = 5 + 3 = 8$ $b_5 = \dfrac{13}{8}$ $a_{11} = 89 + 55 = 144$

119. $a_n = \dfrac{x^n}{n!}$

$a_1 = \dfrac{x}{1} = x$

$a_2 = \dfrac{x^2}{2!} = \dfrac{x^2}{2}$

$a_3 = \dfrac{x^3}{3!} = \dfrac{x^3}{6}$

$a_4 = \dfrac{x^4}{4!} = \dfrac{x^4}{24}$

$a_5 = \dfrac{x^5}{5!} = \dfrac{x^5}{120}$

121. $a_n = \dfrac{(-1)^n x^{2n}}{(2n)!}$

$a_1 = \dfrac{-x^2}{2}$

$a_2 = \dfrac{x^4}{4!} = \dfrac{x^4}{24}$

$a_3 = \dfrac{-x^6}{6!} = \dfrac{-x^6}{720}$

$a_4 = \dfrac{x^8}{8!} = \dfrac{x^8}{40,320}$

$a_5 = \dfrac{-x^{10}}{10!} = \dfrac{-x^{10}}{3,628,800}$

123. (a) $A - B = \begin{bmatrix} 8 & 1 \\ -3 & 7 \end{bmatrix}$

(b) $2B - 3A = \begin{bmatrix} -22 & -7 \\ 3 & -18 \end{bmatrix}$

(c) $AB = \begin{bmatrix} 18 & 9 \\ 18 & 0 \end{bmatrix}$

(d) $BA = \begin{bmatrix} 0 & 6 \\ 27 & 18 \end{bmatrix}$

125. (a) $A - B = \begin{bmatrix} -3 & -7 & 4 \\ 4 & 4 & 1 \\ 1 & 4 & 3 \end{bmatrix}$

(b) $2B - 3A = \begin{bmatrix} 8 & 17 & -14 \\ -12 & -13 & -9 \\ -3 & -15 & -10 \end{bmatrix}$

(c) $AB = \begin{bmatrix} -2 & 7 & -16 \\ 4 & 42 & 45 \\ 1 & 23 & 48 \end{bmatrix}$

(d) $BA = \begin{bmatrix} 16 & 31 & 42 \\ 10 & 47 & 31 \\ 13 & 22 & 25 \end{bmatrix}$

Section 8.2   Arithmetic Sequences and Partial Sums

■   You should be able to recognize an arithmetic sequence, find its common difference, and find its nth term.

■   You should be able to find the nth partial sum of an arithmetic sequence with common difference d using the formula

$$S_n = \frac{n}{2}(a_1 + a_n).$$

Solutions to Odd-Numbered Exercises

1. 10, 8, 6, 4, 2, . . .

Arithmetic sequence, $d = -2$

3. $3, \frac{5}{2}, 2, \frac{3}{2}, 1, \ldots$

Arithmetic sequence, $d = -\frac{1}{2}$

5. $-24, -16, -8, 0, 8$

Arithmetic sequence, $d = 8$

7. 3.7, 4.3, 4.9, 5.5, 6.1, . . .

Arithmetic sequence, $d = 0.6$

9. $a_n = 8 + 13n$

21, 34, 47, 60, 73

Arithmetic sequence, $d = 13$

11. $a_n = \dfrac{1}{n+1}$

$\dfrac{1}{2}, \dfrac{1}{3}, \dfrac{1}{4}, \dfrac{1}{5}, \dfrac{1}{6}$

Not an arithmetic sequence

13. $a_n = 150 - 7n$

143, 136, 129, 122, 115

Arithmetic sequence, $d = -7$

15. $a_n = 3 + \dfrac{(-1)^n 2}{n}$

$1, \ 4, \ \dfrac{7}{3}, \dfrac{7}{2}, \dfrac{13}{5}$

Not an arithmetic sequence

17. $a_1 = 1, \ d = 3$

$a_n = a_1 + (n-1)d = 1 + (n-1)(3) = 3n - 2$

19. $a_1 = 100, \ d = -8$

$a_n = a_1 + (n-1)d = 100 + (n-1)(-8) = 108 - 8n$

21. $4, \frac{3}{2}, -1, -\frac{7}{2}, \ldots$

$d = -\frac{5}{2}$

$a_n = a_1 + (n-1)d = 4 + (n-1)\left(-\frac{5}{2}\right) = \frac{13}{2} - \frac{5}{2}n$

23. $a_1 = 5, \ a_4 = 15$

$a_4 = a_1 + 3d \implies 15 = 5 + 3d \implies d = \frac{10}{3}$

$a_n = a_1 + (n-1)d = 5 + (n-1)\left(\frac{10}{3}\right) = \frac{10}{3}n + \frac{5}{3}$

25. $a_3 = 94,\ a_6 = 85$

$a_6 = a_3 + 3d \Rightarrow 85 = 94 + 3d \Rightarrow d = -3$

$a_1 = a_3 - 2d \Rightarrow a_1 = 94 - 2(-3) = 100$

$a_n = a_1 + (n-1)d = 100 + (n-1)(-3) = 103 - 3n$

27. $a_1 = 5,\ d = 6$

$a_1 = 5$

$a_2 = 5 + 6 = 11$

$a_3 = 11 + 6 = 17$

$a_4 = 17 + 6 = 23$

$a_5 = 23 + 6 = 29$

29. $a_1 = -10,\ \ d = -12$

$a_1 = -10$

$a_2 = -10 - 12 = -22$

$a_3 = -22 - 12 = -34$

$a_4 = -34 - 12 = -46$

$a_5 = -46 - 12 = -58$

31. $a_8 = 26,\ a_{12} = 42$

$26 = a_8 = a_1 + (n-1)d = a_1 + 7d$

$42 = a_{12} = a_1 + (n-1)d = a_1 + 11d$

Answer: $d = 4,\ a_1 = -2$

$a_1 = -2$

$a_2 = -2 + 4 = 2$

$a_3 = 2 + 4 = 6$

$a_4 = 6 + 4 = 10$

$a_5 = 10 + 4 = 14$

33. $a_3 = 19,\ a_{15} = -1.7$

$a_{15} = a_3 + 12d$

$-1.7 = 19 + 12d \Rightarrow d = -1.725$

$a_3 = a_1 + 2d \Rightarrow 19 = a_1 + 2(-1.725)$

$\Rightarrow a_1 = 22.45$

$a_2 = a_1 - 1.725 = 20.725$

$a_3 = 19$

$a_4 = 19 - 1.725 = 17.275$

$a_5 = 17.275 - 1.725 = 15.55$

35. $a_1 = 15,\ \ a_{k+1} = a_k + 4$

$a_2 = a_1 + 4 = 15 + 4 = 19$

$a_3 = 19 + 4 = 23$

$a_4 = 23 + 4 = 27$

$a_5 = 27 + 4 = 31$

$d = 4,\ \ a_n = 11 + 4n$

37. $a_1 = \frac{7}{2},\ a_{k+1} = a_k - \frac{1}{4}$

$a_2 = \frac{7}{2} - \frac{1}{4} = \frac{13}{4}$

$a_3 = \frac{13}{4} - \frac{1}{4} = \frac{12}{4} = 3$

$a_4 = \frac{12}{4} - \frac{1}{4} = \frac{11}{4}$

$a_5 = \frac{11}{4} - \frac{1}{4} = \frac{10}{4} = \frac{5}{2}$

$d = -\frac{1}{4},\ a_n = \frac{15}{4} - \frac{1}{4}n$

39. $a_1 = 5,\ a_2 = 11 \Rightarrow d = 6$

$a_{10} = a_1 + 9d = 5 + 9(6) = 59$

41. $a_1 = 4.2,\ a_2 = 6.6 \Rightarrow d = 2.4$

$a_7 = a_1 + 6d = 4.2 + 6(2.4) = 18.6$

43. $a_n = 15 - \frac{3}{2}n$

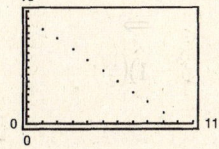

45. $a_n = 0.5n + 4$

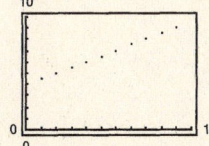

47. $a_n = 4n - 5$

n	1	2	3	4	5	6	7	8	9	10
a_n	-1	3	7	11	15	19	23	27	31	35

49. $a_n = 20 - \frac{3}{4}n$

n	1	2	3	4	5	6	7	8	9	10
a_n	19.25	18.5	17.75	17	16.25	15.5	14.75	14	13.25	12.5

51. $a_n = 1.5 + 0.005n$

n	1	2	3	4	5	6	7	8	9	10
a_n	1.505	1.51	1.515	1.52	1.525	1.53	1.535	1.54	1.545	1.55

53. $8, 20, 32, 44, \ldots n = 10$

$a_1 = 8, a_2 = 20 \implies d = 12$

$a_{10} = a_1 + 9d = 8 + 9(12) = 116$

$S_{10} = \frac{n}{2}[a_1 + a_{10}] = \frac{10}{2}[8 + 116] = 620$

55. $a_1 = 0.5, a_2 = 1.3 \implies d = 0.8$

$a_{10} = a_1 + 9d = 0.5 + 9(0.8) = 7.7$

$S_{10} = \frac{10}{2}(a_1 + a_{10}) = 5(0.5 + 7.7) = 41$

57. $a_1 = 100, a_{25} = 220$

$S_{25} = \frac{25}{2}(a_1 + a_{25}) = 12.5(100 + 220) = 4000$

59. $a_1 = 1, a_{100} = 199, n = 100$

$\displaystyle\sum_{n=1}^{100}(2n - 1) = \frac{100}{2}(1 + 199) = 10,000$

61. $a_1 = 1, a_{50} = 50, n = 50$

$\displaystyle\sum_{n=1}^{50} n = \frac{50}{2}(1 + 50) = 1275$

63. $a_1 = 5, a_{100} = 500, n = 100$

$\displaystyle\sum_{n=1}^{100} 5n = \frac{100}{2}(5 + 500) = 25,250$

65. $\displaystyle\sum_{n=11}^{30} n - \sum_{n=1}^{10} n = \frac{20}{2}(11 + 30) - \frac{10}{2}(1 + 10) = 410 - 55 = 355$

67. $\displaystyle\sum_{n=1}^{500}(n + 8) = \frac{500}{2}[9 + 508] = 129,250$

69. $a_1 = 7, a_{20} = 45, n = 20$

$\displaystyle\sum_{n=1}^{20}(2n + 5) = \frac{20}{2}(7 + 45) = 520$

71. $a_n = \dfrac{n + 4}{2} = 2 + \dfrac{n}{2}$

$a_1 = \dfrac{5}{2}, a_{100} = 52, d = 2$

$S_{100} = \dfrac{100}{2}\left(\dfrac{5}{2} + 52\right) = 2725$

73. $a_1 = \dfrac{742}{3}, a_{60} = 90, n = 60$

$\displaystyle\sum_{i=1}^{60}\left(250 - \dfrac{8}{3}i\right) = \dfrac{60}{2}\left(\dfrac{742}{3} + 90\right) = 10,120$

75. (a) $a_1 = 32,500$, $d = 1500$

$a_6 = a_1 + 5d = 32,500 + 5(1500) = \$40,000$

(b) $S_6 = \frac{6}{2}[32,500 + 40,000] = \$217,500$

(c) 1st year: \$32,500; 2nd year: \$34,000;
3rd year: \$35,500; 4th year: \$37,000;
5th year: \$38,500; 6th year: \$40,000

$32,500 + 34,000 + 35,500 + 37,000 +$

$38,500 + 40,000 = \$217,500$

77. $a_1 = 14$, $a_{18} = 31$

$S_{18} = \frac{18}{2}(14 + 31) = 405$ bricks

79. $a_1 = 25$, $a_2 = 25 + 2 = 27$, etc. $\implies d = 2$ and $n = 15$

$a_{15} = 25 + (15 - 1)(2)$

$S_{15} = \frac{15}{2}(25 + 53) = \frac{15}{2} \cdot 78 = 585$ seats

81. $a_1 = 20,000$

$a_2 = 20,000 + 5000 = 25,000$

$d = 5000$

$a_5 = 20,000 + 4(5000) = 40,000$

$S_5 = \frac{5}{2}(20,000 + 40,000) = 150,000$

83. True. Given a_1 and a_2, you know $d = a_2 - a_1$. Thus, $a_n = a_1 + (n - 1)d$.

85. $a_1 = x$ $a_6 = 11x$

$a_2 = x + 2x = 3x$ $a_7 = 13x$

$a_3 = 3x + 2x = 5x$ $a_8 = 15x$

$a_4 = 7x$ $a_9 = 17x$

$a_5 = 9x$ $a_{10} = 19x$

87. $S_{20} = \frac{20}{2}(a_1 + [a_1 + 19(3)]) = 650$

$650 = 10(2a_1 + 57)$

$65 = 2a_1 + 57$

$a_1 = 4$

89. (a) $1 + 3 = 4$

$1 + 3 + 5 = 9$

$1 + 3 + 5 + 7 = 16$

$1 + 3 + 5 + 7 + 9 = 25$

$1 + 3 + 5 + 7 + 9 + 11 = 36$

(b) $S_n = n^2$

$S_7 = 1 + 3 + 5 + 7 + 9 + 11 + 13 = 49 = 7^2$

(c) $S_n = \frac{n}{2}[1 + (2n - 1)] = \frac{n}{2}(2n) = n^2$

91. $\begin{bmatrix} 2 & -1 & 7 & : & -10 \\ 3 & 2 & -4 & : & 17 \\ 6 & -5 & 1 & : & -20 \end{bmatrix}$ row reduces to $\begin{bmatrix} 1 & 0 & 0 & : & 1 \\ 0 & 1 & 0 & : & 5 \\ 0 & 0 & 1 & : & -1 \end{bmatrix}$

Answer: $(1, 5, -1)$

93. $\begin{vmatrix} 0 & 0 & 1 \\ 4 & -3 & 1 \\ 2 & 6 & 1 \end{vmatrix} = 30$

Area $= \frac{1}{2}(30) = 15$ square units

Section 8.3 Geometric Sequences and Series

■ You should be able to identify a geometric sequence, find its common ratio, and find the nth term.

■ You should be able to find the nth partial sum of a geometric sequence with common ratio r using the formula.

$$S_n = a_1 \left(\frac{1 - r^n}{1 - r} \right)$$

■ You should know that if $|r| < 1$, then

$$\sum_{n=1}^{\infty} a_1 r^{n-1} = \frac{a_1}{1 - r}.$$

Solutions to Odd-Numbered Exercises

1. 5, 15, 45, 135, . . .

Geometric sequence, $r = 3$

3. 6, 18, 30, 42, . . .

Not a geometric sequence

(Note: It is an arithmetic sequence with $d = 12$.)

5. $1, -\frac{1}{2}, \frac{1}{4}, -\frac{1}{8}, \ldots$

Geometric sequence, $r = -\frac{1}{2}$

7. $\frac{1}{8}, \frac{1}{4}, \frac{1}{2}, 1, \ldots$

Geometric sequence, $r = 2$

9. $1, \frac{1}{2}, \frac{1}{3}, \frac{1}{4}, \ldots$

Not a geometric sequence

11. $a_1 = 6, r = 3$

$a_2 = 6(3) = 18$

$a_3 = 18(3) = 54$

$a_4 = 54(3) = 162$

$a_5 = 162(3) = 486$

13. $a_1 = 1, r = \frac{1}{2}$

$a_1 = 1$

$a_2 = 1\left(\frac{1}{2}\right) = \frac{1}{2}$

$a_3 = \frac{1}{2}\left(\frac{1}{2}\right) = \frac{1}{4}$

$a_4 = \frac{1}{4}\left(\frac{1}{2}\right) = \frac{1}{8}$

$a_5 = \frac{1}{8}\left(\frac{1}{2}\right) = \frac{1}{16}$

15. $a_1 = 5, r = -\frac{1}{10}$

$a_1 = 5$

$a_2 = 5\left(-\frac{1}{10}\right) = -\frac{1}{2}$

$a_3 = \left(-\frac{1}{2}\right)\left(-\frac{1}{10}\right) = \frac{1}{20}$

$a_4 = \frac{1}{20}\left(-\frac{1}{10}\right) = -\frac{1}{200}$

$a_5 = \left(-\frac{1}{200}\right)\left(-\frac{1}{10}\right) = \frac{1}{2000}$

17. $a_1 = 1, r = e$

$a_1 = 1$

$a_2 = 1(e) = e$

$a_3 = (e)(e) = e^2$

$a_4 = (e^2)(e) = e^3$

$a_5 = (e^3)(e) = e^4$

19. $a_1 = 64$, $a_{k+1} = \frac{1}{2}a_k$

$a_1 = 64$

$a_2 = \frac{1}{2}(64) = 32$

$a_3 = \frac{1}{2}(32) = 16$

$a_4 = \frac{1}{2}(16) = 8$

$a_5 = \frac{1}{2}(8) = 4$

$r = \frac{1}{2}$, $a_n = 64\left(\frac{1}{2}\right)^{n-1} = 128\left(\frac{1}{2}\right)^n$

21. $a_1 = 9$, $a_{k+1} = 2a_k$

$a_2 = 2(9) = 18$

$a_3 = 2(18) = 36$

$a_4 = 2(36) = 72$

$a_5 = 2(72) = 144$

$r = 2$

$a_n = \left(\frac{9}{2}\right) 2^n = 9(2^{n-1})$

23. $a_1 = 6$, $a_{k+1} = -\frac{3}{2}a_k$

$a_1 = 6$

$a_2 = -\frac{3}{2}(6) = -9$

$a_3 = -\frac{3}{2}(-9) = \frac{27}{2}$

$a_4 = -\frac{3}{2}\left(\frac{27}{2}\right) = -\frac{81}{4}$

$a_5 = -\frac{3}{2}\left(-\frac{81}{4}\right) = \frac{243}{8}$

$r = -\frac{3}{2}$, $a_n = 6\left(-\frac{3}{2}\right)^{n-1}$

$\qquad = -4\left(-\frac{3}{2}\right)^n$

25. $a_1 = 4$, $r = \frac{1}{2}$, $n = 10$

$a_n = a_1 r^{n-1}$

$a_{10} = 4\left(\frac{1}{2}\right)^9 = \left(\frac{1}{2}\right)^7 = \frac{1}{128}$

27. $a_1 = 6$, $r = -\frac{1}{3}$, $n = 12$

$a_n = a_1 r^{n-1}$

$a_{12} = 6\left(-\frac{1}{3}\right)^{11} = \frac{-2}{3^{10}}$

29. $a_1 = 500$, $r = 1.02$, $n = 14$

$a_n = a_1 r^{n-1}$

$a_{14} = 500(1.02)^{13} \approx 646.8$

31. $a_2 = a_1 r = -18 \implies a_1 = \frac{-18}{r}$

$a_5 = a_1 r^4 = (a_1 r)r^3 = -18r^3 = \frac{2}{3} \implies r = -\frac{1}{3}$

$a_1 = \frac{-18}{r} = \frac{-18}{-1/3} = 54$

$a_6 = a_1 r^5 = 54\left(\frac{-1}{3}\right)^5 = \frac{-54}{243} = -\frac{2}{9}$

33. $r = \frac{21}{7} = 3$

$a_9 = a_1 r^{9-1} = 7(3)^8 = 45,927$

35. $r = \frac{30}{5} = 6$

$a_{10} = a_1 r^{10-1} = 5(6)^9 = 50,388,480$

37. $a_n = 12(-0.75)^{n-1}$

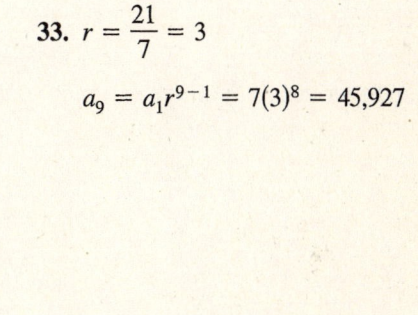

39. $a_n = 2(1.3)^{n-1}$

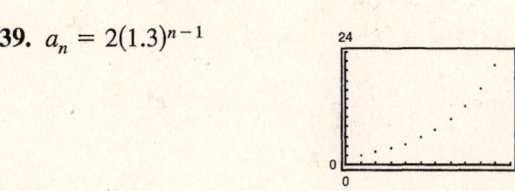

41. 8, -4, 2, -1, $\frac{1}{2}$

$S_1 = 8$

$S_2 = 8 + (-4) = 4$

$S_3 = 8 + (-4) + 2 = 6$

$S_4 = 8 + (-4) + 2 + (-1) = 5$

43. $\displaystyle\sum_{n=1}^{\infty} 16\left(-\frac{1}{2}\right)^{n-1}$

n	1	2	3	4	5	6	7	8	9	10
S_n	16	24	28	30	31	31.5	31.75	31.875	31.9375	31.96875

45. $\displaystyle\sum_{n=1}^{9} 2^{n-1} \Rightarrow a_1 = 1,\ r = 2$

$$S_9 = \frac{1(1 - 2^9)}{1 - 2} = 511$$

47. $\displaystyle\sum_{i=1}^{7} 64\left(-\frac{1}{2}\right)^{i-1} \Rightarrow a_1 = 64,\ r = -\frac{1}{2}$

$$S_7 = 64\left[\frac{1 - (-1/2)^7}{1 - (-1/2)}\right] = \frac{128}{3}\left[1 - \left(-\frac{1}{2}\right)^7\right] = 43$$

49. $\displaystyle\sum_{n=0}^{20} 3\left(\frac{3}{2}\right)^n = \sum_{n=1}^{21} 3\left(\frac{3}{2}\right)^{n-1} \Rightarrow a_1 = 3,\ r = \frac{3}{2}$

$$S_{21} = 3\left[\frac{1 - (3/2)^{21}}{1 - (3/2)}\right] = -6\left[1 - \left(\frac{3}{2}\right)^{21}\right] \approx 29{,}921.31$$

51. $\displaystyle\sum_{i=1}^{10} 8\left(-\frac{1}{4}\right)^{i-1} \Rightarrow a_1 = 8,\ r = -\frac{1}{4}$

$$S_{10} = 8\left[\frac{1 - (-1/4)^{10}}{1 - (-1/4)}\right] = \frac{32}{5}\left[1 - \left(-\frac{1}{4}\right)^{10}\right] \approx 6.4$$

53. $\displaystyle\sum_{n=0}^{5} 300(1.06)^n = \sum_{n=1}^{6} 300(1.06)^{n-1} \Rightarrow a_1 = 300,\ r = 1.06$

$$S_6 = 300\left[\frac{1 - (1.06)^6}{1 - 1.06}\right] \approx 2092.60$$

55. $5 + 15 + 45 + \cdots + 3645$

$r = 3$ and $3645 = 5(3)^{n-1} \Rightarrow n = 7$

Thus, the sum can be written as $\displaystyle\sum_{n=1}^{7} 5(3)^{n-1}$.

57. $2 - \frac{1}{2} + \frac{1}{8} - \cdots + \frac{1}{2048}$

$r = -\frac{1}{4}$ and $\frac{1}{2048} = 2\left(-\frac{1}{4}\right)^{n-1} \Rightarrow n = 7$

$$\sum_{n=1}^{7} 2\left(-\frac{1}{4}\right)^{n-1}$$

59. $a_1 = 1,\ r = \dfrac{1}{2}$

$$\sum_{n=0}^{\infty} \left(\frac{1}{2}\right)^n = \frac{a_1}{1 - r} = \frac{1}{1 - (1/2)} = 2$$

61. $a_1 = 1,\ r = -\dfrac{1}{2}$

$$\sum_{n=0}^{\infty} \left(-\frac{1}{2}\right)^n = \sum_{n=1}^{\infty} \left(-\frac{1}{2}\right)^{n-1} = \frac{a_1}{1 - r} = \frac{1}{1 - (-1/2)} = \frac{2}{3}$$

63. $\displaystyle\sum_{n=1}^{\infty} 2\left(\frac{7}{3}\right)^{n-1}$ does not have a finite sum $\left(\frac{7}{3} > 1\right)$.

65. $a_1 = 1,\ r = 0.4$

$$\sum_{n=0}^{\infty} (0.4)^n = \frac{a_1}{1 - r} = \frac{1}{1 - 0.4} = \frac{1}{0.6} = \frac{10}{6} = \frac{5}{3}$$

67. $a = -3,\ r = 0.9$

$$\sum_{n=0}^{\infty} -3(0.9)^n = \frac{a_1}{1 - r} = \frac{-3}{1 - 0.9} = \frac{-3}{0.1} = -30$$

69. $8 + 6 + \dfrac{9}{2} + \dfrac{27}{8} + \cdots = \displaystyle\sum_{n=0}^{\infty} 8\left(\frac{3}{4}\right)^n$

$$= \frac{8}{1 - 3/4} = 32$$

71. $3 - 1 + \dfrac{1}{3} - \dfrac{1}{9} + \cdots = \displaystyle\sum_{n=0}^{\infty} 3\left(-\dfrac{1}{3}\right)^n = \dfrac{a_1}{1-r} = \dfrac{3}{1-\left(-\dfrac{1}{3}\right)} = 3\left(\dfrac{3}{4}\right) = \dfrac{9}{4}$

73. $0.\overline{36} = \displaystyle\sum_{n=0}^{\infty} 0.36(0.01)^n = \dfrac{0.36}{1-0.01} = \dfrac{0.36}{0.99} = \dfrac{36}{99} = \dfrac{4}{11}$

75. $0.3\overline{18} = 0.3 + \displaystyle\sum_{n=0}^{\infty} 0.018(0.01)^n = \dfrac{3}{10} + \dfrac{0.018}{1-0.01}$

$$= \dfrac{3}{10} + \dfrac{0.018}{0.99} = \dfrac{3}{10} + \dfrac{18}{990} = \dfrac{3}{10} + \dfrac{2}{110}$$

$$= \dfrac{35}{110} = \dfrac{7}{22}$$

77. $A = P\left(1 + \dfrac{r}{n}\right)^{nt} = 1000\left(1 + \dfrac{0.03}{n}\right)^{n(10)}$

    (a) $n = 1$: $A = 1000(1 + 0.03)^{10} \approx 1343.92$

    (b) $n = 2$: $A = 1000\left(1 + \dfrac{0.03}{2}\right)^{2(10)} \approx 1346.86$

    (c) $n = 4$: $A = 1000\left(1 + \dfrac{0.03}{4}\right)^{4(10)} \approx 1348.35$

    (d) $n = 12$: $A = 1000\left(1 + \dfrac{0.03}{12}\right)^{12(10)} \approx 1349.35$

    (e) $n = 365$: $A = 1000\left(1 + \dfrac{0.03}{365}\right)^{365(10)} \approx 1349.84$

79. $A = \displaystyle\sum_{n=1}^{60} 100\left(1 + \dfrac{0.06}{12}\right)^n = 100\left(1 + \dfrac{0.06}{12}\right) \cdot \dfrac{\left[1 - \left(1 + \dfrac{0.06}{12}\right)^{60}\right]}{\left[1 - \left(1 + \dfrac{0.06}{12}\right)\right]} \approx \$7011.89$

81. Let $N = 12t$ be the total number of deposits.

$$A = P\left(1 + \dfrac{r}{12}\right) + P\left(1 + \dfrac{r}{12}\right)^2 + \cdots + P\left(1 + \dfrac{r}{12}\right)^N$$

$$= \left(1 + \dfrac{r}{12}\right)\left[P + P\left(1 + \dfrac{r}{12}\right) + \cdots + P\left(1 + \dfrac{r}{12}\right)^{N-1}\right]$$

$$= P\left(1 + \dfrac{r}{12}\right)\displaystyle\sum_{n=1}^{N}\left(1 + \dfrac{r}{12}\right)^{n-1}$$

$$= P\left(1 + \dfrac{r}{12}\right)\dfrac{1 - \left(1 + \dfrac{r}{12}\right)^N}{1 - \left(1 + \dfrac{r}{12}\right)}$$

$$= P\left(1 + \dfrac{r}{12}\right)\left(-\dfrac{12}{r}\right)\left[1 - \left(1 + \dfrac{r}{12}\right)^N\right]$$

$$= P\left(\dfrac{12}{r} + 1\right)\left[-1 + \left(1 + \dfrac{r}{12}\right)^N\right]$$

$$= P\left[\left(1 + \dfrac{r}{12}\right)^N - 1\right]\left(1 + \dfrac{12}{r}\right)$$

$$= P\left[\left(1 + \dfrac{r}{12}\right)^{12t} - 1\right]\left(1 + \dfrac{12}{r}\right)$$

83. $P = \$50$, $r = 7\%$, $t = 20$ years

    (a) Compounded monthly: $A = 50\left[\left(1 + \dfrac{0.07}{12}\right)^{12(20)} - 1\right]\left(1 + \dfrac{12}{0.07}\right) \approx \$26,198.27$

    (b) Compounded continuously: $A = \dfrac{50e^{0.07/12}(e^{0.07(20)} - 1)}{e^{0.07/12} - 1} \approx \$26,263.88$

85. $P = 100$, $r = 5\% = 0.05$, $t = 40$

    (a) Compounded monthly: $A = 100\left[\left(1 + \dfrac{0.05}{12}\right)^{12(40)} - 1\right]\left(1 + \dfrac{12}{0.05}\right) \approx \$153,237.86$

    (b) Compounded continuously: $A = \dfrac{100e^{0.05/12}(e^{0.05(40)} - 1)}{e^{0.05/12} - 1} \approx \$153,657.02$

87. First shaded area: $\dfrac{16^2}{4}$ Second shaded area: $\dfrac{16^2}{4} + \dfrac{1}{2} \cdot \dfrac{16^2}{4}$

Third shaded area: $\dfrac{16^2}{4} + \dfrac{1}{2}\dfrac{16^2}{4} + \dfrac{1}{4}\dfrac{16^2}{4}$, etc

Total area of shaded region:

$$\dfrac{16^2}{4} \sum_{n=0}^{5} \left(\dfrac{1}{2}\right)^n = 64\left[\dfrac{1 - \left(\dfrac{1}{2}\right)^6}{1 - \dfrac{1}{2}}\right] = 128\left(1 - \left(\dfrac{1}{2}\right)^6\right) = 126 \text{ square units}$$

89. $a_n = 3343(1.013)^n$

    (a) The population is growing at 0.013 or 1.3% per year.

    (b) In 2010, $n = 20$ and $a_{20} \approx 4328$ or 4,328,000 people.

    (c) $a_n = 4100$ when $n = 15.8$ or during 2005.

91. $S_n = \displaystyle\sum_{i=1}^{n} 0.01(2)^{i-1}$ **93.** False. See definition page 516.

    $S_{29} = \$5,368,709.11$

    $S_{30} = \$10,737,418.23$

    $S_{31} = \$21,474,836.47$

95. $a_1 = 3$, $r = \dfrac{x}{2}$ **97.** $a_1 = 100$, $r = e^x$, $n = 9$

    $a_2 = 3\left(\dfrac{x}{2}\right) = \dfrac{3x}{2}$ $a_n = a_1 r^{n-1}$

    $a_3 = \dfrac{3x}{2}\left(\dfrac{x}{2}\right) = \dfrac{3x^2}{4}$ $a_9 = 100(e^x)^8 = 100e^{8x}$

    $a_4 = \dfrac{3x^2}{4}\left(\dfrac{x}{2}\right) = \dfrac{3x^3}{8}$

    $a_5 = \dfrac{3x^3}{8}\left(\dfrac{x}{2}\right) = \dfrac{3x^4}{16}$

99. (a) $f(x) = 6\left[\dfrac{1 - 0.5^x}{1 - 0.5}\right]$

$\displaystyle\sum_{n=0}^{\infty} 6\left(\dfrac{1}{2}\right)^n = \dfrac{6}{1 - \dfrac{1}{2}} = 12$

The horizontal asymptote of $f(x)$ is $y = 12$. This corresponds to the sum of the series.

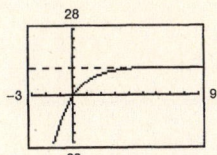

(b) $f(x) = 2\left[\dfrac{1 - 0.8^x}{1 - 0.8}\right]$

$\displaystyle\sum_{n=0}^{\infty} 2\left(\dfrac{4}{5}\right)^n = \dfrac{2}{1 - \dfrac{4}{5}} = 10$

The horizontal asymptote of $f(x)$ is $y = 10$. This corresponds to the sum of the series.

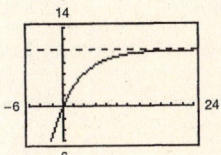

101. To use the first two terms of a geometric series to find the nth term, first divide the second term by the first term to obtain the constant ratio. The nth term is the first term multiplied by the common ratio raised to the $(n - 1)$ power.

$$r = \dfrac{a_2}{a_1}, \; a_n = a_1 r^{n-1}$$

103. $\text{Time} = \dfrac{\text{Distance}}{\text{Speed}} = \dfrac{200}{50} + \dfrac{200}{42} = 200\left[\dfrac{92}{2100}\right] \text{ hours}$

$\text{Speed} = \dfrac{\text{Distance}}{\text{Time}} = \dfrac{400}{200\left[\frac{92}{2100}\right]} = \dfrac{2(2100)}{92} \approx 45.65 \text{ mph}$

105. $\det\begin{bmatrix} 4 & -1 \\ 6 & 2 \end{bmatrix} = 8 - (-6) = 14$

107. $\det\begin{bmatrix} -1 & 3 & 4 \\ -2 & 8 & 0 \\ 2 & 5 & -1 \end{bmatrix}$

$= 4(-10 - 16) - 1(-8 + 6)$

$= -104 + 2 = -102$

Section 8.4   Mathematical Induction

- You should be sure that you understand the principle of mathematical induction. If P_n is a statement involving the positive integer n, where P_1 is true and the truth of P_k implies the truth of P_{k+1}, then P_n is true for all positive integers n.

- You should be able to verify (by induction) the formulas for the sums of powers of integers and be able to use these formulas.

- You should be able to work with finite differences.

Solutions to Odd-Numbered Exercises

1. $P_k = \dfrac{5}{k(k+1)}$

$P_{k+1} = \dfrac{5}{(k+1)[(k+1)+1]} = \dfrac{5}{(k+1)(k+2)}$

3. $P_k = \dfrac{3(2k+1)}{k-1}$

$P_{k+1} = \dfrac{3(2(k+1)+1)}{(k+1)-1} = \dfrac{3(2k+3)}{k}$

5. $P_k = 1 + 6 + 11 + \cdots + [5(k-1) - 4] + [5k - 4]$

$P_{k+1} = 1 + 6 + 11 + \cdots + [5k - 4] + [5(k+1) - 4]$

$\qquad = 1 + 6 + 11 + \cdots + [5k - 4] + [5k + 1]$

7. 1. When $n = 1$, $S_1 = 2 = 1(1 + 1)$.

   2. Assume that

$\qquad S_k = 2 + 4 + 6 + 8 + \cdots + 2k = k(k+1)$.

   Then,

$\qquad S_{k+1} = 2 + 4 + 6 + 8 + \cdots + 2k + 2(k+1)$

$\qquad\qquad = S_k + 2(k+1) = k(k+1) + 2(k+1) = (k+1)(k+2)$.

We conclude by mathematical induction that the formula is valid for all positive integer values of n.

9. 1. When $n = 1$, $S_1 = 3 = \dfrac{1}{2}(5(1) + 1)$

   2. Assume that $S_k = 3 + 8 + 13 + \cdots + (5k - 2) = \dfrac{k}{2}(5k + 1)$

   Then: $S_{k+1} = 3 + 8 + 13 + \cdots + (5k - 2) + [5(k+1) - 2]$

$\qquad\qquad = S_k + [5k + 3] = \dfrac{k}{2}(5k + 1) + 5k + 3$

$\qquad\qquad = \dfrac{1}{2}[5k^2 + 11k + 6] = \dfrac{1}{2}(k + 1)(5k + 6)$

$\qquad\qquad = \dfrac{1}{2}(k + 1)(5(k + 1) + 1)$

We conclude by mathematical induction that the formula is valid for all positive integers n.

11. 1. When $n = 1$, $S_1 = 1 = 2^1 - 1$.

   2. Assume that

   $$S_k = 1 + 2 + 2^2 + 2^3 + \cdots + 2^{k-1} = 2^k - 1.$$

   Then,

   $$S_{k+1} = 1 + 2 + 2^2 + 2^3 + \cdots + 2^{k-1} + 2^k$$
   $$= S_k + 2^k = 2^k - 1 + 2^k = 2(2^k) - 1 = 2^{k+1} - 1.$$

   Therefore, by mathematical induction, the formula is valid for all positive integer values of n.

13. 1. When $n = 1$, $S_1 = 1 = \dfrac{1(1 + 1)}{2}$.

   2. Assume that

   $$S_k = 1 + 2 + 3 + 4 + \cdots + k = \frac{k(k + 1)}{2}.$$

   Then,

   $$S_{k+1} = 1 + 2 + 3 + 4 + \cdots + k + (k + 1)$$
   $$= S_k + (k + 1) = \frac{k(k + 1)}{2} + \frac{2(k + 1)}{2} = \frac{(k + 1)(k + 2)}{2}.$$

   Therefore, we conclude that this formula holds for all positive integer values of n.

15. 1. When $n = 1$,

   $$S_1 = 1^4 = \frac{1(1 + 1)(2 \cdot 1 + 1)(3 \cdot 1^2 + 3 \cdot 1 - 1)}{30}.$$

   2. Assume that $S_k = \displaystyle\sum_{i=1}^{k} i^4 = \frac{k(k + 1)(2k + 1)(3k^2 + 3k - 1)}{30}.$

   Then, $S_{k+1} = S_k + (k + 1)^4$

   $$= \frac{k(k + 1)(2k + 1)(3k^2 + 3k - 1)}{30} + (k + 1)^4 = \frac{k(k + 1)(2k + 1)(3k^2 + 3k - 1) + 30(k + 1)^4}{30}$$

   $$= \frac{(k + 1)[k(2k + 1)(3k^2 + 3k - 1) + 30(k + 1)^3]}{30} = \frac{(k + 1)(6k^4 + 39k^3 + 91k^2 + 89k + 30)}{30}$$

   $$= \frac{(k + 1)(k + 2)(2k + 3)(3k^2 + 9k + 5)}{30} = \frac{(k + 1)(k + 2)(2(k + 1) + 1)(3(k + 1)^2 + 3(k + 1) - 1)}{30}.$$

   Therefore, we conclude that this formula holds for all positive integer values of n.

17. 1. When $n = 1$, $S_1 = 2 = \dfrac{1(2)(3)}{3}$.

2. Assume that $S_k = 1(2) + 2(3) + 3(4) + \cdots + k(k + 1) = \dfrac{k(k + 1)(k + 2)}{3}$.

Then,

$$S_{k+1} = 1(2) + 2(3) + 3(4) + \cdots + k(k + 1) + (k + 1)(k + 2)$$

$$= S_k + (k + 1)(k + 2) = \dfrac{k(k + 1)(k + 2)}{3} + \dfrac{3(k + 1)(k + 2)}{3}$$

$$= \dfrac{(k + 1)(k + 2)(k + 3)}{3}.$$

Thus, this formula is valid for all positive integer values of n.

19. $\displaystyle\sum_{n=1}^{10} n^3 = \dfrac{10^2(10 + 1)^2}{4} = 3025$

21. $\displaystyle\sum_{n=1}^{6} (n^2 - n) = \sum_{n=1}^{6} n^2 - \sum_{n=1}^{6} n$

$$= \dfrac{6(6 + 1)(2(6) + 1)}{6} - \dfrac{6(6 + 1)}{2}$$

$$= 91 - 21 = 70$$

23. 1. When $n = 4$, $4! = 24$ and $2^4 = 16$, thus $4! > 2^4$.

2. Assume $k! > 2^k$, $k > 4$. Then, $(k + 1)! = k!(k + 1) > 2^k(2)$ since $k + 1 > 2$. Thus, $(k + 1)! > 2^{k+1}$.

Therefore, by mathematical induction, the formula is valid for all integers n such that $n \geq 4$.

25. 1. When $n = 2$, $\dfrac{1}{\sqrt{1}} + \dfrac{1}{\sqrt{2}} \approx 1.707$ and $\sqrt{2} \approx 1.414$, thus $\dfrac{1}{\sqrt{1}} + \dfrac{1}{\sqrt{2}} > \sqrt{2}$.

2. Assume $\dfrac{1}{\sqrt{1}} + \dfrac{1}{\sqrt{2}} + \dfrac{1}{\sqrt{3}} + \cdots + \dfrac{1}{\sqrt{k}} > \sqrt{k}$, $k > 2$.

Then, $\dfrac{1}{\sqrt{1}} + \dfrac{1}{\sqrt{2}} + \dfrac{1}{\sqrt{3}} + \cdots + \dfrac{1}{\sqrt{k}} + \dfrac{1}{\sqrt{k + 1}} > \sqrt{k} + \dfrac{1}{\sqrt{k + 1}}$.

Now we need to show that $\sqrt{k} + \dfrac{1}{\sqrt{k + 1}} > \sqrt{k + 1}$, $k > 2$.

This is true because

$$\sqrt{k(k + 1)} > k$$

$$\sqrt{k(k + 1)} + 1 > k + 1$$

$$\dfrac{\sqrt{k(k + 1)} + 1}{\sqrt{k + 1}} > \dfrac{k + 1}{\sqrt{k + 1}}$$

$$\sqrt{k} + \dfrac{1}{\sqrt{k + 1}} > \sqrt{k + 1}.$$

Therefore, $\dfrac{1}{\sqrt{1}} + \dfrac{1}{\sqrt{2}} + \dfrac{1}{\sqrt{3}} + \cdots + \dfrac{1}{\sqrt{k}} + \dfrac{1}{\sqrt{k + 1}} > \sqrt{k + 1}$.

Therefore, by mathematical induction, the formula is valid for all integers n such that $n \geq 2$.

27. 1. When $n = 1, 1 + a \geq a$ since $1 > 0$.

2. Assume $(1 + a)^k \geq ka$

Then $(1 + a)^{k+1} = (1 + a)^k(1 + a) \geq ka(1 + a)$

$$= ka + ka^2 \geq ka + a \quad \text{(because } a > 1)$$

$$= (k + 1)a$$

Therefore, by mathematical induction, the inequality is valid for all integers $n \geq 1$.

29. 1. When $n = 1, (ab)^1 = a^1b^1 = ab$.

2. Assume that $(ab)^k = a^kb^k$.

Then, $(ab)^{k+1} = (ab)^k(ab)$

$$= a^kb^kab$$

$$= a^{k+1}b^{k+1}.$$

Thus, $(ab)^n = a^nb^n$.

31. 1. When $n = 1, (x_1)^{-1} = x_1^{-1}$.

2. Assume that

$(x_1x_2x_3 \cdots x_k)^{-1} = x_1^{-1}x_2^{-1}x_3^{-1} \cdots x_k^{-1}$.

Then,

$(x_1x_2x_3 \cdots x_kx_{k+1})^{-1} = [(x_1x_2x_3 \cdots x_k)x_{k+1}]^{-1}$

$$= (x_1x_2x_3 \cdots x_k)^{-1}x_{k+1}^{-1}$$

$$= x_1^{-1}x_2^{-1}x_3^{-1} \cdots x_k^{-1}x_{k+1}^{-1}.$$

Thus, the formula is valid.

33. 1. When $n = 1, x(y_1) = xy_1$.

2. Assume that $x(y_1 + y_2 + \cdots + y_k) = xy_1 + xy_2 + \cdots + xy_k$.

Then,

$xy_1 + xy_2 + \cdots + xy_k + xy_{k+1} = x(y_1 + y_2 + \cdots + y_k) + xy_{k+1}$

$$= x[(y_1 + y_2 + \cdots + y_k) + y_{k+1}]$$

$$= x(y_1 + y_2 + \cdots + y_k + y_{k+1}).$$

Hence, the formula holds.

35. 1. When $n = 1, [1^3 + 3(1)^2 + 2(1)] = 6$ and 3 is a factor.

2. Assume that 3 is a factor of $(k^3 + 3k^2 + 2k)$. Then,

$[(k + 1)^3 + 3(k + 1)^2 + 2(k + 1)] = k^3 + 3k^2 + 3k + 1 + 3k^2 + 6k + 3 + 2k + 2$

$$= (k^3 + 3k^2 + 2k) + (3k^2 + 9k + 6)$$

$$= (k^3 + 3k^2 + 2k) + 3(k^2 + 3k + 2).$$

Since 3 is a factor of $(k^3 + 3k^2 + 2k)$ by our assumption, and 3 is a factor of $3(k^2 + 3k + 2)$ then 3 is a factor of the whole sum.

Thus, 3 is a factor of $(n^3 + 3n^2 + 2n)$ for every positive integer n.

37. $a_1 = 0, a_n = a_{n-1} + 3$

$a_1 = 0$

$a_2 = a_1 + 3 = 0 + 3 = 3$

$a_3 = a_2 + 3 = 3 + 3 = 6$

$a_4 = a_3 + 3 = 6 + 3 = 9$

$a_5 = a_4 + 3 = 9 + 3 = 12$

a_n:　0　　3　　6　　9　　12

First differences:　　3　　3　　3　　3

Second differences:　　0　　0　　0

Since the first differences are equal, the sequence has a linear model.

39. $a_1 = 3, a_n = a_{n-1} - n$

$a_1 = 3$

$a_2 = a_1 - 2 = 3 - 2 = 1$

$a_3 = a_2 - 3 = 1 - 3 = -2$

$a_4 = a_3 - 4 = -2 - 4 = -6$

$a_5 = a_4 - 5 = -6 - 5 = -11$

a_n:　3　　1　　−2　　−6　　−11

First differences:　　−2　　−3　　−4　　−5

Second differences:　　−1　　−1　　−1

Since the second differences are all the same, the sequence has a quadratic model.

41. $a_0 = 0, a_n = a_{n-1} + n$

$a_0 = 0$

$a_1 = a_0 + 1 = 0 + 1 = 1$

$a_2 = a_1 + 2 = 1 + 2 = 3$

$a_3 = a_2 + 3 = 3 + 3 = 6$

$a_4 = a_3 + 4 = 6 + 4 = 10$

a_n:　0　　1　　3　　6　　10

First differences:　　1　　2　　3　　4

Second differences:　　1　　1　　1

Since the second differences are equal, the sequence has a quadratic model.

43. $a_1 = 2, a_n = a_{n-1} + 2$

$a_1 = 2$

$a_2 = a_1 + 2 = 2 + 2 = 4$

$a_3 = a_2 + 2 = 4 + 2 = 6$

$a_4 = a_3 + 2 = 6 + 2 = 8$

$a_5 = a_4 + 2 = 8 + 2 = 10$

a_n:　2　　4　　6　　8　　10

First differences:　　2　　2　　2　　2

Second differences:　　0　　0　　0

Since the first differences are equal, the sequence has a linear model.

45. $a_0 = 3, a_1 = 3, a_4 = 15$

Let $a_n = an^2 + bn + c$. Thus

$a_0 = a(0)^2 + b(0) + c = 3 \implies c = 3$

$a_1 = a(1)^2 + b(1) + c = 3 \implies a + b + c = 3$

$$a + b \quad\quad = 0$$

$a_4 = a(4)^2 + b(4) + c = 15 \implies 16a + 4b + c = 15$

$$16a + 4b \quad = 12$$

$$4a + b \quad = 3$$

By elimination: $-a - b = 0$

$$\underline{4a + b = 3}$$

$$3a \quad\quad = 3$$

$$a = 1 \implies b = -1$$

Thus, $a_n = n^2 - n + 3$.

47. $a_0 = -3, a_2 = 1, a_4 = 9$

Let $a_n = an^2 + bn + c$. Then

$a_0 = a(0)^2 + b(0) + c = -3 \implies c = -3$

$a_2 = a(2)^2 + b(2) + c = 1 \implies 4a + 2b + c = 1$

$$4a + 2b = 4$$

$$2a + b = 2$$

$a_4 = a(4)^2 + b(4) + c = 9 \implies 16a + 4b + c = 9$

$$16a + 4b = 12$$

$$4a + b = 3$$

By elimination: $-2a - b = -2$

$$\frac{4a + b = \quad 3}{2a \quad\quad = \quad 1}$$

$$a = \tfrac{1}{2} \implies b = 1$$

Thus, $a_n = \tfrac{1}{2}n^2 + n - 3$.

49. (a) 33.3   35.5   38.0   40.1   41.9   44.2   46.1

First difference: 2.2   2.5   2.1   1.8   2.3   1.9

(b) Yes, the data is approximately linear because the first differences are nearly constant.

$a_n = 2.13n + 22.8$

(c)

n	5	6	7	8	9	10	11
a_n	33.5	35.6	37.7	39.8	42.0	44.1	46.2

The model is a good fit.

(d) For 2006, $n = 16$ and $a_{16} \approx 56.9$ or %56,900.

51. False. P_1 might not even be defined.

53. False. It has $n - 2$ second differences.

55. $(2x^2 - 1)^2 = 4x^4 - 4x^2 + 1$

57. $(5 - 4x)^3 = -64x^3 + 240x^2 - 300x + 125$

59.

$$-4 \begin{array}{|rrrr} 1 & 1 & -10 & 8 \\ & -4 & 12 & -8 \\ \hline 1 & -3 & 2 & 0 \end{array}$$

$x^2 - 3x + 2$

61.

$$-5 \begin{array}{|rrrr} 4 & 11 & -43 & 10 \\ & -20 & 45 & -10 \\ \hline 4 & -9 & 2 & 0 \end{array}$$

$4x^2 - 9x + 2$

63. $3\sqrt{-27} - \sqrt{-12} = 3\sqrt{3 \cdot 3 \cdot (-3)} - \sqrt{2 \cdot 2(-3)}$

$$= 9\sqrt{3}i - 2\sqrt{3}i = 7\sqrt{3}i$$

65. $10\left(\sqrt[3]{64} - 2\sqrt[3]{-16}\right) = 10\left(4 - 2^2\sqrt[3]{-2}\right)$

$$= 40 - 40\sqrt[3]{-2}$$

$$= 40\left(1 + \sqrt[3]{2}\right)$$

Section 8.5   The Binomial Theorem

■ You should be able to use the Binomial Theorem

$$(x + y)^n = x^n + nx^{n-1}y + \frac{n(n-1)}{2!}x^{n-2}y^2 + \cdots + {}_nC_r x^{n-r}y^r + \cdots + y^n$$

where ${}_nC_r = \dfrac{n!}{(n-r)!r!}$, to expand $(x + y)^n$.

■ You should be able to use Pascal's Triangle.

Solutions to Odd-Numbered Exercises

1. ${}_7C_5 = \dfrac{7!}{2!5!} = \dfrac{7 \cdot 6 \cdot 5!}{2 \cdot 5!} = \dfrac{42}{2} = 21$

3. $\dbinom{12}{0} = {}_{12}C_0 = \dfrac{12!}{0!12!} = 1$

5. ${}_{20}C_{15} = \dfrac{20!}{15!5!} = \dfrac{20 \cdot 19 \cdot 18 \cdot 17 \cdot 16}{5 \cdot 4 \cdot 3 \cdot 2 \cdot 1} = 15{,}504$

7. ${}_{14}C_1 = \dfrac{14!}{13!1!} = \dfrac{14 \cdot 13!}{13!} = 14$

9. $\dbinom{100}{98} = {}_{100}C_{98} = \dfrac{100!}{98!2!} = \dfrac{100 \cdot 99}{2 \cdot 1} = 4950$

11. ${}_{32}C_{28} = 35{,}960$

13. ${}_{22}C_9 = 497{,}420$

15. ${}_{41}C_{36} = 749{,}398$

17.
```
            1
          1   1
        1   2   1
      1   3   3   1
    1   4   6   4   1
  1   5   10  10  5   1
1   6   15  20  15  6   1
1 7  21  35 (35) 21  7   1
```

$_7C_4 = 35$, the 5th entry in the 7th row.

19.
```
              1
            1   1
          1   2   1
        1   3   3   1
      1   4   6   4   1
    1   5   10  10  5   1
  1   6   15  20  15  6   1
1   7  21  35  35  21  7   1
1 8  28  56  70 (56) 28  8   1
```

$_8C_5 = 56$, the 6th entry in the 8th row.

21. $(x + 2)^4 = {}_4C_0x^4 + {}_4C_1x^3(2) + {}_4C_2x^2(2)^2 + {}_4C_3x(2)^3 + {}_4C_4(2)^4$

$\qquad = x^4 + 8x^3 + 24x^2 + 32x + 16$

23. $(a + 3)^3 = {}_3C_0a^3 + {}_3C_1a^2(3) + {}_3C_2a(3)^2 + {}_3C_3(3)^3$

$\qquad = a^3 + 3a^2(3) + 3a(3)^2 + (3)^3$

$\qquad = a^3 + 9a^2 + 27a + 27$

25. $(y - 2)^4 = {}_4C_0y^4 - {}_4C_1y^3(2) + {}_4C_2y^2(2)^2 - {}_4C_3y(2)^3 + {}_4C_4(2)^4$

$\qquad = y^4 - 4y^3(2) + 6y^2(4) - 4y(8) + 16$

$\qquad = y^4 - 8y^3 + 24y^2 - 32y + 16$

27. $(x + y)^5 = {}_5C_0x^5 + {}_5C_1x^4y + {}_5C_2x^3y^2 + {}_5C_3x^2y^3 + {}_5C_4xy^4 + {}_5C_5y^5$

$\qquad = x^5 + 5x^4y + 10x^3y^2 + 10x^2y^3 + 5xy^4 + y^5$

29. $(3r + 2s)^6 = {}_6C_0(3r)^6 + {}_6C_1(3r)^5(2s) + {}_6C_2(3r)^4(2s)^2 + {}_6C_3(3r)^3(2s)^3 + {}_6C_4(3r)^2(2s)^4 + {}_6C_5(3r)(2s)^5 + {}_6C_6(2s)^6$

$\qquad = 729r^6 + 2916r^5s + 4860r^4s^2 + 4320r^3s^3 + 2160r^2s^4 + 576rs^5 + 64s^6$

31. $(x - y)^5 = {}_5C_0x^5 - {}_5C_1x^4y + {}_5C_2x^3y^2 - {}_5C_3x^2y^3 - {}_5C_4xy^4 - {}_5C_5y^5$

$\qquad = x^5 - 5x^4y + 10x^3y^2 - 10x^2y^3 + 5xy^4 - y^5$

33. $(1 - 4x)^3 = {}_3C_01^3 - {}_3C_11^2(4x) + {}_3C_21(4x)^2 - {}_3C_3(4x)^3$

$\qquad\quad = 1 - 3(4x) + 3(4x)^2 - (4x)^3$

$\qquad\quad = 1 - 12x + 48x^2 - 64x^3$

35. $(x^2 + y^2)^4 = {}_4C_0(x^2)^4 + {}_4C_1(x^2)^3(y^2) + {}_4C_2(x^2)^2(y^2)^2 + {}_4C_3(x^2)(y^2)^3 + {}_4C_4(y^2)^4$

$\qquad\quad = x^8 + 4x^6y^2 + 6x^4y^4 + 4x^2y^6 + y^8$

37. $\left(\dfrac{1}{x} + y\right)^5 = {}_5C_0\left(\dfrac{1}{x}\right)^5 + {}_5C_1\left(\dfrac{1}{x}\right)^4y + {}_5C_2\left(\dfrac{1}{x}\right)^3y^2 + {}_5C_3\left(\dfrac{1}{x}\right)^2y^3 + {}_5C_4\left(\dfrac{1}{x}\right)y^4 + {}_5C_5y^5$

$\qquad\quad = \dfrac{1}{x^5} + \dfrac{5y}{x^4} + \dfrac{10y^2}{x^3} + \dfrac{10y^3}{x^2} + \dfrac{5y^4}{x} + y^5$

39. $2(x - 3)^4 + 5(x - 3)^2 = 2[x^4 - 4(x^3)(3) + 6(x^2)(3^2) - 4(x)(3^3) + 3^4] + 5[x^2 - 2(x)(3) + 3^2]$

$\qquad\qquad\qquad\qquad = 2(x^4 - 12x^3 + 54x^2 - 108x + 81) + 5(x^2 - 6x + 9)$

$\qquad\qquad\qquad\qquad = 2x^4 - 24x^3 + 113x^2 - 246x + 207$

41. $-3(x - 2)^3 - 4(x + 1)^6$

$\qquad = [-3x^3 + 18x^2 - 36x + 24] - [4x^6 + 24x^5 + 60x^4 + 80x^3 + 60x^2 + 24x + 4]$

$\qquad = -4x^6 - 24x^5 - 60x^4 - 83x^3 - 42x^2 - 60x + 20$

43. 5th Row of Pascal's Triangle: 1   5   10   10   5   1

$\qquad (3t - s)^5 = 1(3t)^5 + 5(3t)^4(-s) + 10(3t)^3(-s)^2 + 10(3t)^2(-s)^3 + 5(3t)(-s)^4 + 1(-s)^5$

$\qquad\qquad = 243t^5 - 405t^4s + 270t^3s^2 - 90t^2s^3 + 15ts^4 - s^5$

45. 5th row of Pascal's Triangle: 1   5   10   10   5   1

$\qquad (x + 2y)^5 = (1)x^5 + 5x^42y + 10x^3(2y)^2 + 10x^2(2y)^3 + 5x(2y)^4 + (2y)^5$

$\qquad\qquad = x^5 + 10x^4y + 40x^3y^2 + 80x^2y^3 + 80xy^4 + 32y^5$

47. $(x + 8)^{10}, n = 4$

$\qquad {}_{10}C_3x^{10-3}(8)^3 = 120x^7(512) = 61{,}440x^7$

49. $(x - 6y)^5, n = 3$

$\qquad {}_5C_2x^{5-2}(-6y)^2 = 10x^3(36)y^2 = 360x^3y^2$

51. $(4x + 3y)^9, n = 8$

$$_9C_7(4x)^{9-7}(3y)^7 = 36(16)x^2(3^7)y^7$$
$$= 1,259,712x^2y^7$$

53. $(10x - 3y)^{12}, n = 9$

$$_{12}C_8(10x)^{12-8}(-3y)^8 = 495(10^4)(3^8)x^4y^8$$
$$= 32,476,950,000x^4y^8$$

55. The term involving x^4 in the expansion of $(x + 3)^{12}$ is $_{12}C_8x^4(3)^8 = 495x^4(3)^8 = 3,247,695x^4$. The coefficient is 3,247,695.

57. The term involving x^8y^2 in the expansion of $(x - 2y)^{10}$ is

$$_{10}C_2x^8(-2y)^2 = \frac{10!}{2!8!} \cdot 4x^8y^2 = 180x^8y^2.$$ The coefficient is 180.

59. The term involving x^6y^3 in $(3x - 2y)^9$ is $_9C_3(3x)^6(-2y)^3 = 84(3)^6(-2)^3x^6y^3 = -489,888x^6y^3$. The coefficient is $-489,888$.

61. The coefficient of $x^8y^6 = (x^2)^4y^6$ in the expansion of $(x^2 + y)^{10}$ is $_{10}C_6 = 210$.

63.
$$(\sqrt{x} + 5)^4 = (\sqrt{x})^4 + 4(\sqrt{x})^3(5) + 6(\sqrt{x})^2(5)^2 + 4(\sqrt{x})(5^3) + 5^4$$
$$= x^2 + 20x\sqrt{x} + 150x + 500\sqrt{x} + 625$$
$$= x^2 + 20x^{3/2} + 150x + 500x^{1/2} + 625$$

65.
$$(x^{2/3} - y^{1/3})^3 = (x^{2/3})^3 - 3(x^{2/3})^2(y^{1/3}) + 3(x^{2/3})(y^{1/3})^2 - (y^{1/3})^3$$
$$= x^2 - 3x^{4/3}y^{1/3} + 3x^{2/3}y^{2/3} - y$$

67.
$$\frac{f(x + h) - f(x)}{h} = \frac{(x + h)^3 - x^3}{h}$$
$$= \frac{x^3 + 3x^2h + 3xh^2 + h^3 - x^3}{h}$$
$$= \frac{h(3x^2 + 3xh + h^2)}{h}$$
$$= 3x^2 + 3xh + h^2, \ h \neq 0$$

69.
$$\frac{f(x + h) - f(x)}{h} = \frac{\sqrt{x + h} - \sqrt{x}}{h}$$
$$= \frac{\sqrt{x + h} - \sqrt{x}}{h} \cdot \frac{\sqrt{x + h} + \sqrt{x}}{\sqrt{x + h} + \sqrt{x}}$$
$$= \frac{(x + h) - x}{h[\sqrt{x + h} + \sqrt{x}]}$$
$$= \frac{1}{\sqrt{x + h} + \sqrt{x}}, h \neq 0$$

71.
$$(1 + i)^4 = {}_4C_01^4 + {}_4C_1(1)^3i + {}_4C_2(1)^2i^2 + {}_4C_31 \cdot i^3 + {}_4C_4i^4$$
$$= 1 + 4i - 6 - 4i + 1$$
$$= -4$$

73.
$$(2 - 3i)^6 = {}_6C_02^6 - {}_6C_12^5(3i) + {}_6C_22^4(3i)^2 - {}_6C_32^3(3i)^3 + {}_6C_42^2(3i)^4 - {}_6C_52(3i)^5 + {}_6C_6(3i)^6$$
$$= 64 - 576i - 2160 + 4320i + 4860 - 2916i - 729$$
$$= 2035 + 828i$$

75. $\left(-\dfrac{1}{2} + \dfrac{\sqrt{3}}{2}i\right)^3 = \dfrac{1}{8}\left(-1 + \sqrt{3}i\right)^3$

$$= \dfrac{1}{8}\left[(-1)^3 + 3(-1)^2\left(\sqrt{3}i\right) + 3(-1)\left(\sqrt{3}i\right)^2 + \left(\sqrt{3}i\right)^3\right]$$

$$= \dfrac{1}{8}\left[-1 + 3\sqrt{3}i + 9 - 3\sqrt{3}i\right]$$

$$= 1$$

77. $(1.02)^8 = (1 + 0.02)^8 = 1 + 8(0.02) + 28(0.02)^2 + 56(0.02)^3 + 70(0.02)^4 + 56(0.02)^5$

$$+ 28(0.02)^6 + 8(0.02)^7 + (0.02)^8$$

$$= 1 + 0.16 + 0.0112 + 0.000448 + \cdots \approx 1.172$$

79. $(2.99)^{12} = (3 - 0.01)^{12}$

$$= 3^{12} - 12(3)^{11}(0.01) + 66(3)^{10}(0.01)^2 - 220(3)^9(0.01)^3 + 495(3)^8(0.01)^4$$

$$- 792(3)^7(0.01)^5 + 924(3)^6(0.01)^6 - 792(3)^5(0.01)^7 + 495(3)^4(0.01)^8$$

$$- 220(3)^3(0.01)^9 + 66(3)^2(0.01)^{10} - 12(3)(0.01)^{11} + (0.01)^{12}$$

$$\approx 510{,}568.785$$

81. $f(x) = x^3 - 4x$

$g(x) = f(x + 3)$

$\quad = (x + 3)^3 - 4(x + 3)$

$\quad = x^3 + 9x^2 + 27x + 27 - 4x - 12$

$\quad = x^3 + 9x^2 + 23x + 15$

g is shifted 3 units to the left

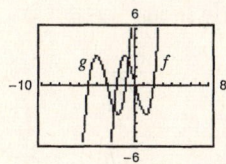

83. $f(x) = (1 - x)^3$

$g(x) = 1 - 3x$

$h(x) = 1 - 3x + 3x^2$

$p(x) = 1 - 3x + 3x^2 - x^3$

Since $p(x)$ is the expansion of $f(x)$, they have the same graph.

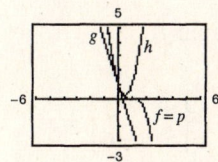

85. $_7C_4\left(\dfrac{1}{2}\right)^4\left(\dfrac{1}{2}\right)^3 = 35\left(\dfrac{1}{16}\right)\left(\dfrac{1}{8}\right) \approx 0.273$

87. $_8C_4\left(\dfrac{1}{3}\right)^4\left(\dfrac{2}{3}\right)^4 = 70\left(\dfrac{1}{81}\right)\left(\dfrac{16}{81}\right) \approx 0.171$

89. $f(t) = 0.018t^2 + 5.15t + 41.6 \quad 5 \le t \le 20$

(a) $g(t) = f(t + 10)$

$\quad = 0.018(t + 10)^2 + 5.15(t + 10) + 41.6 \quad -5 \le t \le 10$

$\quad = 0.018(t^2 + 20t + 100) + 5.15t + 51.5 + 41.6$

$\quad = 0.018t^2 + 5.51t + 94.9$

(b)

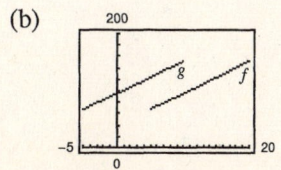

91. False. The x^4y^8 term is

$$_{12}C_4 x^4(-2y)^8 = 495x^4(-2)^8 y^8 = 126{,}720x^4y^8$$

[Note 7920 is the coefficient of x^8y^4]

93. Answers will vary. See page 536.

95. The expansions of $(x + y)^n$ and $(x - y)^n$ are almost the same except that the signs of the terms in the expansion of $(x - y)^n$ alternate from positive to negative.

97.
$$_nC_{n-r} = \frac{n!}{[n - (n - r)]!(n - r)!}$$
$$= \frac{n!}{r!(n - r)!}$$
$$= \frac{n!}{(n - r)!r!}$$
$$= {_nC_r}$$

99.
$$_nC_r + {_nC_{r-1}} = \frac{n!}{(n - r)!r!} + \frac{n!}{(n - r + 1)!(r - 1)!}$$
$$= \frac{n!(n - r + 1)}{(n - r)!r!(n - r + 1)} + \frac{n!}{(n - r + 1)!(r - 1)!} \cdot \frac{r}{r}$$
$$= \frac{n!(n - r + 1)}{(n - r + 1)!r!} + \frac{n!r}{(n - r + 1)!r!}$$
$$= \frac{n!(n - r + 1 + r)}{(n - r + 1)!r!}$$
$$= \frac{n!(n + 1)}{(n - r + 1)!r!}$$
$$= \frac{(n + 1)!}{(n + 1 - r)!r!} = {_{n+1}C_r}$$

101. $g(x) = f(x) + 8$

$g(x)$ is shifted 8 units up from $f(x)$.

103. $g(x) = f(-x)$

$g(x)$ is the reflection of $f(x)$ in the y-axis.

105.
$$\begin{bmatrix} -6 & 5 \\ -5 & 4 \end{bmatrix}^{-1} = \frac{1}{-24 + 25}\begin{bmatrix} 4 & -5 \\ 5 & -6 \end{bmatrix} = \begin{bmatrix} 4 & -5 \\ 5 & -6 \end{bmatrix}$$

Section 8.6   Counting Principles

- ■   You should know The Fundamental Counting Principle.

- ■ $_nP_r = \dfrac{n!}{(n - r)!}$ is the number of permutations of n elements taken r at a time.

- ■   Given a set of n objects that has n_1 of one kind, n_2 of a second kind, and so on, the number of distinguishable permutations is

  $$\frac{n!}{n_1!n_2! \cdots n_k!}.$$

- ■ $_nC_r = \dfrac{n!}{(n - r)!r!}$ is the number of combinations of n elements taken r at a time.

Solutions to Odd-Numbered Exercises

1. Odd integers: 1, 3, 5, 7, 9, 11

6 ways

3. Prime integers: 2, 3, 5, 7, 11

5 ways

5. Divisible by 4: 4, 8, 12

3 ways

7. Sum is 8: $1 + 7, 2 + 6, 3 + 5, 4 + 4, 5 + 3, 6 + 2, 7 + 1$

7 ways

9. Amplifiers: 4 choices

Compact disc players: 6 choices

Speakers: 5 choices

Total: $4 \cdot 6 \cdot 5 = 120$ ways

11. Chemist: 3 choices

Statistician: 8 choices

Total: $3 \cdot 8 = 24$ ways

13. $2^{10} = 1024$ ways

15. 1st Position: 2 choices

2nd Position: 3 choices

3rd Position: 2 choices

4th Position: 1 choice

Total: $2 \cdot 3 \cdot 2 \cdot 1 = 12$ ways

Label the four people A, B, C, and D and suppose that A and B are willing to take the first position. The twelve combinations are as follows.

ABCD   BACD

ABDC   BADC

ACBD   BCAD

ACDB   BCDA

ADBC   BDAC

ADCB   BDCA

17. $10 \cdot 10 \cdot 10 \cdot 26 \cdot 26 \cdot 26 = 17{,}576{,}000$ license plates

19. (a) $9 \cdot 10 \cdot 10 = 900$

(c) $9 \cdot 10 \cdot 2 = 180$

(b) $9 \cdot 9 \cdot 8 = 648$

(d) $10 \cdot 10 \cdot 10 - 400 = 600$

21. $2(8 \cdot 10 \cdot 10)(10 \cdot 10 \cdot 10 \cdot 10) = 16{,}000{,}000$ numbers

23. (a) $6 \cdot 5 \cdot 4 \cdot 3 \cdot 2 \cdot 1 = 720$

(b) $6 \cdot 1 \cdot 4 \cdot 1 \cdot 2 \cdot 1 = 48$

25. $_nP_r = \dfrac{n!}{(n - r)!}$

So, $_4P_4 = \dfrac{4!}{0!} = 4! = 24.$

27. $_8P_3 = \dfrac{8!}{5!} = 8 \cdot 7 \cdot 6 = 336$

29. $_5P_4 = \dfrac{5!}{1!} = 120$

31. $14 \cdot {}_nP_3 = {}_{n+2}P_4$ Note $n \geq 3$ for this to be defined.

$$14\left[\frac{n!}{(n-3)!}\right] = \frac{(n+2)!}{(n-2)!}$$

$14n(n-1)(n-2) = (n+2)(n+1)n(n-1)$ (We can divide here by $n(n-1)$ since $n \neq 0, n \neq 1$.)

$$14n - 28 = n^2 + 3n + 2$$
$$0 = n^2 - 11n + 30$$
$$0 = (n-5)(n-6)$$
$$n = 5 \text{ or } n = 6$$

33. ${}_{20}P_6 = 27,907,200$ **35.** ${}_{120}P_4 = 197,149,680$ **37.** ${}_{20}C_4 = 4845$

39. $5! = 120$ ways   **41.** ${}_{12}P_4 = \dfrac{12!}{8!} = 12 \cdot 11 \cdot 10 \cdot 9 = 11,880$ ways

43. $\dfrac{7!}{2!1!3!1!} = \dfrac{7!}{2!3!} = 420$ **45.** $\dfrac{7!}{2!1!1!1!1!1!} = \dfrac{7!}{2!} = 7 \cdot 6 \cdot 5 \cdot 4 \cdot 3 = 2520$

47. (a) ABCD   BACD   CABD   DABC   (b) ABCD

    ABDC   BADC   CADB   DACB   ACBD

    ACBD   BCAD   CBAD   DBAC   DBCA

    ACDB   BCDA   CBDA   DBCA   DCBA

    ADBC   BDAC   CDAB   DCAB

    ADCB   BDCA   CDBA   DCBA

49. ${}_{20}C_4 = 4845$ groups   **51.** ${}_{49}C_6 = 13,983,816$ ways   **53.** ${}_9C_2 = 36$ lines

55. Select type of card for three of a kind: ${}_{13}C_1$ **57.** (a) ${}_{12}C_4 = 495$ ways

Select three of four cards for three of a kind: ${}_4C_3$ (b) $({}_5C_2)({}_7C_2) = (10)(21) = 210$ ways

Select type of card for pair: ${}_{12}C_1$

Select two of four cards for pair: ${}_4C_2$

${}_{13}C_1 \cdot {}_4C_3 \cdot {}_{12}C_1 \cdot {}_4C_2 = 13 \cdot 4 \cdot 12 \cdot 6 = 3744$
ways to get a full house

59. $({}_7C_1)({}_{12}C_3)({}_{20}C_2) = 7 \cdot 220 \cdot 190$ **61.** ${}_5C_2 - 5 = 10 - 5 = 5$ diagonals

$$= 292,600 \text{ ways}$$

63. ${}_8C_2 - 8 = 28 - 8 = 20$ diagonals   **65.** False. This is an example of a combination.

67. ${}_{100}P_{80} \approx 3.836 \times 10^{139}$.   **69.** ${}_nC_r = {}_nC_{n-r} = \dfrac{n!}{r!(n-r)!}$

This number is too large for some calculators to
evaluate.

71. $_nP_{n-1} = \dfrac{n!}{(n-(n-1))!} = \dfrac{n!}{1!} = \dfrac{n!}{0!} = {_nP_n}$

73. $_nC_{n-1} = \dfrac{n!}{[n-(n-1)]!(n-1)!} = \dfrac{n!}{(1)!(n-1)!} = \dfrac{n!}{(n-1)!1!} = {_nC_1}$

75. From the graph of $y = \sqrt{x-3} - x + 6$, you see that there is one zero, $x \approx 8.303$. Analytically,

$$\sqrt{x-3} = x - 6$$
$$x - 3 = x^2 - 12x + 36$$
$$0 = x^2 - 13x + 39.$$

By the Quadratic Formula, $x = \dfrac{13 \pm \sqrt{(-13)^2 - 4(39)}}{2} = \dfrac{13 \pm \sqrt{13}}{2}$.

Selecting the larger solution, $x = \dfrac{13 + \sqrt{13}}{2} \approx 8.303$. (The other solution is extraneous)

77. $\log_2(x-3) = 5$

$$2^5 = x - 3$$
$$2^5 + 3 = x$$
$$x = 35$$

79. $x = \dfrac{\begin{vmatrix} -14 & 3 \\ 2 & -2 \end{vmatrix}}{\begin{vmatrix} -5 & 3 \\ 7 & -2 \end{vmatrix}} = \dfrac{22}{-11} = -2$

$y = \dfrac{\begin{vmatrix} -5 & -14 \\ 7 & 2 \end{vmatrix}}{\begin{vmatrix} -5 & 3 \\ 7 & -2 \end{vmatrix}} = \dfrac{88}{-11} = -8$

Answer: $(-2, -8)$

81. $x = \dfrac{\begin{vmatrix} -1 & -4 \\ -4 & 5 \end{vmatrix}}{\begin{vmatrix} -3 & -4 \\ 9 & 5 \end{vmatrix}} = \dfrac{-21}{21} = -1$

$y = \dfrac{\begin{vmatrix} -3 & -1 \\ 9 & -4 \end{vmatrix}}{\begin{vmatrix} -3 & -4 \\ 9 & 5 \end{vmatrix}} = \dfrac{21}{21} = 1$

Answer: $(-1, 1)$

Section 8.7   Probability

You should know the following basic principles of probability.

■ If an event E has $n(E)$ equally likely outcomes and its sample space has $n(S)$ equally likely outcomes, then the probability of event E is

$$P(E) = \frac{n(E)}{n(S)}, \text{ where } 0 \le P(E) \le 1.$$

■ If A and B are mutually exclusive events, then $P(A \cup B) = P(A) + P(B)$.

If A and B are not mutually exclusive events, then $P(A \cup B) = P(A) + P(B) - P(A \cap B)$.

■ If A and B are independent events, then the probability that both A and B will occur is $P(A)P(B)$.

■ The probability of the complement of an event A is $P(A') = 1 - P(A)$.

Solutions to Odd-Numbered Exercises

1. $\{(H, 1), (H, 2), (H, 3), (H, 4), (H, 5), (H, 6),$
   $(T, 1), (T, 2), (T, 3), (T, 4), (T, 5), (T, 6)\}$

3. $\{ABC, ACB, BAC, BCA, CAB, CBA\}$

5. $\{(A, B), (A, C), (A, D), (A, E), (B, C), (B, D), (B, E), (C, D), (C, E), (D, E)\}$

7. $E = \{HTT, THT, TTH\}$

$P(E) = \dfrac{n(E)}{n(S)} = \dfrac{3}{8}$

9. $E = \{HHH, HHT, HTH, HTT, THH, THT, TTH\}$

$P(E) = \dfrac{n(E)}{n(S)} = \dfrac{7}{8}$

11. $E = \{K, K, K, K, Q, Q, Q, Q, J, J, J, J\}$

$P(E) = \dfrac{n(E)}{n(S)} = \dfrac{12}{52} = \dfrac{3}{13}$

13. $E = \{K, K, Q, Q, J, J\}$

$P(E) = \dfrac{n(E)}{n(S)} = \dfrac{6}{52} = \dfrac{3}{26}$

15. $E = \{(1, 4), (2, 3), (3, 2), (4, 1)\}$

$P(E) = \dfrac{n(E)}{n(S)} = \dfrac{4}{36} = \dfrac{1}{9}$

17. not $E = \{(5, 6), (6, 5), (6, 6)\}$

$n(E) = n(S) - n(\text{not } E) = 36 - 3 = 33$

$P(E) = \dfrac{n(E)}{n(S)} = \dfrac{33}{36} = \dfrac{11}{12}$

19. $P(E) = \dfrac{_3C_2}{_6C_2} = \dfrac{3}{15} = \dfrac{1}{5}$

21. $P(E) = \dfrac{_4C_2}{_6C_2} = \dfrac{6}{15} = \dfrac{2}{5}$

23. $P(E') = 1 - P(E) = 1 - 0.8 = 0.2$

25. $P(E') = 1 - P(E) = 1 - \frac{1}{3} = \frac{2}{3}$

27. $P(E) = 1 - P(E') = 1 - p = 1 - 0.12 = 0.88$

29. $P(E) = 1 - P(E') = 1 - \frac{13}{20} = \frac{7}{20}$

31. (a) $0.18(6.7) \approx 1.2$ million

(b) $\dfrac{0.42}{1.0} = 0.42$

(c) $\dfrac{0.21}{1.0} = 0.21$

(d) $\dfrac{0.21 + 0.01}{1.0} = 0.22$

33. (a) $\dfrac{34}{100} = 0.34$

(b) $\dfrac{45}{100} = 0.45$

(c) $\dfrac{23}{100} = 0.23$

35. (a) $\dfrac{672}{1254}$

(b) $\dfrac{582}{1254}$

(c) $\dfrac{672 - 124}{1254} = \dfrac{548}{1254}$

37. $p + p + 2p = 1$
   $p = 0.25$

Taylor: $0.50 = \dfrac{1}{2}$

Moore: $0.25 = \dfrac{1}{4}$

Perez: $0.25 = \dfrac{1}{4}$

39. (a) $\dfrac{_{15}C_{10}}{_{20}C_{10}} = \dfrac{3003}{184,756} = \dfrac{21}{1292} \approx 0.016$

(b) $\dfrac{_{15}C_8 \cdot {_5}C_2}{_{20}C_{10}} = \dfrac{64,350}{184,756} = \dfrac{225}{646} \approx 0.348$

(c) $\dfrac{_{15}C_9 \cdot {_5}C_1}{_{20}C_{10}} + \dfrac{_{15}C_{10}}{_{20}C_{10}} = \dfrac{25,025 + 3003}{184,756} = \dfrac{28,028}{184,756} = \dfrac{49}{323} \approx 0.152$

41. (a) $\dfrac{1}{_5P_5} = \dfrac{1}{120}$

(b) $\dfrac{1}{_4P_4} = \dfrac{1}{24}$

43. (a) $\dfrac{20}{52} = \dfrac{5}{13}$

(b) $\dfrac{13 + 13}{52} = \dfrac{1}{2}$

(c) $\dfrac{4 + 12}{52} = \dfrac{4}{13}$

45. (a) $\dfrac{_9C_4}{_{12}C_4} = \dfrac{126}{495} = \dfrac{14}{55}$ (4 good units)

(b) $\dfrac{(_9C_2)\,(_3C_2)}{_{12}C_4} = \dfrac{108}{495} = \dfrac{12}{55}$ (2 good units)

(c) $\dfrac{(_9C_3)(_3C_1)}{_{12}C_4} = \dfrac{252}{495} = \dfrac{28}{55}$ (3 good units)

At least 2 good units: $\dfrac{12}{55} + \dfrac{28}{55} + \dfrac{14}{55} = \dfrac{54}{55}$

47. $(0.32)^2 = 0.1024$

49. (a) $P(SS) = (0.985)^2 \approx 0.9702$

(b) $P(S) = 1 - P(FF) = 1 - (0.015)^2 \approx 0.9998$

(c) $P(FF) = (0.015)^2 \approx 0.0002$

51. (a) $\left(\dfrac{1}{5}\right)^6 = \dfrac{1}{15,625}$

(b) $\left(\dfrac{4}{5}\right)^6 = \dfrac{4096}{15,625} = 0.262144$

(c) $1 - 0.262144 = 0.737856 = \dfrac{11,529}{15,625}$

53. $1 - \dfrac{(45)^2}{(60)^2} = 1 - \left(\dfrac{45}{60}\right)^2 = 1 - \left(\dfrac{3}{4}\right)^2 = 1 - \dfrac{9}{16} = \dfrac{7}{16}$

55. True. $P(E) + P(E') = 1$

57. (a) As you consider successive people with distinct birthdays, the probabilities must decrease to take into account the birth dates already used. Since the birth dates of people are independent events, multiply the respective probabilities of distinct birthdays.

(b) $\dfrac{365}{365} \cdot \dfrac{364}{365} \cdot \dfrac{363}{365} \cdot \dfrac{362}{365}$

(c) $P_1 = \dfrac{365}{365} = 1$

$P_2 = \dfrac{365}{365} \cdot \dfrac{364}{365} = \dfrac{364}{365} P_1 = \dfrac{365 - (2-1)}{365} P_1$

$P_3 = \dfrac{365}{365} \cdot \dfrac{364}{365} \cdot \dfrac{363}{365} = \dfrac{363}{365} P_2 = \dfrac{365 - (3-1)}{365} P_2$

$P_n = \dfrac{365}{365} \cdot \dfrac{364}{365} \cdot \dfrac{363}{365} \cdot \ldots \cdot \dfrac{365 - (n-1)}{365} = \dfrac{365 - (n-1)}{365} P_{n-1}$

(d) Q_n is the probability that the birthdays are *not* distinct which is equivalent to at least 2 people having the same birthday.

(e)

n	10	15	20	23	30	40	50
P_n	0.88	0.75	0.59	0.49	0.29	0.11	0.03
Q_n	0.12	0.25	0.41	(0.51)	0.71	0.89	0.97

(f) 23, See the chart above.

59. $\dfrac{2}{x-5} = 4$

$2 = 4(x-5) = 4x - 20$

$4x = 22$

$x = \dfrac{11}{2}$

61. $\dfrac{3}{x-2} + \dfrac{x}{x+2} = 1$

$3(x+2) + x(x-2) = (x-2)(x+2)$

$3x + 6 + x^2 - 2x = x^2 - 4$

$x = -10$

63. $e^x + 7 = 35$

$e^x = 28$

$x = \ln(28) \approx 3.332$

65. $4 \ln 6x = 16$

$\ln 6x = 4$

$e^4 = 6x$

$x = \tfrac{1}{6} e^4 \approx 9.10$

67. $_5P_3 = \dfrac{5!}{(5-3)!} = \dfrac{120}{2} = 60$

69. $_{11}P_8 = \dfrac{11!}{(11-8)!} = \dfrac{11!}{3!} = 6{,}652{,}800$

71. $_6C_2 = \dfrac{6!}{4!2!} = \dfrac{6 \cdot 5 \cdot 4!}{4!2} = 15$

73. $_{11}C_8 = \dfrac{11!}{8!3!} = \dfrac{11 \cdot 10 \cdot 9 \cdot 8!}{8!6} = 165$

Review Exercises for Chapter 8

Solutions to Odd-Numbered Exercises

1. $a_n = 2 + \dfrac{6}{n}$

$a_1 = 2 + \dfrac{6}{1} = 8$

$a_2 = 2 + \dfrac{6}{2} = 5$

$a_3 = 2 + \dfrac{6}{3} = 4$

$a_4 = 2 + \dfrac{6}{4} = \dfrac{7}{2}$

$a_5 = 2 + \dfrac{6}{5} = \dfrac{16}{5}$

3. $a_n = \dfrac{72}{n!}$

$a_1 = \dfrac{72}{1!} = 72$

$a_2 = \dfrac{72}{2!} = 36$

$a_3 = \dfrac{72}{3!} = 12$

$a_4 = \dfrac{72}{4!} = 3$

$a_5 = \dfrac{72}{5!} = \dfrac{3}{5}$

5. $a_n = \dfrac{3}{2}n$

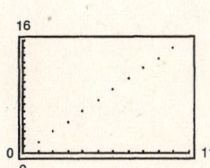

7. $a_n = 4(0.4)^{n-1}$

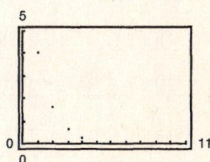

9. $\dfrac{18!}{20!} = \dfrac{18!}{20 \cdot 19 \cdot 18!} = \dfrac{1}{20 \cdot 19} = \dfrac{1}{380}$

11. $\dfrac{2!5!}{6!} = \dfrac{2 \cdot 5!}{6 \cdot 5!} = \dfrac{2}{6} = \dfrac{1}{3}$

13. $\displaystyle\sum_{i=1}^{6} 5 = 6(5) = 30$

15. $\displaystyle\sum_{j=1}^{4} \dfrac{6}{j^2} = \dfrac{6}{1^2} + \dfrac{6}{2^2} + \dfrac{6}{3^2} + \dfrac{6}{4^2} = 6 + \dfrac{3}{2} + \dfrac{2}{3} + \dfrac{3}{8} = \dfrac{205}{24}$

17. $\displaystyle\sum_{k=1}^{10} 2k^3 = 2(1)^3 + 2(2)^3 + 2(3)^3 + \cdots + 2(10)^3 = 6050$

19. $\displaystyle\sum_{n=0}^{10} (n^2 + 3) = \sum_{n=0}^{10} n^2 + \sum_{n=0}^{10} 3 = \dfrac{10(11)(21)}{6} + 11(3) = 418$

21. $\dfrac{1}{2(1)} + \dfrac{1}{2(2)} + \dfrac{1}{2(3)} + \cdots + \dfrac{1}{2(20)} = \displaystyle\sum_{k=1}^{20} \dfrac{1}{2k} \approx 1.799$

23. $\dfrac{1}{2} + \dfrac{2}{3} + \dfrac{3}{4} + \cdots + \dfrac{9}{10} = \displaystyle\sum_{k=1}^{9} \dfrac{k}{k+1} \approx 7.071$

25. (a) $\sum_{k=1}^{4} \frac{5}{10^k} = \frac{5}{10} + \frac{5}{100} + \frac{5}{1000} + \frac{5}{10,000} = .5 + .05 + .005 + .0005 = .5555 = \frac{1111}{2000}$

(b) $\sum_{k=1}^{\infty} \frac{5}{10^k} = \frac{5}{10} \sum_{k=0}^{\infty} \frac{1}{10^k} = \frac{5}{10} \cdot \frac{1}{1 - \frac{1}{10}} = \frac{5}{10} \cdot \frac{10}{9} = \frac{5}{9}$

27. (a) $\sum_{k=1}^{4} 2\left(\frac{1}{100}\right)^k = 2\left(\frac{1}{100}\right) + 2\left(\frac{1}{100}\right)^2 + 2\left(\frac{1}{100}\right)^3 + 2\left(\frac{1}{100}\right)^4$

$= 0.02 + 0.0002 + 0.000002 + 0.00000002$

$= 0.02020202$

(b) $\sum_{k=1}^{\infty} 2\left(\frac{1}{100}\right)^k = \frac{2}{100} \sum_{k=0}^{\infty} \left(\frac{1}{100}\right)^k = \frac{2}{100} \cdot \frac{1}{1 - \left(\frac{1}{100}\right)} = \frac{2}{100} \cdot \frac{100}{99} = \frac{2}{99}$

29. (a) $a_1 = 2500\left(1 + \frac{0.08}{4}\right)^1 = 2500(1.02) = 2550$ **(b)** $a_{40} = 2500(1.02)^{40} = 5520.10$

$a_2 = 2500(1.02)^2 = 2601$

$a_3 = 2500(1.02)^3 = 2653.02$

$a_4 = 2500(1.02)^4 = 2706.08$

$a_5 = 2500(1.02)^5 = 2760.20$

$a_6 = 2500(1.02)^6 = 2815.41$

$a_7 = 2500(1.02)^7 = 2871.71$

$a_8 = 2500(1.02)^8 = 2929.15$

31. Yes. $d = 3 - 5 = -2$ **33.** Yes. $d = 1 - \frac{1}{2} = \frac{1}{2}$

35. $a_1 = 3, d = 4$

$a_1 = 3$

$a_2 = 3 + 4 = 7$

$a_3 = 7 + 4 = 11$

$a_4 = 11 + 4 = 15$

$a_5 = 15 + 4 = 19$

37. $a_4 = 10 \quad a_{10} = 28$

$a_{10} = a_4 + 6d$

$28 = 10 + 6d$

$18 = 6d$

$3 = d$

$a_1 = a_4 - 3d$

$a_1 = 10 - 3(3)$

$a_1 = 1$

$a_2 = 1 + 3 = 4$

$a_3 = 4 + 3 = 7$

$a_4 = 7 + 3 = 10$

$a_5 = 10 + 3 = 13$

39. $a_1 = 35, a_{k+1} = a_k - 3$

$a_1 = 35$

$a_2 = a_1 - 3 = 35 - 3 = 32$

$a_3 = a_2 - 3 = 32 - 3 = 29$

$a_4 = a_3 - 3 = 29 - 3 = 26$

$a_5 = a_4 - 3 = 26 - 3 = 23$

$a_n = 35 + (n - 1)(-3) = 38 - 3n, d = -3$

41. $a_1 = 9, a_{k+1} = a_k + 7$

$a_1 = 9$

$a_2 = a_1 + 7 = 9 + 7 = 16$

$a_3 = a_2 + 7 = 16 + 7 = 23$

$a_4 = a_3 + 7 = 23 + 7 = 30$

$a_5 = a_4 + 7 = 30 + 7 = 37$

$a_n = 9 + (n - 1)(7) = 2 + 7n, d = 7$

43. $a_n = 100 + (n - 1)(-3) = 103 - 3n$

$$\sum_{n=1}^{20} (103 - 3n) = \sum_{n=1}^{20} 103 - 3 \sum_{n=1}^{20} n = 20(103) - 3\left[\frac{(20)(21)}{2}\right] = 1430$$

45. $\displaystyle\sum_{j=1}^{10} (2j - 3) = 2\sum_{j=1}^{10} j - \sum_{j=1}^{10} 3 = 2\left[\frac{10(11)}{2}\right] - 10(3) = 80$

47. $\displaystyle\sum_{k=1}^{11} \left(\frac{2}{3}k + 4\right) = \frac{2}{3}\sum_{k=1}^{11} k + \sum_{k=1}^{11} 4 = \frac{2}{3} \cdot \frac{(11)(12)}{2} + 11(4) = 88$

49. $\displaystyle\sum_{k=1}^{100} 5k = 5\left[\frac{(100)(101)}{2}\right] = 25,250$

51. (a) $34,000 + 4(2250) = \$43,000$

(b) $\displaystyle\sum_{k=1}^{5} [34,000 + (k - 1)(2250)] = \sum_{k=1}^{5} (31,750 + 2250k) = \$192,500$

53. 5, 10, 20, 40

Geometric: $r = 2$

55. $\frac{1}{2}, \frac{2}{3}, \frac{3}{4}, \frac{4}{5}$

Not geometric: $\frac{1}{2}r = \frac{2}{3} \implies r = \frac{4}{3}$

$\qquad\qquad\quad \frac{2}{3}r = \frac{3}{4} \implies r = \frac{9}{8}$

57. $a_1 = 4, \ r = -\frac{1}{4}$

$a_1 = 4$

$a_2 = 4\left(-\frac{1}{4}\right) = -1$

$a_3 = -1\left(-\frac{1}{4}\right) = \frac{1}{4}$

$a_4 = \frac{1}{4}\left(-\frac{1}{4}\right) = -\frac{1}{16}$

$a_5 = -\frac{1}{16}\left(-\frac{1}{4}\right) = \frac{1}{64}$

59. $a_1 = 9, \ a_3 = 4$

$a_3 = a_1 r^2$

$4 = 9r^2$

$\frac{4}{9} = r^2 \implies r = \pm\frac{2}{3}$

$a_1 = 9$ $\qquad\qquad\qquad a_1 = 9$

$a_2 = 9\left(\frac{2}{3}\right) = 6$ $\qquad\quad a_2 = 9\left(-\frac{2}{3}\right) = -6$

$a_3 = 6\left(\frac{2}{3}\right) = 4 \quad$ OR $\quad a_3 = -6\left(-\frac{2}{3}\right) = 4$

$a_4 = 4\left(\frac{2}{3}\right) = \frac{8}{3}$ $\qquad\quad a_4 = 4\left(-\frac{2}{3}\right) = -\frac{8}{3}$

$a_5 = \frac{8}{3}\left(\frac{2}{3}\right) = \frac{16}{9}$ $\qquad a_5 = -\frac{8}{3}\left(-\frac{2}{3}\right) = \frac{16}{9}$

61. $a_1 = 120, a_{k+1} = \frac{1}{3}a_k$

$a_1 = 120$

$a_2 = \frac{1}{3}(120) = 40$

$a_3 = \frac{1}{3}(40) = \frac{40}{3}$

$a_4 = \frac{1}{3}\left(\frac{40}{3}\right) = \frac{40}{9}$

$a_5 = \frac{1}{3}\left(\frac{40}{9}\right) = \frac{40}{27}$

$a_n = 120\left(\frac{1}{3}\right)^{n-1}, r = \frac{1}{3}$

63. $a_1 = 25, \ a_{k+1} = -\frac{3}{5}a_k$

$a_1 = 25$

$a_2 = -\frac{3}{5}(25) = -15$

$a_3 = -\frac{3}{5}(-15) = 9$

$a_4 = -\frac{3}{5}(9) = -\frac{27}{5}$

$a_5 = -\frac{3}{5}\left(-\frac{27}{5}\right) = \frac{81}{25}$

$a_n = 25\left(-\frac{3}{5}\right)^{n-1}, r = -\frac{3}{5}$

65. $a_2 = a_1 r$

$-8 = 16r$

$-\frac{1}{2} = r$

$a_n = 16\left(-\frac{1}{2}\right)^{n-1}$

$\sum_{n=1}^{20} 16\left(-\frac{1}{2}\right)^{n-1} = 16\left[\frac{1 - (-1/2)^{20}}{1 - (-1/2)}\right] \approx 10.67$

67. $a_1 = 100, r = 1.05$

$a_n = 100(1.05)^{n-1}$

$\sum_{n=1}^{20} 100(1.05)^{n-1} = 100\left[\frac{1 - (1.05)^{20}}{1 - 1.05}\right] \approx 3306.60$

69. $\displaystyle\sum_{i=1}^{7} 2^{i-1} = \frac{1 - 2^7}{1 - 2} = 127$

71. $\displaystyle\sum_{n=1}^{7} (-4)^{n-1} = \frac{1 - (-4)^7}{1 - (-4)} = 3277$

73. $\displaystyle\sum_{n=0}^{4} 250(1.02)^n = 250\left(\frac{1 - 1.02^5}{1 - 1.02}\right) = 1301.01004$

75. $\displaystyle\sum_{i=1}^{10} 10\left(\frac{3}{5}\right)^{i-1} \approx 24.849$

77. $\displaystyle\sum_{i=1}^{\infty} \left(\frac{7}{8}\right)^{i-1} = \frac{1}{1 - 7/8} = 8$

79. $\displaystyle\sum_{k=1}^{\infty} 4\left(\frac{2}{3}\right)^{k-1} = \frac{4}{1 - 2/3} = 12$

81. (a) $a_t = 120,000(0.7)^t$

(b) $a_5 = 120,000(0.7)^5 = \$20,168.40$

83. 1. When $n = 1, 2 = \frac{1}{2}(5(1) - 1)$

2. Assume that $S_k = 2 + 7 + \cdots + (5k - 3) = \frac{k}{2}(5k - 1)$

Then, $S_{k+1} = 2 + 7 + \cdots + (5k - 3) + [5(k + 1) - 3]$

$\qquad = S_k + 5k + 2$

$\qquad = \frac{k}{2}(5k - 1) + 5k + 2$

$\qquad = \frac{1}{2}[5k^2 + 9k + 4]$

$\qquad = \frac{1}{2}[(5k + 4)(k + 1)]$

$\qquad = \frac{k + 1}{2}(5(k + 1) - 1)$

Therefore, by mathematical induction, the formula is true for all positive integers n.

85. 1. When $n = 1$, $a = a\left(\dfrac{1 - r}{1 - r}\right)$.

2. Assume that

$$S_k = \sum_{i=0}^{k-1} ar^i = \frac{a(1 - r^k)}{1 - r}.$$

Then,

$$S_{k+1} = \sum_{i=0}^{k} ar^i = \sum_{i=0}^{k-1} ar^i + ar^k = \frac{a(1 - r^k)}{1 - r} + ar^k$$

$$= \frac{a(1 - r^k + r^k - r^{k+1})}{1 - r} = \frac{a(1 - r^{k+1})}{1 - r}.$$

Therefore, by mathematical induction, the formula is valid for all positive integer values of n.

87. $\displaystyle\sum_{n=1}^{30} n = \frac{30(31)}{2} = 465$

89. $\displaystyle\sum_{n=1}^{7} (n^4 - n) = \sum_{n=1}^{7} n^4 - \sum_{n=1}^{7} n$

$$= \frac{7(8)(15)[3(7)^2 + 3(7) - 1]}{30} - \frac{7(8)}{2}$$

$$= \frac{840(167)}{30} - 28$$

$$= 4676 - 28 = 4648$$

91. $a_1 = f(1) = 5$

$a_2 = a_1 + 5 = 5 + 5 = 10$

$a_3 = a_2 + 5 = 15$

$a_4 = a_3 + 5 = 20$

$a_5 = a_4 + 5 = 25$

n:	1	2	3	4	5
a_n:	5	10	15	20	25

First differences: $\quad\quad$ 5 $\quad$ 5 $\quad$ 5 $\quad$ 5

Second difference: $\quad\quad$ 0 $\quad$ 0 $\quad$ 0

Linear model: $a_n = 5n$

93. $a_1 = f(1) = 16$

$a_2 = a_1 - 1 = 16 - 1 = 15$

$a_3 = a_2 - 1 = 15 - 1 = 14$

$a_4 = 14 - 1 = 13$

$a_5 = 13 - 1 = 12$

n:	1	2	3	4	5
a_n:	16	15	14	13	12

First differences: $\quad\quad$ -1 $\quad$ -1 $\quad$ -1 $\quad$ -1

Second difference: $\quad\quad$ 0 $\quad$ 0 $\quad$ 0

Linear model: $a_n = 17 - n$

95. $_{10}C_8 = 45$

97. $\dbinom{9}{4} = {}_9C_4 = 126$

99. 4th number in 6th row is $_6C_3 = 20$

101. 5th number in 8th row is $\dbinom{8}{4} = {}_8C_4 = 70$

103. $(x + 5)^4 = x^4 + 4x^3(5) + 6x^2(5^2) + 4x(5^3) + 5^4$

$$= x^4 + 20x^3 + 150x^2 + 500x + 625$$

105. $(a - 4b)^5 = a^5 - 5a^4(4b) + 10a^3(4b)^2 - 10a^2(4b)^3 + 5a(4b)^4 - (4b)^5$

$$= a^5 - 20a^4b + 160a^3b^2 - 640a^2b^3 + 1280ab^4 - 1024b^5$$

107. $(7 + 2i)^4 = 7^4 + 4(7)^3(2i) + 6(7)^2(2i)^2 + 4(7)(2i)^3 + (2i)^4$

$= 2401 + 2744i - 1176 - 224i + 16$

$= 1241 + 2520i$

109. $E = \{(1, 11), (2, 10), (3, 9), (4, 8), (5, 7), (7, 5), (8, 4), (9, 3), (10, 2), (11, 1)\}$

$n(E) = 10$

111. $(4)(6)(2) = 48$ schedules

113. $\dfrac{8!}{2!2!2!1!1!} = \dfrac{8!}{8} = 7! = 5040$ permutations

115. $10! = 3{,}628{,}800$ ways

117. $_{20}C_{15} = 15{,}504$ ways

119. $\dfrac{10}{10} \cdot \dfrac{1}{9} = \dfrac{1}{9}$

121. (a) $\dfrac{208}{500} = 0.416$

(b) $\dfrac{400}{500} = 0.8$

(c) $\dfrac{37}{500} = 0.074$

123. $P(2 \text{ pairs}) = \dfrac{(_{13}C_2)(_4C_2)(_4C_2)(_{44}C_1)}{(_{52}C_5)} = 0.0475$

125. True.

$\dfrac{(n + 2)!}{n!} = \dfrac{(n + 2)(n + 1)n!}{n!} = (n + 2)(n + 1)$

127. Answers will vary. See pages 507 and 516.

129. (a) arithmetic-linear model

(b) geometric model

131. Answers will vary. See page 509. To define a sequence recursively, you need to be given one or more of the first few terms. All other terms are defined using previous terms.

133. If n is even, the expansion are the same. If n is odd, the expansion of $(-x + y)^n$ is the negative of that of $(x - y)^n$.

Chapter 8   Practice Test

1. Write out the first five terms of the sequence $a_n = \dfrac{2n}{(n+2)!}$.

2. Write an expression for the nth term of the sequence $\left\{\dfrac{4}{3}, \dfrac{5}{9}, \dfrac{6}{27}, \dfrac{7}{81}, \dfrac{8}{243}, \ldots\right\}$.

3. Find the sum $\displaystyle\sum_{i=1}^{6} (2i - 1)$.

4. Write out the first five terms of the arithmetic sequence where $a_1 = 23$ and $d = -2$.

5. Find a_{50} for the arithmetic sequence with $a_1 = 12$, $d = 3$, and $n = 50$.

6. Find the sum of the first 200 positive integers.

7. Write out the first five terms of the geometric sequence with $a_1 = 7$ and $r = 2$.

8. Evaluate $\displaystyle\sum_{n=0}^{9} 6\left(\dfrac{2}{3}\right)^n$.   **9.** Evaluate $\displaystyle\sum_{n=0}^{\infty} (0.03)^n$.

10. Use mathematical induction to prove that $1 + 2 + 3 + 4 + \cdots + n = \dfrac{n(n+1)}{2}$.

11. Use mathematical induction to prove that $n! > 2^n$, $n \geq 4$.

12. Evaluate $_{13}C_4$. Verify with a graphing utility.

13. Expand $(x + 3)^5$.

14. Find the term involving x^7 in $(x - 2)^{12}$.

15. Evaluate $_{30}P_4$.

16. How many ways can six people sit at a table with six chairs?

17. Twelve cars run in a race. How many different ways can they come in first, second, and third place? (Assume that there are no ties.)

18. Two six-sided dice are tossed. Find the probability that the total of the two dice is less than 5.

19. Two cards are selected at random from a deck of 52 playing cards without replacement. Find the probability that the first card is a King and the second card is a black ten.

20. A manufacturer has determined that for every 1000 units it produces, 3 will be faulty. What is the probability that an order of 50 units will have one or more faulty units?

C H A P T E R   9
Topics in Analytic Geometry

CHAPTER 9
Topics in Analytic Geometry

Section 9.1   Introduction to Conics: Parabolas

■   A **parabola** is the set of all points (x, y) that are equidistant from a fixed line (**directrix**) and a fixed point (**focus**) not on the line.

■   The standard equation of a parabola with vertex (h, k) and:
   (a)   Vertical axis $x = h$ and directrix $y = k - p$ is:
      $(x - h)^2 = 4p(y - k), \ p \neq 0$
   (b)   Horizontal axis $y = k$ and directrix $x = h - p$ is:
      $(y - k)^2 = 4p(x - h), \ p \neq 0$

■   The tangent line to a parabola at a point P makes **equal angles** with:
   (a)   the line through P and the focus
   (b)   the axis of the parabola

Solutions to Odd-Numbered Exercises

1. $y^2 = -4x$

Vertex: $(0, 0)$

Opens to the left since p is negative.

Matches graph (e).

3. $x^2 = -8y$

Vertex: $(0, 0)$

Opens downward since p is negative.

Matches graph (d).

5. $(y - 1)^2 = 4(x - 3)$

Vertex: $(3, 1)$

Opens to the right since p is positive.

Matches graph (a).

7. $y = \frac{1}{2}x^2$

$x^2 = 2y = 4\left(\frac{1}{2}\right)y; \ p = \frac{1}{2}$

Vertex: $(0, 0)$

Focus: $\left(0, \frac{1}{2}\right)$

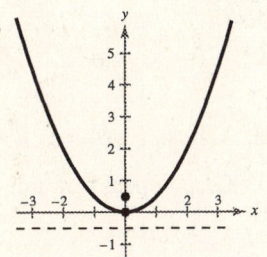

9. $y^2 = -6x$

$y^2 = 4\left(-\frac{3}{2}\right)x; \ p = -\frac{3}{2}$

Vertex: $(0, 0)$

Focus: $\left(-\frac{3}{2}, 0\right)$

Directrix: $x = \frac{3}{2}$

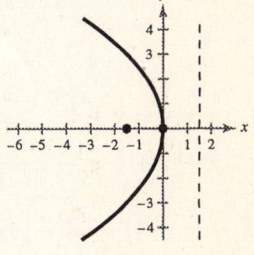

11. $x^2 + 8y = 0$

$x^2 = 4(-2)y; \ p = -2$

Vertex: $(0, 0)$

Focus: $(0, -2)$

Directrix: $y = 2$

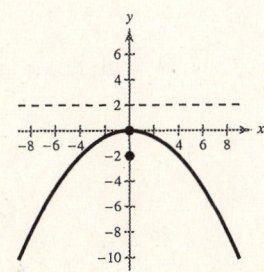

13. $(x + 1)^2 + 8(y + 3) = 0$

$(x + 1)^2 = 4(-2)(y + 3)$

$h = -1, k = -3, p = -2$

Vertex: $(-1, -3)$

Focus: $(-1, -5)$

Directrix: $y = -1$

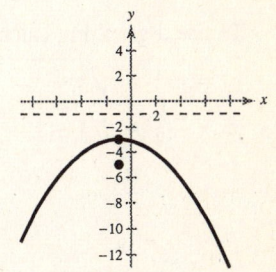

15. $y^2 + 6y + 8x + 25 = 0$

$$(y + 3)^2 = 4(-2)(x + 2); \quad p = -2$$

Vertex: $(-2, -3)$

Focus: $(-4, -3)$

Directrix: $x = 0$

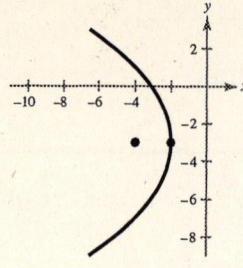

17. $\left(x + \frac{3}{2}\right)^2 = 4(y - 2) \implies h = -\frac{3}{2}, k = 2, p = 1$

Vertex: $\left(-\frac{3}{2}, 2\right)$

Focus: $\left(-\frac{3}{2}, 2 + 1\right) = \left(-\frac{3}{2}, 3\right)$

Directrix: $y = 1$

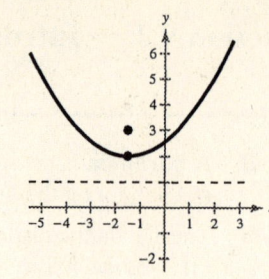

19. $\quad y = \frac{1}{4}(x^2 - 2x + 5)$

$$4y - 4 = (x - 1)^2$$

$$(x - 1)^2 = 4(1)(y - 1)$$

$$h = 1, k = 1, p = 1$$

Vertex: $(1, 1)$

Focus: $(1, 2)$

Directrix: $y = 0$

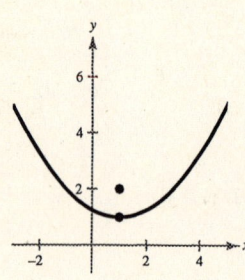

21. $x^2 + 4x + 6y - 2 = 0$

$$x^2 + 4x + 4 = -6y + 2 + 4 = -6y + 6$$

$$(x + 2)^2 = -6(y - 1)$$

$$(x + 2)^2 = 4\left(-\frac{3}{2}\right)(y - 1)$$

Vertex: $(-2, 1)$

Focus: $\left(-2, 1 - \frac{3}{2}\right) = \left(-2, -\frac{1}{2}\right)$

Directrix: $y = \frac{5}{2}$

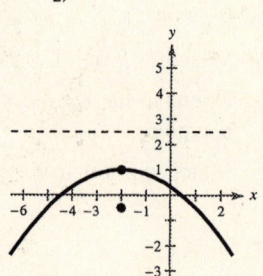

23. $y^2 + x + y = 0$

$$y^2 + y + \frac{1}{4} = -x + \frac{1}{4}$$

$$\left(y + \frac{1}{2}\right)^2 = 4\left(-\frac{1}{4}\right)\left(x - \frac{1}{4}\right)$$

$$h = \frac{1}{4}, k = -\frac{1}{2}, p = -\frac{1}{4}$$

Vertex: $\left(\frac{1}{4}, -\frac{1}{2}\right)$

Focus: $\left(0, -\frac{1}{2}\right)$

Directrix: $x = \frac{1}{2}$

To use a graphing calculator, enter:

$$y_1 = -\frac{1}{2} + \sqrt{\frac{1}{4} - x}$$

$$y_2 = -\frac{1}{2} - \sqrt{\frac{1}{4} - x}$$

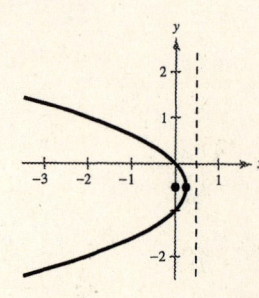

25. Vertex: $(0, 0) \implies h = 0, k = 0$

Graph opens upward.

$$x^2 = 4py$$

Point on graph: $(3, 6)$

$$3^2 = 4p(6)$$

$$9 = 24p$$

$$\frac{3}{8} = p$$

Thus, $x^2 = 4\left(\frac{3}{8}\right)y \implies y = \frac{2}{3}x^2$

$$\implies x^2 = \frac{3}{2}y.$$

27. Vertex: $(0,0) \Rightarrow h = 0, k = 0$

Focus: $\left(0, -\frac{3}{2}\right) \Rightarrow p = -\frac{3}{2}$

$(x - h)^2 = 4p(y - k)$

$x^2 = 4\left(-\frac{3}{2}\right)y$

$x^2 = -6y$

29. Vertex: $(0,0) \Rightarrow h = 0, k = 0$

Focus: $(-2, 0) \Rightarrow p = -2$

$(y - k)^2 = 4p(x - h)$

$y^2 = 4(-2)x$

$y^2 = -8x$

31. Vertex: $(0,0) \Rightarrow h = 0, k = 0$

Directrix: $y = -1 \Rightarrow p = 1$

$(x - h)^2 = 4p(y - k)$

$(x - 0)^2 = 4(1)(y - 0)$

$x^2 = 4y$ or $y = \frac{1}{4}x^2$

33. Vertex: $(0,0) \Rightarrow h = 0, k = 0$

Directrix: $x = 2 \Rightarrow p = -2$

$y^2 = 4px$

$y^2 = -8x$

35. Vertex: $(0,0) \Rightarrow h = 0, k = 0$

Horizontal axis and passes through the point $(4, 6)$

$(y - k)^2 = 4p(x - h)$

$(y - 0)^2 = 4p(x - 0)$

$y^2 = 4px$

$6^2 = 4p(4)$

$36 = 16p \Rightarrow p = \frac{9}{4}$

$y^2 = 4\left(\frac{9}{4}\right)x$

$y^2 = 9x$

37. Vertex: $(3, 1)$ and opens downward.

Passes through $(2, 0)$ and $(4, 0)$.

$y = -(x - 2)(x - 4)$

$= -x^2 + 6x - 8$

$= -(x - 3)^2 + 1$

$(x - 3)^2 = -(y - 1)$

39. Vertex: $(-2, 0)$ and opens to the right.

Focus: $\left(-\frac{3}{2}, 0\right)$

$\frac{1}{2} = p$

$y^2 = 4\left(\frac{1}{2}\right)(x + 2)$

$y^2 = 2(x + 2)$

41. Vertex: $(5, 2)$

Focus: $(3, 2)$

Horizontal axis: $p = 3 - 5 = -2$

$(y - 2)^2 = 4(-2)(x - 5)$

$(y - 2)^2 = -8(x - 5)$

43. Vertex: $(0, 4)$

Directrix: $y = 2$

Vertical axis

$p = 4 - 2 = 2$

$(x - 0)^2 = 4(2)(y - 4)$

$x^2 = 8(y - 4)$

45. Focus: $(2, 2)$

Directrix: $x = -2$

Horizontal axis

Vertex: $(0, 2)$

$p = 2 - 0 = 2$

$(y - 2)^2 = 4(2)(x - 0)$

$(y - 2)^2 = 8x$

47. $y^2 - 8x = 0$ and $x - y + 2 = 0$

$$y^2 = 8x \qquad\qquad y_3 = x + 2$$
$$y_1 = \sqrt{8x}$$
$$y_2 = -\sqrt{8x}$$

The point of tangency is (2, 4).

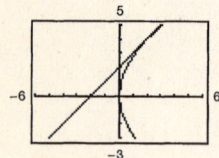

49. $x^2 = 2y$, $(4, 8)$, $p = \dfrac{1}{2}$, focus: $\left(0, \dfrac{1}{2}\right)$

Following Example 4, we find the y-intercept $(0, b)$:

$$d_1 = \frac{1}{2} - b$$

$$d_2 = \sqrt{(4 - 0)^2 + \left(8 - \frac{1}{2}\right)^2} = \frac{17}{2}$$

$$d_1 = d_2 \implies \frac{1}{2} - b = \frac{17}{2} \implies b = -8$$

$$m = \frac{8 - (-8)}{4 - 0} = 4$$

$$y = 4x - 8 \quad \text{Tangent line}$$

Let $y = 0 \implies x = 2 \implies$ x-intercept $(2, 0)$

51. $y = -2x^2 \implies x^2 = -\dfrac{1}{2}y = 4\left(-\dfrac{1}{8}\right)y \implies p = -\dfrac{1}{8}$

Focus: $\left(0, -\dfrac{1}{8}\right)$

Following Example 4, we find the y-intercept $(0, b)$:

$$d_1 = \frac{1}{8} + b$$

$$d_2 = \sqrt{(-1 - 0)^2 + \left(-2 + \frac{1}{8}\right)^2} = \frac{17}{8}$$

$$d_1 = d_2 \implies \frac{1}{8} + b = \frac{17}{8} \implies b = 2$$

$$m = \frac{-2 - 2}{-1 - 0} = 4$$

$$y = 4x + 2$$

Let $y = 0 \implies x = -\dfrac{1}{2} \implies$ x-intercept $\left(-\dfrac{1}{2}, 0\right)$.

53. $R = 375x - \dfrac{3}{2}x^2$

R is a maximum of \$23,437.50 when $x = 125$ televisions.

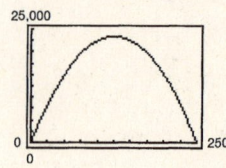

55. (a)

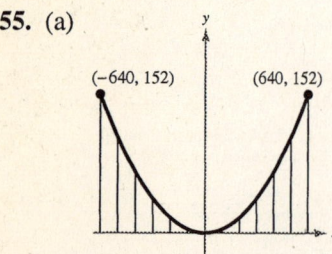

(b) $x^2 = 4py$

$$640^2 = 4p(152)$$
$$p = \frac{12,800}{19}$$
$$y = \frac{19}{51,200}x^2$$

(c)

x	0	200	400	500	600
y	0	14.84	59.38	92.77	133.59

57. Vertex: $(0, 0]$

$$y^2 = 4px$$

Point: $(1000, 800)$

$$800^2 = 4p(1000] \implies p = 160$$
$$y^2 = 4(160)x$$
$$y^2 = 640x$$

59. $-12.5(y - 7.125) = (x - 6.25)^2$

$-12.5y + 89.0625 = x^2 - 12.5x + 39.0625$

$y = -0.08x^2 + x + 4$

(a)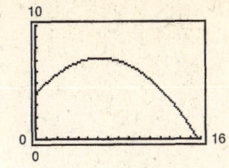

(b) The highest point is at $(6.25, 7.125)$. The distance is the x-intercept of ≈ 15.69 feet.

61. False. A parabola cannot intersect its directrix nor focus.

63. Answers will vary. See the reflective property of parabolas, page 635.

65. $f(x) = 3x^3 - 4x + 2$

Relative maximum: $(-0.67, 3.78)$

Relative minimum: $(0.67, 0.22)$

67. $f(x) = x^4 + 2x + 2$

Relative minimum: $(-0.79, 0.81)$

Section 9.2   Ellipses

- An **ellipse** is the set of all points (x, y) the sum of whose distances from two distinct fixed points (**foci**) is constant.
- The standard equation of an ellipse with center (h, k) and major and minor axes of lengths $2a$ and $2b$ is:

   (a) $\dfrac{(x - h)^2}{a^2} + \dfrac{(y - k)^2}{b^2} = 1$ if the major axis is horizontal.

   (b) $\dfrac{(x - h)^2}{b^2} + \dfrac{(y - k)^2}{a^2} = 1$ if the major axis is vertical.

- $c^2 = a^2 - b^2$ where c is the distance from the center to a focus.
- The eccentricity of an ellipse is $e = \dfrac{c}{a}$.

Solutions to Odd-Numbered Exercises

1. $\dfrac{x^2}{4} + \dfrac{y^2}{9} = 1$

Center: $(0, 0)$

$a = 3, b = 2$

Vertical major axis

Matches graph (b).

3. $\dfrac{x^2}{4} + \dfrac{y^2}{25} = 1$

Center: $(0, 0)$

$a = 5, b = 2$

Vertical major axis

Matches graph (d).

5. $\dfrac{(x - 2)^2}{16} + (y + 1)^2 = 1$

Center: $(2, -1)$

$a = 4, b = 1$

Horizontal major axis

Matches graph (a).

7. $\dfrac{x^2}{5} + \dfrac{y^2}{27} = 1$

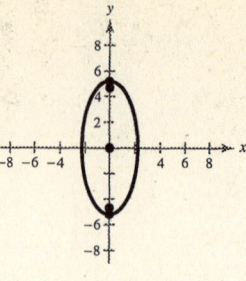

Center: $(0, 0)$

$a = 3\sqrt{3}, b = \sqrt{5}, c = \sqrt{27 - 5} = \sqrt{22}$

Vertices: $\left(0, \pm\sqrt{27}\right)$

Foci: $\left(0, \pm\sqrt{22}\right)$

$e = \dfrac{c}{a} = \dfrac{\sqrt{22}}{3\sqrt{3}} = \dfrac{\sqrt{66}}{9}$

9. $\dfrac{(x - 4)^2}{16} + \dfrac{(y + 1)^2}{25} = 1$

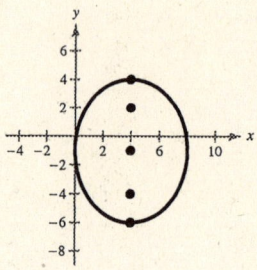

Center: $(4, -1)$

$a = 5, b = 4, c = 3$

Vertices: $(4, -1 \pm 5)$: $(4, -6), (4, 4)$

Foci: $(4, -1 \pm 3)$: $(4, -4), (4, 2)$

$e = \dfrac{c}{a} = \dfrac{3}{5}$

11. $\dfrac{(x + 5)^2}{9/4} + (y - 1)^2 = 1$

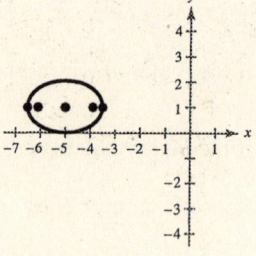

Center: $(-5, 1)$

$a = \dfrac{3}{2}, b = 1, c = \sqrt{\dfrac{9}{4} - 1} = \dfrac{\sqrt{5}}{2}$

Foci: $\left(-5 + \dfrac{\sqrt{5}}{2}, 1\right), \left(-5 - \dfrac{\sqrt{5}}{2}, 1\right)$

Vertices: $\left(-5 + \dfrac{3}{2}, 1\right) = \left(-\dfrac{7}{2}, 1\right), \left(-5 - \dfrac{3}{2}, 1\right) = \left(-\dfrac{13}{2}, 1\right)$

$e = \dfrac{\sqrt{5}/2}{3/2} = \dfrac{\sqrt{5}}{3}$

13. (a) $4x^2 + 9y^2 = 36$

$\qquad \dfrac{x^2}{9} + \dfrac{y^2}{4} = 1$

(c), (d)

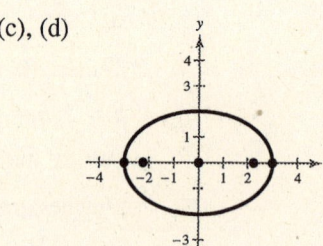

(b) $a = 3, b = 2, c = \sqrt{9 - 4} = \sqrt{5}$

$\quad$ Center: $(0, 0)$

$\quad$ Vertices: $(\pm 3, 0)$

$\quad$ Foci: $\left(\pm\sqrt{5}, 0\right)$

$\quad e = \dfrac{c}{a} = \dfrac{\sqrt{5}}{3}$

15. $9x^2 + 4y^2 + 36x - 24y + 36 = 0$

$9(x^2 + 4x + 4) + 4(y^2 - 6y + 9) = -36 + 36 + 36$

$$\frac{(x+2)^2}{4} + \frac{(y-3)^2}{9} = 1$$

$a = 3, b = 2, c = \sqrt{5}$

Center: $(-2, 3)$

Foci: $\left(-2, 3 \pm \sqrt{5}\right)$

Vertices: $(-2, 6), (-2, 0)$

$e = \dfrac{\sqrt{5}}{3}$

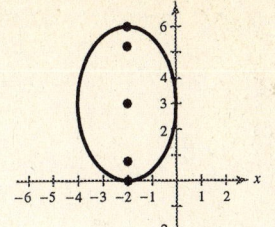

17. $6x^2 + 2y^2 + 18x - 10y + 2 = 0$

$6\left(x^2 + 3x + \dfrac{9}{4}\right) + 2\left(y^2 - 5y + \dfrac{25}{4}\right) = -2 + \dfrac{27}{2} + \dfrac{25}{2}$

$$6\left(x + \frac{3}{2}\right)^2 + 2\left(y - \frac{5}{2}\right)^2 = 24$$

$$\frac{\left(x + \frac{3}{2}\right)^2}{4} + \frac{\left(y - \frac{5}{2}\right)^2}{12} = 1$$

Center: $\left(-\dfrac{3}{2}, \dfrac{5}{2}\right)$

$a = 2\sqrt{3}, b = 2, c = 2\sqrt{2}$

Foci: $\left(-\dfrac{3}{2}, \dfrac{5}{2} \pm 2\sqrt{2}\right)$

Vertices: $\left(-\dfrac{3}{2}, \dfrac{5}{2} \pm 2\sqrt{3}\right)$

$e = \dfrac{\sqrt{2}}{\sqrt{3}} = \dfrac{\sqrt{6}}{3}$

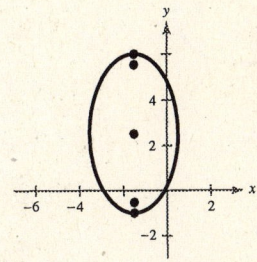

19. $16x^2 + 25y^2 - 32x + 50y + 16 = 0$

$16(x^2 - 2x + 1) + 25(y^2 + 2y + 1) = -16 + 16 + 25$

$$\frac{(x-1)^2}{25/16} + (y+1)^2 = 1$$

$a = \dfrac{5}{4}, b = 1, c = \dfrac{3}{4}$

Center: $(1, -1)$

Foci: $\left(\dfrac{7}{4}, -1\right), \left(\dfrac{1}{4}, -1\right)$

Vertices: $\left(\dfrac{9}{4}, -1\right), \left(-\dfrac{1}{4}, -1\right)$

$e = \dfrac{3}{5}$

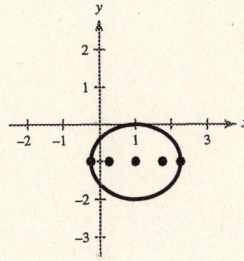

21. (a) $12x^2 + 20y^2 - 12x + 40y - 37 = 0$

$$12\left(x^2 - 1 + \frac{1}{4}\right) + 20(y^2 + 2y + 1) = 37 + 3 + 20$$

$$12\left(x - \frac{1}{2}\right)^2 + 20(y + 1)^2 = 60$$

$$\frac{\left(x - \frac{1}{2}\right)^2}{5} + \frac{(y + 1)^2}{3} = 1$$

(b) Center: $\left(\frac{1}{2}, 1\right)$

$a = \sqrt{5}, b = \sqrt{3}, c = \sqrt{5 - 3} = \sqrt{2}$

Vertices: $\left(\frac{1}{2} \pm \sqrt{5}, -1\right)$

Foci: $\left(\frac{1}{2} \pm \sqrt{2}, -1\right)$

Eccentricity: $\frac{c}{a} = \frac{\sqrt{2}}{\sqrt{5}} = \frac{\sqrt{10}}{5}$

(c), (d)

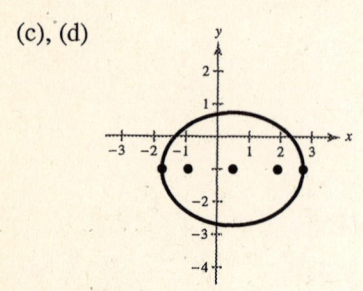

23. Center: $(0, 0)$

$a = 4, b = 2$

Vertical major axis

$$\frac{x^2}{4} + \frac{y^2}{16} = 1$$

25. Center: $(0, 0)$

$a = 3,$

$c = 2 \implies b = \sqrt{9 - 4} = \sqrt{5}$

Horizontal major axis

$$\frac{x^2}{9} + \frac{y^2}{5} = 1$$

27. Center: $(0, 0)$

$c = 4,$

$a = 5 \implies b = \sqrt{25 - 16} = 3$

Horizontal major axis

$$\frac{x^2}{25} + \frac{y^2}{9} = 1$$

29. Vertices: $(0, \pm 5) \implies a = 5$

Center: $(0, 0)$

Vertical major axis

$$\frac{(x - h)^2}{b^2} + \frac{(y - k)^2}{a^2} = 1$$

$$\frac{x^2}{b^2} + \frac{y^2}{25} = 1$$

Point: $(4, 2)$

$$\frac{4^2}{b^2} + \frac{2^2}{25} = 1$$

$$\frac{16}{b^2} = 1 - \frac{4}{25} = \frac{21}{25}$$

$$400 = 21b^2$$

$$\frac{400}{21} = b^2$$

$$\frac{x^2}{400/21} + \frac{y^2}{25} = 1$$

$$\frac{21x^2}{400} + \frac{y^2}{25} = 1$$

31. Center: $(2, 3)$

$a = 3, b = 1$

Vertical major axis

$$\frac{(x - h)^2}{b^2} + \frac{(y - k)^2}{a^2} = 1$$

$$\frac{(x - 2)^2}{1} + \frac{(y - 3)^2}{9} = 1$$

33. Center: $(2, 2)$

$a = 2, b = 1 \implies c = \sqrt{3}$

Horizontal major axis

$$\frac{(x - 2)^2}{4} + \frac{(y - 2)^2}{1} = 1$$

35. Foci: $(0, 0), (0, 8) \implies c = 4$

Major axis of length $16 \implies a = 8$

$b^2 = a^2 - c^2 = 64 - 16 = 48$

Center: $(0, 4) = (h, k)$

$$\frac{(x - h)^2}{b^2} + \frac{(y - k)^2}{a^2} = 1$$

$$\frac{x^2}{48} + \frac{(y - 4)^2}{64} = 1$$

37. Vertices: $(3, 1), (3, 9) \implies a = 4$

Center: $(3, 5)$

Minor axis of length $6 \implies b = 3$

Vertical major axis

$$\frac{(x - h)^2}{b^2} + \frac{(y - k)^2}{a^2} = 1$$

$$\frac{(x - 3)^2}{9} + \frac{(y - 5)^2}{16} = 1$$

39. Center: $(0, 4)$

Vertices: $(-4, 4), (4, 4) \implies a = 4$

$a = 2c \implies 4 = 2c \implies c = 2$

$2^2 = 4^2 - b^2 \implies b^2 = 12$

Horizontal major axis

$$\frac{(x - h)^2}{a^2} + \frac{(y - k)^2}{b^2} = 1$$

$$\frac{x^2}{16} + \frac{(y - 4)^2}{12} = 1$$

41. Vertices: $(\pm 5, 0) \implies a = 5$

Eccentricity: $\frac{3}{5} \implies c = \frac{3}{5}a = 3$

$b^2 = a^2 - c^2 = 25 - 9 = 16$

Center: $(0, 0) = (h, k)$

$$\frac{(x - h)^2}{a^2} + \frac{(y - k)^2}{b^2} = 1$$

$$\frac{x^2}{25} + \frac{y^2}{16} = 1$$

43. (a)

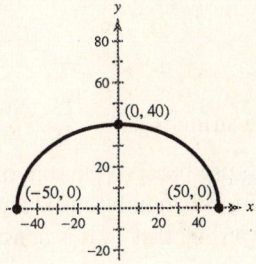

(b) Vertices: $(\pm 50, 0) \implies a = 50$

Height at center: $40 \implies b = 40$

Horizontal major axis

$$\frac{x^2}{a^2} + \frac{y^2}{b^2} = 1$$

$$\frac{x^2}{2500} + \frac{y^2}{1600} = 1, \; y \geq 0$$

(c) For $x = 45, \dfrac{45^2}{2500} + \dfrac{y^2}{1600} = 1.$

$$y^2 = 1600\left(1 - \frac{45^2}{2500}\right)$$

$$y^2 = 304$$

$$y \approx 17.44$$

The height five feet from the edge of the tunnel is approximately 17.44 feet.

45. Area of ellipse = 2(area of circle)

$$\pi a b = 2 \pi r^2$$

$$\pi a (10) = 2 \pi (10)^2$$

$$\pi a (10) = 200$$

$$a = 20$$

Length of major axis: $2a = 2(20) = 40$ units

47. $a + c = 4.08$

$a - c = 0.34$

$2a = 4.42 \Rightarrow a = 2.21 \Rightarrow c = 1.87$

$b^2 = a^2 - c^2 \Rightarrow b^2 = 1.3872$

$$\frac{x^2}{4.8841} + \frac{y^2}{1.3872} = 1$$

49. For $\dfrac{x^2}{a^2} + \dfrac{y^2}{b^2} = 1$, we have $c^2 = a^2 - b^2$.

When $x = c$,

$$\frac{c^2}{a^2} + \frac{y^2}{b^2} = 1 \Rightarrow y^2 = b^2 \left(1 - \frac{a^2 - b^2}{a^2} \right) \Rightarrow y^2 = \frac{b^4}{a^2} \Rightarrow 2y = \frac{2b^2}{a}.$$

51. $\dfrac{x^2}{9} + \dfrac{y^2}{16} = 1$

$a = 4, b = 3, c = \sqrt{7}$

Points on the ellipse: $(\pm 3, 0), (0, \pm 4)$

Length of latus recta: $\dfrac{2b^2}{a} = \dfrac{2(3)^2}{4} = \dfrac{9}{2}$

Additional points: $\left(\pm \dfrac{9}{4}, -\sqrt{7} \right), \left(\pm \dfrac{9}{4}, \sqrt{7} \right)$

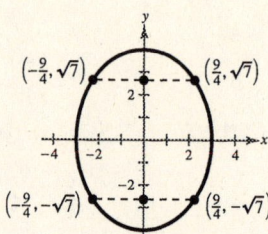

53. $5x^2 + 3y^2 = 15$

$$\frac{x^2}{3} + \frac{y^2}{5} = 1$$

$a = \sqrt{5}, b = \sqrt{3}, c = \sqrt{2}$

Points on the ellipse: $\left(\pm \sqrt{3}, 0 \right), \left(0, \pm \sqrt{5} \right)$

Length of latus recta: $\dfrac{2b^2}{a} = \dfrac{2 \cdot 3}{\sqrt{5}} = \dfrac{6\sqrt{5}}{5}$

Additional points: $\left(\pm \dfrac{3\sqrt{5}}{5}, -\sqrt{2} \right), \left(\pm \dfrac{3\sqrt{5}}{5}, \sqrt{2} \right)$

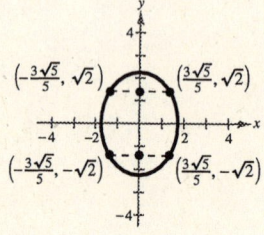

55. True. If $e \approx 1$ then the ellipse is elongated, not circular.

57. (a) The length of the string is $2a$.

(b) The path is an ellipse because the sum of the distances from the two thumbtacks is always the length of the string, that is, it is constant.

59. Arithmetic: $d = -11$　　**61.** Geometric: $r = 2$　　**63.** $\displaystyle\sum_{n=0}^{6} 3^n = 1093$　　**65.** $\displaystyle\sum_{n=1}^{10} 4 \left(\frac{3}{4} \right)^{n-1} \approx 15.099$

Section 9.3   Hyperbolas

■ A **hyperbola** is the set of all points (x, y) the difference of whose distances from two distinct fixed points (**foci**) is constant.

■ The standard equation of a hyperbola with center (h, k) and transverse and conjugate axes of lengths $2a$ and $2b$ is:

(a) $\dfrac{(x - h)^2}{a^2} - \dfrac{(y - k)^2}{b^2} = 1$ if the transverse axis is horizontal.

(b) $\dfrac{(y - k)^2}{a^2} - \dfrac{(x - h)^2}{b^2} = 1$ if the transverse axis is vertical.

■ $c^2 = a^2 + b^2$ where c is the distance from the center to a focus.

■ The asymptotes of a hyperbola are:

(a) $y = k \pm \dfrac{b}{a}(x - h)$ if the transverse axis is horizontal.

(b) $y = k \pm \dfrac{a}{b}(x - h)$ the transverse axis is vertical.

■ The eccentricity of a hyperbola is $e = \dfrac{c}{a}$.

■ To classify a nondegenerate conic from its general equation $Ax^2 + Cy^2 + Dx + Ey + F = 0$:
(a) If $A = C$ $(A \neq 0, C \neq 0)$, then it is a circle.
(b) If $AC = 0$ $(A = 0$ or $C = 0$, but not both), then it is a parabola.
(c) If $AC > 0$, then it is an ellipse.
(d) If $AC < 0$, then it is a hyperbola.

Solutions to Odd-Numbered Exercises

1. Center: $(0, 0)$

$a = 3, b = 5, c = \sqrt{34}$

Vertical transverse axis

Matches graph (b).

3. Center: $(1, 0)$

$a = 4, b = 2$

Horizontal transverse axis

Matches graph (a).

5. $x^2 - y^2 = 1$

$a = 1, b = 1, c = \sqrt{2}$

Center: $(0, 0)$

Vertices: $(\pm 1, 0)$

Foci: $(\pm \sqrt{2}, 0)$

Asymptotes: $y = \pm x$

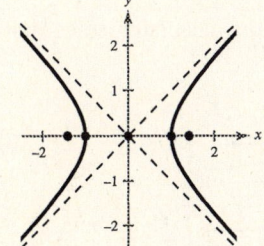

7. $\dfrac{y^2}{1} - \dfrac{x^2}{4} = 1$

$a = 1, b = 2, c = \sqrt{5}$

Center: $(0, 0)$

Vertices: $(0, \pm 1)$

Foci: $(0, \pm \sqrt{5})$

Asymptotes: $y = \pm \dfrac{1}{2}x$

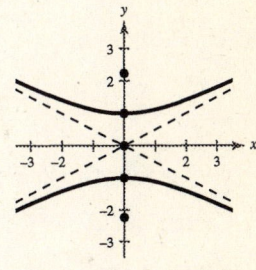

9. $\dfrac{y^2}{25} - \dfrac{x^2}{81} = 1$

$a = 5, b = 9, c = \sqrt{a^2 + b^2} = \sqrt{106}$

Center: $(0, 0)$

Vertices: $(0, \pm 5)$

Foci: $\left(0, \pm \sqrt{106}\right)$

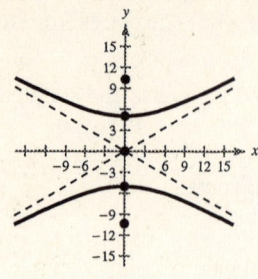

Asymptotes:

$y = \pm\dfrac{a}{b}x = \pm\dfrac{5}{9}x$

11. $\dfrac{(x - 1)^2}{4} - \dfrac{(y + 2)^2}{1} = 1$

$a = 2, b = 1, c = \sqrt{5}$

Center: $(1, -2)$

Vertices:

$(-1, -2), (3, -2)$

Foci: $\left(1 \pm \sqrt{5}, -2\right)$

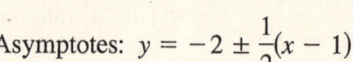

Asymptotes: $y = -2 \pm \dfrac{1}{2}(x - 1)$

13. $\dfrac{(y + 5)^2}{1/9} - \dfrac{(x - 1)^2}{1/4} = 1$

$a = \dfrac{1}{3}, b = \dfrac{1}{2}, c = \sqrt{\dfrac{1}{9} + \dfrac{1}{4}} = \dfrac{\sqrt{13}}{6}$

Center: $(1, -5)$

Vertices: $\left(1, -5 \pm \dfrac{1}{3}\right)$: $\left(1, -\dfrac{16}{3}\right), \left(1, -\dfrac{14}{3}\right)$

Foci: $\left(1, -5 \pm \dfrac{\sqrt{13}}{6}\right)$

Asymptotes: $y = k \pm \dfrac{a}{b}(x - h)$

$y = -5 \pm \dfrac{2}{3}(x - 1)$

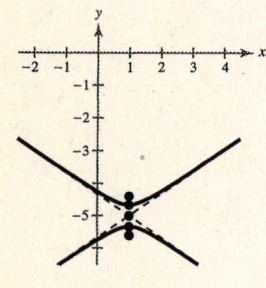

15. (a) $9x^2 - 4y^2 = 36$

$\dfrac{x^2}{4} - \dfrac{y^2}{9} = 1$

(b) Center: $(0, 0)$

$a = 2, b = 3, c = \sqrt{4 + 9} = \sqrt{13}$

Vertices: $(\pm 2, 0)$

Foci: $\left(\pm \sqrt{13}, 0\right)$

Asymptotes: $y = \pm\dfrac{3}{2}x$

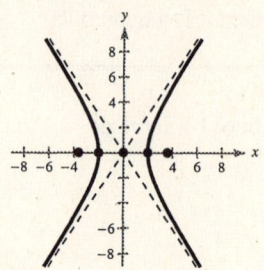

17. $2x^2 - 3y^2 = 6$

$\dfrac{x^2}{3} - \dfrac{y^2}{2} = 1$

$a = \sqrt{3}, b = \sqrt{2}, c = \sqrt{5}$

Center: $(0, 0)$

Vertices: $\left(\pm\sqrt{3}, 0\right)$

Foci: $\left(\pm\sqrt{5}, 0\right)$

Asymptotes: $y = \pm\sqrt{\dfrac{2}{3}}x$

To use a graphing calculator, solve first for y.

$y^2 = \dfrac{2x^2 - 6}{3}$

$y_1 = \sqrt{\dfrac{2x^2 - 6}{3}}$ } Hyperbola

$y_2 = -\sqrt{\dfrac{2x^2 - 6}{3}}$

$y_3 = \sqrt{\dfrac{2}{3}}x$ } Asymptotes

$y_4 = -\sqrt{\dfrac{2}{3}}x$

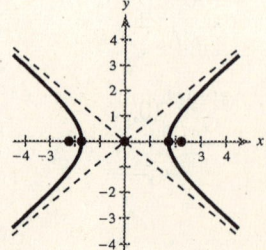

19. $9x^2 - y^2 - 36x - 6y + 18 = 0$

$9(x^2 - 4x + 4) - (y^2 + 6y + 9) = -18 + 36 - 9$

$$\frac{(x-2)^2}{1} - \frac{(y+3)^2}{9} = 1$$

$a = 1, b = 3, c = \sqrt{10}$

Center: $(2, -3)$

Vertices: $(1, -3), (3, -3)$

Foci: $\left(2 \pm \sqrt{10}, -3\right)$

Asymptotes: $y = -3 \pm 3(x - 2)$

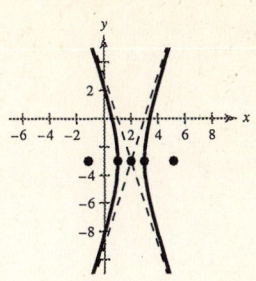

21. $x^2 - 9y^2 + 2x - 54y - 80 = 0$

$(x^2 + 2x + 1) - 9(y^2 + 6y + 9) = 80 + 1 - 81$

$(x + 1)^2 - 9(y + 3)^2 = 0$

$$y + 3 = \pm\frac{1}{3}(x + 1)$$

Degenerate hyperbola is two lines intersecting at $(-1, -3)$.

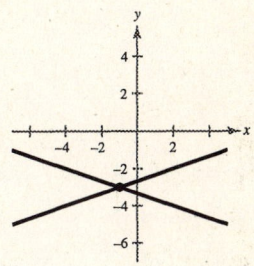

23. $9y^2 - x^2 + 2x + 54y + 62 = 0$

$9(y^2 + 6y + 9) - (x^2 - 2x + 1) = -62 - 1 + 81$

$$\frac{(y+3)^2}{2} - \frac{(x-1)^2}{18} = 1$$

$a = \sqrt{2}, b = 3\sqrt{2}, c = 2\sqrt{5}$

Center: $(1, -3)$

Vertices: $\left(1, -3 \pm \sqrt{2}\right)$

Foci: $\left(1, -3 \pm 2\sqrt{5}\right)$

Asymptotes:

$$y = -3 \pm \frac{1}{3}(x - 1)$$

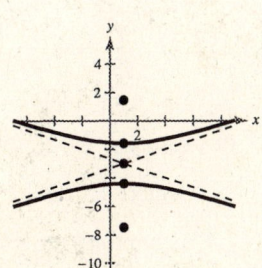

To use a graphing calculator, solve for y first.

$9(y + 3)^2 = 18 + (x - 1)^2$

$$y = -3 \pm \sqrt{\frac{18 + (x-1)^2}{9}}$$

$\left. \begin{array}{l} y_1 = -3 + \dfrac{1}{3}\sqrt{18 + (x-1)^2} \\[2mm] y_2 = -3 - \dfrac{1}{3}\sqrt{18 + (x-1)^2} \end{array} \right\}$ Hyperbola

$\left. \begin{array}{l} y_3 = -3 + \dfrac{1}{3}(x - 1) \\[2mm] y_4 = -3 - \dfrac{1}{3}(x - 1) \end{array} \right\}$ Asymptotes

25. Vertices: $(0, \pm 2) \Rightarrow a = 2$

Foci: $(0, \pm 4) \Rightarrow c = 4$

$b^2 = c^2 - a^2 = 16 - 4 = 12$

Center: $(0, 0) = (h, k)$

$$\frac{(y-k)^2}{a^2} - \frac{(x-h)^2}{b^2} = 1$$

$$\frac{y^2}{4} - \frac{x^2}{12} = 1$$

27. Vertices: $(\pm 1, 0) \Rightarrow a = 1$

Asymptotes: $y = \pm 5x \Rightarrow \dfrac{b}{a} = 5 \Rightarrow b = 5$

Center: $(0, 0)$

$$\frac{x^2}{1} - \frac{y^2}{25} = 1$$

29. Foci: $(0, \pm 8) \implies c = 8$

Asymptotes: $y = \pm 4x \implies \dfrac{a}{b} = 4 \implies a = 4b$

Center: $(0, 0) = (h, k)$

$c^2 = a^2 + b^2 \implies 64 = 16b^2 + b^2$

$\dfrac{64}{17} = b^2 \implies a^2 = \dfrac{1024}{17}$

$\dfrac{(y - k)^2}{a^2} - \dfrac{(x - h)^2}{b^2} = 1$

$\dfrac{y^2}{1024/17} - \dfrac{x^2}{64/17} = 1$

$\dfrac{17y^2}{1024} - \dfrac{17x^2}{64} = 1$

31. Vertices: $(2, 0), (6, 0) \implies a = 2$

Foci: $(0, 0), (8, 0) \implies c = 4$

$b^2 = c^2 - a^2 = 16 - 4 = 12$

Center: $(4, 0) = (h, k)$

$\dfrac{(x - h)^2}{a^2} - \dfrac{(y - k)^2}{b^2} = 1$

$\dfrac{(x - 4)^2}{4} - \dfrac{y^2}{12} = 1$

33. Vertices: $(4, 1), (4, 9) \implies a = 4$

Foci: $(4, 0), (4, 10) \implies c = 5$

$b^2 = c^2 - a^2 = 25 - 16 = 9$

Center: $(4, 5) = (h, k)$

$\dfrac{(y - k)^2}{a^2} - \dfrac{(x - h)^2}{b^2} = 1$

$\dfrac{(y - 5)^2}{16} - \dfrac{(x - 4)^2}{9} = 1$

35. Vertices: $(2, 3), (2, -3) \implies a = 3$

Solution point: $(0, 5)$

Center: $(2, 0) = (h, k)$

$\dfrac{(y - k)^2}{a^2} - \dfrac{(x - h)^2}{b^2} = 1$

$\dfrac{y^2}{9} - \dfrac{(x - 2)^2}{b^2} = 1 \implies$

$b^2 = \dfrac{9(x - 2)^2}{y^2 - 9}$

$= \dfrac{9(-2)^2}{25 - 9} = \dfrac{36}{16} = \dfrac{9}{4}$

$\dfrac{y^2}{9} - \dfrac{(x - 2)^2}{9/4} = 1$

37. Vertices: $(0, 4), (0, 0)$

Center: $(0, 2), a = 2$

$\dfrac{(y - 2)^2}{4} - \dfrac{x^2}{b^2} = 1$

Passes through $\left(\sqrt{5}, -1 \right)$:

$\dfrac{(-1 - 2)^2}{4} - \dfrac{5}{b^2} = 1$

$\dfrac{9}{4} - 1 = \dfrac{5}{b^2}$

$b^2 = 4 \implies b = 2$

$\dfrac{(y - 2)^2}{4} - \dfrac{x^2}{4} = 1$

39. Vertices: $(1, 2), (3, 2) \implies a = 1$

Center: $(2, 2)$

Asymptotes: $y = x, y = 4 - x$

$\dfrac{b}{a} = 1 \implies b = 1$

$\dfrac{(x - 2)^2}{1} - \dfrac{(y - 2)^2}{1} = 1$

41. Vertices: $(0, 2), (6, 2) \implies a = 3$

Asymptotes: $y = \frac{2}{3}x,\ y = 4 - \frac{2}{3}x$

$\dfrac{b}{a} = \dfrac{2}{3} \implies b = 2$

Center: $(3, 2) = (h, k)$

$\dfrac{(x - h)^2}{a^2} - \dfrac{(y - k)^2}{b^2} = 1$

$\dfrac{(x - 3)^2}{9} - \dfrac{(y - 2)^2}{4} = 1$

43. F_1: Friend's location $(-10{,}560, 0)$

F_2: Your location $(10{,}560, 0)$

$P(x, y)$ location of lightning strike

$(1100)(18) = 19{,}800$

$\dfrac{x^2}{a^2} - \dfrac{y^2}{b^2} = 1$

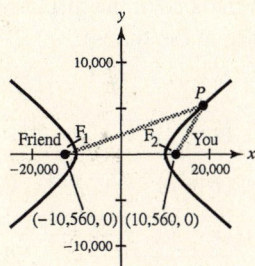

$c = 10{,}560,\ a = \dfrac{19{,}800}{2} = 9900 \implies a^2 = 98{,}010{,}000$

$b^2 = c^2 - a^2 = 13{,}503{,}600$

$\dfrac{x^2}{98{,}010{,}000} - \dfrac{y^2}{13{,}503{,}600}$

45. Foci: $(\pm 150, 0) \implies c = 150$

Center: $(0, 0)$

(a) $d_2 - d_1 = (186{,}000)(0.001)$

$\qquad = 186 \implies 2a = 186 \implies a = 93$

$b^2 = c^2 - a^2 = 150^2 - 93^2 = 13{,}851$

$\dfrac{x^2}{93^2} - \dfrac{y^2}{13{,}851} = 1$

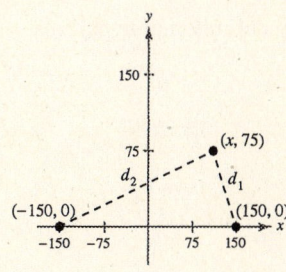

$x^2 = 93^2\left(1 + \dfrac{75^2}{13{,}851}\right) \approx 12{,}161.43$

$x \approx 110.3$ miles

(b) $150 - 93 = 57$ miles

(c) Bay to Station 1: 30 miles

Bay to Station 2: 270 miles

$\dfrac{(270 - 30)}{186{,}000} \approx 0.00129$ second

(d) In this case,

$d_2 - d_1 = 186{,}000(0.00129) \approx 239.94 \implies a \approx 120$

and $b^2 = c^2 - a^2 = 8100$. The hyperbola is

$\dfrac{x^2}{120^2} - \dfrac{y^2}{90^2} = 1.$

For $y = 60$, $x^2 = 20{,}800$ and $x \approx 144.2$.

Position: $(144.2, 60)$

47. $x^2 + y^2 - 6x + 4y + 9 = 0$

$A = 1, C = 1$

$A = C \implies$ Circle

49. $4x^2 - y^2 - 4x - 3 = 0$

$A = 4, C = -1$

$AC = 4(-1)$

$\qquad = -4 < 0 \implies$ Hyperbola

51. $4x^2 + 3y^2 + 8x - 24y + 51 = 0$

$A = 4, C = 3$

$AC = 4(3)$

$\qquad = 12 \implies$ Ellipse

53. $25x^2 - 10x - 200y - 119 = 0$

$A = 25, C = 0$

$AC = 25(0) = 0 \implies$ Parabola

55. $x^2 - 6x - 2y + 7 = 0$

$A = 1, C = 0, D = -6,$

$E = -2, F = 7$

$AC = 0 \implies$ Parabola

57. True. $e = \dfrac{c}{a} = \dfrac{\sqrt{a^2 + b^2}}{a}$

59. Let (x, y) be such that the difference of the distances from $(c, 0)$ and $(-c, 0)$ is $2a$ (again only deriving one of the forms).

$$2a = \left| \sqrt{(x + c)^2 + y^2} - \sqrt{(x - c) + y^2} \right|$$

$$2a + \sqrt{(x - c)^2 + y^2} = \sqrt{(x + c)^2 + y^2}$$

$$4a^2 + 4a\sqrt{(x - c)^2 + y^2} + (x - c)^2 + y^2 = (x + c)^2 + y^2$$

$$4a\sqrt{(x - c)^2 + y^2} = 4cx - 4a^2$$

$$a\sqrt{(x - c)^2 + y^2} = cx - a^2$$

$$a^2(x^2 - 2cx + c^2 + y^2) = c^2x^2 - 2a^2cx + a^4$$

$$a^2(c^2 - a^2) = (c^2 - a^2)x^2 - a^2y^2$$

Let $b^2 = c^2 - a^2$. Then $a^2b^2 = b^2x^2 - a^2y^2 \implies 1 = \dfrac{x^2}{a^2} - \dfrac{y^2}{b^2}$.

61. $|d_2 - d_1| = $ constant by definition of hyperbola

At the point $(a, 0)$,

$|d_2 - d_1| = |(a + c) - (c - a)| = 2a.$

63. $(x^3 - 3x^2) - (6 - 2x - 4x^2) = x^3 + x^2 + 2x - 6$

65. $-2 \begin{array}{|rrrr} 1 & 0 & -3 & 4 \\ & -2 & 4 & -2 \\ \hline 1 & -2 & 1 & 2 \end{array}$

$\dfrac{x^3 - 3x + 4}{x + 2} = x^2 - 2x + 1 + \dfrac{2}{x + 2}$

67. $x^3 - 16x = x(x^2 - 16) = x(x - 4)(x + 4)$

69. $2x^3 - 24x^2 + 72x = 2x(x^2 - 12x + 36)$

$\qquad\qquad\qquad\qquad = 2x(x - 6)^2$

71. $16x^3 + 54 = 2(8x^3 + 27)$

$\qquad\qquad = 2(2x + 3)(4x^2 - 6x + 9)$

73. $f(x) = 8 - 5x^2$

(a) $f(1) = 8 - 5 = 3$

(b) $f(-2) = 8 - 20 = -12$

(c) $f(t) = 8 - 5t^2$

75. $h(x) = \dfrac{x + 5}{x - 12}$

(a) $h(-5) = \dfrac{-5 + 5}{-5 - 12} = 0$

(b) $h(12)$ is undefined

(c) $h(-3) = \dfrac{-3 + 5}{-3 - 12} = -\dfrac{2}{15}$

77. $A = 26°, b = 8$

$$\tan 26° = \frac{a}{b} \implies a = 8 \tan 26° \approx 3.90$$

$$\cos 26° = \frac{b}{c} \implies c = \frac{8}{\cos 26°} \approx 8.90$$

$$B = 90° - 26° = 64°$$

79. $a = 4, b = 7$

$$c = \sqrt{4^2 + 7^2} = \sqrt{65} \approx 8.06$$

$$\tan A = \frac{a}{b} = \frac{4}{7} \implies A = 29.74°$$

$$\tan B = \frac{b}{a} = \frac{7}{4} \implies B \approx 60.26$$

Section 9.4   Rotation and Systems of Quadratic Equations

■ The general second-degree equation $Ax^2 + Bxy + Cy^2 + Dx + Ey + F = 0$ can be rewritten as $A'(x')^2 + C'(y')^2 + D'x' + E'y' + F' = 0$ by rotating the coordinate axes through the angle θ where $\cot 2\theta = (A - C)/B$.

■ $x = x' \cos \theta - y' \sin \theta$
   $y = x' \sin \theta + y' \cos \theta$

■ The graph of the nondegenerate equation $Ax^2 + Bxy + Cy^2 + Dx + Ey + F = 0$ is:

(a) An ellipse or circle if $B^2 - 4AC < 0$.

(b) A parabola if $B^2 - 4AC = 0$.

(c) A hyperbola if $B^2 - 4AC > 0$.

Solutions to Odd-Numbered Exercises

1. $\theta = 90°$; Point: $(0, 4)$

$x = x' \cos \theta - y' \sin \theta$ $\qquad$ $y = x' \sin \theta + y' \cos \theta$

$0 = x' \cos 90° - y' \sin 90°$ $\qquad$ $4 = x' \sin 90° + y' \cos 90°$

$0 = y'$ $\qquad\qquad\qquad\qquad$ $4 = x'$

Thus, $(x', y') = (4, 0)$.

3. $xy + 1 = 0$

$A = 0, B = 1, C = 0$

$$\cot 2\theta = \frac{A - C}{B} = 0 \implies 2\theta = \frac{\pi}{2} \implies \theta = \frac{\pi}{4}$$

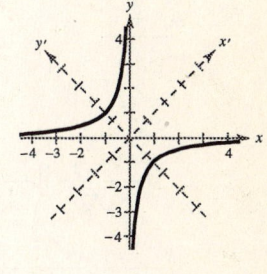

$x = x' \cos \dfrac{\pi}{4} - y' \sin \dfrac{\pi}{4}$ $\qquad\qquad$ $y = x' \sin \dfrac{\pi}{4} + y' \cos \dfrac{\pi}{4}$

$\qquad = x' \left(\dfrac{\sqrt{2}}{2} \right) - y' \left(\dfrac{\sqrt{2}}{2} \right)$ $\qquad\qquad$ $= x' \left(\dfrac{\sqrt{2}}{2} \right) + y' \left(\dfrac{\sqrt{2}}{2} \right)$

$\qquad = \dfrac{x' - y'}{\sqrt{2}}$ $\qquad\qquad\qquad\qquad$ $= \dfrac{x' + y'}{\sqrt{2}}$

$$xy + 1 = 0$$

$$\left(\frac{x' - y'}{\sqrt{2}} \right)\left(\frac{x' + y'}{\sqrt{2}} \right) + 1 = 0$$

$$\frac{(y')^2}{2} - \frac{(x')^2}{2} = 1 \text{ Hyperbola}$$

5. $x^2 - 4xy + y^2 + 1 = 0$

$A = 1, B = -4, C = 1$

$\cot 2\theta = \dfrac{A - C}{B} = 0 \implies 2\theta = \dfrac{\pi}{2} \implies \theta = \dfrac{\pi}{4}$

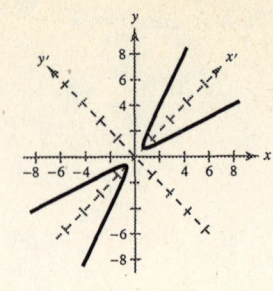

$x = x' \cos \dfrac{\pi}{4} - y' \sin \dfrac{\pi}{4}$ $\qquad\qquad$ $y = x' \sin \dfrac{\pi}{4} + y' \cos \dfrac{\pi}{4}$

$\quad = x'\left(\dfrac{\sqrt{2}}{2}\right) - y'\left(\dfrac{\sqrt{2}}{2}\right)$ $\qquad\qquad$ $= x'\left(\dfrac{\sqrt{2}}{2}\right) + y'\left(\dfrac{\sqrt{2}}{2}\right)$

$\quad = \dfrac{\sqrt{2}}{2}(x' - y')$ $\qquad\qquad\qquad\quad$ $= \dfrac{\sqrt{2}}{2}(x' + y')$

$$x^2 - 4xy + y^2 + 1 = 0$$

$$\left[\dfrac{\sqrt{2}}{2}(x' - y')\right]^2 - 4\left[\dfrac{\sqrt{2}}{2}(x' - y')\dfrac{\sqrt{2}}{2}(x' + y')\right] + \left[\dfrac{\sqrt{2}}{2}(x' + y')\right]^2 + 1 = 0$$

$$\dfrac{1}{2}(x')^2 - x'y' + \dfrac{1}{2}(y')^2 - 2[(x')^2 - (y')^2] + \dfrac{1}{2}(x')^2 + x'y' + \dfrac{1}{2}(y')^2 + 1 = 0$$

$$-(x')^2 + 3(y')^2 = -1$$

$$(x')^2 - \dfrac{(y')^2}{1/3} = 1$$

7. $xy - 2y - 4x = 0$

$A = 0, B = 1, C = 0$

$\cot 2\theta = \dfrac{A - C}{B} = 0 \implies 2\theta = \dfrac{\pi}{2} \implies \theta = \dfrac{\pi}{4}$

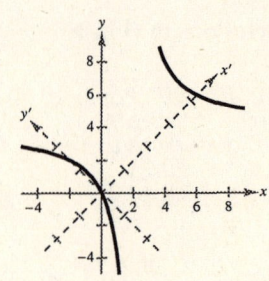

$x = x' \cos \dfrac{\pi}{4} - y' \sin \dfrac{\pi}{4}$ $\qquad\qquad$ $y = x' \sin \dfrac{\pi}{4} + y' \cos \dfrac{\pi}{4}$

$\quad = x'\left(\dfrac{\sqrt{2}}{2}\right) - y'\left(\dfrac{\sqrt{2}}{2}\right)$ $\qquad\qquad$ $= x'\left(\dfrac{\sqrt{2}}{2}\right) + y'\left(\dfrac{\sqrt{2}}{2}\right)$

$\quad = \dfrac{x' - y'}{\sqrt{2}}$ $\qquad\qquad\qquad\qquad$ $= \dfrac{x' + y'}{\sqrt{2}}$

$$xy - 2y - 4x = 0$$

$$\left(\dfrac{x' - y'}{\sqrt{2}}\right)\left(\dfrac{x' + y'}{\sqrt{2}}\right) - 2\left(\dfrac{x' + y'}{\sqrt{2}}\right) - 4\left(\dfrac{x' - y'}{\sqrt{2}}\right) = 0$$

$$\dfrac{(x')^2}{2} - \dfrac{(y')^2}{2} - \sqrt{2}x' - \sqrt{2}y' - 2\sqrt{2}x' + 2\sqrt{2}y' = 0$$

$$\left[(x')^2 - 6\sqrt{2}x' + \left(3\sqrt{2}\right)^2\right] - \left[(y')^2 - 2\sqrt{2}y' + \left(\sqrt{2}\right)^2\right] = 0 + \left(3\sqrt{2}\right)^2 - \left(\sqrt{2}\right)^2$$

$$\left(x' - 3\sqrt{2}\right)^2 - \left(y' - \sqrt{2}\right)^2 = 16$$

$$\dfrac{\left(x' - 3\sqrt{2}\right)^2}{16} - \dfrac{\left(y' - \sqrt{2}\right)^2}{16} = 1 \quad \text{Hyperbola}$$

9. $5x^2 - 6xy + 5y^2 - 12 = 0$

$A = 5, B = -6, C = 5$

$$\cot 2\theta = \frac{A - C}{B} = 0 \implies 2\theta = \frac{\pi}{2} \implies \theta = \frac{\pi}{4}$$

$$x = x'\cos\frac{\pi}{4} - y'\sin\frac{\pi}{4} = \frac{\sqrt{2}}{2}(x' - y')$$

$$y = x'\sin\frac{\pi}{4} + y'\cos\frac{\pi}{4} = \frac{\sqrt{2}}{2}(x' + y')$$

$$5x^2 - 6xy + 5y^2 - 12 = 0$$

$$5\left[\frac{\sqrt{2}}{2}(x' - y')\right]^2 - 6\left[\frac{\sqrt{2}}{2}(x' - y')\frac{\sqrt{2}}{2}(x' + y')\right] + 5\left[\frac{\sqrt{2}}{2}(x' + y')\right]^2 = 12$$

$$\frac{5}{2}(x')^2 - 5x'y' + \frac{5}{2}(y')^2 - 3(x')^2 + 3(y')^2 + \frac{5}{2}(x')^2 + 5x'y' + \frac{5}{2}(y')^2 = 12$$

$$2(x')^2 + 8(y')^2 = 12$$

$$\frac{(x')^2}{6} + \frac{(y')^2}{3/2} = 1 \quad \text{Ellipse}$$

11. $3x^2 - 2\sqrt{3}xy + y^2 + 2x + 2\sqrt{3}y = 0$

$A = 3, B = -2\sqrt{3}, C = 1$

$$\cot 2\theta = \frac{A - C}{B} = -\frac{1}{\sqrt{3}} \implies \theta = 60°$$

$$x = x'\cos 60° - y'\sin 60°$$

$$= x'\left(\frac{1}{2}\right) - y'\left(\frac{\sqrt{3}}{2}\right) = \frac{x' - \sqrt{3}y'}{2}$$

$$y = x'\sin\theta + y'\cos\theta = \frac{\sqrt{3}x' + y'}{2}$$

$$3x^2 - 2\sqrt{3}xy + y^2 + 2x + 2\sqrt{3}y = 0$$

$$3\left(\frac{x' - \sqrt{3}y'}{2}\right)^2 - 2\sqrt{3}\left(\frac{x' - \sqrt{3}y'}{2}\right)\left(\frac{\sqrt{3}x' + y'}{2}\right) + \left(\frac{\sqrt{3}x' + y'}{2}\right)^2 + 2\left(\frac{x' - \sqrt{3}y'}{2}\right)$$

$$+ 2\sqrt{3}\left(\frac{\sqrt{3}x' + y'}{2}\right) = 0$$

$$\frac{3(x')^2}{4} - \frac{6\sqrt{3}x'y'}{4} + \frac{9(y')^2}{4} - \frac{6(x')^2}{4} + \frac{4\sqrt{3}x'y'}{4} + \frac{6(y')^2}{4} + \frac{3(x')^2}{4} + \frac{2\sqrt{3}x'y'}{4} + \frac{(y')^2}{4}$$

$$+ x' - \sqrt{3}y' + 3x' + \sqrt{3}y' = 0$$

$$4(y')^2 + 4x' = 0$$

$$x' = -(y')^2 \quad \text{Parabola}$$

13. $9x^2 + 24xy + 16y^2 + 90x - 130y = 0$

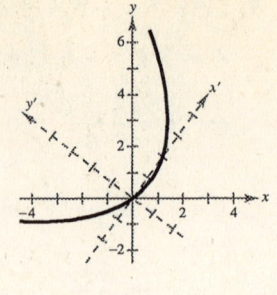

$A = 9, B = 24, C = 16$

$$\cot 2\theta = \frac{A - C}{B} = -\frac{7}{24} \Rightarrow \theta \approx 53.13°$$

$$\cos 2\theta = -\frac{7}{25}$$

$$\sin \theta = \sqrt{\frac{1 - \cos 2\theta}{2}} = \sqrt{\frac{1 - (-7/25)}{2}} = \frac{4}{5}$$

$$\cos \theta = \sqrt{\frac{1 + \cos 2\theta}{2}} = \sqrt{\frac{1 + (-7/25)}{2}} = \frac{3}{5}$$

$x = x' \cos \theta - y' \sin \theta \qquad\qquad y = x' \sin \theta + y' \cos \theta$

$$= x'\left(\frac{3}{5}\right) - y'\left(\frac{4}{5}\right) \qquad\qquad = x'\left(\frac{4}{5}\right) + y'\left(\frac{3}{5}\right)$$

$$= \frac{3x' - 4y'}{5} \qquad\qquad = \frac{4x' + 3y'}{5}$$

$$9x^2 + 24xy + 16y^2 + 90x - 130y = 0$$

$$9\left(\frac{3x' - 4y'}{5}\right)^2 + 24\left(\frac{3x' - 4y'}{5}\right)\left(\frac{4x' + 3y'}{5}\right) + 16\left(\frac{4x' + 3y'}{5}\right)^2 + 90\left(\frac{3x' - 4y'}{5}\right)$$

$$- 130\left(\frac{4x' + 3y'}{5}\right) = 0$$

$$\frac{81(x')^2}{25} - \frac{216x'y'}{25} + \frac{144(y')^2}{25} + \frac{288(x')^2}{25} - \frac{168x'y'}{25} - \frac{288(y')^2}{25} + \frac{256(x')^2}{25} + \frac{384x'y'}{25}$$

$$+ \frac{144(y')^2}{25} + 54x' - 72y' - 104x' - 78y' = 0$$

$$25(x')^2 - 50x' - 150y' = 0$$

$$(x')^2 - 2x' + 1 = 6y' + 1$$

$$y' = \frac{(x')^2}{6} - \frac{x'}{3} \quad \text{Parabola}$$

15. $x^2 + 3xy + y^2 = 20$

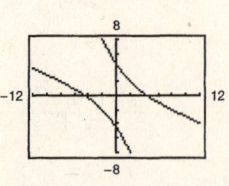

$$\cot 2\theta = \frac{A - C}{B} = \frac{1 - 1}{3} = 0 \Rightarrow \theta = \frac{\pi}{4} = 45°$$

Solve for y in terms of x: Graph:

$$y^2 + 3xy = 20 - x^2$$

$$y_1 = -\frac{3x}{2} + \frac{\sqrt{80 + 5x^2}}{2}$$

$$y^2 + 3xy + \frac{9x^2}{4} = 20 - x^2 + \frac{9x^2}{4}$$

$$y_2 = -\frac{3x}{2} - \frac{\sqrt{80 + 5x^2}}{2}$$

$$\left(y + \frac{3}{2}x\right)^2 = 20 + \frac{5x^2}{4} = \frac{80 + 5x^2}{4}$$

$$y = -\frac{3}{2}x \pm \frac{\sqrt{80 + 5x^2}}{2}$$

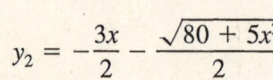

17. $17x^2 + 32xy - 7y^2 = 75$

$$\cot 2\theta = \frac{A - C}{B} = \frac{17 + 7}{32} = \frac{24}{32} = \frac{3}{4} \implies \theta \approx 26.57°$$

Solve for y in terms of x by completing the square.

$$-7y^2 + 32xy = -17x^2 + 75$$

$$y^2 - \frac{32}{7}xy = \frac{17}{7}x^2 - \frac{75}{7}$$

$$y^2 - \frac{32}{7}xy + \frac{256}{49}x^2 = \frac{119}{49}x^2 - \frac{525}{49} + \frac{256}{49}x^2$$

$$\left(y - \frac{16}{7}x\right)^2 = \frac{375x^2 - 525}{49}$$

$$y = \frac{16}{7}x \pm \sqrt{\frac{375x^2 - 525}{49}}$$

$$y = \frac{16x \pm 5\sqrt{15x^2 - 21}}{7}$$

Use $y_1 = \dfrac{16x + 5\sqrt{15x^2 - 21}}{7}$ and

$y_2 = \dfrac{16x - 5\sqrt{15x^2 - 21}}{7}$.

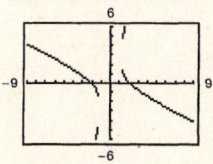

19. $32x^2 + 48xy + 8y^2 = 50$

$$\cot 2\theta = \frac{A - C}{B} = \frac{32 - 8}{48} = \frac{1}{2} \implies \theta \approx 31.72°$$

Solve for y in terms of x:

$$8y^2 + 48xy = -32x^2 + 50$$

$$y^2 + 6xy = -4x^2 + \frac{25}{4}$$

$$y^2 + 6xy + 9x^2 = -4x^2 + \frac{25}{4} + 9x^2$$

$$(y + 3x)^2 = 5x^2 + \frac{25}{4} = \frac{20x^2 + 25}{4}$$

$$y = -3x \pm \frac{\sqrt{20x^2 + 25}}{2}$$

Graph:

$$y_1 = -3x + \frac{\sqrt{20x^2 + 25}}{2}$$

$$y_2 = -3x - \frac{\sqrt{20x^2 + 25}}{2}$$

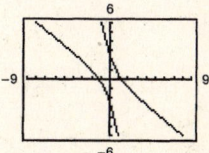

21. $xy + 4 = 0$

$B^2 - 4AC = 1 \implies$ The graph is a hyperbola.

$$\cot 2\theta = \frac{A - C}{B} = 0 \implies \theta = 45°$$

Matches graph (e).

23. $-2x^2 + 3xy + 2y^2 + 3 = 0$

$B^2 - 4AC = (3)^2 - 4(-2)(2)$

$\qquad = 25 \implies$ The graph is a hyperbola.

$$\cot 2\theta = \frac{A - C}{B} = -\frac{4}{3} \implies \theta \approx -18.43°$$

Matches graph (f).

25. $3x^2 + 2xy + y^2 - 10 = 0$

$B^2 - 4AC = (2)^2 - 4(3)(1)$

$\qquad = -8 \implies$ The graph is an ellipse or circle.

$$\cot 2\theta = \frac{A - C}{B} = 1 \implies \theta = 22.5°$$

Matches graph (d).

27. $16x^2 - 24xy + 9y^2 - 30x - 40y = 0$

    (a) $B^2 - 4AC = (-24)^2 - 4(16)(9) = 0 \implies$ Parabola

    (b) $9y^2 - (24x + 40)y + (16x^2 - 30x) = 0$

$$y = \frac{(24x + 40) \pm \sqrt{(24x + 40)^2 - 4(9)(16x^2 - 30x)}}{2(9)}$$

$$= \frac{24x + 40 \pm \sqrt{3000x + 1600}}{18}$$

(c)

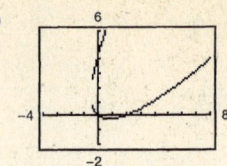

29. $15x^2 - 8xy + 7y^2 - 45 = 0$

    (a) $B^2 - 4AC = (-8)^2 - 4(15)(7) = -356 \implies$ Ellipse or circle

    (b) $7y^2 - 8xy + (15x^2 - 45) = 0$

$$y = \frac{8x \pm \sqrt{(-8x)^2 - 4(7)(15x^2 - 45)}}{14}$$

$$= \frac{8x \pm \sqrt{1260 - 356x^2}}{14}$$

(c)

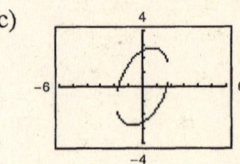

31. $x^2 - 6xy - 5y^2 + 4x - 22 = 0$

    (a) $B^2 - 4AC = (-6)^2 - 4(1)(-5) = 56 \implies$ Hyperbola

    (b) $-5y^2 - 6xy + (x^2 + 4x - 22) = 0$

$$y = \frac{6x \pm \sqrt{(-6x)^2 - 4(-5)(x^2 + 4x - 22)}}{-10}$$

$$= \frac{6x \pm \sqrt{56x^2 + 80x - 440}}{-10}$$

(c)

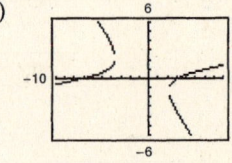

33. $x^2 + 4xy + 4y^2 - 5x - y - 3 = 0$

    (a) $B^2 - 4AC = 4^2 - 4(1)(4) = 0 \implies$ Parabola

    (b) $4y^2 + (4x - 1)y + (x^2 - 5x - 3) = 0$

$$y = \frac{(1 - 4x) \pm \sqrt{(4x - 1)^2 - 4(4)(x^2 - 5x - 3)}}{8} = \frac{1 - 4x \pm \sqrt{72x + 49}}{8}$$

(c)

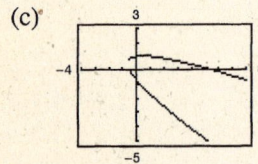

35. $y^2 - 16x^2 = 0$

$$y^2 = 16x^2$$

$$y = \pm 4x$$

Two intersecting lines

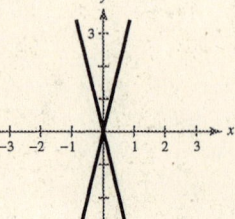

37. $x^2 + 2xy + y^2 - 1 = 0$

$\qquad (x + y)^2 - 1 = 0$

$\qquad (x + y)^2 = 1$

$\qquad x + y = \pm 1$

$\qquad y = -x \pm 1$

Two parallel lines

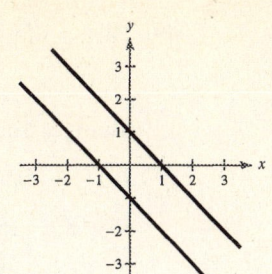

39. $x^2 + y^2 = 4$

$3x - y^2 = 0$

Adding:

$\quad x^2 + 3x - 4 = 0$

$(x + 4)(x - 1) = 0 \implies x = 1, -4$

For $x = 1$, $y = \pm\sqrt{3}$.

$x = -4$ is impossible.

Solutions: $\left(1, \sqrt{3}\right), \left(1, -\sqrt{3}\right)$

41. $-4x^2 - y^2 - 16x + 24y - 16 = 0$

$\underline{4x^2 + y^2 + 40x - 24y + 208 = 0}$

$\qquad 24x \qquad\quad + 192 = 0$

$\qquad\qquad\qquad 24x = -192$

$\qquad\qquad\qquad\quad x = -8$

For $x = -8$:

$-4(64) - y^2 - 16(-8) + 24y - 16 = 0$

$\qquad\qquad -y^2 + 24y - 144 = 0$

$\qquad\qquad\quad y^2 - 24y + 144 = 0$

$\qquad\qquad\qquad (y - 12)^2 = 0$

$\qquad\qquad\qquad\quad \implies y = 12$

Solution: $(-8, 12)$

43. $x^2 - y^2 - 12x + 16y - 64 = 0$

$\underline{x^2 + y^2 - 12x - 16y + 64 = 0}$

$2x^2 \quad\ - 24x \qquad\qquad = 0$

$\qquad\qquad\quad x^2 - 12x = 0$

$\qquad\qquad x(x - 12) = 0 \implies x = 0, 12$

For $x = 0$:

$-y^2 + 16y - 64 = 0$

$\ \ y^2 - 16y + 64 = 0$

$\qquad (y - 8)^2 = 0 \implies y = 8$

For $x = 12$:

$144 - y^2 - 12(12) + 16y - 64 = 0$

$\qquad\quad -y^2 + 16y - 64 = 0 \implies y = 8$

Solutions: $(0, 8), (12, 8)$

45. $-16x^2 - \ y^2 + 24y - 80 = 0$

$\underline{16x^2 + 25y^2 \qquad\quad - 400 = 0}$

$\qquad 24y^2 + 24y - 480 = 0$

$\qquad 24(y + 5)(y - 4) = 0$

$\qquad\qquad y = -5$ or $y = 4$

When $y = -5$:

$16x^2 + 25(-5)^2 - 400 = 0$

$\qquad\qquad\quad 16x^2 = -225$

No real solution

When $y = 4$:

$16x^2 + 25(4)^2 - 400 = 0$

$\qquad\qquad\quad 16x^2 = 0$

$\qquad\qquad\qquad x = 0$

The point of intersection is $(0, 4)$.

In standard form the equations are:

$$\frac{x^2}{4} + \frac{(y - 12)^2}{64} = 1$$

$$\frac{x^2}{25} + \frac{y^2}{16} = 1$$

47. $2x^2 - y^2 + 6 = 0$

$2x + y = 0 \implies y = -2x$

$2x^2 - (-2x)^2 + 6 = 0$

$-2x^2 + 6 = 0$

$x^2 = 3 \implies x = \pm\sqrt{3}$

Two solutions: $\left(\sqrt{3}, -2\sqrt{3}\right), \left(-\sqrt{3}, 2\sqrt{3}\right)$

49. $10x^2 - 25y^2 - 100x + 160 = 0$

$y^2 - 2x + 16 = 0$

Multiply Equation 2 by 25 and add to Equation 1:

$10x^2 - 150x + 560 = 0$

$x^2 - 15x + 56 = 0$

$(x - 8)(x - 7) = 0$

$x = 8 \implies y^2 = 0 \implies (8, 0)$

$x = 7 \implies y^2 = -2$ impossible

One solution: $(8, 0)$

51.

$$xy + x - 2y + 3 = 0 \implies y = \frac{-x - 3}{x - 2}$$

$$x^2 + 4y^2 - 9 = 0$$

$$x^2 + 4\left(\frac{-x - 3}{x - 2}\right)^2 = 9$$

$$x^2(x - 2)^2 + 4(-x - 3)^2 = 9(x - 2)^2$$

$$x^2(x^2 - 4x + 4) + 4(x^2 + 6x + 9) = 9(x^2 - 4x + 4)$$

$$x^4 - 4x^3 + 4x^2 + 4x^2 + 24x + 36 = 9x^2 - 36x + 36$$

$$x^4 - 4x^3 - x^2 + 60x = 0$$

$$x(x + 3)(x^2 - 7x + 20) = 0$$

$$x = 0 \text{ or } x = -3$$

Note: $x^2 - 7x + 20 = 0$ has no real solution.

When $x = 0$: $y = \dfrac{-0 - 3}{0 - 2} = \dfrac{3}{2}$

When $x = -3$: $y = \dfrac{-(-3) - 3}{-3 - 2} = 0$

The points of intersection are $\left(0, \frac{3}{2}\right), (-3, 0)$.

53. True. $B^2 - 4AC = 1 - 4k$

If $k < \frac{1}{4}$, then $B^2 - 4AC > 0$.

55. $g(x) = \dfrac{2}{2 - x}$

Asymptotes: $x = 2, y = 0$

Intercepts: $(0, 1)$

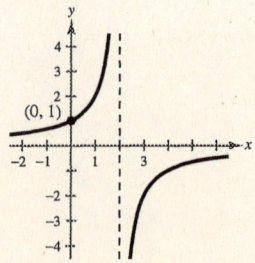

57. $h(t) = \dfrac{t^2}{2 - t} = -t - 2 + \dfrac{4}{2 - t}$

Slant asymptote: $y = -t - 2$

Vertical asymptote: $t = 2$

Intercept: $(0, 0)$

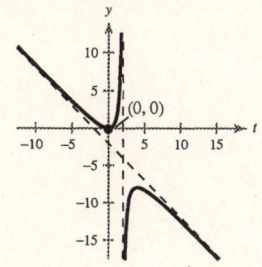

59. (a) $AB = \begin{bmatrix} 1 & -3 \\ 2 & 5 \end{bmatrix}\begin{bmatrix} 0 & 6 \\ 5 & -1 \end{bmatrix} = \begin{bmatrix} -15 & 9 \\ 25 & 7 \end{bmatrix}$

(b) $BA = \begin{bmatrix} 0 & 6 \\ 5 & -1 \end{bmatrix}\begin{bmatrix} 1 & -3 \\ 2 & 5 \end{bmatrix} = \begin{bmatrix} 12 & 30 \\ 3 & -20 \end{bmatrix}$

(c) $A^2 = \begin{bmatrix} 1 & -3 \\ 2 & 5 \end{bmatrix}\begin{bmatrix} 1 & -3 \\ 2 & 5 \end{bmatrix} = \begin{bmatrix} -5 & -18 \\ 12 & 19 \end{bmatrix}$

61. (a) $AB = \begin{bmatrix} 4 & -2 & 5 \end{bmatrix}\begin{bmatrix} 3 \\ -4 \\ 5 \end{bmatrix} = [12 + 8 + 25] = [45]$

(b) $BA = \begin{bmatrix} 3 \\ -4 \\ 5 \end{bmatrix}\begin{bmatrix} 4 & -2 & 5 \end{bmatrix} = \begin{bmatrix} 12 & -6 & 15 \\ -16 & 8 & -20 \\ 20 & -10 & 25 \end{bmatrix}$

(c) A^2 does not exist.

Section 9.5   Parametric Equations

- If f and g are continuous functions of t on an interval I, then the set of ordered pairs $(f(t), g(t))$ is a *plane curve C*. The equations $x = f(t)$ and $y = g(t)$ are *parametric equations* for C and t is the *parameter*.
- You should be able to graph plane curves with your graphing utility.
- To eliminate the parameter:

    Solve for t in one equation and substitute into the second equation.
- You should be able to find the parametric equations for a graph.

Solutions to Odd-Numbered Exercises

1. $x = t$

$y = t + 2$

$y = x + 2$, line

Matches (c).

3. $x = \sqrt{t}$

$y = t$

$y = x^2$, parabola, $x \geq 0$

Matches (b).

5. $x = \ln t \iff t = e^x$

$y = \frac{1}{2}t - 2$

$y = \frac{1}{2}e^x - 2$

Matches (f).

7. $x = \sqrt{t}, y = 2 - t$

(a)

t	0	1	2	3	4
x	0	1	$\sqrt{2}$	$\sqrt{3}$	2
y	2	1	0	-1	-2

(b) Graph by hand. *Note:* $x \geq 0$

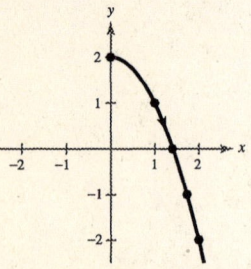

(c)

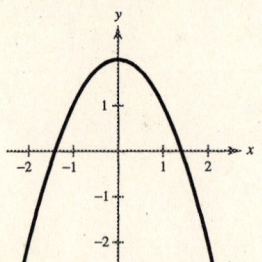

(d) $y = 2 - t = 2 - x^2$, parabola

In part (c), $x \geq 0$.

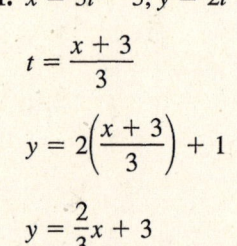

9. $x = t, y = -4t$

$y = -4x$

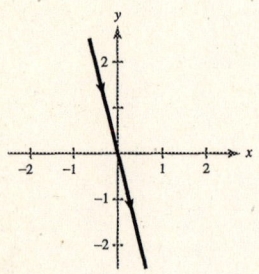

11. $x = 3t - 3, y = 2t + 1$

$t = \dfrac{x + 3}{3}$

$y = 2\left(\dfrac{x + 3}{3}\right) + 1$

$y = \dfrac{2}{3}x + 3$

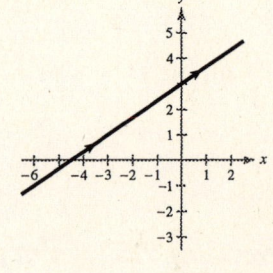

13. $x = \frac{1}{4}t, y = t^2$

$y = (4x)^2$

$y = 16x^2$

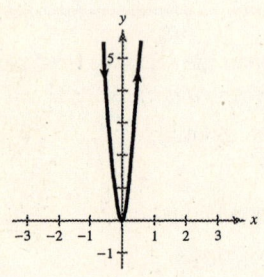

15. $x = t + 2, y = t^2$

$t = x - 2$

$y = (x - 2)^2$

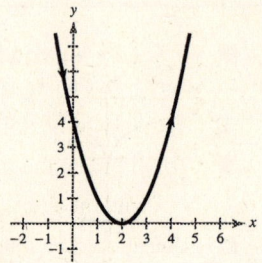

17. $x = 2t, y = |t - 2|$

$t = \dfrac{x}{2} \implies y = |t - 2|$

$= \left|\dfrac{x}{2} - 2\right|$

$= \dfrac{1}{2}|x - 4|$

19. $x = 3 \cos \theta \implies \left(\dfrac{x}{3}\right)^2 = \cos^2 \theta$

$y = 3 \sin \theta \implies \left(\dfrac{y}{3}\right)^2 = \sin^2 \theta$

$\left(\dfrac{x}{3}\right)^2 + \left(\dfrac{y}{3}\right)^2 = 1 \implies x^2 + y^2 = 9$

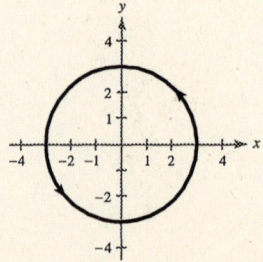

21. $x = e^{-t} \Rightarrow \dfrac{1}{x} = e^t$

$y = e^{3t} \Rightarrow y = (e^t)^3$

$y = \left(\dfrac{1}{x}\right)^3$

$y = \dfrac{1}{x^3}, \ x > 0, \ y > 0$

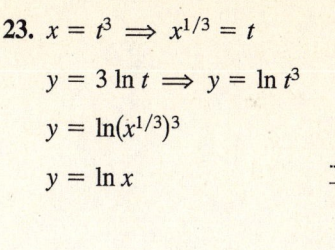

23. $x = t^3 \Rightarrow x^{1/3} = t$

$y = 3 \ln t \Rightarrow y = \ln t^3$

$y = \ln(x^{1/3})^3$

$y = \ln x$

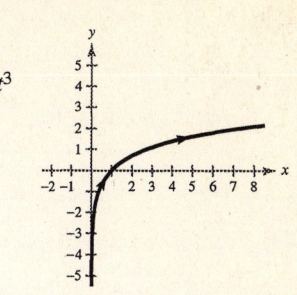

25. $x = 4 + 2\cos\theta \Rightarrow \left(\dfrac{x-4}{2}\right)^2 = \cos^2\theta$

$y = -1 + \sin\theta \Rightarrow (y+1)^2 = \sin^2\theta$

$\left(\dfrac{x-4}{2}\right)^2 + (y+1)^2 = \cos^2\theta + \sin^2\theta$

$\dfrac{(x-4)^2}{4} + (y+1)^2 = 1$

Ellipse

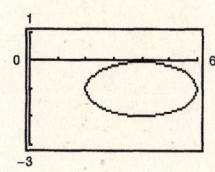

27. $x = 4\sec\theta \Rightarrow \left(\dfrac{x}{4}\right)^2 = \sec^2\theta$

$y = 3\tan\theta \Rightarrow \left(\dfrac{y}{3}\right)^2 = \tan^2\theta$

$\left(\dfrac{x}{4}\right)^2 - \left(\dfrac{y}{3}\right)^2 = \sec^2\theta - \tan^2\theta$

$\dfrac{x^2}{16} - \dfrac{y^2}{9} = 1$

Hyperbola

29. $x = \dfrac{t}{2}$

$y = \ln(t^2 + 1)$

31. By eliminating the parameters in (a)–(d), we get $y = 2x + 1$. They differ from each other in restricted domain and in orientation.

   (a) Domain: $-\infty < x < \infty$
   Orientation: Left to right

   (b) Domain: $-1 \le x \le 1$
   Orientation: Depends on θ

   (c) Domain: $0 < x < \infty$
   Orientation: Right to left

   (d) Domain: $0 < x < \infty$
   Orientation: Left to right

33. $t = \dfrac{(x - x_1)}{(x_2 - x_1)}$

$y = y_1 + \left(\dfrac{x - x_1}{x_2 - x_1}\right)(y_2 - y_1)$

$\Rightarrow y - y_1 = \left(\dfrac{y_2 - y_1}{x_2 - x_1}\right)(x - x_1)$

35. $x = h + a\cos\theta$

$y = k + b\sin\theta$

$\dfrac{x - h}{a} = \cos\theta, \ \dfrac{y - k}{b} = \sin\theta$

$\dfrac{(x - h)^2}{a^2} + \dfrac{(y - k)^2}{b^2} = 1$

37. $x = x_1 + t(x_2 - x_1) = 0 + t(5 - 0)$ $= 5t$

$y = y_1 + t(y_2 - y_1) = 0 + t(-2 - 0) = -2t$

Solution not unique.

39. From Exercise 34:

$x = 2 + 4 \cos \theta$

$y = 1 + 4 \sin \theta$

Solution not unique.

41. From Exercise 35:

$a = 5, c = 4$, and hence, $b = 3$.

$x = 5 \cos \theta$

$y = 3 \sin \theta$

Center: $(0, 0)$

Solution not unique.

43. $y = 4x - 3$

Sample answers:

$x = t, y = 4t - 3$

$x = \frac{1}{4}t + \frac{3}{4}, y = t$

45. $y = \dfrac{1}{x}$

Sample answers:

$x = t, y = \dfrac{1}{t}$

$x = t^3, y = \dfrac{1}{t^3}$

47. $y = x^2 + 4$

Sample answers:

$x = t, y = t^2 + 4$

$x = t^3, y = t^6 + 4$

49. $y = x^3 + 2x$

Sample answers:

$x = t, y = t^3 + 2t$

$x = \dfrac{1}{2}t, y = \dfrac{t^3}{8} + t$

51. $x = 2(\theta - \sin \theta)$

$y = 2(1 - \cos \theta)$

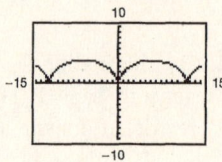

53. $x = 2 \cot \theta, \ y = 2 \sin^2 \theta$

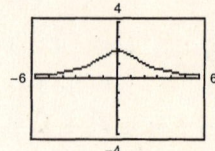

55. Matches graph (b).

57. Matches graph (d).

59. $x = (v_0 \cos \theta)t, \ y = h + (v_0 \sin \theta)t - 16t^2$

(a) $100 \text{ miles/hour} = \dfrac{100 \text{ mi/hr} \cdot 5280 \text{ ft/mi}}{3600 \text{ sec/hr}}$

$= 146.67 \text{ ft/sec}$

$x = (146.67 \cos \theta)t$

$y = 3 + (146.67 \sin \theta)t - 16t^2$

(b) $\theta = 15°$

$x = (146.67 \cos 15°)t = 141.7t$

$y = 3 + (146.67 \sin 15°)t - 16t^2$

$= 3 + 38.0t - 16t^2$

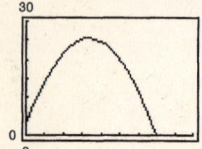

It is not a home run because $y < 10$ when $x = 400$.

(c) $\theta = 23°$

$x = (146.67 \cos 23°)t = 135.0t$

$y = 3 + (146.67 \sin 23°)t - 16t^2$

$= 3 + 57.3t - 16t^2$

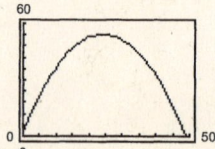

Yes, it is a home run because $y > 10$ when $x = 400$.

(d) $\theta \approx 19.4°$ is the minimum angle.

61. True

$x = t$ first set $x = 3t$ second set

$y = t^2 + 1 = x^2 + 1$ $y = 9t^2 + 1 = (3t)^2 + 1 = x^2 + 1$

63. One possible answer:

$x = \cos\theta$

$y = -2\sin\theta$

65. Answers will vary.

67. $5x^2 + 8 = 0$

$x^2 = -\frac{8}{5}$

$x = \pm\sqrt{\frac{8}{5}}i = \pm\frac{2}{5}\sqrt{10}i$

69. $4x^2 + 4x - 11 = 0$

$$x = \frac{-4 \pm \sqrt{16 + 176}}{8} = -\frac{1}{2} \pm \sqrt{3}$$

71. $\sum_{n=1}^{50} 8n = 8\frac{(50)(51)}{2} = 10{,}200$

73. $\sum_{n=1}^{40}\left(300 - \frac{1}{2}n\right) = 300(40) - \frac{1}{2}\frac{(40)(41)}{2} = 12{,}000 - 410 = 11{,}590$

Section 9.6   Polar Coordinates

- In polar coordinates you do not have unique representation of points. The point (r, θ) can be represented by $(r, \theta \pm 2n\pi)$ or by $(-r, \theta \pm (2n+1)\pi)$ where n is any integer. The pole is represented by $(0, \theta)$ where θ is any angle.
- To convert from polar coordinates to rectangular coordinates, use the following relationships.

  $x = r\cos\theta$

  $y = r\sin\theta$
- To convert from rectangular coordinates to polar coordinates, use the following relationships.

  $r = \pm\sqrt{x^2 + y^2}$

  $\tan\theta = y/x$

  If θ is in the same quadrant as the point (x, y), then r is positive. If θ is in the opposite quadrant as the point (x, y), then r is negative.
- You should be able to convert rectangular equations to polar form and vice versa.

Solutions to Odd-Numbered Exercises

1. Polar coordinates: $\left(4, \frac{\pi}{2}\right)$

$x = 4\cos\left(\frac{\pi}{2}\right) = 0$

$y = 4\sin\left(\frac{\pi}{2}\right) = 4$

Rectangular coordinates: $(0, 4)$

3. Polar coordinates: $\left(-1, \frac{5\pi}{4}\right)$

$x = -1\cos\left(\frac{5\pi}{4}\right) = \frac{\sqrt{2}}{2}$

$y = -1\sin\left(\frac{5\pi}{4}\right) = \frac{\sqrt{2}}{2}$

Rectangular coordinates: $\left(\frac{\sqrt{2}}{2}, \frac{\sqrt{2}}{2}\right)$

5.

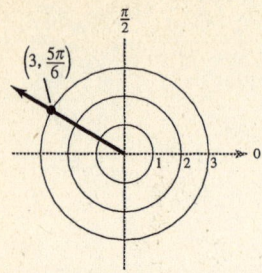

Three additional representations:

$$\left(3, \frac{5\pi}{6} - 2\pi\right) = \left(3, -\frac{7\pi}{6}\right)$$

$$\left(-3, \frac{5\pi}{6} + \pi\right) = \left(-3, \frac{11\pi}{6}\right)$$

$$\left(-3, \frac{5\pi}{6} - \pi\right) = \left(-3, -\frac{\pi}{6}\right)$$

7.

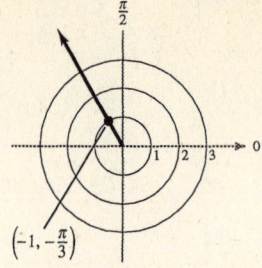

Three additional representations:

$$\left(-1, -\frac{\pi}{3} + 2\pi\right) = \left(-1, \frac{5\pi}{3}\right)$$

$$\left(1, -\frac{\pi}{3} + \pi\right) = \left(1, \frac{2\pi}{3}\right)$$

$$\left(1, -\frac{\pi}{3} - \pi\right) = \left(1, -\frac{4\pi}{3}\right)$$

9.

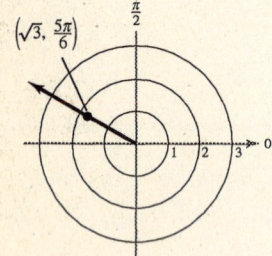

Three additional representations:

$$\left(\sqrt{3}, -\frac{7\pi}{6}\right), \left(-\sqrt{3}, -\frac{\pi}{6}\right), \left(-\sqrt{3}, \frac{11\pi}{6}\right)$$

11.

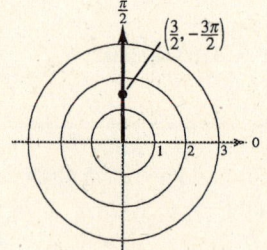

Three additional representations:

$$\left(\frac{3}{2}, \frac{\pi}{2}\right), \left(-\frac{3}{2}, \frac{3\pi}{2}\right), \left(-\frac{3}{2}, -\frac{\pi}{2}\right)$$

13. Polar coordinates: $\left(4, -\frac{\pi}{3}\right)$

$$x = 4\cos\left(-\frac{\pi}{3}\right) = 2$$

$$y = 4\sin\left(-\frac{\pi}{3}\right) = -2\sqrt{3}$$

Rectangular coordinates: $\left(2, -2\sqrt{3}\right)$

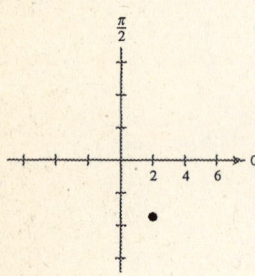

15. Polar coordinates: $\left(-1, \frac{-3\pi}{4}\right)$

$$x = -1\cos\left(\frac{-3\pi}{4}\right) = \frac{\sqrt{2}}{2}$$

$$y = -1\sin\left(\frac{-3\pi}{4}\right) = \frac{\sqrt{2}}{2}$$

Rectangular coordinates: $\left(\frac{\sqrt{2}}{2}, \frac{\sqrt{2}}{2}\right)$

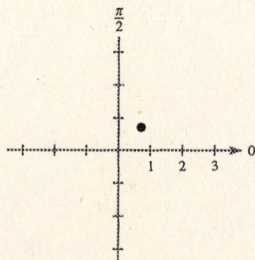

17. Polar coordinates: $\left(0, -\dfrac{7\pi}{6}\right)$ (origin!)

$x = 0\cos\left(-\dfrac{7\pi}{6}\right) = 0$

$y = 0\sin\left(-\dfrac{7\pi}{6}\right) = 0$

Rectangular coordinates: $(0, 0)$

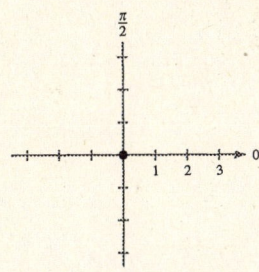

19. Polar coordinates: $\left(\sqrt{2}, 2.36\right)$

$x = \sqrt{2}\cos(2.36) \approx -1.004$

$y = \sqrt{2}\sin(2.36) \approx 0.996$

Rectangular coordinates: $(-1.004, 0.996)$

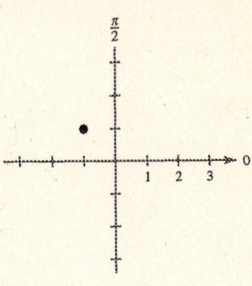

21. $(r, \theta) = \left(2, \dfrac{3\pi}{4}\right) \;\Rightarrow\; (x, y) = (-1.414, 1.414) = \left(-\sqrt{2}, \sqrt{2}\right)$

23. $(r, \theta) = (-4.5, 1.3) \;\Rightarrow\; (x, y) = (-1.204, -4.336)$

25. Rectangular coordinates: $(-7, 0)$

$r = 7, \tan\theta = 0, \theta = 0$

Polar coordinates: $(7, \pi), (-7, 0)$

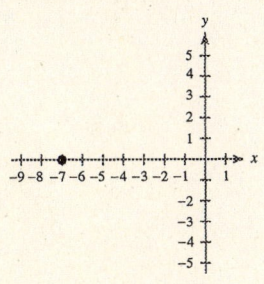

27. Rectangular coordinates: $(1, 1)$

$r = \sqrt{2}, \tan\theta = 1, \theta = \dfrac{\pi}{4}$

Polar coordinates: $\left(\sqrt{2}, \dfrac{\pi}{4}\right), \left(-\sqrt{2}, \dfrac{5\pi}{4}\right)$

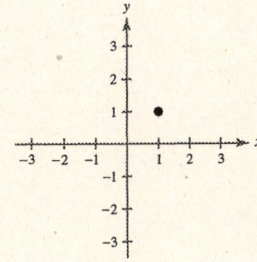

29. Rectangular coordinates: $\left(-\sqrt{3}, -\sqrt{3}\right)$

$r = \sqrt{3+3} = \sqrt{6}$, $\tan\theta = 1$, $\theta = \dfrac{\pi}{4}$

Polar coordinates: $\left(\sqrt{6}, \dfrac{5\pi}{4}\right), \left(-\sqrt{6}, \dfrac{\pi}{4}\right)$

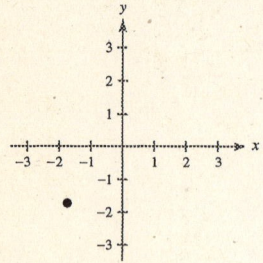

31. $(x, y) = (6, 9)$

$r = \sqrt{6^2 + 9^2} = \sqrt{117} \approx 10.8$

$\tan\theta = \dfrac{9}{6} = \dfrac{3}{2} \implies \theta \approx 0.983$

Polar coordinates: $(10.8, 0.983), (-10.8, 4.124)$

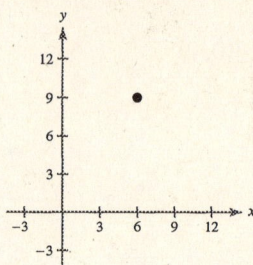

33. $(x, y) = (3, -2) \implies r = \sqrt{3^2 + (-2)^2} = \sqrt{13}$

$\theta = \arctan\left(-\dfrac{2}{3}\right) \approx -0.588$

$(r, \theta) \approx \left(\sqrt{13}, -0.588\right)$

35. $(x, y) = \left(\sqrt{3}, 2\right) \implies r = \sqrt{3 + 2^2} = \sqrt{7}$

$\theta = \arctan\left(\dfrac{2}{\sqrt{3}}\right) \approx 0.857$

$(r, \theta) \approx \left(\sqrt{7}, 0.857\right)$

37. $(x, y) = \left(\dfrac{5}{2}, \dfrac{4}{3}\right) \implies r = \sqrt{\left(\dfrac{5}{2}\right)^2 + \left(\dfrac{4}{3}\right)^2} = \dfrac{17}{6}$

$\theta = \arctan\left(\dfrac{4/3}{5/2}\right) \approx 0.490$

$(r, \theta) \approx \left(\dfrac{17}{6}, 0.490\right)$

39. $x^2 + y^2 = 81$

$r^2 = 81$

$r = 9$

41. $y = 4$

$r\sin\theta = 4$

$r = 4\csc\theta$

43. $x = 8$

$r\cos\theta = 8$

$r = 8\sec\theta$

45. $3x - 6y + 2 = 0$

$3r\cos\theta - 6r\sin\theta = -2$

$r(3\cos\theta - 6\sin\theta) = -2$

$r = \dfrac{2}{6\sin\theta - 3\cos\theta}$

47. $xy = 4$

$(r\cos\theta)(r\sin\theta) = 4$

$r^2\cos\theta\sin\theta = 4$

$r^2(2\cos\theta\sin\theta) = 8$

$r^2\sin 2\theta = 8$

$r^2 = 8\csc 2\theta$

49. $(x^2 + y^2)^2 = 9(x^2 - y^2)$

$(r^2)^2 = 9(r^2\cos^2\theta - r^2\sin^2\theta)$

$r^2 = 9(\cos^2\theta - \sin^2\theta)$

$r^2 = 9\cos(2\theta)$

51. $x^2 + y^2 - 6x = 0$

$r^2 - 6r\cos\theta = 0$

$r^2 = 6r\cos\theta$

$r = 6\cos\theta$

53. $x^2 + y^2 - 2ax = 0$

$r^2 - 2a \cos \theta = 0$

$r(r - 2 \, arcos \, \theta) = 0$

$r = 2a \cos \theta$

55. $y^2 = x^3$

$(r \sin \theta)^2 = (r \cos \theta)^3$

$\sin^2 \theta = r \cos^3 \theta$

$r = \dfrac{\sin^2 \theta}{\cos^3 \theta}$

$= \tan^2 \theta \sec \theta$

57. $r = 6 \sin \theta$

$r^2 = 6r \sin \theta$

$x^2 + y^2 = 6y$

$x^2 + y^2 - 6y = 0$

59. $\theta = \dfrac{4\pi}{3}$

$\tan \theta = \tan \dfrac{4\pi}{3} = \dfrac{y}{x}$

$\sqrt{3} = \dfrac{y}{x}$

$y = \sqrt{3}x$

61. $r = 4$

$r^2 = 16$

$x^2 + y^2 = 16$

63. $r = -3 \csc \theta$

$r \sin \theta = -3$

$y = -3$

65. $r^2 = \cos \theta$

$r^3 = r \cos \theta$

$(x^2 + y^2)^{3/2} = x$

$x^2 + y^2 = x^{2/3}$

$(x^2 + y^2)^3 = x^2$

67. $r = 2 \sin 3\theta$

$r = 2(3 \sin \theta - 4 \sin^3 \theta)$

$r^4 = 6r^3 \sin \theta - 8r^3 \sin^3 \theta$

$(x^2 + y^2)^2 = 6(x^2 + y^2)y - 8y^3$

$(x^2 + y^2)^2 = 6x^2y - 2y^3$

69. $r = \dfrac{1}{1 - \cos \theta}$

$r - r \cos \theta = 1$

$\sqrt{x^2 + y^2} - x = 1$

$x^2 + y^2 = 1 + 2x + x^2$

$y^2 = 2x + 1$

71. $r = \dfrac{6}{2 - 3 \sin \theta}$

$r(2 - 3 \sin \theta) = 6$

$2r = 6 + 3r \sin \theta$

$2\left(\pm \sqrt{x^2 + y^2}\right) = 6 + 3y$

$4(x^2 + y^2) = (6 + 3y)^2$

$4x^2 + 4y^2 = 36 + 36y + 9y^2$

$4x^2 - 5y^2 - 36y - 36 = 0$

73. $r = 7$

$r^2 = 49$

$x^2 + y^2 = 49$

The graph is a circle centered at the origin with radius 7.

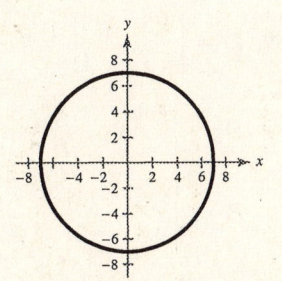

75. $\theta = \dfrac{\pi}{4}$

$\tan \theta = \tan \dfrac{\pi}{4} = 1 = \dfrac{y}{x}$

$y = x$

The graph is the line $y = x$, which makes an angle of $\theta = \pi/4$ with the positive x-axis.

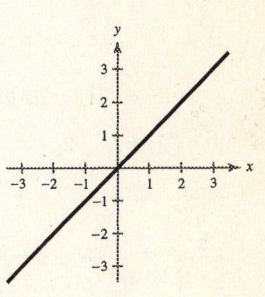

77.
$$r = 3 \sec \theta$$
$$r \cos \theta = 3$$
$$x = 3$$
$$x - 3 = 0$$
Vertical line

79. True, the distances from the origin are the same.

81. (a) $(r_1, \theta_1) = (x_1, y_1)$ where $x_1 = r_1 \cos \theta_1$ and $y_1 = r_1 \sin \theta_1$.

$(r_2, \theta_2) = (x_2, y_2)$ where $x_2 = r_2 \cos \theta_2$ and $y_2 = r_2 \sin \theta_2$.

Then $x_1^2 + y_1^2 = r_1^2 \cos^2 \theta_1 + r_1^2 \sin^2 \theta_1 = r_1^2$ and $x_2^2 + y_2^2 = r_2^2$. Thus,

$$
\begin{aligned}
d &= \sqrt{(x_1 - x_2)^2 + (y_1 - y_2)^2} \\
&= \sqrt{x_1^2 - 2x_1 x_2 + x_2^2 + y_1^2 - 2y_1 y_2 + y_2^2} \\
&= \sqrt{(x_1^2 + y_1^2) + (x_2^2 + y_2^2) - 2(x_1 x_2 + y_1 y_2)} \\
&= \sqrt{r_1^2 + r_2^2 - 2(r_1 r_2 \cos \theta_1 \cos \theta_2 + r_1 r_2 \sin \theta_1 \sin \theta_2)} \\
&= \sqrt{r_1^2 + r_2^2 - 2r_1 r_2 \cos(\theta_1 - \theta_2)}.
\end{aligned}
$$

(b) If $\theta_1 = \theta_2$, the points are on the same line through the origin. In this case,
$$d = \sqrt{r_1^2 + r_2^2 - 2r_1 r_2 \cos(0)} = \sqrt{(r_1 - r_2)^2} = |r_1 - r_2|.$$

(c) If $\theta_1 - \theta_2 = 90°$, $d = \sqrt{r_1^2 + r_2^2}$, the Pythagorean Theorem.

(d) For instance, $\left(3, \dfrac{\pi}{6}\right), \left(4, \dfrac{\pi}{3}\right)$ gives $d \approx 2.053$ and $\left(-3, \dfrac{7\pi}{6}\right), \left(-4, \dfrac{4\pi}{3}\right)$ gives $d \approx 2.053$. (same!)

83. $\cos A = \dfrac{b^2 + c^2 - a^2}{2bc} = \dfrac{19^2 + 25^2 - 13^2}{2(19)(25)} = 0.86$

$A \approx 30.7°$

$\cos B = \dfrac{a^2 + c^2 - b^2}{2ac} = \dfrac{13^2 + 25^2 - 19^2}{2(13)(25)} = 0.66615$

$B \approx 48.2°$

$C \approx 180° - 30.7° - 48.2° \approx 101.1°$

85. $B = 180° - 56° - 38° = 86°$

$\dfrac{a}{\sin A} = \dfrac{c}{\sin C} \Rightarrow a = \dfrac{c \sin A}{\sin C} = \dfrac{12 \sin(56°)}{\sin(38°)} \approx 16.16$

$\dfrac{b}{\sin B} = \dfrac{c}{\sin C} \Rightarrow b = \dfrac{c \sin B}{\sin C} = \dfrac{12 \sin(86°)}{\sin(38°)} \approx 19.44$

87. $c^2 = a^2 + b^2 - 2ab \cos C$

$= 8^2 + 4^2 - 2(8)(4) \cos(35°)$

≈ 27.57

$c \approx 5.25$

$\dfrac{b}{\sin B} = \dfrac{c}{\sin C} \Rightarrow \sin B = \dfrac{b \sin C}{c} = \dfrac{4 \sin(35°)}{5.25}$

$\Rightarrow B \approx 25.9°$

$A = 180° - B - C = 119.1°$

89. $D = \begin{vmatrix} 5 & -7 \\ -3 & 1 \end{vmatrix} = 5 - 21 = -16$

$D_x = \begin{vmatrix} -11 & -7 \\ -3 & 1 \end{vmatrix} = -11 - 21 = -32$

$D_y = \begin{vmatrix} 5 & -11 \\ -3 & -3 \end{vmatrix} = -15 - 33 = -48$

$x = \dfrac{D_x}{D} = \dfrac{-32}{-16} = 2$

$y = \dfrac{D_y}{D} = \dfrac{-48}{-16} = 3$

Solution: $(2, 3)$

91. $D = \begin{vmatrix} 3 & -2 & 1 \\ 2 & 1 & -3 \\ 1 & -3 & 9 \end{vmatrix} = 35$

$D_a = \begin{vmatrix} 0 & -2 & 1 \\ 0 & 1 & -3 \\ 0 & -3 & 9 \end{vmatrix} = 0$

$D_b = \begin{vmatrix} 3 & 0 & 1 \\ 2 & 0 & -3 \\ 1 & 0 & 9 \end{vmatrix} = 0$

$D_c = \begin{vmatrix} 3 & -2 & 0 \\ 2 & 1 & 0 \\ 1 & -3 & 0 \end{vmatrix} = 0$

$a = \dfrac{D_a}{D} = 0$

$b = \dfrac{D_b}{D} = 0$

$c = \dfrac{D_c}{D} = 0$

Solution: $(0, 0, 0)$

Section 9.7   Graphs of Polar Equations

■   When graphing polar equations:

1. Test for symmetry

    (a) $\theta = \pi/2$: Replace (r, θ) by $(r, \pi - \theta)$ or $(-r, -\theta)$.

    (b) Polar axis: Replace (r, θ) by $(r, -\theta)$ or $(-r, \pi - \theta)$.

    (c) Pole: Replace (r, θ) by $(r, \pi + \theta)$ or $(-r, \theta)$.

    (d) $r = f(\sin \theta)$ is symmetric with respect to the line $\theta = \pi/2$.

    (e) $r = f(\cos \theta)$ is symmetric with respect to the polar axis.

2. Find the θ values for which $|r|$ is maximum.

3. Find the θ values for which $r = 0$.

4. Know the different types of polar graphs.

    (a) Limaçons   (b) Rose curves, $n \geq 2$

         $r = a \pm b \cos \theta$ $r = a \cos n\theta$

         $r = a \pm b \sin \theta$ $r = a \sin n\theta$

    (c) Circles   (d) Lemniscates

         $r = a \cos \theta$ $r^2 = a^2 \cos 2\theta$

         $r = a \sin \theta$ $r^2 = a^2 \sin 2\theta$

         $r = a$

■   You should be able to graph polar equations of the form $r = f(\theta)$ with your graphing utility. If your utility does not have a polar mode, use

     $x = f(t) \cos t$

     $y = f(t) \sin t$

in parametric mode.

Solutions to Odd-Numbered Exercises

1. $r = 3 \cos 2\theta$ is a rose curve.   **3.** $r = 3 \cos \theta$ is a circle.   **5.** $r = 6 \sin 2\theta$ is a rose curve.

7. $r = 10 + 4 \cos$

$\theta = \dfrac{\pi}{2}$: $-r = 10 + 4 \cos(-\theta)$

          $-r = 10 + 4 \cos \theta$

          Not an equivalent equation

          $r = 10 + 4 \cos(\pi - \theta)$

          $r = 10 + 4(\cos \pi \cos \theta + \sin \pi \sin \theta)$

          $r = 10 - 4 \cos \theta$

          Not an equivalent equation

Polar axis: $r = 10 + 4 \cos(-\theta)$

               $r = 10 + 4 \cos \theta$

               Equivalent equation

Pole: $-r = 10 + 4 \cos \theta$

            Not an equivalent equation

            $r = 10 + 4 \cos(\pi + \theta)$

               $r = 10 + 4(\cos \pi \cos \theta - \sin \pi \sin \theta)$

               $r = 10 - 4 \cos \theta$

            Not an equivalent equation

Answer: Symmetric with respect to polar axis.

9. $r = \dfrac{4}{1 + \sin \theta}$

$\theta = \dfrac{\pi}{2}$: $r = \dfrac{4}{1 + \sin(\pi - \theta)}$

$\qquad\qquad r = \dfrac{4}{1 + \sin \pi \cos \theta - \cos \pi \sin \theta}$

$\qquad\qquad r = \dfrac{4}{1 + \sin \theta}$

Equivalent equation

Polar axis: $r = \dfrac{4}{1 + \sin(-\theta)}$

$\qquad\qquad r = \dfrac{4}{1 - \sin \theta}$

Not an equivalent equation

$\qquad -r = \dfrac{4}{1 + \sin(\pi - \theta)}$

$\qquad -r = \dfrac{4}{1 + \sin \theta}$

Not an equivalent equation

Pole: $-r = \dfrac{4}{1 + \sin \theta}$

Not an equivalent equation

$\qquad r = \dfrac{4}{1 + \sin(\pi + \theta)}$

$\qquad r = \dfrac{4}{1 - \sin \theta}$

Not an equivalent equation

Answer: Symmetric with respect to $\theta = \dfrac{\pi}{2}$

11. $r = 6 \sin \theta$

$\theta = \dfrac{\pi}{2}$: $-r = 6 \sin(-\theta)$

$\qquad\qquad r = 6 \sin \theta$

Equivalent equation

Polar axis: $r = 6 \sin(-\theta)$

$\qquad\qquad r = -6 \sin \theta$

Not an equivalent equation

$\qquad -r = 6 \sin(\pi - \theta)$

$\qquad -r = 6(\sin \pi \cos \theta - \cos \pi \sin \theta)$

$\qquad -r = 6 \sin \theta$

Not an equivalent equation

Pole: $-r = 6 \sin \theta$

Not an equivalent equation

$\qquad r = 6 \sin(\pi + \theta)$

$\qquad r = -6 \sin \theta$

Not an equivalent equation

Answer: Symmetric with respect to $\theta = \dfrac{\pi}{2}$

13. $r^2 = 25 \sin 2\theta$

$\theta = \dfrac{\pi}{2}$: $(-r)^2 = 25 \sin(2(-\theta))$

$\qquad\qquad r^2 = -25 \sin 2\theta$

Not an equivalent equation

$\qquad\qquad r^2 = 25 \sin(2(\pi - \theta))$

$\qquad\qquad r^2 = 25 \sin(2\pi - 2\theta)$

$\qquad\qquad r^2 = 25(\sin 2\pi \cos 2\theta - \cos 2\pi \sin 2\theta)$

$\qquad\qquad r^2 = -25 \sin 2\theta$

Not an equivalent equation

Polar axis: $r^2 = 25 \sin(2(-\theta))$

$\qquad\qquad r^2 = -25 \sin 2\theta$

Not an equivalent equation

$\qquad\qquad (-r)^2 = 25 \sin(2(\pi - \theta))$

$\qquad\qquad r^2 = -25 \sin 2\theta$

Not an equivalent equation

Pole: $(-r)^2 = 25 \sin(2\theta)$

$\qquad\qquad r^2 = 25 \sin 2\theta$

Equivalent equation

Answer: Symmetric with respect to pole

15. $|r| = |10(1 - \sin \theta)|$

$\qquad = 10|1 - \sin \theta| \le 10(2) = 20$

$|1 - \sin \theta| = 2$

$1 - \sin \theta = 2 \quad$ or $\quad 1 - \sin \theta = -2$

$\qquad \sin \theta = -1 \qquad\qquad \sin \theta = 3$

$\qquad\quad \theta = \dfrac{3\pi}{2} \qquad\quad$ Not possible

Maximum: $|r| = 20$ when $\theta = \dfrac{3\pi}{2}$.

$r = 0$ when $1 - \sin \theta = 0$

$\qquad\qquad\qquad \sin \theta = 1$

$\qquad\qquad\qquad\quad \theta = \dfrac{\pi}{2}$.

17. $|r| = |4 \cos 3\theta| = 4 |\cos 3\theta| \le 4$

$|\cos 3\theta| = 1$

$\cos 3\theta = \pm 1$

$\theta = 0, \dfrac{\pi}{3}, \dfrac{2\pi}{3}, \pi$

Maximum: $|r| = 4$ when $\theta = 0, \dfrac{\pi}{3}, \dfrac{2\pi}{3}, \pi$.

$r = 0$ when $\cos 3\theta = 0$

$\theta = \dfrac{\pi}{6}, \dfrac{\pi}{2}, \dfrac{5\pi}{6}$.

19. Circle: $r = 4$

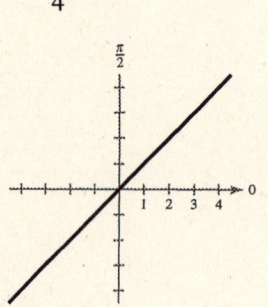

21. $\theta = \dfrac{\pi}{4}$, line

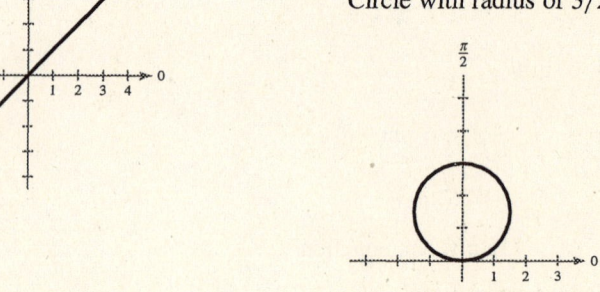

23. $r = 3 \sin \theta$

Symmetric with respect to $\theta = \pi/2$

Circle with radius of $3/2$

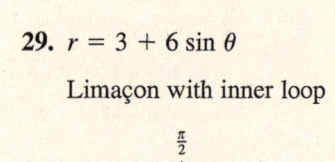

25. $r = 3(1 - \cos \theta)$

Cardioid

27. $r = 3 - 4 \cos \theta$

Limaçon

29. $r = 3 + 6 \sin \theta$

Limaçon with inner loop

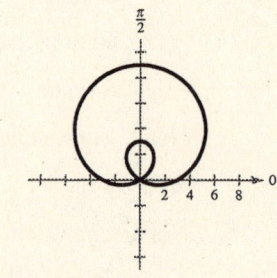

31. $r = 5 \cos 3\theta$

Rose curve

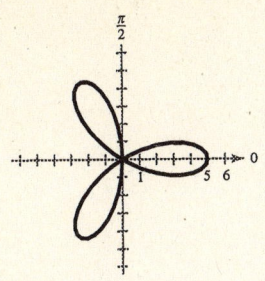

33. $r = 7 \sin 2\theta$

Rose curve, four petals

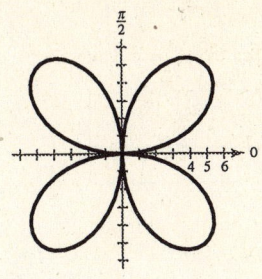

35. $r = 6 \cos 2\theta, \; 0 \le \theta < 2\pi$

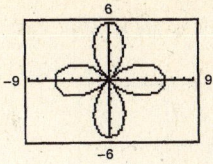

37. $r = 2(3 - \sin \theta), \; 0 \le \theta < 2\pi$

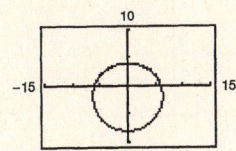

39. $r = 3 - 6 \cos \theta, \; 0 \le \theta \le 2\pi$

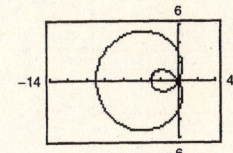

41. $r = \dfrac{3}{\sin \theta - 2 \cos \theta}, \; 0 \le \theta \le \dfrac{\pi}{2}$

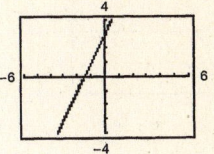

43. $r^2 = 4 \cos 2\theta, \; -2\pi \le \theta \le 2\pi$

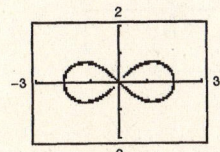

45. $r = 4 \sin \theta \cos^2 \theta, \; 0 \le \theta \le \pi$

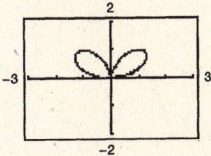

47. $r = 2 \csc \theta + 5, \; 0 < \theta < 2\pi, \; \theta \ne \pi$

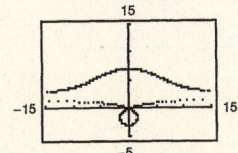

49. $r = 3 - 2 \cos \theta, \; 0 \le \theta < 2\pi$

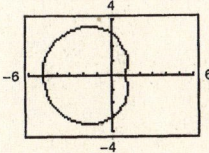

51. $r = 2 \cos\left(\dfrac{3\theta}{2}\right), \; 0 \le \theta < 4\pi$

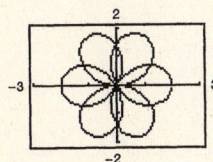

53. $r^2 = \sin 2\theta, \; 0 \le \theta < \dfrac{\pi}{2}$

$\left(\text{Use } r_1 = \sqrt{\sin 2\theta} \text{ and } r_2 = -\sqrt{\sin 2\theta}.\right)$

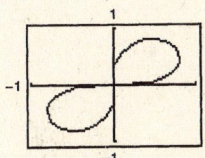

55. $r = 2 - \sec\theta$

$x = -1$ is an asymptote.

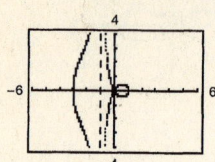

57. $r = \dfrac{2}{\theta}$

$y = 2$ is an asymptote.

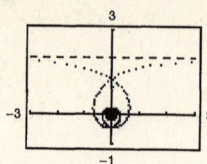

59. True. It has five petals.

61. $r = \cos(5\theta) + n\cos\theta,\ 0 \le \theta < \pi$

Answers will vary.

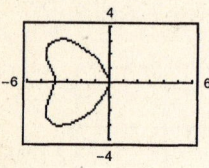

$n = -5$

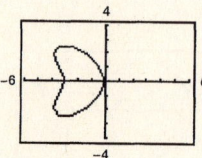

$n = -4$

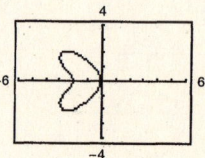

$n = -3$

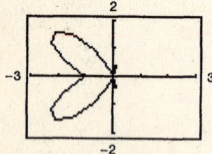

$n = -2$

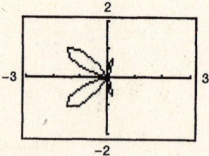

$n = -1$

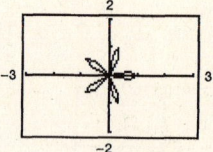

$n = 0$

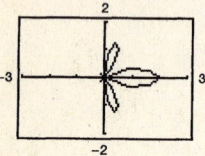

$n = 1$

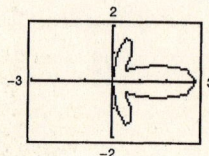

$n = 2$

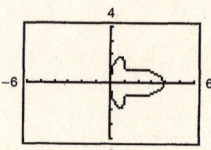

$n = 3$

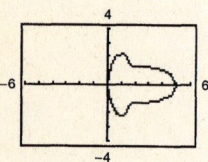

$n = 4$

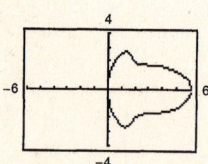

$n = 5$

63. Use the result of Exercise 62.

(a) Rotation: $\phi = \dfrac{\pi}{2}$

Original graph: $r = f(\sin \theta)$

Rotated graph: $r = f\left(\sin\left(\theta - \dfrac{\pi}{2}\right)\right) = f(-\cos \theta)$

(b) Rotation: $\phi = \pi$

Original graph: $r = f(\sin \theta)$

Rotated graph: $r = f(\sin(\theta - \pi)) = f(-\sin \theta)$

(c) Rotation: $\phi = \pi$

Original graph: $r = f(\sin \theta)$

Rotated graph: $r = f\left(\sin\left(\theta - \dfrac{3\pi}{2}\right)\right) = f(\cos \theta)$

65. (a) $r = 2 \sin\left[2\left(\theta - \dfrac{\pi}{6}\right)\right]$

$= 2 \sin\left(2\theta - \dfrac{\pi}{3}\right)$

$= \sin 2\theta - \sqrt{3}\cos 2\theta$

(b) $r = 2 \sin\left[2\left(\theta - \dfrac{\pi}{2}\right)\right]$

$= 2 \sin(2\theta - \pi)$

$= -2 \sin 2\theta$

$= -4 \sin \theta \cos \theta$

(c) $r = 2 \sin\left[2\left(\theta - \dfrac{2\pi}{3}\right)\right]$

$= 2 \sin\left(2\theta - \dfrac{4\pi}{3}\right)$

$= \sqrt{3}\cos 2\theta - \sin 2\theta$

(d) $r = 2 \sin[2(\theta - \pi)]$

$= 2 \sin(2\theta - 2\pi)$

$= 2 \sin 2\theta$

$= 4 \sin \theta \cos \theta$

67. $r = 2 + k \cos \theta$

$k = 0$	$k = 1$	$k = 2$	$k = 3$
Circle	Convex limaçon	Cardioid	Limaçon with inner loop

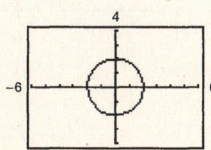   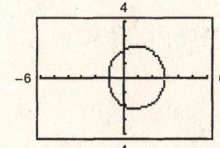   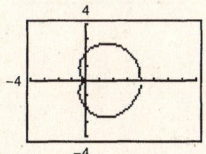   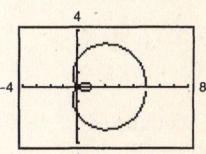

69. $f(x) = \dfrac{1}{2} \sin 2x$

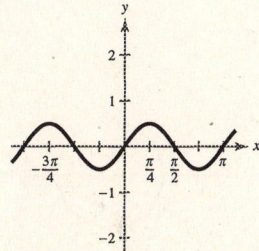

71. $g(x) = \tan \dfrac{x}{3}$

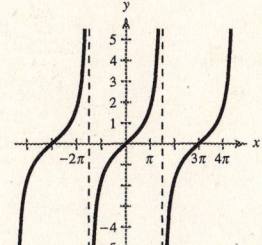

73. $h(x) = \sec\left(x + \dfrac{\pi}{4}\right)$

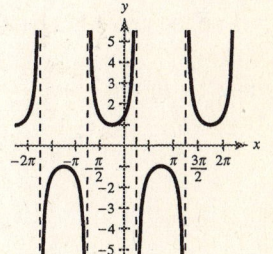

Section 9.8   Polar Equations of Conics

■ The graph of a polar equation of the form

$$r = \frac{ep}{1 \pm e \cos \theta} \quad \text{or} \quad r = \frac{ep}{1 \pm e \sin \theta}$$

is a conic, where $e > 0$ is the eccentricity and $|p|$ is the distance between the focus (pole) and the directrix.

(a) If $e < 1$, the graph is an ellipse.

(b) If $e = 1$, the graph is a parabola.

(c) If $e > 1$, the graph is a hyperbola.

■ Guidelines for finding polar equations of conics:

(a) Horizontal directrix above the pole: $r = \dfrac{ep}{1 + e \sin \theta}$

(b) Horizontal directrix below the pole: $r = \dfrac{ep}{1 - e \sin \theta}$

(c) Vertical directrix to the right of the pole: $r = \dfrac{ep}{1 + e \cos \theta}$

(d) Vertical directrix to the left of the pole: $r = \dfrac{ep}{1 - e \cos \theta}$

Solutions to Odd-Numbered Exercises

1. $r = \dfrac{2e}{1 + e \cos \theta}$

   (a) Parabola

   (b) Ellipse

   (c) Hyperbola

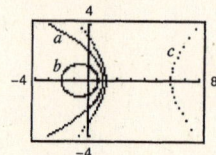

3. $r = \dfrac{2e}{1 - e \sin \theta}$

   (a) Parabola

   (b) Ellipse

   (c) Hyperbola

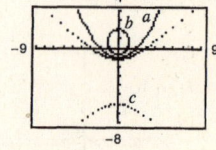

5. Matches (b).

   (Parabola, $e = 1$ and the directrix is vertical)

7. Matches (d).

   (Hyperbola, $e = 2$)

9. $r = \dfrac{3}{1 - \cos \theta}$

   $e = 1$ so the graph is a parabola.

   Vertex: $(r, \theta) = \left(\dfrac{3}{2}, \pi \right)$

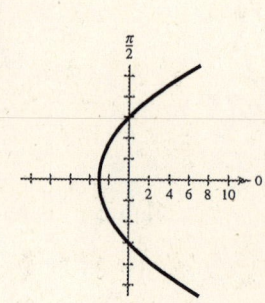

11. $r = \dfrac{4}{4 - \cos \theta} = \dfrac{1}{1 - (1/4) \cos \theta}$

   $e = \dfrac{1}{4}, p = 4$

   Ellipse

   Vertices:

   $(r, \theta) = \left(\dfrac{4}{3}, 0 \right), \left(\dfrac{4}{5}, \pi \right)$

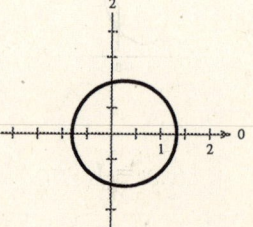

13. $r = \dfrac{8}{4 + 3 \sin \theta} = \dfrac{2}{1 + (3/4) \sin \theta}$

$e = \dfrac{3}{4}$

Ellipse

Vertices:

$(r, \theta) = \left(\dfrac{8}{7}, \dfrac{\pi}{2}\right), \left(8, \dfrac{3\pi}{2}\right)$

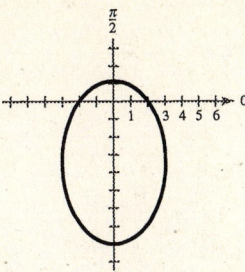

15. $r = \dfrac{6}{2 + \sin \theta} = \dfrac{(1/2)(6)}{1 + (1/2) \sin \theta}$

Ellipse, $\left(\theta = \dfrac{1}{2}\right)$

Vertices:

$\left(2, \dfrac{\pi}{2}\right), \left(6, \dfrac{3\pi}{2}\right)$

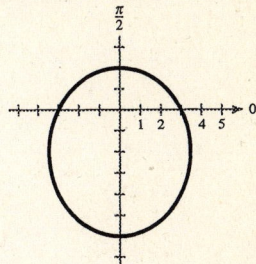

17. $r = \dfrac{3}{4 - 8 \cos \theta} = \dfrac{3/4}{1 - 2 \cos \theta}$

$e = 2$

Hyperbola

Vertices:

$(r, \theta) = \left(-\dfrac{3}{4}, 0\right), \left(\dfrac{1}{4}, \pi\right)$

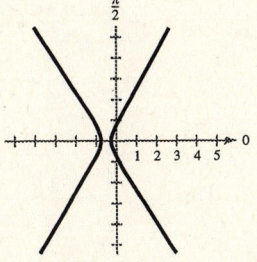

19. $r = \dfrac{-5}{1 - \sin \theta}$

Parabola

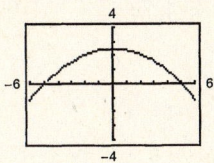

21. $r = \dfrac{14}{14 + 17 \sin \theta} = \dfrac{1}{1 + (17/14) \sin \theta}$

Hyperbola

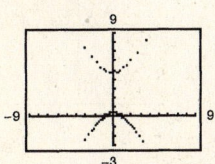

23. $r = \dfrac{3}{1 - \cos[\theta - (\pi/4)]}$

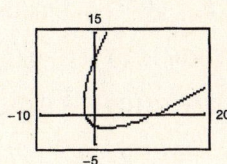

25. $r = \dfrac{8}{4 + 3 \sin[\theta + (\pi/6)]}$

Rotated ellipse

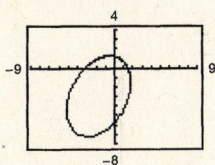

27. $e = 1, x = -1, p = 1$

Vertical directrix to the left of the pole

$r = \dfrac{1(1)}{1 - 1 \cos \theta} = \dfrac{1}{1 - \cos \theta}$

29. $e = \dfrac{1}{2}, y = 1, p = 1$

Horizontal directrix above the pole

$r = \dfrac{(1/2)(1)}{1 + (1/2) \sin \theta} = \dfrac{1}{2 + \sin \theta}$

31. $e = 2, x = 1, p = 1$

Vertical directrix to the right of the pole

$r = \dfrac{2(1)}{1 + 2 \cos \theta} = \dfrac{2}{1 + 2 \cos \theta}$

33. Vertex: $\left(1, -\dfrac{\pi}{2}\right) \Rightarrow e = 1, p = 2$

Horizontal directrix below the pole

$$r = \frac{1(2)}{1 - 1\sin\theta} = \frac{2}{1 - \sin\theta}$$

35. Center: $(4, \pi), c = 4, a = 6, e = \dfrac{2}{3}$

Vertical directrix to the right of the pole

$$r = \frac{(2/3)p}{1 + (2/3)\cos\theta} = \frac{2p}{3 + 2\cos\theta}$$

$$2 = \frac{2p}{3 + 2\cos\theta} = \frac{2p}{5} \Rightarrow p = 5$$

$$r = \frac{10}{3 + 2\cos\theta}$$

37. Center: $\left(5, \dfrac{3\pi}{2}\right); c = 5, a = 4, e = \dfrac{5}{4}$

Horizontal directrix below the pole

$$r = \frac{(5/4)p}{1 - (5/4)\sin\theta} = \frac{5p}{4 - 5\sin\theta}$$

$$1 = \frac{5p}{4 - 5\sin(3\pi/2)}$$

$$p = \frac{9}{5}$$

$$r = \frac{5(9/5)}{4 - 5\sin\theta} = \frac{9}{4 - 5\sin\theta}$$

39. When $\theta = 0, r = c + a = ea + a = a(1 + e)$.

Therefore,

$$a(1 + e) = \frac{ep}{1 - e\cos 0}$$

$$a(1 + e)(1 - e) = ep$$

$$a(1 - e^2) = ep.$$

Thus, $r = \dfrac{ep}{1 - e\cos\theta} = \dfrac{(1 - e^2)a}{1 - e\cos\theta}.$

41. $r = \dfrac{[1 - (0.0167)^2](92.956 \times 10^6)}{1 - 0.0167\cos\theta} \approx \dfrac{9.2930 \times 10^7}{1 - 0.0167\cos\theta}$

Perihelion distance: $r = 92.956 \times 10^6(1 - 0.0167) \approx 9.1404 \times 10^7$

Aphelion distance: $r = 92.956 \times 10^6(1 + 0.0167) \approx 9.4508 \times 10^7$

43. $r = \dfrac{(1 - 0.0484^2)77.841 \times 10^7}{1 - 0.0484\cos\theta} = \dfrac{7.7659 \times 10^8}{1 - 0.0484\cos\theta}$

Perihelion: $r = 77.841 \times 10^7(1 - 0.0484) \approx 7.4073 \times 10^8$ km

Aphelion: $r = 77.841 \times 10^7(1 + 0.0484) \approx 8.1609 \times 10^8$ km

45. $a = 4.498 \times 10^9, e = 0.0086$, Neptune

$a = 5.906 \times 10^9, e = 0.2488$, Pluto

(a) Neptune: $r = \dfrac{(1 - 0.0086^2)4.498 \times 10^9}{1 - 0.0086\cos\theta} = \dfrac{4.4977 \times 10^9}{1 - 0.0086\cos\theta}$

Pluto: $r = \dfrac{(1 - 0.2488^2)5.906 \times 10^9}{1 - 0.2488\cos\theta} = \dfrac{5.5404 \times 10^9}{1 - 0.2488\cos\theta}$

(b) Neptune: Perihelion: $4.498 \times 10^9(1 - 0.0086) \approx 4.4593 \times 10^9$ km

Aphelion: $4.498 \times 10^9(1 + 0.0086) \approx 4.5367 \times 10^9$ km

Pluto: Perihelion: $5.906 \times 10^9(1 - 0.2488) \approx 4.4366 \times 10^9$ km

Aphelion: $5.906 \times 10^9(1 + 0.2488) \approx 7.3754 \times 10^9$ km

—CONTINUED—

45. —CONTINUED—

(c)

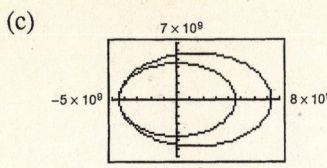

(d) Although the graphs intersect, three orbits do not, and the planets won't collide.

(e) Yes. Pluto is closer to the sun for just a very short time.

47. $r = \dfrac{4}{-3 - 3 \sin \theta} = \dfrac{-4/3}{1 + \sin \theta}$

False. The directrix is below the pole.

49.
$$\frac{x^2}{a^2} + \frac{y^2}{b^2} = 1$$

$$\frac{r^2 \cos^2 \theta}{a^2} + \frac{r^2 \sin^2 \theta}{b^2} = 1$$

$$\frac{r^2 \cos^2 \theta}{a^2} + \frac{r^2(1 - \cos^2 \theta)}{b^2} = 1$$

$$r^2 b^2 \cos^2 \theta + r^2 a^2 - r^2 a^2 \cos^2 \theta = a^2 b^2$$

$$r^2(b^2 - a^2) \cos^2 \theta + r^2 a^2 = a^2 b^2$$

For an ellipse, $b^2 - a^2 = -c^2$. Hence,

$$-r^2 c^2 \cos^2 \theta + r^2 a^2 = a^2 b^2$$

$$-r^2 \left(\frac{c}{a}\right)^2 \cos^2 \theta + r^2 = b^2, \ e = \frac{c}{a}$$

$$-r^2 e^2 \cos^2 \theta + r^2 = b^2$$

$$r^2(1 - e^2 \cos^2 \theta) = b^2$$

$$r^2 = \frac{b^2}{1 - e^2 \cos^2 \theta}.$$

51. $\dfrac{x^2}{169} + \dfrac{y^2}{144} = 1$

$a = 13, b = 12, c = 5, e = \dfrac{5}{13}$

$r^2 = \dfrac{144}{1 - (25/169) \cos^2 \theta} = \dfrac{24{,}336}{169 - 25 \cos^2 \theta}$

53. $r = \dfrac{4}{1 - 0.4 \cos \theta}$

Vertical directrix to left of pole

(a) $e = 0.4 \implies$ ellipse

(c)

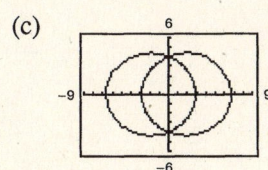 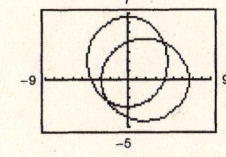

(b) $r = \dfrac{4}{1 + 0.4 \cos \theta}$

Vertical directrix to right of pole

Graph is reflected in line $\theta = \pi/2$.

$r = \dfrac{4}{1 - 0.4 \sin \theta}$

Horizontal directrix below pole

90° rotation counterclockwise

55. Answers will vary.

57. $4\sqrt{3}\tan\theta - 3 = 1$

$$\tan\theta = \frac{1}{\sqrt{3}} = \frac{\sqrt{3}}{3}$$

$$\theta = \frac{\pi}{6} + n\pi$$

59. $12\sin^2\theta = 9$

$$\sin^2\theta = \frac{3}{4}$$

$$\sin\theta = \pm\frac{\sqrt{3}}{2}$$

$$\theta = \frac{\pi}{3} + n\pi, \frac{2\pi}{3} + n\pi$$

For Exercises 61 and 63: $\sin u = -\frac{3}{5}$, $\cos u = \frac{4}{5}$, $\cos v = \frac{1}{\sqrt{2}}$, $\sin v = -\frac{1}{\sqrt{2}}$

61. $\cos(u + v) = \cos u \cos v - \sin u \sin v$

$$= \frac{4}{5}\left(\frac{1}{\sqrt{2}}\right) - \left(-\frac{3}{5}\right)\left(-\frac{1}{\sqrt{2}}\right)$$

$$= \frac{1}{5\sqrt{2}}$$

$$= \frac{\sqrt{2}}{10}$$

63. $\sin(u - v) = \sin u \cos v - \sin v \cos u$

$$= \left(-\frac{3}{5}\right)\left(\frac{1}{\sqrt{2}}\right) - \left(-\frac{1}{\sqrt{2}}\right)\left(\frac{4}{5}\right)$$

$$= \frac{1}{5\sqrt{2}}$$

$$= \frac{\sqrt{2}}{10}$$

65. $_{12}C_9 = 220$

67. $_{10}P_3 = 720$

Review Exercises for Chapter 9

Solutions to Odd-Numbered Exercises

1. Hyperbola

3. Vertex: $(0, 0)$, Focus: $(-6, 0)$

Parabola opens to left.

$$y^2 = 4px$$

$$y^2 = 4(-6)x = -24x$$

5. Vertex: $(0, 2) = (h, k)$

Directrix: $x = -3 \Rightarrow p = 3$

$$(y - k)^2 = 4p(x - h)$$

$$(y - 2)^2 = 12x$$

7. $x^2 = -2y = 4\left(-\frac{1}{2}\right)y, p = -\frac{1}{2}$

Focus: $\left(0, -\frac{1}{2}\right)$

$$d_1 = \frac{1}{2} + b$$

$$d_2 = \sqrt{(2 - 0)^2 + \left(-2 + \frac{1}{2}\right)^2} = \frac{5}{2}$$

$$d_1 = d_2 \Rightarrow \frac{1}{2} + b = \frac{5}{2} \Rightarrow b = 2$$

Slope of tangent line: $\dfrac{b + 2}{0 - 2} = \dfrac{4}{-2} = -2$

Equation: $y + 2 = -2(x - 2)$

$$y = -2x + 2$$

x-intercept: $(1, 0)$

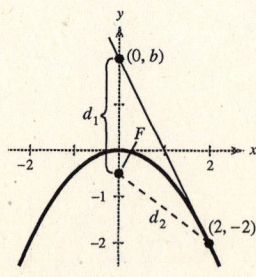

9. $x^2 = 4p(y - 12)$; $(4, 10)$ on curve:

$16 = 4p(10 - 12) = -8p \Rightarrow p = -2$

$x^2 = 4(-2)(y - 12) = -8y + 96$

$y = \dfrac{-x^2 + 96}{8}$

$y = 0$ if $x^2 = 96 \Rightarrow x = 4\sqrt{6}$

$\Rightarrow$ width is $8\sqrt{6}$ meters.

11. Vertices: $(\pm 5, 0)$

Foci: $(\pm 4, 0)$

$a = 5, c = 4 \Rightarrow b = 3$

$\dfrac{x^2}{25} + \dfrac{y^2}{9} = 1$

13. Vertices: $(0, 1), (4, 1)$

Endpoints of minor axis: $(2, 0), (2, 2)$

Horizontal major axis

Center: $(2, 1)$

$a = 2, b = 1$

$\dfrac{(x - h)^2}{a^2} + \dfrac{(y - k)^2}{b^2} = 1$

$\dfrac{(x - 2)^2}{4} + \dfrac{(y - 1)^2}{1} = 1$

15. $a = 5, b = 4, c = \sqrt{a^2 - b^2} = \sqrt{25 - 16} = 3$

The foci should be placed three feet on either side of the center and have the same height as the pillars.

17. $16(x^2 - 2x + 1) + 9(y^2 + 8y + 16) = -16 + 16 + 144$

$16(x - 1)^2 + 9(y + 4)^2 = 144$

$\dfrac{(x - 1)^2}{9} + \dfrac{(y + 4)^2}{16} = 1$

Center: $(1, -4)$

$a = 4, b = 3, c = \sqrt{16 - 9} = \sqrt{7}$

Vertices: $(1, 0), (1, -8)$

Foci: $\left(1, -4 \pm \sqrt{7}\right)$

$e = \dfrac{c}{a} = \dfrac{\sqrt{7}}{4}$

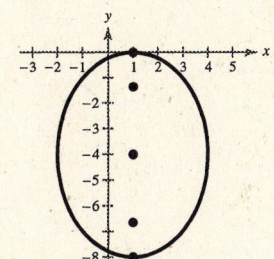

19. (a) $3(x^2 + 4x + 4) + 8(y^2 - 14y + 49) = -403 + 12 + 392$

$$3(x + 2)^2 + 8(y - 7)^2 = 1$$

$$\frac{(x + 2)^2}{1/3} + \frac{(y - 7)^2}{1/8} = 1$$

(b) Center: $(-2, 7)$

(c), (d)

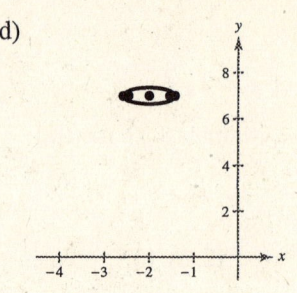

$$a = \frac{\sqrt{3}}{3}, b = \frac{\sqrt{2}}{4}$$

$$c^2 = a^2 - b^2 = \frac{1}{3} - \frac{1}{8} = \frac{5}{24} \implies c = \frac{\sqrt{30}}{12}$$

Vertices: $\left(-2 \pm \frac{\sqrt{3}}{3}, 7\right)$

Foci: $\left(-2 \pm \frac{\sqrt{30}}{12}, 7\right)$

Eccentricity: $\frac{c}{a} = \frac{\sqrt{30}/12}{\sqrt{3}/3} = \frac{\sqrt{10}}{4}$

21. Vertices: $(\pm 4, 0)$, $a = 4$

Foci: $(\pm 6, 0)$, $c = 6$

$b^2 = c^2 - a^2 = 20 \implies b = 2\sqrt{5}$

$$\frac{x^2}{16} - \frac{y^2}{20} = 1$$

23. Foci: $(0, 0)(8, 0) \implies c = 4$

Center: $(4, 0)$

Asymptotes:

$y = \pm 2(x - 4) \implies \frac{b}{a} = 2 \implies b = 2a$

$c^2 = a^2 + b^2$

$16 = a^2 + (2a)^2 = 5a^2 \implies a = \frac{4}{\sqrt{5}}, b = \frac{8}{\sqrt{5}}$

$$\frac{(x - 4)^2}{\frac{16}{5}} - \frac{y^2}{\frac{64}{5}} = 1$$

25. $9(x^2 - 2x + 1) - 16(y^2 + 2y + 1) = 151 + 9 - 16$

$$9(x - 1)^2 - 16(y + 1)^2 = 144$$

$$\frac{(x - 1)^2}{16} - \frac{(y + 1)^2}{9} = 1$$

Center: $(1, -1)$, $a = 4$, $b = 3$, $c = 5$

Vertices: $(5, -1), (-3, -1)$

Foci: $(6, -1), (-4, -1)$

Eccentricity: $\frac{5}{4}$

Asymptotes: $y = -1 \pm \frac{3}{4}(x - 1)$

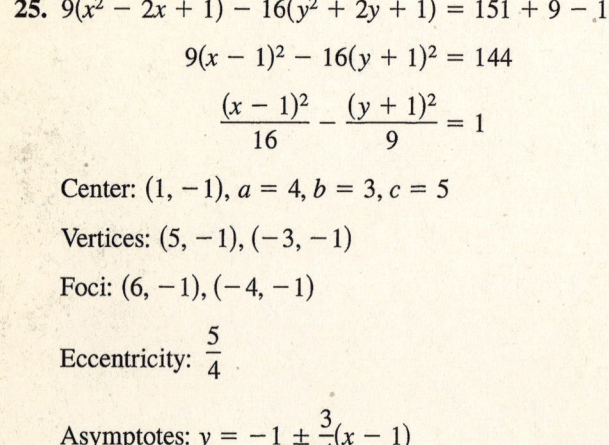

27. (a) $(y^2 - 2y + 1) - 4(x^2 + 12x + 36) = -59 + 1 - 144$

$$(y - 1)^2 - 4(x + 6)^2 = -202$$

$$\frac{(x + 6)^2}{(101/2)} - \frac{(y - 1)^2}{202} = 1$$

(b) Center: $(-6, 1)$

$a^2 = \dfrac{101}{2}, b^2 = 202, c^2 = \dfrac{101}{2} + 202 = \dfrac{505}{2}$

Vertices: $\left(-6 \pm \sqrt{\dfrac{101}{2}}, 1 \right)$

Foci: $\left(-6 \pm \sqrt{\dfrac{505}{2}}, 1 \right)$

Asymptotes: $y = 1 \pm 2(x + 6)$

Eccentricity: $e = \dfrac{c}{a} = \dfrac{\sqrt{505}}{\sqrt{101}} = \sqrt{5}$

(c), (d)

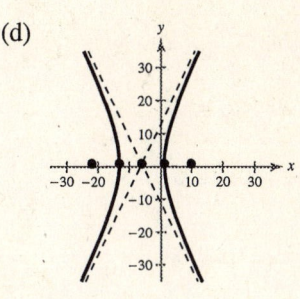

29. $d_2 - d_1 = 186,000(0.0005)$

$2a = 93$

$a = 46.5$

$c = 100$

$b = \sqrt{c^2 - a^2}$

$\dfrac{x^2}{a^2} - \dfrac{y^2}{b^2} = 1$

$x = 60 \implies y^2 = b^2\left(\dfrac{x^2}{a^2} - 1 \right) = (100^2 - 46.5^2)\left(\dfrac{60^2}{46.5^2} - 1 \right) \approx 5211.57 \implies y \approx 72.2$

72.2 miles north

31. $3x^2 + 2y^2 - 12x + 12y + 29 = 0$

$3(x^2 - 4x + 4) + 2(y^2 + 6y + 9) = -29 + 12 + 18$

$$3(x - 2)^2 + 2(y + 3)^2 = 1$$

Ellipse

33. $5x^2 - 2y^2 + 10x - 4y + 17 = 0$

$5(x^2 + 2x + 1) - 2(y^2 + 2y + 1) = -17 + 5 - 2$

$$5(x + 1)^2 - 2(y + 1)^2 = -14$$

$$\frac{(y + 1)^2}{7} - \frac{(x + 1)^2}{(14/5)} = 1$$

Hyperbola

35. $xy - 4 = 0$

$A = 0, B = 1, C = 0$

$\cot 2\theta = \dfrac{A - C}{B} = 0 \implies \theta = \dfrac{\pi}{4}$

$x = \dfrac{\sqrt{2}}{2}(x' - y'), y = \dfrac{\sqrt{2}}{2}(x' + y')$

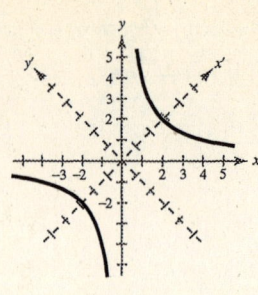

$$xy = 4$$

$$\frac{\sqrt{2}}{2}(x' - y')\frac{\sqrt{2}}{2}(x' + y') = 4$$

$$\frac{1}{2}(x')^2 - \frac{1}{2}(y')^2 = 4$$

$$\frac{(x')^2}{8} - \frac{(y')^2}{8} = 1$$

Hyperbola

37. $5x^2 - 2xy + 5y^2 - 12 = 0$

$A = 5, B = -2, C = 5$

$\cot 2\theta = 0 \implies \theta = \dfrac{\pi}{4}$

$x = \dfrac{\sqrt{2}}{2}(x' - y'), y = \dfrac{\sqrt{2}}{2}(x' + y')$

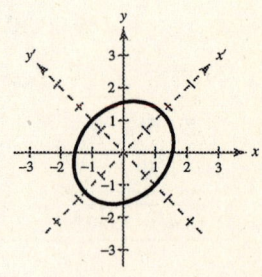

$$5x^2 - 2xy + 5y^2 = 12$$

$$5\left[\frac{\sqrt{2}}{2}(x' - y')\right]^2 - 2\left[\frac{\sqrt{2}}{2}(x' - y')\right]\left[\frac{\sqrt{2}}{2}(x' + y')\right] + 5\left[\frac{\sqrt{2}}{2}(x' + y')\right]^2 = 12$$

$$5\left[\frac{1}{2}(x')^2 - x'y' + \frac{1}{2}(y')^2\right] - (x')^2 + (y')^2 + 5\left[\frac{1}{2}(x')^2 + x'y' + \frac{1}{2}(y')^2\right] = 12$$

$$4(x')^2 + 6(y')^2 = 12$$

$$\frac{(x')^2}{3} + \frac{(y')^2}{2} = 1$$

Ellipse

39. (a) $B^2 - 4AC = (-8)^2 - 4(16)(1) = 0$

Parabola

(b) $y^2 + (5 - 8x)y + (16x^2 - 10x) = 0$

$$y = \frac{(8x - 5) \pm \sqrt{(5 - 8x)^2 - 4(16x^2 - 10x)}}{2}$$

(c)

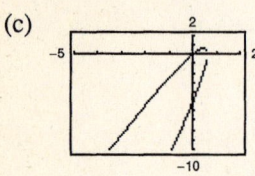

41. (a) $B^2 - 4AC = (2)^2 - 4(1)(1) = 0$

Parabola

(b) $y^2 + (2x - 2\sqrt{2})y + (x^2 + 2\sqrt{2}x + 2) = 0$

$$y = \frac{(2\sqrt{2} - 2x) \pm \sqrt{(2x - 2\sqrt{2})^2 - 4(x^2 + 2\sqrt{2}x + 2)}}{2}$$

(c)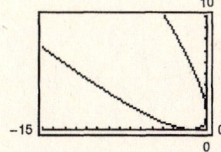

43. Adding the equations, $24x + 240 = 0 \implies x = -10$. Then:

$$4(100) + y^2 - 560 - 24y + 304 = 0$$

$$y^2 - 24y + 144 = 0$$

$$(y - 12)^2 = 0 \implies y = 12$$

Solution: $(-10, 12)$

45.

t	-2	-1	0	1	2	3
x	-8	-5	-2	1	4	7
y	15	11	7	3	-1	-5

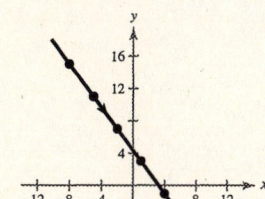

47. $x = 1 + 4t, y = 2 - 3t$

$$t = \frac{x - 1}{4}$$

$$y = 2 - 3\left(\frac{x - 1}{4}\right)$$

$$3x + 4y = 11$$

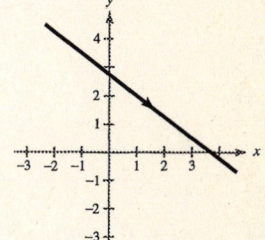

49. $x = \dfrac{1}{t}, y = t^2$

$$t = \frac{1}{x}$$

$$y = \frac{1}{x^2}$$

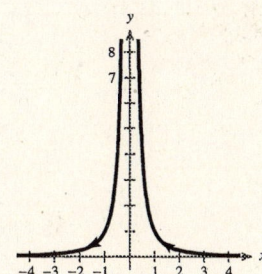

51. $x = 6\cos\theta, y = 6\sin\theta$

$$\cos\theta = \frac{x}{6}, \sin\theta = \frac{y}{6}$$

$$\frac{x^2}{36} + \frac{y^2}{36} = 1$$

$$x^2 + y^2 = 36$$

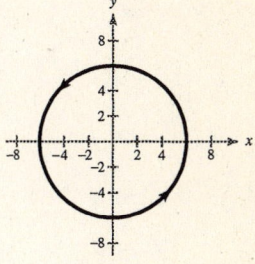

53. $x = \sec\theta, y = \tan\theta$

$$\tan^2\theta + 1 = \sec^2\theta$$

$$y^2 + 1 = x^2$$

$$x^2 - y^2 = 1$$

Hyperbola

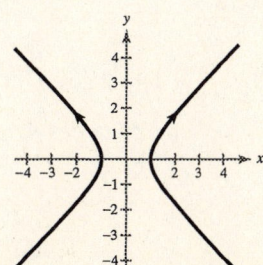

55.

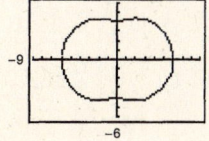

57. Answers will vary.

$x = t, y = t^2 + 2$

$x = 2t, y = 4t^2 + 2$

59. $\left(1, \dfrac{\pi}{4}\right)$

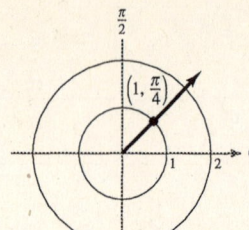

$\left(1, -\dfrac{7\pi}{4}\right), \left(-1, \dfrac{5\pi}{4}\right), \left(-1, -\dfrac{3\pi}{4}\right)$

61. $\left(\sqrt{5}, -\dfrac{2\pi}{3}\right), \left(-\sqrt{5}, \dfrac{\pi}{3}\right),$

$\left(-\sqrt{5}, -\dfrac{5\pi}{3}\right)$

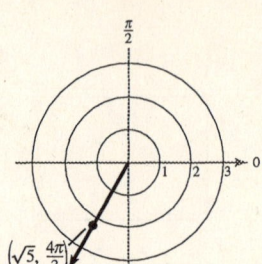

63. $(r, \theta) = \left(5, -\dfrac{7\pi}{6}\right)$

$(x, y) = \left(-\dfrac{5\sqrt{3}}{2}, \dfrac{5}{2}\right)$

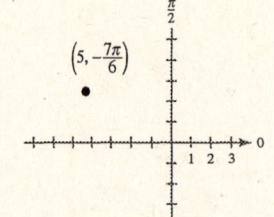

65. $(r, \theta) = \left(3, \dfrac{3\pi}{4}\right)$

$(x, y) = \left(3 \cos \dfrac{3\pi}{4}, 3 \sin \dfrac{3\pi}{4}\right) = \left(\dfrac{-3\sqrt{2}}{2}, \dfrac{3\sqrt{2}}{2}\right)$

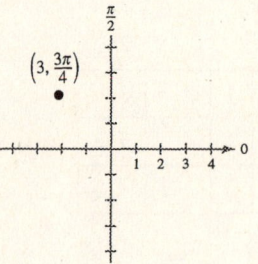

67. $(x, y) = (0, -9)$

$(r, \theta) = \left(9, \dfrac{3\pi}{2}\right), \left(-9, \dfrac{\pi}{2}\right)$

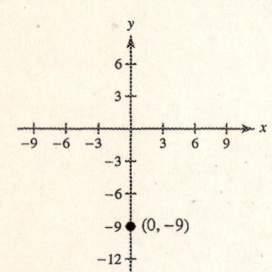

69. $(x, y) = (5, -5)$

$(r, \theta) = \left(5\sqrt{2}, \dfrac{7\pi}{4}\right), \left(-5\sqrt{2}, \dfrac{3\pi}{4}\right)$

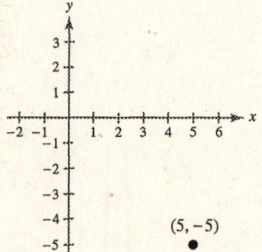

71. $x^2 + y^2 = 9$

$r^2 = 9$

$r = 3$

73. $x^2 + y^2 - 4x = 0$

$r^2 - 4r \cos \theta = 0$

$r = 4 \cos \theta$

75. $xy = 5$

$(r \cos \theta)(r \sin \theta) = 5$

$r^2 = 5 \csc \theta \cdot \sec \theta$

77. $r = 5$

$x^2 + y^2 = 5^2 = 25$

Circle

79. $r = 3 \cos \theta$

$r^2 = 3r \cos \theta$

$x^2 + y^2 = 3x$

81.
$$r^2 = \cos 2\theta$$
$$r^2 = 1 - 2\sin^2\theta$$
$$r^4 = r^2 - 2r^2\sin^2\theta$$
$$(x^2 + y^2)^2 = x^2 + y^2 - 2y^2$$
$$(x^2 + y^2)^2 - x^2 + y^2 = 0$$

83. $r = 5$, circle

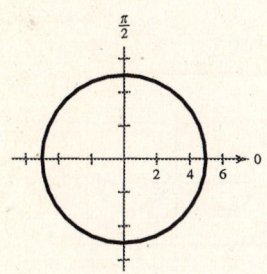

85. $\theta = \dfrac{\pi}{2}$, y-axis

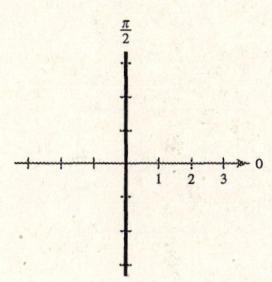

87. $r = 5\cos\theta$, circle

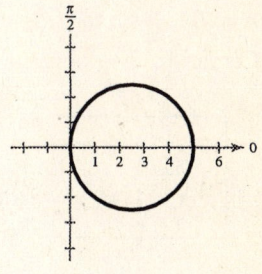

89. $r = 5 + 4\cos\theta$

Dimpled limaçon

Symmetric with respect to polar axis

r is maximum at $\theta = 0$: $(r, \theta) = (9, 0)$

$r \neq 0$ (No zeros)

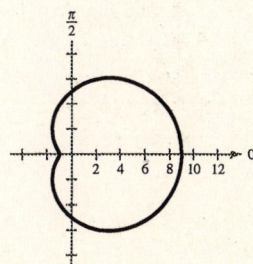

91. $r = 3 - 5\sin\theta$

Limaçon with loop

Symmetry: line $\theta = \dfrac{\pi}{2}$

Maximum $|r|$-value: $|r| = 8$ when $\theta = \dfrac{3\pi}{2}$.

Zeros: $r = 0$ when $\theta \approx 0.6435, 2.4981$ $\left(\sin\theta = \dfrac{3}{5}\right)$.

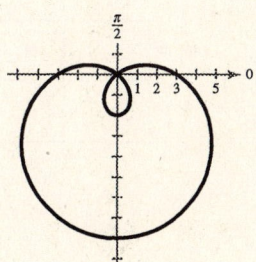

93. $r = -3\cos 2\theta$, $0 \leq \theta \leq 2\pi$

4-leaved rose

Symmetric with respect to $\theta = \dfrac{\pi}{2}$, polar axis, and pole

The value of $|r|$ is a maximum (3) at $\theta = 0, \dfrac{\pi}{2}, \pi, \dfrac{3\pi}{2}$.

$r = 0$ for $\theta = \dfrac{\pi}{4}, \dfrac{3\pi}{4}, \dfrac{5\pi}{4}, \dfrac{7\pi}{4}$

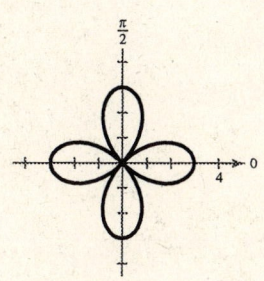

95. $r^2 = 5 \sin 2\theta$

Lemniscate

Symmetry with respect to pole

Maximum $|r|$-value: $\sqrt{5}$ when $\theta = \dfrac{\pi}{4}, \dfrac{5\pi}{4}$.

Zeros: $r = 0$ when $\theta = 0, \dfrac{\pi}{2}, \pi, \dfrac{3\pi}{2}$.

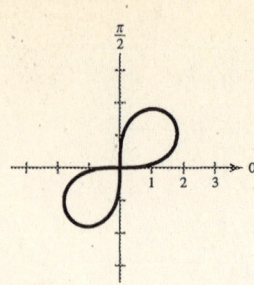

97. $r = 7(1 - \sin \theta)$

Cardioid

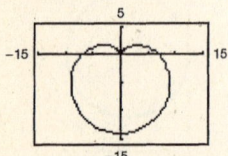

99. $r = 4 \sin 5\theta$

5-leaved rose

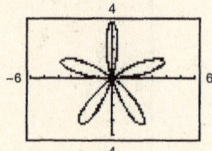

101. $r^2 = 8 \cos 2\theta$

Lemniscate

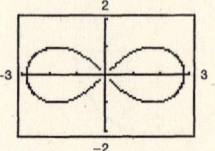

103. $r = \dfrac{2}{1 - \sin \theta}$

$e = 1$

Parabola

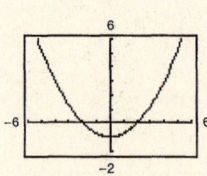

105. $r = \dfrac{4}{5 - 3 \cos \theta}$

$\quad = \dfrac{4/5}{1 - (3/5) \cos \theta}$

$e = \dfrac{3}{5}$

Ellipse

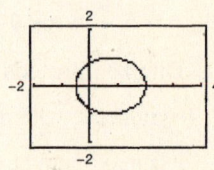

107. $r = \dfrac{5}{6 + 2 \sin \theta}$

$\quad = \dfrac{5/6}{1 + (1/3) \sin \theta}$

$e = \dfrac{1}{3}$

Ellipse

109. $e = 1$

$r = \dfrac{4}{1 - \cos \theta}$

Vertical directrix: $x = -4$

111. Ellipse: $r = \dfrac{ep}{1 - e \cos \theta}$

Vertices: $(5, 0), (1, \pi) \implies a = 3$

One focus: $(0, 0) \implies c = 2$

$e = \dfrac{c}{a} = \dfrac{2}{3}, p = \dfrac{5}{2}$

$r = \dfrac{(2/3)(5/2)}{1 - (2/3) \cos \theta} = \dfrac{5/3}{1 - (2/3) \cos \theta}$

$\quad = \dfrac{5}{3 - 2 \cos \theta}$

113. Hyperbola

Vertices: $(2, 0), (8, 0)$

$a = 3, c = 5, e = \dfrac{5}{3}$

Vertical directrix to right of pole: $p = 2 + \dfrac{6}{5} = \dfrac{16}{5}$

$r = \dfrac{ep}{1 + e \cos \theta} = \dfrac{(5/3)(16/5)}{1 + (5/3) \cos \theta}$

$r = \dfrac{16}{3 + 5 \cos \theta}$

115. $e = 0.093$

Use $r = \dfrac{ep}{1 - e \cos \theta}$.

$2a = \dfrac{0.093p}{1 - 0.093 \cos 0} + \dfrac{0.093p}{1 - 0.093 \cos \pi} = 0.1876p = 3.05 \Rightarrow p \approx 16.258, \; ep \approx 1.512$

$r = \dfrac{1.512}{1 - 0.093 \cos \theta}$

Perihelion: $\dfrac{1.512}{1 + 0.093} \approx 1.383$ astronomical units

Aphelion: $\dfrac{1.512}{1 - 0.093} \approx 1.667$ astronomical units

117. False. The y^4-term is not second degree.

119. (a) Vertical translation

(b) Horizontal translation

(c) Reflection in the y-axis

(d) Parabola opens more slowly

121. The number b must be less than 5. The ellipse becomes more circular and approaches a circle of radius 5.

123. The orientation of the graph would be reversed.

Chapter 9   Practice Test

1. Find the vertex, focus and directrix of the parabola $x^2 - 6x - 4y + 1 = 0$.

2. Find an equation of the parabola with its vertex at $(2, -5)$ and focus at $(2, -6)$.

3. Find the center, foci, vertices, and eccentricity of the ellipse $x^2 + 4y^2 - 2x + 32y + 61 = 0$.

4. Find an equation of the ellipse with vertices $(0, \pm 6)$ and eccentricity $e = \frac{1}{2}$.

5. Find the center, vertices, foci, and asymptotes of the hyperbola $16y^2 - x^2 - 6x - 128y + 231 = 0$.

6. Find an equation of the hyperbola with vertices at $(\pm 3, 2)$ and foci at $(\pm 5, 2)$.

7. Rotate the axes to eliminate the xy-term. Sketch the graph of the resulting equation, showing both sets of axes.

   $5x^2 + 2xy + 5y^2 - 10 = 0$

8. Use the discriminant to determine whether the graph of the equation is a parabola, ellipse, or hyperbola.

   (a) $6x^2 - 2xy + y^2 = 0$ (b) $x^2 + 4xy + 4y^2 - x - y + 17 = 0$

For Exercises 9 and 10, eliminate the parameter and write the corresponding rectangular equation.

9. $x = 3 - 2\sin\theta, y = 1 + 5\cos\theta$ 10. $x = e^{2t}, y = e^{4t}$

11. Convert the polar point $\left(\sqrt{2}, (3\pi)/4\right)$ to rectangular coordinates.

12. Convert the rectangular point $\left(\sqrt{3}, -1\right)$ to polar coordinates.

13. Convert the rectangular equation $4x - 3y = 12$ to polar form.

14. Convert the polar equation $r = 5\cos\theta$ to rectangular form.

15. Sketch the graph of $r = 1 - \cos\theta$.

16. Sketch the graph of $r = 5\sin 2\theta$.

17. Sketch the graph of $r = \dfrac{3}{6 - \cos\theta}$.

18. Find a polar equation of the parabola with its vertex at $(6, \pi/2)$ and focus at $(0, 0)$.

CHAPTER 10
Analytic Geometry in Three Dimensions

CHAPTER 10
Analytic Geometry in Three Dimensions

Section 10.1   The Three-Dimensional Coordinate System

- You should be able to plot points in the three-dimensional coordinate system.
- The distance between the points (x_1, y_1, z_1) and (x_2, y_2, z_2) is
$$d = \sqrt{(x_2 - x_1)^2 + (y_2 - y_1)^2 + (z_2 - z_1)^2}.$$
- The midpoint of the line segment joining the points (x_1, y_1, z_1) and (x_2, y_2, z_2) is
$$\left(\frac{x_1 + x_2}{2}, \frac{y_1 + y_2}{2}, \frac{z_1 + z_2}{2}\right).$$
- The equation of the sphere with center (h, k, j) and radius r is
$$(x - h)^2 + (y - k)^2 + (z - j)^2 = r^2.$$
- You should be able to find the trace of a surface in space.

Solutions to Odd-Numbered Exercises

1. $A(-1, 4, 3), B(1, 3, -2), C(-3, 0, -2)$

3.

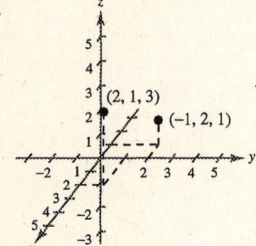

5.

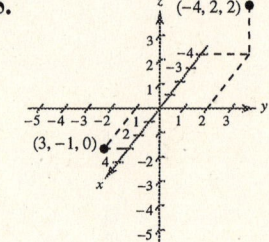

7. $x = -3, y = 3, z = 4$: $(-3, 3, 4)$

9. $y = z = 0, x = 10$: $(10, 0, 0)$

11. Octant IV

13. Octants I, II, III, IV (above the xy-plane)

15. Octants II, IV, VI, VIII

17. $d = \sqrt{(7 - 3)^2 + (4 - 2)^2 + (8 - 5)^2}$

$= \sqrt{4^2 + 2^2 + 3^2}$

$= \sqrt{16 + 4 + 9}$

$= \sqrt{29}$

≈ 5.385

19. $d = \sqrt{[6 - (-1)]^2 + [0 - 4]^2 + [-9 - (-2)]^2}$

$= \sqrt{7^2 + 4^2 + 7^2}$

$= \sqrt{49 + 16 + 49}$

$= \sqrt{114}$

≈ 10.677

21. $d = \sqrt{(1-0)^2 + [0-(-3)]^2 + (-10-0)^2}$

$\quad = \sqrt{1+9+100}$

$\quad = \sqrt{110} \approx 10.488$

23. $d_1 = \sqrt{(-2-0)^2 + (5-0)^2 + (2-2)^2} = \sqrt{4+25} = \sqrt{29}$

$\quad d_2 = \sqrt{(0-0)^2 + (4-0)^2 + (0-2)^2} = \sqrt{16+4} = \sqrt{20} = 2\sqrt{5}$

$\quad d_3 = \sqrt{(0+2)^2 + (4-5)^2 + (0-2)^2} = \sqrt{4+1+4} = \sqrt{9} = 3$

$\quad d_1{}^2 = d_2{}^2 + d_3{}^2 = 29$

25. $d_1 = \sqrt{(2-0)^2 + (2-0)^2 + (1-0)^2} = \sqrt{4+4+1} = \sqrt{9} = 3$

$\quad d_2 = \sqrt{(2-0)^2 + (-4-0)^2 + (4-0)^2} = \sqrt{4+16+16} = \sqrt{36} = 6$

$\quad d_3 = \sqrt{(2-2)^2 + (-4-2)^2 + (4-1)^2} = \sqrt{36+9} = \sqrt{45} = 3\sqrt{5}$

$\quad d_1{}^2 + d_2{}^2 = 9 + 36 = 45 = d_3{}^2$

27. $d_1 = \sqrt{(5-1)^2 + (-1+3)^2 + (2+2)^2} = \sqrt{16+4+16} = \sqrt{36} = 6$

$\quad d_2 = \sqrt{(5+1)^2 + (-1-1)^2 + (2-2)^2} = \sqrt{36+4} = \sqrt{40} = 2\sqrt{10}$

$\quad d_3 = \sqrt{(-1-1)^2 + (1+3)^2 + (2+2)^2} = \sqrt{4+16+16} = \sqrt{36} = 6$

$\quad d_1 = d_3$ Isosceles triangle

29. Midpoint: $\left(\dfrac{3-3}{2}, \dfrac{-6+4}{2}, \dfrac{10+4}{2}\right) = (0, -1, 7)$ **31.** Midpoint: $\left(\dfrac{6-4}{2}, \dfrac{-2+2}{2}, \dfrac{5+6}{2}\right) = \left(1, 0, \dfrac{11}{2}\right)$

33. Midpoint: $\left(\dfrac{-2+7}{2}, \dfrac{8-4}{2}, \dfrac{10+2}{2}\right) = \left(\dfrac{5}{2}, 2, 6\right)$ **35.** $(x-3)^2 + (y-2)^2 + (z-4)^2 = 16$

37. $(x-0)^2 + (y-4)^2 + (z-3)^2 = 3^2$ **39.** Radius $= \dfrac{\text{Diameter}}{2} = 5$

$\qquad x^2 + (y-4)^2 + (z-3)^2 = 9$

$\qquad\qquad\qquad (x+3)^2 + (y-7)^2 + (z-5)^2 = 5^2 = 25$

41. Center: $\left(\dfrac{3+0}{2}, \dfrac{0+0}{2}, \dfrac{0+6}{2}\right) = \left(\dfrac{3}{2}, 0, 3\right)$

$\quad$ Radius: $\sqrt{\left(3-\dfrac{3}{2}\right)^2 + (0-0)^2 + (0-3)^2} = \sqrt{\dfrac{9}{4}+9} = \sqrt{\dfrac{45}{4}}$

$\quad$ Sphere: $\left(x-\dfrac{3}{2}\right)^2 + (y-0)^2 + (z-3)^2 = \dfrac{45}{4}$

43. $\left(x^2 - 5x + \dfrac{25}{4}\right) + y^2 + z^2 = \dfrac{25}{4}$

$\qquad \left(x-\dfrac{5}{2}\right)^2 + y^2 + z^2 = \dfrac{25}{4}$

$\quad$ Center: $\left(\dfrac{5}{2}, 0, 0\right)$

$\quad$ Radius: $\dfrac{5}{2}$

45. $(x^2 - 4x + 4) + (y^2 + 2y + 1) + (z^2 - 6z + 9) = -10 + 4 + 1 + 9$

$$(x - 2)^2 + (y + 1)^2 + (z - 3)^2 = 4$$

Center: $(2, -1, 3)$

Radius: 2

47. $(x^2 + 4x + 4) + y^2 + (z^2 - 8z + 16) = -19 + 4 + 16$

$$(x + 2)^2 + y^2 + (z - 4)^2 = 1$$

Center: $(-2, 0, 4)$

Radius: 1

49.
$$x^2 + y^2 + z^2 - 2x - \tfrac{2}{3}y - 8z = -\tfrac{73}{9}$$

$$(x^2 - 2x + 1) + \left(y^2 - \tfrac{2}{3}y + \tfrac{1}{9}\right) + (z^2 - 8z + 16) = -\tfrac{73}{9} + 1 + \tfrac{1}{9} + 16$$

$$(x - 1)^2 + \left(y - \tfrac{1}{3}\right)^2 + (z - 4)^2 = 9$$

Center: $\left(1, \tfrac{1}{3}, 4\right)$

Radius: 3

51.
$$9x^2 - 6x + 9y^2 + 18y + 9z^2 = -1$$

$$x^2 - \tfrac{2}{3}x + \tfrac{1}{9} + y^2 + 2y + 1 + z^2 = -\tfrac{1}{9} + \tfrac{1}{9} + 1$$

$$\left(x - \tfrac{1}{3}\right)^2 + (y + 1)^2 + z^2 = 1$$

Center: $\left(\tfrac{1}{3}, -1, 0\right)$

Radius: 1

53.

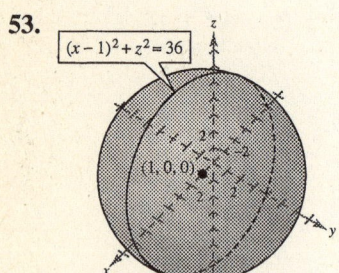

55.

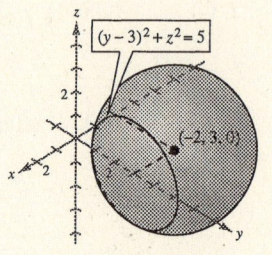

57.

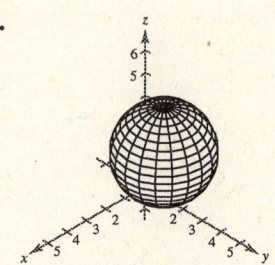

59. The length of each side is 3. Thus, $(x, y, z) = (3, 3, 3)$.

61. $d = 165 \implies r = \tfrac{165}{2} = 82.5$

$$x^2 + y^2 + z^2 = \left(\tfrac{165}{2}\right)^2$$

63. False. x is the directed distance from the yz-plane to P.

65. In the xy-plane, the z-coordinate is 0.
In the xz-plane, the y-coordinate is 0.
In the yz-plane, the x-coordinate is 0.

67. The trace is a circle, or a single point.

69. $x_m = \dfrac{x_2 + x_1}{2} \implies x_2 = 2x_m - x_1$

Similarly for y_2 and z_2,

$(x_2, y_2, z_2) = (2x_m - x_1, 2y_m - y_1, 2z_m - z_1)$.

71. $v^2 + 3v + \dfrac{9}{4} = 2 + \dfrac{9}{4}$

$\left(v + \dfrac{3}{2}\right)^2 = \dfrac{17}{4}$

$v + \dfrac{3}{2} = \pm\dfrac{\sqrt{17}}{2}$

$v = -\dfrac{3}{2} \pm \dfrac{\sqrt{17}}{2}$

73. $x^2 - 5x + \dfrac{25}{4} = -5 + \dfrac{25}{4}$

$\left(x - \dfrac{5}{2}\right)^2 = \dfrac{5}{4}$

$x - \dfrac{5}{2} = \pm\dfrac{\sqrt{5}}{2}$

$x = \dfrac{5}{2} \pm \dfrac{\sqrt{5}}{2}$

75. $4y^2 + 4y = 9$

$y^2 + y + \dfrac{1}{4} = \dfrac{9}{4} + \dfrac{1}{4}$

$\left(y + \dfrac{1}{2}\right)^2 = \dfrac{10}{4}$

$y + \dfrac{1}{2} = \pm\dfrac{\sqrt{10}}{2}$

$y = -\dfrac{1}{2} \pm \dfrac{\sqrt{10}}{2}$

77. $\mathbf{v} = 3\mathbf{i} - 3\mathbf{j}$, Quadrant IV

$\|\mathbf{v}\| = \sqrt{3^2 + (-3)^2}$

$= \sqrt{18}$

$= 3\sqrt{2}$

$\tan\theta = -\dfrac{3}{3} = -1 \implies$

$\theta = -45°$ or $315°$

79. $\mathbf{v} = 4\mathbf{i} + 5\mathbf{j}$, Quadrant I

$\|\mathbf{v}\| = \sqrt{16 + 25} = \sqrt{41}$

$\tan\theta = \dfrac{5}{4} \implies \theta \approx 51.34°$

81. $\mathbf{u} \cdot \mathbf{v} = \langle -4, 1 \rangle \cdot \langle 3, 5 \rangle$

$= -4(3) + 1(5)$

$= -7$

83. $a_0 = 1, a_n = a_{n-1} + n^2$

$a_1 = 1 + 1^2 = 2$

$a_2 = 2 + 2^2 = 6$

$a_3 = 6 + 3^2 = 15$

$a_4 = 15 + 4^2 = 31$

	1		2		6		15		31
First differences:		1		4		9		16	
Second differences:			3		5		7		

Neither model

85. $a_1 = -1, a_n = a_{n-1} + 3$

$a_2 = -1 + 3 = 2$

$a_3 = 2 + 3 = 5$

$a_4 = 5 + 3 = 8$

$a_5 = 8 + 3 = 11$

	−1		2		5		8		11
First differences:		3		3		3		3	
Second differences:			0		0		0		

Linear model

87. $(x + 5)^2 + (y - 1)^2 = 49$

89. $(y - 1)^2 = 4p(x - 4)$, $p = -3$

$(y - 1)^2 = 4(-3)(x - 4)$

$(y - 1)^2 = -12(x - 4)$

91. $a = 3, b = 2$, center: $(3, 3)$, horizontal major axis

$$\frac{(x - 3)^2}{9} + \frac{(y - 3)^2}{4} = 1$$

93. Center: $(6, 0)$, horizontal transverse axis

$$a = 2, c = 6, b^2 = c^2 - a^2 = 36 - 4 = 32$$

$$\frac{(x - 6)^2}{4} - \frac{y^2}{32} = 1$$

Section 10.2   Vectors in Space

- Vectors in space $\mathbf{v} = \langle v_1, v_2, v_3 \rangle$ have many of the same properties as vectors in the plane.
- The dot product of two vectors $\mathbf{u} = \langle u_1, u_2, u_3 \rangle$ and $\mathbf{v} = \langle v_1, v_2, v_3 \rangle$ in space is $\mathbf{u} \cdot \mathbf{v} = u_1 v_1 + u_2 v_2 + u_3 v_3$.
- Two nonzero vectors $\mathbf{u}$ and $\mathbf{v}$ are said to be parallel if there is some scalar c such that $\mathbf{u} = c\mathbf{v}$.
- You should be able to use vectors to solve real life problems.

Solutions to Odd-Numbered Exercises

1. $\mathbf{v} = \langle 0 - 2, 3 - 0, 2 - 1 \rangle = \langle -2, 3, 1 \rangle$

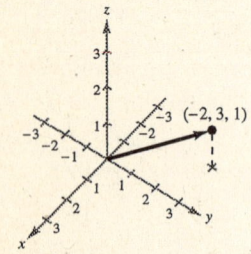

3. (a) $\mathbf{v} = \langle 1 - (-6), -1 - 4, 3 - (-2) \rangle$

$$= \langle 7, -5, 5 \rangle$$

(b) $\|\mathbf{v}\| = \sqrt{7^2 + (-5)^2 + 5^2}$

$$= \sqrt{49 + 25 + 25}$$

$$= \sqrt{99}$$

$$= 3\sqrt{11}$$

(c) $\dfrac{\mathbf{v}}{\|\mathbf{v}\|} = \dfrac{1}{3\sqrt{11}} \langle 7, -5, 5 \rangle = \dfrac{\sqrt{11}}{33} \langle 7, -5, 5 \rangle$

5. (a)

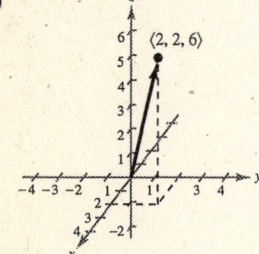

(b)

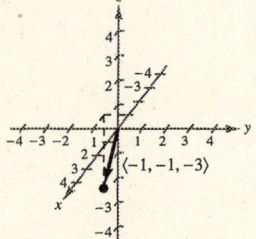

(c)

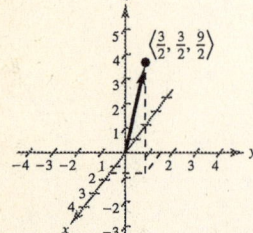

(d)

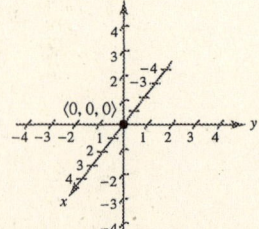

7. $\mathbf{z} = \mathbf{u} - 2\mathbf{v} = \langle -1, 3, 2 \rangle - 2\langle 1, -2, -2 \rangle = \langle -3, 7, 6 \rangle$

9. $2\mathbf{z} - 4\mathbf{u} = \mathbf{w} \implies \mathbf{z} = \frac{1}{2}(4\mathbf{u} + \mathbf{w}) = \frac{1}{2}(4\langle -1, 3, 2 \rangle + \langle 5, 0, -5 \rangle) = \left\langle \frac{1}{2}, 6, \frac{3}{2} \right\rangle$

11. $\|\mathbf{v}\| = \|\langle 7, 8, 7\rangle\|$

$= \sqrt{49 + 64 + 49} = \sqrt{162} = 9\sqrt{2}$

13. $\|\mathbf{v}\| = \sqrt{4^2 + (-3)^2 + (-7)^2}$

$= \sqrt{16 + 9 + 49} = \sqrt{74}$

15. $\mathbf{v} = \langle 1 - 1, 0 - (-3), -1 - 4\rangle = \langle 0, 3, -5\rangle$

$\|\mathbf{v}\| = \sqrt{0 + 3^2 + (-5)^2} = \sqrt{34}$

17. (a) $\dfrac{\mathbf{u}}{\|\mathbf{u}\|} = \dfrac{\langle 8, 3, -1\rangle}{\sqrt{74}}$

$= \dfrac{1}{\sqrt{74}}(8\mathbf{i} + 3\mathbf{j} - \mathbf{k}) = \dfrac{\sqrt{74}}{74}\langle 8, 3, -1\rangle$

(b) $-\dfrac{1}{\sqrt{74}}(8\mathbf{i} + 3\mathbf{j} - \mathbf{k}) = -\dfrac{\sqrt{74}}{74}\langle 8, 3, -1\rangle$

19. $6\mathbf{u} - 4\mathbf{v} = 6\langle -1, 3, 4\rangle - 4\langle 5, 4.5, -6\rangle$

$= \langle -6, 18, 24\rangle + \langle -20, -18, 24\rangle$

$= \langle -26, 0, 48\rangle$

21. $\mathbf{u} + \mathbf{v} = \langle -1, 3, 4\rangle + \langle 5, 4.5, -6\rangle = \langle 4, 7.5, -2\rangle$

$\|\mathbf{u} + \mathbf{v}\| = \sqrt{4^2 + 7.5^2 + (-2)^2} = \frac{1}{2}\sqrt{305} \approx 34.93$

23. $\mathbf{u} \cdot \mathbf{v} = \langle 4, 4, -1\rangle \cdot \langle 2, -5, -8\rangle$

$= 8 - 20 + 8 = -4$

25. $\mathbf{u} \cdot \mathbf{v} = \langle 2, -5, 3\rangle \cdot \langle 9, 3, -1\rangle$

$= 18 - 15 - 3 = 0$

27. $\cos\theta = \dfrac{\mathbf{u} \cdot \mathbf{v}}{\|\mathbf{u}\|\,\|\mathbf{v}\|} = \dfrac{-8}{\sqrt{8}\sqrt{25}} \Rightarrow \theta \approx 124.45°$

29. $\cos\theta = \dfrac{\mathbf{u} \cdot \mathbf{v}}{\|\mathbf{u}\|\,\|\mathbf{v}\|} = \dfrac{-120}{\sqrt{1700}\sqrt{73}} \Rightarrow \theta \approx 109.92°$

31. $-\frac{3}{2}\langle 8, -4, -10\rangle = \langle -12, 6, 15\rangle \Rightarrow$ parallel

33. $\mathbf{u} \cdot \mathbf{v} = 3 - 5 + 2 = 0 \Rightarrow$ orthogonal

35. $\mathbf{v} = \langle 7 - 5, 3 - 4, -1 - 1\rangle = \langle 2, -1, -2\rangle$

$\mathbf{u} = \langle 4 - 7, 5 - 3, 3 - (-1)\rangle = \langle -3, 2, 4\rangle$

Since $\mathbf{u}$ and $\mathbf{v}$ are not parallel, the points are not collinear.

37. $\mathbf{v} = \langle -1 - 1, 2 - 3, 5 - 2\rangle = \langle -2, -1, 3\rangle$

$\mathbf{u} = \langle 3 - (-1), 4 - 2, -1 - 5\rangle = \langle 4, 2, -6\rangle$

Since $\mathbf{u} = -2\mathbf{v}$, the points are collinear.

39. $\mathbf{v} = \langle 2, -4, 7\rangle = \langle q_1 - 1, q_2 - 5, q_3 - 0\rangle \Rightarrow$

$\left.\begin{array}{l} 2 = q_1 - 1 \\ -4 = q_2 - 5 \\ 7 = q_3 \end{array}\right\} \Rightarrow \left.\begin{array}{l} q_1 = 3 \\ q_2 = 1 \\ q_3 = 7 \end{array}\right\} \Rightarrow$

Terminal point is $(3, 1, 7)$.

41. $\mathbf{v} = \left\langle 4, \frac{3}{2}, -\frac{1}{4}\right\rangle = \left\langle q_1 - 2, q_2 - 1, q_3 + \frac{3}{2}\right\rangle$

$4 = q_1 - 2 \Rightarrow q_1 = 6$

$\frac{3}{2} = q_2 - 1 \Rightarrow q_2 = \frac{5}{2}$

$-\frac{1}{4} = q_3 + \frac{3}{2} \Rightarrow q_3 = -\frac{7}{4}$

Terminal point: $\left(6, \frac{5}{2}, -\frac{7}{4}\right)$

43. $c\mathbf{u} = c\mathbf{i} + 2c\mathbf{j} + 3c\mathbf{k}$

$\|c\mathbf{u}\| = \sqrt{c^2 + 4c^2 + 9c^2} = |c|\sqrt{14} = 3 \Rightarrow$

$c = \pm\dfrac{3}{\sqrt{14}} = \pm\dfrac{3\sqrt{14}}{14}$

45. $\mathbf{v} = \langle q_1, q_2, q_3\rangle$

Since $\mathbf{v}$ lies in the yz-plane, $q_1 = 0$. Since $\mathbf{v}$ makes an angle of $45°$, $q_2 = q_3$. Finally, $\|\mathbf{v}\| = 4$ implies that $q_2{}^2 + q_3{}^2 = 16$. Thus, $q_2 = q_3 = 2\sqrt{2}$ and $\mathbf{v} = \left\langle 0, 2\sqrt{2}, 2\sqrt{2}\right\rangle$, or $q_2 = 2\sqrt{2}$ and $q_3 = -2\sqrt{2}$ and $\mathbf{v} = \left\langle 0, 2\sqrt{2}, -2\sqrt{2}\right\rangle$.

47. $\overrightarrow{PQ_1} = \langle 0, -24, -12\sqrt{21} \rangle$

$\overrightarrow{PQ_2} = \langle 12\sqrt{3}, 12, -12\sqrt{21} \rangle$

$\overrightarrow{PQ_3} = \langle -12\sqrt{3}, 12, -12\sqrt{21} \rangle$

Let $\mathbf{F}_1$, $\mathbf{F}_2$, and $\mathbf{F}_3$ be the tension on each wire. Since $\|\mathbf{F}_1\| = \|\mathbf{F}_2\| = \|\mathbf{F}_3\|$, there exists a constant c such that

$\mathbf{F}_1 = c\langle 0, -24, -12\sqrt{21} \rangle$

$\mathbf{F}_2 = c\langle 12\sqrt{3}, 12, -12\sqrt{21} \rangle$

$\mathbf{F}_3 = c\langle -12\sqrt{3}, 12, -12\sqrt{21} \rangle$.

The total force is $-30\mathbf{k} = \mathbf{F}_1 + \mathbf{F}_2 + \mathbf{F}_3 \implies$ the vertical ($\mathbf{k}$) component satisfies

$-10 = -12\sqrt{21}\,c \implies c = \dfrac{5}{6\sqrt{21}}$.

Hence,

$\mathbf{F}_1 = \left\langle 0, \dfrac{-20}{\sqrt{21}}, -10 \right\rangle$

$\mathbf{F}_2 = \left\langle \dfrac{10}{\sqrt{7}}, \dfrac{10}{\sqrt{21}}, -10 \right\rangle$

$\mathbf{F} = \left\langle \dfrac{-10}{\sqrt{7}}, \dfrac{10}{\sqrt{21}}, -10 \right\rangle$

$\|\mathbf{F}_1\| = \|\mathbf{F}_2\| = \|\mathbf{F}_3\| \approx 10.91$ pounds.

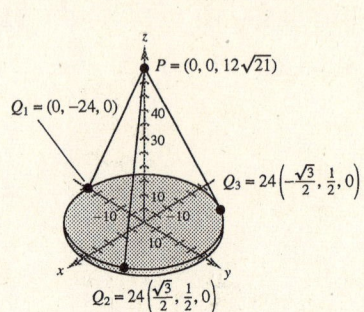

$Q_1 = (0, -24, 0)$

$Q_2 = (20.8, 12, 0)$

$Q_3 = (-20.8, 12, 0)$

$P = (0, 0, 55)$

49. True. $\cos\theta = 0 \implies \theta = 90°$

51. (a)

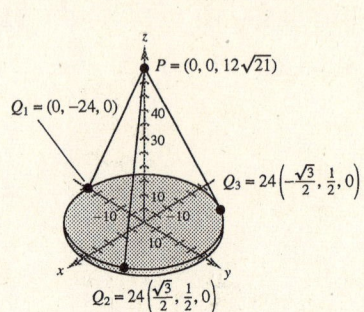

(b) $\mathbf{w} = a\mathbf{u} + b\mathbf{v} = a\langle 1, 1, 0 \rangle + b\langle 0, 1, 1 \rangle$

$\mathbf{0} = \langle a, a + b, b \rangle \implies a = b = 0$

(c) $\mathbf{w} = \langle 1, 2, 1 \rangle = a\langle 1, 1, 0 \rangle + b\langle 0, 1, 1 \rangle$

$1 = a$

$2 = a + b$

$1 = b$

Hence, $a = b = 1$.

(d) $\mathbf{w} = \langle 1, 2, 3 \rangle = a\langle 1, 1, 0 \rangle + b\langle 0, 1, 1 \rangle$

$1 = a$

$2 = a + b$

$3 = b$

Impossible

53. If $\mathbf{u} \cdot \mathbf{v} < 0$, then $\cos\theta < 0$ and the angle between $\mathbf{u}$ and $\mathbf{v}$ is obtuse, $180° > \theta > 90°$.

55. (a) $x = t$

$y = 3t + 2$

(b) $x = t - 1$

$y = 3(t - 1) + 2$

$= 3t - 1$

57. (a) $x = t$

$y = t^2 - 8$

(b) $x = t - 1$

$y = (t - 1)^2 - 8$

$= t^2 - 2t - 7$

Section 10.3   The Cross Product of Two Vectors

■ The cross product of two vectors $\mathbf{u} = u_1\mathbf{i} + u_2\mathbf{j} + u_3\mathbf{k}$ and $\mathbf{v} = v_1\mathbf{i} + v_2\mathbf{j} + v_3\mathbf{k}$ is given by

$$\mathbf{u} \times \mathbf{v} = (u_2v_3 - u_3v_2)\mathbf{i} - (u_1v_3 - u_3v_1)\mathbf{j} + (u_1v_2 - u_2v_1)\mathbf{k}$$

$$= \begin{vmatrix} \mathbf{i} & \mathbf{j} & \mathbf{k} \\ u_1 & u_2 & u_3 \\ v_1 & v_2 & v_3 \end{vmatrix}.$$

■ The cross product satisfies the following algebraic properties.

(a) $\mathbf{u} \times \mathbf{v} = -(\mathbf{v} \times \mathbf{u})$

(b) $\mathbf{u} \times (\mathbf{v} + \mathbf{w}) = (\mathbf{u} \times \mathbf{v}) + (\mathbf{u} \times \mathbf{w})$

(c) $c(\mathbf{u} \times \mathbf{v}) = (c\mathbf{u}) \times \mathbf{v} = \mathbf{u} \times (c\mathbf{v})$

(d) $\mathbf{u} \times \mathbf{0} = \mathbf{0} \times \mathbf{u} = \mathbf{0}$

(e) $\mathbf{u} \times \mathbf{u} = \mathbf{0}$

(f) $\mathbf{u} \cdot (\mathbf{v} \times \mathbf{w}) = (\mathbf{u} \times \mathbf{v}) \cdot \mathbf{w}$

■ The following geometric properties of the cross product are valid, where θ is the angle between the vectors $\mathbf{u}$ and $\mathbf{v}$:

(a) $\mathbf{u} \times \mathbf{v}$ is orthogonal to both $\mathbf{u}$ and $\mathbf{v}$.

(b) $\|\mathbf{u} \times \mathbf{v}\| = \|\mathbf{u}\| \, \|\mathbf{v}\| \sin \theta$

(c) $\mathbf{u} \times \mathbf{v} = \mathbf{0}$ if and only if $\mathbf{u}$ and $\mathbf{v}$ are scalar multiples.

(d) $\|\mathbf{u} \times \mathbf{v}\|$ is the area of the parallelogram having $\mathbf{u}$ and $\mathbf{v}$ as sides.

■ The absolute value of the triple scalar product is the volume of the parallelepiped having $\mathbf{u}$, $\mathbf{v}$, and $\mathbf{w}$ as sides.

$$\mathbf{u} \cdot (\mathbf{v} \times \mathbf{w}) = \begin{vmatrix} u_1 & u_2 & u_3 \\ v_1 & v_2 & v_3 \\ w_1 & w_2 & w_3 \end{vmatrix}$$

Solutions to Odd-Numbered Exercises

1. $\mathbf{j} \times \mathbf{i} = \begin{vmatrix} \mathbf{i} & \mathbf{j} & \mathbf{k} \\ 0 & 1 & 0 \\ 1 & 0 & 0 \end{vmatrix} = -\mathbf{k}$

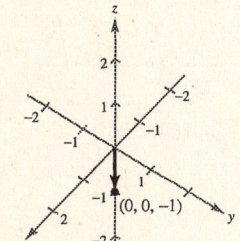

3. $\mathbf{i} \times \mathbf{k} = \begin{vmatrix} \mathbf{i} & \mathbf{j} & \mathbf{k} \\ 1 & 0 & 0 \\ 0 & 0 & 1 \end{vmatrix} = -\mathbf{j}$

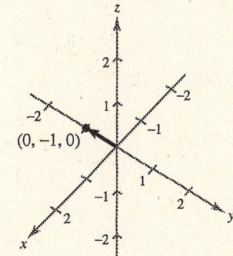

5. $\mathbf{u} \times \mathbf{v} = \begin{vmatrix} \mathbf{i} & \mathbf{j} & \mathbf{k} \\ 3 & -2 & 5 \\ 0 & -1 & 1 \end{vmatrix} = \langle 3, -3, -3 \rangle$

$(\mathbf{u} \times \mathbf{v}) \cdot \mathbf{u} = \langle 3, -3, -3 \rangle \cdot \langle 3, -2, 5 \rangle = 0$

$(\mathbf{u} \times \mathbf{v}) \cdot \mathbf{v} = \langle 3, -3, -3 \rangle \cdot \langle 0, -1, 1 \rangle = 0$

7. $\mathbf{u} \times \mathbf{v} = \begin{vmatrix} \mathbf{i} & \mathbf{j} & \mathbf{k} \\ -10 & 0 & 6 \\ 7 & 0 & 0 \end{vmatrix} = \langle 0, 42, 0 \rangle$

$(\mathbf{u} \times \mathbf{v}) \cdot \mathbf{u} = \langle 0, 42, 0 \rangle \cdot \langle -10, 0, 6 \rangle = 0$

$(\mathbf{u} \times \mathbf{v}) \cdot \mathbf{v} = \langle 0, 42, 0 \rangle \cdot \langle 7, 0, 0 \rangle = 0$

9. $\mathbf{u} \times \mathbf{v} = \begin{vmatrix} \mathbf{i} & \mathbf{j} & \mathbf{k} \\ 6 & 2 & 1 \\ 1 & 3 & -2 \end{vmatrix} = \langle -7, 13, 16 \rangle$

$$= -7\mathbf{i} + 13\mathbf{j} + 16\mathbf{k}$$

11. $\mathbf{u} \times \mathbf{v} = \begin{vmatrix} \mathbf{i} & \mathbf{j} & \mathbf{k} \\ 0 & 0 & 6 \\ -1 & 3 & 1 \end{vmatrix} = \langle -18, -6, 0 \rangle$

$$= -18\mathbf{i} - 6\mathbf{j}$$

13. $\mathbf{u} \times \mathbf{v} = \begin{vmatrix} \mathbf{i} & \mathbf{j} & \mathbf{k} \\ -1 & 0 & 1 \\ 0 & 1 & -2 \end{vmatrix} = \langle -1, -2, -1 \rangle$

$$= -\mathbf{i} - 2\mathbf{j} - \mathbf{k}$$

15. $\mathbf{u} \times \mathbf{v} = \begin{vmatrix} \mathbf{i} & \mathbf{j} & \mathbf{k} \\ 2 & 4 & 3 \\ 0 & -2 & 1 \end{vmatrix} = \langle 10, -2, -4 \rangle$

17. $\mathbf{u} \times \mathbf{v} = \begin{vmatrix} \mathbf{i} & \mathbf{j} & \mathbf{k} \\ 6 & -5 & 1 \\ \frac{1}{2} & -\frac{3}{4} & \frac{2}{10} \end{vmatrix} = \langle -0.25, -0.7, -2 \rangle$

19. $\mathbf{u} \times \mathbf{v} = \begin{vmatrix} \mathbf{i} & \mathbf{j} & \mathbf{k} \\ 3 & 1 & 0 \\ 0 & 1 & 1 \end{vmatrix} = \mathbf{i} - 3\mathbf{j} + 3\mathbf{k}$

$$\|\mathbf{u} \times \mathbf{v}\| = \sqrt{19}$$

$$\text{Unit vector} = \frac{\mathbf{u} \times \mathbf{v}}{\|\mathbf{u} \times \mathbf{v}\|} = \frac{1}{\sqrt{19}}(\mathbf{i} - 3\mathbf{j} + 3\mathbf{k})$$

$$= \frac{\sqrt{19}}{19} \langle 1, -3, 3 \rangle$$

21. $\mathbf{u} \times \mathbf{v} = \begin{vmatrix} \mathbf{i} & \mathbf{j} & \mathbf{k} \\ -3 & 2 & -5 \\ \frac{1}{2} & -\frac{3}{4} & \frac{1}{10} \end{vmatrix} = \left\langle -\frac{71}{20}, -\frac{11}{5}, \frac{5}{4} \right\rangle$

Consider the parallel vector $\langle -71, -44, 25 \rangle = \mathbf{w}$.

$$\|\mathbf{w}\| = \sqrt{71^2 + 44^2 + 25^2} = \sqrt{7602}$$

$$\text{Unit vector} = \frac{1}{\sqrt{7602}} \langle -71, -44, 25 \rangle$$

$$= \frac{\sqrt{7602}}{7602} \langle -71, -44, 25 \rangle$$

23. $\mathbf{u} \times \mathbf{v} = \begin{vmatrix} \mathbf{i} & \mathbf{j} & \mathbf{k} \\ 1 & 1 & -1 \\ 1 & 1 & 1 \end{vmatrix} = 2\mathbf{i} - 2\mathbf{j}$

$$\|\mathbf{u} \times \mathbf{v}\| = 2\sqrt{2}$$

$$\text{Unit vector} = \frac{\mathbf{u} \times \mathbf{v}}{\|\mathbf{u} \times \mathbf{v}\|} = \frac{1}{2\sqrt{2}}(2\mathbf{i} - 2\mathbf{j})$$

$$= \frac{1}{\sqrt{2}}\mathbf{i} - \frac{1}{\sqrt{2}}\mathbf{j}$$

$$= \frac{\sqrt{2}}{2}\mathbf{i} - \frac{\sqrt{2}}{2}\mathbf{j}$$

25. $\mathbf{u} \times \mathbf{v} = \begin{vmatrix} \mathbf{i} & \mathbf{j} & \mathbf{k} \\ 0 & 0 & 1 \\ 1 & 0 & 1 \end{vmatrix} = \mathbf{j}$

$\text{Area} = \|\mathbf{u} \times \mathbf{v}\| = \|\mathbf{j}\| = 1$ square unit

27. $\mathbf{u} \times \mathbf{v} = \begin{vmatrix} \mathbf{i} & \mathbf{j} & \mathbf{k} \\ 3 & 4 & 6 \\ 2 & -1 & 5 \end{vmatrix} = 26\mathbf{i} - 3\mathbf{j} - 11\mathbf{k}$

$\text{Area} = \|\mathbf{u} \times \mathbf{v}\| = \sqrt{26^2 + -(3)^2 + (-11)^2}$

$$= \sqrt{806} \text{ square units}$$

29. $\mathbf{u} \times \mathbf{v} = \begin{vmatrix} \mathbf{i} & \mathbf{j} & \mathbf{k} \\ 2 & 2 & -3 \\ 0 & 2 & 3 \end{vmatrix} = \langle 12, -6, 4 \rangle$

$\text{Area} = \|\mathbf{u} \times \mathbf{v}\| = \sqrt{12^2 + (-6)^2 + 4^2} = 14$ square units

31. (a) $\overrightarrow{AB} = \langle 3 - 2, 1 - (-1), 2 - 4 \rangle = \langle 1, 2, -2 \rangle$
is parallel to

$\overrightarrow{DC} = \langle 0 - (-1), 5 - 3, 6 - 8 \rangle = \langle 1, 2, -2 \rangle.$

$\overrightarrow{AD} = \langle -3, 4, 4 \rangle$ is parallel to $\overrightarrow{BC} = \langle -3, 4, 4 \rangle.$

(c) $\overrightarrow{AB} \cdot \overrightarrow{AD} = \langle 1, 2, -2 \rangle \cdot \langle -3, 4, 4 \rangle$

$\neq 0 \implies$ not a rectangle

(b) $\overrightarrow{AB} \times \overrightarrow{AD} = \begin{vmatrix} \mathbf{i} & \mathbf{j} & \mathbf{k} \\ 1 & 2 & -2 \\ -3 & 4 & 4 \end{vmatrix} = \langle 16, 2, 10 \rangle$

Area $= \|\overrightarrow{AB} \times \overrightarrow{AD}\|$

$= \sqrt{16^2 + 2^2 + 10^2}$

$= \sqrt{360} = 6\sqrt{10}$ square units

33. $\mathbf{u} = \langle 1, 2, 3 \rangle, \mathbf{v} = \langle -3, 0, 0 \rangle$

$\mathbf{u} \times \mathbf{v} = \begin{vmatrix} \mathbf{i} & \mathbf{j} & \mathbf{k} \\ 1 & 2 & 3 \\ -3 & 0 & 0 \end{vmatrix} = \langle 0, -9, 6 \rangle$

Area $= \frac{1}{2}\|\mathbf{u} \times \mathbf{v}\| = \frac{1}{2}\sqrt{81 + 36} = \frac{3}{2}\sqrt{13}$

35. $\mathbf{u} = \langle -2 - 2, -2 - 3, 0 - (-5) \rangle = \langle -4, -5, 5 \rangle$

$\mathbf{v} = \langle 3 - 2, 0 - 3, 6 - (-5) \rangle = \langle 1, -3, 11 \rangle$

$\mathbf{u} \times \mathbf{v} = \begin{vmatrix} \mathbf{i} & \mathbf{j} & \mathbf{k} \\ -4 & -5 & 5 \\ 1 & -3 & 11 \end{vmatrix} = \langle -40, 49, 17 \rangle$

Area $= \frac{1}{2}\|\mathbf{u} \times \mathbf{v}\| = \frac{1}{2}\sqrt{(-40)^2 + 49^2 + 17^2}$

$= \frac{1}{2}\sqrt{4290}$ square units

37. $\mathbf{u} \cdot (\mathbf{v} \times \mathbf{w}) = \begin{vmatrix} 2 & 3 & 3 \\ 4 & 4 & 0 \\ 0 & 0 & 4 \end{vmatrix}$

$= 2(16) - 3(16) + 3(0) = -16$

39. $\mathbf{u} \cdot (\mathbf{v} \times \mathbf{w}) = \begin{vmatrix} 2 & 3 & 1 \\ 1 & -1 & 0 \\ 4 & 3 & 1 \end{vmatrix}$

$= 2(-1) - 3(1) + 1(7) = 2$

41. $\mathbf{u} \cdot (\mathbf{v} \times \mathbf{w}) = \begin{vmatrix} 1 & 1 & 0 \\ 0 & 1 & 1 \\ 1 & 0 & 1 \end{vmatrix} = 1 + 1 = 2$

Volume $= |\mathbf{u} \cdot (\mathbf{v} \times \mathbf{w})| = 2$ cubic units

43. $\mathbf{u} \cdot (\mathbf{v} \times \mathbf{w}) = \begin{vmatrix} 0 & 2 & 2 \\ 0 & 0 & -2 \\ 3 & 0 & 2 \end{vmatrix}$

$= 0 - 2(6) + 2(0) = -12$

Volume $= |\mathbf{u} \cdot (\mathbf{v} \times \mathbf{w})| = 12$ cubic units

45. $\mathbf{u} = \langle 4, 0, 0 \rangle, \mathbf{v} = \langle 0, -2, 3 \rangle, \mathbf{w} = \langle 0, 5, 3 \rangle$

$\mathbf{u} \cdot (\mathbf{v} \times \mathbf{w}) = \begin{vmatrix} 4 & 0 & 0 \\ 0 & -2 & 3 \\ 0 & 5 & 3 \end{vmatrix} = 4(-21) = -84$

Volume $= |-84| = 84$ cubic units

47. $\mathbf{V} \times \mathbf{F} = \begin{vmatrix} \mathbf{i} & \mathbf{j} & \mathbf{k} \\ 0 & \frac{1}{2}\cos 40° & \frac{1}{2}\sin 40° \\ 0 & 0 & -20 \end{vmatrix}$

$= -10\cos 40°\mathbf{i}$

$\|\mathbf{V} \times \mathbf{F}\| = 10\cos 40° \approx 7.66$ foot-pounds

49. True. The cross product is defined for vectors in three-dimensional space.

51. $\mathbf{u} \times \mathbf{u} = \begin{vmatrix} \mathbf{i} & \mathbf{j} & \mathbf{k} \\ u_1 & u_2 & u_3 \\ u_1 & u_2 & u_3 \end{vmatrix} = (u_2 u_3 - u_2 u_3)\mathbf{i} - (u_1 u_3 - u_1 u_3)\mathbf{j} + (u_1 u_2 - u_1 u_2)\mathbf{k} = \mathbf{0}$

53. $c(\mathbf{u} \times \mathbf{v}) = c \begin{vmatrix} \mathbf{i} & \mathbf{j} & \mathbf{k} \\ u_1 & u_2 & u_3 \\ v_1 & v_2 & v_3 \end{vmatrix}$

$= c\langle u_2 v_3 - v_2 u_3, v_1 u_3 - u_1 v_3, u_1 v_2 - u_2 v_1 \rangle$

$= \langle c u_2 v_3 - v_2 c u_3, v_1 c u_3 - c u_1 v_3, c u_1 v_2 - c u_2 v_1 \rangle$

$= \begin{vmatrix} \mathbf{i} & \mathbf{j} & \mathbf{k} \\ c u_1 & c u_2 & c u_3 \\ v_1 & v_2 & v_3 \end{vmatrix}$

$= (c\mathbf{u}) \times \mathbf{v}$

$= \begin{vmatrix} \mathbf{i} & \mathbf{j} & \mathbf{k} \\ u_1 & u_2 & u_3 \\ c v_1 & c v_2 & c v_3 \end{vmatrix}$

$= \mathbf{u} \times (c\mathbf{v})$

55. $\cos 480° = \cos 120° = -\frac{1}{2}$

57. $\sin 690° = \sin 330° = -\frac{1}{2}$

59. $\sin \dfrac{19\pi}{6} = \sin\left(\dfrac{7\pi}{6}\right) = -\dfrac{1}{2}$

61. $\tan \dfrac{15\pi}{4} = \tan \dfrac{7\pi}{4} = -1$

Section 10.4   Lines and Planes in Space

■   The parametric equations of the line in space parallel to the vector $\langle a, b, c \rangle$ and passing through the point (x_1, y_2, z_3) are

$$x = x_1 + at, \quad y = y_1 + bt, \quad z = z_1 + ct.$$

■   The standard equation of the plane in space containing the point (x_1, y_1, z_1) and having normal vector (a, b, c) is

$$a(x - x_1) + b(y - y_1) + c(z - z_1) = 0.$$

■   You should be able to find the angle between two planes by calculating the angle between their normal vectors.

■   You should be able to sketch a plane in space.

■   The distance between a point Q and a plane having normal $\mathbf{n}$ is

$$D = \|\text{proj}_{\mathbf{n}} \overrightarrow{PQ}\| = \frac{| = \overrightarrow{PQ} \cdot \mathbf{n}|}{\|\mathbf{n}\|}$$

where P is a point in the plane.

Solutions to Odd-Numbered Exercises

1. $x = x_1 + at = 0 + t$

     $y = y_1 + bt = 0 + 2t$

     $z = z_1 + ct = 0 + 3t$

   (a) Parametric equations: $x = t, y = 2t, z = 3t$

   (b) Symmetric equations: $\dfrac{x}{1} = \dfrac{y}{2} = \dfrac{z}{3}$

3. $x = x_1 + at = -4 + \frac{1}{2}t, \ y = y_1 + bt = 1 + \frac{4}{3}t, \ z = z_1 + ct = 0 - t$

    (a) Parametric equations: $x = -4 + \frac{1}{2}t, y = 1 + \frac{4}{3}t, z = -t$

       Equivalently: $x = -4 + 3t, y = 1 + 8t, z = -6t$

    (b) Symmetric equations: $\dfrac{x+4}{3} = \dfrac{y-1}{8} = \dfrac{z}{-6}$

5. $x = x_1 + at = 2 + 2t, \ y = y_1 + bt = -3 - 3t, \ z = z_1 + ct = 5 + t$

    (a) Parametric equations: $x = 2 + 2t, y = -3 - 3t, z = 5 + t$

    (b) Symmetric equations: $\dfrac{x-2}{2} = \dfrac{y+3}{-3} = z - 5$

7. (a) $\mathbf{v} = \langle 1 - 2, 4 - 0, -3 - 2 \rangle = \langle -1, 4, -5 \rangle$

      Point: $(2, 0, 2)$

      $x = 2 - t, y = 4t, z = 2 - 5t$

   (b) $\dfrac{x-2}{-1} = \dfrac{y}{4} = \dfrac{z-2}{-5}$

9. (a) $\mathbf{v} = \langle 1 - (-3), -2 - 8, 16 - 15 \rangle = \langle 4, -10, 1 \rangle$

      Point: $(-3, 8, 15)$

      $x = -3 + 4t, y = 8 - 10t, z = 15 + t$

   (b) $\dfrac{x+3}{4} = \dfrac{y-8}{-10} = \dfrac{z-15}{1}$

11.

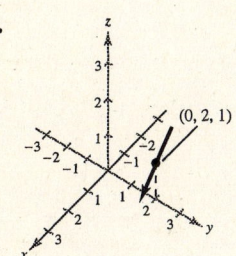

(0, 2, 1)

13. $a(x - x_1) + b(y - y_1) + c(z - z_1) = 0$

    $1(x - 2) + 0(y - 1) + 0(z - 2) = 0$

                       $x - 2 = 0$

15. $-2(x - 5) + 1(y - 6) - 2(z - 3) = 0$

           $-2x + y - 2z + 10 = 0$

17. $\mathbf{n} = \langle -1, -2, 1 \rangle \implies -1(x - 2) - 2(y - 0) + 1(z - 0) = 0$

                               $-x - 2y + z + 2 = 0$

19. $\mathbf{u} = \langle 1 - 0, 2 - 0, 3 - 0 \rangle = \langle 1, 2, 3 \rangle$

    $\mathbf{v} = \langle -2 - 0, 3 - 0, 3 - 0 \rangle = \langle -2, 3, 3 \rangle$

    $\mathbf{n} = \mathbf{u} \times \mathbf{v} = \begin{vmatrix} \mathbf{i} & \mathbf{j} & \mathbf{k} \\ 1 & 2 & 3 \\ -2 & 3 & 3 \end{vmatrix} = \langle -3, -9, 7 \rangle$

    $-3(x - 0) - 9(y - 0) + 7(z - 0) = 0$

               $-3x - 9y + 7z = 0$

                  $3x + 9y - 7z = 0$

21. $\mathbf{u} = \langle 3 - 2, 4 - 3, 2 + 2 \rangle = \langle 1, 1, 4 \rangle$

    $\mathbf{v} = \langle 1 - 2, -1 - 3, 0 + 2 \rangle = \langle -1, -4, 2 \rangle$

    $\mathbf{n} = \mathbf{u} \times \mathbf{v} = \begin{vmatrix} \mathbf{i} & \mathbf{j} & \mathbf{k} \\ 1 & 1 & 4 \\ -1 & -4 & 2 \end{vmatrix} = \langle 18, -6, -3 \rangle$

    $18(x - 2) - 6(y - 3) - 3(z + 2) = 0$

            $18x - 6y - 3z - 24 = 0$

               $6x - 2y - z - 8 = 0$

23. $\mathbf{n} = \mathbf{j}$: $0(x - 2) + 1(y - 5) + 0(z - 3) = 0$

$$y - 5 = 0$$

25. $\mathbf{n}_1 = \langle 5, -3, 1 \rangle$, $\mathbf{n}_2 = \langle 1, 4, 7 \rangle$

$\mathbf{n}_1 \cdot \mathbf{n}_2 = 5 - 12 + 7 = 0$; orthogonal

27. $\mathbf{n}_1 = \langle 2, 0, -1 \rangle$, $\mathbf{n}_2 = \langle 4, 1, 8 \rangle$

$\mathbf{n}_1 \cdot \mathbf{n}_2 = 8 - 8 = 0$; orthogonal

29. (a) $\mathbf{n}_1 = \langle 3, -4, 5 \rangle$, $\mathbf{n}_2 = \langle 1, 1, -1 \rangle$; normal vectors to planes

$$\cos \theta = \frac{|\mathbf{n}_1 \cdot \mathbf{n}_2|}{\|\mathbf{n}_1\| \|\mathbf{n}_2\|} = \frac{|-6|}{\sqrt{50}\sqrt{3}} = \frac{6}{\sqrt{150}} \implies \theta \approx 60.67°$$

(b) $3x - 4y + 5z = 6$ Equation 1

    $x + y - z = 2$ Equation 2

(-3) times Equation 2 added to Equation 1 gives

$-7y + 8z = 0$

$$y = \frac{8}{7}z.$$

Substituting back into Equation 2, $x = 2 - y + z = 2 - \frac{8}{7}z + z = 2 - \frac{1}{7}z$.

Letting $t = z/7$, we obtain $x = 2 - t$, $y = 8t$, $z = 7t$.

31. (a) $\mathbf{n}_1 = \langle 1, 1, -1 \rangle$, $\mathbf{n}_2 = \langle 2, -5, -1 \rangle$; normal vectors to planes

$$\cos \theta = \frac{|\mathbf{n}_1 \cdot \mathbf{n}_2|}{\|\mathbf{n}_1\| \|\mathbf{n}_2\|} = \frac{|-2|}{\sqrt{3}\sqrt{30}} = \frac{2}{\sqrt{90}} \implies \theta \approx 77.83°$$

(b) $x + y - z = 0$ Equation 1

  $2x - 5y - z = 1$ Equation 2

(-2) times Equation 1 added to Equation 2 gives

  $-7y + z = 1$

$$y = \frac{z - 1}{7}.$$

Substituting back into Equation 1, $x = z - y = z - \frac{z - 1}{7} = \frac{6z}{7} + \frac{1}{7} = \frac{1}{7}(6z + 1)$.

Letting $z = t$, $x = \frac{6t + 1}{7}$, $y = \frac{t - 1}{7}$.

Equivalently, let $y = t$, $z = 7t + 1$ and $x = 6t + 1$.

33. $x + 2y + 3z = 6$

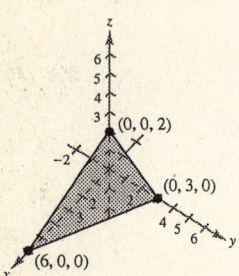

35. $x + 2y = 4$

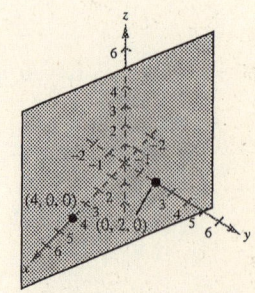

37. $3x + 2y - z = 6$

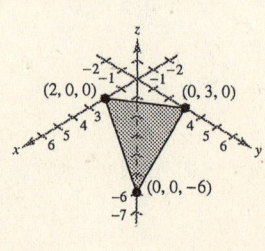

39. $D = \dfrac{|\overrightarrow{PQ} \cdot \mathbf{n}|}{\|\mathbf{n}\|}$

$P = (1, 0, 0)$ on plane, $Q = (0, 0, 0)$,

$\mathbf{n} = \langle 8, -4, 1 \rangle, \overrightarrow{PQ} = \langle -1, 0, 0 \rangle$

$D = \dfrac{|\langle -1, 0, 0 \rangle \cdot \langle 8, -4, 1 \rangle|}{\sqrt{64 + 16 + 1}} = \dfrac{|-8|}{\sqrt{81}} = \dfrac{8}{9}$

41. $D = \dfrac{|\overrightarrow{PQ} \cdot \mathbf{n}|}{\|\mathbf{n}\|}$

$P = (2, 0, 0)$ on plane, $Q = (4, -2, -2)$,

$\mathbf{n} = \langle 2, -1, 1 \rangle, \overrightarrow{PQ} = \langle 2, -2, -2 \rangle$

$D = \dfrac{|\langle 2, -2, -2 \rangle \cdot \langle 2, -1, 1 \rangle|}{\sqrt{6}} = \dfrac{4}{\sqrt{6}} = \dfrac{2\sqrt{6}}{3}$

43. The normal vector to plane containing $(0, 0, 0)$, $(2, 2, 12)$ and $(10, 0, 0)$ is obtained as follows.

$\mathbf{v}_1 = \langle 2, 2, 12 \rangle, \mathbf{v}_2 = \langle 10, 0, 0 \rangle$

$\mathbf{v}_1 \times \mathbf{v}_2 = \begin{vmatrix} \mathbf{i} & \mathbf{j} & \mathbf{k} \\ 2 & 2 & 12 \\ 10 & 0 & 0 \end{vmatrix} = \langle 0, 120, -20 \rangle$

$\mathbf{n}_1 = \langle 0, 6, -1 \rangle$

The normal vector to the plane containing $(0, 0, 0)$, $(2, 2, 12)$ and $(0, 10, 0)$ is obtained as follows.

$\mathbf{u}_1 = \langle 2, 2, 12 \rangle, \mathbf{u}_2 = \langle 0, 10, 0 \rangle$

$\mathbf{u}_1 \times \mathbf{u}_2 = \begin{vmatrix} \mathbf{i} & \mathbf{j} & \mathbf{k} \\ 2 & 2 & 12 \\ 0 & 10 & 0 \end{vmatrix} = \langle -120, 0, 20 \rangle$

$\mathbf{n}_2 = \langle -6, 0, 1 \rangle$

The angle θ between two adjacent sides is given by

$\cos \theta = \dfrac{|\mathbf{n}_1 \cdot \mathbf{n}_2|}{\|\mathbf{n}_1\| \|\mathbf{n}_2\|} = \dfrac{|-1|}{\sqrt{37}\sqrt{37}} = \dfrac{1}{37} \implies \theta \approx 88.45°.$

45. False. They might be skew lines, such as:

$L_1: x = t, y = 0, z = 0$ (x-axis)

and $L_2: x = 0, y = t, z = 1$

47. The lines are parallel:

$-\frac{3}{2}\langle 10, -18, 20 \rangle = \langle -15, 27, -30 \rangle$

49. $x^2 + y^2 = 10^2 = 100$

51. $\begin{aligned} r &= 3 \cos \theta \\ r^2 &= 3r \cos \theta \\ x^2 + y^2 &= 3x \end{aligned}$

53. $\begin{aligned} r^2 &= 49 \\ r &= 7 \end{aligned}$

55. $\begin{aligned} y &= 5 \\ r \sin \theta &= 5 \\ r &= 5 \csc \theta \end{aligned}$

Review Exercises for Chapter 10

Solutions to Odd-Numbered Exercises

1. (a) and (b)

3. $(-5, 4, 0)$

5. $d = \sqrt{(5 - 4)^2 + (2 - 0)^2 + (1 - 7)^2}$

$= \sqrt{1 + 4 + 36}$

$= \sqrt{41}$

7. $d_1 = \sqrt{(3-0)^2 + (-2-3)^2 + (0-2)^2} = \sqrt{9 + 25 + 4} = \sqrt{38}$

$d_2 = \sqrt{(0-0)^2 + (5-3)^2 + (-3-2)^2} = \sqrt{4 + 25} = \sqrt{29}$

$d_3 = \sqrt{(0-3)^2 + (5-(-2))^2 + (-3-0)^2} = \sqrt{9 + 49 + 9} = \sqrt{67}$

$d_1{}^2 + d_2{}^2 = 38 + 29 = 67 = d_3{}^2$

9. Midpoint: $\left(\dfrac{8+5}{2}, \dfrac{-2+6}{2}, \dfrac{3+7}{2}\right) = \left(\dfrac{13}{2}, 2, 5\right)$

11. Midpoint: $\left(\dfrac{10-8}{2}, \dfrac{6-2}{2}, \dfrac{-12-6}{2}\right) = (1, 2, -9)$

13. $(x-2)^2 + (y-3)^2 + (z-5)^2 = 1$

15. Radius: 6

$(x-1)^2 + (y-5)^2 + (z-2)^2 = 36$

17. $(x^2 - 4x + 4) + (y^2 - 6y + 9) + z^2 = -4 + 4 + 9$

$(x-2)^2 + (y-3)^2 + z^2 = 9$

Center: $(2, 3, 0)$

Radius: 3

19. (a) xz-trace $(y = 0)$: $x^2 + z^2 = 7$, circle

(b) yz-trace $(x = 0)$: $(y-3)^2 + z^2 = 16$, circle

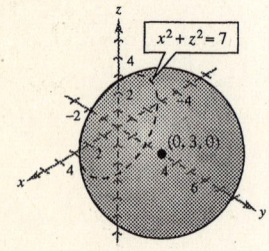

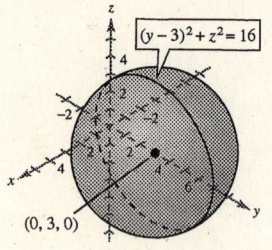

21. $\overrightarrow{PQ} = \langle 3 - 2, 3 - (-1), 0 - 4 \rangle = \langle 1, 4, -4 \rangle$

$\|\overrightarrow{PQ}\| = \sqrt{1^2 + 4^2 + (-4)^2} = \sqrt{33}$

23. $\overrightarrow{PQ} = \langle -3 - 7, 2 - (-4), 10 - 3 \rangle = \langle -10, 6, 7 \rangle$

$\|\overrightarrow{PQ}\| = \sqrt{(-10)^2 + 6^2 + 7^2} = \sqrt{185}$

25. $\mathbf{u} \cdot \mathbf{v} = -1(0) + 4(-6) + 3(5) = -9$

27. $\mathbf{u} \cdot \mathbf{v} = 2(1) - 1(0) + 1(-1) = 1$

29. Since $\mathbf{u} \cdot \mathbf{v} = 0$, the angle is $90°$.

31. Since $-\frac{2}{3}\langle 39, -12, 21 \rangle = \langle -26, 8, -14 \rangle$, the vectors are parallel.

33. First two points: $\mathbf{u} = \langle -3, 4, 1 \rangle$

Last two points: $\mathbf{v} = \langle 0, -2, 6 \rangle$

Since $\mathbf{u} \neq c\mathbf{v}$, the points are not collinear.

35. Let **a**, **b**, and **c** be the three force vectors determined by $A(0, 10, 10)$, $B(-4, -6, 10)$ and $C(4, -6, 10)$.

$$\mathbf{a} = \|\mathbf{a}\| \frac{\langle 0, 10, 10 \rangle}{10\sqrt{2}} = \|\mathbf{a}\| \left\langle 0, \frac{1}{\sqrt{2}}, \frac{1}{\sqrt{2}} \right\rangle$$

$$\mathbf{b} = \|\mathbf{b}\| \frac{\langle -4, -6, 10 \rangle}{\sqrt{152}} = \|\mathbf{b}\| \left\langle \frac{-2}{\sqrt{38}}, \frac{-3}{\sqrt{38}}, \frac{5}{\sqrt{38}} \right\rangle$$

$$\mathbf{c} = \|\mathbf{c}\| \frac{\langle 4, -6, 10 \rangle}{\sqrt{152}} = \|\mathbf{c}\| \left\langle \frac{2}{\sqrt{38}}, \frac{-3}{\sqrt{38}}, \frac{5}{\sqrt{38}} \right\rangle$$

Must have $\mathbf{a} + \mathbf{b} + \mathbf{c} = 300\mathbf{k}$. Thus:

$$\frac{-2}{\sqrt{38}} \|\mathbf{b}\| + \frac{2}{\sqrt{38}} \|\mathbf{c}\| = 0$$

$$\frac{1}{\sqrt{2}} \|\mathbf{a}\| - \frac{3}{\sqrt{38}} \|\mathbf{b}\| - \frac{3}{\sqrt{38}} \|\mathbf{c}\| = 0$$

$$\frac{1}{\sqrt{2}} \|\mathbf{a}\| + \frac{5}{\sqrt{38}} \|\mathbf{b}\| + \frac{5}{\sqrt{38}} \|\mathbf{c}\| = 300.$$

From the first equation $\|\mathbf{b}\| = \|\mathbf{c}\|$. From the second equation, $\dfrac{1}{\sqrt{2}} \|\mathbf{a}\| = \dfrac{6}{\sqrt{38}} \|\mathbf{b}\|$.

From the third equation, $\dfrac{1}{\sqrt{2}} \|\mathbf{a}\| = 300 - \dfrac{10}{\sqrt{38}} \|\mathbf{b}\|$. Thus,

$$\frac{6}{\sqrt{38}} \|\mathbf{b}\| = 300 - \frac{10}{\sqrt{38}} \|\mathbf{b}\| \quad \Rightarrow \quad \frac{16}{\sqrt{38}} \|\mathbf{b}\| = 300 \text{ and } \|\mathbf{b}\| = \|\mathbf{c}\| = \frac{75\sqrt{38}}{4} \approx 115.58.$$

Finally, $\|\mathbf{a}\| = \sqrt{2} \left(\dfrac{6}{\sqrt{38}} \right) \left(\dfrac{75\sqrt{38}}{4} \right) = \dfrac{225\sqrt{2}}{2} \approx 159.10$.

37. $\mathbf{u} \times \mathbf{v} = \begin{vmatrix} \mathbf{i} & \mathbf{j} & \mathbf{k} \\ -2 & 8 & 2 \\ 1 & 1 & -1 \end{vmatrix} = \langle -10, 0, -10 \rangle$

39. $\mathbf{u} \times \mathbf{v} = \begin{vmatrix} \mathbf{i} & \mathbf{j} & \mathbf{k} \\ -3 & 2 & -5 \\ 10 & -15 & 2 \end{vmatrix} = \langle -71, -44, 25 \rangle$

$\|\mathbf{u} \times \mathbf{v}\| = \sqrt{7602}$

Unit vector: $\dfrac{1}{\sqrt{7602}} \langle -71, -44, 25 \rangle$

41. First two points: $\langle 3, 2, 3 \rangle$

Last two points: $\langle 3, 2, 3 \rangle$

First and third points: $\langle -2, 2, 0 \rangle$

$\begin{vmatrix} \mathbf{i} & \mathbf{j} & \mathbf{k} \\ 3 & 2 & 3 \\ -2 & 2 & 0 \end{vmatrix} = \langle -6, -6, 10 \rangle$

Area $= |\langle -6, -6, 10 \rangle|$

$\qquad = \sqrt{36 + 36 + 100}$

$\qquad = \sqrt{172}$

$\qquad = 2\sqrt{43}$ square units

43. The parallelogram is determined by the three vectors with initial point $(0, 0, 0)$.

$\mathbf{u} = \langle 3, 0, 0 \rangle$, $\mathbf{v} = \langle 2, 0, 5 \rangle$, $\mathbf{w} = \langle 0, 5, 1 \rangle$

$\mathbf{u} \cdot (\mathbf{v} \times \mathbf{w}) = \begin{vmatrix} 3 & 0 & 0 \\ 2 & 0 & 5 \\ 0 & 5 & 1 \end{vmatrix} = -75$

Volume $= |-75| = 75$ cubic units

45. $\mathbf{v} = \langle 3 + 1, 6 - 3, -1 - 5 \rangle = \langle 4, 3, -6 \rangle$, point: $(-1, 3, 5)$

(a) Parametric equations: $x = -1 + 4t, y = 3 + 3t, z = 5 - 6t$

(b) Symmetric equations: $\dfrac{x + 1}{4} = \dfrac{y - 3}{3} = \dfrac{z - 5}{-6}$

47. Use $2\mathbf{v} = \langle -4, 5, 2 \rangle$, point: $(0, 0, 0)$.

(a) Parametric equations: $x = -4t, y = 5t, z = 2t$

(b) Symmetric equations: $\dfrac{x}{-4} = \dfrac{y}{5} = \dfrac{z}{2}$

49. $\mathbf{u} = \langle 5, 0, 2 \rangle, \mathbf{v} = \langle 2, 3, 8 \rangle$

$\mathbf{u} \times \mathbf{v} = \begin{vmatrix} \mathbf{i} & \mathbf{j} & \mathbf{k} \\ 5 & 0 & 2 \\ 2 & 3 & 8 \end{vmatrix} = \langle -6, -36, 15 \rangle$

$\mathbf{n} = \langle 2, 12, -5 \rangle$

$a(x - x_0) + b(y - y_0) + c(z - z_0) = 0$

$2(x - 0) + 12(y - 0) - 5(z - 0) = 0$

$2x + 12y - 5z = 0$

51. $\mathbf{n} = \mathbf{k}$, normal vector

Plane: $0(x - 5) + 0(y - 3) + 1(z - 2) = 0$

$z - 2 = 0$

53. $3x - 2y + 3z = 6$

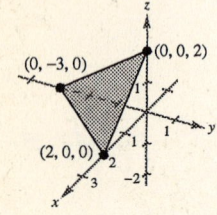

55. $2x - 3z = 6$

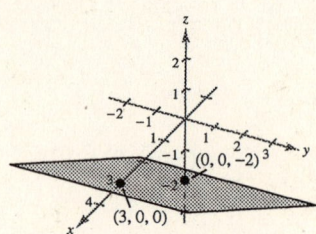

57. $\mathbf{n} = \langle 2, -20, 6 \rangle, P = (0, 0, 1)$ in plane, $Q = (2, 3, 10), \overrightarrow{PQ} = \langle 2, 3, 9 \rangle$

$D = \dfrac{|\overrightarrow{PQ} \cdot \mathbf{n}|}{\|\mathbf{n}\|} = \dfrac{|-2|}{\sqrt{440}} = \dfrac{1}{\sqrt{110}} = \dfrac{\sqrt{110}}{110} \approx 0.0953$

59. $\mathbf{n} = \langle 1, -10, 3 \rangle, P = (2, 0, 0)$ in plane, $Q = (0, 0, 0), \overrightarrow{PQ} = \langle -2, 0, 0 \rangle$

$D = \dfrac{|\overrightarrow{PQ} \cdot \mathbf{n}|}{\|\mathbf{n}\|} = \dfrac{|-2|}{\sqrt{1 + 100 + 9}} = \dfrac{2}{\sqrt{110}} = \dfrac{2\sqrt{110}}{110} = \dfrac{\sqrt{110}}{55} \approx 0.191$

61. False. $\mathbf{a} \times \mathbf{b} = -(\mathbf{b} \times \mathbf{a})$

63. $\mathbf{u} \cdot \mathbf{u} = \langle 3, -2, 1 \rangle \cdot \langle 3, -2, 1 \rangle$

$= 9 + 4 + 1$

$= 14$

$= \|\mathbf{u}\|^2$

65. $\mathbf{u} \cdot (\mathbf{v} + \mathbf{w}) = \langle 3, -2, 1 \rangle \cdot \langle 1, -2, -1 \rangle = 6$

$\mathbf{u} \cdot \mathbf{v} + \mathbf{u} \cdot \mathbf{w} = 11 + (-5) = 6$

Chapter 10   Practice Test

1. Find the lengths of the sides of the triangle with vertices $(0, 0, 0)$, $(1, 2, -4)$, and $(0, -2, -1)$. Show that the triangle is a right triangle.

2. Find the standard form of the equation of a sphere having center $(0, 4, 1)$ and radius 5.

3. Find the center and radius of the sphere $x^2 + y^2 + z^2 + 2x - 4z - 11 = 0$.

4. Find the vector $\mathbf{u} - 3\mathbf{v}$ given $\mathbf{u} = \langle 1, 0, -1 \rangle$ and $\mathbf{v} = \langle 4, 3, -6 \rangle$.

5. Find the length of $\frac{1}{2}\mathbf{v}$ if $\mathbf{v} = \langle 2, 4, -6 \rangle$.

6. Find the dot product of $\mathbf{u} = \langle 2, 1, -3 \rangle$ and $\mathbf{v} = \langle 1, 1, -2 \rangle$.

7. Determine whether $\mathbf{u} = \langle 1, 1, -1 \rangle$ and $\mathbf{v} = \langle -3, -3, 3 \rangle$ are orthogonal, parallel, or neither.

8. Find the cross product of $\mathbf{u} = \langle -1, 0, 2 \rangle$ and $\mathbf{v} = \langle 1, -1, 3 \rangle$. What is $\mathbf{v} \times \mathbf{u}$?

9. Use the triple scalar product to find the volume of the parallelepiped having adjacent edges $\mathbf{u} = \langle 1, 1, 1 \rangle$, $\mathbf{v} = \langle 0, -1, 1 \rangle$, and $\mathbf{w} = \langle 1, 0, 4 \rangle$.

10. Find a set of parametric equations for the line through the points $(0, -3, 3)$ and $(2, -3, 4)$.

11. Find an equation of the plane passing through $(1, 2, 3)$ and perpendicular to the vector $\mathbf{n} = \langle 1, -1, 0 \rangle$.

12. Find an equation of the plane passing through the three points $A = (0, 0, 0)$, $B = (1, 1, 1)$, and $C = (1, 2, 3)$.

13. Determine whether the planes $x + y - z = 12$ and $3x - 4y - z = 9$ are parallel, orthogonal or neither.

14. Find the distance between the point $(1, 1, 1)$ and the plane $x + 2y + z = 6$.

CHAPTER 11
Limits and an Introduction to Calculus

C H A P T E R 1 1
Limits and an Introduction to Calculus

Section 11.1   Introduction to Limits

- ■ If $f(x)$ becomes arbitrarily close to a unique number L as x approaches c from either side, then the limit of $f(x)$ as x approaches c is L:

    $$\lim_{x \to c} f(x) = L.$$

- ■ You should be able to use a calculator to find a limit.
- ■ You should be able to use a graph to find a limit.
- ■ You should understand how limits can fail to exist:
    - (a) $f(x)$ approaches a different number from the right of c than it approaches from the left of c.
    - (b) $f(x)$ increases or decreases without bound as x approaches c.
    - (c) $f(x)$ oscillates between two fixed values as x approaches c.
- ■ You should know and be able to use the elementary properties of limits.

Solutions to Odd-Numbered Exercises

1. (a)

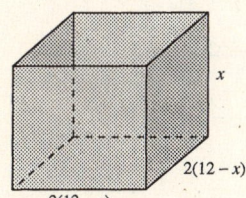

(b) $V = (\text{base})\text{height} = (24 - 2x)^2 x = 4x(12 - x)^2$

(d)

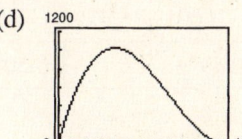

Maximum at $x = 4$

(c) $\lim_{x \to 4} V = 1024$

x	3	3.5	3.9	4	4.1	4.5	5
V	972.0	1011.5	1023.5	1024.0	1023.5	1012.5	980.0

3. $\lim_{x \to 2} (5x + 4) = 14$

x	1.9	1.99	1.999	2	2.001	2.01	2.1
$f(x)$	13.5	13.95	13.995	14	14.005	14.05	14.5

The limit is reached.

5. $\lim\limits_{x \to 3} \dfrac{x-3}{x^2-9} = \dfrac{1}{6}$

x	2.9	2.99	2.999	3	3.001	3.01	3.1
$f(x)$	0.1695	0.1669	0.16669	?	0.16664	0.1664	0.1639

The limit is not reached.

7. $\lim\limits_{x \to 1} \dfrac{x-1}{x^2+2x-3} = \dfrac{1}{4}$

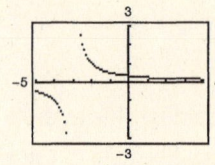

x	0.9	0.99	0.999	1.0	1.001	1.01	1.1
$f(x)$	0.2564	0.2506	0.2501	?	0.2499	0.2494	0.2439

9. $\lim\limits_{x \to 0} \dfrac{\sqrt{x+5}-\sqrt{5}}{x} \approx 0.2236$ $\left(\text{Actual limit is } \dfrac{1}{2\sqrt{5}}.\right)$

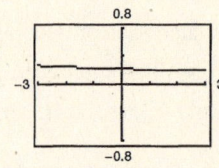

x	-0.1	-0.01	-0.001	0	0.001	0.01	0.1
$f(x)$	0.2247	0.2237	0.2236	?	0.2236	0.2235	0.2225

11. $\lim\limits_{x \to -4} \dfrac{[x/(x+2)]-2}{x+4} = \dfrac{1}{2}$

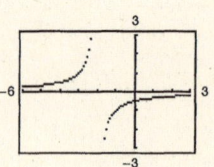

x	-4.1	-4.01	-4.001	-4.0	-3.999	-3.99	-3.9
$f(x)$	0.4762	0.4975	0.4998	?	0.5003	0.5025	0.5263

13. Make sure your calculator is set in radian mode.

$$\lim\limits_{x \to 0} \dfrac{\sin x}{x} = 1$$

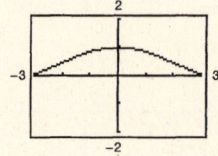

x	-0.1	-0.01	-0.001	0	0.001	0.01	0.1
$f(x)$	0.9983	0.99998	0.9999998	?	0.9999998	0.99998	0.9983

15. $f(x) = \begin{cases} 2x+1, & x < 2 \\ x+3, & x \geq 2 \end{cases}$

The limit exists as x approaches 2:

$$\lim\limits_{x \to 2} f(x) = 5$$

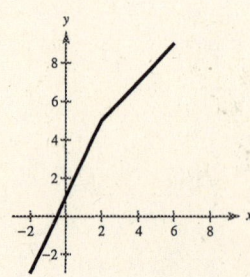

17. $\lim\limits_{x \to -4} (x^2-3) = 13$

19. $\lim\limits_{x \to -2} \dfrac{|x+2|}{x+2}$ does not exist. $f(x) = \dfrac{|x+2|}{x+2}$ equals -1 to the left of -2, and equals 1 to the right of -2.

21. The limit does not exist because $f(x)$ oscillates between 2 and -2.

23.

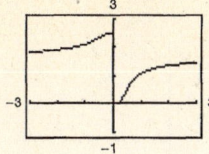

$\displaystyle\lim_{x \to 0} \frac{5}{2 + e^{1/x}}$ does not exist.

25.

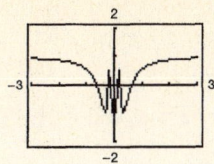

$\displaystyle\lim_{x \to 0} \cos\frac{1}{x}$ does not exist. The graph oscillates between -1 and 1.

27.

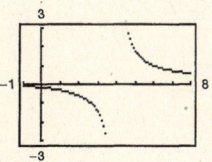

$\displaystyle\lim_{x \to 4} \frac{\sqrt{x + 3} - 1}{x - 4}$ does not exist.

29.

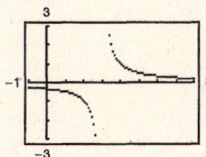

$\displaystyle\lim_{x \to 1} \frac{x - 1}{x^2 - 4x + 3} = -\frac{1}{2}$

31.

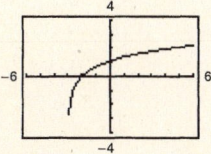

$\displaystyle\lim_{x \to 4} \ln(x + 3) \approx 1.946$ (Exact limit is $\ln 7$.)

33. (a) $\displaystyle\lim_{x \to c} [-2g(x)] = -2(6) = -12$

(b) $\displaystyle\lim_{x \to c} [f(x) + g(x)] = 3 + 6 = 9$

(c) $\displaystyle\lim_{x \to c} \frac{f(x)}{g(x)} = \frac{3}{6} = \frac{1}{2}$

(d) $\displaystyle\lim_{x \to c} \sqrt{f(x)} = \sqrt{3}$

35. (a) $\displaystyle\lim_{x \to 2} f(x) = 2^3 = 8$

(c) $\displaystyle\lim_{x \to 2} [f(x)g(x)] = 8\left(\frac{3}{8}\right) = 3$

(b) $\displaystyle\lim_{x \to 2} g(x) = \frac{\sqrt{2^2 + 5}}{2(2^2)} = \frac{3}{8}$

(d) $\displaystyle\lim_{x \to 2} [g(x) - f(x)] = \frac{3}{8} - 8 = -\frac{61}{8}$

37. $\displaystyle\lim_{x \to 5} (10 - x^2) = 10 - 5^2 = -15$

39. $\displaystyle\lim_{x \to -3} (2x^2 + 4x + 1) = 2(-3)^2 + 4(-3) + 1 = 7$

41. $\displaystyle\lim_{x \to 3} \left(-\frac{9}{x}\right) = -\frac{9}{3} = -3$

43. $\displaystyle\lim_{x \to -3} \frac{3x}{x^2 + 1} = -\frac{9}{10}$

45. $\displaystyle\lim_{x \to -2} \frac{5x + 3}{2x - 9} = \frac{5(-2) + 3}{2(-2) - 9} = \frac{-7}{-13} = \frac{7}{13}$

47. $\displaystyle\lim_{x \to -1} \sqrt{x + 2} = \sqrt{-1 + 2} = 1$

49. $\displaystyle\lim_{x \to 7} \frac{5x}{\sqrt{x + 2}} = \frac{5(7)}{\sqrt{7 + 2}} = \frac{35}{3}$

51. $\displaystyle\lim_{x \to 3} e^x = e^3 \approx 20.0855$

53. $\displaystyle\lim_{x \to \pi} \sin 2x = \sin 2\pi = 0$

55. $\displaystyle\lim_{x \to 1/2} \arcsin x = \arcsin\frac{1}{2} = \frac{\pi}{6} \approx 0.5236$

57. True

59. Answers will vary.

61. (a) No. The limit may or may not exist. And if it does exist, it may not equal 4.

(b) No. $f(2)$ may or may not exist. And if $f(2)$ exists, it may not equal 4.

63. $f(x) = \dfrac{x - 9}{\sqrt{x} - 3}$

$\lim\limits_{x \to 9} f(x) = 6$

Domain: all $x \geq 0$, $x \neq 9$

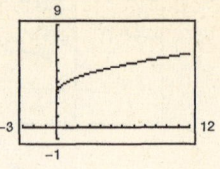

It is difficult to determine the domain from the graph because it is not obvious that the function is undefined at $x = 9$.

65. $\dfrac{5 - x}{3x - 15} = \dfrac{5 - x}{-3(5 - x)} = -\dfrac{1}{3}$, $x \neq 5$

67. $\dfrac{15x^2 + 7x - 4}{15x^2 + x - 2} = \dfrac{(3x - 1)(5x + 4)}{(3x - 1)(5x + 2)}$

$\qquad = \dfrac{5x + 4}{5x + 2}$, $x \neq \dfrac{1}{3}$

69. $\dfrac{x^3 + 27}{x^2 + x - 6} = \dfrac{(x + 3)(x^2 - 3x + 9)}{(x + 3)(x - 2)}$

$\qquad = \dfrac{x^2 - 3x + 9}{x - 2}$, $x \neq -3$

71. $d = \sqrt{(3 - 3)^2 + (2 - 2)^2 + (8 - 7)^2} = 1$

73. $d = \sqrt{(0 - 3)^2 + (5 + 3)^2 + (-5 - 0)^2}$

$\quad = \sqrt{9 + 64 + 25}$

$\quad = \sqrt{98}$

$\quad = 7\sqrt{2}$

Section 11.2   Techniques for Evaluating Limits

> ■   You can use direct substitution to find the limit of a polynomial function $p(x)$:
>
> $\qquad \lim\limits_{x \to c} p(x) = p(c).$
>
> ■   You can use direct substitution to find the limit of a rational function $r(x) = \dfrac{p(x)}{q(x)}$, as long as $q(c) \neq 0$:
>
> $\qquad \lim\limits_{x \to c} r(x) = r(c) = \dfrac{p(c)}{q(c)}$, $q(c) \neq 0.$
>
> ■   You should be able to use cancellation techniques to find a limit.
> ■   You should know how to use rationalization techniques to find a limit.
> ■   You should know how to use technology to find a limit.
> ■   You should be able to calculate one-sided limits.

Solutions to Odd-Numbered Exercises

1. $g(x) = \dfrac{-2x^2 + x}{x}$, $g_2(x) = -2x + 1$

(a) $\lim\limits_{x \to 0} g(x) = 1$ (b) $\lim\limits_{x \to -1} g(x) = 3$

(c) $\lim\limits_{x \to -2} g(x) = 5$

3. $g(x) = \dfrac{x^3 - x}{x - 1}$, $g_2(x) = x^2 + x = x(x + 1)$

(a) $\lim\limits_{x \to 1} g(x) = 2$ (b) $\lim\limits_{x \to -1} g(x) = 0$

(c) $\lim\limits_{x \to 0} g(x) = 0$

5. $\lim\limits_{x\to6}\dfrac{x-6}{x^2-36}=\lim\limits_{x\to6}\dfrac{x-6}{(x-6)(x+6)}$

$\qquad\qquad=\lim\limits_{x\to6}\dfrac{1}{x+6}=\dfrac{1}{12}$

7. $\lim\limits_{x\to-1}\dfrac{1-2x-3x^2}{1+x}=\lim\limits_{x\to-1}\dfrac{(1+x)(1-3x)}{1+x}$

$\qquad\qquad=\lim\limits_{x\to-1}(1-3x)=4$

9. $\lim\limits_{t\to2}\dfrac{t^3-8}{t-2}=\lim\limits_{t\to2}\dfrac{(t-2)(t^2+2t+4)}{t-2}$

$\qquad\qquad=\lim\limits_{t\to2}(t^2+2t+4)$

$\qquad\qquad=4+4+4=12$

11. $\lim\limits_{y\to0}\dfrac{\sqrt{5+y}-\sqrt{5}}{y}=\lim\limits_{y\to0}\dfrac{\sqrt{5+y}-\sqrt{5}}{y}\cdot\dfrac{\sqrt{5+y}+\sqrt{5}}{\sqrt{5+y}+\sqrt{5}}$

$\qquad\qquad=\lim\limits_{y\to0}\dfrac{(5+y)-5}{y\left(\sqrt{5+y}+\sqrt{5}\right)}$

$\qquad\qquad=\lim\limits_{y\to0}\dfrac{1}{\sqrt{5+y}+\sqrt{5}}$

$\qquad\qquad=\dfrac{1}{2\sqrt{5}}$

$\qquad\qquad=\dfrac{\sqrt{5}}{10}$

13. $\lim\limits_{x\to-3}\dfrac{\sqrt{x+7}-2}{x+3}=\lim\limits_{x\to-3}\dfrac{\sqrt{x+7}-2}{x+3}\cdot\dfrac{\sqrt{x+7}+2}{\sqrt{x+7}+2}$

$\qquad\qquad=\lim\limits_{x\to-3}\dfrac{(x+7)-4}{(x+3)\left(\sqrt{x+7}+2\right)}$

$\qquad\qquad=\lim\limits_{x\to-3}\dfrac{1}{\sqrt{x+7}+2}$

$\qquad\qquad=\dfrac{1}{4}$

15. $\lim\limits_{x\to0}\dfrac{1/(1+x)-1}{x}=\lim\limits_{x\to0}\dfrac{1-(1+x)}{(1+x)x}$

$\qquad\qquad=\lim\limits_{x\to0}\dfrac{-1}{1+x}=-1$

17. $\lim\limits_{x\to0}\dfrac{\sec x}{\tan x}=\lim\limits_{x\to0}\dfrac{1}{\cos x}\cdot\dfrac{\cos x}{\sin x}$

$\qquad\qquad=\lim\limits_{x\to0}\dfrac{1}{\sin x},$ does not exist

19. $f(x)=\dfrac{\sqrt{x+3}-\sqrt{3}}{x}$

$\lim\limits_{x\to0}f(x)\approx0.2887$

$\left(\text{Exact limit: }\dfrac{1}{2\sqrt{3}}\right)$

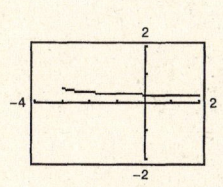

21. $f(x)=\dfrac{\sqrt{2x+1}-1}{x}$

$\lim\limits_{x\to0}f(x)=1$

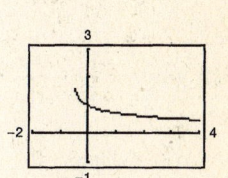

23. $\lim\limits_{x\to 2}\dfrac{x^5-32}{x-2}=80$

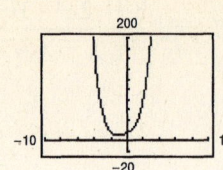

25. $f(x)=\dfrac{1/(x+4)-(1/4)}{x}$

$\lim\limits_{x\to 0}f(x)=-\dfrac{1}{16},\ (-0.0625)$

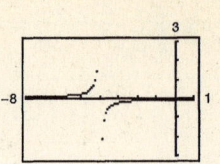

27. $f(x)=\dfrac{e^{2x}-1}{x}$

$\lim\limits_{x\to 0}f(x)=2$

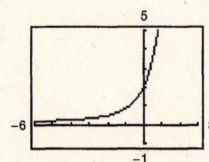

29. $\lim\limits_{x\to 0^+}x\ln x=0$

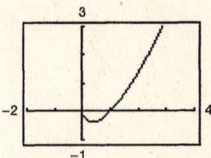

31. $\lim\limits_{x\to 0}\dfrac{\sin 2x}{x}=2$

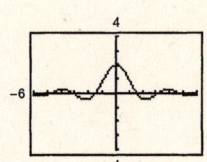

33. $\lim\limits_{x\to 0}\dfrac{\tan x}{x}=1$

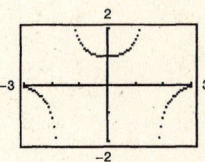

35. $\lim\limits_{x\to 1}\dfrac{1-\sqrt[3]{x}}{1-x}=\dfrac{1}{3}\approx 0.333$

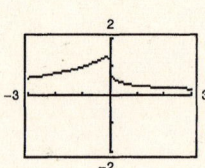

37. $f(x)=(1-x)^{2/x}$

$\lim\limits_{x\to 0}f(x)\approx 0.135$

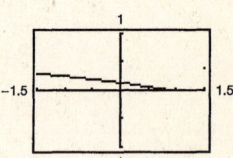

39. $f(x)=\dfrac{x-1}{x^2-1}$

(a) Graphically, $\lim\limits_{x\to 1^-}\dfrac{x-1}{x^2-1}=\dfrac{1}{2}$.

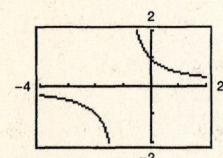

(b)

x	0.5	0.9	0.99	0.999	1
$f(x)$	0.6667	0.5263	0.5025	0.5003	0.5

Numerically, $\lim\limits_{x\to 1^-}\dfrac{x-1}{x^2-1}=\dfrac{1}{2}$.

(c) Algebraically, $\lim\limits_{x\to 1^-}\dfrac{x-1}{x^2-1}=\lim\limits_{x\to 1^-}\dfrac{x-1}{(x-1)(x+1)}=\lim\limits_{x\to 1^-}\dfrac{1}{x+1}=\dfrac{1}{2}$.

41. $f(x)=\dfrac{4-\sqrt{x}}{x-16}$

(a) Graphically, $\lim\limits_{x\to 16^+}\dfrac{4-\sqrt{x}}{x-16}=-\dfrac{1}{8}$.

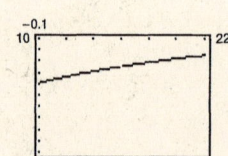

—CONTINUED—

41. —CONTINUED—

(b)

x	16	16.001	16.01	16.1	16.5
$f(x)$	-0.1250	-0.1250	-0.1250	-0.1248	-0.1240

Numerically, $\displaystyle\lim_{x\to 16^+}\frac{4-\sqrt{x}}{x-16}=-0.125.$

(c) Algebraically, $\displaystyle\lim_{x\to 16^+}\frac{4-\sqrt{x}}{x-16}=\lim_{x\to 16^+}\frac{4-\sqrt{x}}{\left(\sqrt{x}-4\right)\left(\sqrt{x}+4\right)}$

$$=\lim_{x\to 16^+}\frac{-1}{\sqrt{x}+4}$$

$$=\frac{-1}{4+4}=-\frac{1}{8}.$$

43. $f(x)=\dfrac{|x-6|}{x-6}$

$\displaystyle\lim_{x\to 6^+}f(x)=1$

$\displaystyle\lim_{x\to 6^-}f(x)=-1$

Limit does not exist.

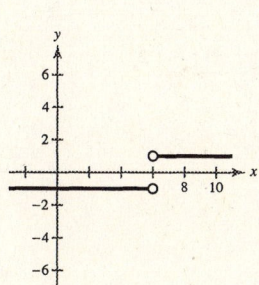

45. $f(x)=\dfrac{1}{x^2+1}$

$\displaystyle\lim_{x\to 1^-}\frac{1}{x^2+1}=\lim_{x\to 1^+}\frac{1}{x^2+1}$

$\displaystyle\qquad=\lim_{x\to 1}\frac{1}{x^2+1}$

$\displaystyle\qquad=\frac{1}{2}$

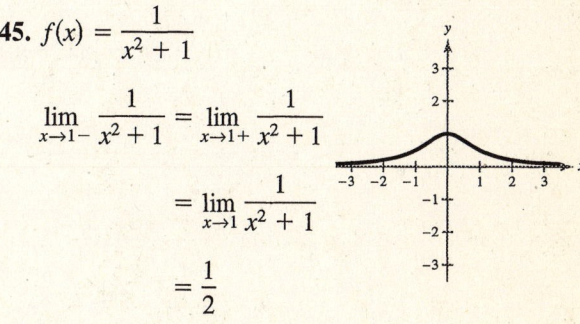

47. $\displaystyle\lim_{x\to 2^-}f(x)=2-1=1$

$\displaystyle\lim_{x\to 2^+}f(x)=2(2)-3=1$

$\displaystyle\lim_{x\to 2}f(x)=1$

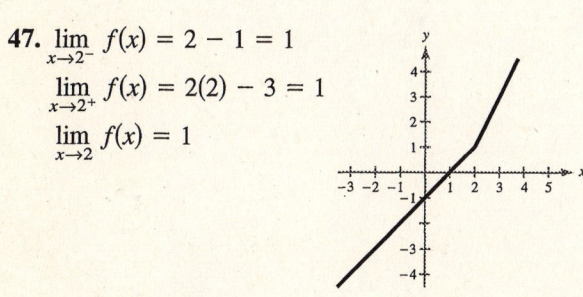

49. $f(x)=\begin{cases}4-x^2, & x\le 1\\ 3-x, & x>1\end{cases}$

$\displaystyle\lim_{x\to 1^-}f(x)=4-1=3$

$\displaystyle\lim_{x\to 1^+}f(x)=3-1=2$

$\displaystyle\lim_{x\to 1}f(x)$ does not exist.

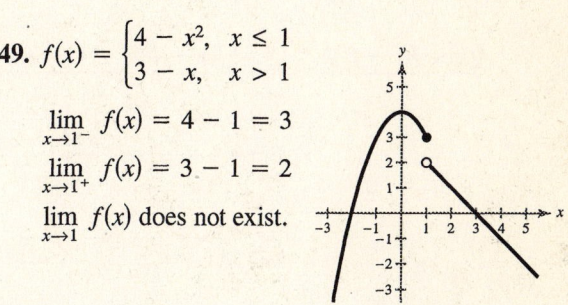

51.

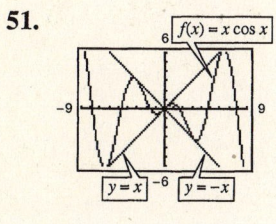

$\displaystyle\lim_{x\to 0}f(x)=0$

53.

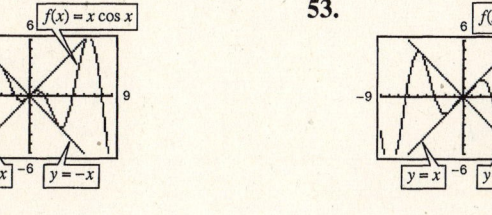

$\displaystyle\lim_{x\to 0}f(x)=0$

55.

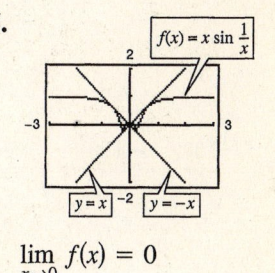

$\displaystyle\lim_{x\to 0}f(x)=0$

57. (a) Can be evaluated by direct substitution: $\displaystyle\lim_{x\to 0}x^2\sin x^2=0^2\sin 0^2=0$

(b) Cannot be evaluated by direct substitution: $\displaystyle\lim_{x\to 0}\frac{\sin x^2}{x^2}=1$

59. $\lim\limits_{h \to 0} \dfrac{f(x+h) - f(x)}{h} = \lim\limits_{h \to 0} \dfrac{3(x+h) - 1 - (3x-1)}{h}$

$= \lim\limits_{h \to 0} \dfrac{3x + 3h - 1 - 3x + 1}{h}$

$= \lim\limits_{h \to 0} \dfrac{3h}{h} = 3$

61. $\lim\limits_{h \to 0} \dfrac{f(x+h) - f(x)}{h} = \lim\limits_{h \to 0} \dfrac{\sqrt{x+h} - \sqrt{x}}{h} \cdot \left(\dfrac{\sqrt{x+h} + \sqrt{x}}{\sqrt{x+h} + \sqrt{x}} \right)$

$= \lim\limits_{h \to 0} \dfrac{(x+h) - x}{h\left(\sqrt{x+h} + \sqrt{x}\right)}$

$= \lim\limits_{h \to 0} \dfrac{1}{\sqrt{x+h} + \sqrt{x}} = \dfrac{1}{2\sqrt{x}}$

63. $\lim\limits_{h \to 0} \dfrac{f(x+h) - f(x)}{h} = \lim\limits_{h \to 0} \dfrac{\left((x+h)^2 - 3(x+h)\right) - (x^2 - 3x)}{h}$

$= \lim\limits_{h \to 0} \dfrac{x^2 + 2xh + h^2 - 3x - 3h - x^2 + 3x}{h}$

$= \lim\limits_{h \to 0} \dfrac{2xh + h^2 - 3h}{h}$

$= \lim\limits_{h \to 0} (2x + h - 3) = 2x - 3$

65. $\lim\limits_{h \to 0} \dfrac{f(x+h) - f(x)}{h} = \lim\limits_{h \to 0} \dfrac{1/(x+h+2) - 1/(x+2)}{h}$

$= \lim\limits_{h \to 0} \dfrac{(x+2) - (x+h+2)}{h(x+h+2)(x+2)}$

$= \lim\limits_{h \to 0} \dfrac{-h}{h(x+h+2)(x+2)}$

$= \lim\limits_{h \to 0} \dfrac{-1}{(x+h+2)(x+2)}$

$= \dfrac{-1}{(x+2)^2}$

67. $\lim\limits_{t \to 1} \dfrac{(-16(1) + 128) - (-16t^2 + 128)}{1 - t} = \lim\limits_{t \to 1} \dfrac{16t^2 - 16}{1 - t}$

$= \lim\limits_{t \to 1} \dfrac{16(t-1)(t+1)}{1 - t}$

$= \lim\limits_{t \to 1} -16(t+1)$

$= -32 \dfrac{\text{ft}}{\text{sec}}$

69. $C(t) = 0.75 - 0.50[\![-(t - 1)]\!]$

(a)

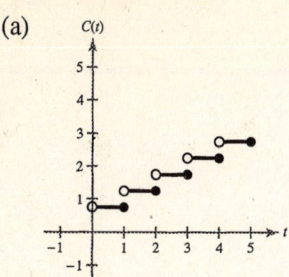

(b)

t	3	3.3	3.4	3.5	3.6	3.7	4
C	1.75	2.25	2.25	2.25	2.25	2.25	2.25

$$\lim_{t \to 3.5} C(t) = 2.25$$

(c)

t	2	2.5	2.9	3	3.1	3.5	4
C	1.25	1.75	1.75	1.75	2.25	2.25	2.25

No, $\lim_{t \to 3} C(t)$ does not exist.

$$\lim_{t \to 3^-} C(t) = 1.75, \lim_{t \to 3^+} C(t) = 2.25$$

71. $\lim_{t \to 2^-} f(t) = 30.80, \lim_{t \to 2^+} f(t) = 33.88$

Thus, the limit of f as $t \to 2$ does not exist.

73. True

75. Many answers possible

(a)

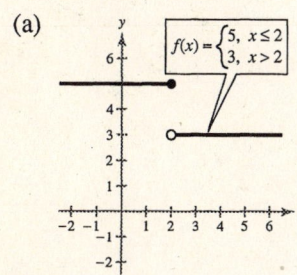

$f(x) = \begin{cases} 5, & x \le 2 \\ 3, & x > 2 \end{cases}$

(b)

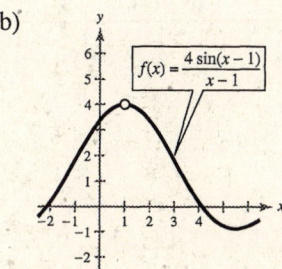

$f(x) = \dfrac{4 \sin(x - 1)}{x - 1}$

77. Slope of line through $(4, -6)$ and $(3, -4)$:

$$\frac{-6 + 4}{4 - 3} = -2$$

Slope of perpendicular line: $\dfrac{1}{2}$

Equation: $y + 10 = \dfrac{1}{2}(x - 6)$

$$2y - x + 26 = 0$$

79. $r = \dfrac{3}{1 + \cos \theta}, e = 1$, Parabola

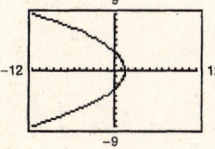

81. $r = \dfrac{9}{2 + 3 \cos \theta} = \dfrac{9/2}{1 + (3/2) \cos \theta}, e = \dfrac{3}{2}$,

Hyperbola

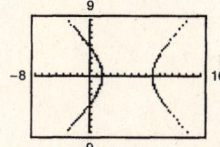

83. $r = \dfrac{5}{1 - \sin \theta}, e = 1$, Parabola

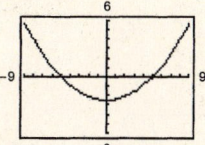

85. $\langle 7, -2, 3 \rangle \cdot \langle -1, 4, 5 \rangle = -7 - 8 + 15$

$$= 0 \implies \text{orthogonal}$$

87. $-3\langle -4, 3, -6 \rangle = \langle 12, -9, 18 \rangle \implies$ parallel

Section 11.3   The Tangent Line Problem

- ■ You should be able to visually approximate the slope of a graph.
- ■ The slope m of the graph of f at the point $(x, f(x))$ is given by

$$m = \lim_{h \to 0} \frac{f(x + h) - f(x)}{h}$$

provided this limit exists.

- ■ You should be able to use the limit definition to find the slope of a graph.
- ■ The derivative of f at x is given by

$$f'(x) = \lim_{h \to 0} \frac{f(x + h) - f(x)}{h}$$

provided this limit exists. Notice that this is the same limit as that for the tangent line slope.

- ■ You should be able to use the limit definition to find the derivative of a function.

Solutions to Odd-Numbered Exercises

1. Slope is 0 at (x, y).

3. Slope is $\frac{1}{2}$ at (x, y).

5. $m_{\text{sec}} = \dfrac{g(3 + h) - g(3)}{h} = \dfrac{(3 + h)^2 - 4(3 + h) - (-3)}{h} = \dfrac{h^2 + 2h}{h}$

$m = \lim_{h \to 0} \dfrac{h^2 + 2h}{h} = \lim_{h \to 0} \dfrac{h(h + 2)}{h} = \lim_{h \to 0} (h + 2) = 2$

7. $m_{\text{sec}} = \dfrac{g(1 + h) - g(1)}{h} = \dfrac{5 - 2(1 + h) - 3}{h} = \dfrac{-2h}{h}$

$m = \lim_{h \to 0} \dfrac{-2h}{h} = -2$

9. $m_{\text{sec}} = \dfrac{g(2 + h) - g(2)}{h} = \dfrac{[4/(2 + h)] - 2}{h} = \dfrac{4 - 2(2 + h)}{(2 + h)h} = \dfrac{-2}{2 + h}, \ h \neq 0$

$m = \lim_{h \to 0} \left(\dfrac{-2}{2 + h} \right) = -1$

11. $m_{\text{sec}} = \dfrac{h(9 + k) - h(9)}{k} = \dfrac{\sqrt{9 + k} - 3}{k} \cdot \dfrac{\sqrt{9 + k} + 3}{\sqrt{9 + k} + 3} = \dfrac{(9 + k) - 9}{k[\sqrt{9 + k} + 3]} = \dfrac{1}{\sqrt{9 + k} + 3}, \ k \neq 0$

$m = \lim_{k \to 0} \dfrac{1}{\sqrt{9 + k} + 3} = \dfrac{1}{6}$

13. $m_{\text{sec}} = \dfrac{g(x + h) - g(x)}{h} = \dfrac{4 - (x + h)^2 - (4 - x^2)}{h} = \dfrac{-2xh - h^2}{h} = -2x - h, \ h \neq 0$

$m = \lim_{h \to 0} (-2x - h) = -2x$

(a) At $(0, 4)$, $m = -2(0) = 0$.

(b) At $(-1, 3)$, $m = -2(-1) = 2$.

15. $m_{sec} = \dfrac{g(x+h) - g(x)}{h} = \dfrac{\dfrac{1}{x+h+4} - \dfrac{1}{x+4}}{h} = \dfrac{(x+4) - (x+4+h)}{(x+h+4)(x+4)(h)}$

$= \dfrac{-h}{(x+h+4)(x+4)h} = \dfrac{-1}{(x+h+4)(x+4)}, \; h \neq 0$

$m = \lim\limits_{h \to 0} \dfrac{-1}{(x+h+4)(x+4)} = \dfrac{-1}{(x+4)^2}$

(a) At $\left(0, \dfrac{1}{4}\right)$, $m = \dfrac{-1}{(0+4)^2} = \dfrac{-1}{16}$.

(b) At $\left(-2, \dfrac{1}{2}\right)$, $m = \dfrac{-1}{(-2+4)^2} = \dfrac{-1}{4}$.

17. $m_{sec} = \dfrac{g(x+h) - g(x)}{h} = \dfrac{\sqrt{x+h-1} - \sqrt{x-1}}{h} \cdot \dfrac{\sqrt{x+h+1} + \sqrt{x-1}}{\sqrt{x+h-1} + \sqrt{x-1}}$

$= \dfrac{(x+h-1) - (x-1)}{h\left(\sqrt{x+h-1} + \sqrt{x-1}\right)} = \dfrac{1}{\sqrt{x+h-1} + \sqrt{x-1}}, \; h \neq 0$

$m = \lim\limits_{h \to 0} \left(\dfrac{1}{\sqrt{x+h-1} + \sqrt{x-1}} \right) = \dfrac{1}{2\sqrt{x-1}}$

(a) At $(5, 2)$, $m = \dfrac{1}{2\sqrt{5} - 1} = \dfrac{1}{4}$.

(b) At $(10, 3)$, $m = \dfrac{1}{2\sqrt{10} - 1} = \dfrac{1}{6}$.

19.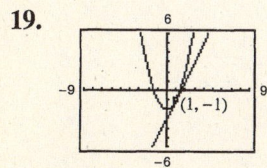
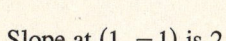

Slope at $(1, -1)$ is 2.

21.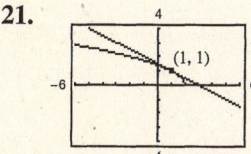
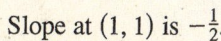

Slope at $(1, 1)$ is $-\dfrac{1}{2}$.

23.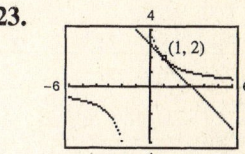

Slope at $(1, 2)$ is -1.

25. $f'(x) = \lim\limits_{h \to 0} \dfrac{f(x+h) - f(x)}{h} = \lim\limits_{h \to 0} \dfrac{5 - 5}{h} = 0$

27. $g'(x) = \lim\limits_{h \to 0} \dfrac{g(x+h) - g(x)}{h} = \lim\limits_{h \to 0} \dfrac{\left[9 - \frac{1}{3}(x+h)\right] - \left[9 - \frac{1}{3}x\right]}{h} = \lim\limits_{h \to 0} \dfrac{-\frac{1}{3}h}{h} = -\dfrac{1}{3}$

29. $f'(x) = \lim\limits_{h \to 0} \dfrac{f(x+h) - f(x)}{h} = \lim\limits_{h \to 0} \dfrac{\left[4 - 3(x+h)^2\right] - (4 - 3x^2)}{h}$

$= \lim\limits_{h \to 0} \dfrac{-3(x^2 + 2xh + h^2) + 3x^2}{h} = \lim\limits_{h \to 0} \dfrac{-6xh - 3h^2}{h} = \lim\limits_{h \to 0}(-6x - 3h) = -6x$

31. $f'(x) = \lim\limits_{h \to 0} \dfrac{f(x+h) - f(x)}{h} = \lim\limits_{h \to 0} \dfrac{\dfrac{1}{(x+h)^2} - \dfrac{1}{x^2}}{h}$

$= \lim\limits_{h \to 0} \dfrac{x^2 - (x^2 + 2xh + h^2)}{(x+h)^2 x^2 h} = \lim\limits_{h \to 0} \dfrac{-2x - h}{(x+h)^2 x^2} = -\dfrac{2x}{x^4} = -\dfrac{2}{x^3}$

33. $f'(x) = \lim\limits_{h \to 0} \dfrac{f(x+h) - f(x)}{h} = \lim\limits_{h \to 0} \dfrac{\dfrac{1}{\sqrt{x+h-9}} - \dfrac{1}{\sqrt{x-9}}}{h} \cdot \dfrac{\dfrac{1}{\sqrt{x+h-9}} + \dfrac{1}{\sqrt{x-9}}}{\dfrac{1}{\sqrt{x+h-9}} + \dfrac{1}{\sqrt{x-9}}}$

$= \lim\limits_{h \to 0} \dfrac{\dfrac{1}{(x+h-9)} - \dfrac{1}{(x-9)}}{h\left[\dfrac{1}{\sqrt{x+h-9}} + \dfrac{1}{\sqrt{x-9}}\right]} = \lim\limits_{h \to 0} \dfrac{(x-9) - (x+h-9)}{h(x+h-9)(x-9)\left[\dfrac{1}{\sqrt{x+h-9}} + \dfrac{1}{\sqrt{x-9}}\right]}$

$= \lim\limits_{h \to 0} \dfrac{-1}{(x+h-9)(x-9)\left[\dfrac{1}{\sqrt{x+h-9}} + \dfrac{1}{\sqrt{x-9}}\right]} = \dfrac{-1}{(x-9)^2\left[\dfrac{2}{\sqrt{x-9}}\right]} = \dfrac{-1}{2(x-9)^{3/2}}$

35. $m_{\text{sec}} = \dfrac{f(2+h) - f(2)}{h} = \dfrac{(2+h)^2 - 1 - 3}{h} = \dfrac{4h + h^2}{h} = 4 + h, h \neq 0$

$m = \lim\limits_{h \to 0}(4 + h) = 4$

Tangent line: $y - 3 = 4(x - 2)$

$\qquad\qquad\qquad y = 4x - 5$

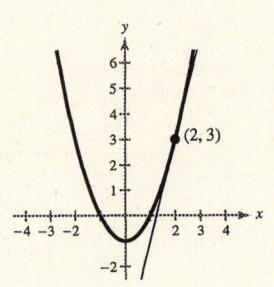

37. $m_{\text{sec}} = \dfrac{f(3+h) - f(3)}{h} = \dfrac{\sqrt{3+h+1} - 2}{h} \cdot \dfrac{\sqrt{4+h} + 2}{\sqrt{4+h} + 2} = \dfrac{(4+h) - 4}{h\left[\sqrt{4+h} + 2\right]} = \dfrac{1}{\sqrt{4+h} + 2}$

$m = \lim\limits_{h \to 0} \dfrac{1}{\sqrt{4+h} + 2} = \dfrac{1}{4}$

Tangent line: $y - 2 = \dfrac{1}{4}(x - 3)$

$\qquad\qquad\qquad 4y = x + 5$

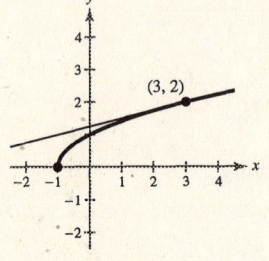

39.

x	-2	-1.5	-1	-0.5	0	0.5	1	1.5	2
$f(x)$	2	1.125	0.5	0.125	0	0.125	0.5	0.125	2
$f'(x)$	-2	-1.5	-1	-0.5	0	0.5	1	1.5	2

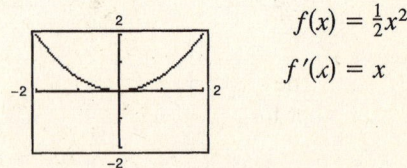

$f(x) = \frac{1}{2}x^2$

$f'(x) = x$

They appear to be the same.

41.

x	-2	-1.5	-1	-0.5	0	0.5	1	1.5	2
$f(x)$	1	1.225	1.414	1.581	1.732	1.871	2	2.121	2.236
$f'(x)$	0.5	0.408	0.354	0.316	0.289	0.267	0.25	0.236	0.224

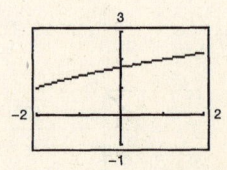

$f(x) = \sqrt{x + 3}$

$f'(x) = \dfrac{1}{2\sqrt{x + 3}}$

They appear to be the same.

43. $f'(x) = \lim_{h \to 0} \dfrac{f(x + h) - f(x)}{h} = \lim_{h \to 0} \dfrac{[(x + h)^2 - 4(x + h) + 3] - [x^2 - 4x + 3]}{h}$

$= \lim_{h \to 0} \dfrac{(x^2 + 2xh + h^2 - 4x - 4h + 3) - (x^2 - 4x + 3)}{h}$

$= \lim_{h \to 0} \dfrac{2xh + h^2 - 4h}{h} = \lim_{h \to 0} 2x + h - 4 = 2x - 4$

$f'(x) = 0 = 2x - 4 \implies x = 2$

f has a horizontal tangent at $(2, -1)$.

45. $f'(x) = \lim_{h \to 0} \dfrac{f(x + h) - f(x)}{h} = \lim_{h \to 0} \dfrac{3(x + h)^3 - 9(x + h) - (3x^3 - 9x)}{h}$

$= \lim_{h \to 0} \dfrac{9x^2h + 9xh^2 + 3h^3 - 9h}{h} = 9x^2 - 9$

$f'(x) = 0 = 9x^2 - 9 \implies x = \pm 1$

f has horizontal tangents at $(1, -6)$ and $(-1, 6)$.

47. (a) $y = 6.679x^2 - 37.74x + 145.0$

Quadratic model

(b)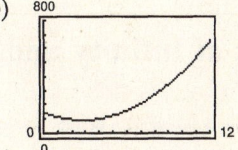

(c)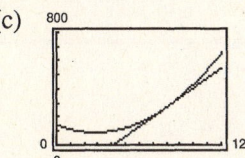

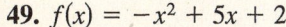

The slopes are the same.

At $x = 8$, slope ≈ 69.1. Sales are increasing at approximately 69.1 million per year at time 1998.

49. $f(x) = -x^2 + 5x + 2$

Using the definition of slope, you obtain $f'(x) = -2x + 5$.

For $0 \le x \le 2$, $f'(x) > 0 \implies$ height increasing.

For $4 \le x \le 6$, $f'(x) < 0 \implies$ height decreasing.

51. True. The slope is $2x$, which is different for all x.

53. Matches (b). (Derivative is always positive, but decreasing.)

55. Matches (d). (Derivative is -1 for $x < 0$, 1 for $x > 0$.)

57. Answers will vary.

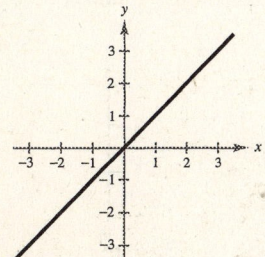

59. Answers will vary.

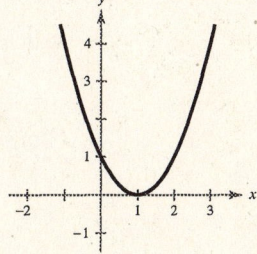

61. $f(x) = \dfrac{1}{x^2 - x - 2} = \dfrac{1}{(x-2)(x+1)}$

Vertical asymptotes: $x = 2, -1$

Horizontal asymptote: $y = 0$

Intercept: $\left(0, -\dfrac{1}{2}\right)$

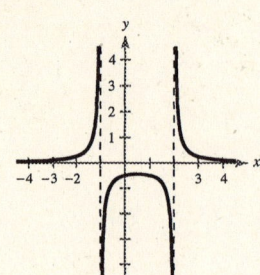

63. $f(x) = \dfrac{x^2 - x - 2}{x - 2}$

$= \dfrac{(x-2)(x+1)}{x-2} = x + 1, \; x \ne 2$

Line with hole at $(2, 3)$

Intercepts: $(0, 1), (-1, 0)$

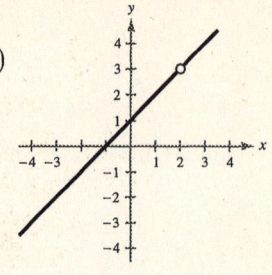

65. $\langle 1, 1, 1 \rangle \times \langle 2, 1, -1 \rangle = \begin{vmatrix} \mathbf{i} & \mathbf{j} & \mathbf{k} \\ 1 & 1 & 1 \\ 2 & 1 & -1 \end{vmatrix}$

$= \langle -2, 3, -1 \rangle$

67. $\langle -4, 10, 0 \rangle \times \langle 4, -1, 0 \rangle = \begin{vmatrix} \mathbf{i} & \mathbf{j} & \mathbf{k} \\ -4 & 10 & 0 \\ 4 & -1 & 0 \end{vmatrix}$

$= \langle 0, 0, -36 \rangle$

Section 11.4   Limits at Infinity and Limits of Sequences

- The limit at infinity
    $$\lim_{x \to \infty} f(x) = L$$
    means that $f(x)$ get arbitrarily close to L as x increases without bound.
- Similarly, the limit at infinity
    $$\lim_{x \to -\infty} f(x) = L$$
    means that $F(x)$ get arbitrarily close to L as x decreases without bound.
- You should be able to calculate limits at infinity, especially those arising from rational functions.
- Limits of functions can be used to evaluate limits of sequences. If f is a function such that $\lim_{x \to \infty} f(x) = L$ and if a_n is a sequence such that $f(n) = a_n$, then $\lim_{n \to \infty} a_n = L$.

Solutions to Odd-Numbered Exercises

1. Intercept: $(0, 0)$

Horizontal asymptote: $y = 4$

Matches (c).

3. Horizontal asymptote: $y = 4$

Vertical asymptote: $x = 0$

Matches (d).

5. $\lim\limits_{x \to \infty} \dfrac{3}{x^2} = 0$

7. $\lim\limits_{x \to \infty} \dfrac{3 + x}{3 - x} = -1$

9. $\lim\limits_{x \to -\infty} \dfrac{4x - 3}{2x + 1} = 2$

11. $\lim\limits_{x \to -\infty} \dfrac{3x^2 - 4}{1 - x^2} = -3$

13. $\lim\limits_{t \to \infty} \dfrac{t^2}{t + 3}$ does not exist.

15. $\lim\limits_{t \to \infty} \dfrac{1 - 2t + 6t^2}{5 + 3t - 4t^2} = \dfrac{6}{-4} = -\dfrac{3}{2}$

17. $\displaystyle\lim_{x\to-\infty}\frac{-(x^2+3)}{(2-x)^2}=\lim_{x\to-\infty}\frac{-x^2-3}{x^2-4x+4}=-1$

19. $\displaystyle\lim_{x\to-\infty}\left[\frac{x}{(x+1)^2}-4\right]=0-4=-4$

21. $\displaystyle\lim_{t\to\infty}\left(\frac{1}{3t^2}-\frac{5t}{t+2}\right)=0-5$
$$=-5$$

23. $y=\dfrac{3x}{1-x}$

Horizontal asymptote:
$y=-3$

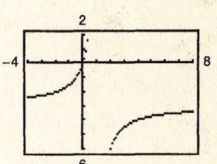

25. Horizontal asymptote:
$y=0$

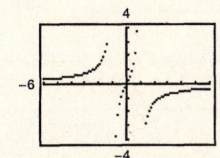

27. $y=1-\dfrac{3}{x^2}$

Horizontal asymptote:
$y=1$

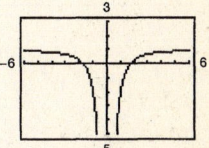

29.

x	10^0	10^1	10^2	10^3	10^4	10^5	10^6
$f(x)$	-0.7321	-0.0995	-0.00999	-0.001	-1×10^{-4}	-1×10^{-5}	-1×10^{-6}

$\displaystyle\lim_{x\to\infty}\left(x-\sqrt{x^2+2}\right)=0$

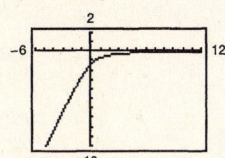

31.

x	10^0	10^1	10^2	10^3	10^4	10^5	10^6
$f(x)$	-0.7082	-0.7454	-0.7495	-0.74995	-0.749995	-0.75	-0.75

$\displaystyle\lim_{x\to\infty}3\left(2x-\sqrt{4x^2+x}\right)=-\frac{3}{4}$

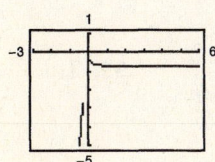

33. $a_n=\dfrac{n+1}{n^2+1}$

$a_1=\dfrac{1+1}{1^2+1}=1$ $a_4=\dfrac{5}{17}$

$a_2=\dfrac{2+1}{2^2+1}=\dfrac{3}{5}$ $a_5=\dfrac{6}{26}=\dfrac{3}{13}$

$a_3=\dfrac{4}{10}=\dfrac{2}{5}$

$\displaystyle\lim_{n\to\infty}a_n=0$

35. $a_n=\dfrac{n}{2n+1}$

$a_1=\dfrac{1}{3}$ $a_4=\dfrac{4}{9}$

$a_2=\dfrac{2}{5}$ $a_5=\dfrac{5}{11}$

$a_3=\dfrac{3}{7}$

$\displaystyle\lim_{n\to\infty}a_n=\dfrac{1}{2}$

37. $\dfrac{1}{5}, \dfrac{1}{2}, \dfrac{9}{11}, \dfrac{8}{7}, \dfrac{25}{17}$

$\displaystyle\lim_{n\to\infty} \dfrac{n^2}{3n+2}$ does not exist.

39. 2, 3, 4, 5, 6

$\displaystyle\lim_{n\to\infty} \dfrac{(n+1)!}{n!} = \lim_{n\to\infty} (n+1)$

does not exist.

41. $-1, \dfrac{1}{2}, -\dfrac{1}{3}, \dfrac{1}{4}, -\dfrac{1}{5}$

$\displaystyle\lim_{n\to\infty} \dfrac{(-1)^n}{n} = 0$

43. $\displaystyle\lim_{n\to\infty} a_n = \lim_{n\to\infty} \left[1 + \dfrac{n(n+1)}{2n^2} \right] = 1 + \dfrac{1}{2} = \dfrac{3}{2}$

n	10^0	10^1	10^2	10^3	10^4	10^5	10^6
a_n	2	1.55	1.505	1.5005	1.50005	1.500005	1.5000005

45. $\displaystyle\lim_{n\to\infty} a_n = \dfrac{16}{1}\left[\dfrac{2}{6}\right] = \dfrac{16}{3}$

n	10^0	10^1	10^2	10^3	10^4	10^5	10^6
a_n	16	6.16	5.4136	5.341336	5.3341	5.33341	5.333341

47. $N = \dfrac{95.1 - 611.40t}{1.0 - 0.52t - 0.005t^2}, \ 5 \le t \le 11$

(a)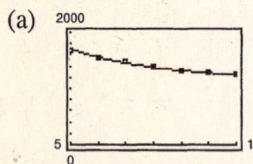

(b) For 2002, $t = 12$ and

$$N = \dfrac{95.1 - 611.40(12)}{1.0 - 0.52(12) - 0.005(12)^2} \approx 1215.05.$$

1,215,000 people

(c) As $t \to \infty, N \to 0$.

49. (a) Average cost $= \overline{C} = \dfrac{C}{x} = 13.50 + \dfrac{45{,}750}{x}$

(b) $\overline{C}(100) = \$471$

$\overline{C}(1000) = \$59.25$

(c) $\displaystyle\lim_{x\to\infty} C(x) = 13.50$

As more units are produced, the fixed costs (45,750) become less dominant.

51. False. $f(x) = \dfrac{x^2 + 1}{1}$ does not have a horizontal asymptote.

53. True

55. For example, let $f(x) = \dfrac{1}{x^2}$ and $g(x) = \dfrac{1}{x^2}$.

Then, $\displaystyle\lim_{x\to 0} \dfrac{1}{x^2}$ increases without bound, but $\displaystyle\lim_{x\to 0} [f(x) - g(x)] = 0$.

57. Converges to 0

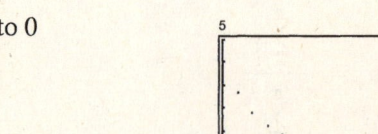

59. Diverges

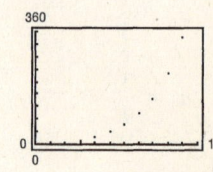

61. $y = x^4$

(a) $f(x) = (x + 3)^4$

(b) $f(x) = x^4 - 1$

(c) $f(x) = -2 + x^4$

(d) $f(x) = \dfrac{1}{2}(x - 4)^4$

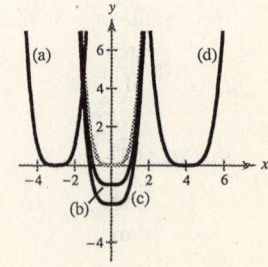

63.

$$
\begin{array}{r}
x^2 + 2x + 1 \\
x^2 - 4\,\overline{)\,x^4 + 2x^3 - 3x^2 - 8x - 4\,} \\
\underline{x^4 \qquad\quad - 4x^2} \\
2x^3 + \ \ x^2 \\
\underline{2x^3 \qquad\quad - 8x} \\
x^2 \qquad - 4
\end{array}
$$

$x^4 + 2x^3 - 3x^2 - 8x - 4 = (x^2 - 4)(x^2 + 2x + 1)$

65.

$$
\begin{array}{r}
x^3 + \ 5x^2 \qquad\quad - 3 \\
3x + 2\,\overline{)\,3x^4 + 17x^3 + 10x^2 - \ 9x - \ 8\,} \\
\underline{3x^4 + \ 2x^3} \\
15x^3 + 10x^2 \\
\underline{15x^3 + 10x^2} \\
-9x - \ 8 \\
\underline{-9x - \ 6} \\
-2
\end{array}
$$

$\dfrac{3x^4 + 17x^3 + 10x^2 - 9x - 8}{3x + 2} = x^3 + 5x^2 - 3 + \dfrac{-2}{3x + 2}$

67. $f(x) = x^4 - x^3 - 20x^2$

$\quad = x^2(x^2 - x - 20)$

$\quad = x^2(x - 5)(x + 4)$

Real zeros: $0, 0, 5, -4$

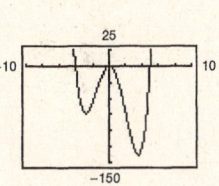

69. $f(x) = x^3 - 3x^2 + 2x - 6$

$\quad = x^2(x - 3) + 2(x - 3)$

$\quad = (x - 3)(x^2 + 2)$

Real zero: 3

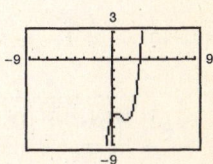

71. $\displaystyle\sum_{i=1}^{6} (2i + 3) = 5 + 7 + 9 + 11 + 13 + 15 = 60$

73. $\displaystyle\sum_{k=1}^{10} 15 = 10(15) = 150$

Section 11.5   The Area Problem

- You should know the following summation formulas and properties.

  (a) $\displaystyle\sum_{i=1}^{n} c = cn$

  (b) $\displaystyle\sum_{i=1}^{n} i = \dfrac{n(n + 1)}{2}$

  (c) $\displaystyle\sum_{i=1}^{n} i^2 = \dfrac{n(n + 1)(2n + 1)}{6}$

  (d) $\displaystyle\sum_{i=1}^{n} i^3 = \dfrac{n^2(n + 1)^2}{4}$

  (e) $\displaystyle\sum_{i=1}^{n} (a_i \pm b_i) = \sum_{i=1}^{n} a_i \pm \sum_{i=1}^{n} b_i$

  (f) $\displaystyle\sum_{i=1}^{n} ka_i = k\sum_{i=1}^{n} a_i$

- You should be able to evaluate a limit of a summation, $\displaystyle\lim_{n\to\infty} S(n)$.

- You should be able to approximate the area of a region using rectangles. By increasing the number of rectangles, the approximation improves.

- The area of a plane region above the x-axis bounded by f between $x = a$ and $x = b$ is the limit of the sum of the approximating rectangles:

  $$A = \lim_{n\to\infty} \sum_{i=1}^{n} f\left(a + \frac{(b - a)i}{n}\right)\left(\frac{b - a}{n}\right)$$

- You should be able to use the limit definition of area to find the area bounded by simple functions in the plane.

Solutions to Odd-Numbered Exercises

1. $\displaystyle\sum_{i=1}^{60} 7 = 7(60) = 420$

3. $\displaystyle\sum_{k=1}^{20} (k^3 + 2) = \frac{20^2(21)^2}{4} + 2(20)$
$= 44{,}100 + 40$
$= 44{,}140$

5. $\displaystyle\sum_{j=1}^{25} (j^2 + j) = \frac{25(26)(51)}{6} + \frac{25(26)}{2}$
$= 5850$

7. (a) $\displaystyle S(n) = \sum_{i=1}^{n} \frac{i^3}{n^4} = \frac{1}{n^4}\left[\frac{n^2(n+1)^2}{4}\right] = \frac{n^2 + 2n + 1}{4n^2}$

(b)

n	10^0	10^1	10^2	10^3	10^4
$S(n)$	1	0.3025	0.255025	0.25050025	0.25005

(c) $\displaystyle\lim_{n\to\infty} S(n) = \frac{1}{4}$

9. (a) $\displaystyle S(n) = \sum_{i=1}^{n} \frac{3}{n^3}(1 + i^2) = \frac{3}{n^3}\left[n + \frac{n(n+1)(2n+1)}{6}\right] = \frac{3}{n^2} + \frac{6n^2 + 9n + 3}{6n^2} = \frac{2n^2 + 3n + 7}{2n^2}$

(b)

n	10^0	10^1	10^2	10^3	10^4
$S(n)$	6	1.185	1.0154	1.0015	1.00015

(c) $\displaystyle\lim_{n\to\infty} S(n) = 1$

11. (a) $\displaystyle S(n) = \sum_{i=1}^{n}\left(\frac{i^2}{n^3} + \frac{2}{n}\right)\left(\frac{1}{n}\right) = \frac{1}{n}\left[\frac{n(n+1)(2n+1)}{6n^3} + \frac{2n}{n}\right] = \frac{1}{6n^3}(2n^2 + 3n + 1) + \frac{2}{n} = \frac{14n^2 + 3n + 1}{6n^3}$

(b)

n	10^0	10^1	10^2	10^3	10^4
$S(n)$	3	0.2385	0.02338	0.00233	0.0002333

(c) $\displaystyle\lim_{n\to\infty} S(n) = 0$

13. (a) $\displaystyle S(n) = \sum_{i=1}^{n}\left[1 - \left(\frac{i}{n}\right)^2\right]\left(\frac{1}{n}\right) = \frac{1}{n}\left[n - \frac{1}{n^2}\left(\frac{n(n+1)(2n+1)}{6}\right)\right] = 1 - \frac{2n^2 + 3n + 1}{6n^2} = \frac{4n^2 - 3n - 1}{6n^2}$

(b)

n	10^0	10^1	10^2	10^3	10^4
$S(n)$	0	0.615	0.66165	0.66617	0.666617

(c) $\displaystyle\lim_{n\to\infty} S(n) = \frac{2}{3}$

15. $f(x) = x + 4$, $[-1, 2]$, $n = 6$, width $= \dfrac{1}{2}$

Area $\approx \dfrac{1}{2}[3.5 + 4 + 4.5 + 5 + 5.5 + 6]$

$= 14.25$ square units

17. The width of each rectangle is $\frac{1}{4}$. The height is obtained by evaluating f at the right-hand endpoint of each interval.

$A \approx \displaystyle\sum_{i=1}^{8} f\left(\frac{i}{4}\right)\left(\frac{1}{4}\right) = \sum_{i=1}^{8} \frac{1}{4}\left(\frac{i}{4}\right)^3\left(\frac{1}{4}\right)$

$= 1.265625$ square units

19. Width of each rectangle is $\dfrac{12}{n}$. The height is $f\left(\dfrac{12}{n}i\right) = -\dfrac{1}{3}\left(\dfrac{12}{n}i\right) + 4$.

$A = \displaystyle\sum_{i=1}^{n}\left[-\frac{1}{3}\left(\frac{12i}{n}\right) + 4\right]\left(\frac{12}{n}\right)$

Note: Exact area is 24.

n	4	8	20	50
Approximate area	18	21	22.8	23.52

21. The width of each rectangle is $\dfrac{3}{n}$. The height is $\dfrac{1}{9}\left(\dfrac{3i}{n}\right)^3$.

n	4	8	20	50
Approximate area	3.52	2.85	2.48	2.34

$$A \approx \sum_{i=1}^{n} \frac{1}{9}\left(\frac{3i}{n}\right)^3\left(\frac{3}{n}\right)$$

23. $A \approx \displaystyle\sum_{i=1}^{n} f\left(\frac{i}{n}\right)\left(\frac{1}{n}\right)$

$= \displaystyle\sum_{i=1}^{n}\left[4\left(\frac{i}{n}\right)+1\right]\left(\frac{1}{n}\right)$

$= \dfrac{1}{n}\displaystyle\sum_{i=1}^{n}\left[\frac{4}{n}i+1\right]$

$= \dfrac{1}{n}\left[\dfrac{4}{n}\dfrac{n(n+1)}{2}+n\right]$

$= \dfrac{1}{n}[2(n+1)+n]$

$= \dfrac{3n+2}{n}$

$A = \displaystyle\lim_{n\to\infty}\dfrac{3n+2}{n} = 3$ square units

25. $A \approx \displaystyle\sum_{i=1}^{n} f\left(\frac{i}{n}\right)\left(\frac{1}{n}\right)$

$= \displaystyle\sum_{i=1}^{n}\left[-2\left(\frac{i}{n}\right)+3\right]\left(\frac{1}{n}\right)$

$= \dfrac{1}{n}\displaystyle\sum_{i=1}^{n}\left[-\frac{2i}{n}+3\right]$

$= \dfrac{1}{n}\left[-\dfrac{2}{n}\dfrac{n(n+1)}{2}+3n\right]$

$= \dfrac{1}{n}[2n-1]$

$A = \displaystyle\lim_{n\to\infty}\dfrac{2n-1}{n} = 2$ square units

27. $A \approx \displaystyle\sum_{i=1}^{n} f\left(-1+\frac{2i}{n}\right)\left(\frac{2}{n}\right) = \sum_{i=1}^{n}\left[2-\left(-1+\frac{2i}{n}\right)^2\right]\frac{2}{n}$

$= \displaystyle\sum_{i=1}^{n}\left[2-1+\frac{4i}{n}-\frac{4i^2}{n^2}\right]\left(\frac{2}{n}\right) = \frac{2}{n}\sum_{i=1}^{n}1+\frac{8}{n^2}\sum_{i=1}^{n}i-\frac{8}{n^3}\sum_{i=1}^{n}i^2$

$= \dfrac{2}{n}(n)+\dfrac{8}{n^2}\dfrac{n(n+1)}{2}-\dfrac{8}{n^3}\dfrac{n(n+1)(2n+1)}{6}$

$A = \displaystyle\lim_{n\to\infty}\left[2+4\frac{n(n+1)}{n^2}-\frac{4}{3}\frac{n(n+1)(2n+1)}{n^3}\right] = 2+4-\frac{8}{3} = \frac{10}{3}$ square units

29. $A \approx \displaystyle\sum_{i=1}^{n} g\left(1+\frac{i}{n}\right)\left(\frac{1}{n}\right)$

$= \displaystyle\sum_{i=1}^{n}\left[8-\left(1+\frac{i}{n}\right)^3\right]\frac{1}{n}$

$= \displaystyle\sum_{i=1}^{n}\left[7-\frac{3i}{n}-\frac{3i^2}{n^2}-\frac{i^3}{n^3}\right]\frac{1}{n}$

$= \dfrac{7}{n}\displaystyle\sum_{i=1}^{n}1-\frac{3}{n^2}\sum_{i=1}^{n}i-\frac{3}{n^3}\sum_{i=1}^{n}i^2-\frac{1}{n^4}\sum_{i=1}^{n}i^3$

$= \dfrac{7}{n}(n)-\dfrac{3}{n^2}\dfrac{n(n+1)}{2}-\dfrac{3}{n^3}\dfrac{n(n+1)(2n+1)}{6}-\dfrac{1}{n^4}\dfrac{n^2(n+1)^2}{4}$

$A = \displaystyle\lim_{n\to\infty}\left[7-\frac{3}{2}\frac{n(n+1)}{n^2}-\frac{1}{2n^3}n(n+1)(2n+1)-\frac{1}{n^4}\frac{n^2(n+1)^2}{4}\right] = 7-\frac{3}{2}-1-\frac{1}{4} = \frac{17}{4}$ square units

31. $A \approx \sum_{i=1}^{n} g\left(\dfrac{i}{n}\right)\left(\dfrac{1}{n}\right)$

$= \sum_{i=1}^{n} \left[2\left(\dfrac{i}{n}\right) - \left(\dfrac{i}{n}\right)^3\right]\left(\dfrac{1}{n}\right)$

$= \dfrac{1}{n}\sum_{i=1}^{n} \left[\dfrac{2}{n}i - \dfrac{1}{n^3}i^3\right]$

$= \dfrac{1}{n}\left[\dfrac{2}{n}\dfrac{n(n+1)}{2} - \dfrac{1}{n^3}\dfrac{n^2(n+1)^2}{4}\right]$

$= \dfrac{n+1}{n} - \dfrac{(n+1)^2}{4n^2}$

$A = \lim_{n\to\infty} \left[\dfrac{n+1}{n} - \dfrac{(n+1)^2}{4n^2}\right]$

$= 1 - \dfrac{1}{4} = \dfrac{3}{4}$ square units

33. $A \approx \sum_{i=1}^{n} f\left(1 + \dfrac{3i}{n}\right)\left(\dfrac{3}{n}\right)$

$= \sum_{i=1}^{n} \left[\dfrac{1}{4}\left(1 + \dfrac{3i}{n}\right)^2 + \left(1 + \dfrac{3i}{n}\right)\right]\left(\dfrac{3}{n}\right)$

$= \sum_{i=1}^{n} \left(\dfrac{1}{4} + \dfrac{3}{2}\dfrac{i}{n} + \dfrac{9}{4}\dfrac{i^2}{n^2} + 1 + \dfrac{3i}{n}\right)\left(\dfrac{3}{n}\right)$

$= \dfrac{15}{4n}\sum_{i=1}^{n} 1 + \dfrac{27}{2n^2}\sum_{i=1}^{n} i + \dfrac{27}{4n^3}\sum_{i=1}^{n} i^2$

$= \dfrac{15}{4n}(n) + \dfrac{27}{2n^2}\left(\dfrac{n(n+1)}{2}\right) + \dfrac{27}{4n^3}\dfrac{n(n+1)(2n+1)}{6}$

$A = \lim_{n\to\infty} \left[\dfrac{15}{4} + \dfrac{27}{4}\dfrac{n(n+1)}{n^2} + \dfrac{9}{8n^3}n(n+1)(2n+1)\right]$

$= \dfrac{15}{4} + \dfrac{27}{4} + \dfrac{9}{4} = \dfrac{51}{4}$ square units

35. $y = (-3.0 \cdot 10^{-6})x^3 + 0.002x^2 - 1.05x + 400$

Note that $y = 0$ when $x = 500$.

Area $\approx 105{,}208.33$ square feet ≈ 2.4153 acres

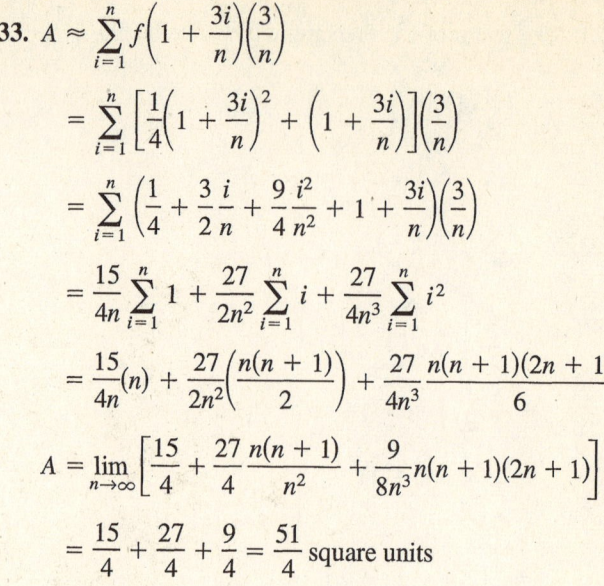

37. True. See Formula 2, page 782.

39. Answers will vary.

41. $\sin 2x - \sqrt{3} \sin x = 0$

$2 \sin x \cos x - \sqrt{3} \sin x = 0$

$\sin x\left(2 \cos x - \sqrt{3}\right) = 0$

$\sin x = 0 \implies x = n\pi$

$\cos x = \dfrac{\sqrt{3}}{2} \implies x = \dfrac{\pi}{6} + 2n\pi, x = \dfrac{11\pi}{6} + 2n\pi$

43. $2 \tan x = \tan 2x = \dfrac{2 \tan x}{1 - \tan^2 x}$

$\tan x = 0 \implies x = n\pi$

45. $2 \cot x = 5 \cos \dfrac{\pi}{2} = 0$

$\cot x = 0 \implies x = \dfrac{\pi}{2} + n\pi$

47. $(\mathbf{u} \cdot \mathbf{v})\mathbf{u} = (\langle 4, -5 \rangle \cdot \langle -1, -2 \rangle)\langle 4, -5 \rangle$

$= 6\langle 4, -5 \rangle$

$= \langle 24, -30 \rangle$

49. $\|\mathbf{v}\| - 2 = \sqrt{5} - 2$

Review Exercises for Chapter 11

Solutions to Odd-Numbered Exercises

1. $\lim_{x\to 3} (6x - 1)$

The limit (17) can be reached.

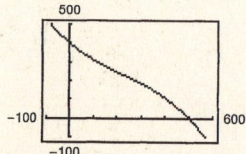

x	2.9	2.99	2.999	3	3.001	3.01	3.1
$f(x)$	16.4	16.94	16.994	17	17.006	17.06	17.6

3. $\lim_{x \to 1} (3 - x) = 2$

5. $\lim_{x \to 1} \dfrac{x^2 - 1}{x - 1} = 2$

7. (a) $\lim_{x \to c} [f(x)]^3 = 4^3 = 64$

    (c) $\lim_{x \to c} [f(x)g(x)] = (4)(5) = 20$

(b) $\lim_{x \to c} [3f(x) - g(x)] = 3(4) - 5 = 7$

(d) $\lim_{x \to c} \dfrac{f(x)}{g(x)} = \dfrac{4}{5}$

9. $\lim_{x \to 4} \left(\dfrac{1}{2}x + 3\right) = \dfrac{1}{2}(4) + 3 = 5$

11. $\lim_{x \to 2} \dfrac{x^2 - 1}{x^3 + 2} = \dfrac{2^2 - 1}{2^3 + 2} = \dfrac{3}{10}$

13. $\lim_{x \to \pi} \sin 3x = \sin 3\pi = 0$

15. $\lim_{x \to 3} (5x - 4) = 5(3) - 4 = 11$

17. $\lim_{x \to 2} (5x - 3)(3x + 5) = (5(2) - 3)(3(2) + 5)$

$$= (7)(11) = 77$$

19. $\lim_{t \to 3} \dfrac{t^2 + 1}{t} = \dfrac{9 + 1}{3} = \dfrac{10}{3}$

21. $\lim_{t \to -2} \dfrac{t + 2}{t^2 - 4} = \lim_{t \to -2} \dfrac{t + 2}{(t + 2)(t - 2)}$

$$= \lim_{t \to -2} \dfrac{1}{t - 2} = -\dfrac{1}{4}$$

23. $\lim_{x \to 5} \dfrac{x - 5}{x^2 + 5x - 50} = \lim_{x \to 5} \dfrac{x - 5}{(x - 5)(x + 10)}$

$$= \lim_{x \to 5} \dfrac{1}{x + 10} = \dfrac{1}{15}$$

25. $\lim_{x \to -2} \dfrac{x^2 - 4}{x^3 + 8} = \lim_{x \to -2} \dfrac{(x + 2)(x - 2)}{(x + 2)(x^2 - 2x + 4)}$

$$= \lim_{x \to -2} \dfrac{x - 2}{x^2 - 2x + 4}$$

$$= \dfrac{-4}{12} = \dfrac{-1}{3}$$

27. $\lim_{x \to -1} \dfrac{1/(x + 2) - 1}{x + 1} = \lim_{x \to -1} \dfrac{1 - (x + 2)}{(x + 2)(x + 1)}$

$$= \lim_{x \to -1} \dfrac{-(x + 1)}{(x + 2)(x + 1)}$$

$$= \lim_{x \to -1} \dfrac{-1}{(x + 2)} = -1$$

29. $\lim_{u \to 0} \dfrac{\sqrt{4 + u} - 2}{u} = \lim_{u \to 0} \dfrac{\sqrt{4 + u} - 2}{u} \cdot \dfrac{\sqrt{4 + u} + 2}{\sqrt{4 + u} + 2}$

$$= \lim_{u \to 0} \dfrac{(4 + u) - 4}{u\left(\sqrt{4 + u} + 2\right)}$$

$$= \lim_{u \to 0} \dfrac{1}{\sqrt{4 + u} + 2} = \dfrac{1}{4}$$

31. $\lim_{x \to 5} \dfrac{\sqrt{x - 1} - 2}{x - 5} = \lim_{x \to 5} \dfrac{\sqrt{x - 1} - 2}{x - 5} \cdot \dfrac{\sqrt{x - 1} + 2}{\sqrt{x - 1} + 2}$

$$= \lim_{x \to 5} \dfrac{(x - 1) - 4}{(x - 5)\left(\sqrt{x - 1} + 2\right)}$$

$$= \lim_{x \to 5} \dfrac{1}{\sqrt{x - 1} + 2} = \dfrac{1}{2 + 2} = \dfrac{1}{4}$$

33. (a)

$$\lim_{x \to 3} \dfrac{x - 3}{x^2 - 9} = \dfrac{1}{6}$$

(b)

x	2.9	2.99	3	3.01	3.1
$f(x)$	0.1695	0.1669	Error	0.1664	0.1639

35. (a)

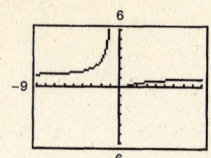

(b) (Answers will vary.)

x	-0.1	-0.01	-0.001	0	0.001	0.01	0.1
y_1	4.85 E 8	7.2 E 86	Error	Error	0	1 E -87	2.1 E -9

$\lim\limits_{x \to 0} e^{-2/x}$ does not exist.

37. (a)

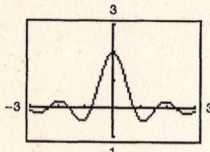

(b)

x	-0.1	-0.01	-0.001	0	0.001	0.01	0.1
y_1	1.9471	1.9995	1.999995	error	1.999995	1.995	1.9471

$\lim\limits_{x \to 0} \dfrac{\sin 4x}{2x} = 2$

39. (a)

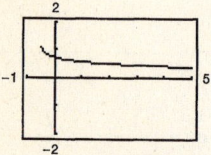

(b)

x	1.1	1.01	1.001	1.0001
$f(x)$	0.5680	0.5764	0.5773	0.5773

$\lim\limits_{x \to 1^+} \dfrac{\sqrt{2x+1} - \sqrt{3}}{x - 1} \approx 0.577$

$\left(\text{Exact value: } \dfrac{\sqrt{3}}{3}\right)$

41. $f(x) = \dfrac{|x - 3|}{x - 3}$

Limit does not exist because

$\lim\limits_{x \to 3^+} f(x) = 1$ and

$\lim\limits_{x \to 3^-} f(x) = -1$.

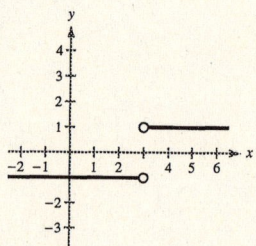

43. $f(x) = \dfrac{2}{x^2 - 4}$

Limit does not exist.

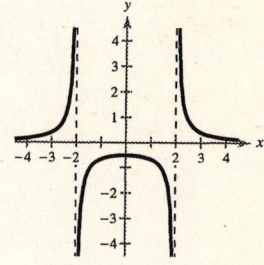

45.

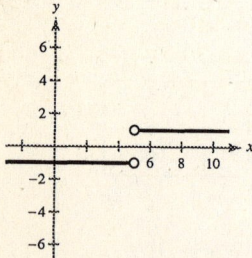

$\lim\limits_{x \to 5} \dfrac{|x - 5|}{x - 5}$ does not exist.

47.

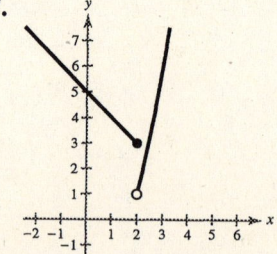

$\lim\limits_{x \to 2} f(x)$ does not exist.

49. $\displaystyle\lim_{h\to 0}\frac{f(x+h)-f(x)}{h} = \lim_{h\to 0}\frac{3(x+h)-(x+h)^2-(3x-x^2)}{h}$

$\displaystyle = \lim_{h\to 0}\frac{3x+3h-x^2-2xh-h^2-3x+x^2}{h} = \lim_{h\to 0}\frac{3h-2xh-h^2}{h}$

$\displaystyle = \lim_{h\to 0}(3-2x-h) = 3-2x$

51. Slope ≈ 2

(Answers will vary.)

53.

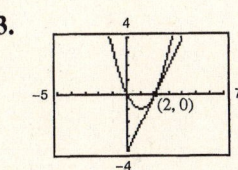

Slope at $(2, f(2))$ is
approximately 2.

55.

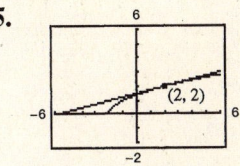

Slope is $\frac{1}{4}$ at $(2, 2)$.

57. $\displaystyle m = \lim_{h\to 0}\frac{g(x+h)-g(x)}{h}$

$\displaystyle = \lim_{h\to 0}\frac{(x+h)^2-4(x+h)-(x^2-4x)}{h}$

$\displaystyle = \lim_{h\to 0}\frac{x^2+2xh+h^2-4x-4h-x^2-4x}{h}$

$\displaystyle = \lim_{h\to 0}\frac{2xh+h^2-4h}{h}$

$\displaystyle = \lim_{h\to 0}(2x+h-4) = 2x-4$

(a) At $(0, 0)$, $m = 2(0)-4 = -4$.

(b) At $(5, 5)$, $m = 2(5)-4 = 6$.

59. $\displaystyle m = \lim_{h\to 0}\frac{g(x+h)-g(x)}{h} = \lim_{h\to 0}\frac{\dfrac{4}{x+h-6}-\dfrac{4}{x-6}}{h}$

$\displaystyle = \lim_{h\to 0}\frac{4(x-6)-4(x+h-6)}{(x+h-6)(x-6)h}$

$\displaystyle = \lim_{h\to 0}\frac{-4h}{(x+h-6)(x-6)h}$

$\displaystyle = \lim_{h\to 0}\frac{-4}{(x+h-6)(x-6)} = \frac{-4}{(x-6)^2}$

(a) At $(7, 4)$, $m = \dfrac{-4}{(7-6)^2} = -4$.

(b) At $(8, 2)$, $m = \dfrac{-4}{(8-6)^2} = -1$.

61. $\displaystyle f'(x) = \lim_{h\to 0}\frac{f(x+h)-f(x)}{h} = \lim_{h\to 0}\frac{5-5}{h} = 0$

63. $\displaystyle h'(x) = \lim_{k\to 0}\frac{h(x+k)-h(x)}{k}$

$\displaystyle = \lim_{k\to 0}\frac{\left[5-\frac{1}{2}(x+k)\right]-\left[5-\frac{1}{2}x\right]}{k}$

$\displaystyle = \lim_{k\to 0}\frac{-\frac{1}{2}k}{k} = -\frac{1}{2}$

65. $\displaystyle g'(x) = \lim_{h\to 0}\frac{g(x+h)-g(x)}{h}$

$\displaystyle = \lim_{h\to 0}\frac{2(x+h)^2-1-(2x^2-1)}{h}$

$\displaystyle = \lim_{h\to 0}\frac{2x^2+4xh+2h^2-2x^2}{h}$

$\displaystyle = \lim_{h\to 0}(4x+2h)$

$\displaystyle = 4x$

67. $f'(t) = \lim\limits_{h \to 0} \dfrac{f(t + h) - f(t)}{h}$

$\quad = \lim\limits_{h \to 0} \dfrac{\sqrt{t + h + 5} - \sqrt{t + 5}}{h} \cdot \dfrac{\sqrt{t + h + 5} + \sqrt{t + 5}}{\sqrt{t + h + 5} + \sqrt{t + 5}}$

$\quad = \lim\limits_{h \to 0} \dfrac{(t + h + 5) - (t + 5)}{h\left(\sqrt{t + h + 5} + \sqrt{t + 5}\right)}$

$\quad = \lim\limits_{h \to 0} \dfrac{1}{\sqrt{t + h + 5} + \sqrt{t + 5}}$

$\quad = \dfrac{1}{2\sqrt{t + 5}}$

69. $g'(s) = \dfrac{g(s + h) - g(s)}{h} = \lim\limits_{h \to 0} \dfrac{\dfrac{4}{s + h + 5} - \dfrac{4}{s + 5}}{h}$

$\quad = \lim\limits_{h \to 0} \dfrac{4s + 20 - 4s - 4h - 20}{(s + h + 5)(s + 5)h}$

$\quad = \lim\limits_{h \to 0} \dfrac{-4h}{(s + h + 5)(s + 5)h}$

$\quad = \lim\limits_{h \to 0} \dfrac{-4}{(s + h + 5)(s + 5)} = \dfrac{-4}{(s + 5)^2}$

71. $\lim\limits_{x \to \infty} \dfrac{4x}{2x - 3} = \dfrac{4}{2} = 2$

73. $\lim\limits_{x \to -\infty} \dfrac{2x}{x^2 - 25} = 0$

75. $\lim\limits_{x \to \infty} \dfrac{x^2}{2x + 3}$ does not exist.

77. $\lim\limits_{x \to \infty} \left[\dfrac{x}{(x - 2)^2} + 3\right] = 0 + 3 = 3$

79. $\dfrac{2}{3}, \dfrac{5}{5} = 1, \dfrac{8}{7}, \dfrac{11}{9}, \dfrac{14}{11}$

$\quad \lim\limits_{n \to \infty} a_n = \dfrac{3}{2}$

81. $a_n = \dfrac{1}{2n^2}[3 - 2n(n + 1)] = \dfrac{3}{2n^2} - \dfrac{n + 1}{n}$

$\quad -0.5, -1.125, -1.16\overline{6}, -1.15625, -1.14$

$\quad \lim\limits_{n \to \infty} a_n = 0 - 1 = -1$

83. (a) $\sum\limits_{i=1}^{n} \left(\dfrac{4i^2}{n^2} - \dfrac{i}{n}\right)\dfrac{1}{n} = \dfrac{4}{n^3}\sum\limits_{i=1}^{n} i^2 - \dfrac{1}{n^2}\sum\limits_{i=1}^{n} i$

$\quad = \dfrac{4}{n^3}\dfrac{n(n + 1)(2n + 1)}{6} - \dfrac{1}{n^2}\dfrac{n(n + 1)}{2}$

$\quad = \dfrac{4n(n + 1)(2n + 1) - 3n^2(n + 1)}{6n^3}$

$\quad = \dfrac{n(n + 1)(8n + 4 - 3n)}{6n^3}$

$\quad = \dfrac{(n + 1)(5n + 4)}{6n^2}$

(b)

n	10^0	10^1	10^2	10^3	10^4
$S(n)$	3	0.99	0.8484	0.8348	0.8335

(c) $\lim\limits_{n \to \infty} S(n) = \dfrac{5}{6}$

85. Area $\approx \frac{1}{2}\left[\frac{7}{2} + 3 + \frac{5}{2} + 2 + \frac{3}{2} + 1\right] = \frac{1}{2}\frac{27}{2} = \frac{27}{4} = 6.75$

87. $f(x) = \frac{1}{4}x^2,\ b - a = 4 - 0 = 4$

n	4	8	20	50
Approximate area	7.5	6.375	5.74	5.4944

(Exact area is $\frac{16}{3} \approx 5.33$.)

$$A \approx \sum_{i=1}^{n} f\left(\frac{4i}{n}\right)\left(\frac{4}{n}\right)$$

$$= \sum_{i=1}^{n} \frac{1}{4}\left(\frac{4i}{n}\right)^2\left(\frac{4}{n}\right)$$

$$= \frac{1}{n}\sum_{i=1}^{n} \frac{16}{n^2}i^2$$

$$= \frac{16}{n^3}\frac{n(n+1)(2n+1)}{6}$$

$$= \frac{8(n+1)(2n+1)}{3n^2}$$

89. $A = \lim_{n\to\infty} \sum_{i=1}^{n}\left(10 - \frac{10i}{n}\right)\left(\frac{10}{n}\right)$

$= \lim_{n\to\infty}\left[\frac{100}{n}\sum_{i=1}^{n} 1 - \frac{100}{n^2}\sum_{i=1}^{n} i\right]$

$= \lim_{n\to\infty}\left[\frac{100}{n}(n) - \frac{100}{n^2}\left(\frac{n(n+1)}{2}\right)\right]$

$= \lim_{n\to\infty}\left[100 - 50\frac{n(n+1)}{n^2}\right]$

$= 100 - 50 = 50$ exact area

91. $A = \lim_{n\to\infty} \sum_{i=1}^{n}\left[\left(-1 + \frac{3i}{n}\right)^2 + 4\right]\left(\frac{3}{n}\right)$

$= \lim_{n\to\infty} \sum_{i=1}^{n}\left[5 - \frac{6i}{n} + \frac{9i^2}{n^2}\right]\frac{3}{n}$

$= \lim_{n\to\infty}\left[\frac{15}{n}\sum_{i=1}^{n} 1 - \frac{18}{n^2}\sum_{i=1}^{n} i + \frac{27}{n^3}\sum_{i=1}^{n} i^2\right]$

$= \lim_{n\to\infty}\left[\frac{15}{n}(n) - \frac{18}{n^2}\frac{n(n+1)}{2} + \frac{27}{n^3}\frac{n(n+1)(2n+1)}{6}\right]$

$= 15 - 9 + 9 = 15$ exact area

93. $A = \lim_{n\to\infty} \sum_{i=1}^{n} 2\left[\left(-1 + \frac{2i}{n}\right)^2 - \left(-1 + \frac{2i}{n}\right)^3\right]\left(\frac{2}{n}\right)$

$= \lim_{n\to\infty} \sum_{i=1}^{n} \frac{4}{n}\left(1 - \frac{4i}{n} + \frac{4i^2}{n^2} - \left(-1 + \frac{6i}{n} - \frac{12i^2}{n^2} + \frac{8i^3}{n^3}\right)\right)$

$= \lim_{n\to\infty} \sum_{i=1}^{n} \frac{4}{n}\left(2 - \frac{10i}{n} + \frac{16i^2}{n^2} - \frac{8i^3}{n^3}\right)$

$= \lim_{n\to\infty}\left[\frac{8}{n}\sum_{i=1}^{n} 1 - \frac{40}{n^2}\sum_{i=1}^{n} i + \frac{64}{n^3}\sum_{i=1}^{n} i^2 - \frac{32}{n^4}\sum_{i=1}^{n} i^3\right]$

$= \lim_{n\to\infty}\left[\frac{8}{n}(n) - \frac{40}{n^2}\frac{n(n+1)}{2} + \frac{64}{n^3}\frac{n(n+1)(2n+1)}{6} - \frac{32}{n^4}\frac{n^2(n+1)^2}{4}\right] = 8 - 20 + \frac{64}{3} - 8 = \frac{4}{3}$ exact area

95. (a) $y = -3.376 \times 10^{-7}x^3 + 3.753 \times 10^{-4}x^2 - 0.168x + 132.168$

(c) Area $\approx$ 88,868 square feet, answers will vary.

(b)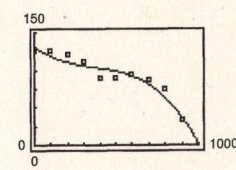

97. False. The limit does not exist.

Chapter 11   Practice Test

1. Use a graphing utility to complete the table and use the result to estimate the limit

$$\lim_{x \to 3} \frac{x - 3}{x^2 - 9}.$$

x	2.9	2.99	3	3.01	3.1
$f(x)$			?		

2. Graph the function

$$f(x) = \frac{\sqrt{x + 4} - 2}{x}$$

and estimate the limit

$$\lim_{x \to 0} \frac{\sqrt{x + 4} - 2}{x}.$$

3. Find the limit $\lim\limits_{x \to 2} e^{x-2}$ by direct substitution.

4. Find the limit $\lim\limits_{x \to 1} \dfrac{x^3 - 1}{x - 1}$ analytically.

5. Use a graphing utility to estimate the limit

$$\lim_{x \to 0} \frac{\sin 5x}{2x}.$$

6. Find the limit

$$\lim_{x \to -2} \frac{|x + 2|}{x + 2}.$$

7. Use the limit process to find the slope of the graph of $f(x) = \sqrt{x}$ at the point $(4, 2)$.

8. Find the derivative of the function $f(x) = 3x - 1$.

9. Find the limits.

(a) $\lim\limits_{x \to \infty} \dfrac{3}{x^4}$

(b) $\lim\limits_{x \to -\infty} \dfrac{x^2}{x^2 + 3}$

(c) $\lim\limits_{x \to \infty} \dfrac{|x|}{1 - x}$

10. Write the first four terms of the sequence $a_n = \dfrac{1 - n^2}{2n^2 + 1}$ and find the limit of the sequence.

11. Find the sum $\sum\limits_{i=1}^{25} (i^2 + i)$.

12. Write the sum $\sum\limits_{i=1}^{n} \dfrac{i^2}{n^3}$ as a rational function $S(n)$, and find $\lim\limits_{n \to \infty} S(n)$.

13. Find the area of the region bounded by $f(x) = 1 - x^2$ over the interval $0 \le x \le 1$.

APPENDICES

APPENDIX B
Review of Graphs, Equations, and Inequalities

Appendix B.1   The Cartesian Plane

- ■ You should be able to plot points.
- ■ You should know that the distance between (x_1, y_1) and (x_2, y_2) in the plane is
  $$d = \sqrt{(x_2 - x_1)^2 + (y_2 - y_1)^2}.$$
- ■ You should know that the midpoint of the line segment joining (x_1, y_1) and (x_2, y_2) is
  $$\left(\frac{x_1 + x_2}{2}, \frac{y_1 + y_2}{2} \right).$$
- ■ You should know the equation of a circle: $(x - h)^2 + (y - k)^2 = r^2$.
- ■ You should be able to translate points in the plane.

Solutions to Odd-Numbered Exercises

1. A: $(2, 6)$, B: $(-6, -2)$, C: $(4, -4)$, D: $(-3, 2)$

3. Plot $(-4, 2)$, $(-3, -6)$, $(0, 5)$, $(1, -4)$

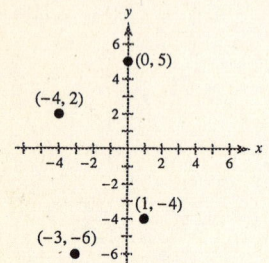

5.

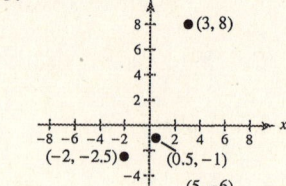

7. The coordinates are $(-5, 4)$

9. The coordinates are $(-6, -6)$

11. $x > 0 \Rightarrow$ The point lies in Quadrant I or in Quadrant IV.

$y < 0 \Rightarrow$ The point lies in Quadrant III or in Quadrant IV.

$x > 0$ and $y < 0 \Rightarrow (x, y)$ lies in Quadrant IV.

13. $x = -4 \Rightarrow x$ is negative $\Rightarrow$ The point lies in Quadrant II or in Quadrant III.

$y > 0 \Rightarrow$ The point lies in Quadrant I or Quadrant II.

$x = -4$ and $y > 0 \Rightarrow (x, y)$ lies in Quadrant II.

15. $y < -5 \Rightarrow y$ is negative $\Rightarrow$ The point lies in either Quadrant III or Quadrant IV.

17. Since $(x, -y)$ is in Quadrant II, we know that $x < 0$ and $-y > 0$. If $-y > 0$, then $y < 0$.

$x < 0 \Rightarrow$ The point lies in Quadrant II or in Quadrant III.

$y < 0 \Rightarrow$ The point lies in Quadrant III or in Quadrant IV.

$x < 0$ and $y < 0 \Rightarrow (x, y)$ lies in Quadrant III.

19. If $xy > 0$, then either x and y are both positive, or both negative. Hence, (x, y) lies in either Quadrant I or Quadrant III.

21.

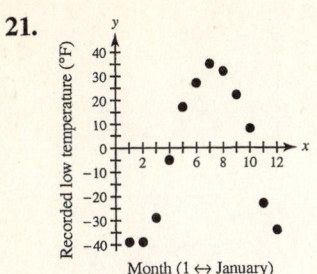

Month (1 ↔ January)

23. $(6, -3), (6, 5)$

$$d = \sqrt{(6 - 6)^2 + (5 - (-3))^2} = \sqrt{64} = 8$$

25. $(-3, -1), (2, -1)$

$$d = \sqrt{(2 - (-3))^2 + (-1 - (-1))^2} = \sqrt{25} = 5$$

27. $d = \sqrt{(3 - (-2))^2 + (-6 - 6)^2} = \sqrt{5^2 + (-12)^2}$

$$= \sqrt{25 + 144} = \sqrt{169} = 13.$$

29. $\left(\frac{1}{2}, \frac{4}{3}\right), (2, -1)$

$$d = \sqrt{\left(\frac{1}{2} - 2\right)^2 + \left(\frac{4}{3} + 1\right)^2}$$
$$= \sqrt{\frac{9}{4} + \frac{49}{9}}$$
$$= \sqrt{\frac{277}{36}} = \sqrt{\frac{277}{6}} \approx 2.77$$

31. $(-4.2, 3.1), (-12.5, 4.8)$

$$d = \sqrt{(-4.2 + 12.5)^2 + (3.1 - 4.8)^2}$$
$$= \sqrt{68.89 + 2.89}$$
$$= \sqrt{71.78} \approx 8.47$$

33. (a) The distance between $(0, 2)$ and $(4, 2)$ is 4.

The distance between $(4, 2)$ and $(4, 5)$ is 3.

The distance between $(0, 2)$ and $(4, 5)$ is

$$\sqrt{(4 - 0)^2 + (5 - 2)^2} = \sqrt{16 + 9} = \sqrt{25} = 5.$$

(b) $4^2 + 3^2 = 16 + 9 = 25 = 5^2$

35. (a) The distance between $(-1, 1)$ and $(9, 1)$ is 10.

The distance between $(9, 1)$ and $(9, 4)$ is 3.

The distance between $(-1, 1)$ and $(9, 4)$ is

$$\sqrt{(9 - (-1))^2 + (4 - 1)^2} = \sqrt{100 + 9} = \sqrt{109}.$$

(b) $10^2 + 3^2 = 109 = \left(\sqrt{109}\right)^2$

37. Find distances between pairs of points.

$$d_1 = \sqrt{(4 - 2)^2 + (0 - 1)^2} = \sqrt{5}$$
$$d_2 = \sqrt{(4 + 1)^2 + (0 + 5)^2} = \sqrt{50}$$
$$d_3 = \sqrt{(2 + 1)^2 + (1 + 5)^2} = \sqrt{45}$$
$$\left(\sqrt{5}\right)^2 + \left(\sqrt{45}\right)^2 = \left(\sqrt{50}\right)^2$$

Because $d_1{}^2 + d_3{}^2 = d_2{}^2$, the triangle is a right triangle.

39. Find distances between pairs of points.

$$d_1 = \sqrt{(0 - 2)^2 + (9 - 5)^2} = \sqrt{4 + 16} = \sqrt{20} = 2\sqrt{5}$$
$$d_2 = \sqrt{(-2 - 0)^2 + (0 - 9)^2} = \sqrt{4 + 81} = \sqrt{85}$$
$$d_3 = \sqrt{(0 - (-2))^2 + (-4 - 0)^2} = \sqrt{4 + 16} = \sqrt{20} = 2\sqrt{5}$$
$$d_4 = \sqrt{(0 - 2)^2 + (-4 - 5)^2} = \sqrt{4 + 81} = \sqrt{85}$$

Opposite sides have equal lengths of $2\sqrt{5}$ and $\sqrt{85}$, so the figure is a parallelogram.

41.

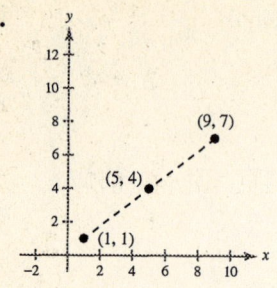

(b) $d = \sqrt{(9-1)^2 + (7-1)^2}$

   $= \sqrt{64 + 36} = 10$

(c) $\left(\dfrac{9+1}{2}, \dfrac{7+1}{2}\right) = (5, 4)$

43.

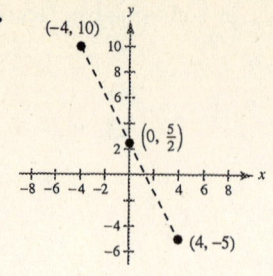

(b) $d = \sqrt{(4+4)^2 + (-5-10)^2}$

   $= \sqrt{64 + 225} = 17$

(c) $\left(\dfrac{4-4}{2}, \dfrac{-5+10}{2}\right) = \left(0, \dfrac{5}{2}\right)$

45.

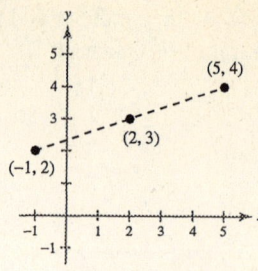

(b) $d = \sqrt{(5+1)^2 + (4-2)^2}$

   $= \sqrt{36 + 4} = \sqrt{40}$

   $= 2\sqrt{10}$

(c) $\left(\dfrac{-1+5}{2}, \dfrac{2+4}{2}\right) = (2, 3)$

47.

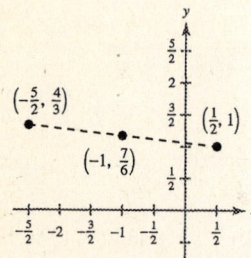

(b) $d = \sqrt{\left(\dfrac{1}{2} + \dfrac{5}{2}\right)^2 + \left(1 - \dfrac{4}{3}\right)^2}$

   $d = \sqrt{9 + \dfrac{1}{9}} = \dfrac{\sqrt{82}}{3}$

(c) $\left(\dfrac{-\frac{5}{2} + \frac{1}{2}}{2}, \dfrac{\frac{4}{3} + 1}{2}\right) = \left(-1, \dfrac{7}{6}\right)$

49.

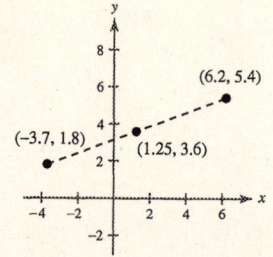

(b) $d = \sqrt{(6.2 + 3.7)^2 + (5.4 - 1.8)^2}$

   $= \sqrt{98.01 + 12.96} = \sqrt{110.97}$

(c) $\left(\dfrac{6.2 - 3.7}{2}, \dfrac{5.4 + 1.8}{2}\right) = (1.25, 3.6)$

51. $\left(\dfrac{1998 + 2002}{2}, \dfrac{839.6 + 1480.0}{2}\right) = (2000, 1159.8)$

The sales in 2000 are $1159.8 million.

53. Since $x_m = \dfrac{x_1 + x_2}{2}$ and $y_m = \dfrac{y_1 + y_2}{2}$ we have:

   $2x_m = x_1 + x_2 \qquad\qquad 2y_m = y_1 + y_2$

   $2x_m - x_1 = x_2 \qquad\qquad 2y_m - y_1 = y_2$

   So, $(x_2, y_2) = (2x_m - x_1, 2y_m - y_1)$.

   (a) $(x_2, y_2) = (2x_m - x_1, 2y_m - y_1) = (2(4) - 1, 2(-1) - (-2)) = (7, 0)$

   (b) $(x_2, y_2) = (2x_m - x_1, 2y_m - y_1) = (2(2) - (-5), 2(4) - 11) = (9, -3)$

55. $(x - 0)^2 + (y - 0)^2 = 3^2$

   $x^2 + y^2 = 9$

57. $(x - 2)^2 + (y + 1)^2 = 4^2$

   $(x - 2)^2 + (y + 1)^2 = 16$

59. $(x + 1)^2 + (y - 2)^2 = r^2$

$(0 + 1)^2 + (0 - 2)^2 = r^2 \implies r^2 = 5$

$(x + 1)^2 + (y - 2)^2 = 5$

61. $r = \dfrac{1}{2}\sqrt{(6 - 0)^2 + (8 - 0)^2} = \dfrac{1}{2}\sqrt{100} = 5$

$\text{Center}\left(\dfrac{0 + 6}{2}, \dfrac{0 + 8}{2}\right) = (3, 4)$

$(x - 3)^2 + (y - 4)^2 = 25$

63. $x^2 + y^2 = 25$

Center: $(0, 0)$

Radius $= 5$

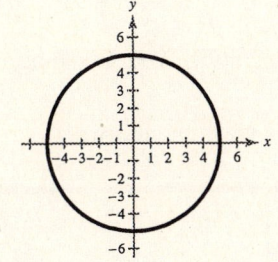

65. Center: $(1, -3)$

Radius $= 2$

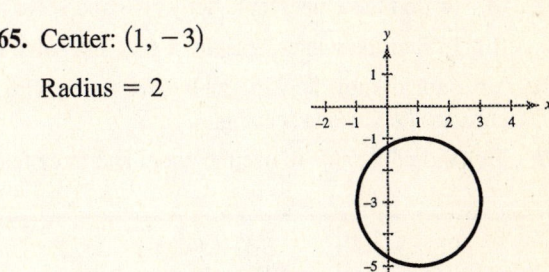

67. Center: $\left(\frac{1}{2}, \frac{1}{2}\right)$

Radius $= \frac{3}{2}$

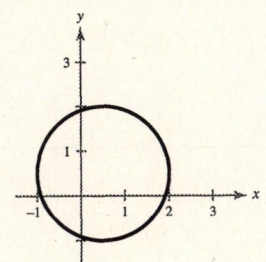

69. The x-coordinates are increased by 2, and the y-coordinates are increased by 5

Old vertex	Shifted vertex
$(-1, -1)$	$(1, 4)$
$(-2, -4)$	$(0, 1)$
$(2, -3)$	$(4, 2)$

71.

Old vertex	Shifted vertex
$(0, 2)$	$(-1, 5)$
$(3, 5)$	$(2, 8)$
$(5, 2)$	$(4, 5)$
$(2, -1)$	$(1, 2)$

73. The point $(65, 83)$ represents an entrance exam score of 65.

75. (a) It appears that the number of artists elected alternates between 6 and 8 per year from 1990 to 2002. If the trend continues, 6, 7 or 8 would be elected in 2005.

(b) Since 1986 and 1987 were the first two years that artists were elected, there was a larger number elected.

77. $d = \sqrt{(45 - 10)^2 + (40 - 15)^2}$

$= \sqrt{35^2 + 25^2}$

$= \sqrt{1850}$

$= 5\sqrt{74}$

≈ 43 yards

79. False. It would be sufficient to use the midpoint formula 15 times.

81. False. The polygon could be a rhombus. For example, consider the points $(4, 0)$, $(0, 6)$, $(-4, 0)$ and $(0, -6)$.

83. No, the scales can be different. The scales depend on the magnitude of the coordinates. See Figure P.13.

Appendix B.2   Graphs of Equations

■ You should be able to use the point-plotting method of graphing.

■ You should be able to find *x*- and *y*-intercepts.

　(a) To find the *x*-intercepts, let $y = 0$ and solve for *x*.

　(b) To find the *y*-intercepts, let $x = 0$ and solve for *y*.

■ You should know how to graph an equation with a graphing utility. You should be able to determine an appropriate viewing rectangle.

■ You should be able to use the zoom and trace features of a graphing utility.

Solutions to Odd-Numbered Exercises

1. $y = \sqrt{x + 4}$

　(a) $(0, 2)$: $2 \overset{?}{=} \sqrt{0 + 4}$

　　　　$2 = 2$ ✓

　　Yes, the point *is* on the graph.

　(b) $(5, 3)$: $3 \overset{?}{=} \sqrt{5 + 4}$

　　　　$3 = \sqrt{9}$ ✓

　　Yes, the point *is* on the graph.

3. $y = 4 - |x - 2|$

　(a) $(1, 5)$: $5 \overset{?}{=} 4 - |1 - 2|$

　　　　　$5 \neq 4 - 1$

　　No, the point *is not* on the graph.

　(b) $(1.2, 3.2)$: $3.2 \overset{?}{=} 4 - |1.2 - 2|$

　　　　　　$3.2 \overset{?}{=} 4 - |-.8|$

　　　　　　$3.2 \overset{?}{=} 4 - .8$

　　　　　　$3.2 \overset{?}{=} 3.2$ ✓

　　Yes, the point *is* on the graph.

5. $x^2 + y^2 = 20$

　(a) $(3, -2)$: $3^2 + (-2)^2 \overset{?}{=} 20$

　　　　　$9 + 4 \overset{?}{=} 20$

　　　　　$13 \neq 20$

　　No, the point *is not* on the graph.

　(b) $(-4, 2)$: $(-4)^2 + 2^2 \overset{?}{=} 20$

　　　　　$16 + 4 \overset{?}{=} 20$

　　　　　$20 = 20$

　　Yes, the point *is* on the graph.

7. $y = -2x + 3$

x	-1	0	1	$\frac{3}{2}$	2
y	5	3	1	0	-1
Solution point	$(-1, 5)$	$(0, 3)$	$(1, 1)$	$\left(\frac{3}{2}, 0\right)$	$(2, -1)$

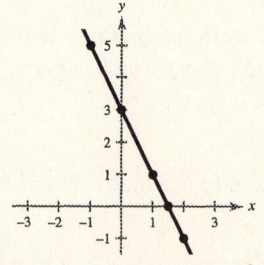

9. $y = x^2 - 2x$

x	-1	0	1	2	3
y	3	0	-1	0	3
Solution point	$(-1, 3)$	$(0, 0)$	$(1, -1)$	$(2, 0)$	$(3, 3)$

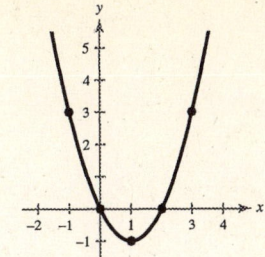

11. (a) $y = \frac{1}{4}x - 3$

x	-2	-1	0	1	2
y	$-\frac{7}{2}$	$-\frac{13}{4}$	-3	$-\frac{11}{4}$	$-\frac{5}{2}$

(c) $y = -\frac{1}{4}x - 3$

x	-2	-1	0	1	2
y	$-\frac{5}{2}$	$-\frac{11}{4}$	-3	$-\frac{13}{4}$	$-\frac{7}{2}$

(b)

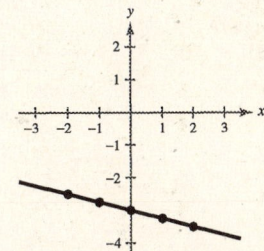

Both graphs are lines. The first graph rises to the right, whereas the second falls. Both pass through $(0, -3)$.

13. $y = 1 - x$ has intercepts $(1, 0)$ and $(0, 1)$.

Matches graph (d).

15. $y = \sqrt{9 - x^2}$ has intercepts $(\pm 3, 0)$ and $(0, 3)$.

Matches graph (f).

17. $y = x^3 - x + 1$ has a y-intercept of $(0, 1)$ and the points $(1, 1)$ and $(-2, -5)$ are on the graph.

Matches (a).

19. $y = -4x + 1$

Slope: -4

y-intercept: $(0, 1)$

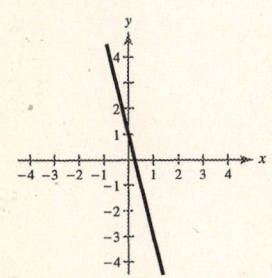

21. $y = 2 - x^2$

Intercepts: $(0, 2), \left(\sqrt{2}, 0\right), \left(-\sqrt{2}, 0\right)$

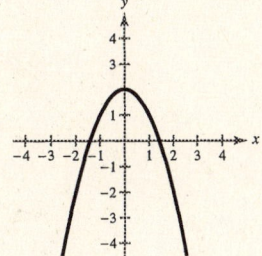

23. $y = x^2 - 3x$

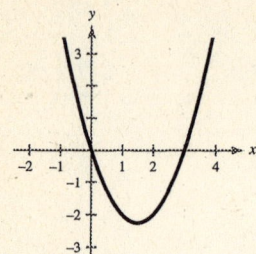

25. $y = x^3 + 2$

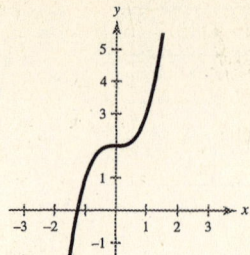

27. $y = \sqrt{x - 3}$

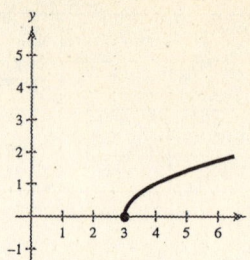

29. $y = |x - 2|$

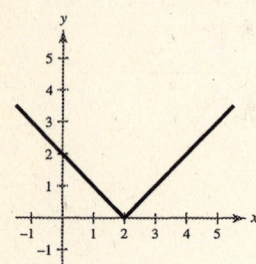

31. $x = y^2 - 1$

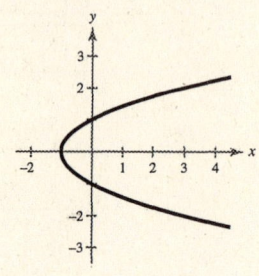

33. $y = x - 7$

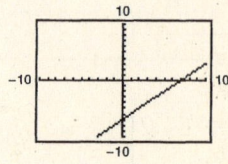

Intercepts: $(0, -7)$, $(7, 0)$

35. $y = 3 - \frac{1}{2}x$

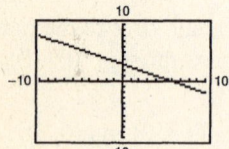

Intercepts: $(6, 0)$, $(0, 3)$

37. $y = x^2 - 4x + 3$

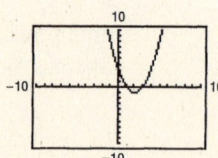

Intercepts: $(3, 0)$, $(1, 0)$, $(0, 3)$

39. $y = x(x - 2)^2$

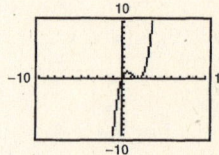

Intercepts: $(0, 0)$, $(2, 0)$

41. $y = \dfrac{2x}{x - 1}$

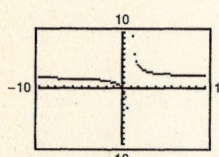

Intercepts: $(0, 0)$

43. $y = x\sqrt{x + 3}$

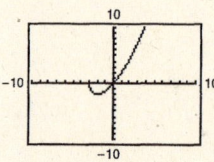

Intercepts: $(0, 0)$, $(-3, 0)$

45. $y = \sqrt[3]{x}$

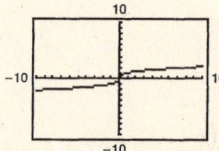

Intercepts: $(0, 0)$

47. $y = \frac{5}{2}x + 5$

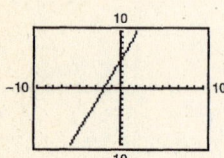

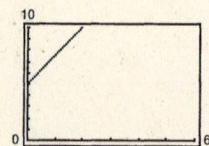

The first setting shows the line and its intercepts. The first setting is better.

The second setting does not show the x-intercept $(-2, 0)$.

49. $y = -x^2 + 10x - 5$

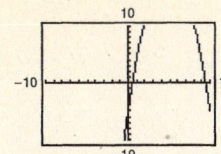 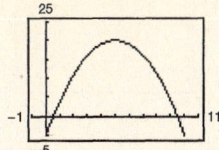

The second viewing window is better because it shows more of the essential features of the function.

51. $y = 4x^2 - 25$

Range/Window

Xmin = -5
Xmax = 5
Xscl = 1
Ymin = -30
Ymax = 10
Yscl = 5

53. $y = |x| + |x - 10|$

Range/Window

Xmin = -30
Xmax = 30
Xscl = 5
Ymin = -10
Ymax = 50
Yscl = 5

55. $x^2 + y^2 = 64$

$$y = \pm\sqrt{64 - x^2}$$

Use: $y_1 = \sqrt{64 - x^2}$

$$y_2 = -\sqrt{64 - x^2}$$

57. $(x - 1)^2 + (y - 2)^2 = 16$

$$(y - 2)^2 = 16 - (x - 1)^2$$

$$y - 2 = \pm\sqrt{16 - (x - 1)^2}$$

$$y = 2 \pm\sqrt{16 - (x - 1)^2}$$

Use $y_1 = 2 + \sqrt{16 - (x - 1)^2}$

$$y_2 = 2 - \sqrt{16 - (x - 1)^2}$$

59. $y_1 = \frac{1}{4}(x^2 - 8)$

$$y_2 = \frac{1}{4}x^2 - 2$$

The graphs are identical.

The Distributive Property is illustrated.

61. $y_1 = \frac{1}{5}[10(x^2 - 1)]$

$$y_2 = 2(x^2 - 1)$$

The graphs are identical.

The Associative Property of Multiplication is illustrated.

63. $y = \sqrt{5 - x}$

(a) $(2, y) \approx (2, 1.73)$

(b) $(x, 3) = (-4, 3)$

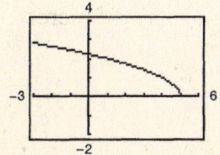

65. $y = x^5 - 5x$

(a) $(-0.5, y) \approx (-0.5, 2.47)$

(b) $(x, -4) = (1, -4)$ or $(x, -4) \approx (-1.65, -4)$

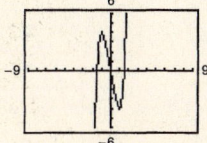

67. (a) $y = 225{,}000 - 20{,}000t$, $0 \le t \le 8$

Window

$X_{min} = 0$

$X_{max} = 8$

$X_{scl} = 1$

$Y_{min} = 60{,}000$

$Y_{max} = 230{,}000$

$Y_{scl} = 10{,}000$

(b)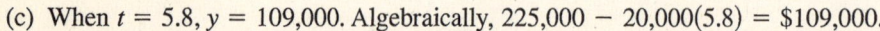

(c) When $t = 5.8$, $y = 109{,}000$. Algebraically, $225{,}000 - 20{,}000(5.8) = \$109{,}000$.

(d) When $t = 2.35$, $y = 178{,}000$. Algebraically, $225{,}000 - 20{,}000(2.35) = \$178{,}000$.

69. (a)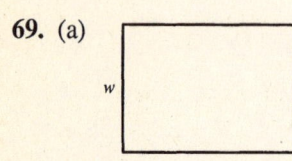

(b) Perimeter: $12 = 2x + 2w$

$$12 = 2(x + w)$$

$$6 = x + w$$

Thus, $w = 6 - x$.

Area: $xw = x(6 - x) \implies A = x(6 - x)$

(c)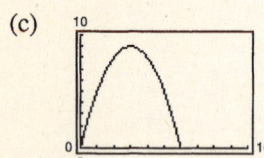

(d) When $w = 4.9$, $x = 1.1$ and Area $= 5.39$ square meters.

Algebraically,
Area $= xw = (1.1)(4.9) = 5.39$ square meters.

(e) The maximum area corresponds to the highest point on the graph, which appears to be $(3, 9)$. Thus, $x \doteq 3$ and $w \doteq 3$, and the rectangle is a square.

71. (a) The y-intercept $(0, 59.97)$ indicates the model's estimate of the life expectancy in 1930. $(t = 0)$

(b) $y = 73.2$ when $t \approx 53.3$, which corresponds to 1983. Algebraically,

$$\frac{59.97 + 0.98t}{1 + 0.01t} = 73.2$$

$$59.97 + 0.98t = 73.2 + 0.732t$$

$$0.248t = 13.23$$

$$t = \frac{13.23}{0.248} \approx 53.3 \text{ or } 1983$$

(c) 1948 corresponds to $t = 18$. Graphically, $y \approx 65.8$ when $t = 18$. Algebraically,

$$\frac{59.97 + 0.98(18)}{1 + 0.01(18)} = \frac{77.61}{1.18} \approx 65.8 \text{ years}$$

(d) 2010 corresponds to $t = 80$:

$$\frac{59.97 + 0.98(80)}{1 + 0.01(80)} = \frac{138.37}{1.8} \approx 76.9 \text{ years}$$

73. False. $y = 1 - x^2$ has two x-intercepts, $(1, 0)$ and $(-1, 0)$. Also, $y = x^2 + 1$ has no x-intercepts.

75. Answers will vary.

77. $7\sqrt{72} - 5\sqrt{18} = 7\sqrt{2(6^2)} - 5\sqrt{2(3^2)}$

$$= 42\sqrt{2} - 15\sqrt{2} = 27\sqrt{2}$$

79. $7^{3/2} \cdot 7^{11/2} = 7^{(3/2 + 11/2)} = 7^7 = 823{,}543$

81. $(9x - 4) + (2x^2 - x + 15) = 2x^2 + 8x + 11$

Appendix B.3   Solving Equations Algebraically and Graphically

- ■ You should know how to solve linear equations: $ax + b = 0$.
- ■ An identity is an equation whose solution consists of every real number in its domain.
- ■ To solve an equation you can:
    - (a) Add or subtract the same quantity from both sides.
    - (b) Multiply or divide both sides by the same nonzero quantity.
- ■ To solve an equation that can be simplified to a linear equation:
    - (a) Remove all symbols of grouping and all fractions.   (b) Combine like terms.
    - (c) Solve by algebra.   (d) Check the answer.
- ■ A "solution" that does not satisfy the original equation is called an extraneous solution.
- ■ You should be able to set up mathematical models to solve problems.
- ■ You should be able to translate key words and phrases.
    - (a) Equality:

        Equals, equal to, is, are, was, will be, represents

    - (b) Addition:

        Sum, plus, greater, increased by, more than, exceeds, total of

    - (c) Subtraction:

        Difference, minus, less than, decreased by, subtracted from, reduced by, the remainder

    - (d) Multiplication:

        Product, multiplied by, twice, times, percent of

    - (e) Division:

        Quotient, divided by, ratio, per

    - (f) Consecutive:

        Next, subsequent

- ■ You should know the following formulas:
    - (a) Perimeter:
        1. Square: $P = 4s$
        2. Rectangle: $P = 2L + 2W$
        3. Circle: $C = 2\pi r$

    - (b) Area:
        1. Square: $A = s^2$
        2. Rectangle: $A = LW$
        3. Circle: $A = \pi r^2$
        4. Triangle: $A = \left(\dfrac{1}{2}\right)bh$

    - (c) Volume
        1. Cube: $V = s^3$
        2. Rectangular solid: $V = LWH$
        3. Cylinder: $V = \pi r^2 h$
        4. Sphere: $V = \left(\dfrac{4}{3}\right)\pi r^3$

    - (d) Simple Interest: $I = Prt$

    - (e) Compound Interest: $A = P\left(1 + \dfrac{r}{n}\right)^{nt}$

    - (f) Distance: $D = r \cdot t$

    - (g) Temperature: $F = \dfrac{9}{5}C + 32$

- ■ You should be able to solve word problems. Study the examples in the text carefully.

Solutions to Odd-Numbered Exercises

1. $\dfrac{5}{2x} - \dfrac{4}{x} = 3$

(a) $\dfrac{5}{2(-1/2)} - \dfrac{4}{(-1/2)} \overset{?}{=} 3$

$\qquad\qquad\qquad 3 = 3$

$\quad x = -\frac{1}{2}$ *is* a solution.

(c) $\dfrac{5}{2(0)} - \dfrac{4}{0}$ is undefined.

$\quad x = 0$ *is not* a solution.

(b) $\dfrac{5}{2(4)} - \dfrac{4}{4} \overset{?}{=} 3$

$\qquad\qquad -\dfrac{3}{8} \neq 3$

$\quad x = 4$ *is not* a solution.

(d) $\dfrac{5}{2(1/4)} - \dfrac{4}{1/4} \overset{?}{=} 3$

$\qquad\qquad\qquad -6 \neq 3$

$\quad x = \frac{1}{4}$ *is not* a solution.

3. $3 + \dfrac{1}{x+2} = 4$

(a) $3 + \dfrac{1}{(-1)+2} \overset{?}{=} 4$

$\qquad\qquad\quad 4 = 4$

$\quad x = -1$ *is* a solution.

(c) $3 + \dfrac{1}{0+2} \overset{?}{=} 4$

$\qquad\qquad \dfrac{7}{2} \neq 4$

$\quad x = 0$ *is not* a solution.

(b) $3 + \dfrac{1}{(-2)+2} = 3 + \dfrac{1}{0}$ is undefined.

$\quad x = -2$ *is not* a solution.

(d) $3 + \dfrac{1}{5+2} \overset{?}{=} 4$

$\qquad\qquad \dfrac{22}{7} = 4$

$\quad x = 5$ *is not* a solution.

5. $\dfrac{\sqrt{x+4}}{6} + 3 = 4$

(a) $\dfrac{\sqrt{-3+4}}{6} + 3 \overset{?}{=} 4$

$\qquad\qquad \dfrac{19}{6} \neq 4$

$\quad x = -3$ *is not* a solution.

(c) $\dfrac{\sqrt{21+4}}{6} + 3 \overset{?}{=} 4$

$\qquad\qquad \dfrac{23}{6} \neq 4$

$\quad x = 21$ *is not* a solution.

(b) $\dfrac{\sqrt{0+4}}{6} + 3 \overset{?}{=} 4$

$\qquad\qquad \dfrac{10}{3} \neq 4$

$\quad x = 0$ *is not* a solution.

(d) $\dfrac{\sqrt{32+4}}{6} + 3 \overset{?}{=} 4$

$\qquad\qquad\quad 4 = 4$

$\quad x = 32$ *is* a solution.

7. $2(x - 1) = 2x - 2$ is an *identity* by the Distributive Property. It is true for all real values of x.

9. $x^2 - 8x + 5 = (x - 4)^2 - 11$ is an *identity* since

$\qquad (x - 4)^2 - 11 = x^2 - 8x + 16 - 11$

$\qquad\qquad\qquad\quad = x^2 - 8x + 5.$

11. $3 + \dfrac{1}{x+1} = \dfrac{4x}{x+1}$ is *conditional*. There are real values of x for which the equation is not true.

13. Method 1:

$$\frac{3x}{8} - \frac{4x}{3} = 4$$

$$\frac{9x - 32x}{24} = 4$$

$$-23x = 96$$

$$x = -\frac{96}{23}$$

Method 2:

Graph

$$y_1 = \frac{3x}{8} - \frac{4x}{3} \text{ and } y_2 = 4$$

in the same viewing window. These lines intersect at $x \approx -4.1739 \approx -\frac{96}{23}$.

15. $\dfrac{x}{5} - \dfrac{x}{2} = 3$

$$\frac{2x - 5x}{10} = 3$$

$$-3x = 30$$

$$x = -10$$

17. $\dfrac{3}{2}(z + 5) - \dfrac{1}{4}(z + 24) = 0$

$$4\left(\frac{3}{2}\right)(z + 5) - 4\left(\frac{1}{4}\right)(z + 24) = 4(0)$$

$$6(z + 5) - (z + 24) = 0$$

$$6z + 30 - z - 24 = 0$$

$$5z = -6$$

$$z = -\frac{6}{5}$$

19. $\dfrac{100 - 4u}{3} = \dfrac{5u + 6}{4} + 6$

$$12\left(\frac{100 - 4u}{3}\right) = 12\left(\frac{5u + 6}{4}\right) + 12(6)$$

$$4(100 - 4u) = 3(5u + 6) + 72$$

$$400 - 16u = 15u + 18 + 72$$

$$-31u = -310$$

$$u = 10$$

21. $\dfrac{5x - 4}{5x + 4} = \dfrac{2}{3}$

$$3(5x - 4) = 2(5x + 4)$$

$$15x - 12 = 10x + 8$$

$$5x = 20$$

$$x = 4$$

23. $\dfrac{1}{x - 3} + \dfrac{1}{x + 3} = \dfrac{10}{x^2 - 9}$

$$\frac{(x + 3) + (x - 3)}{x^2 - 9} = \frac{10}{x^2 - 9}$$

$$2x = 10$$

$$x = 5$$

25. $\dfrac{7}{2x + 1} - \dfrac{8x}{2x - 1} = -4$

$$7(2x - 1) - 8x(2x + 1) = -4(2x + 1)(2x - 1)$$

$$14x - 7 - 16x^2 - 8x = -16x^2 + 4$$

$$6x = 11$$

$$x = \frac{11}{6}$$

27. $\dfrac{1}{x} + \dfrac{2}{x - 5} = 0$

$$1(x - 5) + 2x = 0$$

$$3x - 5 = 0$$

$$3x = 5$$

$$x = \frac{5}{3}$$

29. $\dfrac{3}{x(x-3)} + \dfrac{4}{x} = \dfrac{1}{x-3}$

$3 + 4(x-3) = x$

$3 + 4x - 12 = x$

$3x = 9$

$x = 3$

A check reveals that $x = 3$ is an extraneous solution, so there is no solution.

31. $y = x - 5$

Let $y = 0$:

$0 = x - 5 \implies x = 5 \implies (5, 0)$ x-intercept

Let $x = 0$:

$y = 0 - 5 \implies y = -5 \implies (0, -5)$ y-intercept

33. $y = x^2 + x - 2$

Let $y = 0$:

$(x^2 + x - 2) = (x + 2)(x - 1) = 0 \implies$

$x = -2, 1 \implies$

$(-2, 0), (1, 0)$ x-intercepts

Let $x = 0$:

$y = 0^2 + 0 - 2 = -2 \implies (0, -2)$ y-intercept

35. $y = x\sqrt{x + 2}$

Let $y = 0$: $0 = x\sqrt{x + 2} \implies x = 0, -2 \implies (0, 0), (-2, 0)$ x-intercepts

Let $x = 0$: $y = 0\sqrt{0 + 2} = 0 \implies (0, 0)$ y-intercept

37. $y = |x - 2| - 4$

Let $y = 0$: $|x - 2| - 4 = 0 \implies |x - 2| = 4 \implies x = -2, 6 \implies (-2, 0), (6, 0)$ x-intercepts

Let $x = 0$: $|0 - 2| - 4 = |-2| - 4 = 2 - 4 = -2 = y \implies (0, -2)$ y-intercept

39. $xy - 2y - x + 1 = 0$

Let $y = 0$:

$-x + 1 = 0 \implies x = 1 \implies (1, 0)$ x-intercept

Let $x = 0$:

$-2y + 1 = 0 \implies y = \frac{1}{2} \implies \left(0, \frac{1}{2}\right)$ y-intercept

41. $f(x) = 5(4 - x)$

$5(4 - x) = 0$

$4 - x = 0$

$x = 4$

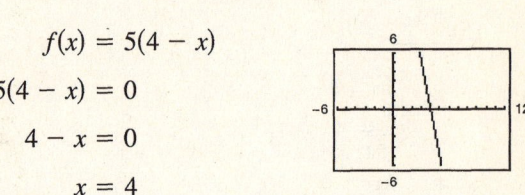

43. $f(x) = x^3 - 6x^2 + 5x$

$x^3 - 6x^2 + 5x = 0$

$x(x^2 - 6x + 5) = 0$

$x(x - 5)(x - 1) = 0$

$x = 0, 5, 1$

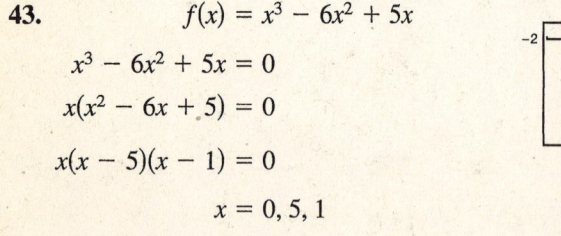

45.
$$f(x) = \frac{x+2}{3} - \frac{x-1}{5} - 1$$

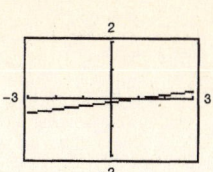

$$\frac{x+2}{3} - \frac{x-1}{5} - 1 = 0$$

$$5(x+2) - 3(x-1) - 15 = 0$$

$$2x = 2$$

$$x = 1$$

47.

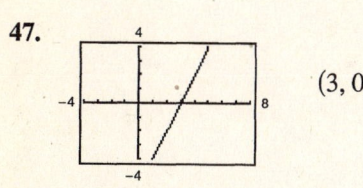

(3, 0)

$$y = 0 = 2(x-1) - 4$$

$$= 2x - 2 - 4$$

$$= 2x - 6 \implies 2x = 6 \implies x = 3$$

49.

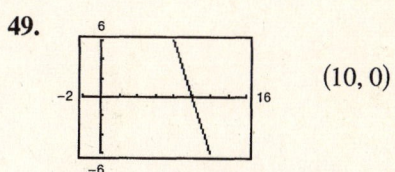

(10, 0)

$$y = 0 = 20 - (3x - 10)$$

$$= 20 - 3x + 10$$

$$= 30 - 3x \implies 3x = 30 \implies x = 10$$

51.
$$2.7x - 0.4x = 1.2$$

$$2.3x = 1.2$$

$$x = \frac{1.2}{2.3} \approx 0.522$$

$$f(x) = 2.7x - 0.4x - 1.2 = 0$$

$$x \approx 0.522$$

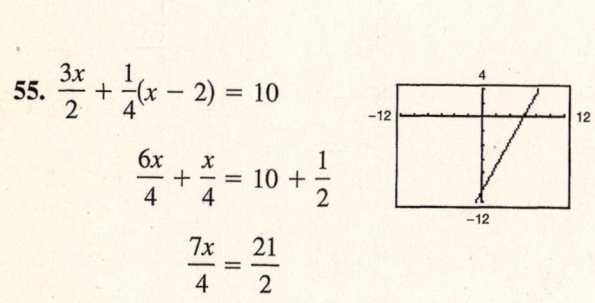

53.
$$25(x-3) = 12(x+2) - 10$$

$$25x - 75 = 12x + 24 - 10$$

$$13x - 89 = 0$$

$$x = \frac{89}{13}$$

$$f(x) = 25(x-3) - 12(x+2) + 10 = 0$$

$$x = 6.846$$

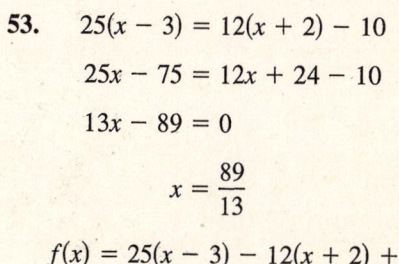

55.
$$\frac{3x}{2} + \frac{1}{4}(x-2) = 10$$

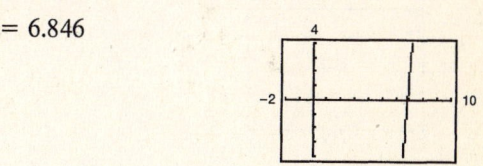

$$\frac{6x}{4} + \frac{x}{4} = 10 + \frac{1}{2}$$

$$\frac{7x}{4} = \frac{21}{2}$$

$$x = 6$$

$$f(x) = \frac{3x}{2} + \frac{1}{4}(x-2) - 10 = 0$$

$$x = 6.0$$

57.
$$\frac{2x}{3} = 10 - \frac{24}{x}$$

$$\frac{2x}{3}(3x) = 10(3x) - \frac{24}{x}(3x)$$

$$2x^2 = 30x - 72$$

$$2x^2 - 30x + 72 = 0$$

$$x^2 - 15x + 36 = 0$$

$$(x-3)(x-12) = 0$$

$$x = 3, 12$$

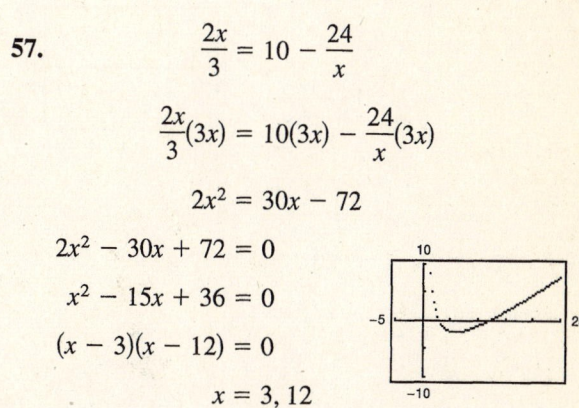

$$f(x) = \frac{2x}{3} - 10 + \frac{24}{x}$$

$$x = 3, 12$$

59.
$$\frac{3}{x+2} - \frac{4}{x-2} = 5$$

$$3(x-2) - 4(x+2) = 5(x+2)(x-2)$$

$$3x - 6 - 4x - 8 = 5(x^2 - 4)$$

$$0 = 5x^2 + x - 6$$

$$0 = (x-1)(5x+6)$$

$$x = 1, -\frac{6}{5}$$

$$f(x) = \frac{3}{x+2} - \frac{4}{x-2} - 5 = 0$$

$$x = 1.0, -1.2$$

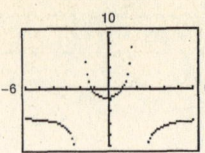

61.
$$(x+2)^2 = x^2 - 6x + 1$$

$$x^2 + 4x + 4 = x^2 - 6x + 1$$

$$10x = -3$$

$$x = -\frac{3}{10}$$

$$f(x) = (x+2)^2 - x^2 + 6x - 1$$

$$x = -\frac{3}{10}$$

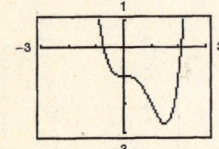

63. $2x^3 - x^2 - 18x + 9 = 0$

$$x = -3.0, 0.5, 3.0$$

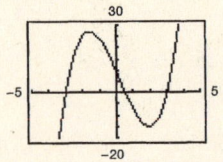

65.
$$x^4 = 2x^3 + 1$$

$$x^4 - 2x^3 - 1 = 0$$

$$x \approx -0.717, 2.107$$

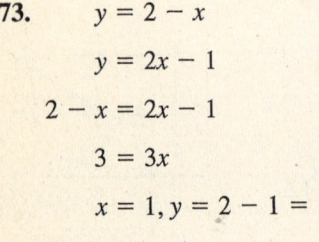

67.
$$\frac{2}{x+2} = 3$$

$$\frac{2}{x+2} - 3 = 0$$

$$x = -\frac{4}{3}$$

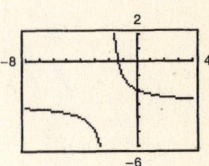

69.
$$|x - 3| = 4$$

$$|x - 3| - 4 = 0$$

$$x = -1, 7$$

71.
$$\sqrt{x-2} = 3$$

$$\sqrt{x-2} - 3 = 0$$

$$x = 11$$

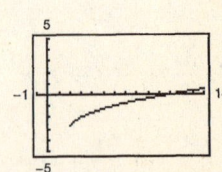

73.
$$y = 2 - x$$

$$y = 2x - 1$$

$$2 - x = 2x - 1$$

$$3 = 3x$$

$$x = 1, y = 2 - 1 = 1$$

$$(x, y) = (1, 1)$$

75.
$$x - y = -4 \implies y = x + 4$$

$$x^2 - y = -2 \implies y = x^2 + 2$$

$$x^2 + 2 = x + 4$$

$$x^2 - x - 2 = 0$$

$$(x - 2)(x + 1) = 0$$

$$x = 2, y = 6$$

$$x = -1, y = 3$$

$$(2, 6), (-1, 3)$$

77. $y = 9 - 2x$

$y = x - 3$

$(4, 1)$

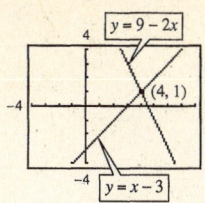

79. $y = 4 - x^2$

$y = 2x - 1$

$(x, y) = (1.449, 1.898), (-3.449, -7.899)$

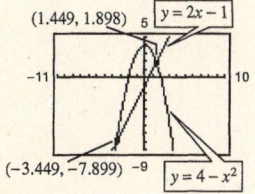

81. $y = 2x^2$

$y = x^4 - 2x^2$

$(x, y) = (0, 0), (2, 8), (-2, 8)$

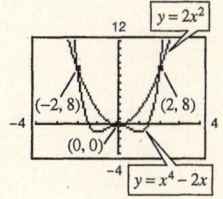

83. $6x^2 + 3x = 0$

$3x(2x + 1) = 0$

$3x = 0$ or $2x + 1 = 0$

$x = 0$ or $x = -\frac{1}{2}$

85. $x^2 - 2x - 8 = 0$

$(x - 4)(x + 2) = 0$

$x - 4 = 0$ or $x + 2 = 0$

$x = 4$ or $x = -2$

87. $3 + 5x - 2x^2 = 0$

$(3 - x)(1 + 2x) = 0$

$3 - x = 0$ or $1 + 2x = 0$

$x = 3$ or $x = -\frac{1}{2}$

89. $x^2 + 4x = 12$

$x^2 + 4x - 12 = 0$

$(x + 6)(x - 2) = 0$

$x + 6 = 0$ or $x - 2 = 0$

$x = -6$ or $x = 2$

91. $(x + a)^2 - b^2 = 0$

$[(x + a) + b][(x + a) - b] = 0$

$x + a + b = 0 \Rightarrow x = -a - b$

$x + a - b = 0 \Rightarrow x = -a + b$

93. $x^2 = 49$

$x = \pm\sqrt{49} = \pm 7$

95. $(x - 12)^2 = 16$

$x - 12 = \pm\sqrt{16} = \pm 4$

$x = 12 \pm 4$

$x = 16, 8$

97. $(2x - 1)^2 = 12$

$2x - 1 = \pm\sqrt{12} = \pm 2\sqrt{3}$

$2x = 1 \pm 2\sqrt{3}$

$x = \frac{1}{2} \pm \sqrt{3}$

$(x = 2.23, -1.23)$

99. $(x - 7)^2 = (x + 3)^2$

$x - 7 = \pm(x + 3)$

$x - 7 = x + 3$ impossible

$x - 7 = -(x + 3) \Rightarrow 2x = 4$

$\Rightarrow x = 2$

101. $x^2 + 4x = 32$

$x^2 + 4x + 4 = 32 + 4$

$(x + 2)^2 = 36$

$x + 2 = \pm 6$

$x = -2 \pm 6$

$x = -8, 4$

103. $x^2 + 6x + 2 = 0$

$x^2 + 6x = -2$

$x^2 + 6x + 3^2 = -2 + 3^2$

$(x + 3)^2 = 7$

$x + 3 = \pm\sqrt{7}$

$x = -3 \pm\sqrt{7}$

105. $9x^2 - 18x + 3 = 0$

$x^2 - 2x + \dfrac{1}{3} = 0$

$x^2 - 2x = -\dfrac{1}{3}$

$x^2 - 2x + 1^2 = -\dfrac{1}{3} + 1^2$

$(x - 1)^2 = \dfrac{2}{3}$

$x - 1 = \pm\sqrt{\dfrac{2}{3}}$

$x = 1 \pm\sqrt{\dfrac{2}{3}}$

$x = 1 \pm\dfrac{\sqrt{6}}{3}$

107. $8 + 4x - x^2 = 0$

$x^2 - 4x = 8$

$x^2 - 4x + 4 = 8 + 4$

$(x - 2)^2 = 12$

$x - 2 = \pm\sqrt{12}$

$= \pm 2\sqrt{3}$

$x = 2 \pm 2\sqrt{3}$

109. $-x^2 + 2x + 2 = 0$

$x = \dfrac{-b \pm \sqrt{b^2 - 4ac}}{2a}$

$= \dfrac{-2 \pm \sqrt{2^2 - 4(-1)(2)}}{2(-1)}$

$= \dfrac{-2 \pm 2\sqrt{3}}{-2} = 1 \pm\sqrt{3}$

111. $x^2 + 8x - 4 = 0$

$x = \dfrac{-b \pm \sqrt{b^2 - 4ac}}{2a}$

$= \dfrac{-8 \pm \sqrt{8^2 - 4(1)(-4)}}{2(1)}$

$= \dfrac{-8 \pm 4\sqrt{5}}{2} = -4 \pm 2\sqrt{5}$

113. $28x - 49x^2 = 4$

$-49x^2 + 28x - 4 = 0$

$x = \dfrac{-b \pm \sqrt{b^2 - 4ac}}{2a}$

$= \dfrac{-28 \pm \sqrt{28^2 - 4(-49)(-4)}}{2(-49)}$

$= \dfrac{-28 \pm 0}{-98} = \dfrac{2}{7}$

115. $4x^2 + 16x + 17 = 0$

$x = \dfrac{-b \pm \sqrt{b^2 - 4ac}}{2a}$

$= \dfrac{-16 \pm \sqrt{16^2 - 4(4)(17)}}{2(4)}$

$= \dfrac{-16 \pm \sqrt{-16}}{8}$

$= \dfrac{-16 \pm 4i}{8}$

$= -2 \pm \dfrac{1}{2}i$

117. $x^2 - 2x - 1 = 0$

$\quad x^2 - 2x = 1$

$\quad x^2 - 2x + 1^2 = 1 + 1^2$

$\quad (x - 1)^2 = 2$

$\quad x - 1 = \pm\sqrt{2}$

$\quad x = 1 \pm \sqrt{2}$

119. $(x + 3)^2 = 81$

$\quad x + 3 = \pm 9$

$\quad x + 3 = 9$ or $x + 3 = -9$

$\quad x = 6$ or $\quad x = -12$

121. $x^2 - x - \frac{11}{4} = 0$

$\quad x^2 - x + \frac{1}{4} = \frac{11}{4} + \frac{1}{4}$

$\quad \left(x - \frac{1}{2}\right)^2 = 3$

$\quad x - \frac{1}{2} = \pm\sqrt{3}$

$\quad x = \frac{1}{2} \pm \sqrt{3}$

$\quad x = \frac{1}{2} + \sqrt{3}, \frac{1}{2} - \sqrt{3}$

123. $4x^4 - 18x^2 = 0$

$\quad 2x^2(2x^2 - 9) = 0$

$\quad 2x^2 = 0 \implies x = 0$

$\quad 2x^2 - 9 = 0 \implies x = \pm\dfrac{3\sqrt{2}}{2}$

125. $x^4 - 4x^2 + 3 = 0$

$\quad (x^2 - 3)(x^2 - 1) = 0$

$\quad \left(x + \sqrt{3}\right)\left(x - \sqrt{3}\right)(x + 1)(x - 1) = 0$

$\quad x + \sqrt{3} = 0 \implies x = -\sqrt{3}$

$\quad x - \sqrt{3} = 0 \implies x = \sqrt{3}$

$\quad x + 1 = 0 \implies x = -1$

$\quad x - 1 = 0 \implies x = 1$

127. $x^3 - 3x^2 - x + 3 = 0$

$\quad x^2(x - 3) - (x - 3) = 0$

$\quad (x - 3)(x^2 - 1) = 0$

$\quad (x - 3)(x + 1)(x - 1) = 0$

$\quad x - 3 = 0 \implies x = 3$

$\quad x + 1 = 0 \implies x = -1$

$\quad x - 1 = 0 \implies x = 1$

129. $4x^4 - 65x^2 + 16 = 0$

$\quad (4x^2 - 1)(x^2 - 16) = 0$

$\quad (2x + 1)(2x - 1)(x + 4)(x - 4) = 0$

$\quad 2x + 1 = 0 \implies x = -\frac{1}{2}$

$\quad 2x - 1 = 0 \implies x = \frac{1}{2}$

$\quad x + 4 = 0 \implies x = -4$

$\quad x - 4 = 0 \implies x = 4$

131. $\dfrac{1}{t^2} + \dfrac{8}{t} + 15 = 0$

$\quad 1 + 8t + 15t^2 = 0$

$\quad (1 + 3t)(1 + 5t) = 0$

$\quad 1 + 3t = 0 \implies t = -\dfrac{1}{3}$

$\quad 1 + 5t = 0 \implies t = -\dfrac{1}{5}$

133. $2x + 9\sqrt{x} - 5 = 0$

$\quad \left(2\sqrt{x} - 1\right)\left(\sqrt{x} + 5\right) = 0$

$\quad \sqrt{x} = \dfrac{1}{2} \implies x = \dfrac{1}{4}$

$(\sqrt{x} = -5$ is not possible.$)$

Note: You can see graphically that there is only one solution.

135. $\sqrt{x - 10} - 4 = 0$

$\quad \sqrt{x - 10} = 4$

$\quad x - 10 = 16$

$\quad x = 26$

137. $\sqrt{x + 1} - 3x = 1$

$\quad \sqrt{x + 1} = 3x + 1$

$\quad x + 1 = 9x^2 + 6x + 1$

$\quad 0 = 9x^2 + 5x$

$\quad 0 = x(9x + 5)$

$\quad x = 0$

$\quad 9x + 5 = 0 \implies x = -\frac{5}{9}$, extraneous

139. $\sqrt[3]{2x + 1} + 8 = 0$

$\qquad \sqrt[3]{2x + 1} = -8$

$\qquad 2x + 1 = -512$

$\qquad 2x = -513$

$\qquad x = -\frac{513}{2} = -256.5$

141. $\sqrt{x} - \sqrt{x - 5} = 1$

$\qquad \sqrt{x} = 1 + \sqrt{x - 5}$

$\qquad \left(\sqrt{x}\right)^2 = \left(1 + \sqrt{x - 5}\right)^2$

$\qquad x = 1 + 2\sqrt{x - 5} + x - 5$

$\qquad 4 = 2\sqrt{x - 5}$

$\qquad 2 = \sqrt{x - 5}$

$\qquad 4 = x - 5$

$\qquad 9 = x$

143. $(x - 5)^{2/3} = 16$

$\qquad x - 5 = \pm 16^{3/2}$

$\qquad x - 5 = \pm 64$

$\qquad x = 69, -59$

145. $3x(x - 1)^{1/2} + 2(x - 1)^{3/2} = 0$

$\qquad (x - 1)^{1/2}[3x + 2(x - 1)] = 0$

$\qquad (x - 1)^{1/2}(5x - 2) = 0$

$\qquad (x - 1)^{1/2} = 0 \implies x - 1 = 0 \implies x = 1$

$\qquad 5x - 2 = 0 \implies x = \frac{2}{5}$ which is extraneous.

147. $\dfrac{1}{x} - \dfrac{1}{x + 1} = 3$

$\qquad x(x + 1)\dfrac{1}{x} - x(x + 1)\dfrac{1}{x + 1} = x(x + 1)(3)$

$\qquad x + 1 - x = 3x(x + 1)$

$\qquad 1 = 3x^2 + 3x$

$\qquad 0 = 3x^2 + 3x - 1;\ a = 3,\ b = 3,\ c = -1$

$\qquad x = \dfrac{-3 \pm \sqrt{(3)^2 - 4(3)(-1)}}{2(3)} = \dfrac{-3 \pm \sqrt{21}}{6}$

149. $x = \dfrac{3}{x} + \dfrac{1}{2}$

$\qquad (2x)(x) = (2x)\left(\dfrac{3}{x}\right) + (2x)\left(\dfrac{1}{2}\right)$

$\qquad 2x^2 = 6 + x$

$\qquad 2x^2 - x - 6 = 0$

$\qquad (2x + 3)(x - 2) = 0$

$\qquad 2x + 3 = 0 \implies x = -\dfrac{3}{2}$

$\qquad x - 2 = 0 \implies x = 2$

151. $|2x - 1| = 5$

$\qquad 2x - 1 = 5 \implies x = 3$

$\qquad -(2x - 1) = 5 \implies x = -2$

153. $|x| = x^2 + x - 3$

$\qquad x = x^2 + x - 3 \qquad$ OR $\qquad -x = x^2 + x - 3$

$x^2 - 3 = 0 \qquad\qquad\qquad x^2 + 2x - 3 = 0$

$\qquad x = \pm\sqrt{3} \qquad\qquad\qquad (x - 1)(x + 3) = 0$

$\qquad\qquad\qquad\qquad\qquad x - 1 = 0 \implies x = 1$

$\qquad\qquad\qquad\qquad\qquad x + 3 = 0 \implies x = -3$

Only $x = \sqrt{3}$, and $x = -3$ are solutions to the original equation. $x = -\sqrt{3}$ and $x = 1$ are extraneous. Note that the graph of $y = x^2 + x - 3 - |x|$ has two x-intercepts.

155. $y = x^3 - 2x^2 - 3x$

(a)

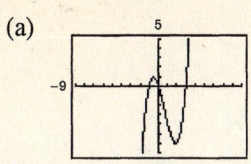

(b) x-intercepts: $(-1, 0), (0, 0), (3, 0)$

(c) $0 = x^3 - 2x^2 - 3x$

$0 = x(x + 1)(x - 3)$

$x = 0$

$x + 1 = 0 \implies x = -1$

$x - 3 = 0 \implies x = \ \ \ 3$

(d) The x-intercepts are the same as the solutions.

157. $y = \sqrt{11x - 30} - x$

(a)

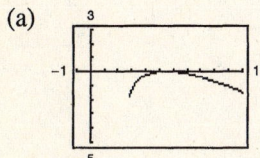

(b) x-intercepts: $(5, 0), (6, 0)$

(d) The x-intercepts and the solutions are the same.

(c)
$$0 = \sqrt{11x - 30} - x$$
$$x = \sqrt{11x - 30}$$
$$x^2 = 11x - 30$$
$$x^2 - 11x + 30 = 0$$
$$(x - 5)(x - 6) = 0$$
$$x - 5 = 0 \implies x = 5$$
$$x - 6 = 0 \implies x = 6$$

159. $y = \dfrac{1}{x} - \dfrac{4}{x - 1} - 1$

(a)

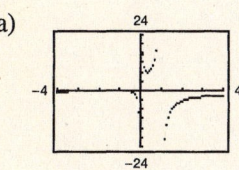

(b) x-intercept: $(-1, 0)$

(d) The x-intercepts and the solutions are the same.

(c) $0 = \dfrac{1}{x} - \dfrac{4}{x - 1} - 1$

$0 = (x - 1) - 4x - x(x - 1)$

$0 = x - 1 - 4x - x^2 + x$

$0 = -x^2 - 2x - 1$

$0 = x^2 + 2x + 1$

$x + 1 = 0 \implies x = -1$

161. $y = |x + 1| - 2$

(a)

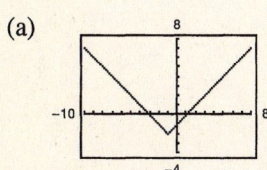

(b) x-intercept: $(1, 0), (-3, 0)$

(d) The x-intercepts and the solutions are the same.

(c) $0 = |x + 1| - 2$

$2 = |x + 1|$

$x + 1 = 2$ or $-(x + 1) = 2$

$x = 1$ or $\ \ -x - 1 = 2$

$-x = 3$

$x = -3$

163. $A = 142.9t + 3729, \quad 5 \le t \le 11$

$M = 52.1t + 4400, \quad 5 \le t \le 11$

(a)

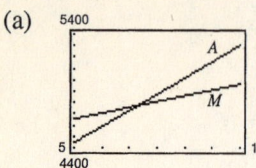

Intersection: $(7.39, 4785.01)$

(b) $142.9t + 3729 = 52.1t + 4400$

$$90.8t = 671$$

$$t = \frac{671}{90.8} = \frac{3355}{454} \approx 7.39$$

This represents the year (1997) when the populations of Arizona and Minnesota were equal.

(c) For 2006, $t = 16$ and $A \approx 6015$ thousand, $M \approx 5234$ thousand.

165. (a) $C = 0.45x^2 - 1.65x + 50.75, \quad 10 \le x \le 25$

(b) If $C = 150$, then $x = 16.797$ degrees.

(c) If the temperature is increased $10°$ to $20°$, then C increases from 79.25 to 197.75, a factor of 2.5.

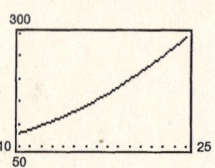

167. False. Two linear equations could have an infinite number of points of intersection. For example, $x + y = 1$ and $2x + 2y = 2$.

169. (a) $ax^2 + bx = 0$

$x(ax + b) = 0$

$x = 0$

$x = -b/a$

(b) $ax^2 - ax = 0$

$ax(x - 1) = 0$

$x = 0$

$x = 1$

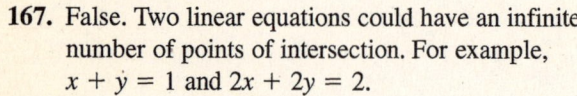

Appendix B.4   Solving Inequalities Algebraically and Graphically

▪ You should know the properties of inequalities.

(a) Transitive: $a < b$ and $b < c$ implies $a < c$.

(b) Addition: $a < b$ and $c < d$ implies $a + c < b + d$.

(c) Adding or Subtracting a Constant: $a \pm c < b \pm c$ if $a < b$.

(d) Multiplying or Dividing by a Constant: For $a < b$,

    1. If $a > 0$, then $ac < bc$ and $\dfrac{a}{c} < \dfrac{b}{c}$.

    2. If $c < 0$, then $ac > bc$ and $\dfrac{a}{c} > \dfrac{b}{c}$.

▪ You should know that

$$|x| = \begin{cases} x & \text{if } x \geq 0 \\ -x & \text{if } x < 0 \end{cases}.$$

▪ You should be able to solve absolute value inequalities.

(a) $|x| < a$ if and only if $-a < x < a$.

(b) $|x| > a$ if and only if $x < -a$ or $x > a$.

▪ You should be able to solve polynomial inequalities.

(a) Find the critical numbers.

    1. Values that make the expression zero

    2. Values that make the expression undefined

(b) Test one value in each interval on the real number line resulting from the critical numbers.

(c) Determine the solution intervals.

▪ You should be able to solve rational and other types of inequalities.

Solutions to Odd-Numbered Exercises

1. $x < 3$

    Matches (d).

3. $-3 < x \leq 4$

    Matches (c).

5. (a) $x = 3$

    $5(3) - 12 \overset{?}{>} 0$

           $3 > 0$

    Yes, $x = 3$ is a solution.

(c) $x = \frac{5}{2}$

    $5\left(\frac{5}{2}\right) - 12 \overset{?}{>} 0$

           $\frac{1}{2} > 0$

    Yes, $x = \frac{5}{2}$ is a solution.

(b) $x = -3$

    $5(-3) - 12 \overset{?}{>} 0$

          $-27 \not> 0$

    No, $x = -3$ is not a solution.

(d) $x = \frac{3}{2}$

    $5\left(\frac{3}{2}\right) - 12 \overset{?}{>} 0$

         $-\frac{9}{2} \not> 0$

    No, $x = \frac{3}{2}$ is not a solution.

7. $-1 < \dfrac{3-x}{2} \le 1$

(a) $x = 0$

$$-1 \overset{?}{<} \frac{3-0}{2} \overset{?}{\le} 1$$

$$-1 \overset{?}{<} \frac{3}{2} \overset{?}{\le} 1$$

No, $x = 0$ is not a solution.

(c) $x = 1$

$$-1 \overset{?}{<} \frac{3-1}{2} \overset{?}{\le} 1$$

$$-1 \overset{?}{<} 1 \overset{?}{\le} 1$$

Yes, $x = 1$ is a solution.

(b) $x = \sqrt{5}$

$$-1 \overset{?}{<} \frac{3-\sqrt{5}}{2} \overset{?}{\le} 1$$

$$-1 \overset{?}{<} 0.382 \overset{?}{\le} 1$$

Yes, $x = \sqrt{5}$ is a solution.

(d) $x = 5$

$$-1 \overset{?}{<} \frac{3-5}{2} \le 1$$

$$-1 \overset{?}{<} -1 \overset{?}{\le} 1$$

No, $x = 5$ is not a solution.

9.

$$-10x < 40$$
$$-\tfrac{1}{10}(-10x) > -\tfrac{1}{10}(40)$$
$$x > -4$$

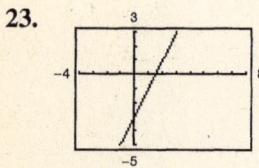

11. $4(x + 1) < 2x + 3$

$$4x + 4 < 2x + 3$$
$$2x < -1$$
$$x < -\tfrac{1}{2}$$

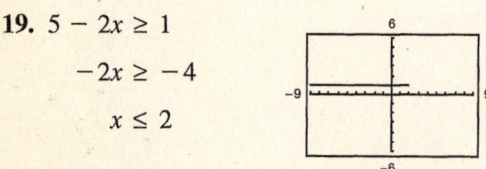

13. $\tfrac{3}{4}x - 6 \le x - 7$

$$1 \le \tfrac{1}{4}x$$
$$4 \le x$$
$$x \ge 4$$

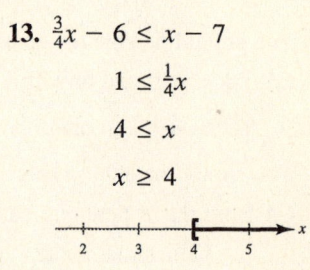

15. $-8 \le 1 - 3(x - 2) < 13$

$$-8 \le 1 - 3x + 6 < 13$$
$$-8 \le -3x + 7 < 13$$
$$-15 \le -3x < 6$$
$$5 \ge x > -2 \;\Rightarrow\; -2 < x \le 5$$

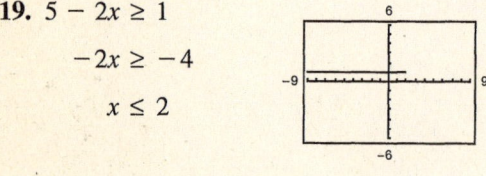

17. $-4 < \dfrac{2x - 3}{3} < 4$

$$-12 < 2x - 3 < 12$$
$$-9 < 2x < 15$$
$$-\frac{9}{2} < x < \frac{15}{2}$$

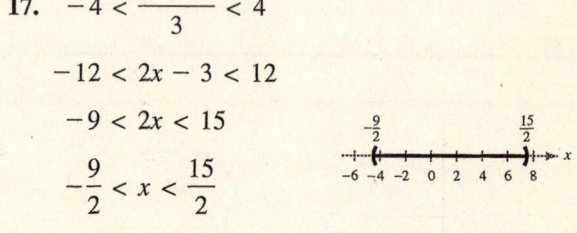

19. $5 - 2x \ge 1$

$$-2x \ge -4$$
$$x \le 2$$

21. $3(x + 1) < x + 7$

$$3x + 3 < x + 7$$
$$2x < 4$$
$$x < 2$$

23.

Algebraically, (a) $y \ge 1$ (b) $y \le 0$

$$2x - 3 \ge 1 \qquad\qquad 2x - 3 \le 0$$
$$2x \ge 4 \qquad\qquad\quad 2x \le 3$$
$$x \ge 2 \qquad\qquad\quad\; x \le \tfrac{3}{2}$$

Using the graph, (a) $y \ge 1$ for $x \ge 2$ and (b) $y \le 0$ for $x \le \tfrac{3}{2}$.

25.

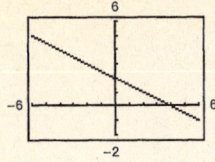

Using the graph, (a) $0 \leq y \leq 3$ for $-2 \leq x \leq 4$
and (b) $y \geq 0$ for $x \leq 4$

Algebraically,

(a) $0 \leq y \leq 3$
$0 \leq -\frac{1}{2}x + 2 \leq 3$
$-2 \leq -\frac{1}{2}x \leq 1$
$4 \geq x \geq -2$

(b) $y \geq 0$
$-\frac{1}{2}x + 2 \geq 0$
$2 \geq \frac{1}{2}x$
$4 \geq x$

27. $|5x| > 10$
$5x < -10$ or $5x > 10$
$x < -2$ or $x > 2$

29. $|x - 7| < 6$
$-6 < x - 7 < 6$
$1 < x < 13$

31. $|x + 14| + 3 > 17$
$|x + 14| > 14$
$x + 14 < -14$ or $x + 14 > 14$
$x < -28$ or $x > 0$

33. $10|1 - 2x| < 5$
$|1 - 2x| < \frac{1}{2}$
$-\frac{1}{2} < 1 - 2x < \frac{1}{2}$
$-\frac{3}{2} < -2x < -\frac{1}{2}$
$\frac{3}{4} > x > \frac{1}{4}$
$\frac{1}{4} < x < \frac{3}{4}$

35. $y = |x - 3|$

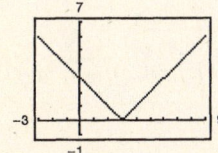

(a) Graphically, $y \leq 2$ for $1 \leq x \leq 5$ and
(b) $y \geq 4$ for $x \leq -1$ or $x \geq 7$

Algebraically,

(a) $y \leq 2$
$|x - 3| \leq 2$
$-2 \leq x - 3 \leq 2$
$1 \leq x \leq 5$

(b) $y \geq 4$
$|x - 3| \geq 4$
$x - 3 \leq -4$ or $x - 3 \geq 4$
$x \leq -1$ or $x \geq 7$

37. The midpoint of the interval $[-3, 3]$ is 0. The interval represents all real numbers x no more than 3 units from 0.
$|x - 0| \leq 3$
$|x| \leq 3$

39. The midpoint of the interval $[-3, 3]$ is 0. The two intervals represent all numbers x more than 3 units from 0.
$|x - 0| > 3$
$|x| > 3$

41. All real numbers within 10 units of 7
$|x - 7| \leq 10$

43. $x^2 - 4x - 5 > 0$

$(x - 5)(x + 1) > 0$

Critical numbers: $-1, 5$

Testing the intervals $(-\infty, -1), (-1, 5)$ and $(5, \infty)$, we have $x^2 - 4x - 5 > 0$ on $(-\infty, -1)$ and $(5, \infty)$

Similarly, $x^2 - 4x - 5 < 0$ on $(-1, 5)$

45. $(x + 2)^2 < 25$

$x^2 + 4x + 4 < 25$

$x^2 + 4x - 21 < 0$

$(x + 7)(x - 3) < 0$

Critical numbers: $x = -7, x = 3$

Test intervals: $(-\infty, -7), (-7, 3), (3, \infty)$

Test: Is $(x + 7)(x - 3) < 0$?

Solution set: $(-7, 3)$

47. $x^2 + 4x + 4 \geq 9$

$x^2 + 4x - 5 \geq 0$

$(x + 5)(x - 1) \geq 0$

Critical numbers: $x = -5, x = 1$

Test intervals: $(-\infty, -5), (-5, 1), (1, \infty)$

Test: Is $(x + 5)(x - 1) \geq 0$?

Solution set: $(-\infty, -5] \cup [1, \infty)$

49. $x^3 - 4x \geq 0$

$x(x + 2)(x - 2) \geq 0$

Critical number: $x = 0, x = \pm 2$

Test intervals: $(-\infty, -2), (-2, 0), (0, 2), (2, \infty)$

Test: Is $x(x + 2)(x - 2) \geq 0$?

Solution set: $[-2, 0] \cup [2, \infty)$

51. $y = -x^2 + 2x + 3$

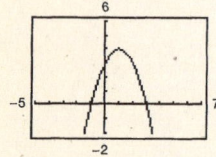

(a) $y \leq 0$ when $x \leq -1$ or $x \geq 3$

(b) $y \geq 3$ when $0 \leq x \leq 2$

Algebraically,

$-x^2 + 2x + 3 \leq 0$

$x^2 - 2x - 3 \geq 0$

$(x - 3)(x + 1) \geq 0$

Critical numbers: $x = -1, x = 3$

Testing the intervals $(-\infty, -1), (-1, 3), (3, \infty)$ you obtain $x \leq -1$ or $x \geq 3$.

$-x^2 + 2x + 3 \geq 3$

$-x^2 + 2x \geq 0$

$x^2 - 2x \leq 0$

$x(x - 2) \leq 0$

Critical numbers: $x = 0, x = 2$

Testing the intervals $(-\infty, 0), (0, 2), (2, \infty)$ you obtain $0 \leq x \leq 2$.

53. $\dfrac{1}{x} - x > 0$

$\dfrac{1 - x^2}{x} > 0$

Critical numbers: $x = 0, x = \pm 1$

Test intervals: $(-\infty, -1), (-1, 0), (0, 1), (1, \infty)$

Test: Is $\dfrac{1 - x^2}{x} > 0$?

Solution set: $(-\infty, -1) \cup (0, 1)$

55. $\dfrac{x+6}{x+1} - 2 < 0$

$\dfrac{x + 6 - 2(x+1)}{x+1} < 0$

$\dfrac{4-x}{x+1} < 0$

Critical numbers: $x = -1, x = 4$

Test intervals: $(-\infty, -1), (-1, 4), (4, \infty)$

Test: Is $\dfrac{4-x}{x+1} < 0$?

Solution set: $(-\infty, -1) \cup (4, \infty)$

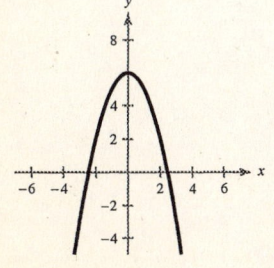

57. $y = \dfrac{3x}{x-2}$

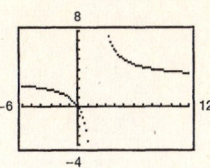

(a) $y \le 0$ when $0 \le x < 2$.

(b) $y \ge 6$ when $2 < x \le 4$.

59. $\sqrt{x-5}$

Need $x - 5 \ge 0$

$x \ge 5$

Domain: $[5, \infty)$

61. $\sqrt[3]{6-x}$

Domain: all real x

63. $\sqrt{x^2 - 4}$

Need $x^2 - 4 \ge 0$

$(x+2)(x-2) \ge 0$

$x \le -2$ or $x \ge 2$

Domain: $(-\infty, -2] \cup [2, \infty)$

65. (a), (b)

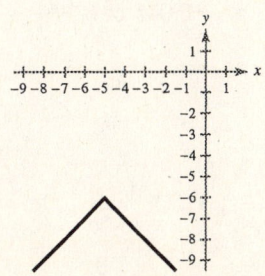

(c) For $y \ge 200$, $x \ge 181.5$ pounds

(d) The model is not accurate. The data is not linear. Other factors include muscle strength, height, physical condition, etc.

67. When $t = 2$, $v \approx 333$ vibrations per second

69. When $200 \le v \le 400$, $1.2 < t < 2.4$

71. False. If $-10 \le x \le 8$, then $10 \ge -x$ and $-x \ge -8$.

73. The polynomial $(x - a)(x - b)$ is zero at $x = a$ and $x = b$.

75. $d = \sqrt{(1 - (-4))^2 + (12 - 2)^2} = \sqrt{25 + 100}$

$= \sqrt{125} = 5\sqrt{5}$

Midpoint: $\left(\dfrac{-4 + 1}{2}, \dfrac{2 + 12}{2} \right) = \left(-\dfrac{3}{2}, 7 \right)$

77. $d = \sqrt{(3 - (-5))^2 + (6 - (-8))^2} = \sqrt{64 + 196}$

$= \sqrt{260} = 2\sqrt{65}$

Midpoint: $\left(\dfrac{3 - 5}{2}, \dfrac{6 - 8}{2} \right) = (-1, -1)$

79. $f(x) = -x^2 + 6$

81. $f(x) = -|x + 5| - 6$

83.
$$y = 12x$$
$$x = 12y$$
$$\frac{x}{12} = y$$
$$f^{-1}(x) = \frac{x}{12}$$

85.
$$y = x^3 + 7$$
$$x = y^3 + 7$$
$$x - 7 = y^3$$
$$\sqrt[3]{x - 7} = y$$
$$f^{-1}(x) = \sqrt[3]{x - 7}$$

Appendix B.5   Exploring Data: Representing Data Graphically

- You should be able to construct line plots.
- You should be able to construct histograms or frequency distributions.
- You should be able to construct bar graphs.
- You should be able to construct line graphs.

Solutions to Odd-Numbered Exercises

1. (a) The price 1.709 occurred with the greatest frequency (6).

   (b) The prices range from 1.599 to 1.789. The range is $1.789 - 1.599 = 0.19$

3.

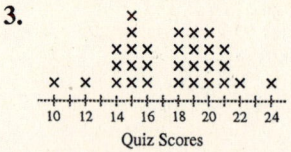

The score of 15 occurred with the greatest frequency.

5.

Interval	Tally
[800, 1000)	$\|\|$
[1000, 1200)	⊦⊦⊦ ⊦⊦⊦ ||||
[1200, 1400)	⊦⊦⊦ ⊦⊦⊦ ⊦⊦⊦ |
[1400, 1600)	⊦⊦⊦ ⊦⊦⊦
[1600, 1800)	⊦⊦⊦ ||
[1800, 2000)	|
[2000, 2200)	|

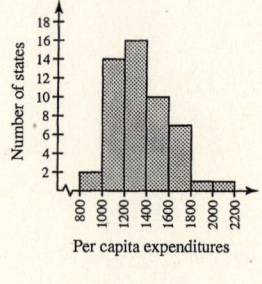

Per capita expenditures

7. 1997 to 2000: $\dfrac{943 - 753}{753} \approx 0.252$ or 25.2%

2000 to 2001: $\dfrac{882 - 943}{943} \approx -0.065$ or 6.5%

decrease

9.

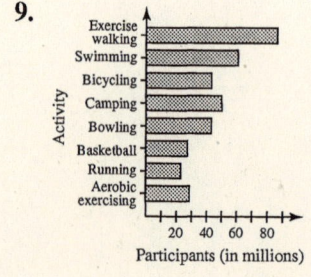

11. The highest price was approximately $2.14 per pound in 2000.

13. 1996 to 2002: $\dfrac{1900 - 1100}{1100} \approx 0.727$ or 72.7%

15.

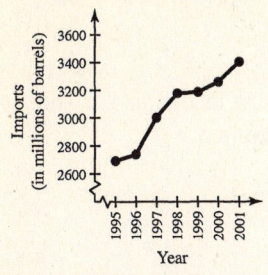

The graph indicates that oil imports increased yearly from 1995 to 2001.

17.

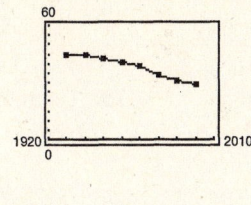

19.

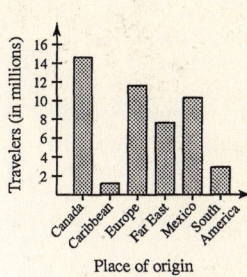

21. A histogram has a portion of the real number line as its horizontal axis, and the bars are not separated by spaces. A bar graph can be either horizontal or vertical. The labels are not necessarily numbers, and the bars are usually separated by spaces.

23. The second graph is misleading because the vertical scale is too small which makes small changes look large. Answers will vary.

APPENDIX D
Concepts in Statistics

Appendix D.1   Measures of Central Tendency and Dispersion

Solutions to Odd-Numbered Exercises

1. Mean $= \dfrac{5 + 12 + 7 + 14 + 8 + 9 + 7}{7} = \dfrac{62}{7} \approx 8.86$

Median: 8

Mode: 7

3. Mean $= \dfrac{5 + 12 + 7 + 24 + 8 + 9 + 7}{7} = \dfrac{72}{7} \approx 10.29$

Median: 8

Mode: 7

5. Mean $= \dfrac{5 + 12 + 7 + 14 + 9 + 7}{6} = \dfrac{54}{6} = 9$

Median: $\dfrac{7 + 9}{2} = 8$

Mode: 7

7. (a) The mean is sensitive to extreme values

(b) Mean: 14.86

Median: 14

Mode: 13

Each is increased by 6.

(c) Each will increase by k.

9. Mean $= \dfrac{410 + 260 + 320 + 320 + 460 + 150}{6} = \dfrac{1920}{6} = 320$

Median: 320

Mode: 320

11. There are many possible answers. For example: {4, 4, 10}

13. The mean is 76.55 and the median is 82. The median is the best description.

15. (a) Mean = 12, $\sigma \approx 2.83$

(b) Mean = 20, $\sigma \approx 2.83$

(c) Mean = 12, $\sigma \approx 1.41$

(d) Mean = 9, $\sigma \approx 1.41$

17. $\bar{x} = 6$

$v = 10$

$\sigma \approx 3.16$

19. $\bar{x} = 2$

$v = \frac{4}{3}$

$\sigma \approx 1.15$

21. $\bar{x} = 4$

$v = 4$

$\sigma \approx 2$

23. $\bar{x} = 47$

$v = 226$

$\sigma \approx 15.03$

25. $\bar{x} = 6$

$$\sigma = \sqrt{\dfrac{2^2 + 4^2 + 6^2 + 6^2 + 13^2 + 5^2}{6} - 6^2}$$

$$= \sqrt{\dfrac{286}{6} - 36}$$

$$= \sqrt{\dfrac{35}{3}} \approx 3.42$$

27. $\bar{x} = 5.8$

$$\sigma = \sqrt{\dfrac{8.1^2 + 6.9^2 + 3.7^2 + 4.2^2 + 6.1^2}{5} - 5.8^2}$$

$$= \sqrt{2.712} \approx 1.65$$

29. $\bar{x} = 12$ and $|x_i - 12| = 8$ for all x_i. Hence, $\sigma = 8$.

31. The mean will increase by 5. The standard deviation will not change.

33. The first histogram has a smaller standard deviation.

35. (a) 12, 13, 13, 14, 14, 15, 20, 23, 23

Median: 14

Lower quartile is median of {12, 13, 13} = 13

Upper quartile is median of {15, 20, 23, 23} = 21.5

(b)

12 13 14   21.5 23

37. (a) 46, 47, 47, 48, 48, 49, 50, 51, 52, 53

Median: $\dfrac{48 + 49}{2} = 48.5$

Lower quartile is median of {46, 47, 47, 48, 48} = 47

Upper quartile is median of {49, 50, 51, 52, 53} = 51

(b)

46   47   48.5   51   53

39.

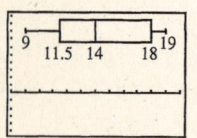

9   11.5   14   18 19

41.

17.3   24.1   34.9   43.4
    21.8

43.

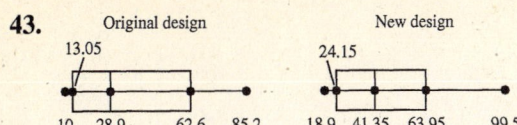

Original design

13.05

10 28.9 62.6 85.2

New design

24.15

18.9 41.35 63.95 99.5

From the plots, you can see that the lifetimes of the units in the new design are greater than the original design. The median increased by over 12 months.

Appendix D.2   Least Squares Regression

1.

x	y	xy	x^2
-4	1	-4	16
-3	3	-9	9
-2	4	-8	4
-1	6	-6	1
Total	-10	14	-27	30

$n = 4$

$4b + (-10)a = 14$

$(-10)b + 30a = -27$

Solving this system, $a = 1.6$ and $b = 7.5$

Answer: $y = 1.6x + 7.5$

3.

x	y	xy	x^2
-3	1	-3	9
-1	2	-2	1
1	2	2	1
4	3	12	16
Total	1	8	9	27

$n = 4$

$4b + a = 8$

$b + 27a = 9$

Solving this system, $a \approx 0.262$ and $b \approx 1.93$

Answer: $y = 0.262x + 1.93$

A P P E N D I X   E
Solving Linear Equations and Inequalities

Solutions to Odd-Numbered Exercises

1. $x + 11 = 15$

$x = 15 - 11$

$x = 4$

3. $x - 2 = 5$

$x = 5 + 2$

$x = 7$

5. $3x = 12$

$x = \frac{12}{3}$

$x = 4$

7. $\frac{x}{5} = 4$

$x = 4(5)$

$x = 20$

9. $8x + 7 = 39$

$8x = 32$

$x = 4$

11. $24 - 7x = 3$

$-7x = -21$

$x = 3$

13. $8x - 5 = 3x + 20$

$5x = 25$

$x = 5$

15. $-2(x + 5) = 10$

$-2x - 10 = 10$

$-2x = 20$

$x = -10$

17. $2x + 3 = 2x - 2$

$3 = -2$

No solution

19. $\frac{3}{2}(x + 5) - \frac{1}{4}(x + 24) = 0$

$\frac{3}{2}(x + 5) = \frac{1}{4}(x + 24)$

$12(x + 5) = 2(x + 24)$

$12x + 60 = 2x + 48$

$10x = -12$

$x = -\frac{12}{10}$

$x = -\frac{6}{5}$

21. $0.25x + 0.75(10 - x) = 3$

$25x + 75(10 - x) = 300$

$25x + 750 - 75x = 300$

$-50x = -450$

$x = 9$

23. $x + 6 < 8$

$x < 8 - 6$

$x < 2$

25. $-x - 8 > -17$

$17 - 8 > x$

$9 > x$

$x < 9$

27. $6 + x \le -8$

$x \le -8 - 6$

$x \le -14$

29. $\frac{4}{5}x > 8$

$x > \frac{5}{4}(8)$

$x > 10$

31. $-\frac{3}{4}x > -3$

$\frac{3}{4}x < 3$

$x < 4$

33. $4x < 12$

$x < 3$

35. $-11x \le -22$

$11x \ge 22$

$x \ge 2$

37. $x - 3(x + 1) \ge 7$

$x - 3x - 3 \ge 7$

$-2x \ge 10$

$x \le -5$

39. $7x - 12 < 4x + 6$

$3x < 18$

$x < 6$

41. $\frac{3}{4}x - 6 \le x - 7$

$1 \le \frac{1}{4}x$

$4 \le x$

$x \ge 4$

43. $3.6x + 11 \ge -3.4$

$3.6x \ge -14.4$

$x \ge \frac{-14.4}{3.6}$

$x \ge -4$

A P P E N D I X   F
Systems of Inequalities

Appendix F.1   Solving Systems of Inequalities

■ You should be able to sketch the graph of an inequality in two variables:

(a) Replace the inequality with an equal sign and graph the equation. Use a dashed line for $<$ or $>$, a solid line for $\le$ or $\ge$.

(b) Test a point in each region formed by the graph. If the point satisfies the inequality, shade the whole region.

Solutions to Odd-Numbered Exercises

1. $x < 2$

Vertical boundary

Matches graph (g).

3. $2x + 3y \ge 6$

$y \ge -\frac{2}{3}x + 2$

Line with negative slope

Matches (a).

5. $x^2 + y^2 < 9$

Circular boundary

Matches (e).

7. $xy > 1$ or $y > \dfrac{1}{x}$

Matches (f).

9. $y < 2 - x^2$

Graph the parabola $y = 2 - x^2$. The region lies below the parabola.

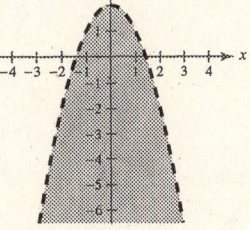

11. $x \geq 4$

Using a solid line, graph the vertical line $x = 4$ and shade to the right of this line.

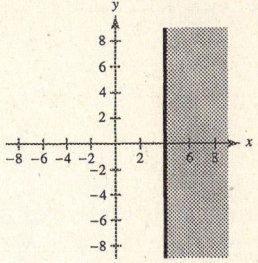

13. $y \geq -1$

Using a solid line, graph the horizontal line $y = -1$ and shade above this line.

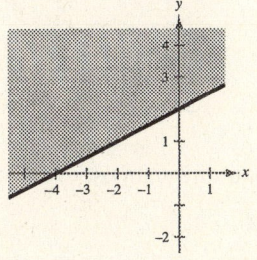

15. $2y - x \geq 4$

Using a solid line, graph $2y - x = 4$, and then shade above the line. (Use $(0, 0)$ as a test point.)

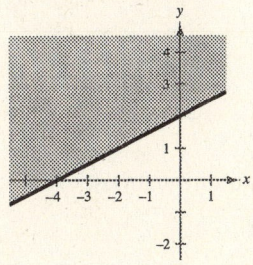

17. $y > 3x^2 + 1$

Sketch the parabola $y = 3x^2 + 1$. The region lies above the parabola.

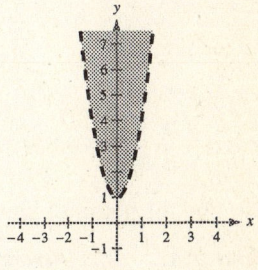

19. $(x + 1)^2 + y^2 < 9$

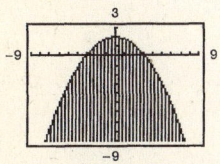

21. $y \geq \dfrac{2}{3}x - 1$

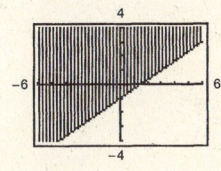

23. $y < -3.8x + 1.1$

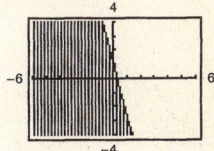

25. $x^2 + 5y - 10 \leq 0$

$$y \leq 2 - \dfrac{x^2}{5}$$

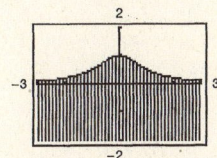

27. $y \leq \dfrac{1}{1 + x^2}$

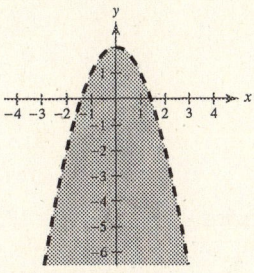

29. $y < \ln x$

Using a dashed line, graph $y = \ln x$, and shade to the right of the curve. (Use $(2, 0)$ as a test point.)

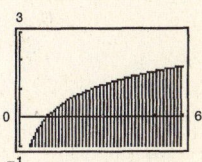

31. $y > 3^{-x-4}$

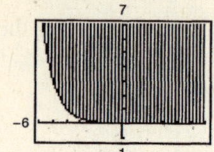

33. The line through $(0, 2)$ and $(3, 0)$ is $y = -\frac{2}{3}x + 2$. For the shaded region above the line, we have:

$$y > -\frac{2}{3}x + 2$$

$$3y > -2x + 6$$

$$2x + 3y > 6$$

$$\frac{x}{3} + \frac{y}{2} > 1$$

35. The circle shown is $x^2 + y^2 = 9$. For the shaded region inside the circle, we have $x^2 + y^2 \le 9$.

37. (a) $(0, 2)$ is a solution: $-2(0) + 5(2) \ge 3$

$$2 < 4$$

$$-4(0) + 2(2) < 7$$

(b) $(-6, 4)$ is not a solution: $4 \not< 4$

(c) $(-8, -2)$ is not a solution:
$-4(-8) + 2(-2) \not< 7$

(d) $(-3, 2)$ is not a solution: $-4(-3) + 2(2) \not< 7$

39. $\begin{cases} x + y \le 1 \\ -x + y \le 1 \\ \quad\quad y \ge 0 \end{cases}$

First, find the points of intersection of each pair of equations.

Vertex A	Vertex B	Vertex C
$\begin{cases} x + y = 1 \\ -x + y = 1 \end{cases}$	$\begin{cases} x + y = 1 \\ \quad\quad y = 0 \end{cases}$	$\begin{cases} -x + y = 1 \\ \quad\quad y = 0 \end{cases}$
$(0, 1)$	$(1, 0)$	$(-1, 0)$

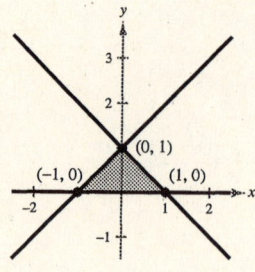

41. $\begin{cases} -3x + 2y < 6 \\ x - 4y > -2 \\ 2x + y < 3 \end{cases}$

First, find the points of intersection of each pair of equations.

Vertex A	Vertex B	Vertex C
$\begin{cases} -3x + 2y = 6 \\ x - 4y = -2 \end{cases}$	$\begin{cases} -3x + 2y = 6 \\ 2x + y = 3 \end{cases}$	$\begin{cases} x - 4y = -2 \\ 2x + y = 3 \end{cases}$
$(-2, 0)$	$(0, 3)$	$\left(\frac{10}{9}, \frac{7}{9}\right)$

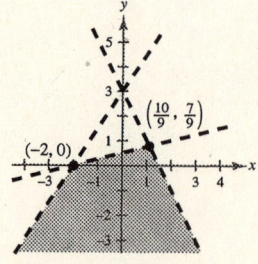

43. $3x + y \leq y^2$

$x - y > 0$

The curves given by $3x + y = y^2$ and $x - y = 0$ intersect as follows:

$3x + x = x^2$

$4x = x^2$

$x = 0, 4$

Intersection points:
$(0, 0), (4, 4)$

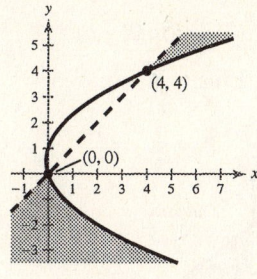

45. $2x + y < 2 \implies y < 2 - 2x$

$x + 3y > 2 \implies y > \frac{1}{3}(2 - x)$

$2 - 2x = \frac{1}{3}(2 - x)$

$6 - 6x = (2 - x)$

$4 = 5x$

$x = \frac{4}{5}$

Intersection: $\left(\frac{4}{5}, \frac{2}{5}\right)$

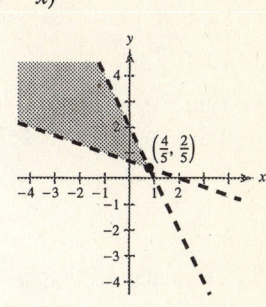

47. $\begin{cases} x < y^2 \\ x > y + 2 \end{cases}$

Points of intersection:

$y^2 = y + 2$

$y^2 - y - 2 = 0$

$(y + 1)(y - 2) = 0$

$y = -1, 2$

$(1, -1), (4, 2)$

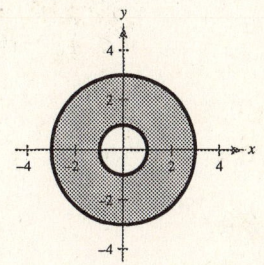

49. $\begin{cases} x^2 + y^2 \leq 9 \\ x^2 + y^2 \geq 1 \end{cases}$

There are no points of intersection. The region in common to both inequalities is the region between the circles.

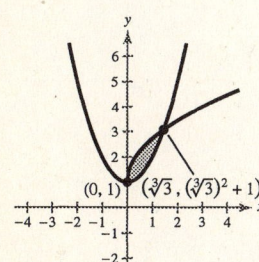

51. $\begin{cases} y \leq \sqrt{3x} + 1 \\ y \geq x^2 + 1 \end{cases}$

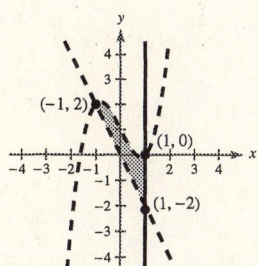

53. $\begin{cases} y < x^3 - 2x + 1 \\ y > -2x \\ x \leq 1 \end{cases}$

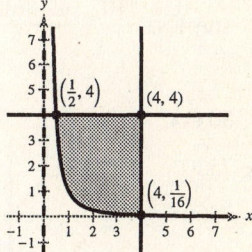

55. $\begin{cases} x^2 y \geq 1 \\ 0 < x \leq 4 \\ y \leq 4 \end{cases}$

57. $\begin{cases} y < -x + 4 \implies \dfrac{x}{4} + \dfrac{y}{4} < 1 \\ x \geq 0 \qquad\qquad x \geq 0 \\ y \geq 0 \qquad\qquad y \geq 0 \end{cases}$

59. $(0, 4), (4, 0)$ Line: $y \leq 4 - x$

$(0, 2), (8, 0)$ Line: $y \leq -\frac{1}{4}x + 2$

$x \geq 0, y \geq 0$

61. $\begin{cases} x \geq 2 \\ x \leq 5 \\ y \geq 1 \\ y \leq 7 \end{cases}$

Thus,
$2 \leq x \leq 5, 1 \leq y \leq 7.$

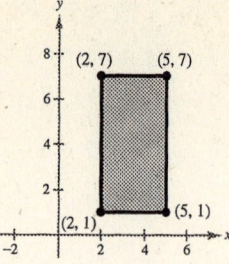

63. $(0, 0), (5, 0)$

Line: $y \geq 0$

$(0, 0), (2, 3)$

Line: $y \leq \frac{3}{2}x$

$(2, 3), (5, 0)$

Line: $y \leq -x + 5$

65.   Demand = Supply

$50 - 0.5x = 0.125x$

$50 = 0.625x$

$x = 80$

$p = 10$

Point of equilibrium: $(80, 10)$

Consumer surplus $= \frac{1}{2}(40)(80) = 1600$

Producer surplus $= \frac{1}{2}(10)(80) = 400$

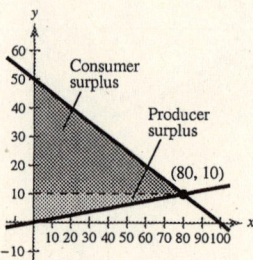

67.   Demand = Supply

$300 - 0.0002x = 225 + 0.0005x$

$75 = 0.0007x$

$x = \dfrac{75}{0.0007} = \dfrac{750,000}{7}$

Equilibrium point: $\left(\dfrac{750,000}{7}, \dfrac{1950}{7}\right) \approx (107,142.86, 278.57)$

Consumer surplus: $\dfrac{(107,142.86)(300 - 278.57)}{2} \approx 1,148,036$

Producer surplus: $\dfrac{(107,142.86)(278.57 - 225)}{2} \approx 2,869,822$

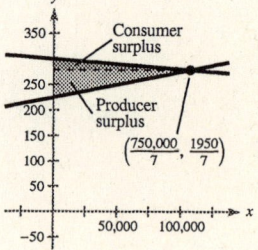

69. $x + y \leq 30,000$

$x \geq 7500$

$y \geq 7500$

$x \geq 2y$

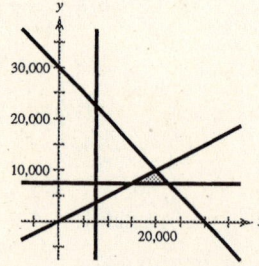

71. (a) Let x = number of ounces of food X

Let y = number of ounces of food Y

Calcium: $20x + 10y \geq 280$

Iron: $15x + 10y \geq 160$

Vitamin B: $10x + 20y \geq 180$

$$x \geq 0$$

$$y \geq 0$$

(b)

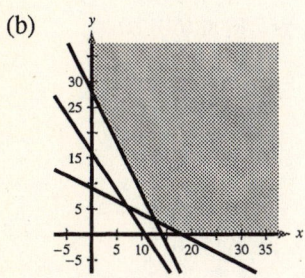

73. (a)

$xy \geq 500$	Body-building space
$2x + \pi y \geq 125$	Track (Two semi-circles and two lengths)
$x \geq 0$	Physical constraint
$y \geq 0$	Physical constraint

(b)

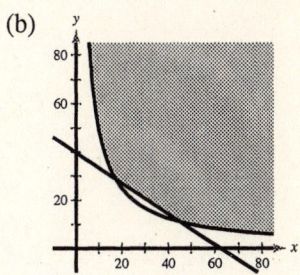

75. Area $= 9 \cdot 11 = 99$ square units

True

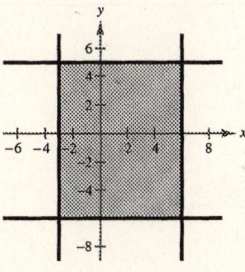

77. Test a point on either side of the boundary.

Appendix F.2   Linear Programming

Solutions to Odd-Numbered Exercises

1. $z = 3x + 5y$

At $(0, 6)$: $z = 3(0) + 5(6) = 30$

At $(0, 0)$: $z = 3(0) + 5(0) = 0$

At $(6, 0)$: $z = 3(6) + 5(0) = 18$

The minimum value is 0 at $(0, 0)$.

The maximum value is 30 at $(0, 6)$.

3. $z = 10x + 7y$

At $(0, 6)$: $z = 10(0) + 7(6) = 42$

At $(0, 0)$: $z = 10(0) + 7(0) = 0$

At $(6, 0)$: $z = 10(6) + 7(0) = 60$

The minimum value is 0 at $(0, 0)$.

The maximum value is 60 at $(6, 0)$.

5. $z = 3x + 2y$

$x + 3y = 15 \implies y = \frac{1}{3}(15 - x)$

$4x + y = 16 \implies y = (16 - 4x)$

$\frac{1}{3}(15 - x) = 16 - 4x$

$(15 - x) = 48 - 12x$

$11x = 33$

$x = 3$

$y = 4$

At $(0, 0)$: $z = 0$

At $(0, 5)$: $z = 10$

At $(4, 0)$: $z = 12$

At $(3, 4)$: $z = 17$

Minimum at $(0, 0)$ is 0.

Maximum at $(3, 4)$ is 17.

7. $z = 5x + 0.5y$

At $(0, 0)$: $z = 0$

At $(0, 5)$: $z = 2.5$

At $(4, 0)$: $z = 20$

At $(3, 4)$: $z = 17$

Minimum at $(0, 0)$ is 0.

Maximum at $(4, 0)$ is 20.

9. $z = 10x + 7y$

At $(0, 45)$: $z = 10(0)$ $+ 7(45) = 315$

At $(30, 45)$: $z = 10(30) + 7(45) = 615$

At $(60, 20)$: $z = 10(60) + 7(20) = 740$

At $(60, 0)$: $z = 10(60) + 7(0)$ $= 600$

At $(0, 0)$: $z = 10(0)$ $+ 7(0) =$ 0

The minimum value is 0 at $(0, 0)$.

The maximum value is 740 at $(60, 20)$.

11. $z = 25x + 30y$

At $(0, 45)$: $z = 25(0)$ $+ 30(45) = 1350$

At $(30, 45)$: $z = 25(30) + 30(45) = 2100$

At $(60, 20)$: $z = 25(60) + 30(20) = 2100$

At $(60, 0)$: $z = 25(60) + 30(0)$ $= 1500$

At $(0, 0)$: $z = 25(0)$ $+ 30(0) =$ 0

The minimum value is 0 at $(0, 0)$.

The maximum value is 2100 at any point along the line segment connecting $(30, 45)$ and $(60, 20)$.

13. $z = 6x + 10y$

At $(0, 2)$: $z = 6(0) + 10(2) = 20$

At $(5, 0)$: $z = 6(5) + 10(0) = 30$

At $(0, 0)$: $z = 6(0) + 10(0) =$ 0

The minimum value is 0 at $(0, 0)$.

The maximum value is 30 at $(5, 0)$.

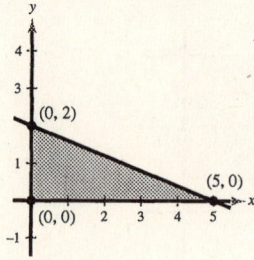

15. $z = 3x + 4y$

At $(0, 0)$: $z = 0$

At $(7, 0)$: $z = 21$

At $(0, 10)$: $z = 40$

At $(5, 8)$: $z = 47$

Minimum at $(0, 0)$ is 0.

Maximum at $(5, 8)$ is 47.

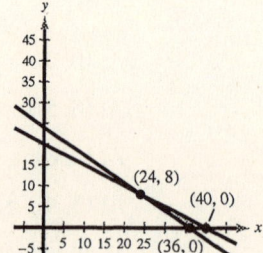

17. $z = 4x + y$

At $(36, 0)$: $z = 4(36) + 0 = 144$

At $(40, 0)$: $z = 4(40) + 0 = 160$

At $(24, 8)$: $z = 4(24) + 8 = 104$

The minimum value is 104 at $(24, 8)$.

The maximum value is 160 at $(40, 0)$.

19. $z = x + 4y$

At $(36, 0)$: $z = 36 + 4(0) = 36$

At $(40, 0)$: $z = 40 + 4(0) = 40$

At $(24, 8)$: $z = 24 + 4(8) = 56$

The minimum value is 36 at $(36, 0)$.

The maximum value is 56 at $(24, 8)$.

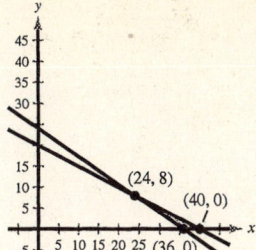

21. $z = 2x + 3y$

At $(36, 0)$: $z = 2(36) + 3(0) = 72$

At $(40, 0)$: $z = 2(40) + 3(0) = 80$

At $(24, 8)$: $z = 2(24) + 3(8) = 72$

Minimum at any point on the line segment joining $(36, 0)$ and $(24, 8)$: 72.

Maximum at $(40, 0)$: 80.

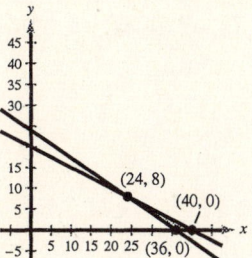

23. $z = 2x + y$

(a), (b)

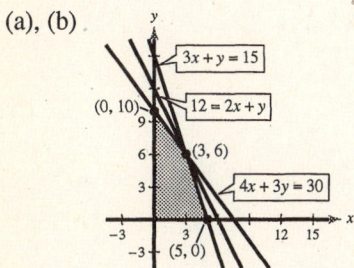

(c) At $(0, 10)$: $z = 2(0) + (10) = 10$

At $(3, 6)$: $z = 2(3) + (6)\ \ = 12$

At $(5, 0)$: $z = 2(5) + (0)\ \ = 10$

At $(0, 0)$: $z = 2(0) + (0)\ \ = 0$

The maximum value is 12 at $(3, 6)$.

25. $z = x + y$

(a), (b)

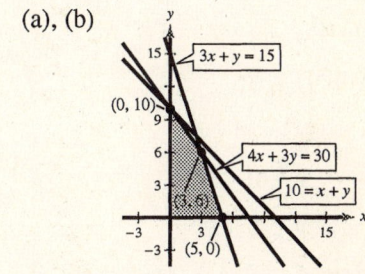

(c) At $(0, 10)$: $z = (0) + (10) = 10$

At $(3, 6)$: $z = (3) + (6)\ \ = 9$

At $(5, 0)$: $z = (5) + (0)\ \ = 5$

At $(0, 0)$: $z = (0) + (0)\ \ = 0$

The maximum value is 10 at $(0, 10)$.

27. $-x + y \leq 1 \implies y \leq x + 1$

$-x + 2y \leq 4 \implies y \leq \frac{1}{2}x + 2$

Intersection: $(2, 3)$

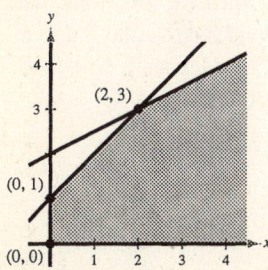

The constraints do not form a closed set of points. Therefore, $z = x + y$ is unbounded.

29. $-x + y \leq 0 \implies y \leq x$

$-3x + y \geq 3 \implies y \geq 3x + 3$

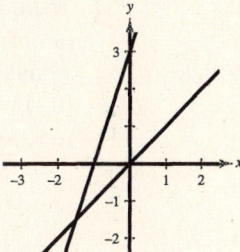

The feasible set is empty.

31. Let x = number of audits

Let y = number of tax returns

Constraints: $100x + 12.5y \leq 800$

$$8x + 2y \leq 96$$

$$x \geq 0$$

$$y \geq 0$$

Objective function: $R = 2000x + 300y$

Vertices of feasible region: $(0, 0), (8, 0), (0, 48), (4, 32)$

At $(0, 0)$: $R = 0$

At $(8, 0)$: $R = 16,000$

At $(0, 48)$: $R = 14,400$

At $(4, 32)$: $R = 17,600$

4 audits, 32 tax returns yields maximum revenue of $17,600.

33. x = number of bags of Brand X
y = number of bags of Brand Y

Constraints: $2x + y \geq 12$

$$2x + 9y \geq 36$$

$$2x + 3y \geq 24$$

$$x \geq 0$$

$$y \geq 0$$

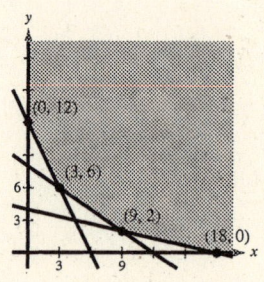

Objective function: $C = 25x + 20y$

Vertices: $(0, 12), (3, 6), (9, 2), (18, 0)$

At $(0, 12)$: $C = 25(0) + 20(12) = 240$

At $(3, 6)$: $C = 25(3) + 20(6) = 195$

At $(9, 2)$: $C = 25(9) + 20(2) = 265$

At $(18, 0)$: $C = 25(18) + 20(0) = 450$

To minimize cost, use three bags of Brand X and six bags of Brand Y for a total cost of $195.

35. True, the maximum value is attained at all points in the segment joining these two vertices.

37. There are an infinite number of objective functions that would have a maximum at $(0, 4)$. One such objective function is $z = x + 5y$.

39. There are an infinite number of objective functions that would have a maximum at $(5, 0)$. One such objective function is $z = 4x + y$.

41. Constraints: $x \geq 0, y \geq 0, x + 3y \leq 15, 4x + y \leq 16$

Vertex	Value of $z = 3x + ty$
$(0, 0)$	$z = 0$
$(0, 5)$	$z = 5t$
$(3, 4)$	$z = 9 + 4t$
$(4, 0)$	$z = 12$

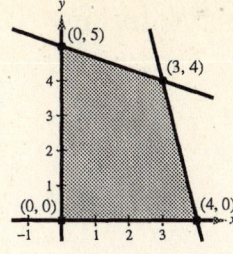

(a) For the maximum value to be at $(0, 5)$, $z = 5t$ must be greater than

$z = 9 + 4t$ and $z = 12$.

$5t > 9 + 4t$ and $5t > 12$

$t > 9$ $\qquad\qquad t > \frac{12}{5}$

Thus, $t > 9$.

(b) For the maximum value to be at $(3, 4)$, $z = 9 + 4t$ must be greater than $z = 5t$ and $z = 12$.

$9 + 4t > 5t$ and $9 + 4t > 12$

$9 > t$ $\qquad\qquad t > 3$

$\qquad\qquad\qquad t > \frac{3}{4}$

Thus, $\frac{3}{4} < t < 9$.

Chapter 1   Practice Test Solutions

1. $m = \dfrac{3 - 2}{1 - (-2)} = \dfrac{1}{3}$

2. Slope $= \dfrac{-2 - (-5)}{3 - 4} = \dfrac{3}{-1} = -3$

$y + 2 = -3(x - 3)$

$y + 2 = -3x + 9$

$y + 3x = 7 \quad \text{or} \quad y = -3x + 7$

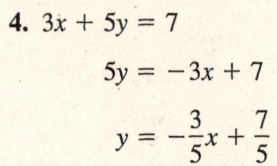

3. $y - 5 = -3(x + 1)$

$y - 5 = -3x - 3$

$y + 3x = 2 \quad \text{or} \quad y = -3x + 2$

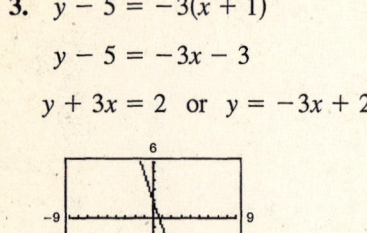

4. $3x + 5y = 7$

$5y = -3x + 7$

$y = -\dfrac{3}{5}x + \dfrac{7}{5}$

Slope of perpendicular line is $m = \dfrac{5}{3}$.

$y - 2 = \dfrac{5}{3}(x + 3)$

$y = \dfrac{5}{3}x + 7$

5. No, y is not a function of x. For example, $(0, 2)$ and $(0, -2)$ both satisfy the equation.

6. $f(0) = \dfrac{|0 - 2|}{(0 - 2)} = \dfrac{2}{-2} = -1$

$f(2)$ is not defined.

$f(4) = \dfrac{|4 - 2|}{(4 - 2)} = \dfrac{2}{2} = 1$

7. The domain of

$f(x) = \dfrac{5}{x^2 - 16}$

is all $x \neq \pm 4$.

8. The domain of $g(t) = \sqrt{4 - t}$ consists of all t satisfying

$4 - t \geq 0 \quad \text{or} \quad t \leq 4.$

9.

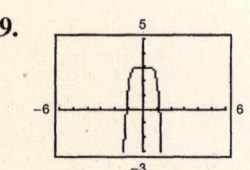

$f(x) = 3 - x^6$ is even.

10. $f(x) = 12x - x^3$

f is increasing or $(-2, 2)$.

11.

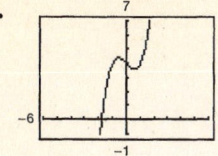

Relative minimum: $(0.577, 3.615)$

Relative maximum: $(-0.577, 4.385)$

12. $f(x) = x^3 - 3$ is a vertical shift of 3 units downward of $y = x^3$.

13. $f(x) = \sqrt{x - 6}$ is a horizontal shift 6 units to the right of $y = \sqrt{x}$.

14. $(g \circ f)(x) = g(f(x))$

$\qquad = g(\sqrt{x}) = (\sqrt{x})^2 - 2 = x - 2$

Domain: $x \geq 0$

15. $\left(\dfrac{f}{g}\right)(x) = \dfrac{f(x)}{g(x)} = \dfrac{3x^2}{16 - x^4}$

The domain is all $x \neq \pm 2$.

16. $(f \circ g)(x) = f\left(\dfrac{x - 1}{3}\right)$

$\qquad = 3\left(\dfrac{x - 1}{3}\right) + 1 = (x - 1) + 1 = x$

$(g \circ f)(x) = g(3x + 1) = \dfrac{(3x + 1) - 1}{3} = \dfrac{3x}{3} = x$

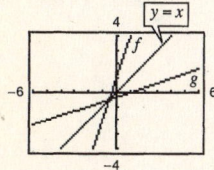

17. $y = \sqrt{9 - x^2}, \quad 0 \leq x \leq 3$

$\qquad x = \sqrt{9 - y^2}$

$\qquad x^2 = 9 - y^2$

$\qquad y^2 = 9 - x^2$

$\qquad y = \sqrt{9 - x^2}$

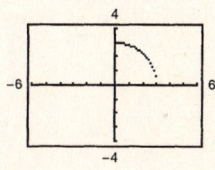

18. $y = 0.882 + 0.912x$

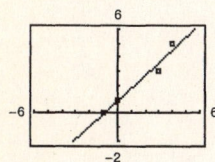

Chapter 2   Practice Test Solutions

1. x-intercepts: $(1, 0), (5, 0)$

y-intercept: $(0, 5)$

Vertex: $(3, -4)$

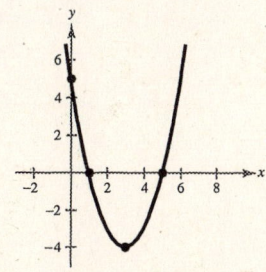

2. $a = 0.01, b = -90$

$\dfrac{-b}{2a} = \dfrac{90}{2(.01)} = 4500$ units

3. Vertex: $(1, 7)$

Opening downward through $(2, 5)$

$y = a(x - 1)^2 + 7$ Standard form

$5 = a(2 - 1)^2 + 7$

$5 = a + 7$

$a = -2$

$y = -2(x - 1)^2 + 7$

$\quad = -2(x^2 - 2x + 1) + 7$

$\quad = -2x^2 + 4x + 5$

4. $y = \pm a(x - 2)(3x - 4)$ where a is any real number.

$y = \pm(3x^2 - 10x + 8)$

5. Leading coefficient: -3

Degree: 5

Moves down to the right and up to the left.

6. $0 = x^5 - 5x^3 + 4x$

$\quad = x(x^4 - 5x^2 + 4)$

$\quad = x(x^2 - 1)(x^2 - 4)$

$\quad = x(x + 1)(x - 1)(x + 2)(x - 2)$

$x = 0, x = \pm 1, x = \pm 2$

7. $f(x) = x(x - 3)(x + 2)$

$\quad = x(x^2 - x - 6)$

$\quad = x^3 - x^2 - 6x$

8. Intercepts: $(0, 0), \left(\pm 2\sqrt{3}, 0\right)$

Moves up to the right.

Moves down to the left.

x	-2	-1	0	1	2
y	16	11	0	-11	-16

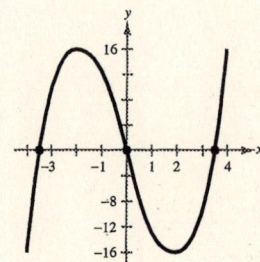

9.
$$3x^3 + 9x^2 + 20x + 62 + \frac{176}{x - 3}$$

$$x - 3 \overline{)\, 3x^4 + 0x^3 - 7x^2 + 2x - 10}$$

$\quad\quad \underline{3x^4 - 9x^3}$

$\quad\quad\quad\quad 9x^3 - 7x^2$

$\quad\quad\quad\quad \underline{9x^3 - 27x^2}$

$\quad\quad\quad\quad\quad\quad 20x^2 + 2x$

$\quad\quad\quad\quad\quad\quad \underline{20x^2 - 60x}$

$\quad\quad\quad\quad\quad\quad\quad\quad 62x - 10$

$\quad\quad\quad\quad\quad\quad\quad\quad \underline{62x - 186}$

$\quad\quad\quad\quad\quad\quad\quad\quad\quad\quad 176$

10.
$$x - 2 + \frac{5x - 13}{x^2 + 2x - 1}$$

$$x^2 + 2x - 1 \overline{)\, x^3 + 0x^2 + 0x - 11}$$

$\quad\quad\quad\quad \underline{x^3 + 2x^2 - x}$

$\quad\quad\quad\quad\quad\quad -2x^2 + x - 11$

$\quad\quad\quad\quad\quad\quad \underline{-2x^2 - 4x + 2}$

$\quad\quad\quad\quad\quad\quad\quad\quad 5x - 13$

11. $-5 \,\big|$

	3	13	0	0	12	-1
		-15	10	-50	250	-1310
	3	-2	10	-50	262	-1311

$$\frac{3x^5 + 13x^4 + 12x - 1}{x + 5} = 3x^4 - 2x^3 + 10x^2 - 50x + 262 - \frac{1311}{x + 5}$$

12. $-6 \,\big|$

	7	40	-12	15
		-42	12	0
	7	-2	0	15

$f(-6) = 15$

13. $0 = x^3 - 19x - 30$

Possible rational roots:

$\pm 1, \pm 2, \pm 3, \pm 5, \pm 6, \pm 10, \pm 15, \pm 30$

$-2 \,\big|$

	1	0	-19	-30
		-2	4	30
	1	-2	-15	0

-2 is a zero.

$0 = (x + 2)(x^2 - 2x - 15)$

$0 = (x + 2)(x + 3)(x - 5)$

Zeros: $x = -2, x = -3, x = 5$

14. $0 = x^4 + x^3 - 8x^2 - 9x - 9$

Possible rational roots: $\pm 1, \pm 3, \pm 9$

$3 \,\big|$

	1	1	-8	-9	-9
		3	12	12	9
	1	4	4	3	0

$x = 3$ is a zero.

$0 = (x - 3)(x^3 + 4x^2 + 4x + 3)$

Possible rational roots of $x^3 + 4x^2 + 4x + 3$: $\pm 1, \pm 3$

$-3 \,\big|$

	1	4	4	3
		-3	-3	-3
	1	1	1	0

$x = -3$ is a zero.

$0 = (x - 3)(x + 3)(x^2 + x + 1)$

The zeros of $x^2 + x + 1$ are $x = \dfrac{-1 \pm \sqrt{3}i}{2}$.

Zeros: $x = 3, x = -3, x = -\dfrac{1}{2} + \dfrac{\sqrt{3}}{2}i, x = -\dfrac{1}{2} - \dfrac{\sqrt{3}}{2}i$

15. $0 = 6x^3 - 5x^2 + 4x - 15$

Possible rational roots: $\pm 1, \pm 3, \pm 5, \pm 15, \pm \frac{1}{2}, \pm \frac{3}{2}, \pm \frac{5}{2}, \pm \frac{15}{2}, \pm \frac{1}{3}, \pm \frac{5}{3}, \pm \frac{1}{6}, \pm \frac{5}{6}$

16. $0 = x^3 - \frac{20}{3}x^2 + 9x - \frac{10}{3}$

$0 = 3x^3 - 20x^2 + 27x - 10$

Possible rational roots:

$\pm 1, \pm 2, \pm 5, \pm 10, \pm \frac{1}{3}, \pm \frac{2}{3}, \pm \frac{5}{3}, \pm \frac{10}{3}$

$$
\begin{array}{r|rrrr}
1 & 3 & -20 & 27 & -10 \\
  &   & 3 & -17 & 10 \\
\hline
  & 3 & -17 & 10 & 0
\end{array}
$$

$x = 1$ is a zero.

$0 = (x - 1)(3x^2 - 17x + 10)$

$0 = (x - 1)(3x - 2)(x - 5)$

Zeros: $x = 1, x = \frac{2}{3}, x = 5$

17. $f(x) = x^4 + x^3 + 3x^2 + 5x - 10$

Possible rational roots: $\pm 1, \pm 2, \pm 5, \pm 10$

$$
\begin{array}{r|rrrrr}
1 & 1 & 1 & 3 & 5 & -10 \\
  &   & 1 & 2 & 5 & 10 \\
\hline
  & 1 & 2 & 5 & 10 & 0
\end{array}
$$

$x = 1$ is a zero.

$$
\begin{array}{r|rrrr}
-2 & 1 & 2 & 5 & 10 \\
   &   & -2 & 0 & -10 \\
\hline
   & 1 & 0 & 5 & 0
\end{array}
$$

$x = -2$ is a zero.

$f(x) = (x - 1)(x + 2)(x^2 + 5)$

$\quad = (x - 1)(x + 2)(x + \sqrt{5}i)(x - \sqrt{5}i)$

18. $\dfrac{2}{1 + i} = \dfrac{2}{1 + i} \cdot \dfrac{1 - i}{1 - i}$

$\quad = \dfrac{2 - 2i}{1 + 1}$

$\quad = 1 - i$

19. $\dfrac{3 + i}{2} - \dfrac{i + 1}{4} = \dfrac{6 + 2i - i - 1}{4} = \dfrac{5}{4} + \dfrac{1}{4}i$

20. $f(x) = (x - 2)[x - (3 + i)][x - (3 - i)][x - (3 - 2i)][x - (3 + 2i)]$

$\quad = (x - 2)[(x - 3)^2 + 1][(x - 3)^2 + 4]$

$\quad = (x - 2)(x^2 - 6x + 10)(x^2 - 6x + 13)$

$\quad = x^5 - 14x^4 + 83x^3 - 256x^2 + 406x - 260$

21.
$$
\begin{array}{r|rrrr}
3i & 1 & 4 & 9 & 36 \\
   &   & 3i & 12i - 9 & -36 \\
\hline
   & 1 & 4 + 3i & 12i & 0
\end{array}
$$

22. $z = \dfrac{kx^2}{\sqrt{y}}$

23. $f(x) = \dfrac{x - 1}{2x}$

Vertical asymptote: $x = 0$

Horizontal asymptote: $y = \dfrac{1}{2}$

x-intercept: $(1, 0)$

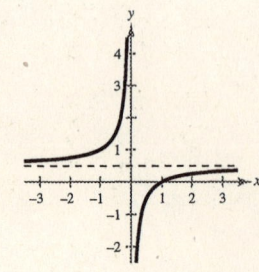

24. $f(x) = \dfrac{3x^2 - 4}{x}$

Vertical asymptote: $x = 0$

Slant asymptote: $y = 3x$

x-intercepts: $\left(\pm \dfrac{2}{\sqrt{3}}, 0 \right)$

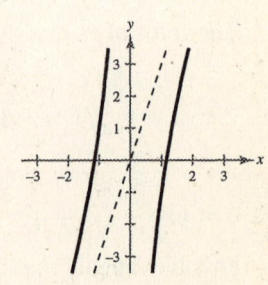

25. $y = 8$ is a horizontal asymptote since the degree of the numerator equals the degree of the denominator. There are no vertical asymptotes.

26. $x = 1$ is a vertical asymptote.

$$\frac{4x^2 - 2x + 7}{x - 1} = 4x + 2 + \frac{9}{x - 1}$$

so $y = 4x + 2$ is a slant asymptote.

27. $f(x) = \dfrac{x - 5}{(x - 5)^2} = \dfrac{1}{x - 5}$

Vertical asymptote: $x = 5$

Horizontal asymptote: $y = 0$

y-intercept: $\left(0, -\dfrac{1}{5}\right)$

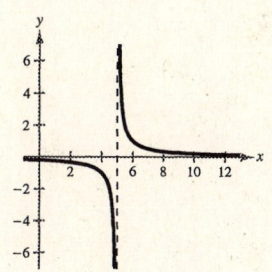

Chapter 3   Practice Test Solutions

1. $x^{3/5} = 8$

$x = 8^{5/3}$

$= \left(\sqrt[3]{8}\right)^5 = 2^5 = 32$

2. $3^{x-1} = \frac{1}{81}$

$3^{x-1} = 3^{-4}$

$x - 1 = -4$

$x = -3$

3. $f(x) = 2^{-x} = \left(\frac{1}{2}\right)^x$

x	-2	-1	0	1	2
$f(x)$	4	2	1	$\frac{1}{2}$	$\frac{1}{4}$

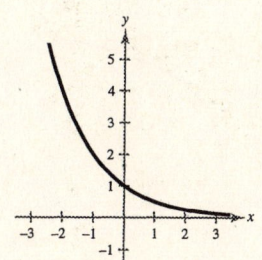

4. $g(x) = e^x + 1$

x	-2	-1	0	1	2
$g(x)$	1.14	1.37	2	3.72	8.39

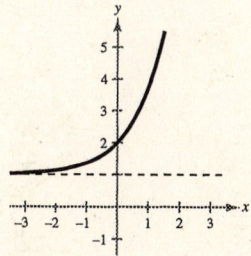

5. $A = P\left(1 + \dfrac{r}{n}\right)^{nt}$

(a) $A = 5000\left(1 + \dfrac{0.09}{12}\right)^{12(3)} \approx \$6543.23$

(b) $A = 5000\left(1 + \dfrac{0.09}{4}\right)^{4(3)} \approx \$6530.25$

(c) $A = 5000e^{(0.09)(3)} \approx \$6549.82$

6. $7^{-2} = \dfrac{1}{49}$

$\log_7 \dfrac{1}{49} = -2$

7. $x - 4 = \log_2 \frac{1}{64}$

$2^{x-4} = \frac{1}{64}$

$2^{x-4} = 2^{-6}$

$x - 4 = -6$

$x = -2$

8. $\log_b \sqrt[4]{\frac{8}{25}} = \frac{1}{4} \log_b \frac{8}{25}$

$= \frac{1}{4}[\log_b 8 - \log_b 25]$

$= \frac{1}{4}[\log_b 2^3 - \log_b 5^2]$

$= \frac{1}{4}[3 \log_b 2 - 2 \log_b 5]$

$= \frac{1}{4}[3(0.3562) - 2(0.8271)]$

$= -0.1464$

9. $5 \ln x - \frac{1}{2} \ln y + 6 \ln z = \ln x^5 - \ln \sqrt{y} + \ln z^6 = \ln\left(\frac{x^5 z^6}{\sqrt{y}}\right)$

10. $\log_9 28 = \dfrac{\log 28}{\log 9} \approx 1.5166$

11. $\log_{10} N = 0.6646$

$N = 10^{0.6646} \approx 4.62$

12.

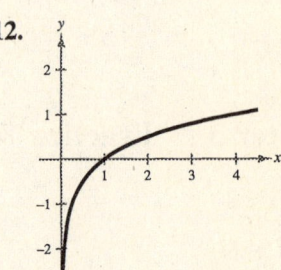

13. Domain:

$x^2 - 9 > 0$

$(x + 3)(x - 3) > 0$

$x < -3 \text{ or } x > 3$

14.

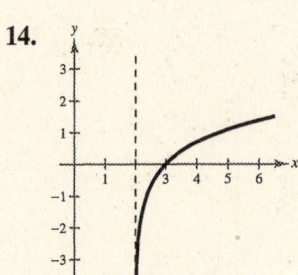

15. $\dfrac{\ln x}{\ln y} \neq \ln(x - y)$ since

$\dfrac{\ln x}{\ln y} = \log_y x.$

16. $5^x = 41$

$x = \log_5 41$

$= \dfrac{\ln 41}{\ln 5}$

≈ 2.3074

17. $x - x^2 = \log_5 \frac{1}{25}$

$5^{x-x^2} = \frac{1}{25}$

$5^{x-x^2} = 5^{-2}$

$x - x^2 = -2$

$0 = x^2 - x - 2$

$0 = (x + 1)(x - 2)$

$x = -1 \text{ or } x = 2$

18. $\log_2 x + \log_2(x - 3) = 2$

$\log_2[x(x - 3)] = 2$

$x(x - 3) = 2^2$

$x^2 - 3x = 4$

$x^2 - 3x - 4 = 0$

$(x + 1)(x - 4) = 0$

$x = 4$

$x = -1$

(extraneous solution)

19. $\dfrac{e^x + e^{-x}}{3} = 4$

$e^x(e^x + e^{-x}) = 12e^x$

$e^{2x} + 1 = 12e^x$

$e^{2x} - 12e^x + 1 = 0$

$e^x = \dfrac{12 \pm \sqrt{144 - 4}}{2}$

$e^x \approx 11.9161$ or $e^x \approx 0.08392$

$x \approx \ln 11.9161$ $x \approx \ln 0.08392$

$x \approx 2.4779$ $x \approx -2.4779$

20. $A = Pe^{rt}$

$12{,}000 = 6000e^{0.13t}$

$2 = e^{0.13t}$

$\ln 2 = 0.13t$

$\dfrac{\ln 2}{0.13} = t$

$t \approx 5.3319 \text{ yr}$ or $5 \text{ yr } 4 \text{ mo}$

21. There are two points of intersection:

(0.0169, −2.983),

(1.731, 1.647)

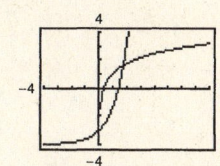

22. $y = 1.0597x^{1.9792}$

Chapter 4   Practice Test Solutions

1. $350° = 350°\left(\dfrac{\pi}{180°}\right) = \dfrac{35\pi}{18}$

2. $\dfrac{5\pi}{9} = \dfrac{5\pi}{9} \cdot \dfrac{180°}{\pi} = 100°$

3. $135° \ 14' \ 12'' = \left(135 + \dfrac{14}{60} + \dfrac{12}{3600}\right)°$

$\approx 135.2367°$

4. $-22.569° = -(22° + 0.569(60)')$

$= -22° \ 34.14'$

$= -(22° \ 34' + 0.14(60)'')$

$\approx -22° \ 34' \ 8''$

5. $\cos \theta = \dfrac{2}{3}$

$x = 2, \ r = 3, \ y = \pm\sqrt{9 - 4} = \pm\sqrt{5}$

$\tan \theta = \dfrac{y}{x} = \pm\dfrac{\sqrt{5}}{2}$

6. $\sin \theta = 0.9063$

$\theta = \arcsin 0.9063$

$\theta \approx 65°$ or $\dfrac{13\pi}{36}$

7. $\tan 20° = \dfrac{35}{x}$

$x = \dfrac{35}{\tan 20°} \approx 96.1617$

8. $\theta = \dfrac{6\pi}{5}$, θ is in Quadrant III.

Reference angle: $\dfrac{6\pi}{5} - \pi = \dfrac{\pi}{5}$ or $36°$

9. $\csc 3.92 = \dfrac{1}{\sin 3.92} \approx -1.4242$

10. $\tan \theta = 6 = \dfrac{6}{1}$, θ lies in Quadrant III.

$y = -6$, $x = -1$, $r = \sqrt{36 + 1} = \sqrt{37}$, so

$\sec \theta = \dfrac{\sqrt{37}}{-1} \approx -6.0828$.

11. Period: 4π

Amplitude: 3

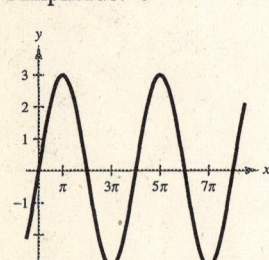

12. Period: 2π

Amplitude: 2

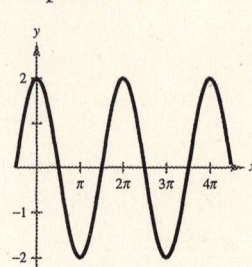

13. Period: $\dfrac{\pi}{2}$

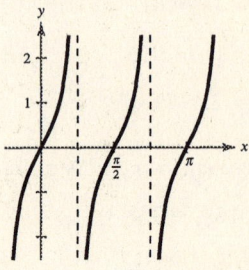

14. Period: 2π

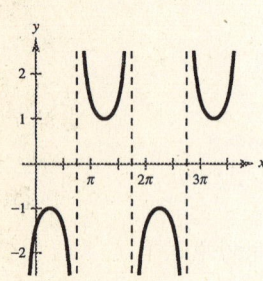

15.

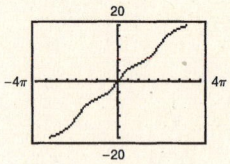

16.

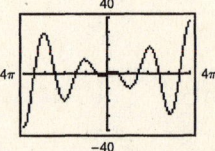

17. $\theta = \arcsin 1$

$\sin \theta = 1$

$\theta = \dfrac{\pi}{2}$

18. $\theta = \arctan(-3)$

$\tan \theta = -3$

$\theta \approx -1.249$ or $-71.565°$

19. $\sin\left(\arccos \dfrac{4}{\sqrt{35}}\right)$

$\sin \theta = \dfrac{\sqrt{19}}{\sqrt{35}} \approx 0.7368$

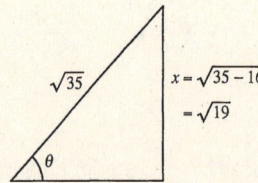

20. $\cos\left(\arcsin \dfrac{x}{4}\right)$

$\cos \theta = \dfrac{\sqrt{16 - x^2}}{4}$

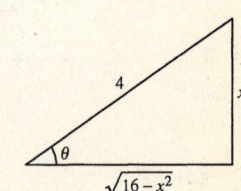

21. Given $A = 40°$, $c = 12$

$B = 90° - 40° = 50°$

$\sin 40° = \dfrac{a}{12}$

$a = 12 \sin 40° \approx 7.713$

$\cos 40° = \dfrac{b}{12}$

$b = 12 \cos 40° \approx 9.193$

22. Given $B = 6.84°$, $a = 21.3$

$A = 90° - 6.84° = 83.16°$

$\sin 83.16° = \dfrac{21.3}{c}$

$c = \dfrac{21.3}{\sin 83.16°} \approx 21.453$

$\tan 83.16° = \dfrac{21.3}{b}$

$b = \dfrac{21.3}{\tan 83.16°} \approx 2.555$

23. Given $a = 5$, $b = 9$

$c = \sqrt{25 + 81} = \sqrt{106}$

$\quad \approx 10.296$

$\tan A = \dfrac{5}{9}$

$A = \arctan \dfrac{5}{9} \approx 29.055°$

$B = 90° - 29.055° = 60.945°$

24. $\sin 67° = \dfrac{x}{20}$

$x = 20 \sin 67° \approx 18.41$ feet

25. $\tan 5° = \dfrac{250}{x}$

$x = \dfrac{250}{\tan 5°}$

$\quad \approx 2857.513$ feet

$\quad \approx 0.541$ mi

Chapter 5   Practice Test Solutions

1. $\tan x = \dfrac{4}{11}$, $\sec x < 0 \implies x$ is in Quadrant III.

$y = -4$, $\bar{x} = -11$, $r = \sqrt{16 + 121} = \sqrt{137}$

$\sin x = -\dfrac{4}{\sqrt{137}} = -\dfrac{4\sqrt{137}}{137} \qquad \csc x = -\dfrac{\sqrt{137}}{4}$

$\cos x = -\dfrac{11}{\sqrt{137}} = -\dfrac{11\sqrt{137}}{137} \qquad \sec x = -\dfrac{\sqrt{137}}{11}$

$\tan x = \dfrac{4}{11} \qquad\qquad\qquad\qquad \cot x = \dfrac{11}{4}$

2. $\dfrac{\sec^2 x + \csc^2 x}{\csc^2 x(1 + \tan^2 x)} = \dfrac{\sec^2 x + \csc^2 x}{\csc^2 x + (\csc^2 x)\tan^2 x}$

$\qquad = \dfrac{\sec^2 x + \csc^2 x}{\csc^2 x + \dfrac{1}{\sin^2 x} \cdot \dfrac{\sin^2 x}{\cos^2 x}}$

$\qquad = \dfrac{\sec^2 x + \csc^2 x}{\csc^2 x + \dfrac{1}{\cos^2 x}}$

$\qquad = \dfrac{\sec^2 x + \csc^2 x}{\csc^2 x + \sec^2 x} = 1$

3. $\ln|\tan \theta| - \ln|\cot \theta| = \ln\dfrac{|\tan \theta|}{|\cot \theta|}$

$\qquad = \ln\left|\dfrac{\sin \theta / \cos \theta}{\cos \theta / \sin \theta}\right| = \ln\left|\dfrac{\sin^2 \theta}{\cos^2 \theta}\right|$

$\qquad = \ln|\tan^2 \theta| = 2\ln|\tan \theta|$

4. $\cos\left(\dfrac{\pi}{2} - x\right) = \dfrac{1}{\csc x}$ is true since

$\cos\left(\dfrac{\pi}{2} - x\right) = \sin x = \dfrac{1}{\csc x}$.

5. $\sin^4 x + (\sin^2 x)\cos^2 x = \sin^2 x(\sin^2 x + \cos^2 x)$

$\qquad\qquad = \sin^2 x(1) = \sin^2 x$

6. $(\csc x + 1)(\csc x - 1) = \csc^2 x - 1 = \cot^2 x$

7. $\dfrac{\cos^2 x}{1 - \sin x} \cdot \dfrac{1 + \sin x}{1 + \sin x} = \dfrac{\cos^2 x(1 + \sin x)}{1 - \sin^2 x} = \dfrac{\cos^2 x(1 + \sin x)}{\cos^2 x} = 1 + \sin x$

8. $\dfrac{1 + \cos \theta}{\sin \theta} + \dfrac{\sin \theta}{1 + \cos \theta} = \dfrac{(1 + \cos \theta)^2 + \sin^2 \theta}{\sin \theta(1 + \cos \theta)}$

$\qquad\qquad\qquad\qquad = \dfrac{1 + 2\cos \theta + \cos^2 \theta + \sin^2 \theta}{\sin \theta(1 + \cos \theta)}$

$\qquad\qquad\qquad\qquad = \dfrac{2 + 2\cos \theta}{\sin \theta(1 + \cos \theta)} = \dfrac{2}{\sin \theta} = 2 \csc \theta$

9. $\tan^4 x + 2\tan^2 x + 1 = (\tan^2 x + 1)^2 = (\sec^2 x)^2 = \sec^4 x$

10. (a) $\sin 105° = \sin(60° + 45°)$

$\qquad\qquad = \sin 60° \cos 45° + \cos 60° \sin 45°$

$\qquad\qquad = \dfrac{\sqrt{3}}{2} \cdot \dfrac{\sqrt{2}}{2} + \dfrac{1}{2} \cdot \dfrac{\sqrt{2}}{2}$

$\qquad\qquad = \dfrac{\sqrt{2}}{4}(\sqrt{3} + 1)$

(b) $\tan 15° = \tan(60° - 45°) = \dfrac{\tan 60° - \tan 45°}{1 + \tan 60° \tan 45°}$

$\qquad\qquad = \dfrac{\sqrt{3} - 1}{1 + \sqrt{3}} \cdot \dfrac{1 - \sqrt{3}}{1 - \sqrt{3}} = \dfrac{2\sqrt{3} - 1 - 3}{1 - 3}$

$\qquad\qquad = \dfrac{2\sqrt{3} - 4}{-2} = 2 - \sqrt{3}$

11. $(\sin 42°) \cos 38° - (\cos 42°) \sin 38° = \sin(42° - 38°) = \sin 4°$

12. $\tan\left(\theta + \dfrac{\pi}{4}\right) = \dfrac{\tan \theta + \tan(\pi/4)}{1 - (\tan \theta) \tan(\pi/4)} = \dfrac{\tan \theta + 1}{1 - \tan \theta(1)} = \dfrac{1 + \tan \theta}{1 - \tan \theta}$

13. $\sin(\arcsin x - \arccos x) = \sin(\arcsin x) \cos(\arccos x) - \cos(\arcsin x) \sin(\arccos x)$

$\qquad\qquad = (x)(x) - \left(\sqrt{1 - x^2}\right)\left(\sqrt{1 - x^2}\right) = x^2 - (1 - x^2) = 2x^2 - 1$

14. (a) $\cos(120°) = \cos[2(60°)] = 2\cos^2 60° - 1 = 2\left(\dfrac{1}{2}\right)^2 - 1 = -\dfrac{1}{2}$

(b) $\tan(300°) = \tan[2(150°)] = \dfrac{2\tan 150°}{1 - \tan^2 150°} = \dfrac{-2\sqrt{3}/3}{1 - (1/3)} = -\sqrt{3}$

15. (a) $\sin 22.5° = \sin \dfrac{45°}{2} = \sqrt{\dfrac{1 - \cos 45°}{2}} = \sqrt{\dfrac{1 - \sqrt{2}/2}{2}} = \dfrac{\sqrt{2 - \sqrt{2}}}{2}$

(b) $\tan \dfrac{\pi}{12} = \tan \dfrac{\pi/6}{2} = \dfrac{\sin(\pi/6)}{1 + \cos(\pi/6)} = \dfrac{1/2}{1 + \sqrt{3}/2} = \dfrac{1}{2 + \sqrt{3}} = 2 - \sqrt{3}$

16. $\sin \theta = \dfrac{4}{5}$, θ lies in Quadrant II $\Rightarrow \cos \theta = -\dfrac{3}{5}$.

$\qquad \cos \dfrac{\theta}{2} = \sqrt{\dfrac{1 + \cos \theta}{2}} = \sqrt{\dfrac{1 - (3/5)}{2}}$

$\qquad\qquad = \sqrt{\dfrac{2}{10}} = \dfrac{1}{\sqrt{5}} = \dfrac{\sqrt{5}}{5}$

17. $(\sin^2 x)\cos^2 x = \dfrac{1 - \cos 2x}{2} \cdot \dfrac{1 + \cos 2x}{2}$

$$= \frac{1}{4}[1 - \cos^2 2x]$$

$$= \frac{1}{4}\left[1 - \frac{1 + \cos 4x}{2}\right]$$

$$= \frac{1}{8}[2 - (1 + \cos 4x)]$$

$$= \frac{1}{8}[1 - \cos 4x]$$

18. $6(\sin 5\theta)\cos 2\theta = 6\left\{\frac{1}{2}[\sin(5\theta + 2\theta) + \sin(5\theta - 2\theta)]\right\} = 3[\sin 7\theta + \sin 3\theta]$

19. $\sin(x + \pi) + \sin(x - \pi) = 2\left(\sin \dfrac{[(x + \pi) + (x - \pi)]}{2}\right)\cos \dfrac{[(x + \pi) - (x - \pi)]}{2} = 2 \sin x \cos \pi = -2 \sin x$

20. $\dfrac{\sin 9x + \sin 5x}{\cos 9x - \cos 5x} = \dfrac{2 \sin 7x \cos 2x}{-2 \sin 7x \sin 2x} = -\dfrac{\cos 2x}{\sin 2x} = -\cot 2x$

21. $\frac{1}{2}[\sin(u + v) - \sin(u - v)] = \frac{1}{2}\{(\sin u)\cos v + (\cos u)\sin v - [(\sin u)\cos v - (\cos u)\sin v]\}$

$$= \frac{1}{2}[2(\cos u)\sin v] = (\cos u)\sin v$$

22. $4 \sin^2 x = 1$

$\sin^2 x = \dfrac{1}{4}$

$\sin x = \pm\dfrac{1}{2}$

$\sin x = \dfrac{1}{2}$ or $\sin x = -\dfrac{1}{2}$

$x = \dfrac{\pi}{6}$ or $\dfrac{5\pi}{6}$ $x = \dfrac{7\pi}{6}$ or $\dfrac{11\pi}{6}$

23. $\tan^2 \theta + \left(\sqrt{3} - 1\right)\tan \theta - \sqrt{3} = 0$

$(\tan \theta - 1)\left(\tan \theta + \sqrt{3}\right) = 0$

$\tan \theta = 1$ or $\tan \theta = -\sqrt{3}$

$\theta = \dfrac{\pi}{4}$ or $\dfrac{5\pi}{4}$ $\theta = \dfrac{2\pi}{3}$ or $\dfrac{5\pi}{3}$

24. $\sin 2x = \cos x$

$2(\sin x)\cos x - \cos x = 0$

$\cos x(2 \sin x - 1) = 0$

$\cos x = 0$ or $\sin x = \dfrac{1}{2}$

$x = \dfrac{\pi}{2}$ or $\dfrac{3\pi}{2}$ $x = \dfrac{\pi}{6}$ or $\dfrac{5\pi}{6}$

25. $\tan^2 x - 6 \tan x + 4 = 0$

$$\tan x = \frac{-(-6) \pm \sqrt{(-6)^2 - 4(1)(4)}}{2(1)}$$

$$\tan x = \frac{6 \pm \sqrt{20}}{2} = 3 \pm \sqrt{5}$$

$\tan x = 3 + \sqrt{5}$ $\qquad$ or $\qquad$ $\tan x = 3 - \sqrt{5}$

$x \approx 1.3821$ or 4.5237 $\qquad$ $x = 0.6524$ or 3.7940

Chapter 6   Practice Test Solutions

1. $C = 180° - (40° + 12°) = 128°$

$a = \sin 40° \left(\dfrac{100}{\sin 12°} \right) \approx 309.164$

$c = \sin 128° \left(\dfrac{100}{\sin 12°} \right) \approx 379.012$

2. $\sin A = 5 \left(\dfrac{\sin 150°}{20} \right) = 0.125$

$A \approx 7.181°$

$B \approx 180° - (150° + 7.181°) = 22.819°$

$b = \sin 22.819° \left(\dfrac{20}{\sin 150°} \right) \approx 15.513$

3. Area $= \frac{1}{2} ab \sin C$

$= \frac{1}{2}(3)(6) \sin 130°$

≈ 6.894 square units

4. $h = b \sin A = 35 \sin 22.5° \approx 13.394$

$a = 10$

Since $a < h$ and A is acute, the triangle has no solution.

5. $\cos A = \dfrac{(53)^2 + (38)^2 - (49)^2}{2(53)(38)} \approx 0.4598$

$A \approx 62.627°$

$\cos B = \dfrac{(49)^2 + (38)^2 - (53)^2}{2(49)(38)} \approx 0.2782$

$B \approx 73.847°$

$C \approx 180° - (62.627° + 73.847°) = 43.526°$

6. $c^2 = (100)^2 + (300)^2 - 2(100)(300) \cos 29°$

$\approx 47{,}522.8176$

$c \approx 218$

$\cos A = \dfrac{(300)^2 + (218)^2 - (100)^2}{2(300)(218)} \approx 0.97495$

$A \approx 12.85°$

$B \approx 180° - (12.85° + 29°) = 138.15°$

7. $s = \dfrac{a + b + c}{2} = \dfrac{4.1 + 6.8 + 5.5}{2} = 8.2$

Area $= \sqrt{s(s - a)(s - b)(s - c)}$

$= \sqrt{8.2(8.2 - 4.1)(8.2 - 6.8)(8.2 - 5.5)}$

$= 11.273$ square units

8. $x^2 = (40)^2 + (70)^2 - 2(40)(70) \cos 168°$

$\approx 11{,}977.6266$

$x \approx 190.442$ miles

9. $\mathbf{w} = 4(3\mathbf{i} + \mathbf{j}) - 7(-\mathbf{i} + 2\mathbf{j}) = 19\mathbf{i} - 10\mathbf{j}$

10. $\dfrac{\mathbf{v}}{\|\mathbf{v}\|} = \dfrac{5\mathbf{i} - 3\mathbf{j}}{\sqrt{25 + 9}} = \dfrac{5}{\sqrt{34}}\mathbf{i} - \dfrac{3}{\sqrt{34}}\mathbf{j}$

$= \dfrac{5\sqrt{34}}{34}\mathbf{i} - \dfrac{3\sqrt{34}}{34}\mathbf{j}$

11. $\mathbf{u} = 6\mathbf{i} + 5\mathbf{j}$, $\mathbf{v} = 2\mathbf{i} - 3\mathbf{j}$

$\mathbf{u} \cdot \mathbf{v} = 6(2) + 5(-3) = -3$

$\|\mathbf{u}\| = \sqrt{61}$, $\|\mathbf{v}\| = \sqrt{13}$

$\cos \theta = \dfrac{-3}{\sqrt{61}\sqrt{13}}$

$\theta \approx 96.116°$

12. $4(\mathbf{i} \cos 30° + \mathbf{j} \sin 30°) = 4\left(\dfrac{\sqrt{3}}{2}\mathbf{i} + \dfrac{1}{2}\mathbf{j}\right) = \langle 4\sqrt{3}, 2 \rangle$

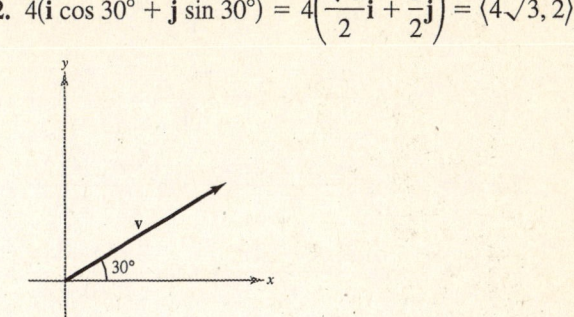

13. $\text{proj}_{\mathbf{v}}\mathbf{u} = \left(\dfrac{\mathbf{u} \cdot \mathbf{v}}{\|\mathbf{v}\|^2}\right)\mathbf{v} = \dfrac{-10}{20}\langle -2, 4 \rangle = \langle 1, -2 \rangle$

14. $r = \sqrt{25 + 25} = \sqrt{50} = 5\sqrt{2}$

$\tan \theta = \dfrac{-5}{5} = -1$

Since z is in Quadrant IV,

$\theta = 315°$

$z = 5\sqrt{2}(\cos 315° + i \sin 315°).$

15. $\cos 225° = -\dfrac{\sqrt{2}}{2}$ $\sin 225° = -\dfrac{\sqrt{2}}{2}$

$z = 6\left(-\dfrac{\sqrt{2}}{2} - i\dfrac{\sqrt{2}}{2}\right)$

$= -3\sqrt{2} - 3\sqrt{2}i$

16. $[7(\cos 23° + i \sin 23°)][4(\cos 7° + i \sin 7°)] = 7(4)[\cos(23° + 7°) + i \sin(23° + 7°)]$

$= 28(\cos 30° + i \sin 30°) = 14\sqrt{3} + 14i$

17. $\dfrac{9\left(\cos \dfrac{5\pi}{4} + i \sin \dfrac{5\pi}{4}\right)}{3(\cos \pi + i \sin \pi)} = \dfrac{9}{3}\left[\cos\left(\dfrac{5\pi}{4} - \pi\right) + i \sin\left(\dfrac{5\pi}{4} - \pi\right)\right] = 3\left(\cos \dfrac{\pi}{4} + i \sin \dfrac{\pi}{4}\right) = \dfrac{3\sqrt{2}}{2} + \dfrac{3\sqrt{2}}{2}i$

18. $(2 + 2i)^8 = [2\sqrt{2}(\cos 45° + i \sin 45°)]^8 = (2\sqrt{2})^8[\cos(8)(45°) + i \sin(8)(45°)]$

$= 4096[\cos 360° + i \sin 360°] = 4096$

19. $z = 8\left(\cos \dfrac{\pi}{3} + i \sin \dfrac{\pi}{3}\right)$, $n = 3$

The cube roots of z are:

For $k = 0$, $\sqrt[3]{8}\left[\cos \dfrac{\pi/3}{3} + i \sin \dfrac{\pi/3}{3}\right] = 2\left(\cos \dfrac{\pi}{9} + i \sin \dfrac{\pi}{9}\right).$

For $k = 1$, $\sqrt[3]{8}\left[\cos \dfrac{\pi/3 + 2\pi}{3} + i \sin \dfrac{\pi/3 + 2\pi}{3}\right] = 2\left(\cos \dfrac{7\pi}{9} + i \sin \dfrac{7\pi}{9}\right).$

For $k = 2$, $\sqrt[3]{8}\left[\cos \dfrac{\pi/3 + 4\pi}{3} + i \sin \dfrac{\pi/3 + 4\pi}{3}\right] = 2\left(\cos \dfrac{13\pi}{9} + i \sin \dfrac{13\pi}{9}\right).$

20. $x^4 = -i = 1\left(\cos\dfrac{3\pi}{2} + i\sin\dfrac{3\pi}{2}\right)$

For $k = 0$, $\cos\dfrac{3\pi/2}{4} + i\sin\dfrac{3\pi/2}{4} = \cos\dfrac{3\pi}{8} + i\sin\dfrac{3\pi}{8}$.

For $k = 1$, $\cos\dfrac{3\pi/2 + 2\pi}{4} + i\sin\dfrac{3\pi/2 + 2\pi}{4} = \cos\dfrac{7\pi}{8} + i\sin\dfrac{7\pi}{8}$.

For $k = 2$, $\cos\dfrac{3\pi/2 + 4\pi}{4} + i\sin\dfrac{3\pi/2 + 4\pi}{4} = \cos\dfrac{11\pi}{8} + i\sin\dfrac{11\pi}{8}$.

For $k = 3$, $\cos\dfrac{3\pi/2 + 6\pi}{4} + i\sin\dfrac{3\pi/2 + 6\pi}{4} = \cos\dfrac{15\pi}{8} + i\sin\dfrac{15\pi}{8}$.

Chapter 7   Practice Test Solutions

1. $\begin{cases} x + y = 1 \\ 3x - y = 15 \end{cases} \implies y = 3x - 15$

$x + (3x - 15) = 1$

$\qquad\qquad 4x = 16$

$\qquad\qquad\ x = 4$

$\qquad\qquad\ y = -3$

2. $\begin{cases} x - 3y = -3 \\ x^2 + 6y = 5 \end{cases} \implies x = 3y - 3$

$\qquad (3y - 3)^2 + 6y = 5$

$\qquad 9y^2 - 18y + 9 + 6y = 5$

$\qquad 9y^2 - 12y + 4 = 0$

$\qquad (3y - 2)^2 = 0$

$\qquad\qquad\qquad\ y = \frac{2}{3}$

$\qquad\qquad\qquad\ x = -1$

3. $\begin{cases} x + y + z = 6 \\ 2x - y + 3z = 0 \\ 5x + 2y - z = -3 \end{cases}$ $\implies z = 6 - x - y$

$2x - y + 3(6 - x - y) = 0 \implies -x - 4y = -18$

$5x + 2y - (6 - x - y) = -3 \implies 6x + 3y = 3$

$\qquad\qquad x = 18 - 4y$

$6(18 - 4y) + 3y = 3$

$\qquad\quad -21y = -105$

$\qquad\qquad\ \ y = 5$

$\qquad\quad x = 18 - 4y = -2$

$\qquad\quad z = 6 - x - y = 3$

4. $\begin{cases} x + y = 110 \\ \ xy = 2800 \end{cases} \implies y = 110 - x$

$x(110 - x) = 2800$

$\qquad 0 = x^2 - 110x + 2800$

$\qquad 0 = (x - 40)(x - 70)$

$\qquad x = 40 \ \text{ or } \ x = 70$

$\qquad y = 70 \qquad\quad y = 40$

5. $\begin{cases} 2x + 2y = 170 \\ \ \ xy = 1500 \end{cases} \implies y = \dfrac{170 - 2x}{2} = 85 - x$

$x(85 - x) = 1500$

$\qquad 0 = x^2 - 85x + 1500$

$\qquad 0 = (x - 25)(x - 60)$

$\qquad x = 25 \ \text{ or } \ x = 60$

$\qquad y = 60 \qquad\quad y = 25$

Dimensions: $60' \times 25'$

6. $\begin{cases} 2x + 15y = 4 \\ x - 3y = 23 \end{cases} \Rightarrow \begin{matrix} 2x + 15y = 4 \\ 5x - 15y = 115 \end{matrix}$

$$\qquad\qquad\qquad\qquad 7x \qquad = 119$$

$$x = 17$$

$$y = \frac{x - 23}{3} = -2$$

7. $\begin{cases} x + y = 2 \\ 38x - 19y = 7 \end{cases} \Rightarrow \begin{matrix} 19x + 19y = 38 \\ 38x - 19y = 7 \end{matrix}$

$$\qquad\qquad\qquad\qquad 57x \qquad = 45$$

$$x = \frac{45}{57} = \frac{15}{19}$$

$$y = 2 - x$$

$$= \frac{38}{19} - \frac{15}{19}$$

$$= \frac{23}{19}$$

8. $y_1 = 2(0.112 - 0.4x)$

$$y_2 = \frac{(0.131 + 0.3x)}{0.7}$$

$\begin{cases} 0.4x + 0.5y = 0.112 \\ 0.3x - 0.7y = -0.131 \end{cases} \Rightarrow \begin{matrix} 0.28x + 0.35y = 0.0784 \\ 0.15x - 0.35y = -0.0655 \end{matrix}$

$$\qquad\qquad\qquad\qquad 0.43x \qquad = 0.0129$$

$$x = \frac{0.0129}{0.43} = 0.03$$

$$y = (2)(0.112 - 0.4x) = 0.20$$

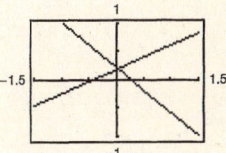

9. Let $x = $ amount in 11% fund and $y = $ amount in 13% fund.

$$\begin{cases} x + y = 17{,}000 \Rightarrow y = 17{,}000 - x \\ 0.11x + 0.13y = 2080 \end{cases}$$

$$0.11x + 0.13(17{,}000 - x) = 2080$$

$$-0.02x = -130$$

$$x = \$6500$$

$$y = \$10{,}500$$

10. Using a graphing utility, you obtain
$y = 0.7857x - 0.1429$. Analytically, $(4, 3)$,
$(1, 1)$, $(-1, -2)$, $(-2, -1)$.

$$n = 4, \sum_{i=1}^{4} x_i = 2, \sum_{i=1}^{4} y_i = 1, \sum_{i=1}^{4} x_i^2 = 22, \sum_{i=1}^{4} x_i y_i = 17$$

$$4b + 2a = 1 \Rightarrow \qquad 4b + 2a = 1$$

$$2b + 22a = 17 \Rightarrow \quad -4b - 44a = -34$$

$$\qquad\qquad\qquad\qquad -42a = -33$$

$$a = \tfrac{33}{42} = \tfrac{11}{14}$$

$$b = \tfrac{1}{4}\left(1 - 2\left(\tfrac{33}{42}\right)\right) = -\tfrac{1}{7}$$

$$y = ax + b = \tfrac{11}{14}x - \tfrac{1}{7}$$

11. $\quad x + y \qquad = -2 \qquad$ Equation 1

$\quad 2x - y + z = 11 \qquad$ Equation 2

$\qquad\quad 4y - 3z = -20 \qquad$ Equation 3

$$\begin{cases} x + y \qquad = -2 \\ \quad -3y + z = 15 \quad -2\text{Eq.1} + \text{Eq.2} \\ \quad\; 4y - 3z = -20 \end{cases}$$

$$\begin{cases} x + y \qquad = -2 \\ \quad -3y + z = 15 \\ \quad -5y \qquad = 25 \quad 3\text{Eq.2} + \text{Eq.3} \end{cases}$$

Answer: $y = -5$

$$x = 3$$

$$z = 0$$

12. $4x - y + 5z = 4$ Equation 1
$2x + y - z = 0$ Equation 2
$2x + 4y + 8z = 0$ Equation 3

$$\begin{cases} 4x - y + 5z = 4 \\ -3y + 7z = 4 \quad \text{Eq.1} - 2\text{Eq.2} \\ 3y + 9z = 0 \quad -\text{Eq.2} + \text{Eq.3} \end{cases}$$

$$\begin{cases} 4x - y + 5z = 4 \\ -3y + 7z = 4 \\ 16z = 4 \quad \text{Eq.2} + \text{Eq.3} \end{cases}$$

Answer: $z = \frac{1}{4}$
$y = -\frac{3}{4}$
$x = \frac{1}{2}$

13. $\begin{cases} 3x + 2y - z = 5 \implies \quad 6x + 4y - 2z = 10 \\ 6x - y + 5z = 2 \implies \quad \underline{-6x + y - 5z = -2} \\ 5y - 7z = 8 \end{cases}$

$$y = \frac{8 + 7z}{5}$$

$3x + 2y - z = 5$
$\underline{12x - 2y + 10z = 4}$
$15x + 9z = 9$

$$x = \frac{9 - 9z}{15} = \frac{3 - 3z}{5}$$

Let $z = a$, then $x = \dfrac{3 - 3a}{5}$ and $y = \dfrac{8 + 7a}{5}$.

14. $y = ax^2 + bx + c$ passes through $(0, -1), (1, 4),$ and $(2, 13)$.

At $(0, -1)$: $-1 = a(0)^2 + b(0) + c \implies c = -1$

At $(1, 4)$: $4 = a(1)^2 + b(1) - 1 \implies 5 = a + b \implies 5 = a + b$

At $(2, 13)$: $13 = a(2)^2 + b(2) - 1 \implies 14 = 4a + 2b \implies \underline{-7 = -2a - b}$
$$-2 = -a$$
$$a = 2$$
$$b = 3$$

Thus, $y = 2x^2 + 3x - 1$.

15. $s = \frac{1}{2}at^2 + v_0 t + s_0$ passes through $(1, 12), (2, 5),$ and $(3, 4)$.

At $(1, 12)$: $12 = \frac{1}{2}a + v_0 + s_0 \implies$ $\begin{cases} \frac{1}{2}a + v_0 + s_0 = 12 \\ -a + s_0 = 19 \quad 2\text{Eq.1} - \text{Eq.2} \\ -3a + s_0 = 7 \quad 3\text{Eq.2} - 2\text{Eq.3} \end{cases}$

At $(2, 5)$: $5 = 2a + 2v_0 + s_0 \implies$

At $(3, 4)$: $4 = \frac{9}{2}a + 3v_0 + s_0 \implies$

$a = 6$
$s_0 = 25$
$v_0 = -16$

$\begin{cases} \frac{1}{2}a + v_0 + s_0 = 12 \\ -a + s_0 = 19 \\ -2a = -12 \quad -\text{Eq.2} + \text{Eq.3} \end{cases}$

Thus,

$s = \frac{1}{2}(6)t^2 - 16t + 25 = 3t^2 - 16t + 25$.

16.
$$\begin{bmatrix} 1 & -2 & 4 \\ 3 & -5 & 9 \end{bmatrix}$$

$-3R_1 + R_2 \rightarrow \begin{bmatrix} 1 & -2 & 4 \\ 0 & 1 & -3 \end{bmatrix}$

$2R_2 + R_1 \rightarrow \begin{bmatrix} 1 & 0 & -2 \\ 0 & 1 & -3 \end{bmatrix}$

17. $3x + 5y = 3$

$2x - y = -11$

$$\begin{bmatrix} 3 & 5 & \vdots & 3 \\ 2 & -1 & \vdots & -11 \end{bmatrix}$$

$$-R_2 + R_1 \rightarrow \begin{bmatrix} 1 & 6 & \vdots & 14 \\ 2 & -1 & \vdots & -11 \end{bmatrix}$$

$$-2R_1 + R_2 \rightarrow \begin{bmatrix} 1 & 6 & \vdots & 14 \\ 0 & -13 & \vdots & -39 \end{bmatrix}$$

$$-\tfrac{1}{13}R_2 \rightarrow \begin{bmatrix} 1 & 6 & \vdots & 14 \\ 0 & 1 & \vdots & 3 \end{bmatrix}$$

$$-6R_2 + R_1 \rightarrow \begin{bmatrix} 1 & 0 & \vdots & -4 \\ 0 & 1 & \vdots & 3 \end{bmatrix}$$

Answer: $x = -4, y = 3$

18. $\begin{cases} 2x + 3y = -3 \\ 3x + 2y = 8 \\ x + y = 1 \end{cases}$

$$\begin{bmatrix} 2 & 3 & \vdots & -3 \\ 3 & 2 & \vdots & 8 \\ 1 & 1 & \vdots & 1 \end{bmatrix}$$

$$\begin{matrix} R_3 \\ \\ R_1 \end{matrix} \begin{bmatrix} 1 & 1 & \vdots & 1 \\ 3 & 2 & \vdots & 8 \\ 2 & 3 & \vdots & -3 \end{bmatrix}$$

$$\begin{matrix} -3R_1 + R_2 \rightarrow \\ -2R_1 + R_3 \rightarrow \end{matrix} \begin{bmatrix} 1 & 1 & \vdots & 1 \\ 0 & -1 & \vdots & 5 \\ 0 & 1 & \vdots & -5 \end{bmatrix}$$

$$\begin{matrix} R_2 + R_1 \rightarrow \\ -R_2 \rightarrow \\ -R_2 + R_3 \rightarrow \end{matrix} \begin{bmatrix} 1 & 0 & \vdots & 6 \\ 0 & 1 & \vdots & -5 \\ 0 & 0 & \vdots & 0 \end{bmatrix}$$

Answer: $x = 6, y = -5$

19. $\begin{cases} x + 3z = -5 \\ 2x + y = 0 \\ 3x + y - z = 3 \end{cases}$

$$\begin{bmatrix} 1 & 0 & 3 & \vdots & -5 \\ 2 & 1 & 0 & \vdots & 0 \\ 3 & 1 & -1 & \vdots & 3 \end{bmatrix}$$

$$\begin{matrix} -2R_1 + R_2 \rightarrow \\ -3R_1 + R_3 \rightarrow \end{matrix} \begin{bmatrix} 1 & 0 & 3 & \vdots & -5 \\ 0 & 1 & -6 & \vdots & 10 \\ 0 & 1 & -10 & \vdots & 18 \end{bmatrix}$$

$$-R_2 + R_3 \rightarrow \begin{bmatrix} 1 & 0 & 3 & \vdots & -5 \\ 0 & 1 & -6 & \vdots & 10 \\ 0 & 0 & -4 & \vdots & 8 \end{bmatrix}$$

$$\begin{matrix} -3R_3 + R_1 \rightarrow \\ 6R_3 + R_2 \rightarrow \\ -\tfrac{1}{4}R_3 \rightarrow \end{matrix} \begin{bmatrix} 1 & 0 & 0 & \vdots & 1 \\ 0 & 1 & 0 & \vdots & -2 \\ 0 & 0 & 1 & \vdots & -2 \end{bmatrix}$$

Answer: $x = 1, y = -2, z = -2$

20. $\begin{bmatrix} 1 & 4 & 5 \\ 2 & 0 & -3 \end{bmatrix} \begin{bmatrix} 1 & 6 \\ 0 & -7 \\ -1 & 2 \end{bmatrix} = \begin{bmatrix} -4 & -12 \\ 5 & 6 \end{bmatrix}$

21. $3A - 5B = 3 \begin{bmatrix} 9 & 1 \\ -4 & 8 \end{bmatrix} - 5 \begin{bmatrix} 6 & -2 \\ 3 & 5 \end{bmatrix}$

$= \begin{bmatrix} -3 & 13 \\ -27 & -1 \end{bmatrix}$

22. $f(A) = \begin{bmatrix} 3 & 0 \\ 7 & 1 \end{bmatrix}^2 - 7\begin{bmatrix} 3 & 0 \\ 7 & 1 \end{bmatrix} + 8\begin{bmatrix} 1 & 0 \\ 0 & 1 \end{bmatrix}$

$\qquad = \begin{bmatrix} 3 & 0 \\ 7 & 1 \end{bmatrix}\begin{bmatrix} 3 & 0 \\ 7 & 1 \end{bmatrix} - \begin{bmatrix} 21 & 0 \\ 49 & 7 \end{bmatrix} + \begin{bmatrix} 8 & 0 \\ 0 & 8 \end{bmatrix}$

$\qquad = \begin{bmatrix} 9 & 0 \\ 28 & 1 \end{bmatrix} - \begin{bmatrix} 21 & 0 \\ 49 & 7 \end{bmatrix} + \begin{bmatrix} 8 & 0 \\ 0 & 8 \end{bmatrix}$

$\qquad = \begin{bmatrix} -4 & 0 \\ -21 & 2 \end{bmatrix}$

23. False.

$\qquad (A + B)(A + 3B) = A(A + 3B) + B(A + 3B)$

$\qquad\qquad\qquad\qquad\quad = A^2 + 3AB + BA + 3B^2$

24.

$\begin{bmatrix} 1 & 2 & \vdots & 1 & 0 \\ 3 & 5 & \vdots & 0 & 1 \end{bmatrix}$

$-3R_1 + R_2 \rightarrow \begin{bmatrix} 1 & 2 & \vdots & 1 & 0 \\ 0 & -1 & \vdots & -3 & 1 \end{bmatrix}$

$\begin{matrix} 2R_2 + R_1 \rightarrow \\ -R_2 \rightarrow \end{matrix} \begin{bmatrix} 1 & 0 & \vdots & -5 & 2 \\ 0 & 1 & \vdots & 3 & -1 \end{bmatrix}$

$A^{-1} = \begin{bmatrix} -5 & 2 \\ 3 & -1 \end{bmatrix}$

25.

$\begin{bmatrix} 1 & 1 & 1 & \vdots & 1 & 0 & 0 \\ 3 & 6 & 5 & \vdots & 0 & 1 & 0 \\ 6 & 10 & 8 & \vdots & 0 & 0 & 1 \end{bmatrix}$

$\begin{matrix} -3R_1 + R_2 \rightarrow \\ -6R_1 + R_3 \rightarrow \end{matrix} \begin{bmatrix} 1 & 1 & 1 & \vdots & 1 & 0 & 0 \\ 0 & 3 & 2 & \vdots & -3 & 1 & 0 \\ 0 & 4 & 2 & \vdots & -6 & 0 & 1 \end{bmatrix}$

$\begin{matrix} -\frac{1}{3}R_2 + R_1 \rightarrow \\ \frac{1}{3}R_2 \rightarrow \\ -4R_2 + R_3 \rightarrow \end{matrix} \begin{bmatrix} 1 & 0 & \frac{1}{3} & \vdots & 2 & -\frac{1}{3} & 0 \\ 0 & 1 & \frac{2}{3} & \vdots & -1 & \frac{1}{3} & 0 \\ 0 & 0 & -\frac{2}{3} & \vdots & -2 & -\frac{4}{3} & 1 \end{bmatrix}$

$\begin{matrix} \frac{1}{2}R_3 + R_1 \rightarrow \\ R_3 + R_2 \rightarrow \\ -\frac{3}{2}R_3 \rightarrow \end{matrix} \begin{bmatrix} 1 & 0 & 0 & \vdots & 1 & -1 & \frac{1}{2} \\ 0 & 1 & 0 & \vdots & -3 & -1 & 1 \\ 0 & 0 & 1 & \vdots & 3 & 2 & -\frac{3}{2} \end{bmatrix}$

$A^{-1} = \begin{bmatrix} 1 & -1 & \frac{1}{2} \\ -3 & -1 & 1 \\ 3 & 2 & -\frac{3}{2} \end{bmatrix}$

26. (a) $x + 2y = 4$

$\qquad 3x + 5y = 1$

$\begin{bmatrix} 1 & 2 & \vdots & 1 & 0 \\ 3 & 5 & \vdots & 0 & 1 \end{bmatrix}$

$-3R_1 + R_2 \rightarrow \begin{bmatrix} 1 & 2 & \vdots & 1 & 0 \\ 0 & -1 & \vdots & -3 & 1 \end{bmatrix}$

$\begin{matrix} -2R_2 + R_1 \rightarrow \\ -R_2 \rightarrow \end{matrix} \begin{bmatrix} 1 & 0 & \vdots & -5 & 2 \\ 0 & 1 & \vdots & 3 & -1 \end{bmatrix}$

$X = A^{-1}B = \begin{bmatrix} -5 & 2 \\ 3 & -1 \end{bmatrix}\begin{bmatrix} 4 \\ 1 \end{bmatrix} = \begin{bmatrix} -18 \\ 11 \end{bmatrix}$

$x = -18, y = 11$

(b) $x + 2y = 3$

$\qquad 3x + 5y = -2$

$X = A^{-1}B - \begin{bmatrix} -5 & 2 \\ 3 & -1 \end{bmatrix}\begin{bmatrix} 3 \\ -2 \end{bmatrix} = \begin{bmatrix} -19 \\ 11 \end{bmatrix}$

$x = -19, y = 11$

27. $\begin{vmatrix} 6 & -1 \\ 3 & 4 \end{vmatrix} = 24 - (-3) = 27$

28. $\begin{vmatrix} 1 & 3 & -1 \\ 5 & 9 & 0 \\ 6 & 2 & -5 \end{vmatrix} = 1(-45) + (-3)(-25) + (-1)(-44)$

$\qquad\qquad\qquad\qquad = 74$

29. $\begin{vmatrix} 1 & 4 & 2 & 3 \\ 0 & 1 & -2 & 0 \\ 3 & 5 & -1 & 1 \\ 2 & 0 & 6 & 1 \end{vmatrix} = -7$

30. $\begin{vmatrix} 6 & 4 & 3 & 0 & 6 \\ 0 & 5 & 1 & 4 & 8 \\ 0 & 0 & 2 & 7 & 3 \\ 0 & 0 & 0 & 9 & 2 \\ 0 & 0 & 0 & 0 & 1 \end{vmatrix} = 6(5)(2)(9)(1) = 540$

31. $\text{Area} = \dfrac{1}{2} \begin{vmatrix} 0 & 7 & 1 \\ 5 & 0 & 1 \\ 3 & 9 & 1 \end{vmatrix} = \dfrac{1}{2}(31)$

$= 15.5 \text{ square units}$

32. $\begin{vmatrix} x & y & 1 \\ 2 & 7 & 1 \\ -1 & 4 & 1 \end{vmatrix} = 3x - 3y + 15 = 0$

or $x - y + 5 = 0$

33. $x = \dfrac{\begin{vmatrix} 4 & -7 \\ 11 & 5 \end{vmatrix}}{\begin{vmatrix} 6 & -7 \\ 2 & 5 \end{vmatrix}} = \dfrac{97}{44}$

34. $z = \dfrac{\begin{vmatrix} 3 & 0 & 1 \\ 0 & 1 & 3 \\ 1 & -1 & 2 \end{vmatrix}}{\begin{vmatrix} 3 & 0 & 1 \\ 0 & 1 & 4 \\ 1 & -1 & 0 \end{vmatrix}} = \dfrac{14}{11}$

35. $y = \dfrac{\begin{vmatrix} 721.4 & 33.77 \\ 45.9 & 19.85 \end{vmatrix}}{\begin{vmatrix} 721.4 & -29.1 \\ 45.9 & 105.6 \end{vmatrix}}$

$= \dfrac{12,769.747}{77,515.530} \approx 0.1647$

Chapter 8   Practice Test Solutions

1. $a_n = \dfrac{2n}{(n+2)!}$

$a_1 = \dfrac{2(1)}{3!} = \dfrac{2}{6} = \dfrac{1}{3}$

$a_2 = \dfrac{2(2)}{4!} = \dfrac{4}{24} = \dfrac{1}{6}$

$a_3 = \dfrac{2(3)}{5!} = \dfrac{6}{120} = \dfrac{1}{20}$

$a_4 = \dfrac{2(4)}{6!} = \dfrac{8}{720} = \dfrac{1}{90}$

$a_5 = \dfrac{2(5)}{7!} = \dfrac{10}{5040} = \dfrac{1}{504}$

Terms: $\dfrac{1}{3}, \dfrac{1}{6}, \dfrac{1}{20}, \dfrac{1}{90}, \dfrac{1}{504}$

2. $a_n = \dfrac{n+3}{3^n}$

3. $\sum_{i=1}^{6}(2i-1) = 1 + 3 + 5 + 7 + 9 + 11 = 36$

4. $a_1 = 23,\ d = -2$

$a_2 = a_1 + d = 21$

$a_3 = a_2 + d = 19$

$a_4 = a_3 + d = 17$

$a_5 = a_4 + d = 15$

Terms: 23, 21, 19, 17, 15

5. $a_1 = 12,\ d = 3,\ n = 50$

$a_n = a_1 + (n - 1)d$

$a_{50} = 12 + (50 - 1)3 = 159$

6. $a_1 = 1$

$a_{200} = 200$

$S_n = \dfrac{n}{2}(a_1 + a_n)$

$S_{200} = \dfrac{200}{2}(1 + 200) = 20{,}100$

7. $a_1 = 7,\ r = 2$

$a_2 = a_1 r = 14$

$a_3 = a_1 r^2 = 28$

$a_4 = a_1 r^3 = 56$

$a_5 = a_1 r^4 = 112$

Terms: 7, 14, 28, 56, 112

8. $\displaystyle\sum_{n=0}^{9} 6\left(\frac{2}{3}\right)^n,\ a_1 = 6,\ r = \frac{2}{3},\ n = 10$

$S_n = \dfrac{a_1(1 - r^n)}{1 - r} = \dfrac{6\left(1 - (2/3)^{10}\right)}{1 - (2/3)} \approx 17.6879$

9. $\displaystyle\sum_{n=0}^{\infty} (0.03)^n,\ a_1 = 1,\ r = 0.03$

$S = \dfrac{a_1}{1 - r} = \dfrac{1}{1 - 0.03} = \dfrac{1}{0.97} = \dfrac{100}{97} \approx 1.0309$

10. For $n = 1,\ 1 = \dfrac{1(1 + 1)}{2}$. Assume that $1 + 2 + 3 = 4 + \cdots + k = \dfrac{k(k + 1)}{2}$. Now for $n = k + 1$,

$$1 + 2 + 3 + 4 + \cdots + k + (k + 1) = \frac{k(k + 1)}{2} + k + 1$$

$$= \frac{k(k + 1)}{2} + \frac{2(k + 1)}{2}$$

$$= \frac{(k + 1)(k + 2)}{2}.$$

Thus, $1 + 2 + 3 + 4 + \cdots + n = \dfrac{n(n + 1)}{2}$ for all integers $n \geq 1$.

11. For $n = 4,\ 4! > 2^4$. Assume that $k! > 2^k$. Then

$(k + 1)! = (k + 1)(k!) > (k + 1)2^k > 2 \cdot 2^k$

$= 2^{k+1}$.

Thus, $n! > 2^n$ for all integers $n \geq 4$.

12. $_{13}C_4 = \dfrac{13!}{(13 - 4)!4!} = 715$

13. $(x + 3)^5 = x^5 + 5x^4(3) + 10x^3(3)^2 + 10x^2(3)^3 + 5x(3)^4 + (3)^5$

$= x^5 + 15x^4 + 90x^3 + 270x^2 + 405x + 243$

14. $_{12}C_5 x^7(-2)^5 = -25{,}344x^7$

15. $_{30}P_4 = \dfrac{30!}{(30 - 4)!} = 657{,}720$

16. $6! = 720$ ways

17. $_{12}P_3 = 1320$

18. $P(2) + P(3) + P(4) = \frac{1}{36} + \frac{2}{36} + \frac{3}{36}$
$$= \frac{6}{36} = \frac{1}{6}$$

19. $P(K, B10) = \frac{4}{52} \cdot \frac{2}{51} = \frac{2}{663}$

20. Let A = probability of no faulty units.

$P(A) = \left(\frac{997}{1000}\right)^{50} \approx 0.8605$

$P(A') = 1 - P(A) \approx 0.1395$

Chapter 9   Practice Test Solutions

1. $x^2 - 6x - 4y + 1 = 0$

$x^2 - 6x + 9 = 4y - 1 + 9$

$(x - 3)^2 = 4y + 8$

$(x - 3)^2 = 4(1)(y + 2) \implies p = 1$

Vertex: $(3, -2)$

Focus: $(3, -1)$

Directrix: $y = -3$

2. Vertex: $(2, -5)$

Focus: $(2, -6)$

Vertical axis; opens downward with $p = -1$

$(x - h)^2 = 4p(y - k)$

$(x - 2)^2 = 4(-1)(y + 5)$

$x^2 - 4x + 4 = -4y - 20$

$x^2 - 4x + 4y + 24 = 0$

3. $x^2 + 4y^2 - 2x + 32y + 61 = 0$

$(x^2 - 2x + 1) + 4(y^2 + 8y + 16) = -61 + 1 + 64$

$(x - 1)^2 + 4(y + 4)^2 = 4$

$$\frac{(x - 1)^2}{4} + \frac{(y + 4)^2}{1} = 1$$

$a = 2, b = 1, c = \sqrt{3}$

Horizontal major axis

Center: $(1, -4)$

Foci: $\left(1 \pm \sqrt{3}, -4\right)$

Vertices: $(3, -4), (-1, -4)$

Eccentricity: $e = \dfrac{\sqrt{3}}{2}$

4. Vertices: $(0, \pm 6)$

Eccentricity: $e = \dfrac{1}{2}$

Center: $(0, 0)$

Vertical major axis

$a = 6, e = \dfrac{c}{a} = \dfrac{c}{6} = \dfrac{1}{2} \implies c = 3$

$b^2 = (6)^2 - (3)^2 = 27$

$\dfrac{x^2}{27} + \dfrac{y^2}{36} = 1$

5.
$$16y^2 - x^2 - 6x - 128y + 231 = 0$$

$$16(y^2 - 8y + 16) - (x^2 + 6x + 9) = -231 + 256 - 9$$

$$16(y - 4)^2 - (x + 3)^2 = 16$$

$$\frac{(y - 4)^2}{1} - \frac{(x + 3)^2}{16} = 1$$

$a = 1, b = 4, c = \sqrt{17}$

Center: $(-3, 4)$

Vertical transverse axis

Vertices: $(-3, 5), (-3, 3)$

Foci: $\left(-3, 4 \pm \sqrt{17}\right)$

Asymptotes: $y = 4 \pm \dfrac{1}{4}(x + 3)$

6. Vertices: $(\pm 3, 2)$ $a = 3, c = 5, b = 4$

Foci: $(\pm 5, 2)$ $\dfrac{(x - 0)^2}{9} - \dfrac{(y - 2)^2}{16} = 1$

Center: $(0, 2)$

Horizontal transverse axis $\dfrac{x^2}{9} - \dfrac{(y - 2)^2}{16} = 1$

7. $5x^2 + 2xy + 5y^2 - 10 = 0$

$A = 5, B = 2, C = 5$

$\cot 2\theta = \dfrac{5 - 5}{2} = 0$

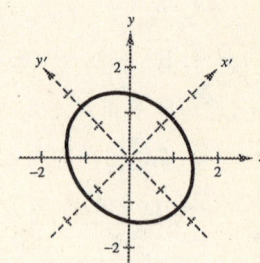

$2\theta = \dfrac{\pi}{2} \Rightarrow \theta = \dfrac{\pi}{4}$

$x = x' \cos \dfrac{\pi}{4} - y' \sin \dfrac{\pi}{4} = \dfrac{x' - y'}{\sqrt{2}}$

$y = x' \sin \dfrac{\pi}{4} + y' \cos \dfrac{\pi}{4} = \dfrac{x' + y'}{\sqrt{2}}$

$$5\left(\frac{x' - y'}{\sqrt{2}}\right)^2 + 2\left(\frac{x' - y'}{\sqrt{2}}\right)\left(\frac{x' + y'}{\sqrt{2}}\right) + 5\left(\frac{x' + y'}{\sqrt{2}}\right)^2 - 10 = 0$$

$$\frac{5(x')^2}{2} - \frac{10x'y'}{2} + \frac{5(y')^2}{2} + (x')^2 - (y')^2 + \frac{5(x')^2}{2} + \frac{10x'y'}{2} + \frac{5(y')^2}{2} - 10 = 0$$

$$6(x')^2 + 4(y')^2 - 10 = 0$$

$$\frac{3(x')^2}{5} + \frac{2(y')^2}{5} = 1$$

$$\frac{(x')^2}{5/3} + \frac{(y')^2}{5/2} = 1$$

Ellipse centered at the origin

8. (a) $6x^2 - 2xy + y^2 = 0$

$A = 6, B = -2, C = 1$

$B^2 - 4AC = (-2)^2 - 4(6)(1) = -20 < 0$

Ellipse

(b) $x^2 + 4xy + 4y^2 - x - y + 17 = 0$

$A = 1, B = 4, C = 4$

$B^2 - 4AC = (4)^2 - 4(1)(4) = 0$

Parabola

9. $x = 3 - 2\sin\theta, y = 1 + 5\cos\theta$

$\dfrac{x-3}{-2} = \sin\theta, \dfrac{y-1}{5} = \cos\theta$

$\left(\dfrac{x-3}{-2}\right)^2 + \left(\dfrac{y-1}{5}\right)^2 = 1$

$\dfrac{(x-3)^2}{4} + \dfrac{(y-1)^2}{25} = 1$

10. $x = e^{2t}, y = e^{4t}$

$x > 0, y > 0$

$y = (e^{2t})^2 = (x)^2 = x^2, x > 0, y > 0$

11. Polar: $\left(\sqrt{2}, \dfrac{3\pi}{4}\right)$

$x = \sqrt{2}\cos\dfrac{3\pi}{4} = \sqrt{2}\left(-\dfrac{1}{\sqrt{2}}\right) = -1$

$y = \sqrt{2}\sin\dfrac{3\pi}{4} = \sqrt{2}\left(\dfrac{1}{\sqrt{2}}\right) = 1$

Rectangular: $(-1, 1)$

12. Rectangular: $\left(\sqrt{3}, -1\right)$

$r = \pm\sqrt{(\sqrt{3})^2 + (-1)^2} = \pm 2$

$\tan\theta = \dfrac{-1}{\sqrt{3}} = -\dfrac{\sqrt{3}}{3}$

$\theta = \dfrac{5\pi}{6}$ or $\theta = \dfrac{11\pi}{6}$

Polar: $\left(-2, \dfrac{5\pi}{6}\right)$ or $\left(2, \dfrac{11\pi}{6}\right)$

13. Rectangular: $4x - 3y = 12$

Polar: $4r\cos\theta - 3r\sin\theta = 12$

$r(4\cos\theta - 3\sin\theta) = 12$

$r = \dfrac{12}{4\cos\theta - 3\sin\theta}$

14. Polar: $r = 5\cos\theta$

$r^2 = 5r\cos\theta$

Rectangular: $x^2 + y^2 = 5x$

$x^2 + y^2 - 5x = 0$

15. $r = 1 - \cos\theta$

Cardioid

Symmetry: Polar axis

Maximum value of $|r|$: $r = 2$ when $\theta = \pi$.

Zero of r: $r = 0$ when $\theta = 0$.

θ	0	$\dfrac{\pi}{2}$	π	$\dfrac{3\pi}{2}$
r	0	1	2	1

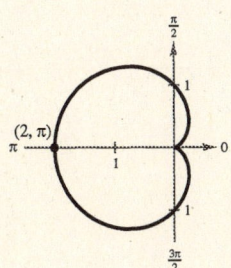

16. $r = 5 \sin 2\theta$

Rose curve with four petals

Symmetry: Polar axis, $\theta = \dfrac{\pi}{2}$, and pole

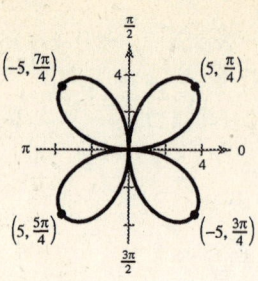

Maximum value of $|r|$: $|r| = 5$ when $\theta = \dfrac{\pi}{4}, \dfrac{3\pi}{4}, \dfrac{5\pi}{4}, \dfrac{7\pi}{4}$.

Zeros of r: $r = 0$ when $\theta = 0, \dfrac{\pi}{2}, \pi, \dfrac{3\pi}{2}$.

17. $r = \dfrac{3}{6 - \cos \theta}$

$r = \dfrac{1/2}{1 - (1/6) \cos \theta}$

$e = \dfrac{1}{6} < 1$, so the graph is an ellipse.

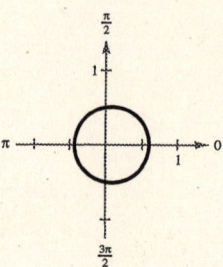

θ	0	$\dfrac{\pi}{2}$	π	$\dfrac{3\pi}{2}$
r	$\dfrac{3}{5}$	$\dfrac{1}{2}$	$\dfrac{3}{7}$	$\dfrac{1}{2}$

18. Parabola

Vertex: $\left(6, \dfrac{\pi}{2}\right)$

Focus: $(0, 0)$

$e = 1$

$r = \dfrac{ep}{1 + e \sin \theta}$

$r = \dfrac{p}{1 + \sin \theta}$

$6 = \dfrac{p}{1 + \sin(\pi/2)}$

$6 = \dfrac{p}{2}$

$12 = p$

$r = \dfrac{12}{1 + \sin \theta}$

Chapter 10   Practice Test Solutions

1. Let $A = (0, 0, 0)$, $B = (1, 2, -4)$, $C = (0, -2, -1)$.

Side AB: $\sqrt{1^2 + 2^2 + 4^2} = \sqrt{21}$

Side AC: $\sqrt{0^2 + 2^2 + 1^2} = \sqrt{5}$

Side BC: $\sqrt{(-1)^2 + (-2 - 2)^2 + (-1 + 4)^2} = \sqrt{1 + 16 + 9} = \sqrt{26}$

$BC^2 = AB^2 + AC^2$

$26 = 21 + 5$

2. $(x - 0)^2 + (y - 4)^2 + (z - 1)^2 = 5^2$

$\quad x^2 + (y - 4)^2 + (z - 1)^2 = 25$

3. $(x^2 + 2x + 1) + y^2 + (z^2 - 4z + 4) = 1 + 4 + 11$

$\quad (x + 1)^2 + y^2 + (z - 2)^2 = 16$

Center: $(-1, 0, 2)$

Radius: 4

4. $\mathbf{u} - 3\mathbf{v} = \langle 1, 0, -1 \rangle - 3 \langle 4, 3, -6 \rangle$

$\quad\quad\quad = \langle 1, 0, -1 \rangle - \langle 12, 9, -18 \rangle$

$\quad\quad\quad = \langle -11, -9, 17 \rangle$

5. $\frac{1}{2}\mathbf{v} = \frac{1}{2} \langle 2, 4, -6 \rangle = \langle 1, 2, -3 \rangle$

$\left\| \frac{1}{2}\mathbf{v} \right\| = \sqrt{1^2 + 2^2 + (-3)^2} = \sqrt{14}$

6. $\mathbf{u} \cdot \mathbf{v} = \langle 2, 1, -3 \rangle \cdot \langle 1, 1, -2 \rangle$

$\quad\quad\quad = 2 + 1 + 6 = 9$

7. Because $\mathbf{v} = \langle -3, -3, 3 \rangle = -3 \langle 1, 1, -1 \rangle = -3\mathbf{u}$, $\mathbf{u}$ and $\mathbf{v}$ are parallel.

8. $\mathbf{u} \times \mathbf{v} = \begin{vmatrix} \mathbf{i} & \mathbf{j} & \mathbf{k} \\ -1 & 0 & 2 \\ 1 & -1 & 3 \end{vmatrix} = \langle 2, 5, 1 \rangle$

$\mathbf{v} \times \mathbf{u} = -(\mathbf{u} \times \mathbf{v}) = \langle -2, -5, -1 \rangle$

9. $\mathbf{u} \cdot (\mathbf{v} \times \mathbf{w}) = \begin{vmatrix} 1 & 1 & 1 \\ 0 & -1 & 1 \\ 1 & 0 & 4 \end{vmatrix}$

$\quad\quad\quad = 1(-4) - 1(-1) + 1(1)$

$\quad\quad\quad = -4 + 1 + 1 = -2$

Volume $= |\mathbf{u} \cdot (\mathbf{v} \times \mathbf{w})| = |-2| = 2$

10. $\mathbf{v} = \langle (2 - 0), -3 - (-3), 4 - 3 \rangle = \langle 2, 0, 1 \rangle$

$x = 2 + 2t, y = -3, z = 4 + t$

11. $1(x - 1) - 1(y - 2) + 0(z - 3) = 0$

$\quad\quad\quad x - 1 - y + 2 = 0$

$\quad\quad\quad\quad x - y + 1 = 0$

12. $\overrightarrow{AB} = \langle 1, 1, 1 \rangle$, $\overrightarrow{AC} = \langle 1, 2, 3 \rangle$

$\mathbf{n} = \overrightarrow{AB} \times \overrightarrow{AC} = \begin{vmatrix} \mathbf{i} & \mathbf{j} & \mathbf{k} \\ 1 & 1 & 1 \\ 1 & 2 & 3 \end{vmatrix} = \langle 1, -2, 1 \rangle$

Plane: $1(x - 0) - 2(y - 0) + (z - 0) = 0$

$\quad\quad\quad x - 2y + z = 0$

13. $\mathbf{n}_1 = \langle 1, 1, -1 \rangle$, $\mathbf{n}_2 = \langle 3, -4, -1 \rangle$

$\mathbf{n}_1 \cdot \mathbf{n}_2 = 3 - 4 + 1 = 0 \implies$ Orthogonal planes

14. $\mathbf{n} = \langle 1, 2, 1 \rangle$, $Q = (1, 1, 1)$, $P = (0, 0, 6)$ on plane, $\overrightarrow{PQ} = \langle 1, 1, -5 \rangle$

$D = \dfrac{|\overrightarrow{PQ} \cdot \mathbf{n}|}{\|\mathbf{n}\|} = \dfrac{|1 + 2 - 5|}{\sqrt{1 + 4 + 1}} = \dfrac{2}{\sqrt{6}} = \dfrac{\sqrt{6}}{3}$

Chapter 11   Practice Test Solutions

1.

x	2.9	2.99	3	3.01	3.1
$f(x)$	0.1695	0.1669	?	0.1664	0.1639

$$\lim_{x \to 3} \frac{x-3}{x^2-9} \approx 0.1667$$

2. $\lim\limits_{x \to 0} \dfrac{\sqrt{x+4}-2}{x} \approx \dfrac{1}{4}$

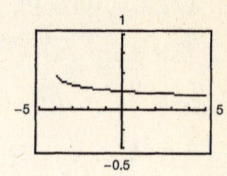

3. $\lim\limits_{x \to 2} e^{x-2} = e^{2-2} = e^0 = 1$

4. $\lim\limits_{x \to 1} \dfrac{x^3-1}{x-1} = \lim\limits_{x \to 1} \dfrac{(x-1)(x^2+x+1)}{x-1}$

$$= \lim_{x \to 1} (x^2 + x + 1) = 3$$

5. $\lim\limits_{x \to 0} \dfrac{\sin 5x}{2x} \approx 2.5$

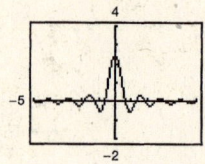

6. The limit does not exist. If

$$f(x) = \frac{|x+2|}{x+2},$$

then $f(x) = 1$ for $x > -2$, and $f(x) = -1$ for $x < -2$.

7. $m_{\text{sec}} = \dfrac{f(4+h) - f(4)}{h}$

$$= \frac{\sqrt{4+h}-2}{h}$$

$$= \frac{\sqrt{4+h}-2}{h} \cdot \frac{\sqrt{4+h}+2}{\sqrt{4+h}+2}$$

$$= \frac{(4+h)-4}{h\left[\sqrt{4+h}+2\right]}$$

$$= \frac{h}{h\left[\sqrt{4+h}+2\right]}$$

$$= \frac{1}{\sqrt{4+h}+2}, \quad h \neq 0$$

$$m = \lim_{h \to 0} \frac{1}{\sqrt{4+h}+2} = \frac{1}{\sqrt{4}+2} = \frac{1}{4}$$

8. $f'(x) = \lim\limits_{h \to 0} \dfrac{f(x+h) - f(x)}{h}$

$$= \lim_{h \to 0} \frac{[3(x+h)-1] - [3x-1]}{h}$$

$$= \lim_{h \to 0} \frac{3x + 3h - 1 - 3x + 1}{h}$$

$$= \lim_{h \to 0} \frac{3h}{h} = \lim_{h \to 0} 3 = 3$$

9. (a) $\lim\limits_{x \to \infty} \dfrac{3}{x^4} = 0$

(b) $\lim\limits_{x \to -\infty} \dfrac{x^2}{x^2+3} = 1$

(c) $\lim\limits_{x \to \infty} \dfrac{|x|}{1-x} = -1$

10. $a_1 = 0, \; a_2 = \dfrac{1-4}{8+1} = -\dfrac{1}{3}, \; a_3 = \dfrac{1-9}{18+1} = -\dfrac{8}{19},$

$$a_4 = \frac{1-16}{33} = -\frac{15}{33}$$

$$\lim_{n \to \infty} a_n = \lim_{n \to \infty} \frac{1-n^2}{2n^2+1} = -\frac{1}{2}$$

11. $\displaystyle\sum_{i=1}^{25} i^2 + \sum_{i=1}^{25} i = \frac{25(26)(51)}{6} + \frac{25(26)}{2} = \frac{25(26)}{6}[51 + 3] = \frac{25(26)(54)}{6} = 5850$

12. $\displaystyle\sum_{i=1}^{n} \frac{i^2}{n^3} = \frac{1}{n^3} \sum_{i=1}^{n} i^2 = \frac{1}{n^3}\left[\frac{n(n+1)(2n+1)}{6}\right] = \frac{2n^2 + 3n + 1}{6n^2} = S(n)$

$\displaystyle\lim_{n\to\infty} S(n) = \frac{1}{3}$

13. Width of rectangles: $\dfrac{b-a}{n} = \dfrac{1}{n}$

Height: $f\left(a + \dfrac{(b-a)i}{n}\right) = f\left(\dfrac{i}{n}\right) = 1 - \left(\dfrac{i}{n}\right)^2$

$\displaystyle A_n \approx \sum_{i=1}^{n} \left[1 - \frac{i^2}{n^2}\right]\frac{1}{n} = \sum_{i=1}^{n} \frac{1}{n} - \sum_{i=1}^{n} \frac{i^2}{n^3} = 1 - \frac{1}{n^2}\frac{n(n+1)(2n+1)}{6}$

$\displaystyle A = \lim_{n\to\infty} A_n = 1 - \frac{1}{3} = \frac{2}{3}$

PART II

Chapter 1   Chapter Test

1. $y - (-1) = \frac{3}{2}(x - 3) = \frac{3}{2}x - \frac{9}{2}$

$\qquad y = \frac{3}{2}x - \frac{11}{2}$

Additional points: $(1, -4)$, $\left(4, \frac{1}{2}\right)$, $(5, 2)$

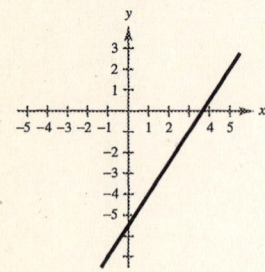

2. $5x + 2y = 3$

$\qquad 2y = -5x + 3$

$\qquad y = -\frac{5}{2}x + \frac{3}{2}$ slope $= -\frac{5}{2}$

(a) parallel line: $\qquad y - 4 = -\frac{5}{2}(x - 0)$

$\qquad\qquad\qquad\qquad y = -\frac{5}{2}x + 4$

$\qquad\qquad 5x + 2y - 8 = 0$

(b) perpendicular line: $\qquad y - 4 = \frac{2}{5}(x - 0)$

$\qquad\qquad\qquad\qquad\qquad y = \frac{2}{5}x + 4$

$\qquad\qquad 2x - 5y + 20 = 0$

3. No, for some x there corresponds more than one value of y. For instance, if $x = 1$, $y = \pm 1/\sqrt{3}$.

4. $f(x) = |x + 2| - 15$

(a) $f(-8) = |-8 + 2| - 15 = 6 - 15 = -9$

(b) $f(14) = |14 + 2| - 15 = 16 - 15 = 1$

(c) $f(t - 6) = |t - 6 + 2| - 15 = |t - 4| - 15$

5. $3 - x \geq 0 \implies$ domain is all $x \leq 3$.

6. $C = 5.60x + 24,000$

$P = R - C = 99.50x - (5.60x + 24,000) = 93.9x - 24,000$

7. $h(x) = \frac{1}{4}x^4 - 2x^2 = \frac{1}{4}x^2(x^2 - 8)$

By graphing h, you see that the graph is increasing on $(-2, 0)$ and $(2, \infty)$ and decreasing on $(-\infty, -2)$ and $(0, 2)$.

8. $g(t) = |t + 2| - |t - 2|$.

By graphing g, you see that the graph is increasing on $(-2, 2)$, and constant on $(-\infty, -2)$ and $(2, \infty)$.

9. Relative minimum: $(-3.33, -6.52)$.

Relative maximum: $(0, 12)$

10. Relative minimum: $(0.77, 1.81)$.

Relative maximum: $(-0.77, 2.19)$

11. (a) Common function $f(x) = x^3$

(b) g is obtained from f by a horizontal shift 5 units to the right, a vertical stretch of 2, a reflection in the x-axis, and a vertical shift 3 units upward.

(c)

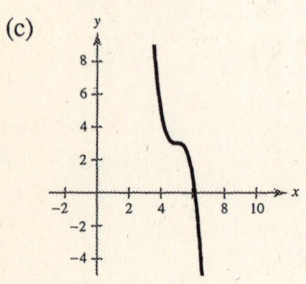

12. (a) Common function $f(x) = \sqrt{x}$.

(b) g is obtained from f by a reflection in the y-axis, and a horizontal shift 7 units to the left.

(c)

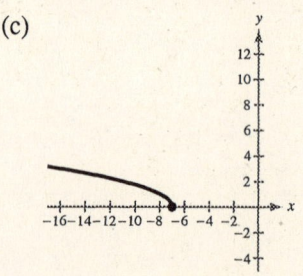

13. (a) Common function $f(x) = |x|$.

(b) $g(x) = 4|-x| - 7 = 4|x| - 7$ is obtained from f by a vertical stretch of 4 followed by a vertical shift 7 units downward.

(c)

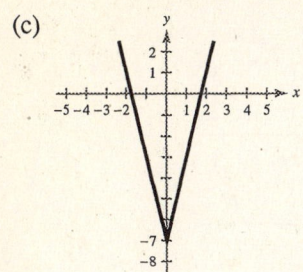

14. (a) $(f - g)(x) = x^2 - \sqrt{2 - x}$, Domain: $x \le 2$

(b) $\left(\dfrac{f}{g}\right)(x) = \dfrac{x^2}{\sqrt{2 - x}}$, Domain: $x < 2$

(c) $(f \circ g)(x) = f\left(\sqrt{2 - x}\right) = 2 - x$, Domain $x \le 2$

(d) $(g \circ f)x = g(x^2) = \sqrt{2 - x^2}$ Domain:
$$-\sqrt{2} \le x \le \sqrt{2}$$

15. $f(x) = x^3 + 8$

Yes, f is one-to-one and has an inverse function.

$y = x^3 + 8$

$x = y^3 + 8$

$x - 8 = y^3$

$\sqrt[3]{x - 8} = y$

$f^{-1}(x) = \sqrt[3]{x - 8}$

16. $f(x) = x^2 + 6$

No, f is not one-to-one, and does not have an inverse function.

17. $f(x) = \dfrac{3x\sqrt{x}}{8}$

Yes, f is one-to-one and has an inverse function.

$y = \frac{3}{8}x^{3/2}$, $x \ge 0$, $y \ge 0$

$x = \frac{3}{8}y^{3/2}$, $y \ge 0$, $x \ge 0$

$\frac{8}{3}x = y^{3/2}$

$\left(\frac{8}{3}x\right)^{2/3} = y$

$f^{-1}(x) = \left(\frac{8}{3}x\right)^{2/3}$, $x \ge 0$

18. $L = 14.8t + 92.4$

$L > 300$

$14.8t + 92.4 > 300$

$14.8t > 207.6$

$t > 14$

The number of lines will exceed 300 in 2004.

Chapter 2   Chapter Test

1. (a) $g(x) = 6 - x^2$ is a reflection in the x-axis followed by a vertical shift 6 units upward.

(b) $g(x) = \left(x - \frac{3}{2}\right)^2$ is a horizontal shift $\frac{3}{2}$ units to the right.

2. $y = x^2 + 4x + 3 = x^2 + 4x + 4 - 1 = (x + 2)^2 - 1$

Vertex: $(-2, -1)$

$x = 0 \implies y = 3$

$y = 0 \implies x^2 + 4x + 3 = 0 \implies (x + 3)(x + 1) = 0 \implies x = -1, -3$

Intercepts: $(0, 3)$, $(-1, 0)$, $(-3, 0)$

3. Let $y = a(x - h)^2 + k$. The vertex $(3, -6)$ implies that $y = a(x - 3)^2 - 6$. For $(0, 3)$ you obtain

$$3 = a(0 - 3)^2 - 6 = 9a - 6 \implies a = 1.$$

Thus, $y = (x - 3)^2 - 6 = x^2 - 6x + 3$.

4.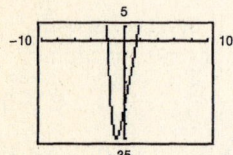

$$3x + \dfrac{x - 1}{x^2 + 1}$$

5.
$$
\begin{array}{r|rrrrr}
2 & 2 & 0 & -5 & 0 & -3 \\
  &   & 4 & 8 & 6 & 12 \\
\hline
  & 2 & 4 & 3 & 6 & 9
\end{array}
$$

$$2x^3 + 4x^2 + 3x + 6 + \dfrac{9}{x - 2}$$

6. Possible rational zeros:

$\pm 24, \pm 12, \pm 8, \pm 6, \pm 4, \pm 3, \pm 2, \pm 1, \pm\frac{3}{2}, \pm\frac{1}{2}$

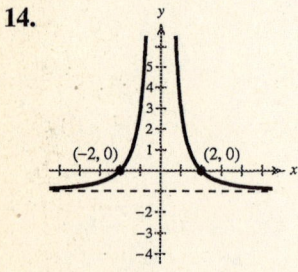

Rational zeros: $-2, \frac{3}{2}$

7. Possible rational zeros:

$\pm 2, \pm 1, \pm\frac{2}{3}, \pm\frac{1}{3}$

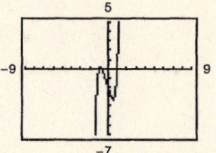

Rational zeros: $\pm 1, -\frac{2}{3}$

8. $(-8 - 3i) + (-i - 15i) = -9 - 18i$

9. $(2 + i)(6 - i) = 12 + 6i - 2i + 1 = 13 + 4i$

10. $(4 + 3i) - (5 + i) = -1 + 2i$

11. $\dfrac{3i}{7 + i} \cdot \dfrac{7 - i}{7 - i} = \dfrac{21i + 3}{49 + 1} = \dfrac{3}{50} + \dfrac{21}{50}i$

12. Real zeros: $1.380, -0.819$

13. Real zeros: $-1.414, -0.667, 1.414$

14.

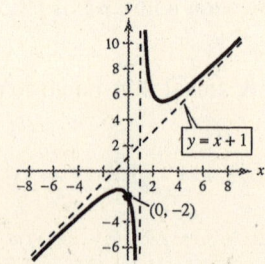

Vertical asymptote: $x = 0$

Intercepts: $(2, 0), (-2, 0)$

Symmetry: y-axis

Horizontal asymptote: $y = -1$

15. $g(x) = \dfrac{x^2 + 2}{x - 1} = x + 1 + \dfrac{3}{x - 1}$

Vertical asymptote: $x = 1$

Intercept: $(0, -2)$

Slant asymptote: $y = x + 1$

16. $f(x) = \dfrac{2x^2 + 9}{5x^2 + 2}$

Horizontal asymptote: $y = \dfrac{2}{5}$

y-axis symmetry

Intercept: $\left(0, \dfrac{9}{2}\right)$

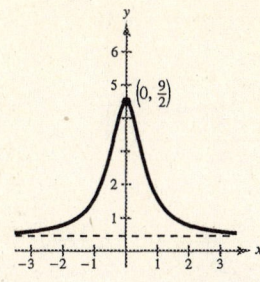

17. (a)

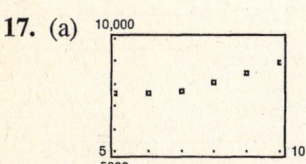

(c)

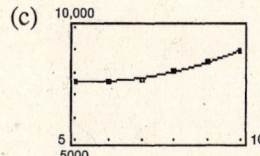

(b) $C = 62.5536t^2 - 654.875t + 9269.1429$

(d) $C = 15{,}000$ when $t \approx 16.1$, or 2006.

Chapter 3   Chapter Test

1. $12.4^{2.79} \approx 1123.690$

2. $4^{3\pi/2} \approx 687.291$

3. $e^{-7/10} \approx 0.497$

4. $e^{3.1} \approx 22.198$

5. $f(x) = 10^{-x}$

x	-3	-1	0	1	2	4
$f(x)$	1000	10	1	0.1	0.01	0.0001

6. $f(x) = -6^{x-2}$

x	-1	0	1	2	3	4	5
$f(x)$	-0.0046	-0.0278	$-.1667$	-1	-6	-36	-216

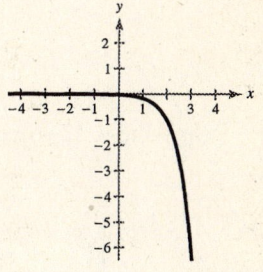

7. $f(x) = 1 - e^{2x}$

x	-3	-1	0	1	2
$f(x)$	0.9975	0.8647	0	-6.389	-53.6

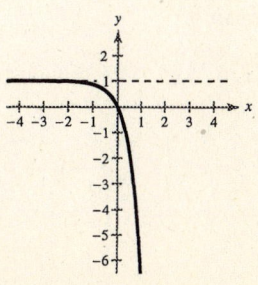

8. (a) $\log_7 7^{-0.89} = -0.89 \log_7 7 = -0.89$

(b) $4.6 \ln e^2 = 4.6(2)\ln e = 9.2$

9. $f(x) = -\log_{10} x - 6$

Domain: $x > 0$

Vertical asymptote: $x = 0$

x-intercept: $(10^{-6}, 0) \approx (0, 0)$

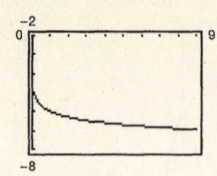

10. $f(x) = \ln(x - 4)$

Domain: $x > 4$

Vertical asymptote: $x = 4$

x-intercept: $(5, 0)$

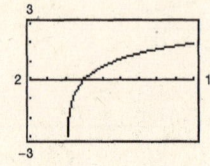

11. $f(x) = 1 + \ln(x + 6)$

Domain: $x > -6$

Vertical asymptote: $x = -6$

x-intercept: $(-5.632, 0)$

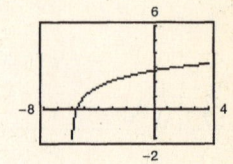

12. $\log_7 44 = \dfrac{\ln 44}{\ln 7} \approx 1.945$

13. $\log_{2/5}(0.9) = \dfrac{\ln(0.9)}{\ln\left(\frac{2}{5}\right)} \approx 0.115$

14. $\log_{24} 68 = \dfrac{\ln 68}{\ln 24} \approx 1.328$

15. $\log_2 3a^4 = \log_2 3 + \log_2 a^4 = \log_2 3 + 4 \log_2 a$

16. $\ln \dfrac{5\sqrt{x}}{6} = \ln 5 + \ln \sqrt{x} - \ln 6 = \ln 5 + \frac{1}{2} \ln x - \ln 6$

17. $\log_3 13 + \log_3 y = \log_3(13y)$

18. $4 \ln x - 4 \ln y = \ln x^4 - \ln y^4 = \ln\left(\dfrac{x^4}{y^4}\right) = \ln\left(\dfrac{x}{y}\right)^4$

19. $\dfrac{1025}{8 + e^{4x}} = 5$

$1025 = 40 + 5e^{4x}$

$985 = 5e^{4x}$

$e^{4x} = 197$

$4x = \ln(197)$

$x = \dfrac{1}{4} \ln(197) \approx 1.321$

20. $\log_{10} - \log_{10}(8 - 5x) = 2$

$\log_{10}\left(\dfrac{x}{8 - 5x}\right) = 2$

$10^2 = \dfrac{x}{8 - 5x}$

$800 - 500x = x$

$800 = 501x$

$x = \dfrac{800}{501} \approx 1.597$

21. $\frac{1}{2} = 1e^{k(22)}$ (half-life is 22 years)

$\ln \frac{1}{2} = 22k$

$k = \frac{1}{22} \ln \frac{1}{2} = -\frac{1}{22} \ln 2 \approx -0.03151$

$A = e^{-0.03151(19)}$

$= e^{-0.03151(19)}$

≈ 0.54953 or 55% remains

22. (a) $y_1 = -0.03095x^2 + 2.3667x + 41.3714$

$y_2 = 44.863(1.0328)^x = 44.8613 \, e^{0.03228x}$

$y_3 = 35.06298 \, x^{0.24661}$

(b)

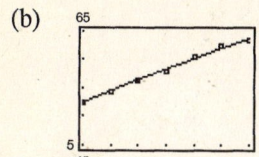 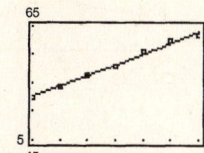 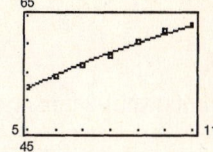

(c) y_1 fits best, although all are good fit

(d) For 2007, $x = 17$ and $y_1 \approx 72.7$ billion dollars

Chapters 1–3   Cumulative Test

1. (a) $y - 8 = \dfrac{-6 - 8}{\frac{1}{2} + 5}(x + 5) = \dfrac{-14}{\frac{11}{2}}(x + 5) = -\dfrac{28}{11}(x + 5)$

$11y - 88 = -28x - 140$

$28x + 11y + 52 = 0$

(b) Three other points: $\left(0, -\frac{52}{11}\right), \left(-\frac{52}{28}, 0\right), \left(1, -\frac{80}{11}\right)$

2. (a) $y - 1 = -2\left(x + \frac{1}{2}\right)$

$y - 1 = -2x - 1$

$y = -2x$

$y + 2x = 0$

(b) Three additional points:

$(0, 0), (1, -2). (2, -4)$

3. (a) Vertical line: $x = -\frac{3}{7}$ or $x + \frac{3}{7} = 0$

(b) Three additional points;

$\left(-\frac{3}{7}, 0\right), \left(-\frac{3}{7}, 1\right), \left(-\frac{3}{7}, 2\right)$

4. (a) $f(6) = \dfrac{6}{6 - 2} = \dfrac{3}{2}$

(b) $f(2)$ is undefined (division by zero).

(c) $f(s + 2) = \dfrac{s + 2}{(s + 2) - 2} = \dfrac{s + 2}{s}$

5. (a) $f\left(-\frac{5}{3}\right) = 3\left(-\frac{5}{3}\right) - 8 = -13$

(b) $f(-1) = 3(-1)^2 + 9(-1) - 8$

$= 3 - 9 - 8 = -14$

(c) $f(0) = 3(0)^2 + 9(0) - 8 = -8$

6. No, for some x there corresponds two values of y.

7.

Decreasing on $(-\infty, 5)$, increasing on $(5, \infty)$

8. (a) $r(x) = \frac{1}{2}\sqrt[3]{x}$ is a vertical shrink of $y = \sqrt[3]{x}$

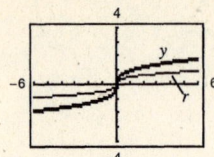

         (b) $h(x) = \sqrt[3]{x} + 2$ is a vertical shift 2 units upward.

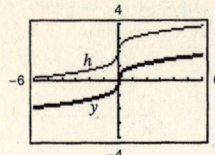

  (c) $g(x) = \sqrt[3]{x + 2}$ is a horizontal shift 2 units to the left.

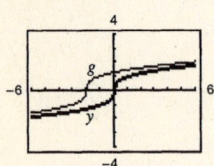

9. $(f + g)(-4) = f(-4) + g(-4) = [-(-4)^2 + 3(-4) - 10] + [4(-4) + 1]$

$$= -38 - 15 = -53$$

10. $(g - f)\left(\frac{3}{4}\right) = \left[4\left(\frac{3}{4}\right) + 1\right] - \left[-\left(\frac{3}{4}\right)^2 + 3\left(\frac{3}{4}\right) - 10\right]$

$$= 4 - (-8.3125) = 12.3125 = \frac{197}{16}$$

11. $(g \circ f)(-2) = g(f(-2)) = g(-20) = 4(-20) + 1 = -79$

12. $(fg)(-1) = f(-1)g(-1) = (-14)(-3) = 42$

13. Yes, $h(x) = 5x - 2$ has an inverse function

$$y = 5x - 2$$

$$x = 5y - 2$$

$$x + 2 = 5y$$

$$\frac{x + 2}{5} = y$$

$$h^{-1}(x) = \frac{x + 2}{5}$$

14. $f(x) = -\frac{1}{2}(x^2 + 4x)$

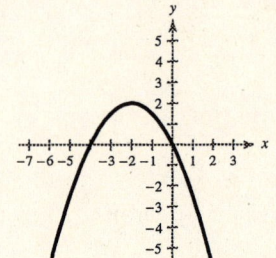

15. $f(x) = \frac{1}{4}x(x - 2)^2$

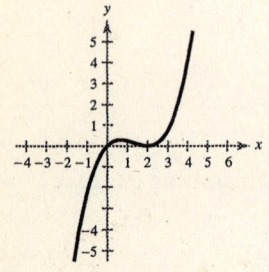

16.

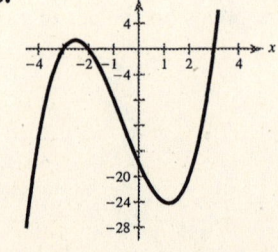

17. $x^3 + 2x^2 + 4x + 8 = (x + 2)(x^2 + 4)$

Zeros: $-2, \pm 2i$

18. Using a graphing utility, $x \approx 1.424$

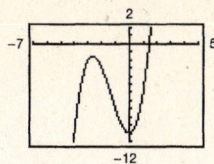

19.

$$
\begin{array}{r}
4x + 2 \\
x + 3 \overline{)\, 4x^2 + 14x - 9} \\
\underline{4x^2 + 12x} \\
2x - 9 \\
\underline{2x + 6} \\
-15
\end{array}
$$

$$\frac{4x^2 + 14x - 9}{x + 3} = 4x + 2 - \frac{15}{x + 3}$$

20.

$$
\begin{array}{r|rrrr}
6 & 2 & -5 & 6 & -20 \\
  &   & 12 & 42 & 288 \\
\hline
  & 2 & 7 & 48 & 268
\end{array}
$$

$$\frac{2x^3 - 5x^2 + 6x - 20}{x - 6} = 2x^2 + 7x + 48 + \frac{268}{x - 6}$$

21.

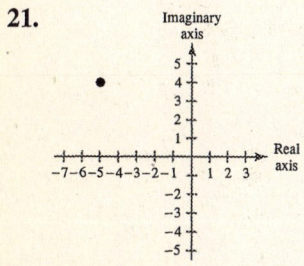

22. $f(x) = (x - 0)(x + 3)\big[x - \big(1 + \sqrt{5}i\big)\big]\big[x - \big(1 - \sqrt{5}i\big)\big]$

$ = x(x + 3)[(x - 1)^2 + 5]$

$ = (x^2 + 3x)(x^2 - 2x + 6)$

$ = x^4 + x^3 + 18x$

23. $f(x) = \dfrac{2x}{x - 3}$

Vertical asymptote: $x = 3$

Horizontal asymptote: $y = 2$

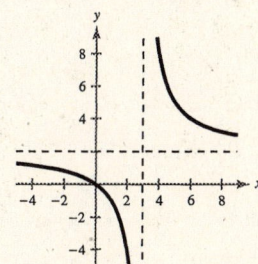

24. $f(x) = \dfrac{5x}{x^2 + x - 6} = \dfrac{5x}{(x + 3)(x - 2)}$

Vertical asymptotes: $x = -3, 2$

Horizontal asymptote: $y = 0$

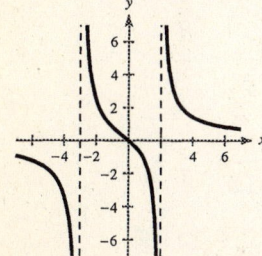

25. $f(x) = \dfrac{x^2 - 3x + 8}{x - 2} = x - 1 + \dfrac{6}{x - 2}$

Vertical asymptote: $x = 2$

Slant asymptote: $y = x - 1$

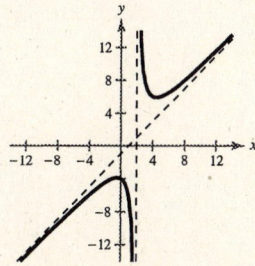

26. $(1.85)^{3.1} \approx 6.733$

27. $58^{\sqrt{5}} \approx 8772.934$

28. $e^{-20/11} \approx 0.162$

29. $4e^{2.56} \approx 51.743$

30. $f(x) = -3^{x+4} - 5$

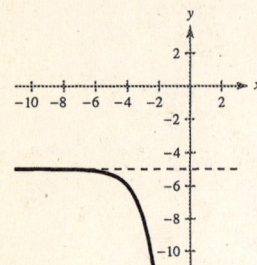

31. $f(x) = -\left(\frac{1}{2}\right)^{-x} - 3$

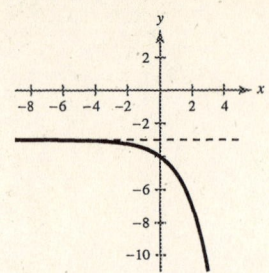

32. $f(x) = 4 + \log_{10}(x - 3)$

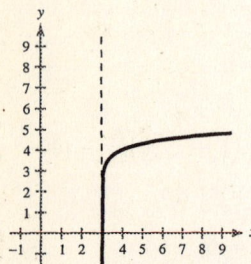

33. $f(x) = \ln(4 - x)$

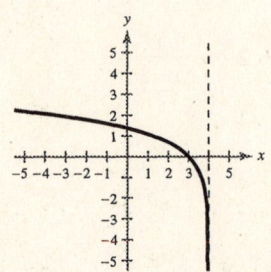

34. $\log_5 21 = \dfrac{\ln 21}{\ln 5} \approx 1.892$

35. $\log_9 6.8 = \dfrac{\ln 6.8}{\ln 9} \approx 0.872$

36. $\log_2\left(\dfrac{3}{2}\right) = \dfrac{\ln\left(\frac{3}{2}\right)}{\ln 2} \approx 0.585$

37. $\log_5\left(\dfrac{x^2 - 16}{x^4}\right) = \log_5[(x - 4)(x + 4)] - \log_5 x^4$

$$= \log_5(x - 4) + \log_5(x - 4) - 4\log_5 x, \quad x > 4$$

38. $2\ln x - \dfrac{1}{2}\ln(x + 5) = \ln x^2 - \ln\sqrt{x + 5}$

$$= \ln\dfrac{x^2}{\sqrt{x + 5}}, \quad x > 0$$

39. $6e^{2x} = 72$

$e^{2x} = 12$

$2x = \ln 12$

$x = \frac{1}{2}\ln 12 \approx 1.242$

40. $4^{x-5} + 21 = 30$

$4^{x-5} = 9$

$(x - 5)\ln 4 = \ln 9$

$x - 5 = \ln 9/\ln 4$

$x = 5 + \dfrac{\ln 9}{\ln 4} \approx 6.585$

41. $\log_2 x + \log_2 5 = 6$

$\log_2 5x = 6$

$5x = 2^6 = 64$

$x = \frac{64}{5} = 12.8$

42. (a) Let x and y be the lengths of the sides $2x + 2y = 546 \implies y = 273 - x$

$A = xy = x(273 - x)$

(b)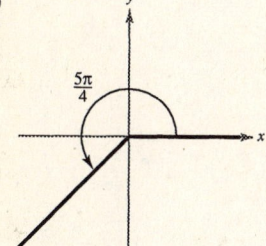

Domain: $0 < x < 273$

(c) If $A = 15000$, then $x = 76.23$ or 196.77

Dimensions in feet:

76.23×196.77 or 196.77×76.23

43. (a) $y = 0.248x^2 - 0.28x + 50.3$

$y = 42.62(1.05)^x = 42.62e^{0.0515x}$

$y = 32.45x^{0.3329}$

(b)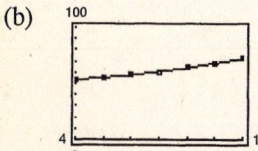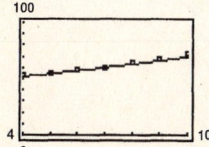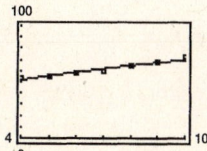

(c) The quadratic model fits best.

(d) In 2006, $x = 16$ and $y \approx 109$ or 109,000 pilots and copilots.

Chapter 4   Chapter Test

1. (a)

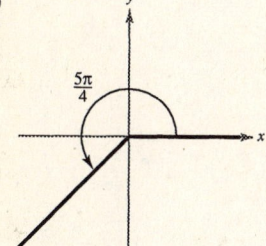

(b) $\dfrac{5\pi}{4} + 2\pi = \dfrac{13\pi}{4}$; $\dfrac{5\pi}{4} - 2\pi = -\dfrac{3\pi}{4}$

(c) $\dfrac{5\pi}{4} \cdot \dfrac{180}{\pi} = 225°$

2. $(90{,}000 \text{ meters/hr})\left(\dfrac{1}{60} \text{ hr/min}\right)\left(\dfrac{2\pi \text{ rad}}{2\pi(1/2) \text{ meters}}\right) = 3000 \text{ rad/min}$

3. $\sin\theta = \dfrac{4}{\sqrt{17}} = \dfrac{4\sqrt{17}}{17}$ $\csc\theta = \dfrac{\sqrt{17}}{4}$

$\cos\theta = -\dfrac{1}{\sqrt{17}} = -\dfrac{\sqrt{17}}{17}$ $\sec\theta = -\sqrt{17}$

$\tan\theta = -4$ $\cot\theta = -\dfrac{1}{4}$

4. $\tan \theta = \dfrac{6}{5}$

$\sin \theta = \pm \dfrac{6}{\sqrt{61}} = \pm \dfrac{6\sqrt{61}}{61}$

$\cos \theta = \pm \dfrac{5}{\sqrt{61}} = \pm \dfrac{5\sqrt{61}}{61}$

$\csc \theta = \pm \dfrac{\sqrt{61}}{6}$

$\sec \theta = \pm \dfrac{\sqrt{61}}{5}$ $\cot \theta = \dfrac{5}{6}$

5. $\theta = 255°$

$\theta' = 255° - 180° = 75°$

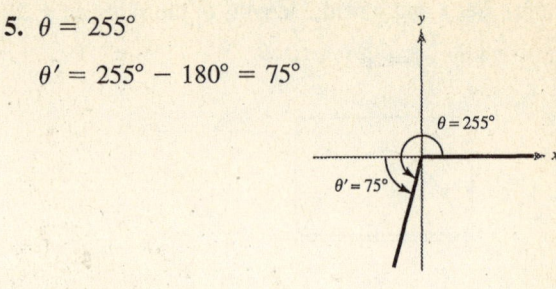

6. $\sec \theta = \dfrac{1}{\cos \theta} < 0 \implies$ Quadrants II or III

$\tan \theta > 0 \implies$ Quadrants I or III

Hence, Quadrant III.

7. If $\cos \theta = -\dfrac{\sqrt{3}}{2}$, then θ is in Quadrant II or III.

$\theta = 150°, 210°$

8. $\csc \theta = \dfrac{1}{\sin \theta} = 1.030 \implies \sin \theta = \dfrac{1}{1.030}$ and θ in

Quadrant I or II. Using a calculator, $\theta = 1.33$, 1.81 radians.

9. $\cos \theta = -\dfrac{3}{5}$, $\sin \theta > 0$, Quadrant II

$\sin \theta = \dfrac{4}{5}$ $\tan \theta = -\dfrac{4}{3}$

$\sec \theta = -\dfrac{5}{3}$ $\csc \theta = \dfrac{5}{4}$

$\cot \theta = -\dfrac{3}{4}$

10. $g(x) = -2 \sin\left(x - \dfrac{\pi}{4}\right)$

Amplitude: 2, shifted $\pi/4$ to the right

11.

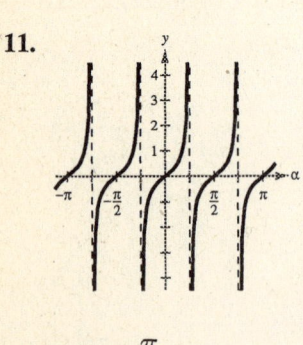

Period: $\dfrac{\pi}{2}$

12. $f(x) = \dfrac{1}{2} \sec(x - \pi)$ is $y = \dfrac{1}{2} \sec x$

shifted π to the right

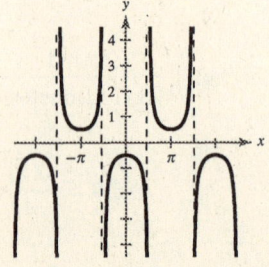

13. $f(x) = 2\cos(\pi - 2x) + 3 = 2\cos(2x - \pi) + 3$

Amplitude: 2

Shifted $\dfrac{\pi}{2}$ to the right, period π

Shifted vertically upward 3

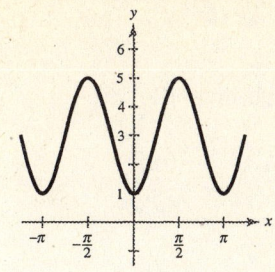

14. $f(x) = 2\csc\left(x + \dfrac{\pi}{2}\right)$

Shifted $\dfrac{\pi}{2}$ to the left

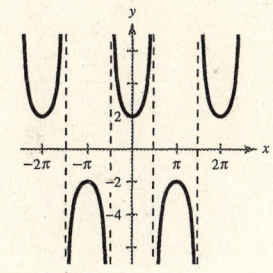

15. $f(x) = \dfrac{1}{4}\cot\left(x - \dfrac{\pi}{2}\right)$

Shifted $\dfrac{\pi}{2}$ to the right

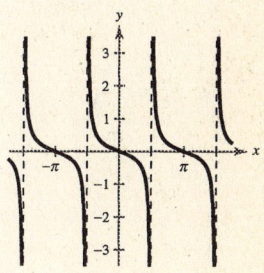

16.

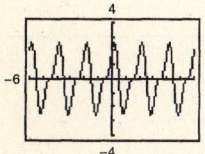

Period is 2.

17.

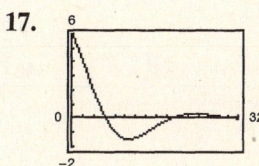

Not periodic

18. Amplitude: 2

Reflected in x-axis $\implies a = -2$

Period 4π and shifted to the right:

$y = -2\sin\left(\dfrac{x}{2} - \dfrac{\pi}{4}\right)$

19. Let $u = \arccos\frac{2}{3} \implies \cos u = \frac{2}{3}$. Then

$$\tan\left(\arccos\frac{2}{3}\right) = \tan u = \frac{\sqrt{5}}{2}.$$

20.

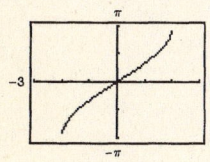

21. $f(x) = 2\arccos x$

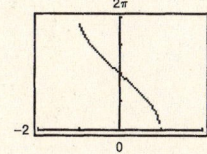

22. $f(x) = \arctan\left(\dfrac{x}{2}\right)$

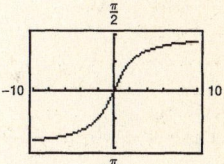

23. $\tan\theta = \dfrac{110}{160} \implies \theta \approx 34.5°$

Bearing: S 34.5° W

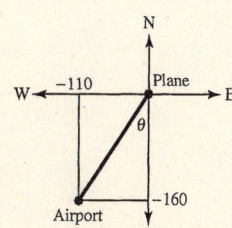

Chapter 5   Chapter Test

1. $\tan \theta = \dfrac{6}{5}, \cos \theta < 0 \implies$ Quadrant III

$\tan^2 \theta + 1 = \sec^2 \theta \implies \dfrac{36}{25} + 1 = \dfrac{61}{25} = \sec^2 \theta \implies \sec \theta = -\dfrac{\sqrt{61}}{5}$

$\cos \theta = \dfrac{-5}{\sqrt{61}} = \dfrac{-5\sqrt{61}}{61}$

$\sin \theta = \tan \theta \cos\theta = \dfrac{6}{5}\left(\dfrac{-5}{\sqrt{61}}\right) = \dfrac{-6}{\sqrt{61}} = \dfrac{-6\sqrt{61}}{61}$

$\csc \theta = \dfrac{\sqrt{61}}{-6}$

$\cot \theta = \dfrac{5}{6}$

2. $\csc^2 \beta(1 - \cos^2 \beta) = \dfrac{1}{\sin^2 \beta} \cdot \sin^2 \beta = 1$

3. $\dfrac{\sec^4 x - \tan^4 x}{\sec^2 x + \tan^2 x} = \dfrac{[(\sec^2 x) + (\tan^2 x)][\sec^2 x - \tan^2 x]}{\sec^2 x + \tan^2 x} = \sec^2 x - \tan^2 x = 1$

4. $\dfrac{\cos \theta}{\sin \theta} + \dfrac{\sin \theta}{\cos \theta} = \dfrac{\cos^2 \theta + \sin^2 \theta}{\sin \theta \cos \theta} = \dfrac{1}{\sin \theta \cos \theta} = \csc \theta \sec \theta$

5. Since $\tan^2 \theta = \sec^2 \theta - 1$ for all θ, then $\tan \theta = -\sqrt{\sec^2 \theta - 1}$ in Quadrants II and IV. Thus, $\dfrac{\pi}{2} < \theta \leq \pi$ and $3\pi/2 < \theta < 2\pi$.

6. The graph appears equal.

Analytically:

$y_1 = \cos x + \sin x \tan x$

$= \cos x + \sin x\left(\dfrac{\sin x}{\cos x}\right)$

$= \dfrac{\cos^2 x + \sin^2 x}{\cos x}$

$= \dfrac{1}{\cos x} = \sec x = y_2$

7. $\sin \theta \cdot \sec \theta = \sin \theta \dfrac{1}{\cos \theta} = \tan \theta$

8. $\sec^2 x \tan^2 x + \sec^2 x = \sec^2 x(\tan^2 x + 1) = \sec^4 x$

9. $\dfrac{\csc \alpha + \sec \alpha}{\sin \alpha + \cos \alpha} = \dfrac{\dfrac{1}{\sin \alpha} + \dfrac{1}{\cos \alpha}}{\sin \alpha + \cos \alpha} = \dfrac{\dfrac{\cos \alpha + \sin \alpha}{\sin \alpha \cdot \cos \alpha}}{(\sin \alpha + \cos \alpha)}$

$= \dfrac{1}{\sin \alpha \cos \alpha} = \dfrac{\cos^2 \alpha + \sin^2 \alpha}{\sin \alpha \cos \alpha}$

$= \dfrac{\cos \alpha}{\sin \alpha} + \dfrac{\sin \alpha}{\cos \alpha} = \cot \alpha + \tan \alpha$

10. $\cos\left(x + \dfrac{\pi}{2}\right) = \cos x \cos \dfrac{\pi}{2} - \sin x \sin \dfrac{\pi}{2}$

$= 0 - \sin x = -\sin x$

11. $\cos(\pi - \theta) + \sin\left(\dfrac{\pi}{2} + \theta\right) = (\cos\pi\cos\theta + \sin\pi\sin\theta) + \left(\sin\dfrac{\pi}{2}\cos\theta + \sin\theta\cos\dfrac{\pi}{2}\right)$

$$= -\cos\theta + \cos\theta = 0$$

12. $(\sin x + \cos x)^2 = \sin^2 x + \cos^2 x + 2\sin x\cos x$

$$= 1 + \sin 2x$$

13. $\tan 105° = \tan(45° + 60°)$

$$= \dfrac{\tan 45° + \tan 60°}{1 - \tan 45°\tan 60°}$$

$$= \dfrac{1 + \sqrt{3}}{1 - \sqrt{3}} = -2 - \sqrt{3}$$

14. $\sin^4 x \tan^2 x = \dfrac{\sin^6 x}{\cos^2 x}$

$$= \dfrac{1}{32} \cdot \dfrac{(10 - 15\cos 2x + 6\cos 4x - \cos 6x)}{(1 + \cos 2x)/2}$$

$$= \dfrac{1}{16}\left[\dfrac{10 - 15\cos 2x + 6\cos 4x - \cos 6x}{1 + \cos 2x}\right]$$

15. $\dfrac{\sin 4\theta}{1 + \cos 4\theta} = \tan\dfrac{4\theta}{2} = \tan 2\theta$

16. $6\sin 4\theta\sin 6\theta = 6\left[\tfrac{1}{2}(\cos(4\theta - 6\theta) - \cos(4\theta + 6\theta)\right]$

$$= 3[\cos(-2\theta) - \cos 10\theta]$$

$$= 3\cos 2\theta - 3\cos 10\theta$$

17. $\cos 5\theta + \cos 3\theta = 2\left(\cos\dfrac{5\theta + 3\theta}{2}\cos\dfrac{5\theta - 3\theta}{2}\right) = 2\cos 4\theta\cos\theta$

18. $\tan^2 x + \tan x = 0$

$\tan x(\tan x + 1) = 0$

$\tan x = 0 \implies x = 0, \pi$

$\tan x + 1 = 0 \implies \tan x = -1 \implies x = \dfrac{3\pi}{4}, \dfrac{7\pi}{4}$

19. $\sin 2\alpha - \cos\alpha = 0$

$2\sin\alpha\cos\alpha - \cos\alpha = 0$

$\cos\alpha(2\sin\alpha - 1) = 0$

$\cos\alpha = 0 \implies a = \dfrac{\pi}{2}, \dfrac{3\pi}{2}$

$2\sin\alpha - 1 = 0 \implies \sin a = \dfrac{1}{2} \implies \alpha = \dfrac{\pi}{6}, \dfrac{5\pi}{6}$

20. $4\cos^2 x - 3 = 0$

$\cos^2 x = \dfrac{3}{4}$

$\cos x = \pm\dfrac{\sqrt{3}}{2}$

$x = \dfrac{\pi}{6}, \dfrac{5\pi}{6}, \dfrac{7\pi}{6}, \dfrac{11\pi}{6}$

21. $\csc^2 x - \csc x - 2 = 0$

$(\csc x - 2)(\csc x + 1) = 0$

$\csc x - 2 = 0 \implies \csc x = 2 \implies \sin x = \dfrac{1}{2} \implies x = \dfrac{\pi}{6}, \dfrac{5\pi}{6}$

$\csc x + 1 = 0 \implies \csc x = -1 \implies \sin x = -1 \implies x = \dfrac{3\pi}{2}$

22. Let $y = 5\cos x - x$ on $[0, 2\pi)$.

The zero is $x \approx 1.306$

23. $\sin 2u = 2\sin u \cos u = 2\dfrac{2}{\sqrt{5}} \cdot \dfrac{1}{\sqrt{5}} = \dfrac{4}{5}$

$\tan 2u = \dfrac{2\tan u}{1 - \tan^2 u} = \dfrac{2(2)}{1 - 2^2} = -\dfrac{4}{3}$

$\cos 2u = 1 - 2\sin^2 u = 1 - 2\left(\dfrac{2}{\sqrt{5}}\right)^2 = -\dfrac{3}{5}$

24. $n = \dfrac{\sin\left(\dfrac{\theta}{2} + \dfrac{\alpha}{2}\right)}{\sin\left(\dfrac{\theta}{2}\right)}$

$\dfrac{3}{2} = \dfrac{\sin\left(\dfrac{\theta}{2} + 30°\right)}{\sin\left(\dfrac{\theta}{2}\right)}$

$3\sin\dfrac{\theta}{2} = 2\left[\sin\dfrac{\theta}{2}\cos 30° + \cos\dfrac{\theta}{2}\sin 30°\right]$

$3\sin\dfrac{\theta}{2} = \sqrt{3}\sin\dfrac{\theta}{2} + \cos\dfrac{\theta}{2}$

$\left(3 - \sqrt{3}\right)\sin\dfrac{\theta}{2} = \cos\dfrac{\theta}{2}$

$\tan\dfrac{\theta}{2} = \dfrac{1}{3 - \sqrt{3}}$

$\dfrac{\theta}{2} = \arctan\left(\dfrac{1}{3 - \sqrt{3}}\right) \approx 38.26°$

$\theta \approx 76.52°$

Chapter 6   Chapter Test

1. Given: $A = 36°, B = 98°, c = 18$

$C = 180° - 36° - 98° = 46°$

$a = \dfrac{c}{\sin C}\sin A = \dfrac{18}{\sin 46°}\sin 36° \approx 14.71$

$b = \dfrac{c}{\sin C}\sin B = \dfrac{18}{\sin 46°}\sin 98° \approx 24.78$

2. Given: $a = 4, b = 7, c = 9$

$$\cos A = \frac{b^2 + c^2 - a^2}{2bc} = \frac{7^2 + 9^2 - 4^2}{2(7)(9)} \approx 0.90476 \Rightarrow A \approx 25.2°$$

$$\sin C = \frac{\sin A}{a}c = \frac{\sin 25.2°}{4}(9) \approx 0.9580 \Rightarrow C \approx 106.6°$$

$$\sin B = \frac{\sin A}{a}b = \frac{\sin 25.2°}{4}(7) \approx 0.7451 \Rightarrow B \approx 48.2°$$

3. Given: $A = 35°, b = 8, c = 11$

$$a^2 = b^2 + c^2 - 2bc \cos A = 8^2 + 11^2 - 2(8)(11) \cos 35° \approx 40.8292 \Rightarrow a \approx 6.39$$

$$\sin B = \frac{\sin A}{a}b = \frac{\sin 35°}{6.39}(8) \approx 0.7181 \Rightarrow B \approx 45.9°$$

$$C = 180° - A - B \approx 99.1°$$

4. Given: $A = 25°, b = 28, a = 15$

$$h = b \sin A \approx 11.83 < a < b \Rightarrow \text{Two possible triangles}$$

$$\sin B = \frac{\sin A}{a}(b) = 0.7889 \Rightarrow B = 52.1° \text{ or } 127.9°$$

For $B = 52.1°$: $C = 180° - 52.1° - 25° = 102.9°$

$$c = \frac{a}{\sin A}\sin C = \frac{15}{\sin 25°}\sin 102.9° \approx 34.59$$

For $B = 127.9°$: $C = 180° - 127.9° - 25° = 27.1°$

$$c = \frac{a}{\sin A}\sin C = \frac{15}{\sin 25°}\sin 27.1° \approx 16.16$$

5. No triangle possible $(5.2 \leq 10.1)$

6. $\sin B = \dfrac{\sin A}{a}b = \dfrac{\sin 150°}{9.4}4.8$

$\qquad \approx 0.2553 \Rightarrow B \approx 14.8°$

$C = 180° - A - B = 15.2$

$c = \dfrac{a}{\sin A}\sin C \approx 4.9$

7. Law of Cosines:

$$a^2 = b^2 + c^2 - 2bc \cos \theta$$
$$= 565^2 + 480^2 - 2(565)(480) \cos 80°$$
$$= 455{,}438.2 \Rightarrow a \approx 674.9 \text{ ft}$$

8. $s = \dfrac{a + b + c}{2} = \dfrac{55 + 85 + 100}{2} = 120$

$A = \sqrt{s(s - a)(s - b)(s - c)}$

$\quad = \sqrt{120(65)(35)(20)}$

$\quad \approx 2336.7$ square meters

9. $\mathbf{w} = \langle 4 - (-8), 1 - (-12) \rangle = \langle 12, 13 \rangle$

$\|\mathbf{w}\| = \sqrt{12^2 + 13^2} = \sqrt{313} \approx 17.69$

10. (a) $2\mathbf{v} + \mathbf{u} = 2\langle -2, 4 \rangle + \langle 0, -4 \rangle = \langle -4, 4 \rangle$

(b) $\mathbf{u} - 3\mathbf{v} = \langle 0, -4 \rangle - 3\langle -2, 4 \rangle = \langle 6, -16 \rangle$

(c) $5\mathbf{u} - \mathbf{v} = 5\langle 0, -4 \rangle - \langle -2, 4 \rangle = \langle 2, -24 \rangle$

11. (a) $2\mathbf{v} + \mathbf{u} = 2\langle 1, 5 \rangle + \langle -2, -3 \rangle = \langle 0, 7 \rangle$

(b) $\mathbf{u} - 3\mathbf{v} = \langle -2, -3 \rangle - 3\langle 1, 5 \rangle = \langle -5, -18 \rangle$

(c) $5\mathbf{u} - \mathbf{v} = 5\langle -2, -3 \rangle - \langle 1, 5 \rangle = \langle -11, -20 \rangle$

12. (a) $2\mathbf{v} + \mathbf{u} = 2(6\mathbf{i} + 9\mathbf{j}) + (\mathbf{i} - \mathbf{j}) = 13\mathbf{i} + 17\mathbf{j}$

(b) $\mathbf{u} - 3\mathbf{v} = (\mathbf{i} - \mathbf{j}) - 3(6\mathbf{i} + 9\mathbf{j}) = -17\mathbf{i} - 28\mathbf{j}$

(c) $5\mathbf{u} - \mathbf{v} = 5(\mathbf{i} - \mathbf{j}) - (6\mathbf{i} + 9\mathbf{j}) = -\mathbf{i} - 14\mathbf{j}$

13. (a) $2\mathbf{v} + \mathbf{u} = 2(-\mathbf{i} - 2\mathbf{j}) + (2\mathbf{i} + 3\mathbf{j}) = -\mathbf{j}$

(b) $\mathbf{u} - 3\mathbf{v} = (2\mathbf{i} + 3\mathbf{j}) - 3(-\mathbf{i} - 2\mathbf{j}) = 5\mathbf{i} + 9\mathbf{j}$

(c) $5\mathbf{u} - \mathbf{v} = 5(2\mathbf{i} + 3\mathbf{j}) - (-\mathbf{i} - 2\mathbf{j}) = 11\mathbf{i} + 17\mathbf{j}$

14. Unit vector $= \dfrac{\mathbf{v}}{\|\mathbf{v}\|} = \dfrac{1}{\sqrt{49 + 16}}\langle 7, 4 \rangle$

$= \left\langle \dfrac{7}{\sqrt{65}}, \dfrac{4}{\sqrt{65}} \right\rangle = \dfrac{\sqrt{65}}{65}\langle 7, 4 \rangle$

15. $12\dfrac{\langle 3, -5 \rangle}{\|\langle 3, -5 \rangle\|} = \dfrac{12}{\sqrt{34}}\langle 3, -5 \rangle = \left\langle \dfrac{36}{\sqrt{34}}, \dfrac{-60}{\sqrt{34}} \right\rangle$

$= \left\langle \dfrac{18\sqrt{34}}{17}, \dfrac{-30\sqrt{34}}{17} \right\rangle$

16. $250(\cos 45°\mathbf{i} + \sin 45°\mathbf{j})$ first force

$130(\cos(-60°)\mathbf{i} + \sin(-60°)\mathbf{j})$ second force

Resultant: $\left[250\left(\dfrac{\sqrt{2}}{2}\right) + 130\left(\dfrac{1}{2}\right) \right]\mathbf{i} + \left[250\left(\dfrac{\sqrt{2}}{2}\right) + 130\left(-\dfrac{\sqrt{3}}{2}\right) \right]\mathbf{j}$

$= \left(125\sqrt{2} + 65\right)\mathbf{i} + \left(125\sqrt{2} - 65\sqrt{3}\right)\mathbf{j}$

Magnitude: $\sqrt{\left(125\sqrt{2} + 65\right)^2 + \left(125\sqrt{2} - 65\sqrt{3}\right)^2} \approx 250.15$

Direction: $\theta = \arctan\left(\dfrac{125\sqrt{2} - 65\sqrt{3}}{125\sqrt{2} + 65}\right) \Rightarrow \theta \approx 14.9°$

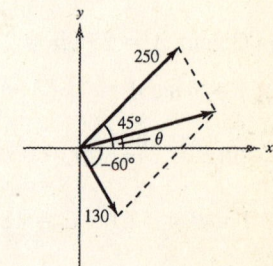

17. $\mathbf{u} \cdot \mathbf{v} = \langle -9, 4 \rangle \cdot \langle 1, 3 \rangle = -9 + 12 = 3$

18. $\cos \theta = \dfrac{\mathbf{u} \cdot \mathbf{v}}{\|\mathbf{u}\| \, \|\mathbf{v}\|} = \dfrac{-8}{\sqrt{53}(4)} \Rightarrow \theta \approx 105.9°$

19. No, the dot product is 24, not 0.

20. $\text{proj}_\mathbf{v}\mathbf{u} = \dfrac{-37}{26}\langle -5, -1 \rangle = \left\langle \dfrac{185}{26}, \dfrac{37}{26} \right\rangle = \mathbf{w}_1$

$\mathbf{w}_2 = \mathbf{u} - \mathbf{w}_1 = \langle 6, 7 \rangle - \left\langle \dfrac{185}{26}, \dfrac{37}{26} \right\rangle = \left\langle \dfrac{-29}{26}, \dfrac{145}{26} \right\rangle$

$\mathbf{u} = \mathbf{w}_1 + \mathbf{w}_2$

21. $|z| = 2\sqrt{2}$

$z = 2\sqrt{2}\left(\cos \dfrac{3\pi}{4} + i \sin \dfrac{3\pi}{4}\right)$

22. $100(\cos 240° + i \sin 240°) = -50 - 50\sqrt{3}i$

23. $\left[3\left(\cos\dfrac{5\pi}{6} + i \sin\dfrac{5\pi}{6}\right) \right]^8 = 3^8\left(\cos\dfrac{40\pi}{6} + i \sin\dfrac{40\pi}{6}\right)$

$= 3^8\left(-\dfrac{1}{2} + \dfrac{\sqrt{3}}{2}i\right)$

$= -3280.5 + 3280.5\sqrt{3}i = -\dfrac{6561}{2} + \dfrac{6561}{2}\sqrt{3}i$

24. $(3 - 3i)^6 = \left[3\sqrt{2}\left(\cos\frac{7\pi}{4} + i\sin\frac{7\pi}{4} \right) \right]^6$

$$= 5832\left(\cos\frac{42\pi}{4} + i\sin\frac{42\pi}{4} \right) = 5832i$$

25. $128\left(1 + \sqrt{3}i\right) = 256\left(\frac{1}{2} + \frac{\sqrt{3}}{2}i\right) = 256\left(\cos\frac{\pi}{3} + i\sin\frac{\pi}{3} \right).$

4th roots: $\sqrt[4]{256}\left(\cos\dfrac{\frac{\pi}{3} + 2\pi k}{4} + i\sin\dfrac{\frac{\pi}{3} + 2k\pi}{4} \right), k = 0, 1, 2, 3$

4 roots are: $4\left(\cos\frac{\pi}{12} + i\sin\frac{\pi}{12} \right) \approx 3.8637 + 1.0353i$

$4\left(\cos\frac{7\pi}{12} + i\sin\frac{7\pi}{12} \right) \approx -1.0353 + 3.8637i$

$4\left(\cos\frac{13\pi}{12} + i\sin\frac{13\pi}{12} \right) \approx -3.8637 - 1.0353i$

$4\left(\cos\frac{19\pi}{12} + i\sin\frac{19\pi}{12} \right) \approx 1.0353 - 3.8637i$

26. $x^4 = 625i.$ Fourth roots of $625i = 625\left(\cos\frac{\pi}{2} + i\sin\frac{\pi}{2} \right)$

$\sqrt[4]{625}\left(\cos\left(\dfrac{\frac{\pi}{2} + 2\pi k}{4}\right) + i\sin\left(\dfrac{\frac{\pi}{2} + 2\pi k}{4}\right) \right), k = 0, 1, 2, 3$

4 roots are $5\left(\cos\frac{\pi}{8} + i\sin\frac{\pi}{8} \right)$

$5\left(\cos\frac{5\pi}{8} + i\sin\frac{5\pi}{8} \right)$

$5\left(\cos\frac{9\pi}{8} + i\sin\frac{9\pi}{8} \right)$

$5\left(\cos\frac{13\pi}{8} + i\sin\frac{13\pi}{8} \right)$

Chapters 4–6   Cumulative Test

1. (a)

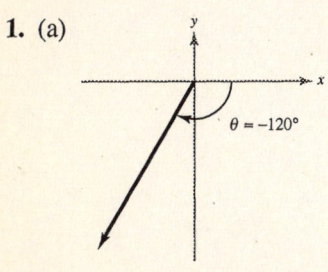

(b) $-120° + 360° = 240°$

(c) $-120° \cdot \dfrac{\pi}{180}\dfrac{\text{rad}}{\text{deg}} = -\dfrac{2\pi}{3}$

(d) $\theta' = 60°$

(e) $\sin\theta = -\dfrac{\sqrt{3}}{2}, \quad \cos\theta = -\dfrac{1}{2}, \quad \tan\theta = \sqrt{3}$

$\csc\theta = -\dfrac{2}{\sqrt{3}} = -\dfrac{2\sqrt{3}}{3}, \quad \sec\theta = -2,$

$\cot\theta = \dfrac{1}{\sqrt{3}} = \dfrac{\sqrt{3}}{3}$

2. $2.35 \text{ radians}\left(\dfrac{180}{\pi}\right) \approx 134.6$

3. $\tan \theta = -\dfrac{4}{3} \implies \sec^2 \theta = \tan^2 \theta + 1 = \dfrac{16}{9} + 1 = \dfrac{25}{9}$

$\implies \sec \theta = -\dfrac{5}{3} \text{ (Quadrant II)} \implies \cos \theta = -\dfrac{3}{5}$

4. $f(x) = 3 - 2 \sin \pi x$

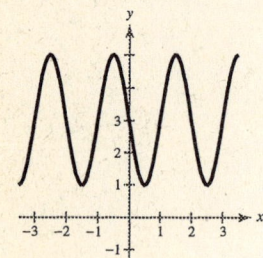

5. $f(x) = \tan(3x)$

Period: $\dfrac{\pi}{3}$

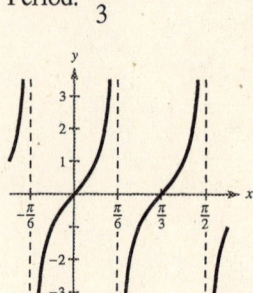

6. $f(x) = \dfrac{1}{2} \sec(x + \pi)$

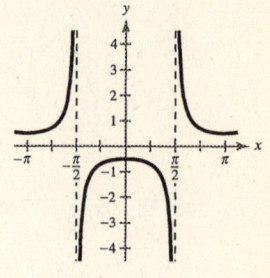

7. Amplitude: 3

Cosine curve reflected about the x-axis

Period: $2 \implies h(x) = -3 \cos(\pi x)$

Answer: $a = -3, b = \pi, c = 0$

8. $\tan(\arctan 6.7) = 6.7$

9. $\tan\left(\arcsin \dfrac{3}{5}\right) = \tan y = \dfrac{3}{4}$

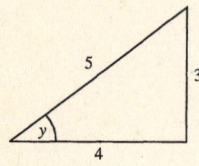

10. Let $u = \arccos 2x \implies \cos u = 2x$. Then

$\sin(\arccos 2x) = \sin u = \sqrt{1 - 4x^2}$.

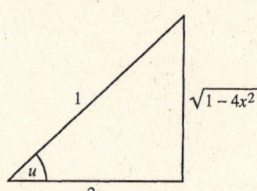

11. $\dfrac{\sin \theta - 1}{\cos \theta} - \dfrac{\cos \theta}{\sin \theta - 1} = \dfrac{\sin^2 \theta - 2 \sin \theta + 1 - \cos^2 \theta}{\cos \theta(\sin \theta - 1)}$

$= \dfrac{\sin^2 \theta - 2 \sin \theta + \sin^2 \theta}{\cos \theta(\sin \theta - 1)}$

$= \dfrac{2 \sin \theta(\sin \theta - 1)}{\cos \theta(\sin \theta - 1)} = 2 \tan \theta$

12. $\cot^2 \alpha(\sec^2 \alpha - 1) = \cot^2 \alpha(\tan^2 \alpha) = 1$

13. $\sin(x + y) \sin(x - y) = [\sin x \cos y + \cos x \sin y][\sin x \cos y - \sin y \cos x]$

$= \sin^2 x \cos^2 y - \sin^2 y \cos^2 x$

$= \sin^2 x(1 - \sin^2 y) - \sin^2 y(1 - \sin^2 x)$

$= \sin^2 x - \sin^2 x \sin^2 y - \sin^2 y + \sin^2 y \sin^2 x$

$= \sin^2 x - \sin^2 y$

14. $\sin^2 x \cos^2 x = \dfrac{1}{4}(2 \sin x \cos x)^2$

$\qquad\qquad = \dfrac{1}{4}(\sin 2x)^2$

$\qquad\qquad = \dfrac{1}{4} \cdot \dfrac{1 - \cos 4x}{2}$

$\qquad\qquad = \dfrac{1}{8}(1 - \cos 4x)$

15. $\sin^2 x + 2 \sin x + 1 = 0$

$\qquad\quad (\sin x + 1)^2 = 0$

$\qquad\qquad \sin x + 1 = 0$

$\qquad\qquad\quad \sin x = -1$

$\qquad\qquad\qquad x = \dfrac{3\pi}{2} + 2n\pi$

16. $3 \tan \theta - \cot \theta = 0$

$\quad 3 \tan \theta - \dfrac{1}{\tan \theta} = 0$

$\quad 3 \tan^2 \theta - 1 = 0$

$\qquad \tan \theta = \pm\dfrac{1}{\sqrt{3}} \implies \theta = \dfrac{\pi}{6} + n\pi, \dfrac{5\pi}{6} + n\pi$

17. Graph $y = \cos^2 x - 5 \cos x - 1$ on $[0, 2\pi)$.

$\quad$ Roots are $x \approx 1.7646, 4.5186$

18.

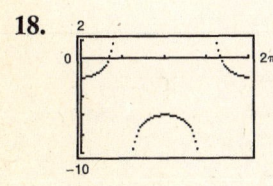

Zeros: $x \approx 1.047, 5.236$

Algebraically, $\qquad \dfrac{1 + \sin x}{\cos x} + \dfrac{\cos x}{1 + \sin x} = 4$

$$\dfrac{1 + 2 \sin x + \sin^2 x + \cos^2 x}{\cos x(1 + \sin x)} = 4$$

$$\dfrac{2 + 2 \sin x}{\cos x(1 + \sin x)} = 4$$

$$\dfrac{2}{\cos x} = 4$$

$$\cos x = \dfrac{1}{2}$$

$$x = \dfrac{\pi}{3}, \dfrac{5\pi}{3}$$

19.

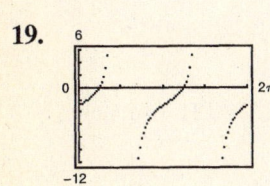

Zeros: $x \approx 0.785, 3.927$

Algebraically, $\qquad \tan^3 x - \tan^2 x + 3 \tan x - 3 = 0$

$$\tan^2 x(\tan x - 1) + 3(\tan x - 1) = 0$$

$$(\tan^2 x + 3)(\tan x - 1) = 0$$

$$\tan x = 1 \implies x = \dfrac{\pi}{4}, \dfrac{5\pi}{4}$$

20. $\sin u = \dfrac{12}{13} \implies \cos u = \dfrac{5}{13} \implies \tan u = \dfrac{12}{5}$

$\quad \cos v = \dfrac{3}{5} \implies \sin v = \dfrac{4}{5} \implies \tan v = \dfrac{4}{3}$

$\quad \tan(u - v) = \dfrac{\tan u - \tan v}{1 + \tan u \tan v} = \dfrac{(12/5) - (4/3)}{1 + (12/5)(4/3)} = \dfrac{16}{63}$

21. $\tan(2\theta) = \dfrac{2 \tan \theta}{1 - \tan^2 \theta} = \dfrac{2(1/2)}{1 - (1/4)} = \dfrac{4}{3}$

22. $\sec^2 \theta = \tan^2 \theta + 1 = \dfrac{25}{9} \implies \sec \theta = -\dfrac{5}{3} \implies \cos \theta = -\dfrac{3}{5}$ (Quadrant III)

$\sin \dfrac{\theta}{2} = \pm \sqrt{\dfrac{1 - \cos \theta}{2}} = \pm \dfrac{2}{\sqrt{5}} = \pm \dfrac{2\sqrt{5}}{5}$

$\dfrac{\theta}{2}$ in Quadrant II: $\sin \dfrac{\theta}{2} = \dfrac{2\sqrt{5}}{5}$

23. $\cos 8x + \cos 4x = 2 \cos\left(\dfrac{8x + 4x}{2}\right) \cos\left(\dfrac{8x - 4x}{2}\right)$

$= 2 \cos 6x \cos 2x$

24. $\tan x(1 - \sin^2 x) = \dfrac{\sin x}{\cos x} \cos^2 x = \sin x \cos x$

$= \dfrac{1}{2}(2 \sin x \cos x) = \dfrac{1}{2} \sin 2x$

25. $\sin 3\theta \sin \theta = \dfrac{1}{2}[\cos(3\theta - \theta) - \cos[3\theta + \theta]]$

$= \dfrac{1}{2}(\cos 2\theta - \cos 4\theta)$

26. $\sin 3x \cos 2x = \dfrac{1}{2}(\sin(3x + 2x) + \sin(3x - 2x))$

$= \dfrac{1}{2}(\sin 5x + \sin x)$

27. $\dfrac{2 \cos 3x}{\sin 4x - \sin 2x} = \dfrac{2 \cos 3x}{2 \cos 3x \cdot \sin x} = \dfrac{1}{\sin x} = \csc x$

28. $\sin B = \dfrac{\sin A}{a} b = 0.2569 \implies B \approx 14.9°$

$C = 180° - 46° - 14.9° = 119.1°$

$c = \dfrac{a}{\sin A}(\sin C) \approx 17.0$

29. $a^2 = b^2 + c^2 - 2bc \cos A = 25.436 \implies a \approx 5.04$

$\sin B = \dfrac{\sin A}{a} \cdot b = 0.7936 \implies B \approx 52.5°$

$C = 180° - 52.5° - 30° = 97.5°$

30. $B = 180° - 24° - 101° = 55°$

$b = \dfrac{a}{\sin A} \sin B \approx 20.14$

$c = \dfrac{a}{\sin A} \sin C \approx 24.13$

31. $\cos A = \dfrac{b^2 + c^2 - a^2}{2bc} = 0.8982 \implies A \approx 26.1°$

$\cos B = \dfrac{a^2 + c^2 - b^2}{2ac} = 0.8355 \implies B \approx 33.3°$

$C = 180° - 26.1° - 33.3° = 120.6°$

32. $A = \dfrac{1}{2}bh = \dfrac{1}{2} 19 \cdot 14 \sin 82° \approx 131.7$ sq inches

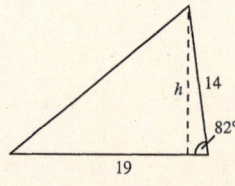

33. $s = \dfrac{a + b + c}{2} = 22$

Area $= \sqrt{s(s - a)(s - b)(s - c)}$

$= \sqrt{22(11)(6)(5)}$

≈ 85.2 square inches

34. $\mathbf{u} = \langle 3, 5 \rangle = 3\mathbf{i} + 5\mathbf{j}$

35. $\|\mathbf{v}\| = \sqrt{2}$

Unit vector: $\left\langle \dfrac{1}{\sqrt{2}}, \dfrac{1}{\sqrt{2}} \right\rangle = \left\langle \dfrac{\sqrt{2}}{2}, \dfrac{\sqrt{2}}{2} \right\rangle$

36. $\mathbf{u} \cdot \mathbf{v} = 3(1) + 4(-2) = -5$

37. $\text{proj}_{\mathbf{v}}\mathbf{u} = \dfrac{8 - 10}{26}\langle 1, 5 \rangle = \left\langle -\dfrac{1}{13}, -\dfrac{5}{13} \right\rangle = \mathbf{w}_1$

$\mathbf{w}_2 = \mathbf{u} - \mathbf{w}_1 = \langle 8, -2 \rangle - \left\langle -\dfrac{1}{13}, -\dfrac{5}{13} \right\rangle = \left\langle \dfrac{105}{13}, -\dfrac{21}{13} \right\rangle$

$\mathbf{u} = \mathbf{w}_1 + \mathbf{w}_2$

38. $|z| = 3\sqrt{2}, \; \theta = \dfrac{3\pi}{4}: \; 3\sqrt{2}\left(\cos \dfrac{3\pi}{4} + i \sin \dfrac{3\pi}{4} \right)$

39. $8\left(-\dfrac{\sqrt{3}}{2} + \dfrac{1}{2}i \right) = -4\sqrt{3} + 4i$

40. $[4(\cos 30° + i \sin 30°)][6(\cos 120° + i \sin 120°)] = 24(\cos(30 + 120°) + i \sin(30° + 120°))$

$$= 24(\cos 150° + i \sin 150°)$$

$$= 24\left(-\dfrac{\sqrt{3}}{2} + i\dfrac{1}{2} \right) = -12\sqrt{3} + 12i$$

41. $1 = 1(\cos 0 + i \sin 0)$

$\cos\left(\dfrac{0 + 2\pi k}{3} \right) + i \sin\left(\dfrac{0 + 2\pi k}{3} \right), \quad k = 0, 1, 2$

$k = 0: \cos 0 + i \sin 0 = 1$

$k = 1: \cos \dfrac{2\pi}{3} + i \sin \dfrac{2\pi}{3} = -\dfrac{1}{2} + \dfrac{\sqrt{3}}{2}i$

$k = 2: \cos \dfrac{4\pi}{3} + i \sin \dfrac{4\pi}{3} = -\dfrac{1}{2} - \dfrac{\sqrt{3}}{2}i$

42. $x^5 = -243$

Five fifth roots of $-243 = 243(\cos \pi + i \sin \pi)$ are:

$\sqrt[5]{243}\left(\cos \dfrac{\pi + 2\pi k}{5} + i \sin \dfrac{\pi + 2\pi k}{5} \right); \; k = 0, 1, 2, 3, 4$

$3\left(\cos \dfrac{\pi}{5} + i \sin \dfrac{\pi}{5} \right)$

$3\left(\cos \dfrac{3\pi}{5} + i \sin \dfrac{3\pi}{5} \right)$

$3\left(\cos \dfrac{5\pi}{5} + i \sin \dfrac{5\pi}{5} \right) = -3$

$3\left(\cos \dfrac{7\pi}{5} + i \sin \dfrac{7\pi}{5} \right)$

$3\left(\cos \dfrac{9\pi}{5} + i \sin \dfrac{9\pi}{5} \right)$

43. $\tan 18° = \dfrac{h}{200}$

$\tan 16° \, 45' = \dfrac{k}{200}$

Hence,

$f = h - k = 200 \tan 18° - 200 \tan 16° \, 45'$

$$\approx 4.8 \approx 5 \text{ feet.}$$

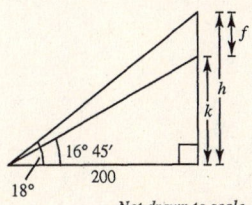

Not drawn to scale

44. $y = 4 \cos\left(\dfrac{\pi}{4}t\right)$ or $y = 4 \sin\left(\dfrac{\pi}{4}t\right)$

Amplitude: 4

Period $\dfrac{2\pi}{(\pi/4)} = 8$

45. Add the two vectors:

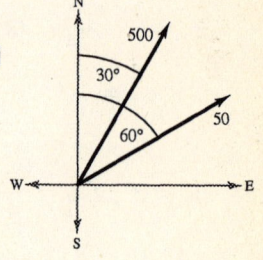

$500(\cos 60°\mathbf{i} + \sin 60°\mathbf{j}) + 50(\cos 30°\mathbf{i} + \sin 30°\mathbf{j}) = \left(250 + 25\sqrt{3}\right)\mathbf{i} + \left(250\sqrt{3} + 25\right)\mathbf{j}$

$\tan \theta = \dfrac{250\sqrt{3} + 25}{250 + 25\sqrt{3}} \approx 1.56 \Rightarrow \theta \approx 57.4°$

Direction: N 32.6° E or 32.6° in airplane navigation

Speed $= \sqrt{\left(250 + 25\sqrt{3}\right)^2 + \left(250\sqrt{3} + 25\right)^2} \approx 543.9$ km/hr

46. $\cos A = \dfrac{60^2 + 125^2 - 100^2}{2(60)(125)} = 0.615 \Rightarrow A \approx 52.05°$

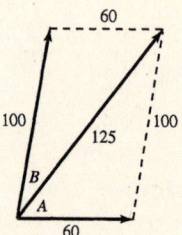

$\cos B = \dfrac{100^2 + 125^2 - 60^2}{2(100)(125)} = 0.881 \Rightarrow B \approx 28.24$

Angle between vectors $= A + B \approx 80.3°$

Chapter 7   Chapter Test

1. $x - y = 6 \Rightarrow y = x - 6$. Then $3x + 5(x - 6) = 2 \Rightarrow 8x = 32 \Rightarrow x = 4, y = 4 - 6 = -2$.

Answer: $(4, -2)$

2. $y = x - 1 = (x - 1)^3 \Rightarrow x = 1$ or $1 = (x - 1)^2 = x^2 - 2x + 1 \Rightarrow x^2 - 2x = 0$.

Thus, $x = 1$ or $x(x - 2) = 0 \Rightarrow x = 0, 1, 2$.

Answer: $(0, -1), (1, 0), (2, 1)$

3. $x - y = 3 \Rightarrow y = x - 3 \Rightarrow$

$$4x - (x - 3)^2 = 7$$
$$4x - (x^2 - 6x + 9) = 7$$
$$x^2 - 10x + 16 = 0$$
$$(x - 2)(x - 8) = 0$$
$$x = 2, 8$$

Answer: $(2, -1), (8, 5)$

4. $\begin{cases} 2x + 5y = -11 & \text{Equation 1} \\ 5x - y = 19 & \text{Equation 2} \end{cases}$

$-\frac{5}{2}$ times Eq. 1 added to Eq. 2 produces

$-\frac{27}{2}y = \frac{93}{2} \implies y = -\frac{31}{9}.$

Then $2x + 5\left(-\frac{31}{9}\right) = -11 \implies x = \frac{28}{9}.$

Answer: $\left(\frac{28}{9}, -\frac{31}{9}\right)$

5. $\begin{cases} x - 2y + 3z = -5 & \text{Equation 1} \\ 2x - z = -4 & \text{Equation 2} \\ 3y + z = 17 & \text{Equation 3} \end{cases}$

$\begin{cases} x - 2y + 3z = -5 \\ 4y - 7z = 6 \\ 3y + z = 17 \end{cases} \quad (-2)\,\text{Eq. 1} + \text{Eq. 2}$

$\begin{cases} x - 2y + 3z = -5 \\ 4y - 7z = 6 \\ \frac{25}{4}z = \frac{50}{4} \end{cases} \quad \left(-\frac{3}{4}\right)\text{Eq. 2} + \text{Eq. 3}$

$z = 2 \implies 4y - 7(2) = 6 \implies y = 5$

$z = 2, y = 5 \implies x - 2(5) + 3(2) = -5 \implies$
$x = -1$

Answer: $(-1, 5, 2)$

6. $\begin{cases} 5x + 5y - z = 0 & \text{Equation 1} \\ 10x + 5y + 2z = 0 & \text{Equation 2} \\ 5x + 15y - 9z = 0 & \text{Equation 3} \end{cases}$

$\begin{cases} 5x + 5y - z = 0 \\ -5y + 4z = 0 \\ 10y - 8z = 0 \end{cases} \quad \begin{array}{l} (-2)\text{Eq. 1} + \text{Eq. 2} \\ (-1)\text{Eq. 1} + \text{Eq. 3} \end{array}$

$\begin{cases} 5x + 5y - z = 0 \\ -5y + 4z = 0 \\ 0 = 0 \end{cases} \quad (2)\text{Eq. 2} + \text{Eq. 3}$

Infinite number of solutions. They are all of the
form $(-3a, 4a, 5a)$ or $\left(-\frac{3}{5}a, \frac{4}{5}a, a\right)$ where a is any
real number.

7. $6 = a(0)^2 + b(0) + c \implies c = 6$

$2 = a(-2)^2 + b(-2) + c$

$\frac{9}{2} = a(3)^2 + b(3) + c$

Hence, $\begin{cases} 4a - 2b + 6 = 2 & \text{or } 2a - b = -2 \\ 9a + 3b + 6 = \frac{9}{2} & \text{or } 9a + 3b = -\frac{3}{2}. \end{cases}$

Solving this system for a and b, you obtain

$a = -\frac{1}{2}, b = 1.$ Thus, $y = -\frac{1}{2}x^2 + x + 6.$

8. $\dfrac{5x - 2}{(x - 1)^2} = \dfrac{A}{x - 1} + \dfrac{B}{(x - 1)^2}$

$5x - 2 = A(x - 1) + B = Ax + (-A + B)$

$\begin{cases} A = 5 \\ -A + B = -2 \implies B = 3 \end{cases}$

$\dfrac{5x - 2}{(x - 1)^2} = \dfrac{5}{x - 1} + \dfrac{3}{(x - 1)^2}$

9. $\begin{bmatrix} 2 & 1 & 2 & \vdots & 4 \\ 2 & 2 & 0 & \vdots & 5 \\ 2 & -1 & 6 & \vdots & 2 \end{bmatrix}$ row reduces to $\begin{bmatrix} 1 & 0 & 2 & \vdots & 1.5 \\ 0 & 1 & -2 & \vdots & 1 \\ 0 & 0 & 0 & \vdots & 0 \end{bmatrix}$

Infinite number of solutions. Let $z = a, y = 2a + 1, x = 1.5 - 2a.$

Answer: $(1.5 - 2a, 1 + 2a, a)$, where a is any real number

10. $\begin{bmatrix} 2 & 3 & 1 & \vdots & 10 \\ 2 & -3 & -3 & \vdots & 22 \\ 4 & -2 & 3 & \vdots & -2 \end{bmatrix}$ row reduces to $\begin{bmatrix} 1 & 0 & 0 & \vdots & 5 \\ 0 & 1 & 0 & \vdots & 2 \\ 0 & 0 & 1 & \vdots & -6 \end{bmatrix}$

Answer: $(5, 2, -6)$

11. (a) $A - B = \begin{bmatrix} 1 & 0 & 4 \\ -7 & -6 & -1 \\ 0 & 4 & 0 \end{bmatrix}$ (b) $3A = \begin{bmatrix} 15 & 12 & 12 \\ -12 & -12 & 0 \\ 3 & 6 & 0 \end{bmatrix}$

(c) $3A - 2B = \begin{bmatrix} 7 & 4 & 12 \\ -18 & -16 & -2 \\ 1 & 10 & 0 \end{bmatrix}$ (d) $AB = \begin{bmatrix} 36 & 20 & 4 \\ -28 & -24 & -4 \\ 10 & 8 & 2 \end{bmatrix}$

12. $A^{-1} = \dfrac{1}{ad - bc} \begin{bmatrix} d & -b \\ -c & a \end{bmatrix} = \dfrac{1}{30 - 40} \begin{bmatrix} -5 & -4 \\ -10 & -6 \end{bmatrix} = \dfrac{1}{10} \begin{bmatrix} 5 & 4 \\ 10 & 6 \end{bmatrix} = \begin{bmatrix} \frac{1}{2} & \frac{2}{5} \\ 1 & \frac{3}{5} \end{bmatrix}$

$X = A^{-1}B = \begin{bmatrix} \frac{1}{2} & \frac{2}{5} \\ 1 & \frac{3}{5} \end{bmatrix} \begin{bmatrix} 10 \\ 20 \end{bmatrix} = \begin{bmatrix} 13 \\ 22 \end{bmatrix} \Rightarrow (x, y) = (13, 22)$

13. $\begin{vmatrix} -25 & 18 \\ 6 & -7 \end{vmatrix} = (-25)(-7) - 6(18) = 67$

14. $\det(A) = \begin{vmatrix} 4 & 0 & 3 \\ 1 & -8 & 2 \\ 3 & 2 & 2 \end{vmatrix} = 4(-16 - 4) - 0 + 3(2 + 24)$

$= -80 + 78 = -2$

15. $\begin{vmatrix} -5 & 0 & 1 \\ 3 & 2 & 1 \\ 4 & 4 & 1 \end{vmatrix} = -5(2 - 4) - 0 + 1(12 - 8) = 10 + 4 = 14$

Area $= \frac{1}{2}(14) = 7$

16. $x = \dfrac{\begin{vmatrix} 11 & 8 \\ 21 & -24 \end{vmatrix}}{\begin{vmatrix} 20 & 8 \\ 12 & -24 \end{vmatrix}} = \dfrac{-432}{-576} = \dfrac{3}{4}$ $y = \dfrac{\begin{vmatrix} 20 & 11 \\ 12 & 21 \end{vmatrix}}{\begin{vmatrix} 20 & 8 \\ 12 & -24 \end{vmatrix}} = \dfrac{288}{-576} = -\dfrac{1}{2}$

Answer: $\left(\frac{3}{4}, -\frac{1}{2}\right)$

17. Upper left: $400 + x_2 = x_1$

Upper right: $x_1 + x_3 = x_4 + 600$

Lower left: $300 = x_2 + x_3 + x_5$

Lower right: $x_5 + x_4 = 100$

$$\begin{cases} x_1 - x_2 & = 400 \\ x_1 \quad\quad + x_3 - x_4 & = 600 \\ \quad x_2 + x_3 \quad\quad + x_5 = 300 \\ \quad\quad\quad\quad x_4 + x_5 = 100 \end{cases}$$

Solving this system,

$$\begin{bmatrix} 1 & -1 & 0 & 0 & 0 & \vdots & 400 \\ 1 & 0 & 1 & -1 & 0 & \vdots & 600 \\ 0 & 1 & 1 & 0 & 1 & \vdots & 300 \\ 0 & 0 & 0 & 1 & 1 & \vdots & 100 \end{bmatrix} \rightarrow \begin{bmatrix} 1 & 0 & 1 & 0 & 1 & \vdots & 700 \\ 0 & 1 & 1 & 0 & 1 & \vdots & 300 \\ 0 & 0 & 0 & 1 & 1 & \vdots & 100 \\ 0 & 0 & 0 & 0 & 0 & \vdots & 0 \end{bmatrix}$$

Letting $x_3 = a$ and $x_5 = b$ be real numbers, we have

$x_5 = b$

$x_4 = 100 - b$

$x_3 = a$

$x_2 = 300 - a - b$

$x_1 = 700 - b - a$

Chapter 8   Chapter Test

1. $a_n = \left(-\frac{2}{3}\right)^{n-1}$ $a_1 = \left(-\frac{2}{3}\right)^{1-1} = \left(-\frac{2}{3}\right)^0 = 1$ **2.** $a_1 = 12,\ a_{k+1} = a_k + 4$

$\qquad\qquad\qquad\qquad\quad a_2 = -\frac{2}{3}$ $\qquad\qquad\qquad a_2 = 12 + 4 = 16$

$\qquad\qquad\qquad\qquad\quad a_3 = \left(-\frac{2}{3}\right)^2 = \frac{4}{9}$ $\qquad\qquad a_3 = 16 + 4 = 20$

$\qquad\qquad\qquad\qquad\quad a_4 = \left(-\frac{2}{3}\right)^3 = -\frac{8}{27}$ $\qquad\qquad a_4 = 20 + 4 = 24$

$\qquad\qquad\qquad\qquad\quad a_5 = \left(-\frac{2}{3}\right)^4 = \frac{16}{81}$ $\qquad\qquad a_5 = 24 + 4 = 28$

3. $\dfrac{11!\,4!}{4!\,7!} = \dfrac{11!}{7!} = \dfrac{11 \cdot 10 \cdot 9 \cdot 8 \cdot 7!}{7!} = 11 \cdot 10 \cdot 9 \cdot 8 = 7920$

4. $a_n = dn + c,\ c = a_1 - d = 5000 - (-100) = 5100$

$\Rightarrow a_n = -100n + 5100 = 5000 - 100(n - 1)$

5. $a_n = a_1 r^{n-1},\ a_1 = 4,\ r = \dfrac{1}{2} \Rightarrow a_n = 4\left(\dfrac{1}{2}\right)^{n-1}$ **6.** $\displaystyle\sum_{n=1}^{12} \dfrac{2}{3n + 1}$

7. $\displaystyle\sum_{n=1}^{7} (8n - 5) = 8\left(\dfrac{7(8)}{2}\right) - 5(7) = 224 - 35 = 189$

8. $\sum_{n=1}^{8} 24\left(\frac{1}{6}\right)^{n-1} = 24\left(\frac{1-\left(\frac{1}{6}\right)^8}{1-\frac{1}{6}}\right) = 24\left(\frac{6}{5}\right)\left(1-\left(\frac{1}{6}\right)^8\right) \approx 28.79998 \approx 28.80$

9. $\sum_{n=1}^{\infty} 5\left(\frac{1}{10}\right)^{n-1} = 5[1 + 0.1 + 0.01 + \ldots] = 5[1.1111\ldots] = 5.555\ldots = \frac{50}{9}$

10. (1) For $n = 1, 3 = \dfrac{3(1)(1+1)}{2}$

(2) Assume $S_k = 3 + 6 + \cdots + 3k = \dfrac{3k(k+1)}{2}$.

Then $S_{k+1} = 3 + 6 + \cdots + 3k + 3(k+1)$

$\qquad\qquad = S_k + 3(k+1)$

$\qquad\qquad = \dfrac{3k(k+1)}{2} + 3(k+1)$

$\qquad\qquad = \dfrac{k+1}{2}[3k+6]$

$\qquad\qquad = \dfrac{3(k+1)(k+2)}{2}$

Therefore, the formula is true for all positive integers n.

11. $(2a - 5b)^4 = (2a)^4 - 4(2a)^3(5b) + 6(2a)^2(5b)^2 - 4(2a)(5b)^3 + (5b)^4$

$\qquad\qquad = 16a^4 - 160a^3b + 600a^2b^2 - 1000ab^3 + 625b^4$

12. $_8C_3(3)^3(2)^5 = 56(27)(32) = 48,384$ **13.** $_9C_3 = 84$

14. $_{20}C_3 = 1140$

15. $_9P_2 = \dfrac{9!}{7!} = 9 \cdot 8 = 72$ **16.** $_{70}P_3 = \dfrac{70!}{67!} = 70 \cdot 69 \cdot 68$

$\qquad\qquad\qquad\qquad\qquad\qquad\qquad\qquad\qquad\qquad\qquad\qquad = 328,440$

17. $26 \cdot 10 \cdot 10 \cdot 10 = 26,000$ ways   **18.** $_{25}C_4 = \dfrac{25!}{21!\,4!} = \dfrac{25 \cdot 24 \cdot 23 \cdot 22}{24} = 12,650$ ways

19. There are 6 red face cards $\Rightarrow$ probability $= \frac{6}{52} = \frac{3}{26}$.

20. There are $_4C_2 = 6$ ways to select 2 spark plugs. Only one way corresponds to both being selected. Probability $= \frac{1}{6}$.

21. (a) $\left(\frac{30}{60}\right)\left(\frac{30}{60}\right) = \frac{1}{2} \cdot \frac{1}{2} = \frac{1}{4}$ (b) $\frac{11}{60} \cdot \frac{11}{60} = \frac{121}{3600} \approx 0.0336$

    (c) $\frac{1}{60} \approx 0.0167$

22. $1 - 0.75 = 0.25 = 25\%$

Chapter 9   Chapter Test

1. $y^2 = 8x = 2(4)x$

Parabola

Vertex: $(0, 0)$

Focus: $(2, 0)$

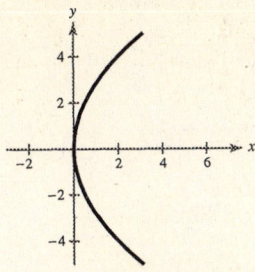

2. Center: $(3, 2)$

$a = 10, b = 9, c = \sqrt{19}$

Vertices: $(3, 12), (3, -8)$

Foci: $\left(3, 2 \pm \sqrt{19}\right)$

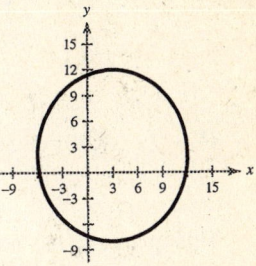

3. $(x^2 - 6x + 9) = -2y - 9 + 9$

$(x - 3)^2 = -2y = 4\left(-\dfrac{1}{2}\right)y, p = -\dfrac{1}{2}$

Vertex: $(3, 0)$

Focus: $\left(3, -\dfrac{1}{2}\right)$

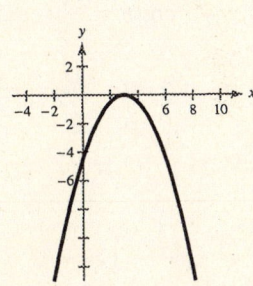

4. $x^2 - 4y^2 - 4x = 0$

$x^2 - 4x + 4 - 4y^2 = 4$

$(x - 2)^2 - 4y^2 = 4$

$\dfrac{(x - 2)^2}{4} - y^2 = 1$

Hyperbola

Center $(2, 0)$

$a = 2, b = 1, c = \sqrt{5}$

Vertices: $(0, 0), (4, 0)$

Foci: $\left(2 \pm \sqrt{5}, 0\right)$

5. Vertex: $(6, -2), p = 2$

$(y + 2)^2 = 4(2)(x - 6)$

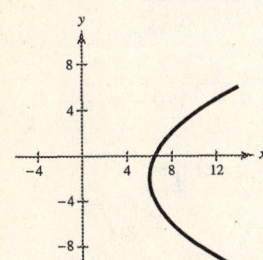

6. Center: $(-6, 3)$

$a = 7, b = 4$

$\dfrac{(x + 6)^2}{16} + \dfrac{(y - 3)^2}{49} = 1$

7. $a = 3, \dfrac{3}{2} = \dfrac{a}{b} \implies b = 2$

$\dfrac{y^2}{9} - \dfrac{x^2}{4} = 1$

8. (a) $\cot 2\theta = \dfrac{A - C}{B} = \dfrac{1 - 1}{6} = 0 \implies \theta = \dfrac{\pi}{4}$ or $45°$

(b) $B^2 - 4AC = 36 - 4 = 32 > 0 \implies$ Hyperbola

$y^2 + 6xy + (x^2 - 6) = 0$

$y = \dfrac{-6x \pm \sqrt{36x^2 - 4(x^2 - 6)}}{2}$

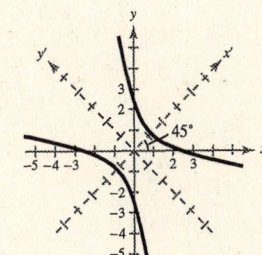

9. $x^2 + 2y^2 - 4x + 6y - 5 = 0$

$$x + y + 5 = 0$$

$y = -x - 5$:

$$x^2 + 2(-x - 5)^2 - 4x + 6(-x - 5) - 5 = 0$$

$$x^2 + 2x^2 + 20x + 50 - 4x - 6x - 30 - 5 = 0$$

$$3x^2 + 10x + 15 = 0$$

This quadratic has no real solutions. Therefore, no solution.

10. $x = 2 + 3 \cos \theta$

$y = 2 \sin \theta$

$$\left(\frac{x - 2}{3}\right)^2 + \left(\frac{y}{2}\right)^2 = 1$$

$$\frac{(x - 2)^2}{9} + \frac{y^2}{4} = 1$$

Ellipse

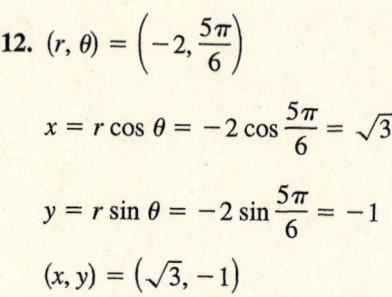

11. $x = t, \qquad y = \frac{1}{4}t - 5$

$$x = -t, \quad y = -\frac{1}{4}t - 5$$

12. $(r, \theta) = \left(-2, \dfrac{5\pi}{6}\right)$

$$x = r \cos \theta = -2 \cos \frac{5\pi}{6} = \sqrt{3}$$

$$y = r \sin \theta = -2 \sin \frac{5\pi}{6} = -1$$

$$(x, y) = \left(\sqrt{3}, -1\right)$$

13. $(x, y) = (2, -2), r = \sqrt{8} = 2\sqrt{2}, \theta = \dfrac{7\pi}{4}$

$$(r, \theta) = \left(2\sqrt{2}, \frac{7\pi}{4}\right) = \left(2\sqrt{2}, -\frac{\pi}{4}\right) = \left(-2\sqrt{2}, \frac{3\pi}{4}\right)$$

14. $x^2 + y^2 - 12y = 0$

$$r^2 - 12r \sin \theta = 0$$

$$r^2 = 12r \sin \theta$$

$$r = 12 \sin \theta$$

15. $r = 2 + 3 \sin \theta$

Limaçon

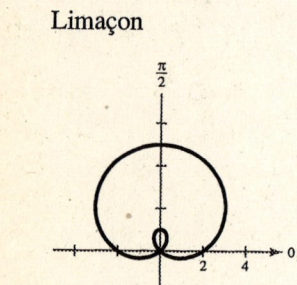

16. $r = 8 \cos 3\theta$

Rose curve

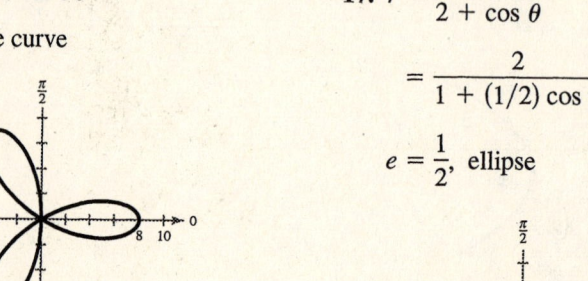

17. $r = \dfrac{4}{2 + \cos \theta}$

$$= \frac{2}{1 + (1/2) \cos \theta}$$

$e = \dfrac{1}{2}$, ellipse

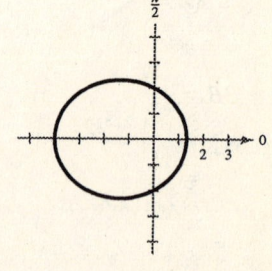

18. $r = \dfrac{ep}{1 + e \sin \theta} = \dfrac{(1/4)(4)}{1 + (1/4) \sin \theta}$

$$r = \frac{4}{4 + \sin \theta} = \frac{1}{1 + (1/4) \sin \theta}$$

19. $r = \dfrac{ep}{1 + e \sin \theta}$, $e = 0.2056$

$$72,000,000 = 2a = \dfrac{ep}{1 + e} + \dfrac{ep}{1 - e} = \left(\dfrac{0.2056}{1.2056} + \dfrac{0.2056}{0.7944}\right)p$$

$$\Rightarrow \quad p = 167,695,676.3 \text{ and } ep \approx 34,478,231$$

$$r = \dfrac{34,478,231}{1 + 0.2056 \sin \theta}$$

Perihelion: $\left(\theta = \dfrac{\pi}{2}\right)$ 28,598,400 miles

Aphelion: $\left(\theta = \dfrac{3\pi}{2}\right)$ 43,401,600 miles

Chapters 7–9   Cumulative Test

1. $\begin{bmatrix} -1 & -3 & \vdots & 5 \\ 4 & 2 & \vdots & 10 \end{bmatrix}$ row reduces to

$\begin{bmatrix} 1 & 0 & \vdots & 4 \\ 0 & 1 & \vdots & -3 \end{bmatrix}$.

Answer: $(4, -3)$

2. $3x + 7y = -10$

$-5x + 7y = 12$

Subtracting:

$8x = -22$

$x = -\dfrac{11}{4}$

$y = \dfrac{1}{7}\left(-10 - 3\left(-\dfrac{11}{4}\right)\right) = \dfrac{1}{7}\left(-\dfrac{7}{4}\right) = -\dfrac{1}{4}$

Answer: $\left(-\dfrac{11}{4}, -\dfrac{1}{4}\right)$

3. $\begin{bmatrix} 2 & -3 & 1 & \vdots & 13 \\ -4 & 1 & -2 & \vdots & -6 \\ 1 & -3 & 3 & \vdots & 12 \end{bmatrix}$ row reduces to

$\begin{bmatrix} 1 & 0 & 0 & \vdots & \frac{3}{5} \\ 0 & 1 & 0 & \vdots & -4 \\ 0 & 0 & 1 & \vdots & -\frac{1}{5} \end{bmatrix}$.

Answer: $\left(\dfrac{3}{5}, -4, -\dfrac{1}{5}\right)$

4. $3A - 2B = 3\begin{bmatrix} -3 & 0 \\ 2 & 4 \end{bmatrix} - 2\begin{bmatrix} -1 & 5 \\ 6 & -3 \end{bmatrix}$

$= \begin{bmatrix} -7 & -10 \\ -6 & 18 \end{bmatrix}$

5. $5A + 3B = 5\begin{bmatrix} -3 & 0 \\ 2 & 4 \end{bmatrix} + 3\begin{bmatrix} -1 & 5 \\ 6 & -3 \end{bmatrix}$

$= \begin{bmatrix} -18 & 15 \\ 28 & 11 \end{bmatrix}$

6. $AB = \begin{bmatrix} -3 & 0 \\ 2 & 4 \end{bmatrix}\begin{bmatrix} -1 & 5 \\ 6 & -3 \end{bmatrix} = \begin{bmatrix} 3 & -15 \\ 22 & -2 \end{bmatrix}$

7. $BA = \begin{bmatrix} -1 & 5 \\ 6 & -3 \end{bmatrix}\begin{bmatrix} -3 & 0 \\ 2 & 4 \end{bmatrix} = \begin{bmatrix} 13 & 20 \\ -24 & -12 \end{bmatrix}$

8. $\begin{vmatrix} 0 & 0 & 1 \\ 6 & 2 & 1 \\ 8 & 10 & 1 \end{vmatrix} = 44 \Rightarrow \text{Area} = \frac{1}{2}(44) = 22$ sq units

9. $\displaystyle\sum_{k=1}^{6}(7k - 2) = \dfrac{7(6)(7)}{2} - 2(6) = 135$

10. $\displaystyle\sum_{k=1}^{4}\dfrac{2}{k^2 + 4} = \dfrac{2}{1 + 4} + \dfrac{2}{4 + 4} + \dfrac{2}{9 + 4} + \dfrac{2}{16 + 4}$

≈ 0.9038

11. $\sum_{n=0}^{10} 9\left(\frac{3}{4}\right)^n = 9\left(\frac{1 - (3/4)^{11}}{1 - (3/4)}\right) \approx 34.4795$

12. $\sum_{n=1}^{\infty} 8(0.9)^{n-1} = 8\left(\frac{1}{1 - 0.9}\right) = 80$

13. For $n = 1$, $3 = 1(2 + 1)$ and the formula is true. Assume true for k, and consider

$$3 + 7 + \cdots + (4k - 1) + (4(k + 1) - 1) = 3 + 7 + \cdots + (4k - 1) + (4k + 3)$$

$$= k(2k + 1) + (4k + 3)$$

$$= 2k^2 + 5k + 3$$

$$= (k + 1)(2k + 3)$$

$$= (k + 1)(2(k + 1) + 1)$$

which shows that the formula is true for $k + 1$.

14. $(2x - 1)^8 = 256x^8 - 1024x^7 + 1792x^6 - 1792x^5 + 1120x^4 - 448x^3 + 112x^2 - 16x + 1$

15. $\dfrac{10!}{3!2!2!} = 151{,}200$

16. Ellipse with center $(2, -1)$

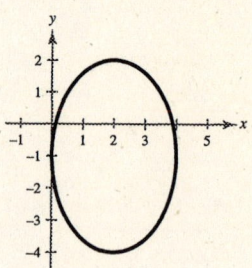

17. $(x^2 - 2x + 1) + (y^2 - 4y + 4) = -1 + 1 + 4$

$$(x - 1)^2 + (y - 2)^2 = 4$$

Circle

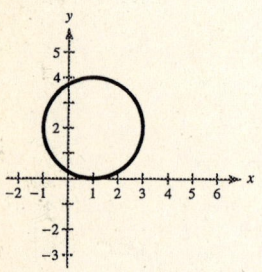

18. Foci: $(0, 0), (0, 4) \Rightarrow c = 2$

Center: $(0, 2)$

Asymptotes: $y = \pm\frac{1}{2}x + 2 \Rightarrow \frac{a}{b} = \frac{1}{2} \Rightarrow 2a = b$

$c^2 = a^2 + b^2$

$4 = a^2 + (2a)^2 = 5a^2 \Rightarrow a = \dfrac{2}{\sqrt{5}}, b = \dfrac{4}{\sqrt{5}}$

$\dfrac{(y - 2)^2}{\frac{4}{5}} - \dfrac{x^2}{\frac{16}{5}} = 1$

19. $B^2 - 4AC = 16 - 8 = 8 \implies$ Hyperbola

$\cot 2\theta = \dfrac{1 - 2}{-4} = \dfrac{1}{4} \implies \theta \approx 38°$

Graph as:

$2y^2 - 4xy + (x^2 - 6) = 0$

$$y = \frac{4x \pm \sqrt{16x^2 - 8(x^2 - 6)}}{4}$$

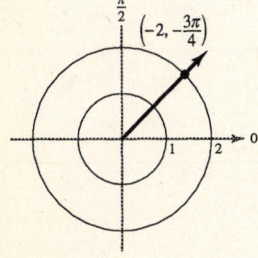

20. $x = 4 \ln t \implies t = e^{x/4}$

$y = \dfrac{1}{2}t^2$

$y = \dfrac{1}{2}(e^{x/4})^2 = \dfrac{1}{2}e^{x/2}$

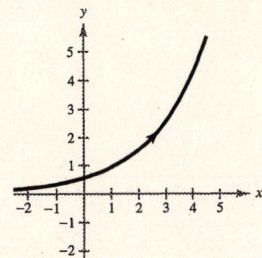

21.

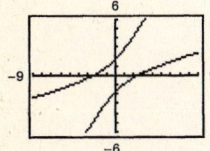

$\left(-2, \dfrac{5\pi}{4}\right), \left(2, \dfrac{\pi}{4}\right), \left(2, -\dfrac{7\pi}{4}\right)$

22.
$$-8x - 3y + 5 = 0$$
$$-8r\cos\theta - 3r\sin\theta + 5 = 0$$
$$r(8\cos\theta + 3\sin\theta) = 5$$
$$r = \frac{5}{8\cos\theta + 3\sin\theta}$$

23.
$$r = \frac{2}{4 - 5\cos\theta}$$
$$4r - 5r\cos\theta = 2$$
$$4(x^2 + y^2)^{1/2} - 5x = 2$$
$$16(x^2 + y^2) = (5x + 2)^2 = 25x^2 + 20x + 4$$
$$16y^2 - 9x^2 - 20x = 4$$

24. $r = -\dfrac{\pi}{6}$, circle

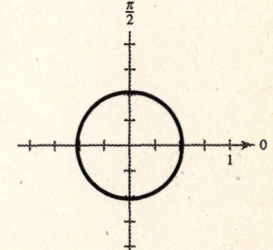

25. $r = 3 - 2\sin\theta$

Limaçon

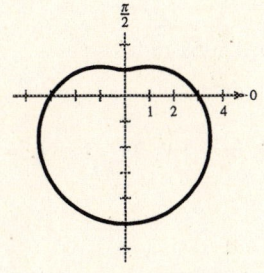

26. $r = 2 + 5\cos\theta$

Limaçon

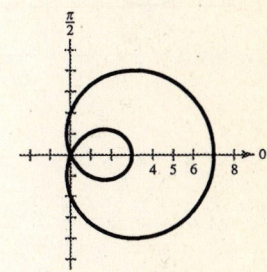

Chapter 10   Chapter Test

1.

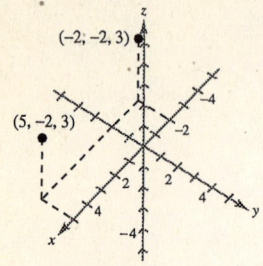

2. $AB = \sqrt{(8-6)^2 + (-2-4)^2 + (5+1)^2} = \sqrt{76}$

$AC = \sqrt{(8+4)^2 + (-2-3)^2 + (5-0)^2} = \sqrt{144 + 25 + 25} = \sqrt{194}$

$BC = \sqrt{(6+4)^2 + (4-3)^2 + (-1-0)^2} = \sqrt{100 + 1 + 1} = \sqrt{102}$

No. $\left(\sqrt{76}\right)^2 + \left(\sqrt{102}\right)^2 \neq \left(\sqrt{194}\right)^2$

3. Midpoint $= \left(\dfrac{8+6}{2}, \dfrac{-2+4}{2}, \dfrac{5-1}{2}\right) = (7, 1, 2)$

4. Diameter $= \sqrt{(8-6)^2 + (-2-4)^2 + (5+1)^2}$

$\qquad\qquad = \sqrt{4 + 36 + 36} = \sqrt{76}$

Radius $= \sqrt{19}$

$(x-7)^2 + (y-1)^2 + (z-2)^2 = 19$

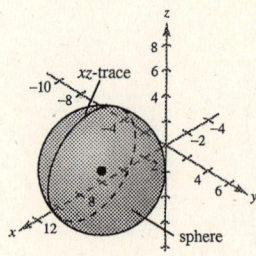

5. $\mathbf{u} = \langle 6-8, 4-(-2), -1-5 \rangle = \langle -2, 6, -6 \rangle$

$\mathbf{v} = \langle -4-8, 3-(-2), 0-5 \rangle = \langle -12, 5, -5 \rangle$

6. (a) $\|\mathbf{v}\| = \sqrt{(-12)^2 + 5^2 + (-5)^2} = \sqrt{194}$

(b) $\mathbf{u} \cdot \mathbf{v} = (-2)(-12) + 6(5) + (-6)(-5) = 84$

(c) $\mathbf{u} \times \mathbf{v} = \begin{vmatrix} \mathbf{i} & \mathbf{j} & \mathbf{k} \\ -2 & 6 & -6 \\ -12 & 5 & -5 \end{vmatrix} = \langle 0, 62, 62 \rangle$

7. $\cos \theta = \dfrac{\mathbf{u} \cdot \mathbf{v}}{\|\mathbf{u}\| \, \|\mathbf{v}\|} = \dfrac{84}{\sqrt{76}\sqrt{194}} \approx 0.6918 \implies \theta \approx 46.23°$ or 0.8068 radians

8. (a) $x = 8 - 2t, y = -2 + 6t, z = 5 - 6t$

(b) $\dfrac{x-8}{-2} = \dfrac{y+2}{6} = \dfrac{z-5}{-6}$

9. $\mathbf{u} \cdot \mathbf{v} = 0 - 2 - 6 \neq 0$ and $\mathbf{u} \neq c\mathbf{v} \implies$ neither

10. $\mathbf{u} \cdot \mathbf{v} = -2 + 3 - 1 = 0 \implies$ orthogonal

11. First two points: $\mathbf{v} = \langle 4, 8, -2 \rangle$

Last two points: $\mathbf{w} = \langle 4, 8, -2 \rangle$

Opposite sides are parallel and equal length.

Adjacent sides: $\mathbf{v}$ and $\mathbf{u} = \langle 1, -3, 3 \rangle$

Area $= \| \mathbf{u} \times \mathbf{v} \|$

$$\mathbf{u} \times \mathbf{v} = \begin{vmatrix} \mathbf{i} & \mathbf{j} & \mathbf{k} \\ 1 & -3 & 3 \\ 4 & 8 & -2 \end{vmatrix} = \langle -18, 14, 20 \rangle$$

$\| \mathbf{u} \times \mathbf{v} \| = \sqrt{18^2 + 14^2 + 20^2} = 2\sqrt{230} \approx 30.33$ square units

12. $\mathbf{u} = \langle 0, 8, -1 \rangle, \mathbf{v} = \langle 4, 5, -4 \rangle$

$$\mathbf{n} = \mathbf{u} \times \mathbf{v} = \begin{vmatrix} \mathbf{i} & \mathbf{j} & \mathbf{k} \\ 0 & 8 & -1 \\ 4 & 5 & -4 \end{vmatrix} = \langle -27, -4, -32 \rangle$$

Plane: $-27(x + 3) - 4(y + 4) - 32(z - 2) = 0$

$$-27x - 4y - 32z - 33 = 0$$

$$27x + 4y + 32z + 33 = 0$$

13. Let $A(0, 0, 5)$ be the vertex.

$\mathbf{u} = \overrightarrow{AD} = \langle 4, 0, 0 \rangle, \mathbf{v} = \overrightarrow{AB} = \langle 0, 10, 0 \rangle,$

$\mathbf{w} = \overrightarrow{AE} = \langle 0, 1, -5 \rangle$

$$\mathbf{u} \cdot (\mathbf{v} \times \mathbf{w}) = \begin{vmatrix} 4 & 0 & 0 \\ 0 & 10 & 0 \\ 0 & 1 & -5 \end{vmatrix} = 4(-50) = -200$$

Volume $= |-200| = 200$ cubic units

14. $2x + 3y + 4z = 12$

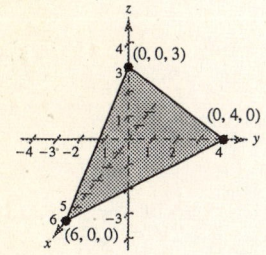

15. $5x - y - 2z = 10$

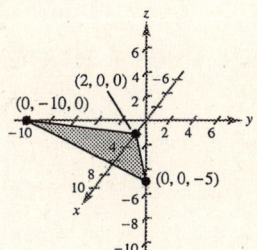

16. $\mathbf{n} = \langle 3, -2, 1 \rangle, Q = (2, -1, 6), P = (0, 0, 6)$ in plane, $\overrightarrow{PQ} = \langle 2, -1, 0 \rangle$

$$D = \frac{|\overrightarrow{PQ} \cdot \mathbf{n}|}{\| \mathbf{n} \|} = \frac{|8|}{\sqrt{14}} = \frac{4\sqrt{14}}{7}$$

Chapter 11   Chapter Test

1. $\displaystyle\lim_{x \to -2} \frac{x^2 - 1}{2x} = \frac{(-2)^2 - 1}{2(-2)} = -\frac{3}{4}$

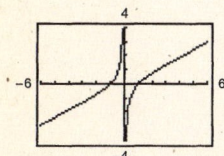

Limit is -0.75.

2. $\displaystyle\lim_{x \to -1} \frac{2x^2 - x - 3}{x + 1} = \lim_{x \to -1} \frac{(x + 1)(2x - 3)}{x + 1}$

$$= \lim_{x \to -1} (2x - 3) = -5$$

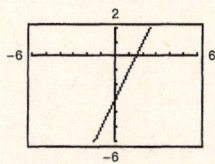

Limit is -5.

3.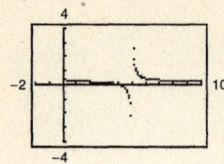

$$\lim_{x \to 5} \frac{\sqrt{x} - 2}{x - 5} \text{ does not exist.}$$

4.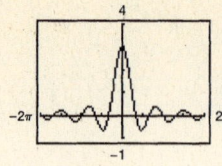

$$\lim_{x \to 0} \frac{\sin 3x}{x} = 3$$

$$f(x) = \frac{\sin 3x}{x}$$

5.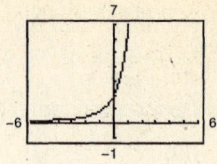

$$\lim_{x \to 0} \frac{e^{2x} - 1}{x} = 2$$

$$f(x) = \frac{e^{2x} - 1}{x}$$

6. (a)
$$\frac{f(x + h) - f(x)}{h} = \frac{3(x + h)^2 - 5(x + h) - 2 - (3x^2 - 5x - 2)}{h}$$

$$= \frac{3x^2 + 6xh + 3h^2 - 5h - 3x^2}{h}$$

$$= 6x + 3h - 5$$

$$f'(x) = \lim_{h \to 0} [6x + 3h - 5] = 6x - 5$$

$$f'(2) = 6(2) - 5 = 7$$

(b)
$$\frac{f(x + h) - f(x)}{h} = \frac{[2(x + h)^3 + 6(x + h)] - [2x^3 + 6x]}{h}$$

$$= \frac{2x^3 + 6x^2h + 6xh^2 + 2h^3 + 6x + 6h - 2x^3 - 6x}{h}$$

$$= \frac{6x^2h + 6xh^2 + 2h^3 + 6h}{h}$$

$$= 6x^2 + 6xh + 2h^2 + 6, \ h \neq 0$$

$$f'(x) = \lim_{h \to 0} [6x^2 + 6xh + 2h^2 + 6] = 6x^2 + 6$$

$$f'(-1) = 6(-1)^2 + 6 = 12$$

7.
$$f'(x) = \lim_{h \to 0} \frac{f(x + h) - f(x)}{h}$$

$$= \lim_{h \to 0} \frac{4 - (3/4)(x + h) - [4 - (3/4)x]}{h}$$

$$= \lim_{h \to 0} \frac{-(3/4)h}{h} = -\frac{3}{4}$$

8.
$$f'(x) = \lim_{h \to 0} \frac{f(x + h) - f(x)}{h}$$

$$= \lim_{h \to 0} \frac{2(x + h)^2 + 4(x + h) - 1 - [2x^2 + 4x - 1]}{h}$$

$$= \lim_{h \to 0} \frac{2x^2 + 4xh + 2h^2 + 4h - 2x^2}{h}$$

$$= \lim_{h \to 0} (4x + 2h + 4) = 4x + 4$$

9. $f'(x) = \lim\limits_{h \to 0} \dfrac{f(x+h) - f(x)}{h}$

$\quad = \lim\limits_{h \to 0} \dfrac{\dfrac{1}{x+3+h} - \dfrac{1}{x+3}}{h}$

$\quad = \lim\limits_{h \to 0} \dfrac{(x+3) - (x+3+h)}{h(x+3+h)(x+3)}$

$\quad = \lim\limits_{h \to 0} \dfrac{-1}{(x+3+h)(x+3)}$

$\quad = \dfrac{-1}{(x+3)^2}$

10. $\lim\limits_{x \to \infty} \dfrac{6}{5x-1} = 0$

11. $\lim\limits_{x \to \infty} \dfrac{1-3x^2}{x^2-5} = -3$

12. $\lim\limits_{x \to -\infty} \dfrac{3x^3}{x+2}$ does not exist.

13. $0, \frac{3}{4}, \frac{14}{19}, \frac{12}{17}, \frac{36}{53}$

$\quad \lim\limits_{n \to \infty} a_n = \frac{1}{2}$

14. $0, 1, 0, \frac{1}{2}, 0$

$\quad \lim\limits_{n \to \infty} a_n = 0$

15. Width of each rectangle: $\frac{1}{2}$

$\quad$ Heights: $8, \frac{15}{2}, 6, \frac{7}{2}$

$\quad$ Area $\approx \frac{1}{2}\left[8 + \frac{15}{2} + 6 + \frac{7}{2}\right] = \frac{25}{2}$

16. Width: $\dfrac{4}{n}$, Height: $f\left(-2 + \dfrac{4i}{n}\right) = \left(-2 + \dfrac{4i}{n}\right) + 2 = \dfrac{4i}{n}$

$A \approx \sum\limits_{i=1}^{n} \left(\dfrac{4i}{n}\right)\left(\dfrac{4}{n}\right) = \dfrac{16}{n^2} \sum\limits_{i=1}^{n} i = \dfrac{16}{n^2} \dfrac{n(n+1)}{2}$

$A = \lim\limits_{n \to \infty} \dfrac{16}{n^2} \cdot \dfrac{n(n+1)}{2} = 8$

17. Width: $\dfrac{1}{n}$, Height: $f\left(\dfrac{i}{n}\right) = 1 - \dfrac{i^3}{n^3}$

$A \approx \sum\limits_{i=1}^{n} \left(1 - \dfrac{i^3}{n^3}\right)\left(\dfrac{1}{n}\right) = \sum\limits_{i=1}^{n} \left(\dfrac{1}{n} - \dfrac{i^3}{n^4}\right)$

$\quad = \dfrac{1}{n} \sum\limits_{i=1}^{n} 1 - \dfrac{1}{n^4} \sum\limits_{i=1}^{n} i^3$

$\quad = \dfrac{1}{n}(n) - \dfrac{1}{n^4}\left(\dfrac{n^2(n+1)^2}{4}\right)$

$\quad = 1 - \dfrac{(n+1)^2}{4n^2}$

$A = \lim\limits_{n \to \infty}\left(1 - \dfrac{(n+1)^2}{4n^2}\right) = 1 - \dfrac{1}{4} = \dfrac{3}{4}$

18. (a) $y = 8.79x^2 - 6.2x - 0.4$

$\quad$ (b) Velocity = Derivative = $17.58x - 6.2$

$\qquad$ At $x = 5$, velocity ≈ 81.7 ft/sec.

Chapters 10–11 Cumulative Test

1. $(-6, 1, 3)$

2. $(0, -4, 0)$

3. $d = \sqrt{(4 - (-2))^2 + (-5 - 3)^2 + (1 - (-6))^2}$

$= \sqrt{36 + 64 + 49}$

$= \sqrt{149}$

4. $d_1 = 3, d_2 = 4, d_3 = \sqrt{4^2 + 3^2} = 5$

$d_1^2 + d_2^2 = d_3^2$

5. Midpoint: $\left(\dfrac{3 - 5}{2}, \dfrac{4 + 0}{2}, \dfrac{-1 + 2}{2}\right) = \left(-1, 2, \dfrac{1}{2}\right)$

6. Center $= (2, 2, 4)$

Radius $= \sqrt{2^2 + 2^2 + 4^2} = \sqrt{24}$

$(x - 2)^2 + (y - 2)^2 + (z - 4)^2 = 24$

7. xy-trace: $(z = 0)$

$(x - 2)^2 + (y + 1)^2 = 4$, Circle

yz-trace: $(x = 0)$

$4 + (y + 1)^2 + z^2 = 4$ or $(y + 1)^2 + z^2 = 0$, Point

$(0, -1, 0)$, Point

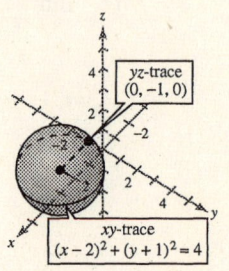

8. $\mathbf{u} \cdot \mathbf{v} = \langle 2, -6, 0 \rangle \cdot \langle -4, 5, 3 \rangle$

$= -8 - 30 = -38$

$\mathbf{u} \times \mathbf{v} = \begin{vmatrix} \mathbf{i} & \mathbf{j} & \mathbf{k} \\ 2 & -6 & 0 \\ -4 & 5 & 3 \end{vmatrix} = \langle -18, -6, -14 \rangle$

9. $\mathbf{u} \cdot \mathbf{v} \neq 0, \mathbf{u} \neq c\mathbf{v} \Rightarrow$ neither

10. $\mathbf{u} \cdot \mathbf{v} = -8 - 12 + 20 = 0 \Rightarrow$ orthogonal

11. $3\mathbf{u} = \langle -3, 18, -9 \rangle = -\mathbf{v} \Rightarrow$ parallel

12. $\overrightarrow{DA} = \langle 0, -2, 0 \rangle, \overrightarrow{DC} = \langle 2, 1, 0 \rangle, \overrightarrow{DH} = \langle 0, 0, 3 \rangle$

$\begin{vmatrix} 0 & -2 & 0 \\ 2 & 1 & 0 \\ 0 & 0 & 3 \end{vmatrix} = 12$ cubic units

13. (a) Vector is $\langle 5 + 2, 8 - 3, 25 - 0 \rangle = \langle 7, 5, 25 \rangle$.

$x = -2 + 7t, y = 3 + 5t, z = 25t$

(b) $\dfrac{x + 2}{7} = \dfrac{y - 3}{5} = \dfrac{z}{25}$

14. $\mathbf{u} = \langle -2, 3, 0 \rangle, \mathbf{v} = \langle 5, 8, 25 \rangle$

$\mathbf{u} \times \mathbf{v} = \begin{vmatrix} \mathbf{i} & \mathbf{j} & \mathbf{k} \\ -2 & 3 & 0 \\ 5 & 8 & 25 \end{vmatrix} = \langle 75, 50, -31 \rangle$

Normal to plane

Plane: $75x + 50y - 31z = 0$

15.

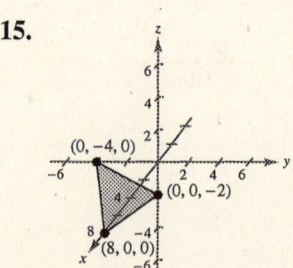

16. $\mathbf{n} = \langle 2, -5, 1 \rangle$, $Q = (0, 0, 25)$, $P = (0, 0, 10)$ in plane, $\overrightarrow{PQ} = \langle 0, 0, 15 \rangle$

$$D = \frac{|\overrightarrow{PQ} \cdot \mathbf{n}|}{\|\mathbf{n}\|} = \frac{15}{\sqrt{30}} = \frac{\sqrt{30}}{2} \approx 2.74$$

17. Normal to plane containing: $(-1, -1, 3)$, $(0, 0, 0)$ and $(2, 0, 0)$ is

$$\langle -1, -1, 3 \rangle \times \langle 2, 0, 0 \rangle = \begin{vmatrix} \mathbf{i} & \mathbf{j} & \mathbf{k} \\ -1 & -1 & 3 \\ 2 & 0 & 0 \end{vmatrix} = \langle 0, 6, 2 \rangle \text{ or } \mathbf{n}_1 = \langle 0, 3, 1 \rangle$$

Normal to front face is: $\langle 1, -1, 3 \rangle \times \langle 0, 2, 0 \rangle = \begin{vmatrix} \mathbf{i} & \mathbf{j} & \mathbf{k} \\ 1 & -1 & 3 \\ 0 & 2 & 0 \end{vmatrix} = \langle -6, 0, 2 \rangle \text{ or } \mathbf{n}_2 = \langle -3, 0, 1 \rangle$

Angle between sides: $\cos\theta = \dfrac{|\mathbf{n}_1 \cdot \mathbf{n}_2|}{\|\mathbf{n}_1\| \|\mathbf{n}_2\|} = \dfrac{1}{\sqrt{10}\sqrt{10}} = \dfrac{1}{10} \implies \theta \approx 84.26°$

18. $\displaystyle\lim_{x\to4} (5x - x^2) = 5(4) - 4^2 = 4$

19. $\displaystyle\lim_{x\to-2^+} \frac{x + 2}{(x + 2)(x - 1)} = \lim_{x\to-2^+} \frac{1}{x - 1} = -\frac{1}{3}$

20. $\displaystyle\lim_{x\to7} \frac{x - 7}{(x - 7)(x + 7)} = \lim_{x\to7} \frac{1}{x + 7} = \frac{1}{14}$

21. $\displaystyle\lim_{x\to0} \frac{\sqrt{x + 4} - 2}{x} \cdot \frac{\sqrt{x + 4} + 2}{\sqrt{x + 4} + 2} = \lim_{x\to0} \frac{(x + 4) - 4}{x(\sqrt{x + 4} + 2)} = \lim_{x\to0} \frac{1}{\sqrt{x + 4} + 2} = \frac{1}{2 + 2} = \frac{1}{4}$

22. $\displaystyle\lim_{x\to4^-} \frac{|x - 4|}{x - 4} = -1$

23. $\displaystyle\lim_{x\to0} \sin\left(\frac{\pi}{x}\right)$ does not exist.

24. $f(x) = 3 - x^2$

$$m = \lim_{h\to0} \frac{f(x + h) - f(x)}{h}$$

$$= \lim_{h\to0} \frac{3 - (x + h)^2 - (3 - x^2)}{h} = \lim_{h\to0} \frac{-2xh - h^2}{h} = \lim_{h\to0} (-2x - h) = -2x$$

At $(1, 2)$, $m = -2$.

25. $f(x) = \sqrt{x + 3}$

$$m = \lim_{h\to0} \frac{f(x + h) - f(x)}{h}$$

$$= \lim_{h\to0} \frac{\sqrt{x + h + 3} - \sqrt{x + 3}}{h} \cdot \frac{\sqrt{x + h + 3} + \sqrt{x + 3}}{\sqrt{x + h + 3} + \sqrt{x + 3}}$$

$$= \lim_{h\to0} \frac{(x + h + 3) - (x + 3)}{h[\sqrt{x + h + 3} + \sqrt{x + 3}]}$$

$$= \lim_{h\to0} \frac{1}{\sqrt{x + h + 3} + \sqrt{x + 3}} = \frac{1}{2\sqrt{x + 3}}$$

At $(-2, 1)$, $m = \frac{1}{2}$.

26. $f(x) = \dfrac{1}{x + 3}$

$m = \lim\limits_{h \to 0} \dfrac{f(x + h) - f(x)}{h}$

$= \lim\limits_{h \to 0} \dfrac{\dfrac{1}{x + h + 3} - \dfrac{1}{x + 3}}{h}$

$= \lim\limits_{h \to 0} \dfrac{(x + 3) - (x + h + 3)}{h(x + h + 3)(x + 3)}$

$= \lim\limits_{h \to 0} \dfrac{-1}{(x + h + 3)(x + 3)}$

$= \dfrac{-1}{(x + 3)^2}$

At $\left(1, \dfrac{1}{4}\right)$, $m = \dfrac{-1}{16}$.

27. $f(x) = x^4$

$m = \lim\limits_{h \to 0} \dfrac{f(x + h) - f(x)}{h}$

$= \lim\limits_{h \to 0} \dfrac{(x + h)^4 - x^4}{h}$

$= \lim\limits_{h \to 0} \dfrac{4x^3h + 6x^2h^2 + 4xh^3 + h^4}{h}$

$= \lim\limits_{h \to 0} \left[4x^3 + 6x^2h + 4xh^2 + h^3\right]$

$= 4x^3$

At $(-1, 1)$, $m = -4$.

28. $\lim\limits_{x \to \infty} \dfrac{2x^4 - x^3 + 4}{x^2 - 9}$

Does not exist

29. $\lim\limits_{x \to \infty} \dfrac{3 - 7x}{x + 4} = -7$

30. $\lim\limits_{x \to \infty} \dfrac{3x^2 + 1}{x^2 + 4} = 3$

31. $\lim\limits_{x \to \infty} \dfrac{2x}{x^2 + 3x - 2} = 0$

32. $\sum\limits_{i=1}^{50}(1 - i^2) = 50 - \dfrac{50(51)(101)}{6} = -42{,}875$

33. $\sum\limits_{k=1}^{20}(3k^2 - 2k) = 3\,\dfrac{20(21)(41)}{6} - 2\,\dfrac{20(21)}{2}$

$= 8610 - 420 = 8190$

34. $\sum\limits_{i=1}^{40}(12 + i^3) = 12(40) + \dfrac{40^2(41)^2}{4}$

$= 480 + 672{,}400 = 672{,}880$

35. Area $\approx \dfrac{1}{2}[1 + 2 + 3 + 4 + 5 + 6]$

$= \dfrac{21}{2}$ square units

36. Area $\approx \dfrac{1}{2}[4.875 + 4.5 + 3.875 + 3]$

$= 8.125$ square units

37. Area $\approx \dfrac{1}{2}\left[\dfrac{9}{16} + 1 + \dfrac{25}{16} + \dfrac{9}{4}\right] = \dfrac{43}{16} = 2.6875$

38. Area $\approx \dfrac{1}{4}\left[\dfrac{1}{1 + \left(-\frac{3}{4}\right)^2} + \dfrac{1}{1 + \left(-\frac{1}{2}\right)^2} + \dfrac{1}{1 + \left(-\frac{1}{4}\right)^2} + \dfrac{1}{1 + 0} + \dfrac{1}{1 + \left(\frac{1}{4}\right)^2} + \dfrac{1}{1 + \left(\frac{1}{2}\right)^2} + \dfrac{1}{1 + \left(\frac{3}{4}\right)^2} + \dfrac{1}{1 + 1^2}\right]$

$= \dfrac{1}{4}\left[2(0.64) + 2(0.8) + 2(0.941176) + 1 + \dfrac{1}{2}\right]$

≈ 1.566 square units

39. Width: $\dfrac{1}{n}$, Height: $f\left(\dfrac{i}{n}\right) = 1 - \left(\dfrac{1}{n}\right)^3$

$A \approx \sum\limits_{i=1}^{n}\left(1 - \left(\dfrac{i}{n}\right)^3\right)\left(\dfrac{1}{n}\right) = \dfrac{1}{n}\sum\limits_{i=1}^{n}1 - \dfrac{1}{n^4}\sum\limits_{i=1}^{n}i^3 = \dfrac{1}{n}(n) - \dfrac{1}{n^4}\left[\dfrac{n^2(n + 1)^2}{4}\right]$

$A = \lim\limits_{n \to \infty}\left[1 - \dfrac{1}{n^4}\left(\dfrac{n^2(n + 1)^2}{4}\right)\right] = 1 - \dfrac{1}{4} = \dfrac{3}{4}$

40. Width: $\dfrac{6}{n}$, Height: $f\left(-3 + \dfrac{6i}{n}\right) = \left(-3 + \dfrac{6i}{n}\right) + 3 = \dfrac{6i}{n}$

$A \approx \displaystyle\sum_{i=1}^{n} \left(\dfrac{6i}{n}\right)\left(\dfrac{6}{n}\right) = \dfrac{36}{n^2} \sum_{i=1}^{n} i = \dfrac{36}{n^2} \dfrac{n(n+1)}{2} = \dfrac{18(n+1)}{n}$

$A = \displaystyle\lim_{n\to\infty} \left[\dfrac{18(n+1)}{n}\right] = 18$

41. Width: $\dfrac{2}{n}$, Height: $f\left(-1 + \dfrac{2i}{n}\right) = \left(-1 + \dfrac{2i}{n}\right)^2 = 1 - \dfrac{4i}{n} + \dfrac{4i^2}{n^2}$

$A \approx \displaystyle\sum_{i=1}^{n} \left[1 - \dfrac{4i}{n} + \dfrac{4i^2}{n^2}\right]\left(\dfrac{2}{n}\right)$

$= \dfrac{2}{n} \displaystyle\sum_{i=1}^{n} 1 - \dfrac{8}{n^2} \sum_{i=1}^{n} i + \dfrac{8}{n^3} \sum_{i=1}^{n} i^2$

$= \dfrac{2}{n}(n) - \dfrac{8}{n^2} \dfrac{n(n+1)}{2} + \dfrac{8}{n^3} \dfrac{n(n+1)(2n+1)}{6}$

$= 2 - \dfrac{4(n+1)}{n} + \dfrac{4(n+1)(2n+1)}{3n^2}$

$A = \displaystyle\lim_{n\to\infty} \left[2 - \dfrac{4(n+1)}{n} + \dfrac{4(n+1)(2n+1)}{3n^2}\right] = 2 - 4 + \dfrac{8}{3} = \dfrac{2}{3}$